HOW PROTEINS WORK

HOW PROTEINS WORK

Mike Williamson

NEW YORK AND LONDON

Garland Science
Vice President: Denise Schanck
Editor: Summers Scholl
Editorial Assistant: Kelly O'Connor
Production Editor and Layout: EJ Publishing Services
Illustrator and Cover Design: Matthew McClements, Blink Studio, Ltd.
Copyeditor: Bruce Goatly
Proofreader: Sally Livitt
Indexer: Medical Indexing Limited

ISBN 978-0-8153-4446-9

Library of Congress Cataloging-in-Publication Data

Williamson, Mike (Michael Paul)
How proteins work / Mike Williamson.
p. cm.
Includes bibliographical references.
ISBN 978-0-8153-4446-9
1. Proteins. 2. Proteins--Analysis. 3. Biochemistry. 4. Biophysics. I. Title.
QD431.W536 2012
572'.636--dc23
201101758

Published by Garland Science, Taylor & Francis Group, LLC, an informa business, 711 Third Avenue, New York, NY 10017, USA, and 2 Park Square, Milton Park, Abingdon, OX14 4RN, UK.

Printed in the United Kingdom

15 14 13 12 11 10 9 8 7 6 5 4 3 2

Visit our website at http://www.garlandscience.com

Preface

I have been occupied for some time with the study of the most essential substances of the animal kingdom: fibrin, albumin and gelatin. I conclude that the organic substance which is present in all constituents of the animal body, also as we shall soon see in the plant kingdom, could be named protein *from* πρωτειος *[proteios],* primarius, *which has the composition* $C_{400}H_{620}N_{100}O_{120}$...

Gerhardus Johannes Mulder (1802–1880)

Proteins are endlessly fascinating. They carry out almost all the catalytic functions in the cell, as well as directing and forming most of the structural framework. They catalyze reactions many orders of magnitude faster than any system that humans can devise under comparable conditions. Proteins are also much larger than most human-designed catalysts. They make many interactions of widely varying strength and duration. Crucial to any understanding of protein function is their structure: although many of the principles governing how proteins work were understood many years ago, it is not until we have the structural details that we can really appreciate exactly what the proteins are doing. This is a major reason why we don't understand membrane proteins as well as we do globular proteins. However, the structure is merely the detail that enables us to reach toward the concepts that really explain how proteins work. I have therefore endeavored to look beyond the structural detail to understand the underlying principles.

This textbook grew out of my courses for intermediate and advanced undergraduates, and is inspired by the idea that proteins are a functional part of living and evolving systems. They have a certain form and function because it works, not necessarily as the most perfect solution to the biological problem, but certainly as a viable and successful solution. Advanced undergraduate and graduate students, as well as practitioners, interested in proteins should find the book useful. A basic foundation in chemistry and biology, as supplied by introductory undergraduate courses in biochemistry, should suffice. Students may have a background in chemistry, biology or physics, but I have tried to write the text so that it is accessible to all.

The book is written in a style that I like to read. This means the text is discursive; occasionally it goes off at a tangent; it has analogies and examples liberally scattered around; it simplifies systems as far as possible in an attempt to see the forest for the trees; and it places more emphasis on principles than on the experiments that were used to derive the principles.

There is considerable discussion of the role of evolution in tinkering with proteins to create something with a desired function. I am particularly interested in how proteins solve 'difficult' biological problems, such as catalysis, movement, and signaling. In the same way that you cannot really understand a foreign country without having some idea of its history, I believe that you cannot understand proteins unless you have some idea of how they got to their present form. The use of everyday analogies and emphasis of the physical environment around the protein enable the reader to understand proteins as well as merely know the facts about them.

Quantitative calculations are used to understand how proteins work. I strongly believe that the field of biochemistry in general, and protein science in particular, will need to place more emphasis on quantitative measurements as they mature. A holistic view—integrating structural, chemical, and biological data to try to understand how proteins help the cell to function properly—is key to this text. We are moving into a new era of biological science, where we have a good idea of many of the pieces, and we are starting to see how the pieces work together to achieve a functional whole (the idea behind Systems Biology). This book is an attempt to do exactly that.

I do not attempt to be comprehensive in the coverage of proteins. There is little coverage of medical aspects of proteins, though they are certainly described where relevant, as in signaling. I have often skipped over the experimental evidence for many of the facts presented, because I do not want to obscure the principles of how proteins work by inclusion of too much experimental detail.

Chapters 1-4 and 6 present the physical constraints that have resulted in proteins looking and working the way they do. These limitations include the structures and properties of amino acids and the forces that hold proteins together, which are discussed in Chapter 1, along with a detailed discussion of the way evolution shapes proteins. Chapter 2 discusses the domain, the fundamental structural and evolutionary building block of proteins, while Chapter 3 considers how domains associate together into oligomeric proteins; it also discusses consequences of oligomerization such as allostery and cooperativity. Chapter 4 covers an important topic that is not often discussed in textbooks, namely the cellular environment and how this influences proteins. It describes the crowded environment of the cell, how proteins bind rapidly and yet specifically to their targets, and natively unstructured proteins, as well as post-translational modifications and protein folding. Finally, Chapter 6 discusses the developing area of internal mobility within proteins.

The second half of the book, Chapters 5 and 7–10, covers various biological functions of proteins, and considers how they carry out these functions, and how their structure enables them to do so. These are enzyme catalysis in Chapter 5, movement and translocation in Chapter 7, signaling in Chapter 8, regulation (by the formation of complexes) in Chapter 9, and coordination of sequential reactions by multi-enzyme complexes in Chapter 10. Additionally, Chapter 9 looks at the results emerging from high-throughput technology. Finally, Chapter 11 discusses the techniques used in studying proteins, both experimental and theoretical.

The main text is augmented with boxes referred to by numbered asterisks (*) that provide more details on select topics, brief biographies of prominent scientists, and pedagogical analogies for further elucidation of concepts. There is also a glossary containing definitions to words that appear in bold throughout the main text.

Online Resources

Accessible from www.garlandscience.com, the Student and Instructor Resources websites provide learning and teaching tools created for *How Proteins Work*.The Student Resources Site is open to everyone, and users have the option to register in order to use book-marking and note-taking tools. The Instructor's Resources Site requires registration and access is available to instructors who have assigned the book to their course. To access the Instructor Resource Site please contact your local sales representative or email science@garland.com. Below is an overview of the resources available for this book. On the website, the resources may be browsed by individual chapters and there is a search engine. You can also access the resources available for other Garland Science titles.

For Students

Animations and Videos

The animations and videos dynamically illustrate important concepts from the book, and make many of the more difficult topics accessible.

Flashcards

Each chapter contains a set of flashcards, built into the website, that allow students to review key terms from the text.

Glossary

The complete glossary from the book is available on the website and can be searched and browsed as a whole or sorted by chapter.

Hints

The hints provide strategies and clues for solving some of the more difficult end-of-chapter problems.

Solutions to Problems

Solutions to the odd-numbered problems are provided for self-testing.

For Instructors

Figures

The images from the book are available in two convenient formats: PowerPoint® and JPEG. They have been optimized for display on a computer. Figures are searchable by figure number, figure name, or by keywords used in the figure legend from the book.

Animations and Videos

The animations and videos that are available to students are also available on the Instructor's website in two formats. The WMV formatted movies are created for instructors who wish to use the movies in PowerPoint presentations on Windows® computers; the QuickTime formatted movies are for use in PowerPoint for Apple computers or Keynote® presentations. The movies can easily be downloaded to your computer using the "download" button on the movie preview page.

Power Point Presentations

The PowerPoint presentations contain the figures and micrographs from the book. There is one presentation for each chapter.

Solutions Manual

A complete solutions manual is provided for all problems in the text.

Acknowledgments

I need to thank the many people who have helped in one way or another. These include my supervisors and mentors Dudley Williams and Kurt Wüthrich, as well as colleagues who have provided much needed insight. For advice and corrections: Pete Artymiuk and Per Bullough. For suggestions particularly on the problems: Abaigael Keegan, Hugh Dannatt, Rebecca Hill, Vicki Kent, Tacita Nye and Muhammed Qureshi. And of course the production team at Garland Science, particularly Summers Scholl who has nurtured the book through its gestation; Emma Jeffcock, who has ably managed the production process; Matt McClements, who has turned my sketches into awesome images; and Bruce Goatly, whose spot-on comments and corrections were appreciated.

Mike Williamson

Contents

Detailed Contents

CHAPTER 1
Protein Structure and Evolution

The basic laws of physics can usually be expressed in exact mathematical form, and they are probably the same throughout the universe. The "laws" of biology, by contrast, are often only broad generalizations, since they describe rather elaborate chemical mechanisms that natural selection has evolved over billions of years.

Francis Crick (1988), [1]

Structural biology has had an enormous influence on biochemistry in general, and on the study of proteins in particular. It can almost be said that unless we know a protein's three-dimensional structure we cannot understand how it functions. However, when the crystallographer John Kendrew determined the structure of the first protein to be described in detail (myoglobin, in 1958), the most striking feature was its irregularity and complexity (or, as Max Perutz wrote, a "hideous and visceral-looking object"—**Figure 1.1** [2]). It soon became clear that proteins require this level of complexity to bind ligands and catalyze reactions specifically. But as soon as we start looking in detail at proteins, we see that there are regular patterns to the way in which proteins fold up, patterns that are determined by the underlying structures of amino acids and by the forces that dictate how they pack together. When we look at the human body, we can identify a hierarchy of structural and functional units, each dependent on the next: limbs, organs, cells, and cellular components. The same is true of proteins—each level of structure (quaternary, tertiary, secondary, and primary) depends on the one below.

Even more importantly, the structure and function of proteins are a product of evolution. This is again true of the human body: we cannot hope to understand its functions, malfunctions, and development without understanding something about the evolutionary processes that shaped it. This is why an evolutionary viewpoint pervades this book, and why a considerable part of the first chapter has been set aside to consider the implications of evolution.

Chapter 1 lays down a framework and sets the scene for the rest of the book. It is, however, far from being just an introduction, and contains some advanced material.

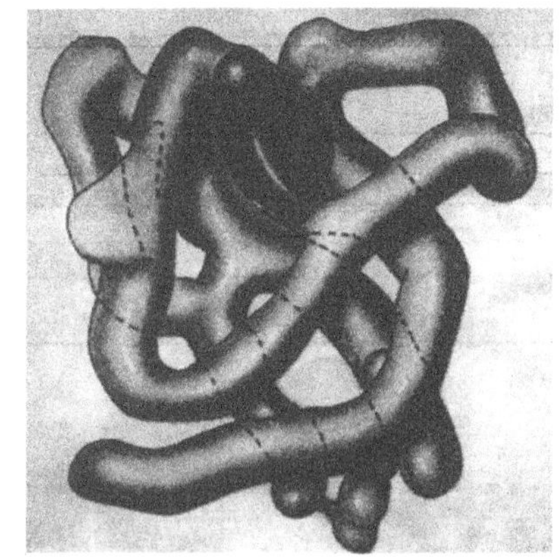

FIGURE 1.1
The first view of a protein structure was Kendrew's "hideous and visceral-looking object": the low-resolution crystal structure of myoglobin, obtained in 1958. At this resolution it is only possible to see the course of the peptide chain, much of which is in the form of α helices. Although the internal structure of an α helix is regular, the rest of the protein (the tertiary structure) is strikingly irregular. The darker region near the top is the heme, which should of course be almost completely flat; in higher-resolution structures it is indeed flat. (From J.C. Kendrew et al., *Nature* 181:662–666, 1958. With permission from Macmillan Publishers Ltd.)

1.1 STRUCTURES OF AMINO ACIDS AND PEPTIDES

1.1.1 Proteins are composed of amino acids

There are 20 common amino acids coded for by DNA and translated into proteins from mRNA on ribosomes, as listed in **Table 1.1**. These are all **L-amino acids (*1.1)**. In addition, selenocysteine is coded for by UGA, the umber codon, which is normally a termination codon; an extra nucleotide sequence slightly downstream in the mRNA directs the cell to insert selenocysteine here. Bacteria can also produce D-amino acids and unusual amino acids by using nonribosomal synthesis, which does not concern us here but is discussed further in Chapter 10. The amino acids are known both by their three-letter abbreviations and also by one-letter codes, which match the three-letter name where possible (see Table 1.1).

An amino acid consists of a carboxylic acid, which is attached to a carbon atom called the α-carbon because it is adjacent to the carboxylate. In turn, the α-carbon is attached to an amine (hence the name *amino acid*). In the smallest amino acid, glycine, this is all there is. In all the others, the α-carbon is attached to a β-carbon, which in turn is often attached to further atoms. These are given succeeding letters from the Greek alphabet: γ, δ, etc. The carbonyl, Cα and amine are called the **backbone (*1.3)**, the other atoms being the **side chain (*1.4)**.

TABLE 1.1 The 20 common amino acids plus selenocysteine

Name	Three-letter code	One-letter code[b]	Side-chain structure[c]	pK_a of side chain	Range of pK_a in proteins	Comments
Alanine	Ala	A	CH_3			Hydrophobic, small
Arginine	Arg	R	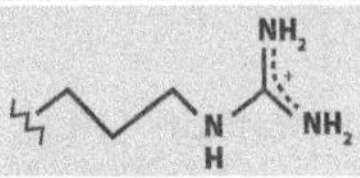	12.5		Hydrophobic in middle, basic at end
Asparagine[a]	Asn	N	CH_2-$CONH_2$			Polar
Aspartic acid[a]	Asp	D	CH_2-CO_2^-	3.9	2.0–6.7	Acidic
Cystine/cysteine	Cys	C	CH_2-S-; CH_2-SH	8.3	2.9–10.5	Hydrophobic Reduced (SH) is called cysteine; oxidized (S–S) is called cystine.
Phenylalanine	Phe	F	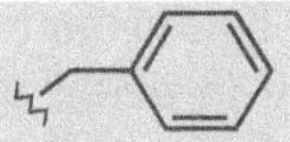			Hydrophobic, aromatic
Glutamine[a]	Gln	Q	CH_2-CH_2-$CONH_2$			Polar
Glutamic acid[a]	Glu	E	CH_2-CH_2-CO_2^-	3.2	2.0–6.7	Acidic
Glycine	Gly	G	H			Hydrophobic
Histidine	His	H	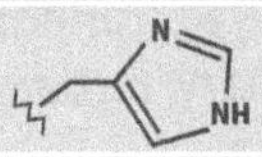	6.0	2.3–9.2	Basic, aromatic
Isoleucine	Ile	I	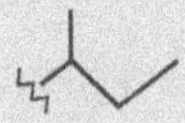			Hydrophobic
Leucine	Leu	L	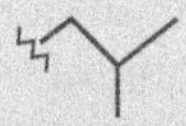			Hydrophobic
Lysine	Lys	K	CH_2-CH_2-CH_2-CH_2-NH_3^+	10.5	6.0	Basic
Methionine	Met	M	CH_2-CH_2-S-CH_3			Hydrophobic
Proline	Pro	P	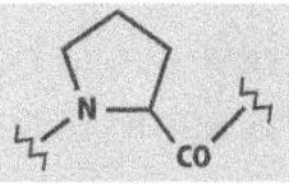			Hydrophobic and hydrophilic[d]
Serine	Ser	S	CH_2-OH	14.0		Polar
Threonine	Thr	T	CH(OH)-CH_3	15.0		Polar
Tryptophan	Trp	W	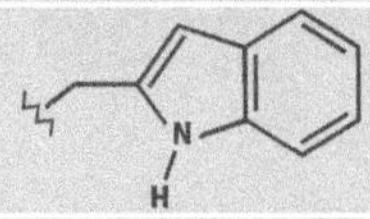			Hydrophobic, aromatic
Tyrosine	Tyr	Y	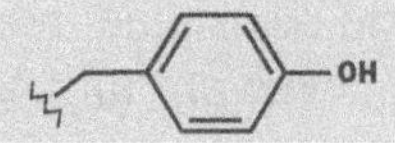	9.7	6.1	Aromatic
Valine	Val	V	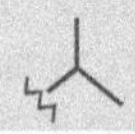			Hydrophobic
Selenocysteine	-	-	CH_2-SeH			Hydrophobic[e]

Amino acids have the common structure ^+H_3N–CH(R)–CO_2^-, where R is the **side chain** and the rest is the **backbone**. The table gives the structure of R.

[a]In addition, Asp and Asn are collectively called Asx with one-letter code B, and Glu and Gln are called Glx with code Z.

[b]The one-letter code for any of the 20 amino acids is usually X. The one-letter code matches the first letter of the amino acid where this is unique (C, H, I, M, S, V). Where more than one amino acid starts with the same letter, the code is assigned to the more common amino acid (A, G, L, P, T). The rest are phonetic where possible (Fenylalanine, asparagiNe, aRginine, Qtamine, tYrosine). Tryptophan has a double ring (double-u or W), and the others have a letter somewhere near the letter that the amino acid starts with (Asp D, Glu E, Lys K).

[c]The backbone CH carbon is the alpha carbon Cα, and its attached proton is Hα. The side-chain atoms are given succeeding letters from the Greek alphabet: β (beta), γ (gamma), δ (delta), ϵ (epsilon), ζ (zeta), η (eta). In computer files such as coordinate files, these labels are given in capital letters: A, B, G, D, E, Z, H. Where there is more than one heavy atom the same distance out from Cα, they are numbered 1 and 2; so for example the two methyl groups of a leucine are called Cδ1 and Cδ2. The dihedral angles along the side chain are called χ_1 (chi-1, pronounced kai, the angle formed by the four atoms N, Cα, Cβ, and Cγ), χ_2, and so on.

[d]The entire amino acid is drawn here. Strictly, proline is not an amino acid but an imino acid because it has an NH group, not an NH_2 group. As discussed in Chapter 4, the ring is hydrophobic but the main chain is unusually hydrophilic, making polyproline, for example, soluble in water.

[e]Selenocysteine is not normally counted as one of the standard amino acids (see the text).

*1.1 L-Amino acid

An L-amino acid is an α-amino acid with L chirality at the α carbon (**Figure 1.1.1**). The prefix L stands for levo and means that the related compound L-glyceraldehyde rotates polarized light to the left.

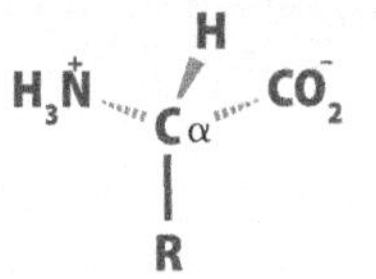

FIGURE 1.1.1
The Cα carbon of amino acids is chiral. This figure shows an L-amino acid.

A D-amino acid (**Figure 1.1.2**) has the opposite **chirality (*1.2)**: D-glyceraldehyde rotates polarized light to the right (dextro).

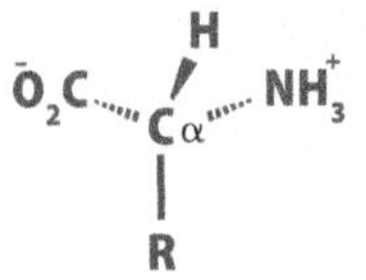

FIGURE 1.1.2
A D-amino acid.

*1.2 Chirality

Any molecule whose reflection in a mirror cannot be superimposed is asymmetric or *chiral*. The two mirror images are called *enantiomers*, or more generally but less specifically *isomers*. Their physical and chemical properties are identical, except that one rotates plane-polarized light to the left and the other rotates it to the right. The most common origin of chirality is carbon atoms that have four nonidentical groups attached to them: for example Cα carbons in amino acids (see Figure 1.1.1). [The exception is glycine, which is not chiral because the Cα has two hydrogens attached and is therefore symmetrical.] The two enantiomers are called L and D. The formal definition of L is as follows: view the Cα with the Hα toward you. If C=O, side chain, N go in a clockwise direction, the amino acid is L, whereas if they are anticlockwise it is D. This nomenclature is related to the organic chemistry (Cahn–Ingold–Prelog) definitions of *S* and *R*: all L-amino acids except cystine are also *S*.

The 20 amino acids are conveniently divided into groups. Four (Asp, Glu, Arg, and Lys) carry a charge at neutral pH: two are positive (basic: Arg and Lys) and two negative (acidic: Asp and Glu). Seven are **hydrophobic** (eight if we include glycine), and the remaining eight have polar groups. Of these, histidine is noteworthy because its pK_a is close to 7. Therefore in a protein at neutral pH it can be either protonated or not, depending on its local environment. Cysteine is also "special" because the side chain is easily oxidized to form the S–S disulfide form, where it is known as cystine. In an extracellular environment, including in the blood, cysteine is usually oxidized to cystine. However, the intracellular environment is normally sufficiently reducing that the dominant form is cysteine. Therefore one commonly finds extracellular proteins that are stabilized by disulfide bridges, whereas disulfides are not usually found in intracellular proteins. (In intracellular proteins, a similar stabilizing role is played by zinc, which binds to a combination of four cysteine or histidine side chains, forming a variety of "zinc finger" structures.) Cysteine also has a fairly low pK_a, making it a good nucleophile (*5.7). It is therefore often found in enzyme active sites.

*1.3 Backbone

This is generally taken to mean the N, Cα and carbonyl CO groups in a protein (**Figure 1.3.1**).

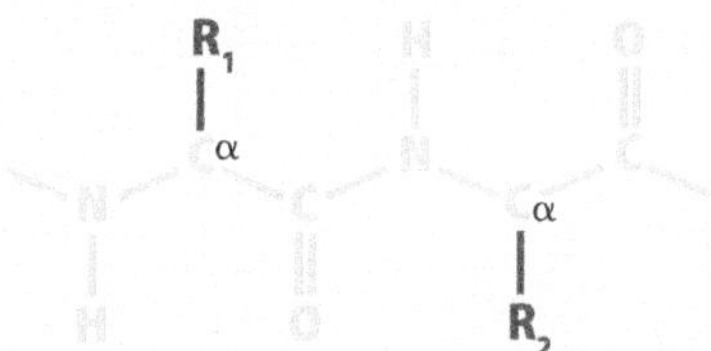

FIGURE 1.3.1
A protein backbone (green).

*1.4 Side chain

The side chain is those parts of a protein that are not the backbone (**Figure 1.4.1**). Each amino acid except glycine has a side chain (Table 1.1).

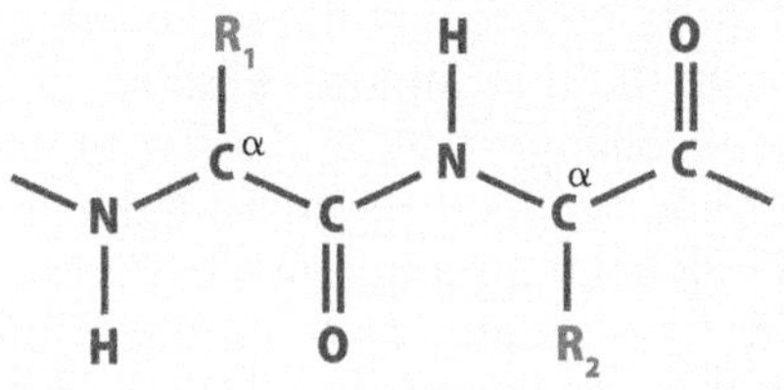

FIGURE 1.4.1
Protein side chains.

TABLE 1.2 Frequency of occurrence of amino acids in proteins

Amino acid	Frequency in intracellular proteins (%)	Frequency in membrane proteins (%)	Number of codons
Ala	7.9	8.1	4
Arg	4.9	4.6	6
Asp	5.5	3.8	2
Asn	4.0	3.7	2
Cys	1.9	2.0	2
Glu	7.1	4.6	2
Gln	4.4	3.1	2
Gly	7.1	7.0	4
His	2.1	2.0	2
Ile	5.2	6.7	3
Leu	8.6	11.0	6
Lys	6.7	4.4	2
Met	2.4	2.8	1
Phe	3.9	5.6	2
Pro	5.3	4.7	4
Ser	6.6	7.3	6
Thr	5.3	5.6	4
Trp	1.2	1.8	1
Tyr	3.1	3.3	2
Val	6.8	7.7	4

(Data taken from J. Cedano et al., *J. Mol. Biol.* 266:594–600, 1997.)

The 20 amino acids differ in their physical and chemical properties, and also in the metabolic cost of making and degrading them. This means that they occur in very different proportions in different proteins (**Table 1.2**). Table 1.2 illustrates that they also are coded for by different numbers of codons, these two facts being related by the very ancient mechanisms that led to just these 20 amino acids being selected as the ones coded for by DNA.

1.1.2 Amino acids have only a few allowed conformations

In proteins, the amino acids are linked together by **peptide bonds** (**Figure 1.2**) and are always drawn with the amino-terminal or N-terminal residue on the left and the carboxy-terminal or C-terminal residue on the right. This is also the order in which they are assembled on the ribosome (but not in a chemical synthesizer, where they are made starting at the C terminus). A short series of amino acids is usually called a peptide or oligopeptide, and a longer one is called a protein. There is no set dividing line between a peptide and a protein, the division being partly structural (a globular structure makes it a protein not a peptide) and partly length (anything short enough to be chemically synthesized is usually a peptide), implying that any peptide longer than about

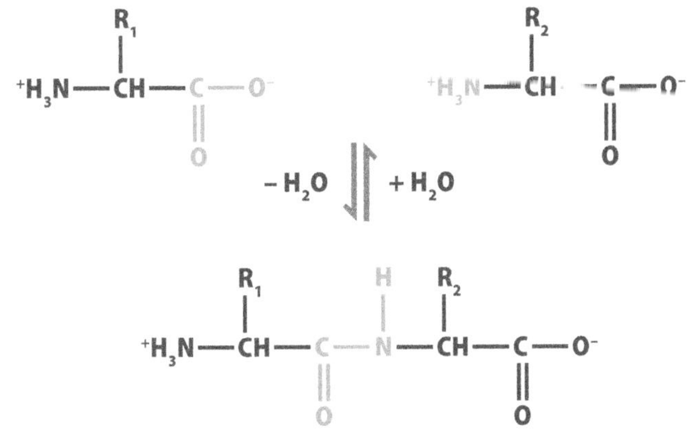

FIGURE 1.2
Formation of a peptide bond. The free amino acids have charges at each end, whereas the peptide bond (shaded part) has no charge. Formation of the bond involves the loss of a water molecule, which is called *condensation*. Both making and breaking peptide bonds are extremely slow reactions under normal conditions unless catalyzed by an enzyme.

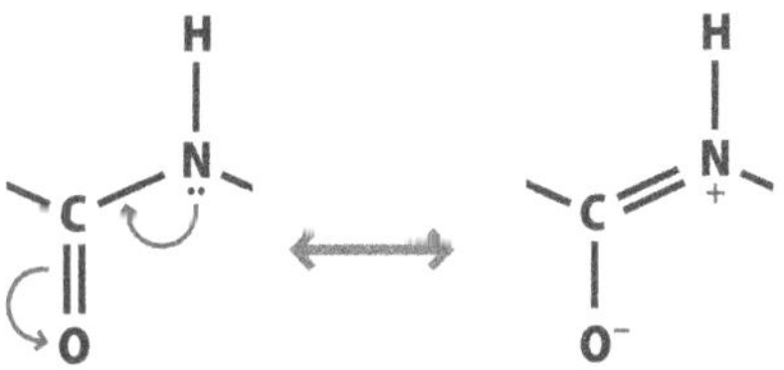

FIGURE 1.3
A peptide bond has partial double bond character. This means that rotation around the peptide (C–N) bond is slow, and that the C, O, N, and H atoms are all in the same plane, as are the Cα atoms at either side. It also tends to make the peptide nitrogen slightly positively charged, and the carbonyl oxygen slightly negatively charged. The double-headed arrow used here has a special meaning in chemistry. It means that the movement of electrons shown here by the curly arrows (*5.10) is not a *reaction* but a phenomenon called *resonance*. That is, the electrons are (on average) somewhere between the two extremes shown here; therefore the C–N bond is somewhere between a single bond and a double bond, and the oxygen is a bit negative while the nitrogen is a bit positive.

40 amino acids is usually called a protein. Amino acids joined together into a polypeptide tend to be called residues.

Of the hydrophobic amino acid residues, proline is unusual. It does not have an NH proton (making it in fact an imino acid rather than an amino acid), which means that it does not fit well into standard regular secondary structure motifs because it cannot form the hydrogen bonds that are normally present. Although the side chain is indeed hydrophobic, the carbonyl oxygen carries a higher partial charge than in other amino acids because the side chain pushes more electron density into it than it can in normal amino acids. This gives proline a hydrophilic character, and proline-containing peptides are well hydrated and soluble in water. This makes proline an important and interesting amino acid, as we shall see in Chapter 4.

The peptide bond has partial double bond character (**Figure 1.3**) and is flat. It is *trans* (the angle ω in **Figure 1.4** is 180°) 99.97% of the time, *except* that the amide bond preceding a proline has a relatively weak preference for *trans* and is found *cis* ($\omega = 0$) roughly 5% of the time in proteins [3] (**Figure 1.5**). This is because the proline ring is uncomfortably close to the side chain of the preceding amino acid in a *trans* amide, destabilizing it. This makes proline an interesting amino acid, because it is almost unique in being able to adopt *cis* conformations.

The rigidity of the amide bond means that there are only two freely rotating single bonds per amino acid in a peptide backbone, which are characterized by the **dihedral angles (*1.5)** ϕ and ψ, pronounced phi and psi respectively (see Figure 1.4). In turn, this means that one can very conveniently represent an amino acid residue's backbone geometry as a single point on a (ϕ, ψ)

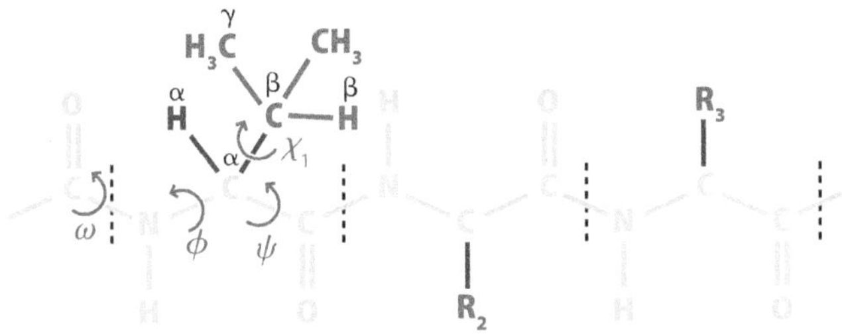

FIGURE 1.4
Part of a protein chain. Dashed lines separate one amino acid from another. The first amino acid in this diagram is a valine. The backbone atoms are shaded in green. The dihedral angles (*1.5) along the backbone are called ϕ (phi, the angle between N and Cα), ψ (psi, the angle between Cα and the carbonyl carbon C′) and ω (omega, the angle between C′ and N). The ω angle is usually 180°, making the peptide bond flat. Side-chain atoms, shaded red, are named by successive letters of the Greek alphabet: β, γ, δ (beta, gamma, delta), and so on, and the dihedral angles along the side chain are called χ_1, χ_2 (chi-1, chi-2), and so on. Side chains are often shown simply with the letter R.

*1.5 Dihedral angle

A dihedral angle is the angle defined by four atoms, as shown in **Figure 1.5.1**. The ψ and ϕ angles along a protein backbone are dihedral angles, as are the side-chain χ angles. The figure also indicates that a dihedral angle has a sign, defined by convention. Dihedral angles adopt preferred positions because of steric interactions between substituents at the two ends.

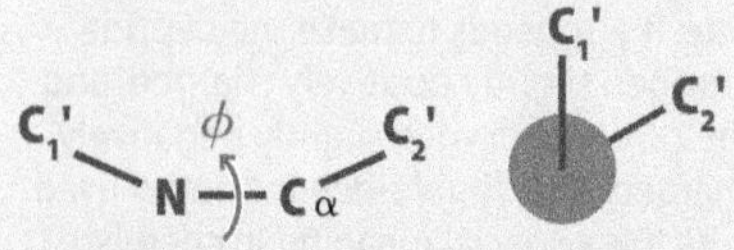

FIGURE 1.5.1
A dihedral angle is the angle defined by four atoms. For example, the backbone dihedral angle ϕ is defined by the four sequential atoms C′, N, Cα, C′ (or equivalently the angle between the planes defined by the first three and the last three atoms). Looking down the N–Cα bond, if the two carbonyl carbons are eclipsed (that is, superimposed), ϕ is zero. If the back one is rotated in a clockwise direction, as shown here, ϕ is positive: here it is about +60°.

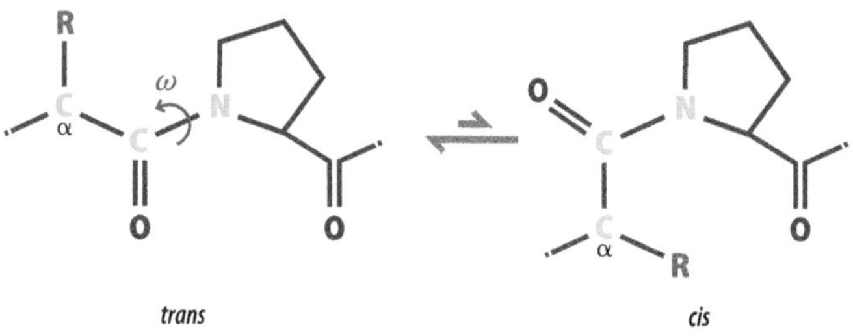

trans *cis*

FIGURE 1.5
The peptide bond preceding a proline (the ω angle) is *cis* approximately 5% of the time, because of steric interactions between the proline ring and the side chain (R) of the preceding amino acid. The *cis/trans* isomerization is slow and results in a change in the direction of the peptide backbone.

surface, and a whole protein as a set of points on the surface, often called a **Ramachandran plot** (**Figure 1.6**) [4]. Amino acid conformations are grouped into two main clusters in this plot, as will be discussed shortly. These derive from steric interactions between backbone and side-chain atoms, which mean that certain combinations of ϕ and ψ are very high in energy because they result in a steric clash.

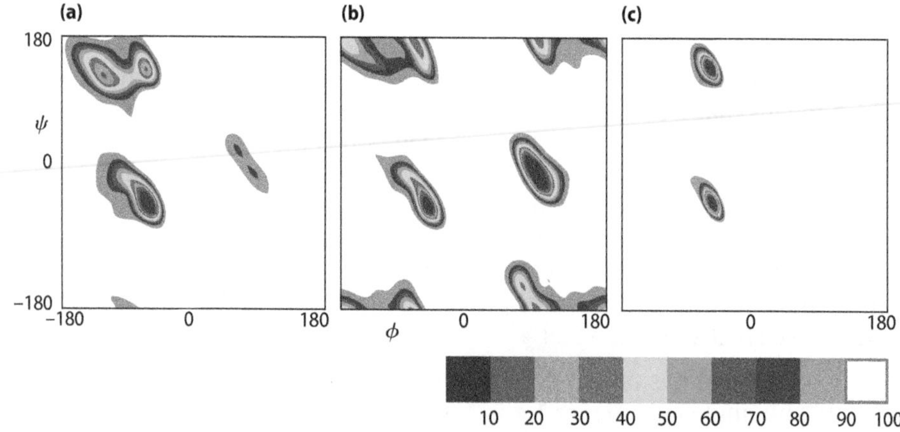

FIGURE 1.6
Examples of Ramachandran plots, showing the energies of different combinations of ϕ and ψ backbone dihedral angles. Areas are color-coded to show the percentage of the total in each area, as shown in the key. (a) Energetically allowed regions for all amino acids. There are two main allowed regions. The lower one, with a minimum at around (ϕ, ψ) = (−64, −41), contains about 40% of all amino acids and is characteristic of α-helical conformation and helical turns. "True" α helices are very close to the energy minimum, but other helical residues have a wide spread in angles. The upper region is characteristic of β-sheet and extended conformations. Residues in a β sheet very largely occupy the left-hand lobe centered at (ϕ, ψ) = (−121, +128), whereas extended regions and polyproline II conformation (PPII) occupy the right-hand lobe (ϕ, ψ) = (−66, +137). The small region centered on about (ϕ, ψ) = (+60, +60) is often called the α_L or left-handed helix region, but in fact it contains almost exclusively residues in γ turns and so is better called the turn region. This region is relatively much less stable than the other two and is therefore populated by only a small number of amino acids. It is noteworthy that this plot is based on experimental data from protein crystal structures and has a rather different shape from many plots found in textbooks based on the original steric calculations of Ramachandran. (b) Energetically allowed regions for glycine. Because glycine has no side chain, it has far fewer steric clashes with neighboring amino acids and can therefore take up a much wider range of conformations. (c) Energetically allowed regions for proline. The side chain is cyclized back to the backbone amide, which restricts the ϕ angle to about −65°. [These figures are taken from data presented in S. Hovmöller, T. Zhou and T. Ohlson, *Acta Cryst. D* 58:768–776, 2002; courtesy of Professor S. Hovmöller (Stockholm University). It should be added that the original paper describing the use of (ϕ, ϕ) plots was written by Sasisekharan, Ramakrishnan and Ramachandran, implying that it should really be called the Sasisekharan–Ramakrishnan–Ramachandran plot.]

***1.6 Ramachandran**

G.N. Ramachandran (**Figure 1.6.1**) was one of India's most distinguished scientists, who unusually chose to work for all of his career in India, except for his (second) PhD in Cambridge, UK. Apart from the Ramachandran plot he also worked out the triple-helical structure of collagen and developed algorithms for back projection in X-ray tomography. He founded two biophysics research centers and was also interested in traditional Indian music, philosophy, and poetry.

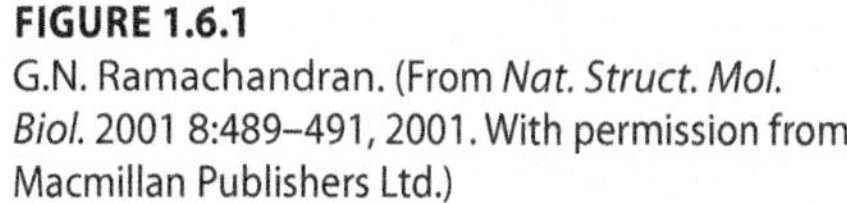

FIGURE 1.6.1
G.N. Ramachandran. (From *Nat. Struct. Mol. Biol.* 2001 8:489–491, 2001. With permission from Macmillan Publishers Ltd.)

It should be noted that this very useful representation relies on ω being 180°. In fact in real residues it typically differs from 180° by 5° or so, meaning that the **Ramachandran (*1.6)** plot should be thought of as having slightly movable angles, depending on the exact value of ω. The Ramachandran plots for all amino acids are approximately the same, except for two. *Glycine* can access a much larger range of backbone dihedral angles because it lacks a side chain (Figure 1.6b), and *proline* can only access ϕ angles of about −65° (Figure 1.6c) because of its cyclized side chain. For the other amino acid residues, propensity to be in the α-helical or β-sheet region varies by up to a factor of 3, and correlates fairly well with their actual occurrence in helix or sheet [5], implying that the natural conformational preferences of the unfolded peptide have a fairly large role in determining their conformations in the folded protein.

The planarity of the peptide bonds means that one can consider the structure of a protein backbone as consisting of a series of planes, each plane being defined by the atoms in the peptide bond together with the Cα atoms on each side (**Figure 1.7**), with the hinge at the Cα atoms. One can therefore represent a protein rather effectively using just the Cα atoms (**Figure 1.8**) (***1.7**).

Amino acid side chains also have preferred conformations. The geometry around a saturated carbon atom is tetrahedral, implying that if one views an amino acid side chain (which is usually made up of saturated carbons) along a carbon-carbon bond, the substituents point in the directions of an equilateral triangle, 120° apart (**Figure 1.9a**). Side chains prefer to have all their atoms as far apart as possible to minimize steric overlap. This implies that neighboring carbons tend to be in a "staggered" conformation (Figure 1.9b). There are three possible staggered rotamers for each side-chain dihedral angle, differing by 120°. Analysis of high-resolution protein crystal structures shows that side chains do tend to be very close to these angles [6]. Often, one of the staggered rotamers is preferred because it is the best way of keeping all the big atoms as far apart as possible. This means, for example, that long side chains such as lysine or methionine have a preference for being extended in a zigzag conformation (Figure 1.9c). However, the energy differences between different orientations are small, meaning that there are many instances of side chains adopting nonstaggered positions, particularly where protein function requires a particular orientation. It should also be pointed out that in lower-resolution crystal structures, the orientations of side chains on the surface are often disordered. Therefore the side-chain structures of surface-exposed side chains in such structures are determined by modeling rather than experimentally.

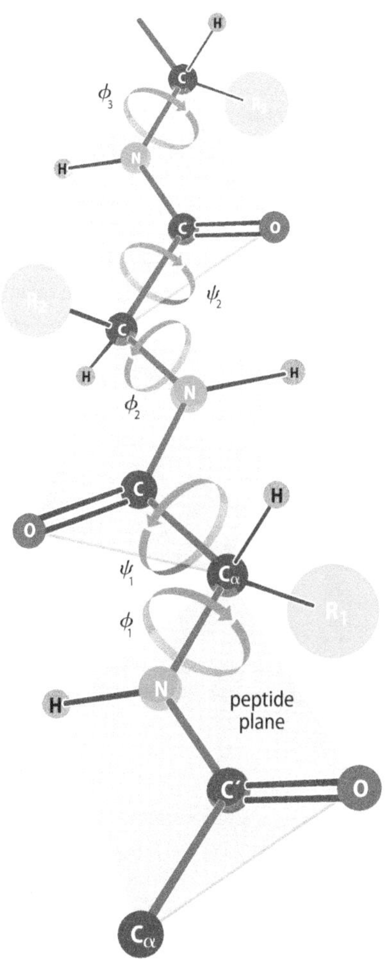

FIGURE 1.7
A protein backbone structure can be thought of as a series of planes, hinged at Cα. (From C. Branden and J. Tooze, Introduction to Protein Structure, 2nd ed. New York: Garland Science, 1999.)

*1.7 Representation of protein structures

This book seeks to clarify the principles by which proteins function, without getting too bogged down in the details. The drawings of protein structures are thus simplified as much as possible, leaving them often as a single round blob. The example of adenylate kinase in Chapter 2 is a good example. **Figure 1.7.1** represents the enzyme in a very simplified manner and implies that the conformations of the three domains are unchanged between free and bound conformations, the only changes being in the loops connecting them. In fact, one of the domains changes its conformation considerably, as can be seen from the more detailed structure in **Figure 1.7.2.** I suggest, however, that the simplified diagram of Figure 1.7.1 is more helpful, even if less "accurate."

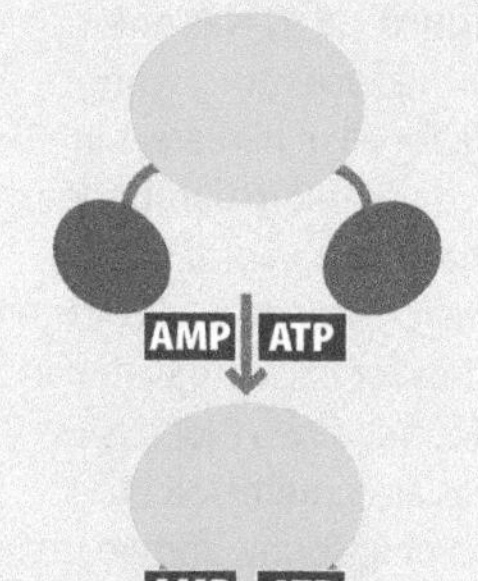

FIGURE 1.7.1 Domain motions in adenylate kinase on binding to substrate, shown in schematic form.

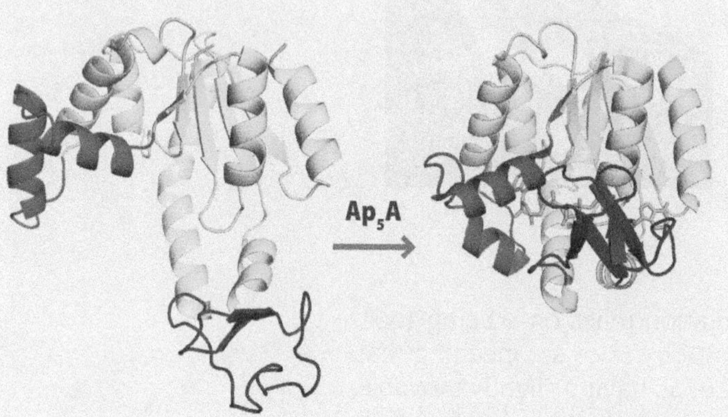

FIGURE 1.7.2 (left) The detailed structures of adenylate kinase, free and bound to P^1,P^5-bis(adenosine-5′) pentaphosphate, Ap_5A, a bisubstrate analog. The same color schemes are used as in Figure 1.7.1: the nucleotide monophosphate-binding domain is shown in purple, the nucleotide triphosphate-binding domain in green, and the lid in brown. Ap_5A is in cyan. (The PDB structures are 4ake and 1ake, and the domain boundaries are as used in C.W. Müller et al. *Structure* 4:147–156, 1996.)

We are used to seeing protein structures as secondary structure cartoons. These are very helpful in guiding our viewing of what are very complex objects. They do, however, have some problems, one of which is that they imply that the loops connecting the regular secondary structure elements are less significant, because these are drawn thinner. Indeed, many software packages, such as Pymol, which I am using for these structures, have an option to draw smooth curves through the loops, which makes them prettier but significantly distorts their appearance. Compare, for example, **Figure 1.7.3**a and Figure 1.7.3b. Both of these representations give the impression that there is lots of empty space in protein structures. They do this because they do not show the side chains. Including the side chains (Figure 1.7.3c) makes the diagram much messier and harder to interpret, but does at least show that proteins are densely packed. This can be seen even more clearly by showing the atoms with their true volumes (Figure 1.7.3d): such a figure comes closest to showing what a protein "actually looks like" but is not very helpful for understanding how it works. A surface representation (Figure 1.7.3e) can be useful for seeing how proteins interact with other molecules.

The best way to appreciate this is to sit down with a graphics program and experiment (see Problem 3 at the end of Chapter 1).

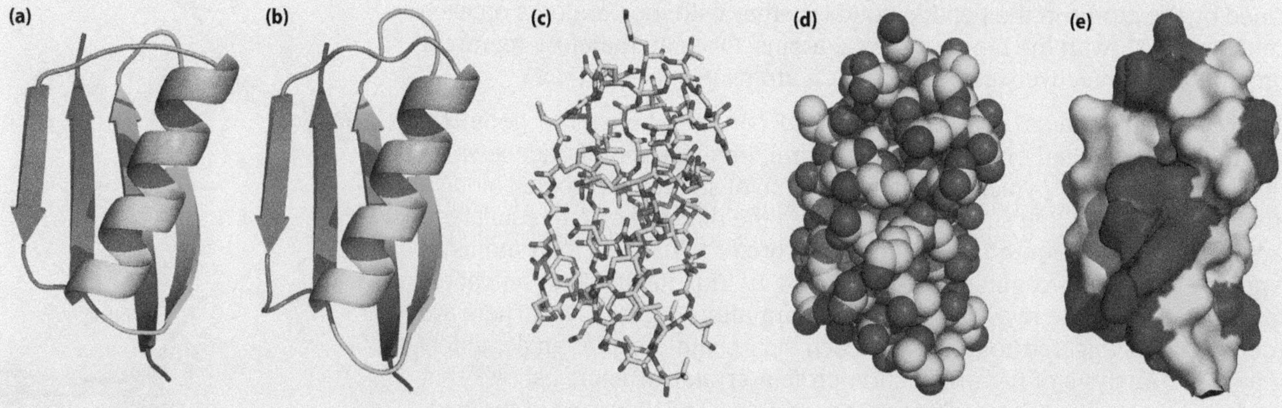

FIGURE 1.7.3 These five figures show the same protein in the same orientation but with different representations; it is the B1 domain of streptococcal protein G. (a) A standard representation of a protein structure, colored from blue at the N terminus to red at the C terminus. Note the marked twist in the β sheet. (b) The same except that the loops are shown closer to their true positions: they are longer and less smooth. (c) All the (non-hydrogen) atoms shown in a standard color scheme (carbon green, oxygen red, nitrogen blue). This is more of a 'true' representation of the protein structure but is rather less helpful in understanding the structure. One can see the side chains and just about make out for example the helical backbone, but it is not at all easy. (d) The atoms at their van der Waals radii. This is closer to a true structure, but it is the least helpful because it is almost impossible to work out how the protein is folded. It does, however, show a nice example of a salt bridge toward the bottom of the figure, between the aspartate in the bottom left corner (compare part (c), where it is slightly easier to see) and a neighboring lysine. (e) Finally, the surface in a Connolly or solvent-excluded surface. This is obtained by rolling a sphere the same size as a water molecule around the protein, and shows the volume not available to water. Acidic side chains (Asp and Glu) are shown in blue, and basic residues (Lys and Arg) in red. It therefore gives a feel of what the protein looks like when approached from a distance in water.

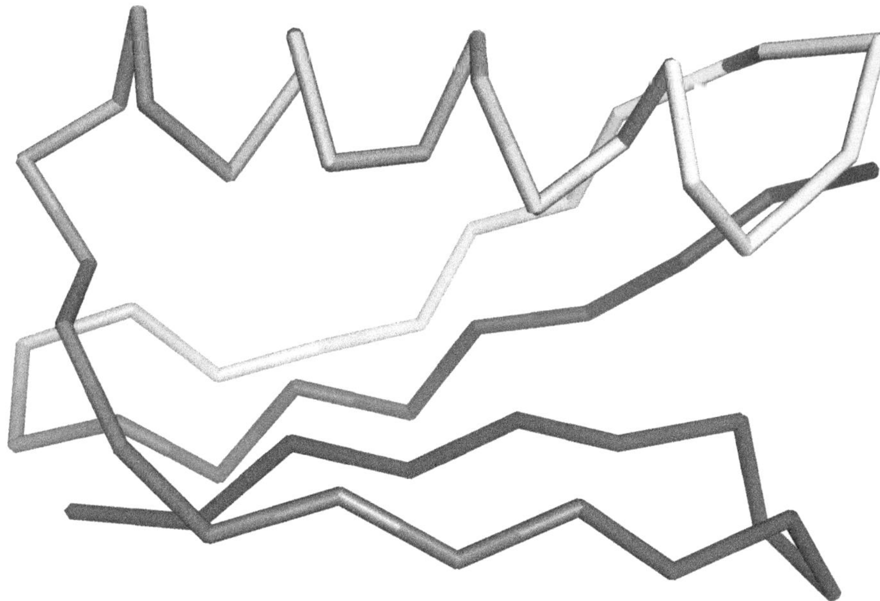

FIGURE 1.8
Because a protein can be represented by a series of planes (Figure 1.7), one can usefully represent a protein's three-dimensional structure simply using its Cα atoms, linked together by pseudo-bonds. The figure shows staphylococcal protein G, colored from blue at the N terminus to red at the C terminus.

1.1.3 The most populated conformation is in the β-sheet region

Because of steric limitations (that is, atoms that get too close together) there are only three broad regions that are energetically allowed, which was the original point of Ramachandran plots. The most populated of these is the β-sheet region, the region containing residues that are in extended conformations including β sheets. These sheets are formed from several single polypeptide chains or **β strands** that line up side by side. Neighboring strands can be either parallel or antiparallel (**Figure 1.10**). *Antiparallel* strands are more common: small β sheets are almost always antiparallel, and sheets composed only of parallel strands are very rare. Antiparallel sheets have a similar local geometry and energy to those of parallel sheets but they are easier to form simply by a single strand folding in the middle (**Figure 1.11**). Antiparallel sheets typically have a marked twist (**Figure 1.12**), which probably arises as a natural consequence of the way in which amino acids fold. The β-sheet region of

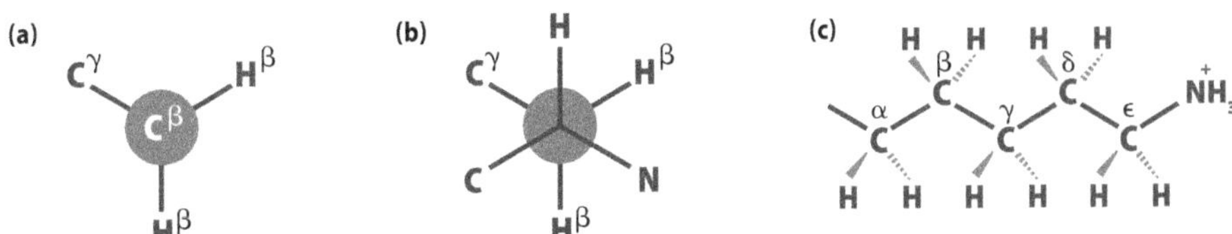

FIGURE 1.9
Staggered rotamers in an amino acid side chain. (a) A long side chain (for example Met or Lys) as viewed from Cα toward Cβ. (b) The same view, now including the Cα and its attached atoms (for an L-amino acid). The substituents on Cα and Cβ are in one of the three possible staggered rotamers. (c) A lysine side chain in its energetically preferred conformation, with all the carbons staggered and minimizing steric overlap. The carbon atoms are in the plane of the paper, and the hydrogens are in front and behind.

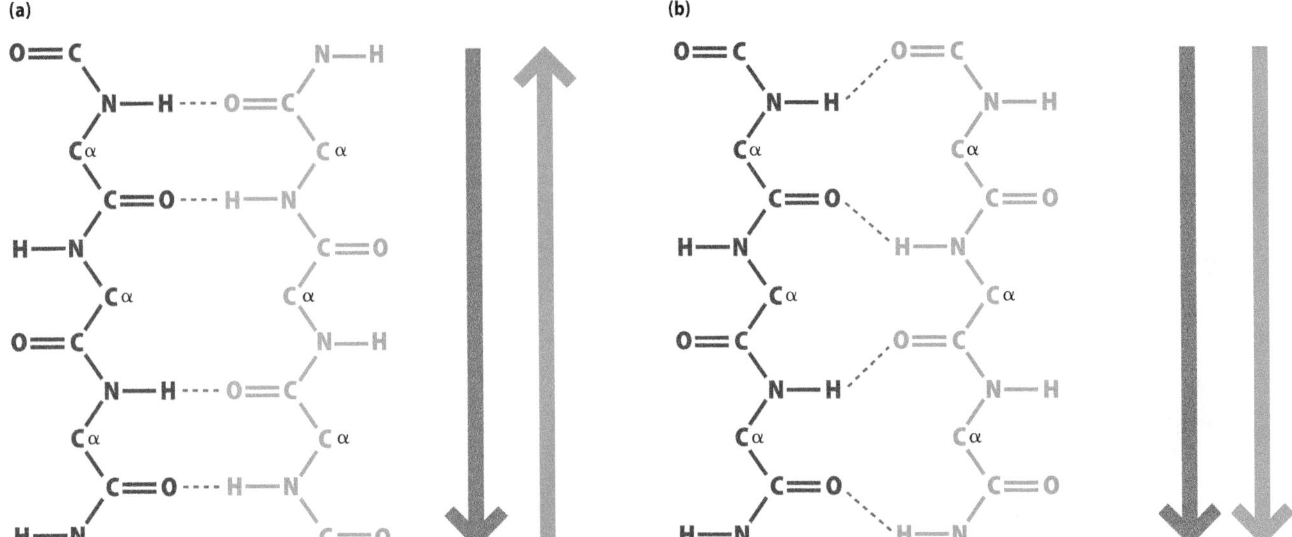

FIGURE 1.10
Two types of β sheet. (a) An antiparallel β sheet. Hydrogen bonds are marked by red dashes and run approximately perpendicular to the sheet direction. Sheets are often drawn simply as arrows. (b) A parallel β sheet. The hydrogen bonds are more asymmetrical.

the Ramachandran plot also contains residues in *polyproline II* helices (**Figure 1.13**) and extended strands. All of these structures have a similar length per amino acid and are close to being as extended as possible. It is not possible to draw fixed boundaries around these different conformation types: in real β sheets, individual residues can have conformations differing quite widely from the "standard" Ramachandran angles. Sometimes β sheets are described as *pleated* because the backbone atoms in a sheet form a zigzag arrangement (**Figure 1.14**).

1.1.4 The other main conformations are the α helix and the "random coil"

The other main populated region in the Ramachandran plot is the α-helical region. An α helix is right-handed (the same direction of twist as a normal screw, which needs rotating clockwise to screw in), and has 3.6 residues per turn (**Figure 1.15**). Because of the requirement for the carbonyl oxygen of residue *i* to hydrogen bond to the amide nitrogen of residue *i* + 4, the structure of an α helix is rather precisely fixed in its geometry, more so than a β sheet. The other striking feature of an α helix is that all the amide NH groups point toward the N-terminal end of the helix, whereas all the carbonyl groups point toward the C-terminal end. Amide N–H bonds are polarized, with the H positive and N negative; carbonyls are also polarized, with the C positive and O negative. This gives the α helix a macroscopic **dipole**: it tends to be positive at the N-terminal end and negative at the C-terminal end, as discussed further in Section 1.3.2.

Most α helices have characteristic patterns of amino acids at the N-terminal end, generally consisting of a hydrophilic amino acid such as Thr, Ser, Asp, or Asn, whose side chain makes a stabilizing hydrogen bond to the last "exposed"

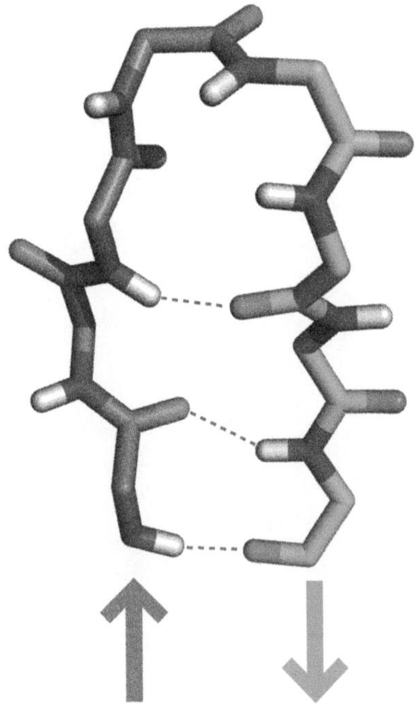

FIGURE 1.11
An antiparallel β sheet composing a β hairpin, in which the two strands form a β turn at the end. Atoms are denoted by gray or cyan = carbon, blue = nitrogen, red = oxygen, white = hydrogen. Only the amide hydrogens are shown. There are two major types of turn; this is a type I turn, which is roughly helical and is the most common type. In a type II turn, the peptide plane of the central peptide bond is rotated by about 180°. This creates a steric clash involving the side chain of the second residue in the turn, which is consequently nearly always glycine.

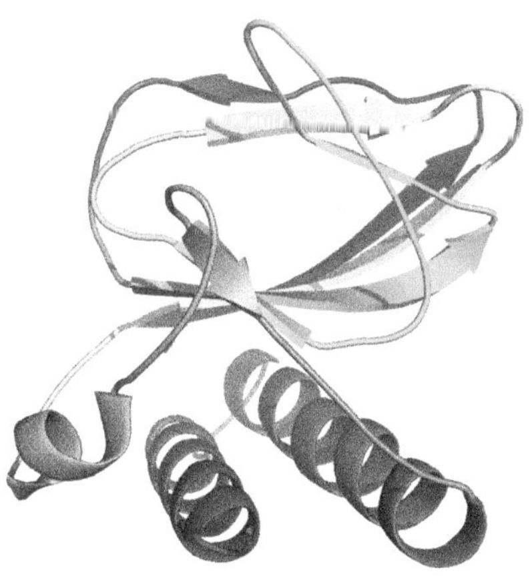

FIGURE 1.12
A twisted sheet formed from β strands. This elegant example is the arabinose-binding protein araC (PDB file 2ara).

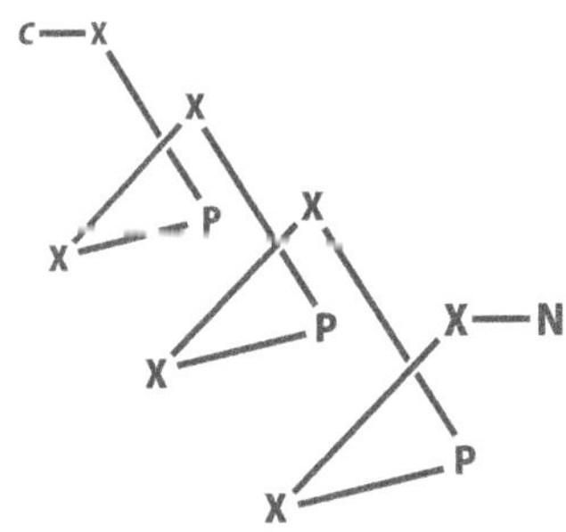

FIGURE 1.13
A schematic polyproline II helix, which has a 120° rotation from one amino acid to the next and therefore three amino acids per turn. In the sequence PXXP this puts the two prolines on the same face. This view is "end on": the helix is actually quite extended. A polyproline I helix involves repeating *cis* prolines and in practice is therefore never found in proteins.

amide group before the helix starts. This is called the N-cap and helps to define where a helix starts. There is not such a clear signal for the C-terminal end of a helix, which in general seems less well defined [7].

It is theoretically possible for amino acids to form a left-handed helix, but this structure is less stable: left-handed helices are not observed in proteins, although individual residues do adopt this conformation sometimes, particularly in turns. An unstructured peptide, such as an unfolded or natively unstructured protein, generally adopts backbone conformations that populate the (ϕ, ψ) surface according to the distribution shown in Figure 1.6a. (Strictly speaking, each amino acid has its own individual (ϕ, ψ) distribution because of the structure of its side chain; in detail this makes the random coil structure sequence-specific [8].) Thus, the predominant conformation is in the extended or β-sheet region, whereas residues occasionally adopt the α-helical conformation and very rarely the left-handed helix conformation. Therefore unfolded or **random-coil** proteins are essentially extended.

Proteins contain a mixture of hydrophobic and **hydrophilic**/polar groups. As we shall see below, the key determinant of their structure is that the

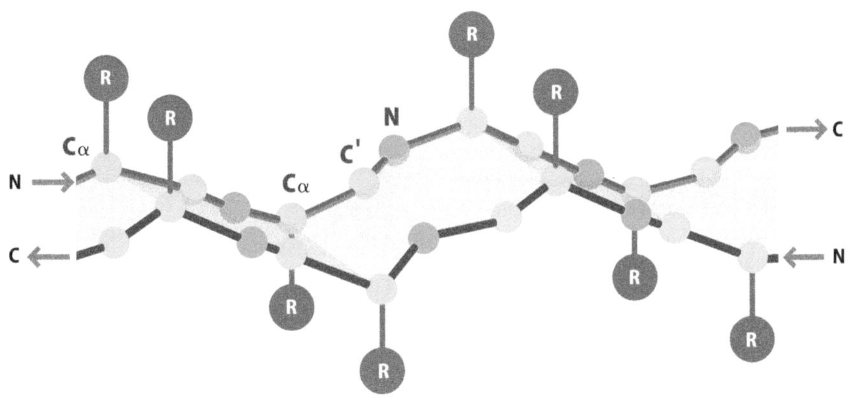

FIGURE 1.14
An antiparallel β sheet emphasizing the "pleat," which gives a zigzag structure to the sheet. Note that the side chains alternate up and down, and that side chains on adjacent strands are close together. (From C. Branden and J. Tooze, Introduction to Protein Structure, 2nd ed. New York: Garland Science, 1999.)

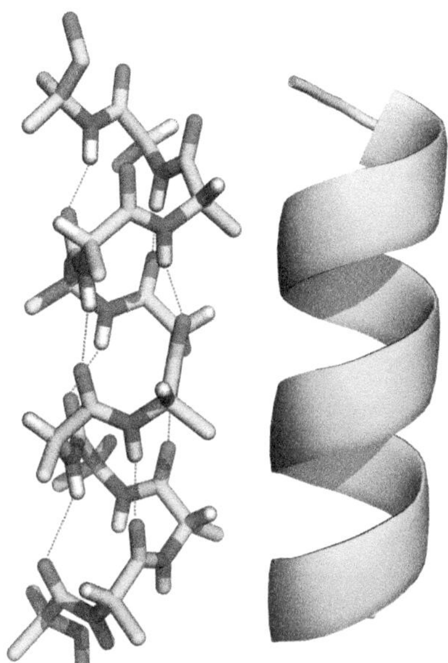

FIGURE 1.15
An α helix. Note that all N–H bonds point roughly back down the helix, whereas all C=O bonds point up the helix: this is what gives the helix its dipole moment. It is also striking how nonlinear many of the hydrogen bonds are. Side chains are omitted for clarity.

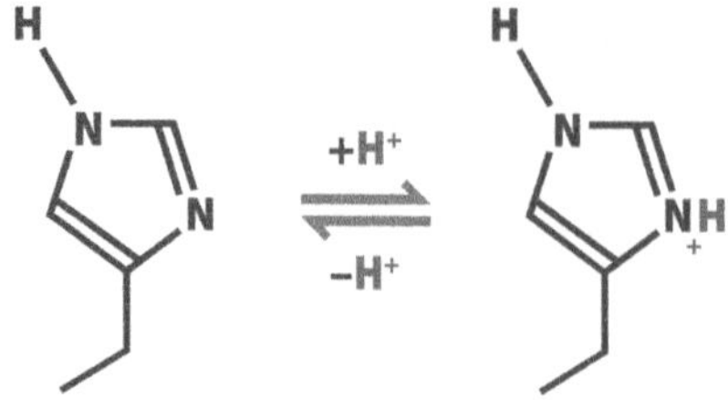

FIGURE 1.16
The protonation equilibrium for histidine. At the pH corresponding to histidine's pK_a it has 50% in each form.

hydrophobic groups should associate together, and charged and polar groups should be able to satisfy their hydrogen bonding requirements. In particular, because every amino acid has a polar peptide bond it is important that the peptide bonds should be able to form hydrogen bonds, either to each other or to water. The regular secondary structures formed by proteins are determined by the need to find structures within the allowed regions of the Ramachandran plot that also allow the peptide groups in the backbone to hydrogen bond together. This was the basis of the original proposal by **Linus Pauling (*1.8)** that proteins should fold up into the two regular structures that he called α and β: these are the only two regular structures in which backbone hydrogen bonds involving all backbone amides and carbonyls are possible.

1.1.5 The pK_a value describes the protonation behavior of side chains

The pK_a is the pH value at which a group is 50% protonated and 50% deprotonated. Thus for example, an exposed histidine side chain has a pK_a of 6.0 (see Table 1.1). This means that at pH 6.0, it is 50% neutral His and 50% protonated His.H^+ (**Figure 1.16**). At other pH values, the proportions of neutral and protonated species are easily calculated by the **Henderson–Hasselbalch equation**:

$$pH = pK_a + \log([\text{base}]/[\text{acid}])$$

(where the neutral species is the base and the protonated species is the acid), implying that at pH 7.0 it is about 90% neutral and only 10% protonated.

***1.8 Linus Pauling**

Linus Pauling (**Figure 1.8.1**) developed several new fields of chemistry. He worked in quantum chemistry, describing the nature of chemical bonds, the tetravalency of carbon, and bond hybridization; developed the concept of electronegativity; correctly described the nature of aromaticity and resonance; studied peptide structure and proposed two new potential structures for peptides, the α helix and β sheet, well before any crystal structures were obtained; studied enzyme catalysis and was the first to suggest clearly that an enzyme works by stabilizing the transition state, and also suggested that enzymes could be studied by looking at transition-state analogs; and explained the molecular basis of sickle-cell anemia, and thereby started the study of genetic diseases at a molecular level. He is one of a very select group of people to have been awarded two Nobel Prizes (the others being Marie Curie, John Bardeen, and Fred Sanger): one for his work on the chemical bond, and the other for peace (he led a campaign against nuclear testing and nuclear weapons, one of the consequences being that the McCarthy-influenced Senate Internal Security Subcommittee called him "the number one scientific name in virtually every major activity of the Communist peace offensive in this country"). The magazine *New Scientist* included him among the 20 greatest scientists of all time. He also campaigned vigorously and controversially for large doses of vitamin C as a health benefit (and died at the age of 93, so maybe there is some truth in it).

FIGURE 1.8.1
Linus Pauling. (Courtesy of US Library of Congress, public domain.)

However, in a real protein environment, the pK_a can be altered by several pH units from those given in Table 1.1. This makes histidine a very useful amino acid side chain in any situations where protons need to be provided or removed. Proteins can alter the pK_a of side chains in two main ways. First, if a side chain is in a hydrophobic environment, it will prefer to be uncharged. This raises the pK_a of acidic side chains (that is, a lower proton concentration is required to protonate them) and lowers the pK_a of basic side chains. Second, putting an acidic side chain close to a positively charged group will make it prefer to be deprotonated (that is, it will lower the pK_a), whereas putting it close to a negatively charged group will increase the pK_a.

Free Asp, Glu, Lys, and Arg have pK_a values of 3.9, 4.1, 10.8, and 12.5 respectively, which are so far from 7 that only one form is expected at pH 7. In proteins, their pK_a values can be shifted to pH 7, but this is less common. Because the large majority of catalytic reactions involve proton transfer, histidine is very commonly found in enzyme active sites, and it is indeed the most common catalytic side chain [9]. However, Asp, Glu, Cys, Lys, and Tyr can all have their pK_a altered to be close to 7 in enzyme active sites if necessary (see Table 1.1).

The overall charge on a protein is just the sum of all the individual charges, and it can be estimated fairly well from the amino acid composition, on the basis of their expected pK_a values. There are many websites that perform such calculations, including some listed at the end of the chapter (for example the NCBI site). The pH at which a protein has zero net charge is called the p*I*; this is a useful thing to know because solubility is lowest at the p*I*, and it dictates how the protein is to be purified by ion-exchange chromatography or separated by two-dimensional SDS-PAGE gels.

Free amino acids have charges on both the backbone amine and carboxylate (*1.1). These are positive and negative respectively, meaning that free amino acids are **zwitterionic**. Polypeptides do not have such charges along the backbone, except that they retain a positively charged amine at the N terminus and a carboxylate at the C terminus (see Figure 1.2).

1.2 THE FORCES HOLDING PROTEINS TOGETHER

By far the strongest bonds in proteins are the covalent bonds, which have strengths of about 200 kJ mol^{-1}. These do not get broken during the lifetime of a protein. The only exception is the disulfide bond, which can be broken by reduction. In extracellular proteins, disulfide bonds clearly do stabilize proteins markedly: for example, if the four disulfide bonds in ribonuclease A are broken, the protein unfolds completely. By contrast, the noncovalent forces discussed below are only a few kJ mol^{-1} each, after one takes into account how water weakens many of them. The stability of proteins under physiological conditions is about 20–60 kJ mol^{-1}, as discussed in Chapter 2, which is much less than the strength of a single covalent bond and is only equal to a small number of noncovalent bonds. The stability of a protein is a delicate balance between large total energies.

1.2.1 Electrostatic forces can be strong

The forces between molecules are all essentially charge–charge interactions. The most obvious of these is the force between permanently charged ions; that is, a coulombic or *electrostatic force* such as that responsible for a **salt bridge**. In water, charges are well solvated by water or by ions (**Figure 1.17**), and thus one very rarely encounters a genuine "naked" charge except possibly in metalloproteins, in which buried metals are quite common. On the surface of a protein, charges are effectively weakened by their interactions

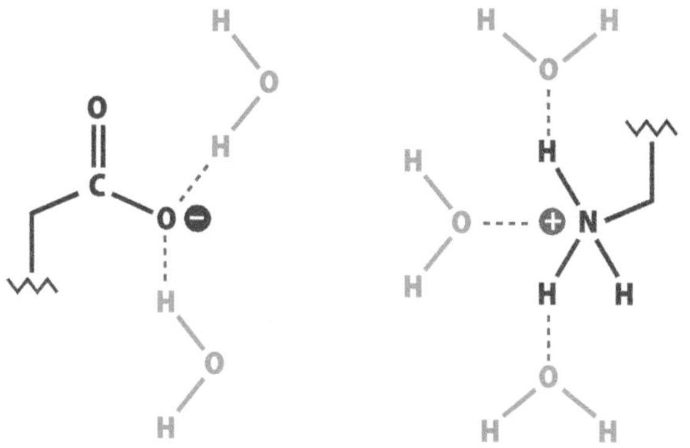

FIGURE 1.17
In water, charged groups are solvated by the water molecules, which act to shield charges from each other.

with water. Nevertheless, they are the only force that is capable of acting over a significant distance. A force between two charged ions only falls off with distance as r^{-2} and can therefore extend over a significant distance. It is, however, weakened dramatically by the large dipole moment of water, which acts to screen the charges from each other and makes it much less significant than one might expect. As discussed in Chapter 4, electrostatic forces are important for attracting and orienting molecules to each other in solution.

Many atoms in proteins have partial charges. Electronegative atoms such as N and O have a partial negative charge, whereas electropositive atoms such as amide protons and carbonyl carbons have a partial positive charge. The typical charges calculated for N, O, C, and H in a peptide bond are respectively about –0.4, –0.5, +0.5, and +0.2, so the partial charges are far from small. These charges also attract and repel. The forces are weaker because the charges are smaller, but because there tend to be many more of these atoms than there are of atoms carrying a full charge, the cumulative effect of these forces can be very significant. It is remarkable that for example the sulfate ion in the bacterial sulfate-binding protein, which has two negative charges, is bound entirely inside the hydrophobic interior of the protein, and stabilized mainly by hydrogen bonds to the partial charges on backbone atoms [10].

1.2.2 Hydrogen bonds are formed by electrostatic dipoles

Hydrogen bonds are formed from two electronegative atoms with a hydrogen in between, bonded to one of them. They are usually regarded as being electrostatic in character, with the N–H and C=O dipoles forming a strong interaction (**Figure 1.18**). As noted in Chapter 4, the energy of a peptide–peptide hydrogen bond is about 15–20 kJ mol^{-1}, but so is the energy for a peptide–water hydrogen bond. Therefore, although a peptide group that is not hydrogen bonded at all is genuinely unfavorable by 15–20 kJ mol^{-1}, the net energy of formation of a hydrogen bond between two peptide groups in water is very small, because the peptide group is usually hydrated (**Figure 1.19**). On the outside of a protein, peptide groups can be hydrogen bonded to water, but inside the protein essentially all peptide groups hydrogen bond either to another peptide group or occasionally to a buried water or a side chain.

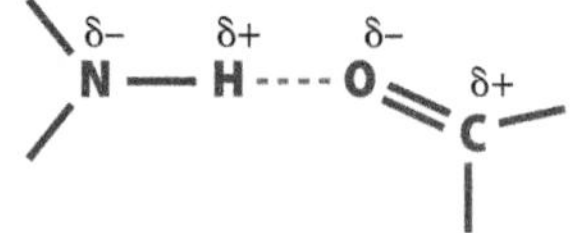

FIGURE 1.18
A hydrogen bond formed by amide NH and CO groups. Each of these forms a dipole, with partial positive and negative charges on adjacent atoms. N, H, O and C have charges of approximately –0.4, +0.2, –0.5 and +0.5, respectively. The optimum geometry is a linear N–H···O angle and a C=O···H angle between 120° and 180° (see Figure 1.23). NMR experiments have shown that there is a small covalent contribution to the bond, because *J*-coupling can be observed across the hydrogen bond (see Chapter 11).

Hydrogen bonds are the most interesting of the noncovalent bonds in proteins because they are so highly directional. They are discussed in more detail below.

1.2.3 Van der Waals forces are individually weak but collectively strong

All atoms consist of a positively charged nucleus surrounded by negatively charged electrons. The nuclei are fixed but the electrons can move around very rapidly. In particular, if an atom with a partial charge is close to an uncharged

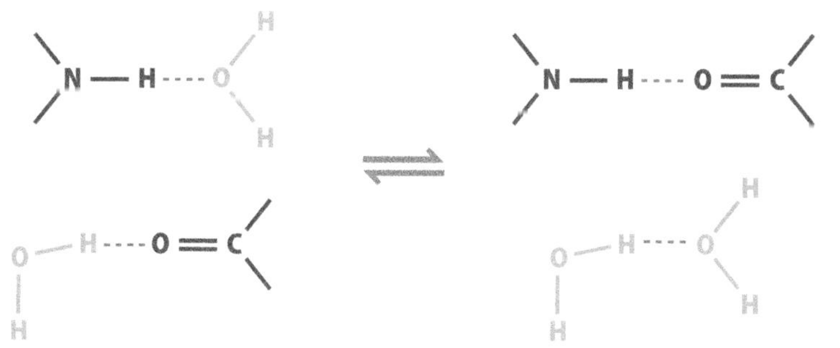

FIGURE 1.19
Peptide groups are normally hydrated, so that formation of a hydrogen bond between two peptide groups in water does not involve a net change in the number of hydrogen bonds.

atom, the charge will produce a redistribution of the electron density in its neighbor, which results in a weak attractive interaction. This is true even for two completely neutral atoms: a transient nonuniform charge produced in one atom will in turn generate a transient redistribution of charge in the other, again resulting in a weak attraction. Thus, all atoms can attract each other: this is usually called a *van der Waals attraction*, although the attraction described in this paragraph is also called a London dispersive force. This force is weaker than the direct electrostatic force, but because all atoms can attract each other it can add up to be very powerful (in contrast with interactions between dipoles, which can be attractive or repulsive, and therefore do not add up constructively). It falls off approximately as r^{-6}, so it is a much shorter-range interaction than an electrostatic interaction. Effectively it only becomes significant when molecules are almost in contact. Different atoms are more or less effective. The strongest forces involve the atoms in which the electrons are most easily moved around, or in other words the more *polarizable* atoms. These tend to be the atoms of higher atomic weight; for example, sulfur is more polarizable than oxygen and therefore has very significantly stronger van der Waals attraction.

The attractive forces also have corresponding repulsive forces. Charge–charge repulsions are again the only repulsive forces that can act over any distance, and they are important for orienting molecules as they approach each other, as discussed in more detail in Chapter 4. *Van der Waals repulsions* are extremely strong when atoms get too close together, increasing approximately as r^{-12} (**Figure 1.20**).

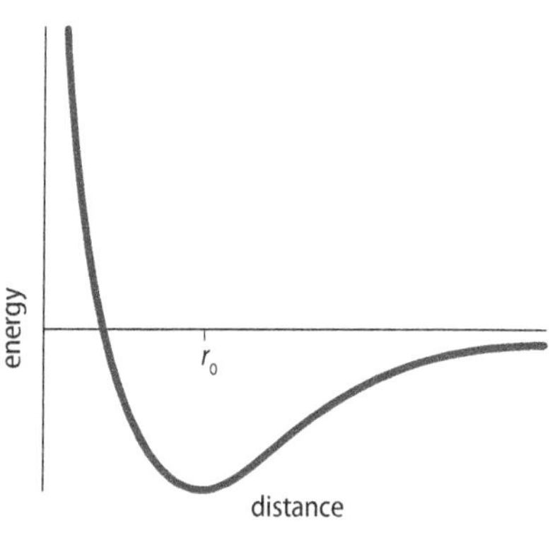

FIGURE 1.20
The van der Waals energy between two atoms, plotted as a function of distance. There is an attractive (negative) energy, which goes as r^{-6}, and a repulsive (positive) energy, which goes roughly as r^{-12} and only becomes significant when atoms get close together (within the sum of their van der Waals radii). The sum of these two energies has a minimum at r_0, the equilibrium distance.

1.2.4 The hydrophobic interaction is entropic in origin

The final force to be discussed is usually known as the hydrophobic force, or *hydrophobic interaction*, in recognition of the fact that it is not a physical force in the same way as the others that have been discussed so far. It is manifested in the general rule that hydrophobic groups, such as the side chains of Leu, Ile, Val, and Phe, tend to cluster together in the protein interior and exclude water. It is different because the driving force is not enthalpy, as it is for the other forces discussed so far, but entropy (*3.1).

A hydrophilic group in **water (*1.9)** has a favorable (enthalpic) interaction with the water: it forms hydrogen bonds. By contrast, a hydrophobic group cannot form hydrogen bonds. Therefore hydrophobic groups in water lead to a decrease in the number of hydrogen bonds that nearby water molecules can form, which puts the waters into an energetically unfavorable state. Water molecules respond by becoming more ordered around hydrophobic groups, which allows them to hydrogen bond to each other better (**Figure 1.21**). This means that the addition of hydrophobic groups to water tends to result in some increase in enthalpy and a significant loss of entropy; on both counts this is therefore energetically unfavorable.

*1.9 Water

We tend to forget what an unusual molecule water is. If, for example, you plot the melting or boiling point of hydrides of different elements (**Figure 1.9.1**), water stands out as having a remarkably high boiling point (along with HF, for similar reasons). Water expands when it freezes; it has a very high surface tension; it has a high heat capacity; and it dissolves many ionic materials that are almost insoluble in many other solvents. All this is of course due to its capacity for **hydrogen bonding.** A hydrogen bond is essentially an electrostatic interaction: the partial negative charge on the oxygen attracts the partial positive charge on the hydrogen. This attraction tends to make water molecules assemble into ice-like structures (**Figure 1.9.2**), in which every hydrogen atom forms a hydrogen bond and every oxygen forms two, and thus every water molecule makes four hydrogen bonds in total. However, such an arrangement is entropically (*3.1) unfavorable, and the result is what has been picturesquely called a "fluctuating iceberg," in which this ordered arrangement is constantly being broken up by thermal motion (see Figure 1.9.2) [133].

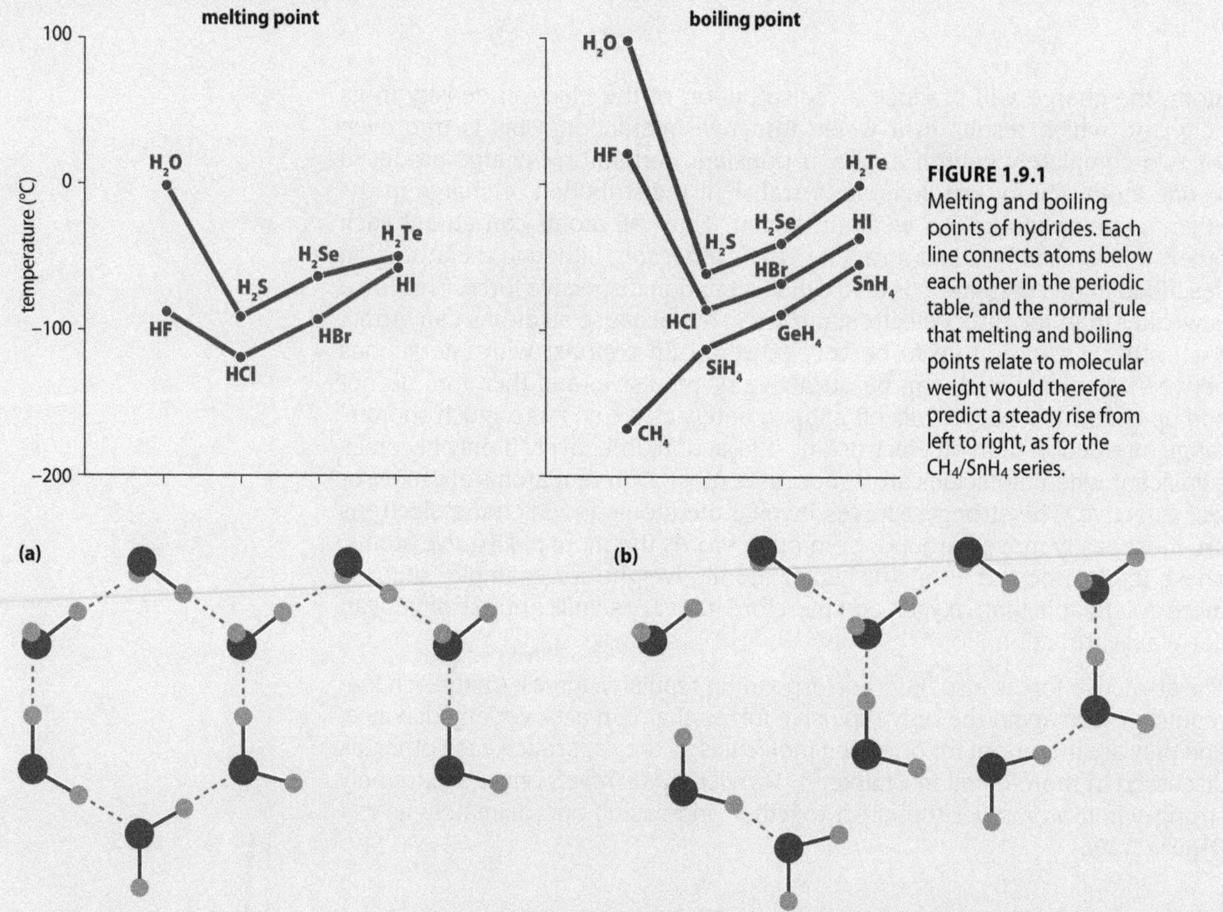

FIGURE 1.9.1
Melting and boiling points of hydrides. Each line connects atoms below each other in the periodic table, and the normal rule that melting and boiling points relate to molecular weight would therefore predict a steady rise from left to right, as for the CH_4/SnH_4 series.

FIGURE 1.9.2
The difference in structure between ice and water. (a) The structure of ice: a three-dimensional lattice, in which every water molecule forms two hydrogen bonds from each oxygen and one from each hydrogen. This is enthalpically good but entropically bad, which is why this structure is favored at low temperature. (b) The structure of water. Some ice-like regions remain, but on average the number of hydrogen bonds per water molecule is reduced. This makes the enthalpy worse but the entropy better. Water molecules make and break hydrogen bonds rapidly.

The hydrophobic force is a result of the system trying to minimize this loss in energy. The hydrophobic groups come together, and thus minimize the hydrophobic area exposed to water. The result is to decrease the effects just described. Water molecules can re-form hydrogen bonds and become more mobile: there is therefore a decrease in enthalpy, and also an increase in entropy. Because $\Delta G = \Delta H - T\Delta S$ (see *3.1), both of these make the overall change in free energy favorable. It is this effect that leads to oil drops coming together from an oil/water mixture. The driving force is not the attraction of the oil molecules for each other; it occurs because the water molecules would

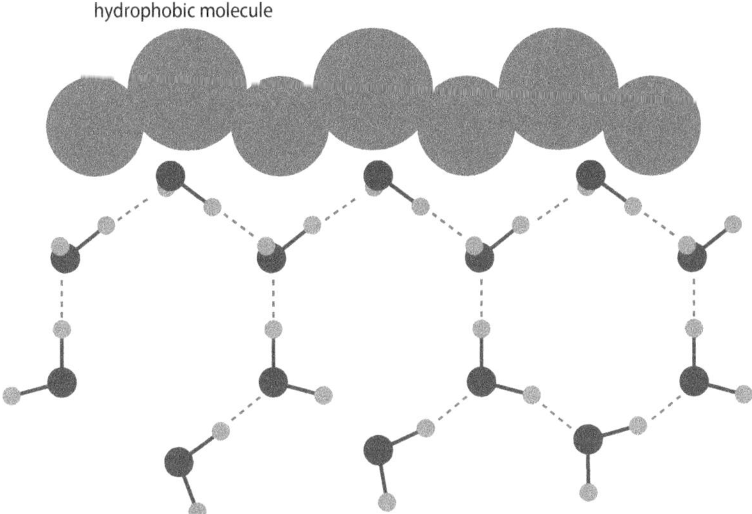

FIGURE 1.21
Introduction of a hydrophobic molecule into water leads to a more ice-like structure around it (compare the structure of water shown in *1.9).

rather associate with each other than with the oil molecules. Therefore essentially the hydrophobic force is not a force at all, but rather a decrease in an unfavorable energy. Hydrophobic effects are proportional to the hydrophobic surface area.

This explanation is valid at normal biological temperatures. However, at high temperature the balance of entropy and enthalpy is different, and it is more sensible to explain the hydrophobic force as originating from the unfavorable enthalpy of interaction of water with hydrophobic solutes.

1.2.5 Hydrogen bonds are uniquely directional

All of the forces described here are directional. However, they differ markedly in the degree of the directionality. Van der Waals attraction occurs between any two atoms that are appropriately close. This means that two molecular surfaces will attract each other, whatever their detailed shapes, because there will always be some atoms close together (**Figure 1.22**). Van der Waals forces are therefore not in practice strongly directional. The same is true for coulombic forces: although the forces are directed toward or away from the charge, small displacements of atoms only have a minor effect on the coulombic forces. However, hydrogen bonds are very different, because their strength depends markedly on the direction of the two dipoles involved (**Figure 1.23**). In this case, a small movement of for example N–H relative to C=O can make a big difference to the strength of the hydrogen bond. In this sense, hydrogen bonds are by far the most directional of the forces considered here. The consequence of this is that we can often consider that the *affinity* of an intermolecular interaction is determined mainly by hydrophobic and van der Waals interactions, whereas the *specificity* is determined by hydrogen bonds (because the bond energy becomes much weaker if the hydrogen bond is absent or has a poor geometry). In a similar way, the initial folding of proteins into a globular shape is generally considered to be due mainly to hydrophobic interactions, whereas the final folding to a stable specific tertiary structure (as opposed to a molten globule) is determined by hydrogen bonding.

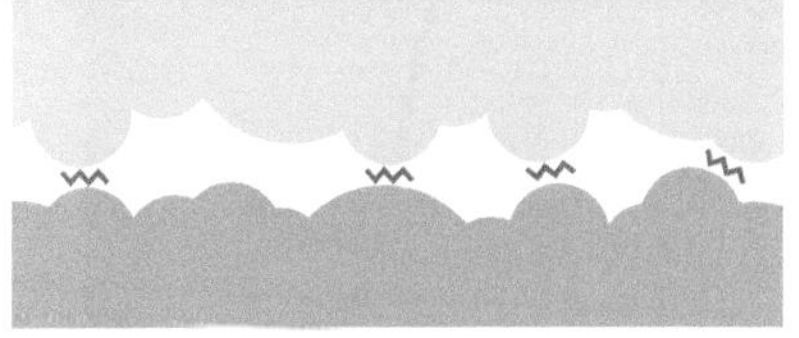

FIGURE 1.22
A close contact between two rough surfaces results in several strong van der Waals interactions, with a range of geometries.

1.2.6 Cooperativity is a feature of large systems

Cooperativity is a striking feature of proteins. **Figure 1.24** shows a typical graph of protein stability against temperature, and displays a highly cooperative transition. Very similar graphs are obtained for stability against

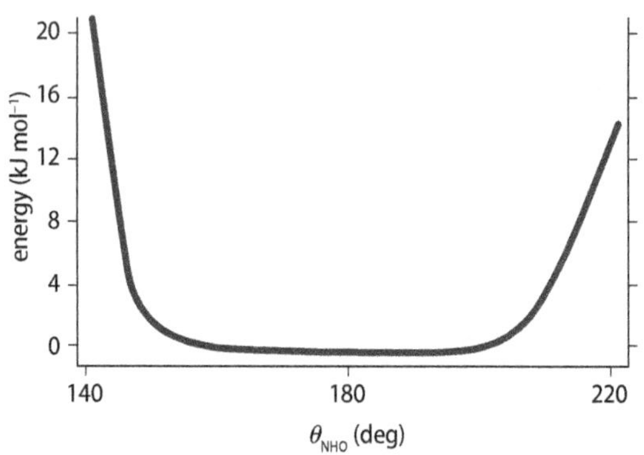

FIGURE 1.23
The energy of an amide hydrogen bond as a function of the N–H···O angle. Data were calculated for $r_{HO} = 2.1$ Å, $\theta_{COH} = 160°$. (Redrawn from H. Adalsteinsson, A.H. Maulitz and T.C. Bruice, *J. Am. Chem. Soc.* 118: 7689–7693, 1996. With permission from the American Chemical Society.)

denaturant (*1.10) concentration. Quite a lot of this is due to the regular secondary structure elements in proteins. Polymers of amino acids will undergo a **helix-coil transition**, being random at high temperature but helical at low temperature (when they are soluble; somewhat surprisingly, most amino acid polymers are insoluble in water). However, *in vitro* it requires a rather long sequence to produce a helix, of the order of 50–100 amino acids, whereas the average helix length in globular proteins is about 10 residues. Some of the cooperativity in proteins is therefore due to the tertiary structure. There have been several attempts at engineering synthetic proteins *de novo*, with designed sequences. Most such attempts produce proteins in which the secondary structure (typically helical) is formed, but the different secondary structure elements are not well packed together: such a structure is typically described as a **molten globule**. These proteins have much less cooperative folding than real proteins. The cooperativity is therefore a property of the entire protein and not merely the secondary structure elements.

Cooperativity is essentially a feature of large systems. One of the most obvious cooperative events is the melting of ice. Any piece of ice that we can easily see (for example, anything bigger than 1 mm^3 or 1 mg) is composed of more than 10^{19} molecules, so on a molecular scale it is a very large system. When one gets down to sections only a few molecules thick, the melting is less cooperative (that is, the melting point is not sharp) and occurs at a much lower temperature, of the order of –90°C [11]. Therefore the sharp melting transition of proteins is one among many indicators that proteins behave as highly cooperative systems. This has a major effect on the way in which proteins behave.

1.2.7 The formation of a β hairpin is cooperative

Regular secondary structure (for example β sheet) in proteins is composed of an array of hydrogen bonds. Each individual hydrogen bond is weak and of marginal stability. However, together the bonds provide very considerable stability and rigidity. This can be understood by thinking about the

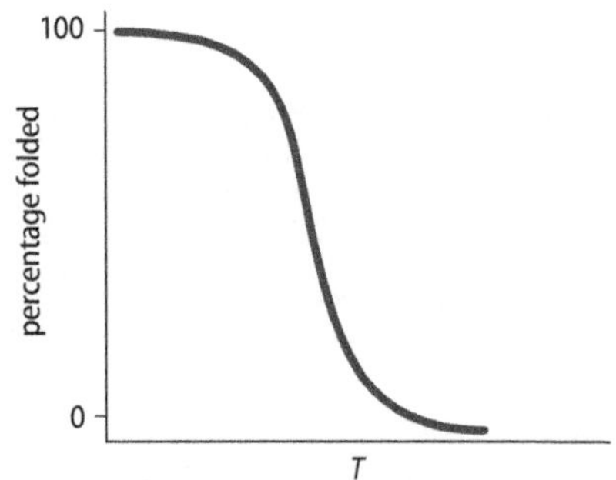

FIGURE 1.24
A typical cooperative thermal unfolding curve.

> ***1.10 Denaturant**
> Denaturants destabilize proteins and cause them to unfold. In other words, they make the unfolded state more favorable by comparison with the folded state. It is not agreed how they work, and it is quite possible that different denaturants work in different ways. However, it is likely that many work by having a favorable binding interaction with backbone amide groups. In this way they stabilize the unfolded protein, which has a greater number of accessible backbone amides. This explanation is supported by experimental data that demonstrate that urea-unfolded ubiquitin adopts conformations that are more extended than expected for a random coil peptide [135]: in other words, urea binds to the unfolded backbone and leads to it adopting a different distribution of conformations.

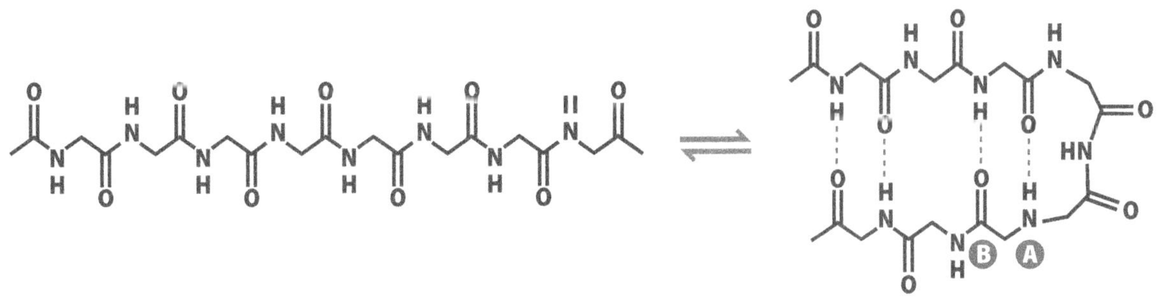

cooperativity between them. Imagine a peptide that exists in an equilibrium between unfolded and β hairpin (**Figure 1.25**). It is not really known how such peptides fold up in solution, although it is likely that hydrophobic clusters stabilize a partly folded structure, which is then further stabilized by hydrogen bonds [12]. We shall ignore the hydrophobic effects, however, and just concentrate on the hydrogen bonds.

FIGURE 1.25
Model system for β hairpin formation.

The first hydrogen bond to form is unfavorable. A single hydrogen bond is (net) not a strong interaction, and is outweighed by the substantial loss in entropy (*3.1) required to fold up the chain and keep the hydrogen-bonding partners together. This does not mean that it does not happen; it just means that its probability is low, and most of the time the hydrogen bond is not present. The most likely hydrogen bond to form is the one marked A in Figure 1.25, because these two amide groups are closest in the sequence and therefore require the smallest number of backbone dihedral angles to be correctly oriented simultaneously to allow them to bond.

However, once the first hydrogen bond has formed, then the next one (B in Figure 1.25) is already in a good geometry to bind and is therefore much more favorable than the first. In fact, formation of this bond is probably energetically favorable by a small amount. The *overall* stability of the hairpin is, however, still unfavorable because the energy of the favorable second hydrogen bond cannot overcome the unfavorable first one. The overall equilibrium constant for formation of the hairpin is the product of the equilibrium constant for the first bond and the equilibrium constant for the second: because the first bond was unfavorable, the second makes the overall formation less unfavorable but not yet favorable (**Figure 1.26**). However, with each new hydrogen bond added, the hairpin becomes more favorable, until with a sufficient number (in Figure 1.26 this number is six) the hairpin is stable.

1.2.8 Hydrogen bond networks are cooperative

As described above, cooperativity means that each hydrogen bond "helps" its neighbor to be stronger. However, there is a second way in which it does this, which is equally important. The strength of a hydrogen bond is very strongly dependent on geometry. This means that a hydrogen bond in a flexible molecule is inherently weak because it is being continually bent and stretched. As more hydrogen bonds are added to the network of hydrogen bonds that form the β hairpin, the network becomes less flexible, hydrogen bonds on average have a better geometry, and thus *all* the hydrogen bonds become stronger. That is, not only does the existing network significantly strengthen a new hydrogen bond (see Figure 1.26), but the new hydrogen bond also strengthens all the hydrogen bonds in the existing network. In other words, a cooperative hydrogen bond network means what it says—all the hydrogen bonds are affected by the presence of each other: remove one and the whole network gets weaker; add one and the whole network gets stronger.

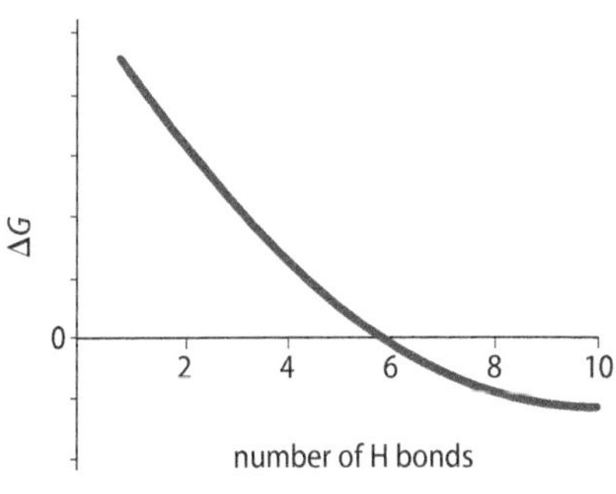

FIGURE 1.26
Schematic free energy for formation of hydrogen bonds in a β hairpin.

One implication of this analysis is that the binding of a ligand to a protein is likely to have long-range effects. In particular, the hydrogen bonds between ligand and protein will affect the strengths of their neighbors, so that hydrogen bond strengths all over the protein are likely to be affected. This is a common observation, either in crystal structures or in nuclear magnetic resonance studies, where for example after the addition of a ligand the exchange rates of amide protons with solvent are affected all over the protein and not just at the binding site: typically they all get slower, although there are a significant number of cases in which some also get faster. The category "ligand" also includes protons: a change of pH can therefore also cause long-range effects in the structure. Similarly, a mutation to a protein usually causes minor rearrangements in structure that propagate a considerable distance through the protein. This makes any detailed analysis or prediction of mutational effects almost impossibly difficult.

1.2.9 Proteins require a layer of water for their function

Water is essential for protein function. However, proteins require remarkably few water molecules. There is some evidence that a completely dehydrated enzyme retains enzyme activity, although the activity is several orders of magnitude less than that of the hydrated enzyme [13]. Experiments involving drying protein solutions and then gradually rehydrating them show that enzymes typically require only about 50 water molecules for the onset of function; these water molecules hydrate essential buried water sites inside the protein (discussed further in Chapter 6) as well as charged groups on the surface. As more water molecules are added, they start to form into clusters and therefore facilitate the transport of solutes and protons around the surface. Full function can be regained by a few hundred water molecules, which corresponds to rather less than a single layer of water molecules surrounding the protein [14, 15]. This upper limit represents a fully active and functional protein, and implies that outside this single layer of water the solution is effectively just bulk water. Indeed, as we shall see in Chapter 4, this surface layer is all the water that there is in crowded solutions *in vivo*, particularly inside mitochondria, where the mass of water is no more than 40% of the mass of proteins. This corresponds approximately to a single-molecule layer; mitochondria therefore have a very similar water content to many protein crystals. Bacterial cells are only slightly more hydrated, with about 50% water:protein by weight [16]. It is estimated that in cells as much as 15% of the water is this surface or hydration layer, the remaining 85% being "bulk." The essential water is required to maintain the correct structure and polarity of the active site; it is less mobile than bulk water [15, 17], not least because it is usually located in pits in the protein surface. Although most of the water on the surface of proteins has a mobility roughly half that of bulk water, a small fraction located in pits is much less mobile [18]. The result is that the mobility of the hydration layer is *on average* 10–15 times less than that of bulk water, most of this coming from the small numbers of molecules in pits. Not surprisingly, this "hydration layer" is somewhat different from bulk water. For example, it is slightly more densely packed and less compressible [19], as well as having rather lower heat capacity and enthalpy [15], and can be viewed essentially as an integral part of the protein. The thickness of the hydration layer has been much discussed [20], but the accepted view now is that it is no more than two water molecules thick and for most purposes, as stated above, only one. It is undoubtedly true that the hydration layer is more structured around charged groups and is therefore unevenly distributed, being for example probably thicker around DNA than proteins, because of the greater charge density on the surface of DNA.

Surveys of protein crystal structures show that most proteins contain water molecules buried inside the protein, completely inaccessible to bulk water (**Figure 1.27**). Homologous proteins tend to have buried waters in the same places; these waters can therefore be regarded as part of the structure of the protein [21, 22]. They are often functionally important, as we shall see in Chapter 6, for example typically being found in interfaces or close to the active site [23].

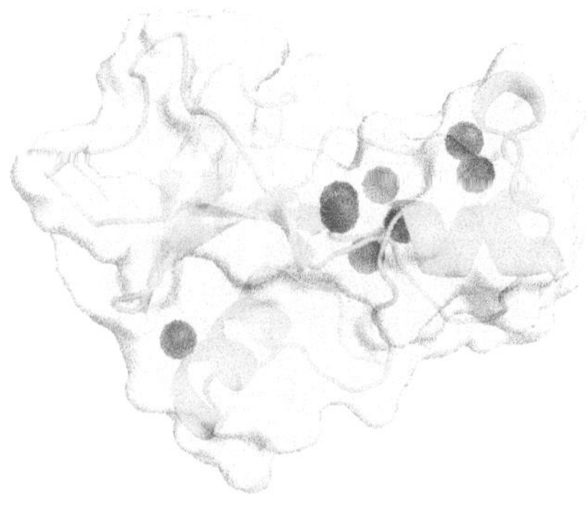

FIGURE 1.27
Buried waters. This figure shows the ribonuclease enzyme barnase in cartoon form, enclosed by the protein surface. There are eight buried waters in this structure (PDB file 1brn).

1.2.10 Entropy and enthalpy tend to change in compensatory directions

It is usually straightforward to rationalize (if not completely to predict) the free energy of interaction between two molecules, for example a protein and a ligand. Add a group that makes a favorable interaction and the free energy becomes more favorable; add one that interferes with the binding and the free energy becomes less favorable. However, once one tries to go further and understand the numerical value of the change, or (even more so) the changes in entropy and enthalpy (*3.1), it gets very much harder. Enthalpy and entropy changes are easily measured by techniques such as isothermal titration calorimetry (ITC, Chapter 11), which has led to large numbers of publications that discuss such issues, often with little enlightenment resulting. It does not help that the other way to measure entropy change is to measure the change in free energy with temperature. This only works if the enthalpy is independent of temperature, which is the same as saying that the **heat capacity** change is zero. Because this is almost never true, a fair proportion of such measurements in the past are probably of low reliability. In addition, the limited range of temperatures and affinities that are experimentally accessible tends to produce entropy/enthalpy compensation automatically [24]. And thirdly, the magnitude of an entropy change is not an absolute value, but changes with the units that are used to measure concentration, with the consequence that it is meaningless to talk of any binding event or reaction as being "entropy driven" or "enthalpy driven" [25].

Experimentally, the changes in entropy and enthalpy are much larger than the change in free energy; the change in free energy is thus a relatively small difference between two much larger effects. And these large effects often do not do what one might expect: they can vary quite widely for rather similar interactions. There is a widely discussed (and poorly understood) phenomenon known as **entropy/enthalpy compensation**, which is the observation that many binding reactions, including protein folding, ligand binding, and salt effects, cause very large changes in entropy and enthalpy that mostly cancel out to give small changes in free energy (**Figure 1.28**). This phenomenon has been so widely observed that many authors have ascribed it to the special features of water molecules. It is suggested that binding reactions tend to involve

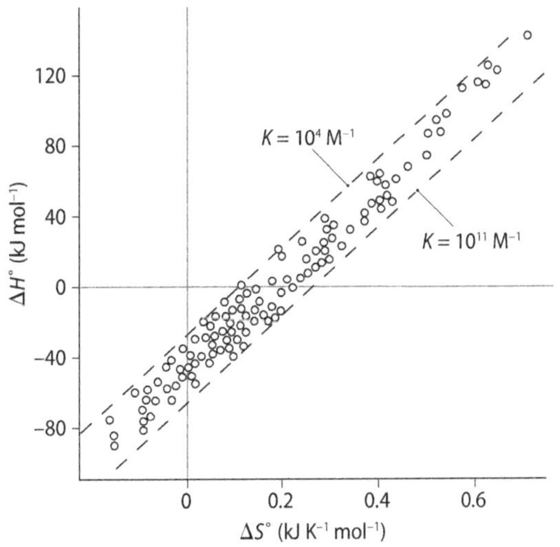

FIGURE 1.28
Entropy/enthalpy compensation. The figure shows experimental data for the binding of ligands to macromolecules. In this case, much of the apparent compensation is probably artifactual, as indicated by the dashed lines, which show association constants of 10^4 and 10^{11} M^{-1}. The large majority of binding data are restricted to this range, because values outside the range are either too weak or too strong (respectively) to be easily measurable. (Adapted from P. Gilli et al., *J. Phys. Chem.* 98:1515–1518, 1994. With permission from the American Chemical Society.)

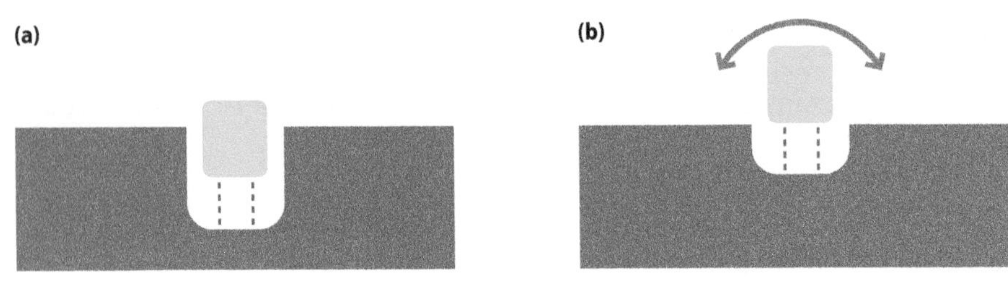

FIGURE 1.29
Hydrogen bonds can be favored either enthalpically or entropically. (a) If a bound ligand is fixed in place, it can form highly directional and strong (enthalpically favorable) hydrogen bonds. However, it has very restricted motion and therefore loses a lot of entropy on binding. (b) In contrast, a ligand that is bound less rigidly has weaker hydrogen bonds but loses less entropy on binding. The two ligands can achieve similar overall free energy but with different combinations of entropy and enthalpy.

the release of a large number of water molecules, and it is the enthalpy and entropy associated with the waters that lead to such large effects.

Alternatively, and more helpfully, it has been noted [26] that entropy/enthalpy compensation is observed equally in nonaqueous systems, implying that it cannot be a feature peculiar to water and that it is telling us something important about the behavior of cooperative systems. A bimolecular system, such as a ligand bound to a protein, will settle into the state of lowest free energy, which as always is a balance between enthalpy and entropy. A favorable *enthalpy* state is one with strong hydrogen bonds; because a hydrogen bond is highly directional, a necessary implication of a strong bond is that it is also fixed and rigid and therefore has an unfavorable *entropy*. Alternatively, the bond can have a favorable entropy, and thus be rather mobile in its geometry, with consequently weak hydrogen bonds (**Figure 1.29**). The optimum position of this balance varies from one ligand to another, and hence the balance between enthalpy and entropy varies. More conformationally rigid complexes (suggested to be typical of agonist interactions, because of the exact fit of hydrogen bond geometry) have favorable enthalpy, whereas more mobile ones have favorable entropy.

The views expressed in the previous two paragraphs cannot both be correct at the same time. Which is correct? As so often happens, the true picture seems to be a mixture of both: sometimes entropy and enthalpy are dominated by water molecules, and sometimes they are dominated by the mobility of the protein and ligand. When two similar complexes are being compared, water can usually be ignored [27], whereas water is very important for explaining the overall entropy and enthalpy change in ligand binding. It is a difficult and largely unsolved problem to work out at which end of the spectrum one is.

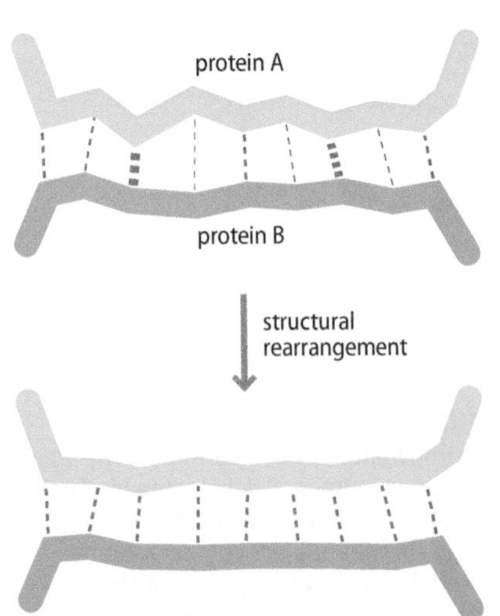

FIGURE 1.30
In an interface between two proteins, composed of a set of roughly equivalent hydrogen bonds, the protein structures are most likely to rearrange themselves so as to make all the bonds of roughly equal enthalpy, as indicated by the widths of the lines.

It is therefore true to say that measurements of entropy and enthalpy changes are seldom useful or interpretable. Fortunately, some clarity is now emerging. Weber [28] noted that most protein interactions (for example protein–protein binding or folding) tend to involve the simultaneous formation of a large number of weak and cooperative interactions. These interactions tend to get distributed roughly evenly between the different hydrogen bonds in the interface, as noted earlier. In essence, the reason for this is that maximum entropy (*3.1) means that the system adopts the most *likely* arrangement (**Figure 1.30**). The most likely arrangement is one in which all bonds are roughly equal in strength. He suggested that on formation of a protein/protein interface from two previous protein/water interfaces, several strong protein–water bonds are converted into weaker protein–protein bonds, and that this loss of enthalpic stabilization, and the consequent rearrangement of the hydrogen bond network to give a more even spread of hydrogen bonds, results in an increase in the entropy of the system. From this he concluded that the increase in entropy seen in protein–protein association (and for that matter in protein folding) was a result not of water molecules released from the hydrophobic surface but of the redistribution of hydrogen bond energies to a more uniform and mobile state. This argument was developed (independently) by Williams [26, 29], who adduced a wide range of experimental data to illustrate the balance between enthalpic and entropic stabilization, as argued above.

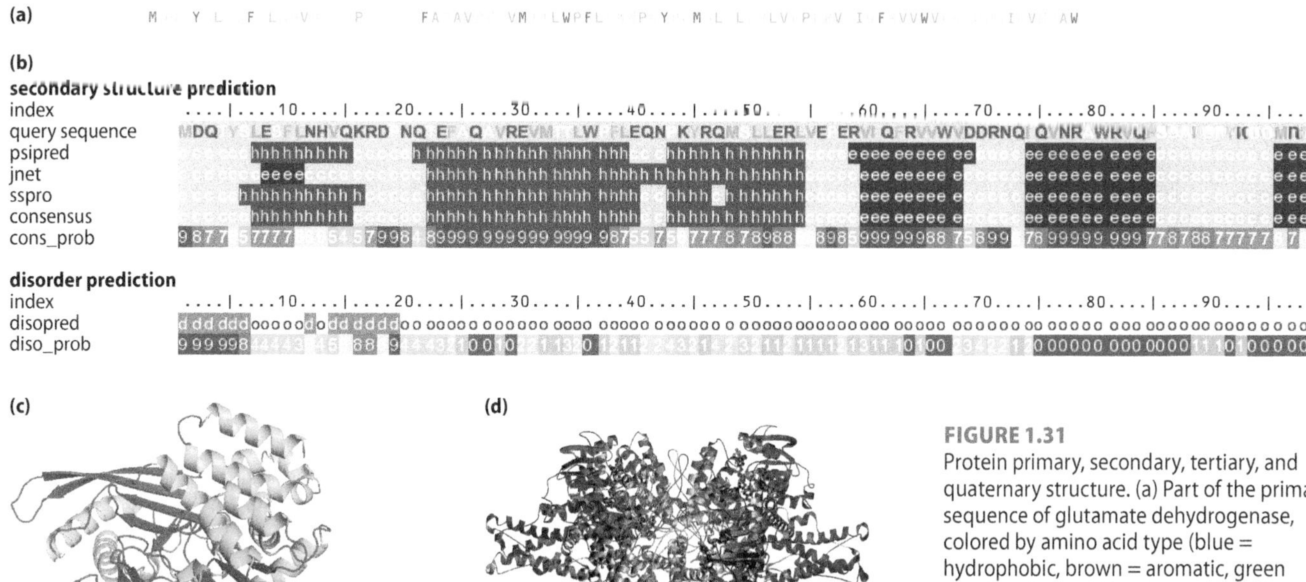

FIGURE 1.31
Protein primary, secondary, tertiary, and quaternary structure. (a) Part of the primary sequence of glutamate dehydrogenase, colored by amino acid type (blue = hydrophobic, brown = aromatic, green = polar, olive = sulfur-containing). (b) Secondary structure of glutamate dehydrogenase. This is the output of a secondary structure prediction by the program phyre (http://www.sbg.bio.ic.ac.uk/phyre/), which also predicts disordered regions. (c) Tertiary structure of glutamate dehydrogenase, colored by the secondary structure elements present. (d) Quaternary structure of glutamate dehydrogenase. This enzyme is a hexamer, composed of a dimer of trimers. The dimer interface is down the center of the diagram.

However, the most convincing proof of the argument has come from Alan Cooper. He has pointed out that an unfolded protein is an uncorrelated system with many degrees of freedom [24, 30]. Therefore an increased energy can be redistributed into a large number of motional modes; in other words, an unfolded protein has a large heat capacity. By contrast, a folded protein is a much more correlated system, and therefore energy has fewer places to go and results in heating: a folded protein has a small heat capacity. This is experimentally true but is usually explained as being due to the large number of water molecules released from the hydrophobic surfaces when a protein folds. However, Cooper's explanation emphasizes that it can also be explained as a simple consequence of the more cooperative nature of a folded protein. The change in heat capacity on protein folding means that the enthalpy of unfolding increases with temperature, and this quantitatively provides an explanation for the entropy/enthalpy compensation usually observed in such cases. Similar arguments apply to ligand binding.

In summary, although most textbooks and research papers state or implicitly assume that the changes in entropy, enthalpy, and heat capacity on ligand binding or folding are due to the release of water molecules from hydrophobic surfaces on the formation of hydrophobic clusters, at least some and possibly most of these thermodynamic changes are due to the increased cooperativity that is a characteristic of larger systems (such as protein dimers or folded proteins, respectively). It is therefore likely that the thermodynamic effects of water release are not as large as normally assumed, and that thermodynamic changes on binding or folding can be more straightforwardly explained in terms of changes in bond strength and internal mobility.

1.3 THE STRUCTURE OF PROTEINS

1.3.1 Proteins are composed of primary, secondary, tertiary, and quaternary structure

Textbooks usually define four levels of protein structure, although not always consistently (**Figure 1.31**). *Primary* structure is the covalent structure: the sequence plus any covalent modifications including disulfide bonds,

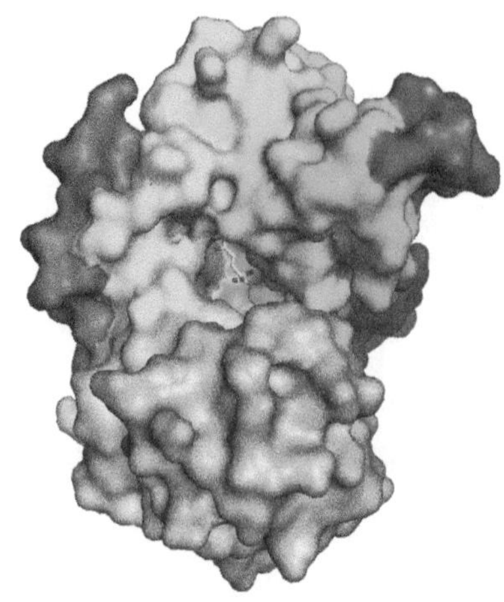

FIGURE 1.32
The active site of glutamate dehydrogenase, which is in a very deep pocket (the glutamate substrate is shown as sticks). The coloring shows the different chains of the quaternary structure.

phosphorylation, and so on. *Secondary* structure is the elements of regular backbone geometry, such as α helices and β sheets, to which one could add β turns, polyproline helices, and 3_{10} helices. It is also often taken to include "coil," which means everything else. Secondary structure is predicted fairly well by primary structure, mainly because different amino acids have different preferences for being in α-helix or β-sheet structures. Proline is not easily accommodated in either an α helix or a β sheet because of its side chain; it therefore often acts as a breaker of regular secondary structure. There are many algorithms for predicting secondary structure, of which the best achieve accuracies of around 75–80%. The errors arise from residues whose secondary structures are determined by tertiary interactions and are therefore much harder to predict. A related observation is that many proteins seem to fold by forming elements of secondary structure first, which then aggregate together and form tertiary interactions. In general, the intermediates have secondary structures that match those found in the folded proteins, although there are some clear exceptions where the initially formed secondary structures have to be broken in the folded protein. Clearly in these cases, prediction of secondary structure is unlikely to work.

Tertiary structure is the full three-dimensional fold. Approximately, therefore, secondary structure describes the backbone arrangement whereas tertiary structure includes also side-chain structure. The tertiary structure of a protein is usually divided into **domains** (see Chapter 2). The tertiary structure of a protein also contains one or more **active sites** on the surface. If the function of the active site is to bind a small molecule, it is always different from the rest of the protein surface. Such sites form concave pits in the surface; they are almost always more hydrophobic than the rest of the protein surface (**Figure 1.32**) and can therefore often be identified in this way. If the active site forms an interface with a second protein or a dimerization interface, it is often not significantly different from the rest of the protein surface.

Finally, *quaternary* structure is the way in which different polypeptide chains pack together into oligomers; this is discussed in more detail in Chapters 2 and 3. Quaternary structure is less well conserved during evolution than secondary and tertiary structure [31].

***1.11 EF hand**
The oddly named EF hand is a motif that binds calcium; it is found in most calcium-binding proteins such as calmodulin and troponin C, which each contain four copies. It consists of two helices connected by a loop, with the calcium ligands typically formed by the side chains of two Asp, one Glu, and one Asn, and a backbone carbonyl, within the loop. It is named after its first observation in parvalbumin, in which it is formed by helices E and F and the loop between them, which hold the calcium rather like the index finger, middle finger, and thumb of a right hand (helix E, loop, and helix F, respectively).

1.3.2 Secondary structures pack together in structure motifs

In between the secondary and tertiary levels, secondary structures often pack into supersecondary structures or **structure motifs**, which include the antiparallel helix bundle, helix–turn–helix (HTH, see Chapter 3), calcium-binding **EF hand (*1.11)**, coiled coil, Greek key, β-α-β, and β hairpin motifs (**Figure 1.33**). These are particularly stable (or easily folded) arrangements of secondary structures, and many proteins are usefully analyzed in this way. These in turn pack together into larger units, of which the most common is the **β barrel (*1.12)** (**Figure 1.34**) , such as that found in the enzyme triosephosphate isomerase (TIM). Because this barrel was first identified in TIM, it is often known as the TIM fold or TIM barrel.

Some of these structure motifs have very characteristic packing arrangements. The clearest is probably the way in which α helices pack against each other [32]. When viewed from the side, an α helix is composed of ridges and grooves. The ridges can be looked at in two ways. If we connect together residues four apart

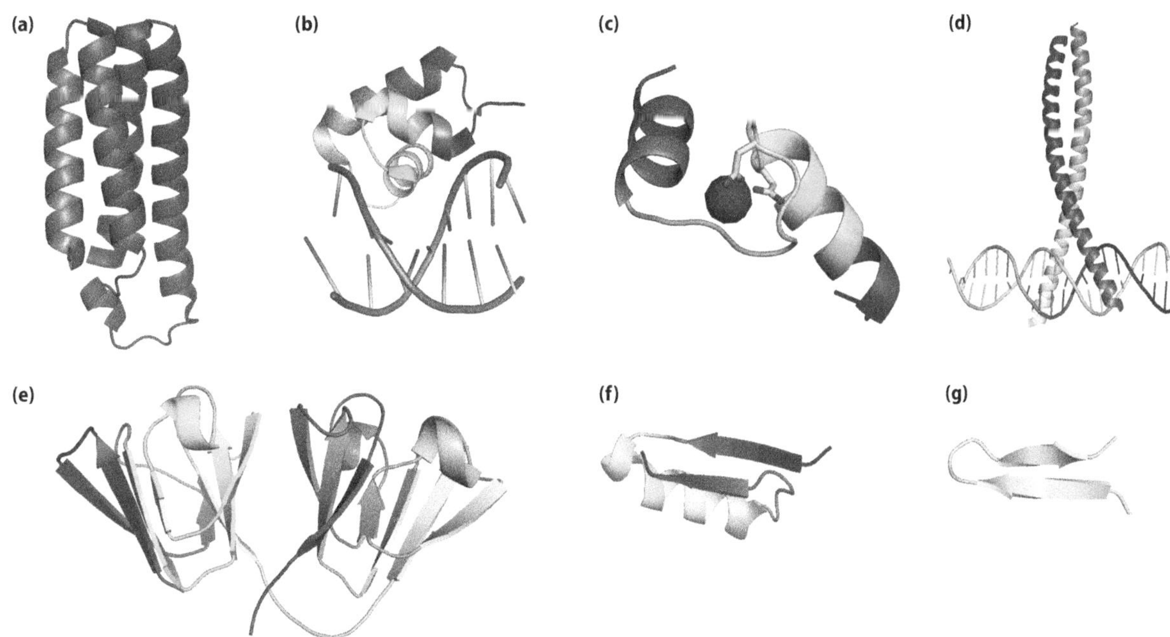

FIGURE 1.33
Examples of some structure motifs. (a) A four-helix bundle; *E. coli* cytochrome *b*562 (PDB file 3c63). (b) A helix–turn–helix (HTH) motif; the *lac* repressor headpiece complexed to DNA (PDB file 2pe5). The recognition helix is in green. As mentioned in Chapter 3, in several eukaryotic transcription factors that form the b-HTH-Zip family such as Max and Myc, the HTH motifs are used not for binding but as dimerization domains, where they form four-helix bundles. (c) An EF hand; one of the four EF hands of calmodulin, showing the two aspartate side chains that form part of the binding site for calcium (PDB file 1a29). (d) A coiled coil; the Fos–Jun dimer binding to DNA (PDB file 1fos). (e) Four Greek keys; human γb-crystallin (PDB file 2jdf). The Greek key motif is probably formed by an antiparallel β hairpin folding in half and doubling back on itself. (f) A β-α-β motif; one of the many such sequential motifs in the triosephosphate isomerase β barrel (PDB file 3gvg). (g) A β hairpin from lysozyme (PDB file 2lym).

*1.12 The β barrel

This is a structure formed by a series of β strands, which typically pack in an alternating up-and-down manner to make a barrel shape (see also Figure 1.34). It is one of the most common folds in enzymes (the "TIM barrel") and has a very versatile function. It is also found in transmembrane proteins, where it typically forms channels. The barrel can also be squashed to make a two-layered sheet (**Figure 1.12.1**) that lacks the hydrogen bonds between the edge strands. This fold is often used for binding and transporting hydrophobic molecules, for example retinol, and also the small molecules present in rodent urine bound to urinary proteins, which are deposited on surfaces and liberated once the urine has dried, producing a range of biological stimuli.

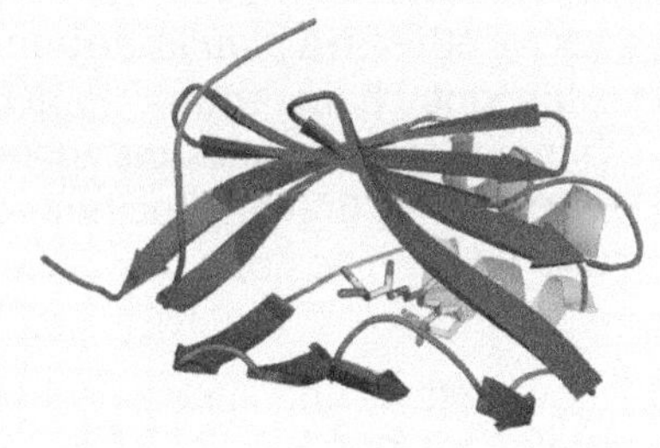

FIGURE 1.12.1
Human cellular retinol binding protein II, with retinol (green) bound (PDB file 2rct). Note how the β barrel is incomplete: the two nearest strands (bottom right) do not form interstrand hydrogen bonds.

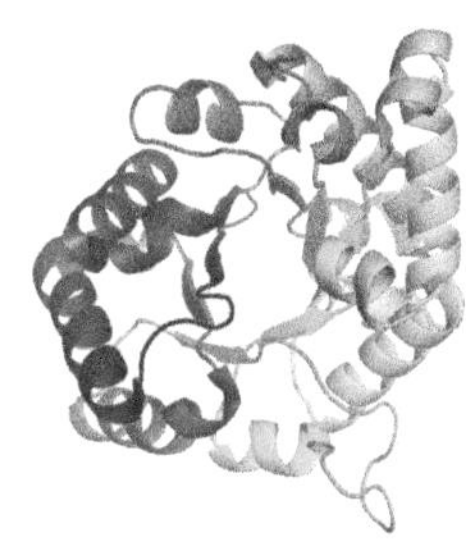

FIGURE 1.34
A β barrel: the structure of triosephosphate isomerase, from which the β-α-β motif in the previous figure was taken. The view is looking down almost directly into the top of the barrel, which is the active site in all β barrels.

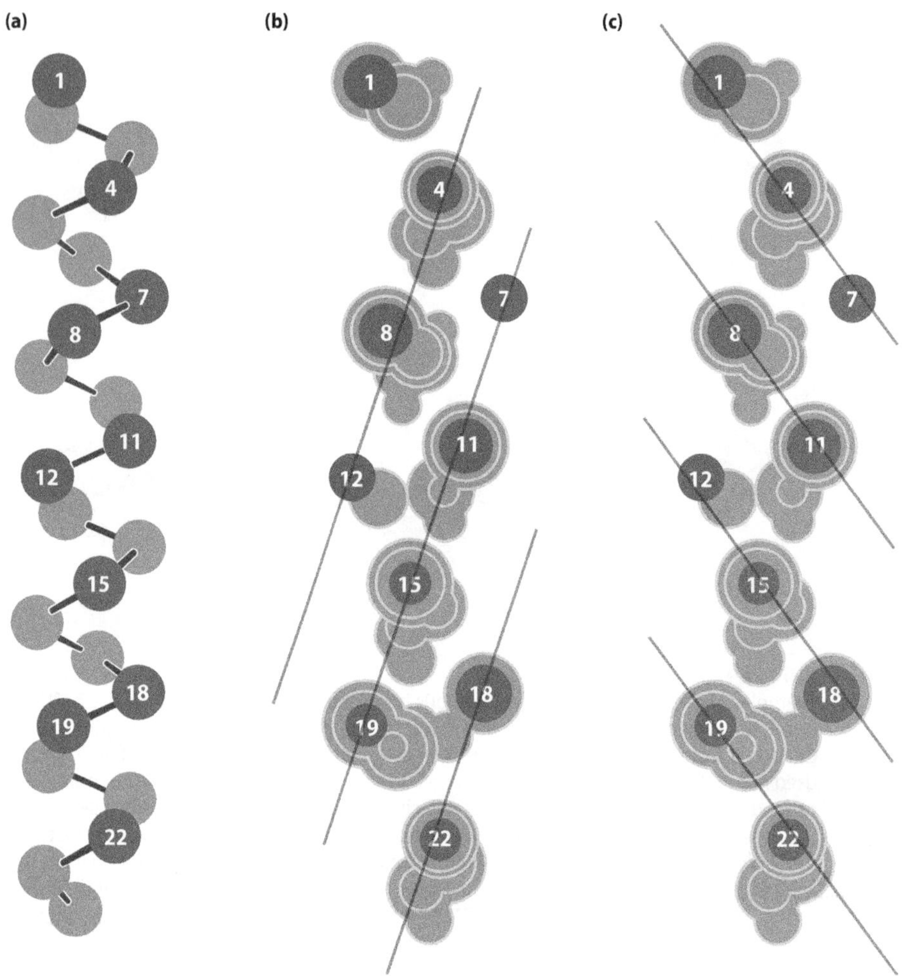

FIGURE 1.35
Rationale for the packing of α helices together. The "knobs" formed by the side chains of α helices (a) run in parallel rows, which can be thought of either as running at 25° to the axis, considering every fourth residue (b), or at 45° to the axis, considering every third residue (c). In (b) and (c), the circles indicate the approximate position and size of the side chains. (From C. Branden and J. Tooze, Introduction to Protein Structure, 2nd ed. New York: Garland Science, 1999.)

in the sequence, the ridges run at an angle of 25° to the helix axis, whereas if we connect together residues three apart in the sequence, the ridges form an angle of about 45°, but in the opposite direction (**Figure 1.35**). It turns out that there are only two geometries that allow two helices to pack together such that the ridges of one helix can fit into the grooves of another. In one, the 25° ridges of one helix fit directly into the 25° grooves of another. This requires the two helices to be antiparallel, with a 50° relative tilt (**Figure 1.36**). In the alternative arrangement, the helices are again antiparallel, but now the 25° ridges of one fit into the 45° grooves of the other, which requires a 20° (= 45 – 25) relative tilt (see Figure 1.36). To pack two helices together in a *parallel* arrangement, the commonest way is to arrange that the side chains from one helix fit into the gaps between the side chains of its neighbor: a *"knobs into holes"* arrangement, as described by **Francis Crick (*1.13)** [33]. By drawing the helix as a flat representation (**Figure 1.37**) we can see that this can be done most easily by tilting the two helices by 18°.

One can make some useful generalizations about helices and sheets. Helices are relatively rigid and straight, although helices on the outside of a protein often bend around the core. Such helices are usually easily identified by

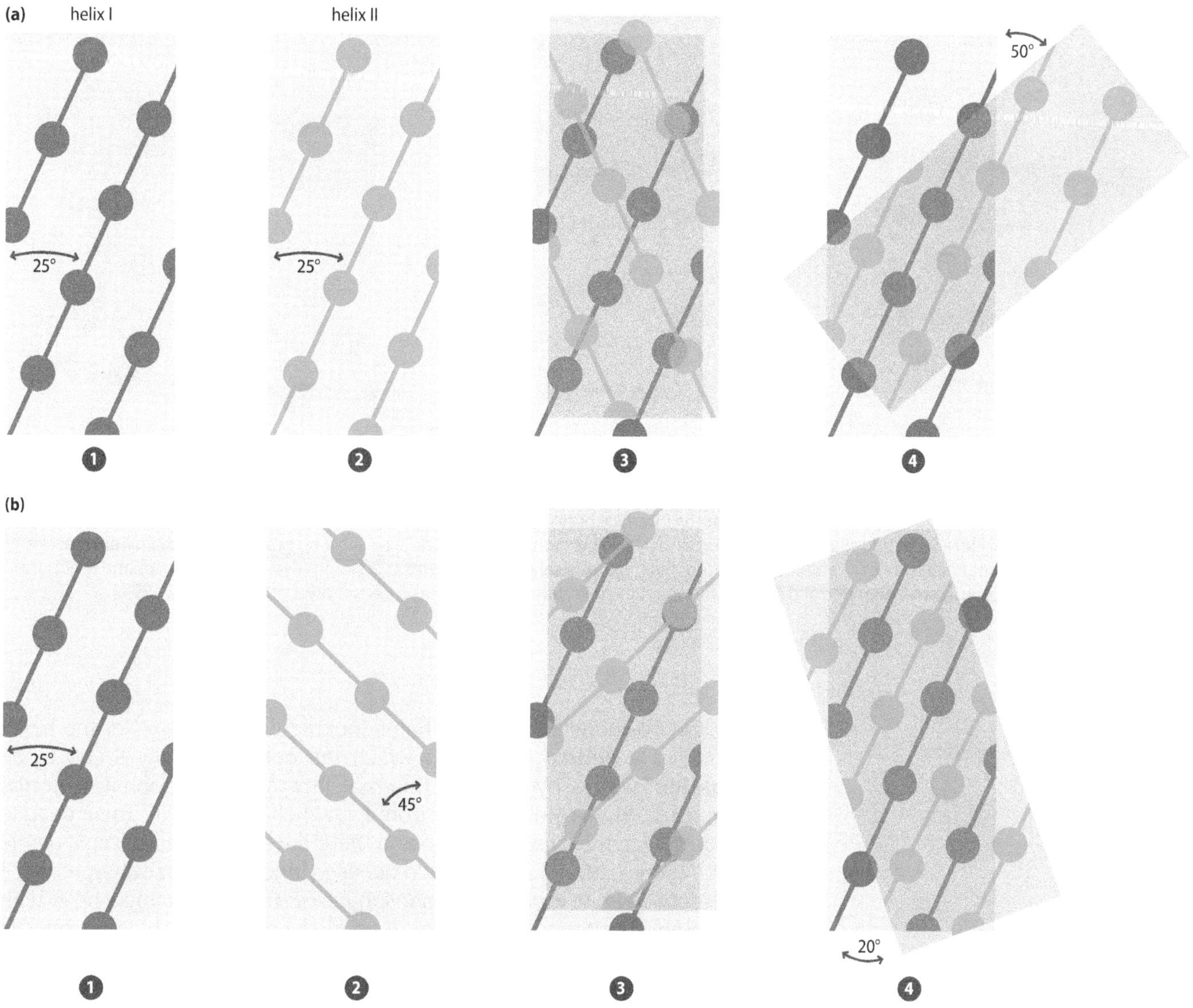

FIGURE 1.36
Two ways of packing helices together antiparallel, knobs into holes. (a) We use the 25° rows discussed above, and consider two helices, I (red) and II (blue). Helix II can be matched with helix I by first turning it to be antiparallel to helix I and then placing it against helix I (panel 3), and then by rotating it by 50°, so that the two rows run in parallel. (b) This time we use the 25° rows in helix I, but the 45° rows in helix II. Now when helix II is made antiparallel and placed against helix I (panel 3), there is only a 20° rotation required to get the rows parallel. (From C. Branden and J. Tooze, Introduction to Protein Structure, 2nd ed. New York: Garland Science, 1999. The figure was based on an idea in C. Chothia et al., *Proc. Natl. Acad. Sci. USA* 74:4130–4134, 1977.)

*1.13 Francis Crick

Francis Crick (**Figure 1.13.1**) is of course best known for his discovery of the double-helical structure of DNA in 1953, which won him the Nobel Prize in Physiology or Medicine in 1962. He made several other important contributions, such as proposing the "central dogma," investigations on protein synthesis and the genetic code, working out the theory of X-ray diffraction by helices, and suggesting the "knobs into holes" model for coiled coils. From 1975 until his death in 2004 he worked on neurobiology and human consciousness at the Salk Institute in California. He had strong views on the relationship between science and religion, being on the whole an agnostic "with a strong inclination towards atheism."

FIGURE 1.13.1
Francis Crick. (Courtesy of Marc Lieberman 2004. Creative Commons Attribution 2.5 Generic license.)

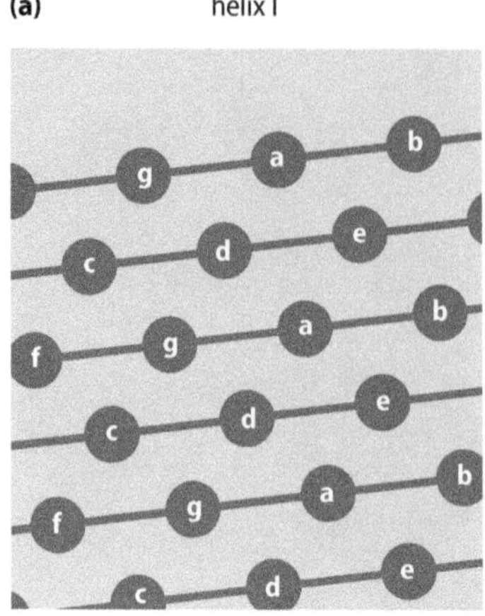

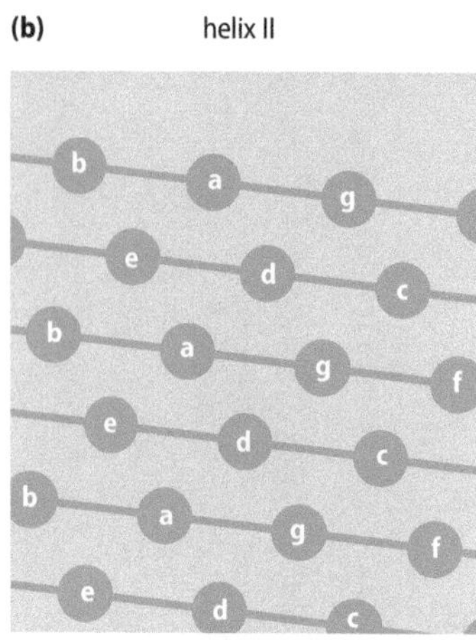

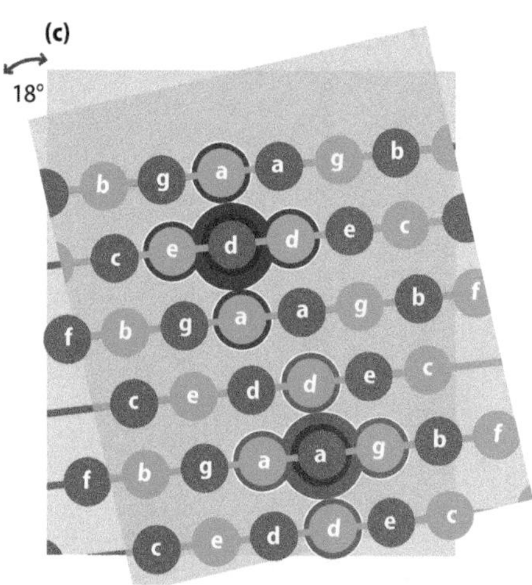

FIGURE 1.37
How to pack two parallel helices together. In parts (a) and (b), the two helices are flattened out onto a plane by "cutting" them down one vertical face and opening the helix out flat. The two helices can be fitted onto each other, knobs into holes, by rotating the helices by 18° (c). Note that the interface between the two helices has alternating pairs of "d" side chains (usually leucine) and "a" side chains (usually hydrophobic). This is highlighted in part (c) by showing how a "d" side chain from helix I is surrounded by two "a," a "d," and an "e" from helix II; conversely, an "a" from helix I is surrounded by an "a," two "d," and a "g" from helix II. (From C. Branden and J. Tooze, Introduction to Protein Structure, 2nd ed. New York: Garland Science, 1999.)

drawing the positions of the side chains looking down the axis of the helix, usually called a **helical wheel**, in which the helix is usually found to be **amphipathic**, with a hydrophobic internal face and a hydrophilic external face (**Figure 1.38**). As noted in Section 1.1.4, helices have all their peptide bonds pointing in the same direction and therefore have a macroscopic dipole moment (**Figure 1.39**). Where a protein needs to bind or stabilize charges, such as in the active site of kinases and phosphatases, there is usually a helix with its N-terminal (positive) end pointing toward the phosphate-binding site. By contrast, sheets are easily twisted and deformed. (This also implies that there are no simple rules for packing sheets together as there are for helices.) It has been noted that where a protein requires the transmission of mechanical movement, such as in proteins that transmit signals across membranes, they use helices. A particularly simple example is that in several "unconventional" myosins (that is, not the standard myosin II as found in muscle; this is discussed further in Chapter 7), the "strut" linking the motor to the cargo seems to consist of a single α helix, stabilized by charge–charge interactions along the length of the helix [34]. By contrast, where a protein needs to deform to bind to a ligand, it usually does this using β sheets [35]. Thus, antibodies and the very common TIM barrel (Figure 1.34) bind their ligands by using the loops linking together different sheets. One of the most rigid structural motifs is the **coiled coil**. Extreme rigidity, as required for example in muscle proteins and in the central spindle of F_0F_1-ATPase (Chapter 7), is usually achieved with coiled coils. Coiled coils therefore are also common in proteins whose function requires them to be very long, such as a number of fibrous proteins (*1.14).

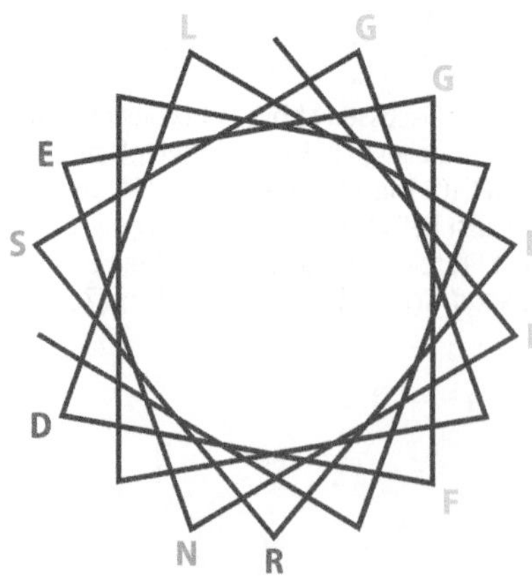

FIGURE 1.38
A helical wheel. Amino acids are plotted every 100° round a circle, showing where they occur on the helix surface. In this example of an 11-residue helix from alcohol dehydrogenase, green is hydrophobic, blue is polar, and brown is charged, clearly showing an amphipathic helix, for which one would expect to find the right side pointing in to the protein interior in a globular protein.

Distinctions are often made between *globular* and **fibrous proteins (*1.14)**. This can indeed be a useful distinction, and the fibrous proteins such as collagen and silk are indeed different from normal globular proteins. But the difference stems not so much from their *structure* as from their *function*. In particular, collagen and silk are extracellular proteins that are built to last. Once made, they are not intended to be degraded or modified. Almost every other protein binds, releases, moves, undergoes reversible covalent modification, and is *regulated*. A good example of such a regulated, but fibrous, protein is actin. However, these fibers are continually formed and broken up, modified and decorated with other proteins. The fiber is therefore formed not from long, continuous, and rather featureless secondary structure, but from smaller monomeric globular building blocks that can be added or removed as required, as discussed in Chapter 7. In other words, *function* drives *structure*.

1.3.3 Membrane proteins are different from globular proteins

We have seen in Section 1.2.2 that an unpaired hydrogen-bonding group is energetically very unfavorable. In globular (that is, water-soluble) proteins, any backbone amide NH or CO that is not hydrogen bonded to some other group on the protein can usually be hydrogen bonded to water. Because the interiors of globular proteins are less accessible to water, they are usually made up from regular secondary structures, which form hydrogen bonds within the structure. However, the outside of a globular protein consists mainly of loops in which many of the hydrogen bonds are to water.

A membrane protein does not have this option, and in particular it requires that all the backbone amides on the transmembrane face should form hydrogen bonds. This means that most transmembrane proteins are made from α helices, in which every amide group is hydrogen bonded (**Figure 1.40**). The ends of the helices are in the aqueous environment outside the membrane and are therefore free to form whatever loops and turns they want.

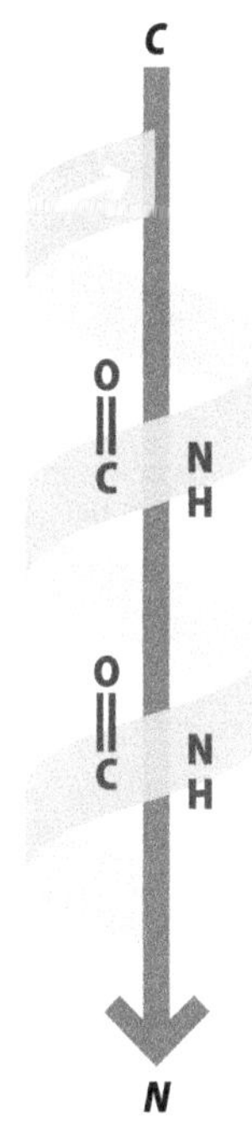

FIGURE 1.39
An α helix has all its NH groups pointing down the helix (toward the N terminus) and all its CO groups pointing up the helix; consequently, it has a macroscopic dipole moment.

***1.14 Fibrous proteins**

Permanent fibrous proteins (that is, not including those proteins such as actin and tubulin that form fibers but are not themselves fibrous) can be either α-helical or β-sheet. Helical proteins include keratin (as found in skin, hair, and feathers) and other intermediate filaments mentioned in Chapter 7. These are **coiled-coil** proteins, which are usually further bundled up into larger fibers. They also include collagen, which is the main protein of connective tissue (skin, tendons, cartilage, and bone) and is the most abundant protein in mammals, composing one-quarter of their total weight. Collagen is formed from the consensus sequence Gly-Pro-Hypro, where Hypro is 4-hydroxyproline, an example of posttranslational modification that requires vitamin C as a cofactor. This is a parallel triple helix (**Figure 1.14.1**), in which glycine is necessary because it comes in the middle of the helix: larger amino acids would disrupt the structure. Each individual peptide chain forms a **polyproline II helix**.

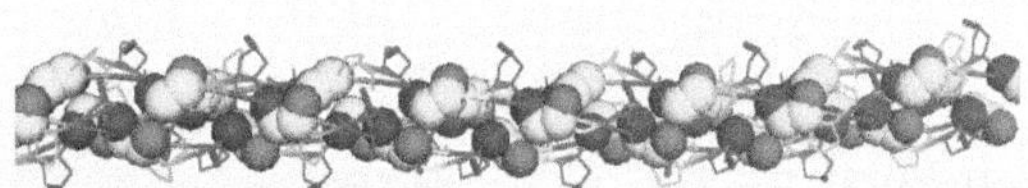

FIGURE 1.14.1
The structure of collagen. Collagen is a parallel triple helix, made of repeating -(Pro-Hypro-Gly)- units, where Hypro is 4-hydroxyproline. The glycines are shown as space-filling spheres and pack in the center of the helix. The prolines (orange) and hydroxyprolines (magenta) are arranged around the outside. The three helices wind around each other. Each strand forms a polyproline II helix.

The β-sheet fibrous proteins are typified by silk. Silk from the commercial silkworm consists of long Gly-Ala repeats, which form an antiparallel β sheet. By contrast, spider silk contains poly-Ala blocks interspersed with sequences that contain a high proportion of glycine. The poly-Ala blocks form crystalline β-sheet regions, and the glycine-rich regions are largely amorphous [136]; the strength of the silk seems to derive largely from the cross-linking effect of the crystalline regions, whereas the amorphous regions give the silk its flexibility.

It is interesting to note that collagen and silk both constitute low-complexity proteins, which are generally natively unstructured.

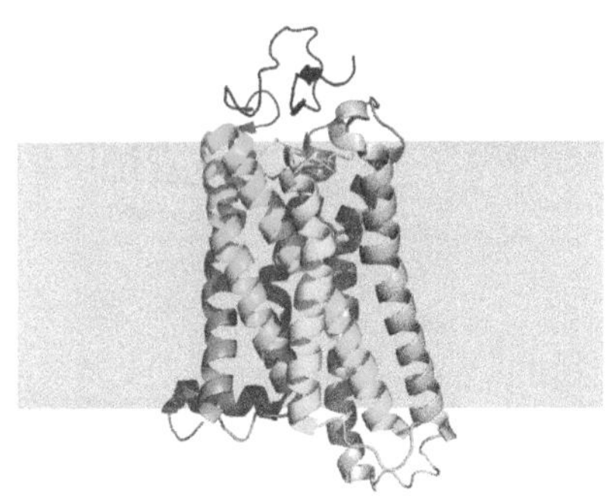

FIGURE 1.40
The structure of a membrane protein, the seven-transmembrane helix protein rhodopsin. The approximate positions of the membrane surfaces are indicated. The retinal ligand is shown in red sticks. All residues within the membrane are helical, except for a few at the top center, which are accessible to water. (From PDB file 3c9l.)

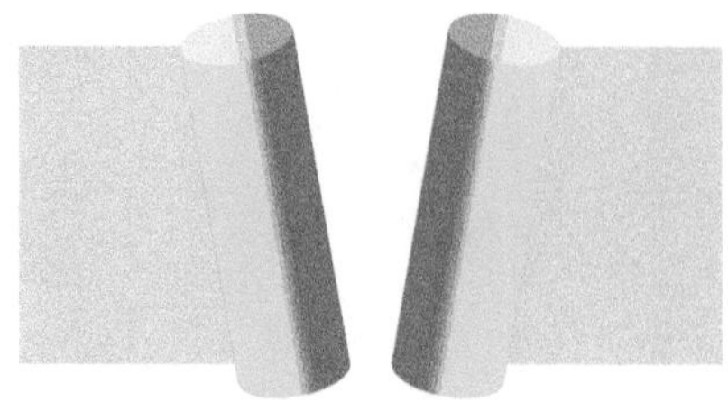

FIGURE 1.41
Many membrane proteins are helical and have hydrophobic outside faces (green) but hydrophilic inside faces (red).

There is still very little information on the structure and dynamics of membrane proteins. It seems that (unlike the case in soluble globular proteins) the interactions between one transmembrane helix and another are often not strong, allowing considerable relative movement of one helix against another. This is likely to contribute significantly to the function of many transmembrane proteins; an example is the integrins, which are important adhesion receptors, and in which the ready separation of transmembrane helices caused by phosphorylation and/or binding to intracellular ligands is probably crucial for their function [36]. As noted in the previous section, many helices are amphipathic. In globular proteins, they are hydrophilic on the outside face and hydrophobic on the inside; in membrane proteins, the opposite is often true, meaning that many helical membrane proteins are able to bind, and sometimes transport, hydrophilic molecules through a channel defined by the hydrophilic faces of their helices (**Figure 1.41**). This can be illustrated by the potassium channel from *Streptomyces lividans*, which has a channel formed by four identical subunits (**Figure 1.42**) [37, 38]. At the inner cytoplasmic end, the channel has a negatively charged hydrophilic face, appropriate for binding a positively charged ion. In the center of the membrane is a water-filled cavity that has four helices pointing their C-terminal ends toward it (**Figure 1.43**). The effect of this is to stabilize positive charges in the cavity, which therefore acts as a holding area for potassium ions ready to pass through the second half of the channel. The crucial selectivity filter for potassium rather than sodium (a factor of 10^4!) is at the other end of the channel: it consists of a series of backbone carbonyls, placed so that the potassium ions have to be dehydrated to pass through and can only go through in single file. It is believed that the free energy required to dehydrate the potassium ions is well matched by the free energy provided by binding of the carbonyl groups, making this a low-energy process, whereas the smaller sodium ions are much less well bound by the carbonyls and are therefore prevented from passing through the channel by the high energetic cost of doing so. Despite this rather restrictive pore, potassium ions are able to move through at up to 10^8 ions per second, a remarkable speed.

A relatively small number of transmembrane proteins cross the membrane as β sheets. A normal β sheet has unsatisfied hydrogen bonds at the edges of the sheet, which would not be possible in a transmembrane protein, and therefore all β-sheet transmembrane proteins form β barrels (*1.12), in which the sheet curls round onto itself to make a continuous hydrogen-bonding pattern

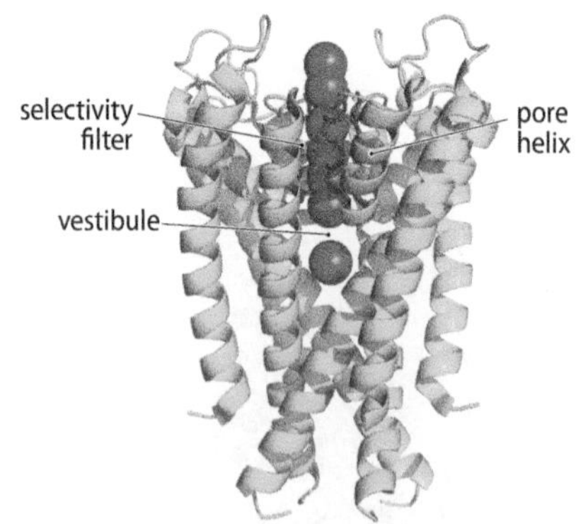

FIGURE 1.42
The potassium channel from *Streptomyces lividans* (PDB file 1k4c). The channel is formed from four identical subunits. The cytoplasmic end is at the bottom, and the purple spheres are potassium ions. The lowest potassium ion is inside the water-filled cavity and is stabilized by the negative dipole of the C termini of the four pore helices. The selectivity filter is formed from the four regions colored red, which bind the potassium ions via their carbonyl oxygens (pointing in toward the potassium ions).

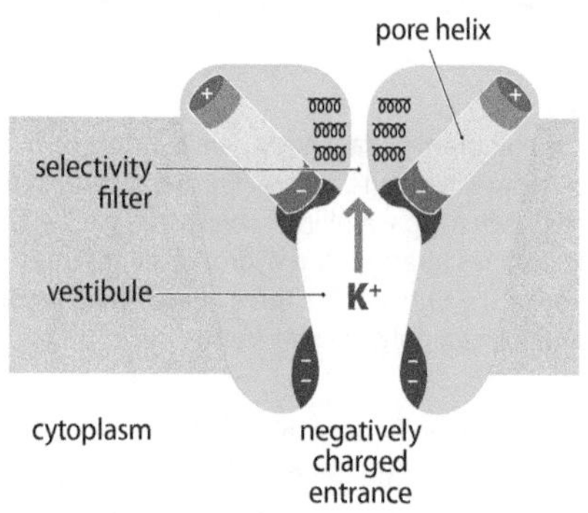

FIGURE 1.43
Diagram of the potassium channel. The selectivity filter is "spring tensioned" by aromatic–hydrophobic interactions, to keep it at exactly the correct dimensions to allow potassium through but prevent sodium. The pore helices have their negative ends pointing toward the vestibule, and they therefore stabilize potassium ions.

(**Figure 1.44**). Almost all transmembrane β barrels have an even number of strands, which means that they can form continuous antiparallel sheets. Many β-barrel proteins act as channels.

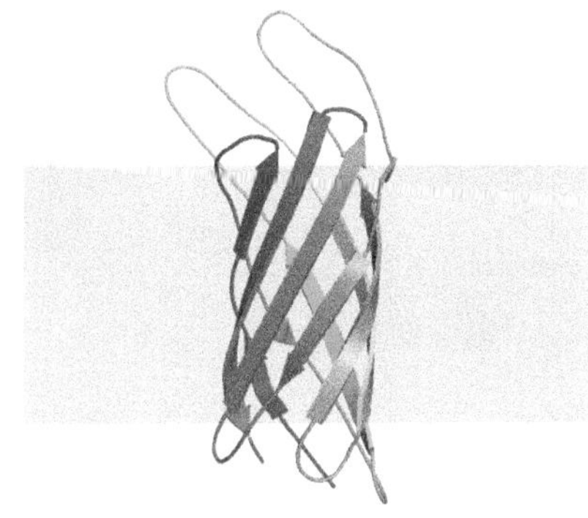

FIGURE 1.44
The NMR structure of the bacterial outer membrane protein OmpX, which is an eight-stranded β barrel (from PDB file 1q9f). Many such proteins are channels, but this protein is used to attach the bacterium to mammalian cells; it is important for generating pathogenicity and is therefore called a *virulence factor*.

1.3.4 The structure of a protein is (more or less) determined by its sequence

A fundamental truth about proteins, first clearly stated by Christian Anfinsen in the 1960s [39], is that the tertiary **native structure** or fold or a protein is determined only by its primary sequence. This observation has two big implications. One is that it should be possible to predict a protein's structure from its sequence. This has been the goal of a large number of scientists for a very long time; we are certainly getting better at it, although we are nowhere near being able to do it properly yet. One of the currently most successful programs for generating structure from sequence (*"structure prediction"*) is called Rosetta [40]. This program is based as far as possible on forces and energies that have a good theoretical basis, such as hydrophobic burial and interactions between pairs of amino acids. However, to produce good predictions it also needs to use relationships derived from analyzing how proteins fold, such as the way in which strands and helices pack together, and it needs to do a lot of sorting of the results to eliminate unlikely-looking protein folds and to cluster results together—which clearly is not something that real proteins do when they fold. This methodology is typical and reflects the fact that although we now know a lot about the forces governing protein structure, we still do not know enough.

The other implication is that the tertiary fold of a protein is the global energy minimum, or in other words the most stable state of the protein. This is often represented on a conformational energy diagram (**Figure 1.45**), in which the energy is on the vertical axis and the horizontal scale represents in a very schematic way the range of conformations available to a protein. This type of diagram is often described as a protein **energy landscape**; it is discussed in more detail in Chapter 6. The landscape drawn in Figure 1.45a has smooth sides, meaning that if the structure is perturbed it will always relax back to the native state. In real life, landscapes are probably not as smooth as this. On a large scale (Figure 1.45b) there are probably "hollows" on the landscape, so that proteins can adopt a range of **metastable** conformations [41]. On a small scale (Figure 1.45c) the landscape is probably rough: proteins can adopt a very large number of similar conformations, differing for example in side-chain conformation or small details of backbone conformation that have similar energies and small energy barriers between them, implying that the exchange between one conformation and another is very rapid.

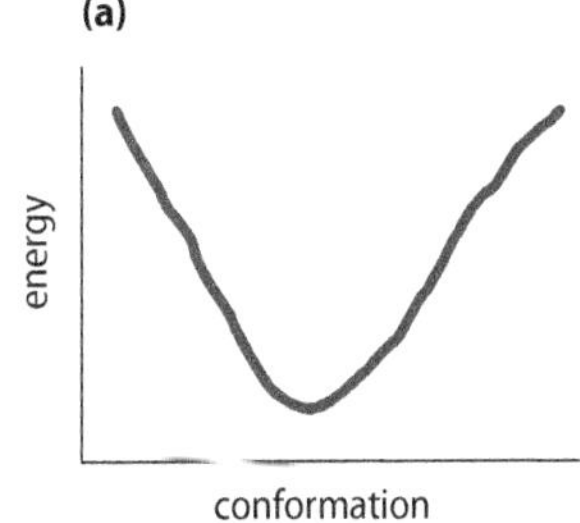

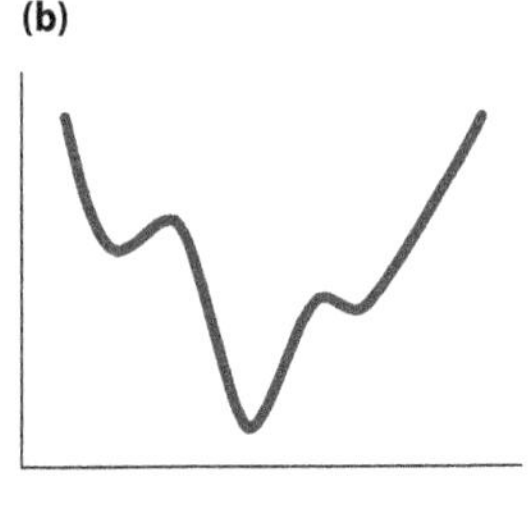

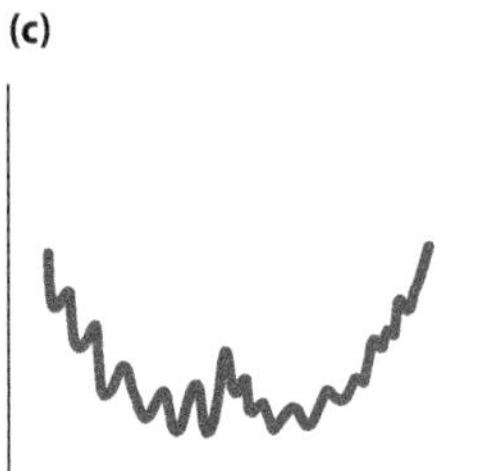

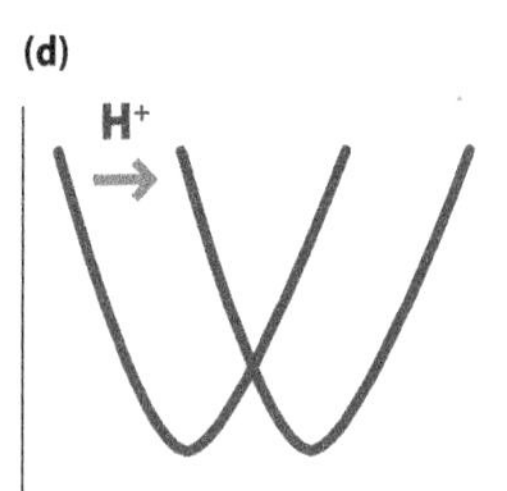

FIGURE 1.45
Protein conformational energy landscapes. The vertical axis is energy (roughly free energy, but not always well defined), and the horizontal axis is a one-dimensional representation of a very multidimensional surface. (a) A typical protein has a single well-defined energy minimum, so that it folds up smoothly to the global minimum. (b) Some (possibly many) proteins have higher-energy metastable states, which usually have low-energy barriers to the global minimum and therefore are very short-lived. (c) Locally, the surface of the landscape is probably very rough. However, most of the energy barriers are low enough for proteins at room temperature to be able to cross them rapidly. (d) A change in conditions, in this case the binding of a proton (that is, a lowering of the pH) causes a change in the structure of the global minimum.

FIGURE 1.46
The formation of an endosome. A protein (in this case hemagglutinin on the surface of a virus particle) binds to a protein on the cell surface. This binding stimulates the formation of an endosome, which is pulled into the cell interior and becomes acidified, normally releasing the protein from the receptor.

We have seen that the conformation of a protein is a balance between a large number of weak forces. A (literally!) vitally important point about this analysis is that the shape of the landscape and the position of the global energy minimum will change as solution conditions change and alter the forces acting on the protein. For example, if the pH is changed, the distribution of charges on the protein will change, leading to conformational rearrangement and in almost all cases eventually to unfolding of the protein. Similarly, if the temperature or pressure is increased, or if a range of denaturants (*1.10) are added, the protein will unfold. Equally significantly, if a ligand is added, its binding will contribute a set of new forces, which are easily strong enough to change the protein conformation by a small but often very significant amount (Figure 1.45d). Ligand binding causes conformational changes to enzymes to make the activated enzyme (Chapter 3) and to regulatory proteins such as GTPases to act as a "switch," flicking the conformation between one of two states depending on the structure of the bound ligand (Chapter 7).

1.3.5 Some proteins form metastable structures

In most cases these conformational changes are small, although they are large enough to have important biological effects. In a few cases evolution has developed the conformational changes to lead to much more dramatic changes in structure. One example is provided by hemagglutinin from influenza virus [42, 43]. This is a protein on the surface of the virus that recognizes human cells and attaches the virus to a glycoprotein on the cell surface, and then uses the cell's response to fuse the viral membrane with the cell membrane and introduce the viral RNA into the cell. The binding of hemagglutinin stimulates the cell to form an endosome (**Figure 1.46**). This is a common response to foreign material: the endosome becomes acidified, weakening the interaction between cell surface protein and ligand; the ligand dissociates and can be digested or taken up by the cell, while the cell membrane and integral proteins get recycled back to the cell surface. However, in this case, the lower pH leads to a marked change in the structure of the viral protein (**Figure 1.47**), in which one hinge opens and another closes, leading to the uncovering of

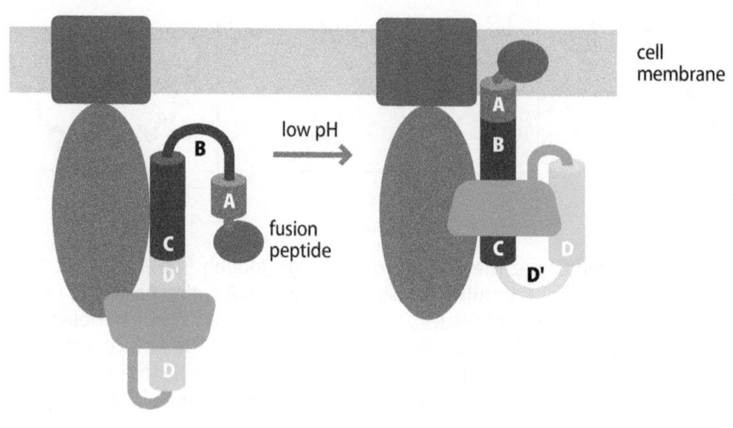

FIGURE 1.47
The conformational change in hemagglutinin at low pH. Hemagglutinin is part of the viral coat and is adjacent to the ochre region that makes the initial binding interaction with the target cell. At low pH, an extended chain (B) becomes helical and forms a continuous helix with C and A, thereby projecting upward and pushing the fusion peptide into the cell membrane, where it inserts to bind the hemagglutinin (and therefore the virus) and prevent it from dissociating. At the same time, part of helix D (D′) unwinds to pull the C-terminal region upward, closer to the cell membrane. This C-terminal region is linked to the virus coat membrane, and the combined effect is to bring the viral membrane close to the cell membrane and (by an as yet unknown mechanism) stimulate membrane fusion. (Adapted from P.A. Bullough et al., *Nature* 371:37–43, 1994. With permission from Macmillan Publishers Ltd.)

a hydrophobic "fusion peptide" that attaches itself to the cell membrane and locks the virus onto the cell membrane, preventing the normal detachment process. There is then a second major conformational change in which the protein segment that links hemagglutinin to the viral membrane zips up the side of the newly uncovered protein, a change that brings the viral membrane close to the cell membrane. Finally, in a process that is still not well understood, the two membranes fuse. A related mechanism is likely to be employed by the proteins used by the cell for membrane fusion, many of which also have pH-regulated conformations [44]. The other major group of proteins showing this behavior is the serpins (see Problem 7).

Another significant example of proteins having quite different conformational states in different conditions is amyloid, discussed at the end of Chapter 4. Three-dimensional domain swapping, discussed in Chapter 2, provides a further example of proteins that can adopt more than one stable conformation. In both these cases, the "different conditions" relate to the protein concentration: at low concentration the protein adopts its normal conformation, but at high concentration it is more stable in the amyloid or domain-swapped conformation. Several other examples are discussed by Murzin [45], who notes that the metastable form is often stabilized by the formation of asymmetric dimers with the "normal" monomer, and argues that these may be mutational accidents that have been seized on and developed by natural selection.

1.3.6 Structure is conserved more than sequence

DNA mutates at a roughly constant rate as a result of copying errors and chemical degradation, although like almost everything else this rate is under evolutionary control and can be increased, for example when an organism is subject to unusually strong selection pressures. This means that proteins also mutate at a roughly constant rate. However, at the protein (and therefore the phenotypic) level, selection operates much more strongly. This means in particular that proteins that interact with a large number of partners (for example interactome core proteins, as discussed in Chapter 9), and therefore need to keep their surfaces constant, have a low mutation rate. The classic example is the **histone** proteins, which interact both with DNA and with numerous proteins that regulate DNA expression: these have very low mutation rates. By contrast, proteins under high evolutionary pressure, such as immunoglobulins or secreted proteins, tend to evolve much faster (**Table 1.3**). Mutation rates of different types of protein have been observed to vary by a factor of 1000, although a factor of 30 is more common. A good example of evolutionary pressure altering the mutation rate is provided by lysozyme proteins in langurs, in which a novel evolutionary pressure (a change in diet requiring ruminant behavior) has led to a mutation rate that is 2.5 times higher than that in the related baboons [46].

TABLE 1.3 Evolutionary rates of proteins

Protein	Rate[a]
Histone H4	0.0025
Histone H1	0.13
Cytochrome *c*	0.07
Insulin	0.07
Triosephosphate isomerase	0.05
Hemoglobin β	0.3
Snake short neurotoxin	1.3
Immunoglobulin IgG (V)	1.4

[a]The rate is quoted as amino acids per 100 residues altered per million years. (Data taken from A.C. Wilson, S.C. Carlson and T.J. White, *Annu. Rev. Biochem.* 44:573–639, 1977.)

1.3.7 Structural homology can be used to identify function

The other major constraint on protein evolution is provided by structure and function: amino acids that are required for the function or structure of the protein change more slowly than others. This is the basis for a wide range of bioinformatics methods, discussed more fully in Chapter 11. Amino acid sequences from homologous proteins can be aligned rapidly and easily. Locations that show greater **conservation** are likely to be important, either for function or for structure (**Figure 1.48**). The active-site residues of an enzyme can often be identified in this way; an example is the Ser-Asp-His **catalytic triad (*1.15)** of serine proteases. Regions of the protein required for function can also be identified. Thus, for example, many proteins that bind nucleotides do so using a domain known as the Rossmann fold (*2.1),

```
score =  108 bits (271),  expect = 8e-29, method: compositional matrix adjust
identities = 84/239 (35%), positives = 120/239 (50%), gaps = 24/239 (10%)

Query  1    VGGEDAIPHSWPWQISLQYLRDNTWRHTCGGTLITPNHVLTAAHCISNTLTYRVA--LGK  58
            VGG  A PH+WP+ +SLQ LR     H CG TLI PN V++AAHC++N     V   LG
Sbjct  1    VGGRRARPHAWPFMVSLQ-LRGG---HFCGATLIAPNFVMSAAHCVANVNVRAVRVVLGA  56

Query  59   NNLEVEDEAGSLYVGVDTIFVHEKWNSFLVRNDIALIKLAETVELSDTIQVACLPEEGSL  118
            +NL   +    ++  V  IF    ++   + NDI +++L  +  ++  +QVA LP +G
Sbjct  57   HNLSRREPTRQVFA-VQRIF-ENGYDPVNLLNDIVILQLNGSATINANVQVAQLPAQGRR  114

Query  119  LPQDYPCFVTGWGRLYTNGPIAAELQQGLQPVVDYATCSQRDWWGTTVKETMVCAGGDGV  178
            L     C   GWG L  N  IA+ LQ+ L   V  + C + +        T+V     GV
Sbjct  115  LGNGVQCLAMGWGLLGRNRGIASVLQE-LNVTVVTSLCRRSNVC------TLVRGRQAGV  167

Query  179  ISACNGDSGGPLNCQAENGNWDVRGIVSFGSGLSCNTFKKPTVFTRVSAYIDWINQKLQ   237
               C GDSG PL C     N  + GI SF  G  C +   P  F  V+ +++WI+  +Q
Sbjct  168  ---CFGDSGSPLVC-----NGLIHGIASFVRG-GCASGLYPDAFAPVAQFVNWIDSIIQ   217
```

FIGURE 1.48
A sequence alignment for bovine chymotrypsin (top) and human neutrophil elastase (bottom), performed with BLAST (http://blast.ncbi.nlm.nih.gov/Blast.cgi). The two sequences are 34% identical and are therefore clearly homologous, with very probably identical folds and probably similar function. The conserved residues are given in the center of the alignment and include the catalytic triad (*1.15) (S, D, H, highlighted) and not surprisingly most of the cysteines, many of the prolines, half of the glycines, and many hydrophobic residues.

*1.15 Catalytic triad

All serine proteases contain three amino acids in their active site that are crucial for enzyme activity, namely serine, histidine, and aspartate. The serine acts as a nucleophile and attacks the peptide bond to be cleaved (**Figure 1.15.1**). As it attacks, the histidine acts as a general base and removes the proton from the serine OH group, making it a much better nucleophile. The aspartate hydrogen bonds to the histidine and stabilizes the positive charge that forms on it as it abstracts the proton. The three-dimensional arrangement of the catalytic triad is closely similar in chymotrypsin and subtilisin, despite their having completely different structures otherwise, and is the classic example of **convergent evolution**. There are related families of proteases in which some of the catalytic triad residues are missing or different, such as Ser-Lys.

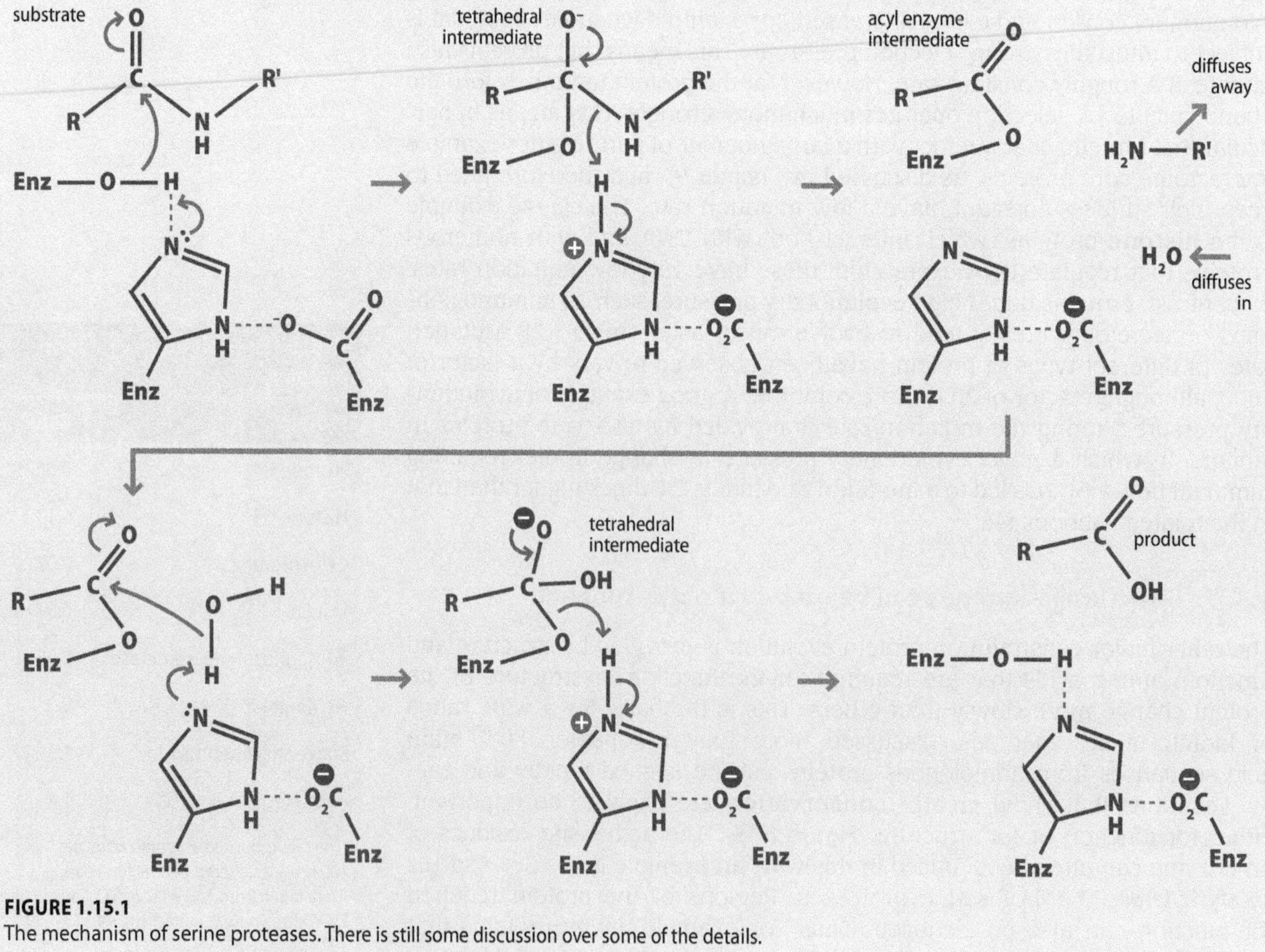

FIGURE 1.15.1
The mechanism of serine proteases. There is still some discussion over some of the details.

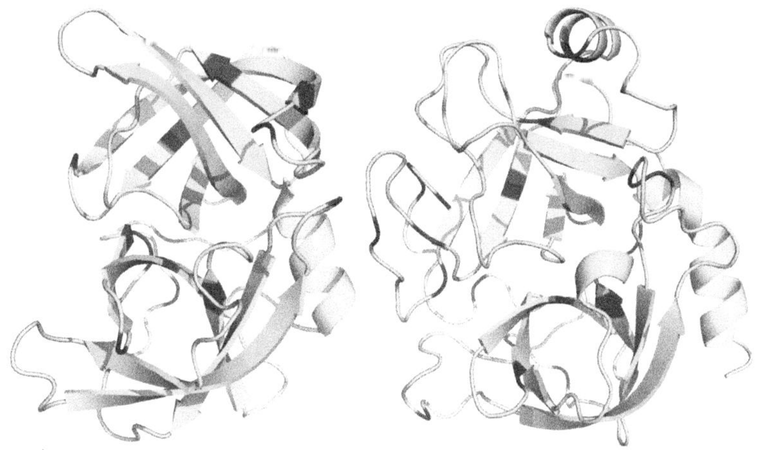

FIGURE 1.49
The structures of *Streptomyces griseus* protease B (left) and bovine β-trypsin (right). The two structures are clearly related, despite the presence for example of an extra helix at the top of β-trypsin. However, the residues with sequence identity (red) are essentially random and of no functional significance: many are in regions of the structure where the structures have diverged significantly and are therefore not "real" matches.

discussed in Chapter 2. This fold has a conserved loop that winds tightly round the nucleotide, with the sequence GXGXXG (G = glycine, X = any other amino acid), where the glycines are required for the nucleotide-binding function because their small size allows them both to approach the nucleotide closely and to adopt an unusual backbone geometry and thus a position in the Ramachandran plot that cannot easily be adopted by other amino acids. This short sequence is called a **sequence motif**. Many motifs are known; these are collected into databases on websites, as listed at the end of this chapter. Indeed, glycines are often fairly highly conserved because they have a unique role in protein function, and the same is true for proline and cysteine.

The requirement to preserve protein structure means that structure is much more highly conserved than sequence in homologous proteins. For example, **Figure 1.49** shows the backbones of two proteins that have diverged a long time ago, and consequently only have less than 15% sequence identity. Nevertheless, their structures are very similar. The sequences matched by BLAST are in different parts of the sequence and are merely fortuitous, whereas the "real" sequence identities (shown in red) have no significance. Over the course of time, protein sequences in homologous proteins have diverged, making it difficult to identify homologous proteins from their sequences. Very similar protein sequences can be assigned with confidence as being homologs, meaning that two proteins with similar sequence have the same fold and very probably the same function. As the sequences become more different, this gets harder: once the protein sequences are less than about 30% identical, it starts to become difficult to make alignments with any confidence, and once it is less than 20% it is virtually impossible. This region between 30% and 20% identity has therefore been called the "twilight zone" [47] (**Figure 1.50**). However, the *structures* of homologous proteins are very largely conserved, and one can find proteins with only 5% sequence identity that still have very similar structures.

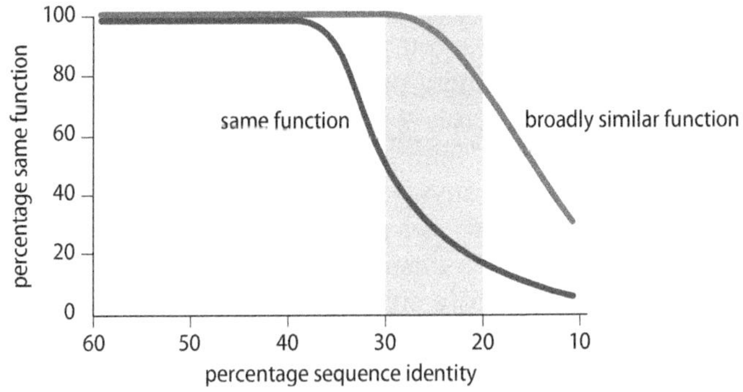

FIGURE 1.50
The twilight zone of sequence similarity, where sequence identity is in the range 20–30%. If the sequence identity is greater than about 30%, pairwise sequence alignments are reasonably straightforward and one can be fairly confident that the two proteins will have the same fold and similar, if not identical, functions. However, by the time the sequence identity is less than 20%, it is no longer possible to make pairwise alignments and very difficult to make any useful alignment, and it is far less likely that the two proteins will have the same function. They are, however, likely to have similar folds.

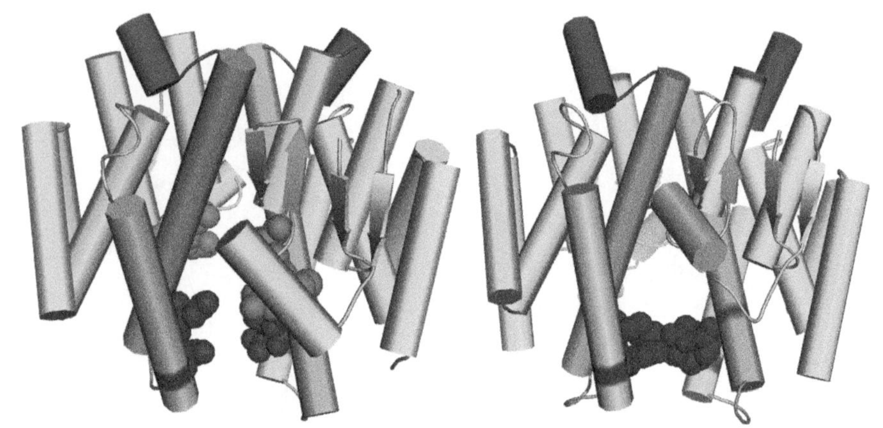

FIGURE 1.51
An example of an indel. There are several isoforms of the dimeric enzyme glutathione S-transferase from the mosquito *Anopheles dirus* species B, which differ by alternative splicing of their mRNAs. Isoform 1-4 (right) contains six extra amino acids compared with isoform 1-3 (left), which have extended the length of a small helix on the surface (shown in purple). The effect of this addition is to alter the packing of the helices, which means that at the other end of a long lever helix (yellow), the residues at the entrance to the binding site (brown) are pushed closer together. Consequently, access to the binding site for the substrate glutathione (orange) is more restricted. Note that the active site is in the interface between the two monomer domains. In this figure, helices are shown as cylinders for clarity. It is common for alternative splicing to lead to indels [133]. (From PDB files 1jlv and 1jlw.)

This level of sequence identity corresponds to the identity between any two proteins chosen at random. [At each position there is a choice of 20 possible amino acids, which would imply a 5% probability. However, different amino acids occur with very different frequencies in proteins (see Table 1.2), which increases the chances of sequence identity, implying that random matches are actually more frequent than 5%.] Several programs are available for comparing tertiary structures, of which a common one is DALI.

In the protein alignments described above, it is normal to allow the sequences to contain a certain number of **insertions and deletions** (indels). Almost all indels occur in loops rather than in regular secondary structure (**Figure 1.51**). Therefore the presence of indels serves both as a check that sequence alignments have been done correctly (if the structure is known) and as an indication of the location of secondary structure (if the structure is not known). Indels can be very long: sometimes a complete extra domain can be inserted. The classic example is immunoglobulins, in which the basic fold is highly conserved and essentially all the variation required for recognition of antigens comes from loops, which undergo extensive insertions and deletions. This topic is discussed in more detail in Section 5.1.3.

1.4 THE EVOLUTION OF PROTEINS

1.4.1 What are the purposes of proteins?

The first part of this chapter dealt with protein *structure*. We turn now to protein *evolution*.

Life got to look like it does now by a process of evolution, or in other words by **natural selection**. If some biochemical change was advantageous in helping its host organism to generate more offspring, then it survived. This section is provocatively called "What are the *purposes* of proteins?"; I hope it is agreed that the "purpose" of a protein is to perform some function that helps the host propagate its offspring (that is, to improve its fitness). It has no other function that can properly be described a "purpose." *However*, this book will occasionally describe the evolution of new protein function as the protein "trying to achieve some aim." This is merely a convenient shorthand. The protein does not "know" anything; it is not "aiming" at anything. *Nature is not teleological*: that is, it is not seeking to get anywhere. Organisms survive and produce offspring, and the more successful ones are those that ultimately prevail. If a new protein, or a new control mechanism, helps the organism, it will be passed on to offspring and will therefore survive. Therefore, when we try to understand what the *function* of a protein is, we have to be careful. As far as real life ("nature") is concerned, we can legitimately say that the function is to make the organism fitter. We can further try to describe how it does this by catalyzing a reaction, regulating a process, increasing selectivity, and so on: these are useful attempts to understand the primary Darwinian function because they explore how an individual protein's properties help to make the host fitter. Therefore it is useful to describe the function of a protein as, for example, being an enzyme or a regulatory protein and to analyze how it does this—how it is that the structure of the protein enables it to perform this function better. To that extent (and only to that extent) it is legitimate to say that the reason *why* a protein has the structure or properties that it does is because it enables it to perform some particular function efficiently. In other words, if we ask *why* a protein has the properties that it does (which this book will do frequently!), we are assuming that the function makes the host fitter. This is the traditional reductionist view, that if individual components work better or faster, then the overall process also does. The new science of *Systems Biology* (Chapter 11) is trying to see whether and how this is really true, and I hope that some of the things described in this book may help generate ideas on some directions in which Systems Biology can develop. *How* a protein performs its biological function is a very relevant and powerful question: *why* it does it like that can only be justified as "because it works better than anything else that has been evolved."

In this context it is perhaps helpful to repeat a story told by Richard Perham in an old but still excellent review on pyruvate dehydrogenase [48]. In the early nineteenth century, the two leading diplomats in Europe were Talleyrand, who was Napoleon's foreign minister but remarkably survived Napoleon's downfall, and Metternich, an Austrian prince. They had frequent diplomatic scuffles, leading them to be respectful of each other but suspicious of their devious motives. It is reported that, on hearing of Tallyrand's death, Metternich said, "I wonder what he *meant* by that?" The moral is that there are some phenomena, even in the world of proteins, that have no "ulterior motive," they just happen, and we should beware of trying to read too much into every observation.

This lengthy and slightly apologetic introduction is important, because *why* is often a very useful question, if understood correctly. In particular, to ask *why* properly, we must always have the random character of evolution at the back of our minds. We must remember that evolution just happens, with no goal or plan. Sometimes this makes it arrive at a dead end, or end up in places that are not actually helpful for further development. It is therefore always possible that there is no good answer to "why" except "it evolved that way." This book will often refer back to evolutionary processes when trying to explain "why," because this is the only way to answer the question.

1.4.2 Evolution is a tinker

In his wonderful book Chance and Necessity [49], Jacques Monod describes nature as a tinker. A tinker is now sadly a dying breed, but these are people who take your old pots and mend them, or perhaps turn them into something else that is more useful to you, the traditional "make do and mend." Nature does exactly the same: it takes whatever is there and turns it into something more useful. Therefore, if a new enzyme is needed, it is converted from an existing one rather than being made anew. Moreover, as we shall see numerous times, nature tends to use a closely related protein as its starting point, mainly because it is easily available and presumably already has a rudimentary desired activity, a characteristic sometimes called *opportunism* [50]. For example, the widespread use of penicillin antibiotics since the 1950s created an urgent need (in the evolutionary sense described above) for bacteria to acquire β-lactamases—enzymes that would break down penicillin. It is now clear that some bacteria already had β-lactamases; these were then transferred from one bacterium to another by horizontal transfer, this being far simpler and

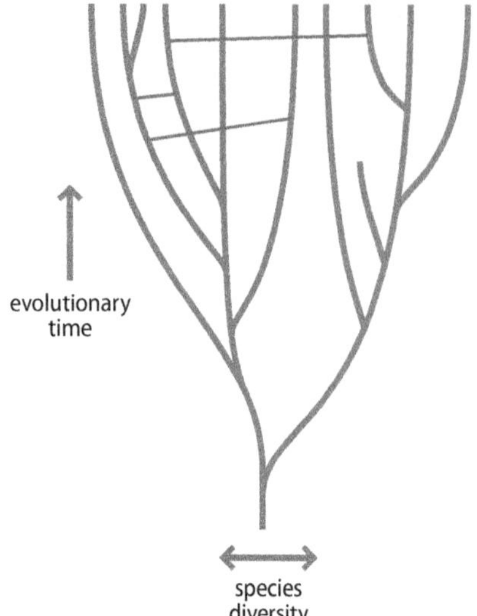

FIGURE 1.52
Horizontal gene transfer. Most evolution is a branching tree: species diverge and differentiate over long periods of historical time. However, horizontal gene transfer involves the direct transfer of DNA from one species to another; it effectively bridges the gaps between different branches of the tree. It is particularly common between bacterial species.

quicker than any other evolutionary process. The original β-lactamases may well have been penicillin-binding proteins, implying that bacteria took proteins whose function was to recognize and bind to penicillin, and modified them by adding catalytic groups in suitable positions—again, a much simpler process than evolving a completely new protein. Indeed, it is now clear that horizontal transfer has been very significant: *Escherichia coli* is thought to have picked up 10–15% of its genome like this recently [51–53]. This makes evolution a far from linear process (**Figure 1.52**).

Such a transfer can of course occur even if there is an existing enzyme. In this case, evolution will try out both enzymes and usually discard the less successful one. In comparing genomes of different organisms, it is common to find that an enzyme with a clearly defined function in one organism is completely unrelated to the same enzyme in other (sometimes closely related) organisms but is similar to an enzyme with a related function. This is called **nonorthologous gene displacement** and represents a second independent evolution of the same function [54]. The unrelated but functionally equivalent enzymes have been called **analogous**. It is a widespread phenomenon that has often been observed as a consequence of evolutionary pressure. Evolution has clearly tinkered in this way repeatedly, a good example being the essential and fundamental proteins that make up the transfer RNA synthetases [55], which fall into two classes representing quite different solutions. Surprisingly, the replacement of enzymes even in very basic pathways seems to be common [51, 56].

1.4.3 Many proteins arose by gene duplication

If a penicillin-binding protein is mutated to create a β-lactamase, then the original and presumably useful penicillin-binding function is lost. Unless the mutated source is a **cryptic (*1.16)** or pseudo gene, the mutation therefore requires a previous **gene duplication**, in which one copy retains its original function and the other is free to acquire a different function. There are several mechanisms for gene duplication [57], one of the most common in eukaryotes being **homologous recombination (*1.17)** between retrovirus-derived mobile genetic elements such as the Alu **pseudogene (*1.18)** sequence in humans. It has been estimated that at least one-third of the genes in *Haemophilus influenzae* have arisen from gene duplications [58], as well as 88% of the *Saccharomyces cerevisiae* genome and 98% of the human genome [59, 60]: the rate of gene

***1.16 Cryptic genes**

Genome sequencing has revealed a very large number of "cryptic genes": genes that are closely related to proteins with known biological function but seem to have lost that function by random mutations (of the gene or of its promoter) and are therefore not expressed or are nonfunctional. There is an evolutionary pressure to remove these genes, but particularly in eukaryotes the pressure is not strong because eukaryotes have no great need to replicate the genome rapidly and are quite happy to carry large amounts of "junk DNA." Therefore these genes can remain for a long time. And indeed it can be useful for the organism to keep cryptic genes: they can be converted back into a functional protein. For example, if there is some sudden environmental change, the organism may need to evolve rapidly if it is not to die: the presence of cryptic genes may give it a "reserve" of potential genes that it can call on quickly if necessary. (This makes it sound as though this were a deliberate choice by the organism, but the evidence clearly shows that like everything else the rate of removal of cryptic genes is under evolutionary control, and it happens like this because it has been selected like this. The "explanation" above is merely a reasonable rationalization.) Alternatively, it provides a "junkroom" of proteins that can be raided and tinkered with to produce possible new functions, as discussed later in Chapter 1.

***1.17 Homologous recombination**

This is the exchange of DNA between two double-stranded DNA molecules, by base pairing between identical or closely similar sequences in the two molecules (**Figure 1.17.1**). It is typically used by organisms for repairing damaged DNA but can also be used for introducing or removing new genetic material.

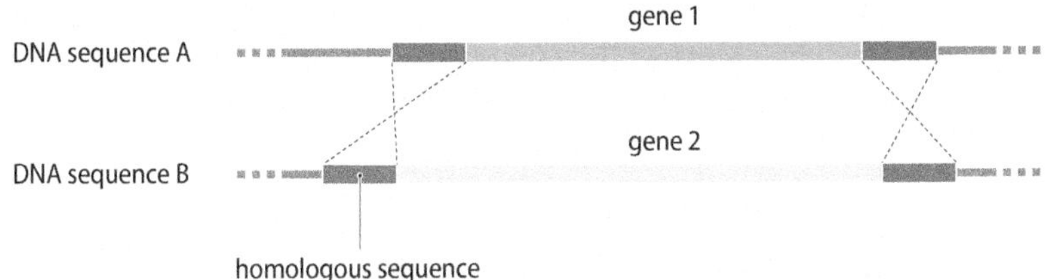

FIGURE 1.17.1
Homologous recombination. In sequence A there are two short DNA sequences that are homologous (that is, identical or closely related) to sequences in B. Homologous recombination leads to an exchange of one sequence for the other, and therefore the replacement of gene 1 by gene 2. This mechanism is used widely both naturally in DNA replication and in molecular biology.

duplication is roughly the same as the rate of mutation per nucleotide site [61]. Another common mechanism is duplication of the entire genome, which has occurred numerous times in evolution and has for example been suggested as the impetus that allowed vertebrates to develop (by two successive duplications) [62, 63]; this allowed *S. cerevisiae* to diverge from *Kluyveromyces* (specifically, to be much more efficient at the anaerobic fermentation of sugar) [64] and also allowed some plants to survive the last mass extinction that killed off the dinosaurs [65].

There is an obvious difficulty with gene duplication: it creates extra copies of genes that require regulating and controlling differently, and it creates a dilemma of how to prevent duplicated genes from interfering with each other. In sexually reproducing species, duplication of the sex chromosomes tends to lead to a loss of fertility [66]. Gene duplication is therefore usually followed by silencing or deletion of a high percentage of unwanted copies [61]. For example, the genome duplication of *Saccharomyces* described above is thought to have been followed by a loss of 85% of the new genes so created [64]. It is also commonly observed that regulatory genes undergo more change than genes coding for catalytic proteins [52, 67], a reflection of the importance of regulating the newly duplicated function correctly.

Occasionally, gene duplication is used just to provide additional copies of the gene and therefore a higher number of protein molecules in the cell. The classic example of this is histone proteins, for which the copy number in *Drosophila* for example is about 100 [57].

It is worth noting that although wholesale genome duplication has clearly occurred many times and has been crucial in the development of new genes, it is unlikely to occur now in eukaryotes at least—except possibly in plants, in which it seems to be much easier than in other eukaryotes. In humans, accidental duplication of a single chromosome occurs reasonably frequently at fertilization, giving three copies of the chromosome. This is known as a trisomy,

***1.18 Pseudogene**

A pseudogene is a DNA sequence that looks like a gene but does not express a protein. It is therefore more or less the same as a cryptic gene (*1.16). Often the creation of a pseudogene has occurred as a result of various mutations and frameshifts. Pseudogenes are therefore nonfunctional, but can be used by evolution as a way of generating new proteins. An important example is the Alu family in primates (a member of the *short interspersed elements* of DNA, or SINEs), which are derived from a gene for cytoplasmic 7SL RNA, a component of the signal recognition particle (*1.20). These are retrotransposons; in other words, they are capable of cutting themselves out and inserting themselves elsewhere in the genome. They therefore form important transposable elements and have often been found associated with gene duplications and insertions. It is estimated that the human genome has more than 20,000 pseudogenes, implying that the number of pseudogenes is similar to the number of genuine genes.

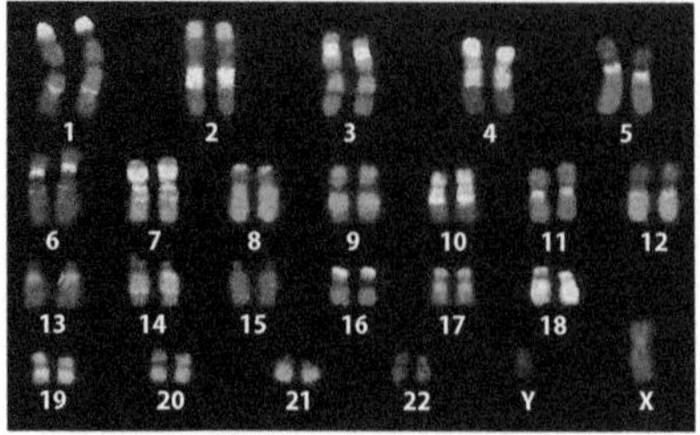

FIGURE 1.53
The 23 pairs of human chromosomes. These are stained with fluorescent probes, a technique known as FISH (fluorescence *in situ* hybridization). The colors are computer-generated by using changes in fluorescence wavelength. Note that the numbering goes in order of size. (Courtesy of Steven M. Carr, after original by Genetix.)

and the most common is trisomy 16 (that is, an extra copy of chromosome 16), which occurs in about 1% of pregnancies but always results in miscarriage early in the pregnancy. In fact, most trisomies are never observed because they cannot produce viable fetuses: the only trisomies that survive birth are of chromosomes 8, 9, 12, 13, 18, 21, and the sex chromosomes. Because human chromosomes are numbered from 1 to 23 from largest to smallest (**Figure 1.53**), this implies that the duplication of large chromosomes always produces nonviable fetuses. Most of the trisomies listed above have severe defects and do not survive, except trisomy 21 (Down syndrome) and trisomy of the sex chromosomes.

1.4.4 Most new proteins arise by modification of duplicated genes

A nice example of this principle can be found in the evolution of the **tricarboxylic acid** (**TCA**) cycle, also known as the **Krebs (*1.19)** cycle. The early atmosphere had little oxygen and was a reducing environment. In its current form in most organisms, the TCA cycle has two main functions: to generate metabolic intermediates and to produce energy in the form of reducing equivalents such as NADH. It is generally believed, however, that in early life on Earth this second function was not required and that it is a relatively late development. In fact, the early Krebs cycle was not a cycle at all but was basically a way of regenerating NAD^+ and also of fixing CO_2, by going from pyruvate to succinate in the "reverse" direction (**Figure 1.54**) [68, 69]. This view is supported, for example, by the fact that the "last" part of the modern-day Krebs cycle is the most highly conserved: a review in 1999 found that in most species whose genomes had been determined the cycle was not in fact a complete cycle [70]. The first steps in the conventional Krebs cycle were also present, as a way of producing the biosynthetic precursor α-ketoglutarate (used to make glutamate) [68]. At some point, the existing enzyme pyruvate dehydrogenase was duplicated and modified to create α-ketoglutarate dehydrogenase, which performs very similar chemistry, therefore permitting the biosynthesis of succinyl-CoA from α-ketoglutarate. Succinyl-CoA is in itself a useful biosynthetic precursor, being used to make porphyrins and therefore heme and chlorophyll.

***1.19 Hans Krebs**

Hans Krebs (**Figure 1.19.1**) was born in Germany in 1900. He was Jewish, and left Germany in 1933 to come to Cambridge, UK, subsequently moving on to Sheffield, UK, where he became the first professor of biochemistry there. He is noted for discovering the tricarboxylic acid cycle, often also called the Krebs cycle, although he also discovered a second cycle, the urea cycle, five years earlier. He was awarded the Nobel Prize in Physiology or Medicine in 1953.

FIGURE 1.19.1
Hans Krebs. (With permission from the Krebs family.)

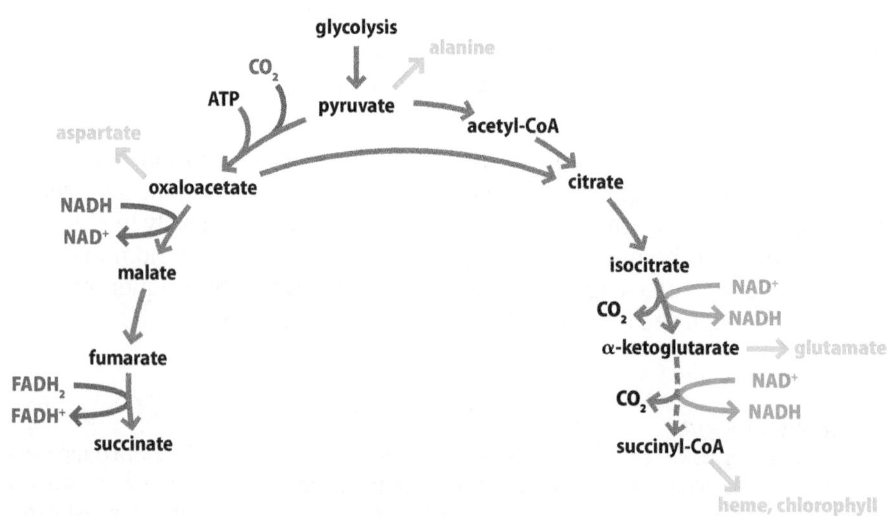

FIGURE 1.54
A probable early form of the Krebs "cycle." The left arm is used for fixing CO_2 and also for regenerating NAD^+—in a reductive early environment, this is a vital function. The right arm is used to produce α-ketoglutarate (the precursor for glutamate). As an extension of the right arm (dashed, using the enzyme α-ketoglutarate dehydrogenase), it goes on to make succinyl-CoA, which is the precursor for heme and chlorophyll.

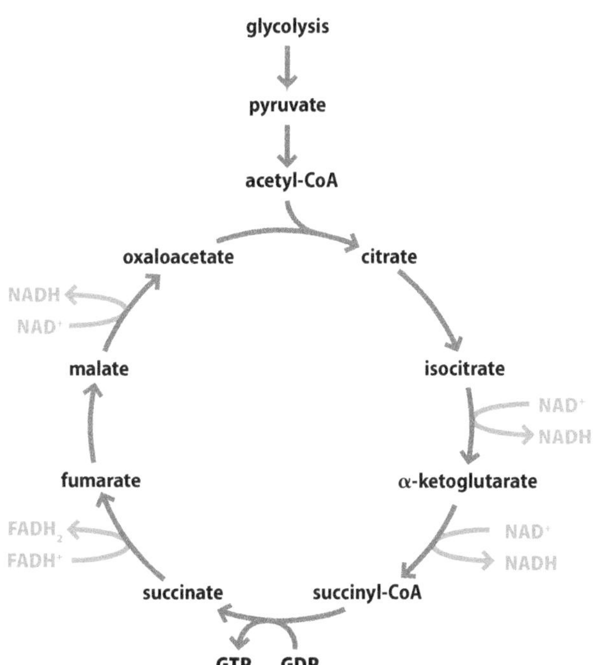

FIGURE 1.55
Modern form of the Krebs cycle. It is worth noting that this is the form we are familiar with in humans, and it works entirely in the direction to create reduced cofactors (NADH and $FADH_2$). However, in many bacteria, particularly photosynthetic ones or bacteria that live in a reducing environment, it still looks much more like that shown in Figure 1.54, with one or more enzymes "missing," and much of it runs "backwards."

As can be seen from **Figure 1.55**, these two arms form most of the enzymes in the current cycle, leaving only one "new" enzyme to be created. In fact in many prokaryotes the Krebs cycle remains in this incomplete form: in many photosynthetic organisms, which again do not need the reducing equivalents but do need the biosynthetic precursors, this "backward" mode of the cycle is very common. The final enzyme needed is now called succinyl-CoA synthetase, and is a common fold, being closely related to acetyl-CoA synthetase and many other acyl-CoA synthetases. It would therefore not be difficult to supply this enzyme by gene duplication and modification.

A somewhat different example is provided by antifreeze proteins. Many fish live in seawater that is below 0°C and therefore need to prevent the formation of ice crystals inside their bodies. They do this by producing antifreeze proteins. This is a relatively recent evolutionary need, and solutions seem to have arisen independently at least four times. The antifreeze from the most common group of Antarctic fish arose from a pancreatic trypsinogen gene. The untranslated 5′ and 3′ ends of the gene provided a secretory signal and a 3′ untranslated region, respectively, and the main bulk of the protein came from **repeat expansions** of a nine-nucleotide section of the trypsinogen gene coding for the amino acids Thr-Ala-Ala, on the boundary between the first intron and second exon [71, 72] (**Figure 1.56**). The scientists who discovered this speculated that the gene may have arisen first in the pancreas, because it was first needed to protect intestinal fluid from freezing, and was subsequently expressed more widely. The rate at which genes and proteins change provides a molecular clock to date the emergence of new proteins [73], which allows the dating of the creation of this antifreeze protein to somewhere in the range

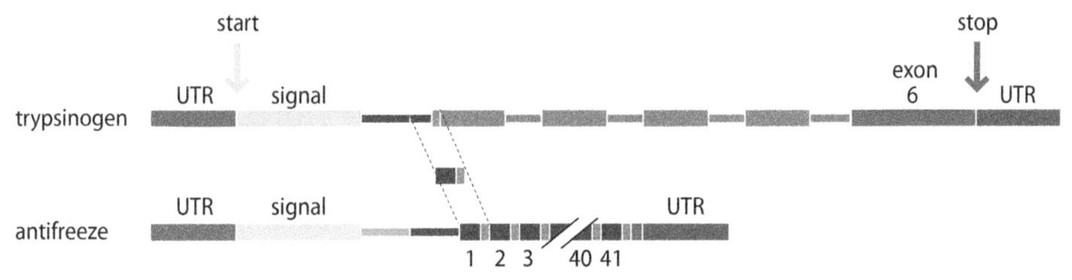

FIGURE 1.56
Suggested evolution of antifreeze protein in Antarctic notothenioid fish. Parts of an ancestral trypsinogen gene were used to create an antifreeze gene. In particular, part of the first intron and the start of the second exon (brown and beige), coding for the tripeptide TAA, were duplicated to give a total of 41 copies, while a frameshift mutation in exon 6 shortened the coding region.

5–14 million years ago; by comparison, Antarctic freezing occurred 10–14 million years ago, supporting the idea that the gene arose as a product of strong selection pressure. By contrast, the antifreeze protein from Arctic cod, which also consists largely of Thr-Ala-Ala repeats, has been shown to have arisen by a completely different route and is thus a good example of **convergent evolution** [74], as discussed below. It also suggests that a very good place to look for recognizable examples of evolutionary mechanisms is in very recent proteins.

Repeat expansions seem to be a good and rapid way for evolution to experiment with new proteins. As a result, such proteins will tend to have very limited sequence variation and will tend to be **low-complexity proteins**, which are surprisingly common, presumably for this reason. They are therefore often natively unstructured [75] (Chapter 4).

1.4.5 Evolutionary tinkering leaves its fingerprints behind

The essence of tinkering, at least as practiced by evolution, is that you take something and alter it at random. If the result is not useful, you discard it; if it has some use, even if that use is quite different from the function of the original object, then you keep it and develop it further. We shall see repeated examples throughout this book of this phenomenon, which has been called embellishment: the nature of evolution is to take what was originally a very simple structure or mechanism and add extra refinements until the result looks quite different from the original.

When nature picks up a protein and starts tinkering with it, it has a tendency to pick up the nearest thing at hand. Most obviously, internal duplication (that is, copying of a domain and attaching it to the original) is extremely common; and the most common type of protein interaction is of a domain interacting with another closely related or **paralogous** one [76]. It is also common to find that a regulatory or modulating protein turns out to be structurally or functionally similar to the protein it regulates. (It is worth reading a remarkably insightful 1968 review on enzymes by Koshland and Neet, based on almost no structural information but some well-informed speculation, which concluded quite correctly with an expectation that "regulatory sites are in most cases mutated active sites" [77].) Thus, proteases are often activated or regulated by proteolysis; kinases by phosphorylation; acetylases by acetylation [78], and so on. We shall see several examples in this book of regulatory proteins that were derived by gene duplication from the protein that they regulate. Another is provided by PyrR, the transcriptional attenuator that modulates the expression of pyrimidine nucleotide genes in *Bacillus subtilis*, which probably evolved from a ribosyl transferase enzyme, keeping the binding functions but losing the catalytic ones [79]. Signaling systems reuse the same limited repertoire of binding domains.

Evolution is thus at the same time both stupid and clever. It is stupid because it is unable to scrap a rather clumsy system and start again; and it is clever because it can keep tinkering with a system until it ends up with something that works very nicely, even if when we investigate the engine under the hood the mechanism is rather like something dreamed up by Heath Robinson or Rube Goldberg (**Figure 1.57**).

1.4.6 New proteins can arise by gene sharing

How does an organism acquire proteins with new functions? Answer: like a tinker, it takes whatever is to hand and adapts it in the easiest way possible.

The simplest way is not to make any change at all, but simply to use an old protein for a new task, an approach sometimes described as **gene sharing**

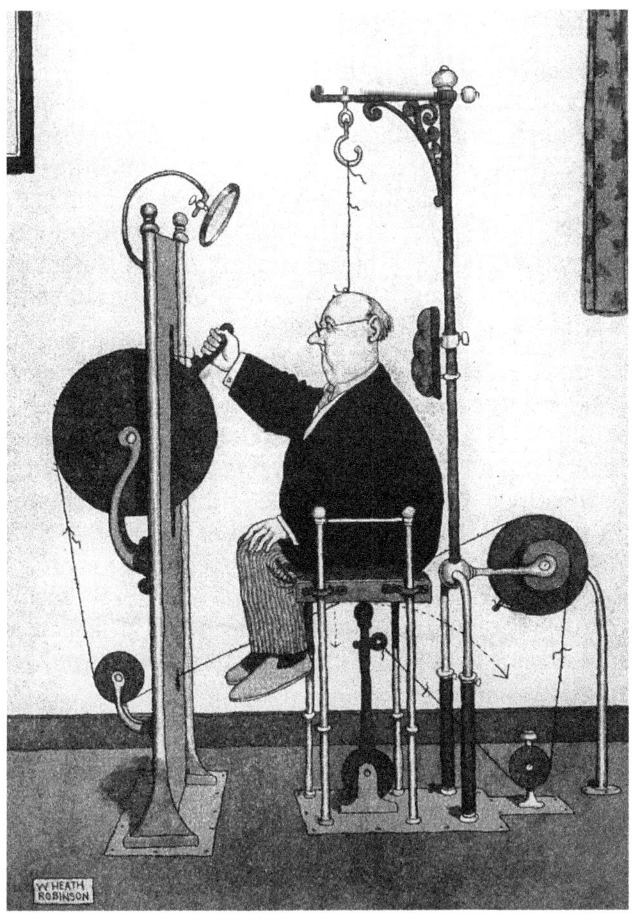

FIGURE 1.57
A drawing by Heath Robinson: The wart chair. The contraption works, but it is a tinkered job, being put together from randomly chosen elements and not necessarily the most efficient construction for the job. Evolution works in much the same way. (From W.H. Robinson, Absurdities. London: Hutchinson, 1934. With permission from Pollinger Ltd and the Estate of J.C. Robinson.)

[57]. The classic example of this is eye lens crystallins [80]. As vision became more important, organisms needed to evolve a lens in the eye to refract light and focus it on the retina. The lens proteins need to have three properties: they should be soluble enough to create a solution of high refractive index, they should remain soluble and transparent throughout the life of the animal, and they should be nontoxic in their new location and concentration. (Because humans now live much longer than they were ever evolved to do, the second of these properties is often not met, resulting in cataracts in older people; this is a good example of the suggestion that many modern diseases are simply biological phenomena that we were never evolved to handle, such as obesity and dialysis-related amyloidosis [81].) Therefore many different proteins could be and were used in lenses. In crocodiles and some birds, ε crystallin is identical to lactate dehydrogenase and even retains enzyme activity. δ2 crystallin, found in all birds and reptiles, is identical to argininosuccinate lyase. These proteins use the identical gene to the normal metabolic one, and just express it at high levels in the eye. Many other lens proteins have minor mutations that make them nonfunctional enzymes but otherwise almost identical to functional enzymes: δ1 crystallin, for example, represents a gene duplication of the δ2 crystallin/argininosuccinate lyase gene with mutations to make it inactive as an enzyme. This gene duplication and subsequent mutations may well be an evolutionary response to the regulatory difficulties created by having one protein with two quite different functions (Section 1.4.3). It has been suggested [80, 82] that δ crystallin represents one of the clearest examples of the evolutionary adaptation of a moonlighting protein (see below) in which the development of a new function generates adaptive conflict and therefore generates a pressure for gene duplication and divergence.

1.4.7 Evolution usually retains chemistry and alters binding

However, the most common way to create a new enzyme is to adapt an old one. Enzymes have two main requirements: they should be able to bind the substrates in the correct orientation, and they should be able to perform some kind of chemistry to catalyze a reaction. Therefore, to generate a new enzyme, nature has two choices. It can either use an enzyme that binds the substrates and change the chemistry or it can use an enzyme that does the right chemistry and alter the binding. (For almost the last time, I should point out that nature does not make choices: it simply does things at random and whatever works best is what emerges.) Which does it do?

Actually, both have been observed [83–85]. However, conservation of the chemistry is much more common, possibly a reflection of the fact that the structural requirements for catalysis are much more demanding than those simply for binding [86–89]: a detailed study concluded that metabolic enzymes "usually conserve their catalytic or cofactor binding properties; substrate recognition is rarely conserved" [90]. This latter study strongly supports the opportunist tinkering picture of evolutionary selection, in that it concludes that there is no pattern or order in the domains used for metabolic pathways: domains were just picked up more or less at random. This is nicely indicated by the word *patchwork* to describe this random putting together of elements from different places [89].

1.4.8 Convergent and divergent evolution are difficult to distinguish

The discussion in the previous two sections implies that evolutionary tinkering typically takes an existing protein, duplicates the gene, and tinkers with one copy. If the result turns out to be useful, it is kept; if not, it is discarded. Because random evolutionary events usually preserve secondary structure, the resulting protein typically has a different sequence but similar tertiary structure to the "original." This process is described as **divergent evolution**: a protein and its ortholog become gradually more different as a result of evolution but typically retain the same tertiary structure. Almost always, the function is related, although not identical [91]. There are very many examples of this.

Less common is **convergent evolution**: when evolution starts off from two quite different proteins and ends up with two proteins that look very similar. Usually this similarity is a consequence of functional requirements: catalysis of the same reaction requires the same arrangement of the protein. We are familiar with this phenomenon on a large scale, in that animals that live in similar evolutionary niches tend to look very similar, although starting from quite different shapes—for example sharks and dolphins. The classic example of this in proteins is the serine proteases. These enzymes digest proteins by using three key residues that form a catalytic triad (*1.15) (**Figure 1.58**) composed of serine,

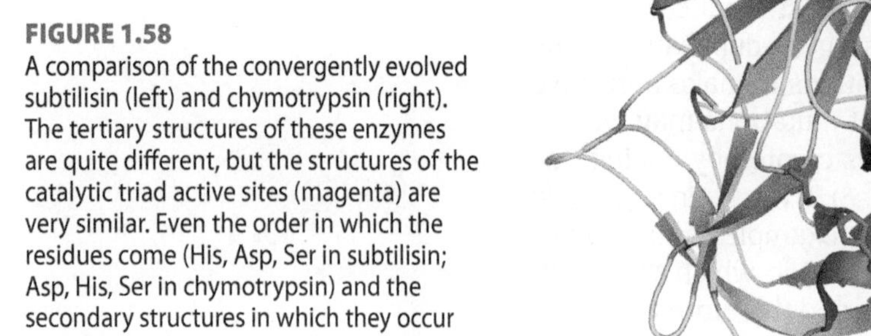

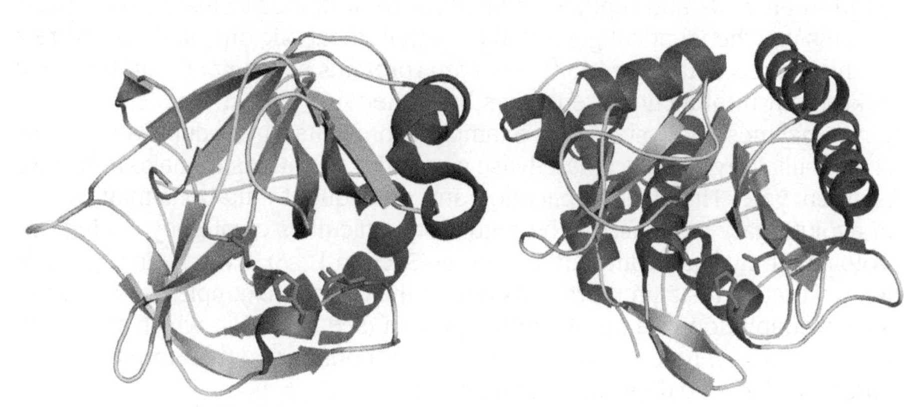

FIGURE 1.58
A comparison of the convergently evolved subtilisin (left) and chymotrypsin (right). The tertiary structures of these enzymes are quite different, but the structures of the catalytic triad active sites (magenta) are very similar. Even the order in which the residues come (His, Asp, Ser in subtilisin; Asp, His, Ser in chymotrypsin) and the secondary structures in which they occur are different.

aspartate, and histidine, positioned in a hydrogen-bonding arrangement that optimizes the catalysis. This arrangement of three residues has evolved twice, quite independently: once in prokaryotes, in subtilisin and its relatives, and once in eukaryotes, in chymotrypsin and its relatives. The geometry of the catalytic triad is very similar, but their locations within the structure, local secondary structure, and even the order of the three residues within the sequence, are all quite different. Although convergent evolution is less common than divergent evolution, it is still remarkably common [92], possibly implying that the optimal arrangement of protein side chains making up an enzyme's active site is actually rather specific.

Convergent evolution is not limited to side-chain arrangement. It is clear that some protein structure motifs are much more stable and "evolvable" than others. In particular, the β-α-β motif seems to be easily transferable, and structures based on it are therefore fairly likely to be arrived at by convergent evolution. The most versatile enzyme fold is the TIM barrel (*1.12), which is formed out of a repeated β-α-β motif (see Figure 1.34). This fold is found in a very wide range of enzymes, catalyzing many different types of reaction. The above argument implies that structural similarity does not necessarily imply divergent evolution; it could equally well be a consequence of convergent evolution. There are many examples, particularly among TIM barrels, where it is still unclear which is the true explanation, although many cases that were previously thought to represent convergent evolution are now believed to be examples of very distant divergent evolution [88, 93]. Nevertheless, there probably are some genuine examples of convergent evolution of the same protein fold [94].

1.4.9 New functions may often develop from promiscuous or moonlighting precursors

The enzymes that exist now are in the main very efficient at what they do. In other words, they catalyze one reaction very specifically and are poor catalysts for any other reaction. This means that to generate a new enzyme from an old one, one has not only to gain new function, but also to lose the old function. And remember that this has to be done "blindly," by selective pressure. One can imagine how selective pressure could operate to lose an unwanted function, but how could it "start" a new function? One suggestion is that, in some instances at least, the new function may be there already: that is, the protein may already be able to catalyze a second, different, reaction, albeit poorly. This phenomenon has been called **moonlighting** and seems to be quite common. Enzymes that catalyze two different reactions have also been called **promiscuous** [86, 96], this term often being restricted to enzymes that use the same active site for catalyzing both reactions.

Promiscuous enzymes are relatively common: a short list is given in **Table 1.4**. It has been persuasively argued that active sites have a structure that makes them inherently good at catalysis—for example, being partly buried, often with charged groups held at strategic positions [95]. This makes enzyme promiscuity particularly likely. It should also be noted that almost all the enzymes in Table 1.4 are hydrolytic enzymes, for which one can easily imagine how the attack of water on an activated substrate could be used for more than one type of substrate. One can therefore make a subtle but important distinction between two possible mechanisms for the evolution of a new enzyme function. In one (the classic model), a gene is duplicated, one copy is "kept" for its original function, and the second is then free for evolution to tinker with it and see if it can do anything useful. In the second, the gene product that is duplicated is already promiscuous—that is, it already has some weak secondary activity—and the duplication event allows this to be developed independently of the primary activity. Evolution therefore does not have so far to go.

TABLE 1.4 Promiscuous enzymes

Enzyme	Normal reaction	Promiscuous reaction
α-Chymotrypsin	Peptide hydrolysis	Hydrolysis of esters and phosphate triesters
Alkaline phosphatase	Phosphate monoester hydrolysis	Hydrolysis of phosphate diesters, phosphonate monoesters, and sulfate esters; phosphate oxidation
Arylsulfatase A	Sulfate ester hydrolysis	Hydrolysis of phosphate diesters
Phosphonate monoester hydrolase	Phosphonate monoester hydrolysis	Hydrolysis of phosphate monoesters, sulfate monoesters, phosphate diesters, and sulfonates
Exonuclease III	DNA phosphate diester hydrolysis	DNA phosphate monoester hydrolysis
Phosphotriesterase	Phosphate triester hydrolysis	Hydrolysis of phosphate diesters, esters, and lactones
Urease	Urea hydrolysis	Phosphoramidite hydrolysis
Dihydroorotase	Hydrolysis of dihydroorotate	Phosphate triester hydrolysis

(Taken from Table 3.2 in A.J. Kirby and F. Hollfelder, From Enzyme Models to Model Enzymes. Cambridge, UK: Royal Society of Chemistry, 2009.)

Because evolution is characteristically opportunistic, this second mechanism sounds more like the kind of thing that it does. These two mechanisms are two positions on a continuous spectrum of possibilities and so are not drastically different, but the second is possibly a more helpful way to think of evolution operating and it has gradually been gaining ground as the "standard" way of viewing the evolution of new enzymes [84].

A plausible example of this "promiscuous" route is illustrated by azurocidin [57]. This protein is homologous to a number of serine proteases that are involved in protecting the cell against attacking bacteria, but it has lost its protease activity and instead binds to lipopolysaccharide on the bacterial cell wall, forming a bridge that activates recognition and attack by the immune system [96]. One can imagine a very reasonable evolutionary scenario in which the protein's original role was as a protease that was also able to recognize lipopolysaccharide; evolution then developed this ability as an alternative protective mechanism, and on the way it dropped the original activity.

There are a surprisingly large number of moonlighting proteins that have two or more quite different functions: some are listed in **Tables 1.5 and 1.6** [97–99]. In most cases, the second function is not enzymatic. The adoption of one protein for a second function is a typical "tinkering" evolutionary strategy. However, the argument of Section 1.4.4 would suggest that in most cases, once a second function has been established, the relevant gene would be duplicated, thereby permitting differentiation not only of the protein but also of its regulation. The "surprise" of moonlighting proteins is not that a protein can have two functions but that evolution has not separated them. Genuine moonlighting cases are therefore presumably either very recent or else represent related functions. This latter is certainly true in some cases. For example, the enzyme aconitase (**Figure 1.59**) uses iron in its active site as an Fe_4S_4 cluster. There is therefore a relevance in its also being used to bind to the mRNA that codes for the iron storage protein ferritin and inhibit its synthesis. Furthermore, aconitase only binds to the iron-responsive element of the mRNA when it has lost one or more iron atoms from the cluster, and therefore iron storage by ferritin is inhibited only when the iron concentration in the cell is too low to replenish the iron atom(s) lost from the aconitase cluster. However, for some of the other moonlighting proteins there seems to be no connection at all between the two

TABLE 1.5 Enzymes with a moonlighting nonenzymatic function

Enzyme	Function	Second function
Cytochrome *c*	Electron carrier	Apoptosis
PutA	Proline dehydrogenase	Transcriptional repressor
Phosphoglucose isomerase	Glycolytic enzyme	Cytokine
Phosphoglycerate kinase	Glycolytic enzyme	Cytokine
Thymidine phosphorylase	Biosynthetic enzyme	Endothelial growth factor
Aconitase	TCA cycle enzyme	Iron-responsive element binding protein
Thymidylate synthase	Biosynthetic enzyme	Translation inhibitor
Band 3 ion exchanger	Cl^-/HCO_3^- antiporter	Regulator of glycolysis
FtsH	Metalloprotease	Chaperone
birA	Biotin synthetase	Bio operon repressor
Lactate dehydrogenase	Glycolytic enzyme	Eye lens crystallin
ARGONAUTE4	RNAse to make siRNA	Chromatin remodeling
Cyclophilin	Peptidyl proline *cis/trans* isomerase	Regulator of calcineurin
Enolase	Glycolytic enzyme	Mitochondrial import
Enolase	Glycolytic enzyme	Transcriptional regulation [131]
Hexokinase	Glycolytic enzyme	Apoptosis repressor [132]
Galectin-1	Lectin (saccharide binding)	Protein:protein interaction

TABLE 1.6 Enzymes with a moonlighting enzymatic function

Enzyme	Moonlighting function
Uracil-DNA glycosylase	Glyceraldehyde-3-phosphate dehydrogenase
Thioredoxin	Subunit of phage T7 DNA polymerase
Ferredoxin-dependent glutamate synthase	Subunit of UDP-sulfoquinovose synthase

functions. It is striking that several of the moonlighting enzymes are glycolytic or TCA-cycle enzymes; that is, they are abundant housekeeping enzymes. It is therefore likely (although so far unproven) that these enzymes were used merely because they were available and abundant. In a similar manner, several ribosomal proteins (which again are very abundant in the cell) have been found to moonlight as translational regulators, DNA repair factors, or regulators of development.

A related phenomenon is seen in a small number of enzymes that catalyze *two sequential reactions* within one active site. The TCA-cycle enzyme isocitrate dehydrogenase is one such [100]. Its "main" reaction is the oxidation of

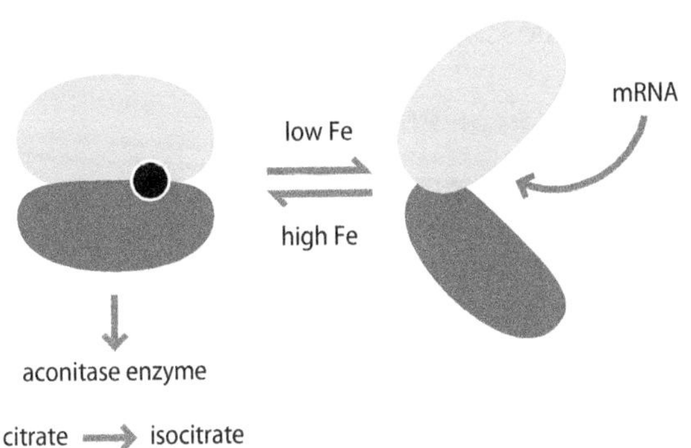

FIGURE 1.59
Aconitase has two functions. When iron levels in the cell are high, it forms a 4Fe–4S cluster (black circle) and is an active enzyme in the Krebs cycle, converting citrate to isocitrate. However, when iron levels are low, the iron–sulfur cluster is disrupted, and the protein opens out to create a binding site for mRNA stem loops. The protein is then also known as iron regulatory protein 1 (IRP1). In particular, it binds to an upstream regulatory sequence known as the iron responsive element (IRE) and alters the transcription of proteins involved in the uptake, storage, or utilization of iron. For example, it blocks transcription of the iron storage protein ferritin but increases transcription of the iron uptake transferrin receptor.

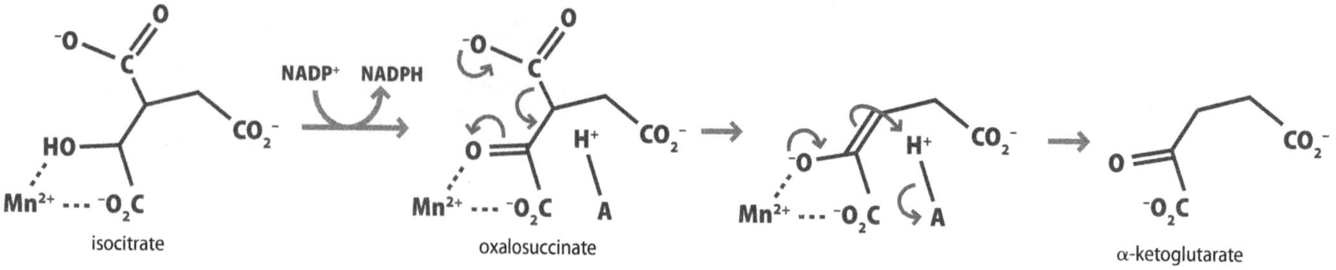

FIGURE 1.60
The reactions catalyzed by isocitrate dehydrogenase. The difficult reaction is the oxidation of isocitrate by NAD^+, but there is also a second reaction: the decarboxylation of oxalosuccinate to give α-ketoglutarate.

isocitrate to oxalosuccinate (**Figure 1.60**). The second reaction is a chemically very straightforward reaction, which requires electrophilic catalysis (Chapter 5) to pull electrons away from the carbonyl carbon: a suitable group is already present as a catalyst for the first reaction in the form of a Mn^{2+} ion. It also requires a general acid (Chapter 5) to provide a proton (AH^+ in Figure 1.60). This is easily provided by a suitable side chain, although in the *E. coli* enzyme it is apparently provided by a rather unusual side chain, either a tyrosine or a lysine [100]. Therefore in this case the development of a second catalytic activity within the same active site is easily achieved. And because the reaction is inherently fairly fast, and the "main" reaction is relatively slow, the catalytic enhancement does not need to be large to make the second reaction go at the same rate as the first.

1.4.10 Retrograde evolution is not common

In 1945, Horowitz looked at the limited sequence information then available and noticed that there were several instances in which enzymes in the same metabolic pathway were clearly homologous [101]. He proposed that these had evolved by *retrograde evolution*. That is, he proposed that in early evolution a metabolic pathway used some metabolite A that was readily available in the surrounding medium, and evolved an enzyme (B synthase) that took A and converted it to a useful product B (**Figure 1.61**). However, once B synthase had evolved to become more efficient, the supply of A got used up, so the organism had to duplicate and adapt B synthase to make a second enzyme, A synthase (see Figure 1.61).

This is an example of an idea that is logical, elegant, and explains the facts, but is unfortunately quite wrong, at least as a general mechanism [88, 90, 102]. We shall meet several of these throughout this book, and it is a pitfall that is particularly likely to occur to those who are trying to rationalize and explain biological observations (as I do throughout the book). Actually very few, if any, enzymes evolved in this way, although it is certainly true that there are several instances of homologous enzymes within the same pathway. It looks much more likely that in pathways containing homologous enzymes, the original enzyme was a rather nonspecific and inefficient enzyme that catalyzed both reactions; it was then duplicated and the two copies evolved to catalyze the two individual reactions [89, 103]. However, it is much more common to find that the different enzymes in a pathway came from quite different sources, a mechanism sometimes described as *mosaic* or *patchwork*.

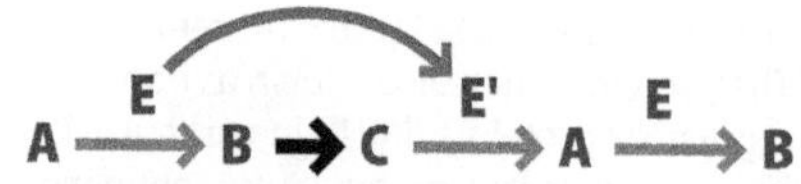

FIGURE 1.61
The retrograde evolution theory. An enzyme E evolves to convert A into B. However, this uses up the available supplies of A, and therefore E duplicates and evolves to make a new enzyme E′, which catalyzes the synthesis of A from C.

1.4.11 Proteins began in an RNA world

There is no clear agreement on how proteins began. It is generally agreed, however, that early life was an RNA world, in which RNA formed both the genetic message and the catalysts [104]. It was also probably a hot world, composed of small patches of water in hot rocks, because for example the most ancient proteins most closely resemble modern thermophilic proteins [105, 106]. We can imagine the first proteins as being roughly like the proteins we can see now in ribosomes; indeed, the ribosome is a good model for typical structures found at an early stage in life, because both the basic structure and the catalytic function in the ribosome come from RNA [107]. The protein component of the ribosome is basically a coating on the RNA (**Figure 1.62**): it helps support it and shape it (and therefore no doubt increases the specificity) and helps it to assemble, but some of it would have no structure or function independently of the RNA. There are some ribosomal proteins that do have an independent existence; these tend to be peripheral proteins, and one can speculate that they have been added on later to enhance the function. One can then envisage that protein or peptide side chains could have gradually replaced RNA in catalytic roles because of their greater range of chemical reactivity, thus becoming more and more a key part of the structure [108]. A possible "snapshot" of a slightly later form of protein evolution is the **signal recognition particle (*1.20)** (**Figure 1.63**), which is common to all forms of life and controls an essential function, namely the insertion of proteins into or through the membrane [109]. This again consists mainly of RNA, onto which are bound several proteins. However, the key conserved proteins fold independently of the RNA although they interact closely with it, and the functions of the particle depend on protein:protein interactions rather than protein:RNA interactions. Thus, one can imagine a steady progression from proteins as adjuncts to the RNA, to proteins as independent catalysts.

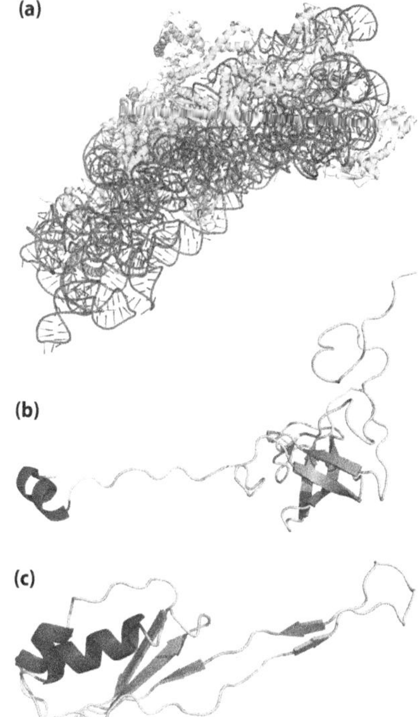

FIGURE 1.62
The 30S subunit of the *Thermus thermophilus* ribosome (PDB file 1j5e). (a) The entire ribosome. The RNA is shown in brown and protein in green: it is clear that the core of the particle is RNA, and the protein mainly decorates the surface. (b) The S12 protein. (c) The S10 protein. Although many of the ribosomal proteins resemble typical globular proteins, some (such as S12 and S10) clearly can only adopt a defined structure in the presence of the rest of the ribosome.

Proteins are much more versatile and effective as catalysts than RNA, because they have a larger array of functional groups and can adopt a wider range of structures. Therefore, in much the same way that DNA took over the role of genetic store from RNA because its double-helical structure is so much more suitable for this purpose than RNA, so proteins took over the roles of catalysts and structural elements. The early proteins must have been rather unstructured, molten-globule-like things with an ill-defined structure that relied on the accompanying RNA to support them. They must also have been extremely poor and nonspecific enzymes. It is likely that, like some of the natively unstructured proteins we shall consider in Chapter 4, they only adopted anything like a folded structure when bound to their substrates [110]. It is therefore not unreasonable to suppose that early enzymes were structurally ill-defined, genetically mobile, and nonspecific catalysts [111] and that they were therefore subject to extremely rapid genetic change: that is, that the rate of improvement in enzyme specificity and rate in early evolution must have been very great [112]. In other words, evolution has slowed down very significantly since the early heady days: a concept that even has a word describing it, namely hypobradytelism ([113], quoted in [114]).

This viewpoint provides a useful insight into the stability of current proteins. Evolutionary pressure has acted in general to lead to proteins adopting a single stable folded structure; this makes sense, as discussed above, because unfolded or poorly folded proteins will in general be more likely to aggregate or degrade and will also be less effective as catalysts, ligands, and so

***1.20 Signal recognition particle**
A signal recognition particle is a ribonucleoprotein (RNA–protein complex) whose function is to recognize proteins coming off the ribosome that should be targeted to membranes (endoplasmic reticulum in eukaryotes, or cell membrane in prokaryotes); see Figure 1.63. It halts translation until the ribosome is docked onto a receptor on the membrane, after which it dissociates. It has to recognize a very wide range of signal sequences, whose main common element is hydrophobicity. It does this partly by having a large number of methionine side chains in the recognition site, which are hydrophobic but also capable of adapting to multiple targets. The widely found calcium signaling protein calmodulin has a similar recognition problem, and a similar solution.

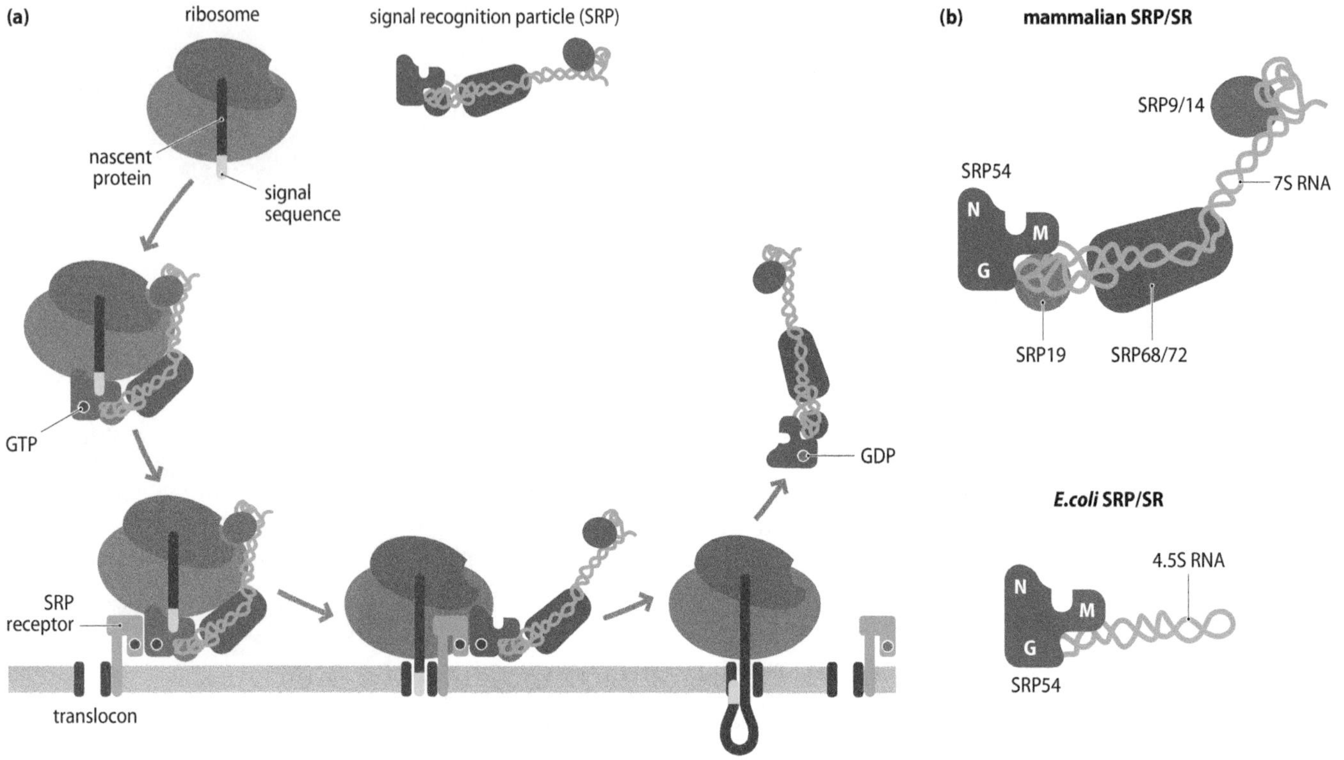

FIGURE 1.63
Structure and function of the signal recognition particle. It recognizes the N-terminal signal sequence as it emerges from the ribosome (green) and binds to it, therefore stalling translation. The mammalian SRP clamps onto the ribosome as a further way of preventing further translation. It then binds to a receptor on the endoplasmic reticulum membrane, a process that requires GTP. Hydrolysis of a receptor-bound GTP moves the receptor to the Sec61 membrane channel, after which further hydrolysis of SRP-bound GTP leads to dissociation of the SRP and continuation of translation through the Sec61 channel. (b) Comparison of the structures of mammalian and bacterial SRP, which are in both cases RNA with globular proteins domains stuck on. G indicates the GTPase domain of SRP54, N the N-terminal domain, and M the methionine-rich domain that binds to the signal sequence. (Adapted from H. Halic and R. Beckmann, *Curr. Opin. Struct. Biol.* 15:116–125, 2005. With permission from Elsevier.)

on. However, once a stable fold has evolved, there is no further evolutionary pressure. That is, once a protein has evolved to be *stable enough*, there is no pressure to make it more stable. Observation of current proteins suggests that the meaning of *stable enough* varies from protein to protein, but for example it normally means that the protein does not unfold enough to generate significant aggregation or misfolding over the life of the protein. This idea has two implications: (1) proteins that have very short lifetimes, such as those involved in regulating the cell cycle, are often much less stable than housekeeping proteins, for example; and (2) because we now live much longer than we were evolved to do, problems involving protein unfolding and aggregation ("post-evolutionary diseases") are becoming more common [80]. Cataracts (misfolding of crystallins in the eye), amyloid diseases, and a range of arthritic diseases all fall within this category.

1.4.12 Most evolutionary innovation happened very early

A comparative genome analysis [115] has shown that most current protein functions, and many protein folds, were already in place in the **Last Universal Common Ancestor** (that is, the original organism or community of organisms from which prokaryotes, eukaryotes, and archaea diverged), with this ancestor containing genes for DNA replication and repair, transcription, translation, protein folding, primitive signaling, and metabolic pathways for sugar, nucleotide and lipid metabolism, glycolysis (although possibly only the bottom half [116]), and parts of the TCA cycle and oxidative phosphorylation. It is suggested that the earliest enzymes to evolve were mainly concerned with phosphate and nucleotide chemistry [117]. That is, that by an early stage in evolution, enzymes had already started to "fossilize" into the structures that we see today, and most "recent" enzymes are recruited from existing ones [118]. It is further suggested that even in the Last Universal Common Ancestor, proteins were already **mosaic**, being composed of multiple modules copied from elsewhere in the genome [90]. This is not too surprising, and had already been suggested 30 years earlier in a very readable and insightful paper [118],

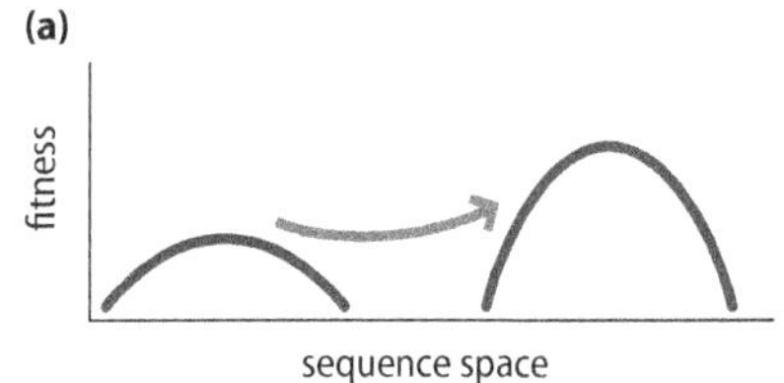

FIGURE 1.64
Evolution selects the fittest genotype by selecting proteins that do the required task. Normally this implies moving only "uphill," or at least not too steeply downhill. (a) Therefore in early evolution, when the existing proteins are not very effective, "downhill" evolution is possible, and searches of sequence space can take place. (b) However, later in evolution, when proteins are more effective, downhill evolution implies such a loss of fitness that it cannot occur except in a nonessential duplicated gene, and even local maxima are sufficient to block further progress.

on the basis that experimentation with new structures and functions was much easier when proteins were relatively "young" and inefficient. If we picture protein sequence space as an enormously large and complex surface, where the height at any point represents the ability of that sequence to catalyze a desired reaction (**Figure 1.64**), then the evolution of enzyme function is roughly a process of exploring this surface to find the highest point[1]. Evolution has two basic methods by which new protein sequences can be sampled: mutation, and splicing or transposition. Mutation normally consists of changing one amino acid at a time, and it corresponds essentially to a slow stepping from one position to the next across the surface. If the step leads to a higher level of activity, or at least one that is not too much lower, it is accepted or tolerated. In the early stages of evolution, all sequences are poor catalysts, and nature is effectively wandering around the foothills. Therefore most changes can at least be tolerated, and exploration is rapid. However, once nature has found a "good" sequence (that is, it has started climbing a taller and steeper mountain) then downhill steps are tolerated less well, and the climb becomes increasingly limited to a simple uphill climb (Figure 1.64b). Thus, mutations can reach the local maximum; however, if this is not the global maximum, mutation cannot provide a good mechanism to "climb down" again and start looking elsewhere.

This analysis implies that the present-day enzymes that have evolved may well not be the best possible for their task: they are just the best that could be achieved by searching close to whatever the original starting point was [119]. An observation that supports this conclusion is that of nonorthologous gene displacement discussed earlier. A survey of enzymes in the GenBank database found 105 reactions that were catalyzed by at least two nonorthologous enzymes [54]. This shows that, at the least, there are several completely different proteins that can do the same job roughly equally well; slightly more speculatively, it suggests that many of the enzymes used currently could be improved by replacing them by something else quite different.

If a completely different function is required, mutation can achieve this, but only if a small number of mutations can significantly degrade the original activity (otherwise the original activity is likely to interfere with the desired new function). However, this is not difficult to achieve, of course! And in fact the common occurrence of pseudogene products provides a set of nonfunctional protein templates that can act as starting points for this process.

There is evidence to suggest that for any given protein sequence, only very roughly 2% of amino acid substitutions are nondeleterious [120]. However, the effect of a mutation depends on the neighboring residues: in the context of a different neighbor, a mutation that might have been impossible for steric reasons may become possible, for example. More than 90% of sites in a protein can eventually accept a substitution, given the right combination of amino acids at other sites, implying that evolution is as yet nowhere near the limit of divergent evolution: sequences are continuing to diverge.

The alternative search mechanism of splicing and transposition is essentially a jump from one position to another, and potentially a very large jump to a completely new position. In most cases this is a random jump; in most cases the result is therefore not useful. If there is a mechanism to direct the jump

[1]Actually a point high enough that there is no further evolutionary pressure to improve further. In practice this means one of two things: either an enzyme activity that is at least as fast as others in the same pathway, so that the enzyme had a negligible effect on the overall metabolic flux through the pathway, or an enzyme activity that is limited by diffusion rates, so that increases in the enzyme turnover do not lead to increases in rate, as for example with triosephosphate isomerase (Chapter 5).

***1.21 Genetic algorithm**

A genetic algorithm (GA) is a computer method for optimizing complex problems in which there are many variables that require optimizing jointly to achieve a correct solution. It is designed to work in the same way as evolution. The problem is represented as a series of genes, each representing one parameter, that are organized into a linear chromosome. (For example, if the GA is used to calculate a protein structure, each gene could represent a backbone dihedral angle; the entire chromosome is then the list of all the backbone dihedral angles.) Each chromosome gives rise to a phenotype (for example, the set of dihedrals produces a three-dimensional structure), and a vital part of the GA is the *fitness function*, which calculates how close the phenotype is to the optimal properties. Almost anything can be put into the fitness function as long as it can be computed. To start the GA, a population of chromosomes is created, for example using randomly chosen genes. The number of individuals depends on the complexity of the problem but is typically a few hundred. The initial collection of individuals represents the first generation, and a fitness is calculated for each individual. This generation is used to produce a second generation, a set of daughter chromosomes. Modeled on evolutionary processes, the probability that a first-generation individual will be selected for reproduction depends on its fitness: the most fit are more likely to be selected than the least fit. Reproduction can be asexual or sexual. In asexual reproduction, a single chromosome generates a daughter chromosome by mutations. In sexual reproduction, two chromosomes generate a daughter by a combination of genetic crossover (that is, part of the chromosome of one individual replaces the equivalent part in the other) and mutation. The fitter offspring replace their less fit parents, and the process continues through many generations (usually many thousand generations) until there is no further improvement. In an interesting parallel with evolution, crossover is more effective near the start but is deleterious toward the end. The crossover and mutation rates, the size of the population and the selection pressures (that is, by how much a fitter parent is preferred) have been found to depend on the nature and complexity of the problem, and for optimal solutions these need to vary throughout the calculation. This is of course also true for evolution. As in evolution, a GA often results in a dead end: a line that is stuck in a low fitness and cannot evolve out of it. These individuals gradually get replaced by their fitter cousins [137].

only in the more promising directions, nature stands some chance of getting somewhere useful. Just such a mechanism is provided by the splicing together of two poor but functional sequences, as discussed further in Chapter 2: this is now almost the only way in which nature can come up with a significantly new function when one is required, because most proteins cannot now be improved much by mutation. It is interesting to compare "real" evolution with the computational optimization technique known as the **genetic algorithm or GA (*1.21)**. GAs tend to work in the same way that is described here: they make rapid early progress, largely as a result of genetic crossovers, they tend to "crystallize" in correct or partly correct solutions, and they refine their successful results by mutations.

However, by far the biggest example of recent evolutionary development is in signaling, where, as discussed in later chapters, nature has again just taken existing proteins and joined them together in different ways. The big difference between enzymes in mammals and in prokaryotes is that mammalian enzymes have a massive expansion in the number of enzymes related to signaling; at the same time they have a big increase in the number of extra domains, as considered in Chapter 2 [121].

1.5 SUMMARY

Proteins consist of a linear chain of amino acid residues, selected from 20 candidates, linked by peptide bonds. A protein therefore consists of a regularly repeating backbone to which is attached a specific series of side chains, which can be charged, polar, or hydrophobic. Each residue has a limited range of conformations available, because of steric clashes. The most energetically favorable conformation gives the peptide an extended structure; the other main conformation gives it a helical structure. Within a protein, the amino acids interact, causing the protein to fold up into its native state. Hydrophobic interactions result from the preference of water to form hydrogen bonds, and lead to the burial of hydrophobic side chains. This hydrophobic collapse also produces good van der Waals interactions, because these interactions depend on close steric contact between two surfaces. Charged groups are almost always on the outside of a protein and are solvated by water.

Hydrogen bonding of one amino acid to another competes with hydrogen bonding to water. Thus the net energy from hydrogen bonds is small, but an unsatisfied hydrogen bond is very unfavorable. This means that in the interior of a protein, almost all backbone amide groups are hydrogen bonded together in regular structure, forming either α helices or β sheets: sheets can be either parallel or antiparallel. This is particularly true for membrane proteins, because within the membrane there are no water molecules with which to form hydrogen bonds. The weak interactions in a protein stabilize each other cooperatively, particularly for hydrogen bonds, because of their strong directionality. Proteins interact with their solvating water molecules and require roughly a monolayer of waters if they are to be fully functional.

The structure of a protein can be analyzed in terms of primary structure (covalent structure), regular secondary structures (sheets and helices), tertiary structure (three-dimensional fold), and quaternary structure (the packing of one polypeptide chain against another). Different secondary structures also pack together in well-defined ways, creating supersecondary structures and structure motifs.

The tertiary structure of a protein is determined by its primary sequence. There are, however, a few proteins that can adopt a metastable state, usually for functional reasons.

Evolution is a tinker: it takes what is close at hand and modifies it. Often this leads to no functional improvement, so that most of the changes between one ortholog and another are random. Evolutionary changes preserve the three-dimensional structure much more than sequence. Thus, most of the variation that we see in related proteins is probably due to divergent rather than convergent evolution. Sequence comparisons identify conserved residues, which are often important for function or structure. However, evolution has frequently replaced one enzyme by a functionally equivalent but evolutionarily unrelated one, leading to a patchwork of different structures, both within and between species.

The most common mechanism for the evolution of new proteins is gene duplication followed by differentiation. It is likely that this was most successful when the duplicated protein was already moonlighting, with a weak ability to undertake its "new" function. In most cases, this represents an ability to catalyze the new chemical reaction rather than to bind new substrate.

The first form of life was probably an RNA world, in which the main role of proteins was to support and protect the RNA. However, proteins have greater catalytic and binding versatility than RNA and therefore gradually supplanted it. Evolutionary changes happened very rapidly in early life. The Last Universal Common Ancestor probably already had most of the protein folds and functions that we see now. Once a protein has reached a certain functional level it starts to "fossilize" into an evolutionary dead end: it is therefore likely that many enzymes are not the best possible for their role, merely the best that have been found by evolution so far.

1.6 FURTHER READING

General introductions to proteins and their role in biochemistry can be found in any biochemistry textbook. I particularly like Berg et al. [122]. There are many others that are equally good. My other favorite is Voet and Voet [123].

Introduction to Protein Structure, by Branden and Tooze [43], is a wonderful description of protein structure and of the rules that govern it.

Lesk's Introduction to Protein Science [124] also presents an excellent account of protein structure and evolution.

Proteins: Structures and Molecular Properties, by Creighton [125], is excellent for the more biophysical aspects discussed here. It is a little old, which makes it less useful for the biology and techniques.

Molecular Evolution, by Li [126], is a very readable account of protein evolution.

1.7 WEBSITES

http://agadir.crg.es/ *Prediction of helix.*

http://ekhidna.biocenter.helsinki.fi/dali_server/ *DALI/FSSP* [129, 130].

http://robetta.bakerlab.org/ *A resource for structure predictors including ROSETTA.*

http://scop.mrc-lmb.cam.ac.uk/scop/ *SCOP* [127].

http://www.biochem.ucl.ac.uk/bsm/sidechains/ *Atlas of side-chain interactions.*

http://www.biophys.uni-duesseldorf.de/BioNet/Pedro/research_tools.html *A large collection of links to biomolecular research websites.*

http://www.bioscience.org/urllists/protdb.htm *An index of protein classification programs.*

http://www.cathdb.info/ *CATH* [128].

http://www.geneinfinity.org/sp_proteinsecondstruct.html *A nice collection of links to programs for predicting secondary structure, transmembrane regions, disorder, etc.*

http://www.genome.jp/kegg/ *The KEGG metabolic pathways site.*

http://www.ncbi.nlm.nih.gov/sites/entrez?db=protein *ENTREZ: top page of NCBI's protein bioinformatics.*

http://www.ncbi.nlm.nih.gov/Structure/VAST/vast.shtml *VAST (vector alignment search tool).*

http://www.piqsi.org *Quaternary structure.*

http://www.proteopedia.org/ *An interesting website that presents moving three-dimensional images of proteins with accompanying information.*

http://www.rcsb.org/pdb/home/home.do *The Protein Data Bank, containing structure coordinates for proteins and nucleic acids: all the detailed structures shown in this book use coordinates taken from the PDB.*

http://www.sbg.bio.ic.ac.uk/~phyre/ *Another secondary structure prediction; also predicts function from structure.*

Some molecular graphics programs

http://jena3d.fli-leibniz.de/ *Links to programs that can display protein structures.*

http://jmol.sourceforge.net/ *Jmol.*

http://rasmol.org/ *Rasmol.*

http://www.liv.ac.uk/Chemistry/Links/refmodl.html *Links to programs that can display protein structures.*

http://www.pdbj.org/jv/index.html *jV.*

http://www.pymol.org/ *Pymol (all the structural diagrams in this book were prepared using pymol).*

http://www.rcsb.org/pdb/static.do?p=software/software_links/molecular_graphics.html *Links to programs that can display protein structures.*

1.8 PROBLEMS

Hints for some of these problems can be found on the book's website.

1. In what form would you normally expect to find the side chains of Asp, Glu, Lys, Arg, His, Ser, Thr, and Tyr at pH 7? If a Glu has a pK_a of 4, what proportion will be protonated at pH 7?

2. Describe the charged groups you would expect to see on the peptide ACEQLRYTFS.

3. Get hold of (or download) a computer program for visualizing protein structures. Two recommended ones are Rasmol (for example http://www.umass.edu/microbio/rasmol/—simple to use but a little basic in some areas) or Pymol (for example http://www.pymol.org/—more sophisticated but correspondingly harder to learn how to use properly). Download a protein coordinate file from the PDB (http://www.rcsb.org/pdb). Play with the different options for displaying the structure: backbone only, Cα only, cartoon, all atoms, surface, ball-and-stick, space-filling. If the program has such options, try varying the default setting, for example loops and helices. Try different coloring schemes. In particular, try to get a feel for how "solid" proteins are: compare the amount of empty space in the cartoon and space-filling representations. Of all of these, which most closely represents what a protein "really" looks like? Print out a cartoon, a space-filling view and a surface.

4. Sketch a diagram of the tripeptide Ser-Arg-Phe. Mark in the backbone, the backbone dihedral angles ϕ and ψ, any side-chain angles, and any acidic protons.

5. (a) What structural features of Pro and Gly explain the shape of the Ramachandran plots in Figure 1.6? (b) The left panel of Figure 1.6 shows that there are actually four distinct minima in the Ramachandran plot. What are these, and where do they occur in proteins?

6. By using the websites listed at the end of this chapter, find the amino acid sequences of human lysozyme and α-lactalbumin. Using BLAST or an equivalent program, perform a sequence alignment of these two proteins, and comment on the results.

7. Serpins are inhibitors of serine proteases and have a remarkable structural change when they inhibit their target proteases. Such a dramatic structural change must have been evolutionarily selected and must therefore have a benefit to the cell. However, it also clearly causes problems, because misfolded serpins are quite common and give rise to a range of diseases including antiprotease deficiency and cell death resulting from polymerization. Use the Internet to obtain information about the conformational change in serpins, and discuss potential advantages for this large structural change.

8. In Section 1.1.1, it is suggested that zinc has a structural role in intracellular proteins similar to that of disulfide bridges in extracellular proteins, by forming interactions that stabilize small domains by cross-linking cysteine and histidine side chains. In evolution, which would have come first? Is it possible that one could have evolved into the other? You will probably need to do some research on the occurrence of such cross-linking zincs in intracellular proteins.

TABLE 1.7 Metabolic cost of amino acids (number of ATP molecules)

Amino acid	Cost
Ala	20
Arg	44
Asp	21
Asn	22
Cys	19
Glu	30
Gln	31
Gly	12
His	42
Ile	55
Leu	47
Lys	51
Met	44
Phe	65
Pro	39
Ser	18
Thr	31
Trp	78
Tyr	62
Val	39

9. In Section 1.4.2, it is mentioned that there are two different kinds of transfer RNA synthetase. Which amino acids do they recognize? Is there any logic in this?

10. There are many examples of pairs of proteins that have the same function but completely different structures, sometimes called *analogous* proteins (including some elsewhere in this book). Describe one such pair.

1.9 NUMERICAL PROBLEMS

N1. If a protein has a histidine side chain with a pK_a of 6.5, what fraction of the time will that histidine be protonated at pH 7.0?

N2. High-resolution structures in the Protein Data Bank were searched for *cis*-peptide bonds. If the average protein length is 270 residues, the frequency of prolines within proteins is 4.8%, and the frequency of *cis* Xaa-Pro (as a percentage of all the peptide bonds before proline) is 5.2% and of *cis* peptides for amino acids other than proline is 0.029%, roughly how many *cis* Xaa-Pro might you expect to find in each protein (or, equivalently, how many proteins would you need to look at before you found one containing a *cis* Xaa-Pro)? And how about for other *cis* peptides not containing proline?

N3. Table 1.2 lists the observed frequencies of amino acids in proteins as well as the number of codons used to code for each amino acid. If we assume that any of the 64 codons is equally likely to occur by chance, but that three of the codons do not code for amino acids, then the chance of a random codon coding for alanine, for example, should be $(4/64) \times (64/61)$ or $4/61 = 6.6\%$. Put the data of Table 1.2 into a spreadsheet and use this to calculate the expected frequencies of all the amino acids. Plot this on the x axis of a graph, with the observed frequencies along the y axis. Use the spreadsheet functions to calculate the correlation coefficient. What do you deduce from this?

N4. Actually the four different codons do not have identical frequencies of occurrence: the actual frequencies of bases are U 22%, A 30.3%, C 21.7%, G 26%. Using these frequencies plus the calculations done for N3 calculate the expected frequency of an amino acid occurring at random. Again you should correct for the three stop codons. What can you deduce from this? Comment on any amino acids with unusual or extreme behavior.

N5. The metabolic cost of different amino acids is difficult to define. One estimate, based on the number of ATP molecules required to make the amino acid from glycolytic intermediates, is listed in **Table 1.7**. Plot the occurrence of the amino acids against their metabolic cost. What do you deduce from this? Comment on any outliers.

1.10 REFERENCES

1. FHC Crick (1988) What Mad Pursuit: A Personal View of Scientific Discovery. New York: Basic Books.

2. MF Perutz (1964) The hemoglobin molecule. *Sci. Am.* 211:64–76.

3. A Jabs, MS Weiss & R Hilgenfeld (1999) Non-proline *cis* peptide bonds in proteins. *J. Mol. Biol.* 286:291–304.

4. S Hovmöller, T Zhou & T Ohlson (2002) Conformations of amino acids in proteins. *Acta Cryst. D* 58:768–776.

5. MB Swindells, MW MacArthur & JM Thornton (1995) Intrinsic ϕ, ψ propensities of amino acids, derived from the coil regions of known structures. *Nature Struct. Biol.* 2:596–603.

6. RA Laskowski, MW MacArthur, DS Moss & JM Thornton (1993) PROCHECK—a program to check the stereochemical quality of protein structures. *J. Appl. Cryst.* 26:283–291.

7. AJ Doig & RL Baldwin (1995) N- and C-capping preferences for all 20 amino acids in α-helical peptides. *Prot. Sci.* 4:1325–1336.

8. P Bernádo, L Blanchard, P Timmins et al. (2005) A structural model for unfolded proteins from residual dipolar couplings and small-angle x-ray scattering. *Proc. Natl. Acad. Sci. USA* 102:17002–17007.

9. GL Holliday, DE Almonacid, JBO Mitchell & JM Thornton (2007) The chemistry of protein catalysis. *J. Mol. Biol.* 372:1261–1277.

10. JW Pflugrath & FA Quiocho (1985) Sulfate sequestered in the sulfate-binding protein of *Salmonella typhimurium* is bound solely by hydrogen bonds. *Nature* 314:257–260.

11. Q Jiang, LH Liang & M Zhao (2001) Modelling of the melting temperature of nano-ice in MCM-41 pores. *J. Phys. Condensed Matter* 13:L397–L401.

12. MS Searle (2001) Peptide models of protein β-sheets: design, folding and insights into stabilising weak interactions. *J. Chem. Soc. Perkin Trans. 2* 1011–1020.

13. M Lopez, V Kurbal-Siebert, RV Dunn et al. (2010) Activity and dynamics of an enzyme, pig liver esterase, in near-anhydrous conditions. *Biophys. J.* 99:L62–L64.

14. AM Klibanov (1989) Enzymatic catalysis in anhydrous organic solvents. *Trends Biochem. Sci.* 14:141–144.

15. JA Rupley & G Careri (1991) Protein hydration and function. *Adv. Protein Chem.* 41:37–172.

16. S Cayley, BA Lewis, HJ Guttman & MT Record (1991) Characterization of the cytoplasm of *Escherichia coli* K-12 as a function of external osmolarity: implications for protein–DNA interactions *in vivo*. *J. Mol. Biol.* 222:281–300.

17. AJ Lapthorn & NC Price (2008) Enzyme action: lock up your waters? *Biochemist* 30:4–9.

18. K Modig, E Liepinsh, G Otting & B Halle (2004) Dynamics of protein and peptide hydration. *J. Am. Chem. Soc.* 126:102–114.

19. DJ Wilton, R Kitahara, K Akasaka et al. (2009) Pressure-dependent structure changes in barnase on ligand binding reveal intermediate rate fluctuations. *Biophys. J.* 97:1482–1490.

20. P Mentré (1995) L'eau Dans la Cellule [Water in the Cell]. Paris: Masson.

21. CA Bottoms, TA White & JJ Tanner (2006) Exploring structurally conserved solvent sites in protein families. *Proteins Struct. Funct. Bioinf.* 64:404–421.

22. U Sreenivasan & PH Axelsen (1992) Buried water in homologous serine proteases. *Biochemistry* 31:12785–12791.

23. S Shaltiel, S Cox & SS Taylor (1998) Conserved water molecules contribute to the extensive network of interactions at the active site of protein kinase A. *Proc. Natl. Acad. Sci. USA* 95:484–491.

24. A Cooper, CM Johnson, JH Lakey & M Nollmann (2001) Heat does not come in different colours: entropy–enthalpy compensation, free energy windows, quantum confinement, pressure perturbation calorimetry, solvation and the multiple causes of heat capacity effects in biomolecular interactions. *Biophys. Chem.* 93:215–230.

25. H-X Zhou & MK Gilson (2009) Theory of free energy and entropy in noncovalent binding. *Chem. Rev.* 109:4092–4107.

26. DH Williams, E Stephens, DP O'Brien & M Zhou (2004) Understanding noncovalent interactions: ligand binding energy and catalytic efficiency from ligand-induced reductions in motion within receptors and enzymes. *Angew. Chem. Int. Ed.* 43:6596–6616.

27. HF Xie, DN Bolam, T Nagy et al. (2001) Role of hydrogen bonding in the interaction between a xylan binding module and xylan. *Biochemistry* 40:5700–5707.

28. G Weber (1993) Thermodynamics of the association and the pressure dissociation of oligomeric proteins. *J. Phys. Chem.* 97:7108–7115.

29. DH Williams, E Stephens & M Zhou (2003) Ligand binding energy and catalytic efficiency from improved packing within receptors and enzymes. *J. Mol. Biol.* 329:389–399.

30. A Cooper (2000) Heat capacity of hydrogen-bonded networks: an alternative view of protein folding thermodynamics. *Biophys. Chem.* 85:25–39.

31. J Janin, RP Bahadur & P Chakrabarti (2008) Protein–protein interaction and quaternary structure. *Q. Rev. Biophys.* 41:133–180.

32. C Chothia, M Levitt & D Richardson (1977) Structure of proteins: packing of α-helices and pleated sheets. *Proc. Natl. Acad. Sci. USA* 74:4130–4134.

33. FHC Crick (1953) The packing of α-helices: simple coiled coils. *Acta Cryst.* 6:689–697.

34. M Peckham & PJ Knight (2009) When a predicted coiled coil is really a single α helix, in myosins and other proteins. *Soft Matter* 5:2493–2503.

35. RJP Williams (1993) Are enzymes mechanical devices? *Trends Biochem. Sci.* 18:115–117.

36. C Ader, S Frey, W Maas et al. (2010) Amyloid-like interactions within nucleoporin FG hydrogels. *Proc. Natl. Acad. Sci. USA* 107:6281–6285.

37. DA Doyle, JM Cabral, RA Pfuetzner et al. (1998) The structure of the potassium channel: molecular basis of K^+ conduction and selectivity. *Science* 280:69–77.

38. Y Zhou, JH Morais-Cabral, A Kaufman & R MacKinnon (2001) Chemistry of ion coordination and hydration revealed by a K^+ channel–Fab complex at 2.0 Å resolution. *Nature* 414:43–48.

39. CB Anfinsen (1973) Principles that govern the folding of protein chains. *Science* 181:223–230.

40. R Bonneau & D Baker (2001) *Ab initio* protein structure prediction: progress and prospects. *Annu. Rev. Biophys. Biomol. Struct.* 30:173–189.

41. RB Tunnicliffe, JL Waby, RJ Williams & MP Williamson (2005) An experimental investigation of conformational fluctuations in proteins G and L. *Structure* 13:1677–1684.

42. PA Bullough, FM Hughson, JJ Skehel & DC Wiley (1994) Structure of influenza haemagglutinin at the pH of membrane fusion. *Nature* 371:37–43.

43. C Branden & J Tooze (1999) Introduction to Protein Structure, 2nd ed. New York: Garland Science.

44. LK Tamm, J Crane & V Kiessling (2003) Membrane fusion: a structural perspective on the interplay of lipids and proteins. *Curr. Opin. Struct. Biol.* 13:453–466.

45. A Murzin (2008) Metamorphic proteins. *Science* 320:1725–1726.

46. CB Stewart & AC Wilson (1987) Sequence convergence and functional adaptation of stomach lysozymes from foregut fermenters. *Cold Spring Harb. Lab. Symp. Quant. Biol.* 52:891–899.

47. DF Feng & RF Doolittle (1996) Progressive alignment of amino acid sequences and construction of phylogenetic trees from them. *Methods Enzymol.* 266:368–382.

48. RN Perham (1975) Self-assembly of biological macromolecules. *Phil. Trans. R. Soc. Lond. B* 272:123–136.

49. J Monod (1971) Chance and Necessity: An Essay on the Natural Philosophy of Modern Biology. New York: Alfred A Knopf.

50. E Meléndez-Hevia, TG Waddell & M Cascante (1996) The puzzle of the Krebs citric acid cycle: assembling the pieces of chemically feasible reactions, and opportunism in the design of metabolic pathways during evolution. *J. Mol. Evol.* 43:293–303.

51. P Bork, T Dandekar, Y Diaz-Lazcoz et al. (1998) Predicting function: from genes to genomes and back. *J. Mol. Biol.* 283:707–725.

52. MA Huynen & P Bork (1998) Measuring genome evolution. *Proc. Natl. Acad. Sci. USA* 95:5849–5856.

53. JG Lawrence & H Ochman (1997) Amelioration of bacterial genomes: rates of change and exchange. *J. Mol. Evol.* 44:383–397.

54. MY Galperin, DR Walker & EV Koonin (1998) Analogous enzymes: independent inventions in enzyme evolution. *Genome Res.* 8:779–790.

55. LR de Pouplana & P Schimmel (2001) Aminoacyl-tRNA synthetases: potential markers of genetic code development. *Trends Biochem. Sci.* 26:591–596.

56. RL Tatusov, EV Koonin & DJ Lipman (1997) A genomic perspective on protein families. *Science* 278:631–637.

57. L Patthy (1999) Protein Evolution. Oxford: Blackwell.
58. SE Brenner, T Hubbard, A Murzin & C Chothia (1995) Gene duplications in *H. influenzae*. *Nature* 378:140.
59. SA Teichmann, C Chothia & M Gerstein (1999) Advances in structural genomics. *Curr. Opin. Struct. Biol.* 9:390–399.
60. A Müller, RM MacCallum & MJE Sternberg (2002) Structural characterization of the human proteome. *Genome Res.* 12:1625–1641.
61. M Lynch & JS Conery (2000) The evolutionary fate and consequences of duplicate genes. *Science* 290:1151–1155.
62. J Spring (1997) Vertebrate evolution by interspecific hybridisation—are we polyploid? *FEBS Lett.* 400:2–8.
63. A Sidow (1996) Gen(om)e duplications in the evolution of early vertebrates. *Curr. Opin. Genet. Dev.* 6:715–722.
64. KH Wolfe & DC Shields (1997) Molecular evidence for an ancient duplication of the entire yeast genome. *Nature* 387:708–713.
65. JA Fawcett, S Maere & Y van de Peer (2009) Plants with double genomes might have had a better chance to survive the Cretaceous–Tertiary extinction event. *Proc. Natl. Acad. Sci. USA* 106:5737–5742.
66. S Ohno (1970) Evolution by Gene Duplication. Berlin: Springer-Verlag.
67. TS Mikkelsen, MJ Wakefield, B Aken et al. (2007) Genome of the marsupial *Monodelphis domestica* reveals innovation in non-coding sequences. *Nature* 447:167-U1.
68. H Gest (1987) Evolutionary roots of the citric acid cycle in prokaryotes. *Biochem. Soc. Symp.* 54:3–16.
69. G Zubay (2000) Origins of Life on the Earth and in the Cosmos, 2nd ed. San Diego: Academic Press.
70. M Huynen, T Dandekar & P Bork (1999) Variation and evolution of the citric-acid cycle: a genomic perspective. *Trends Microbiol.* 7:281–291.
71. LB Chen, AL DeVries & CHC Cheng (1997) Evolution of antifreeze glycoprotein gene from a trypsinogen gene in Antarctic notothenioid fish. *Proc. Natl. Acad. Sci. USA* 94:3811–3816.
72. JM Logsdon & WF Doolittle (1997) Origin of antifreeze protein genes: a cool tale in molecular evolution. *Proc. Natl. Acad. Sci. USA* 94:3485–3487.
73. AC Wilson, SC Carlson & TJ White (1977) Biochemical evolution. *Annu. Rev. Biochem.* 44:573–639.
74. LB Chen, AL DeVries & CHC Cheng (1997) Convergent evolution of antifreeze glycoproteins in Antarctic notothenioid fish and Arctic cod. *Proc. Natl. Acad. Sci. USA* 94:3817–3822.
75. P Tompa (2003) Intrinsically unstructured proteins evolve by repeat expansion. *BioEssays* 25:847–855.
76. J Park, M Lappe & SA Teichmann (2001) Mapping protein family interactions: intramolecular and intermolecular protein family interaction repertoires in the PDB and yeast. *J. Mol. Biol.* 307:929–938.
77. DE Koshland & KE Neet (1968) The catalytic and regulatory properties of enzymes. *Annu. Rev. Biochem.* 37:359–410.
78. BF Pugh (2004) Is acetylation the key to opening locked gates? *Nature Struct. Mol. Biol.* 11:298–300.
79. DR Tomchick, RJ Turner, RL Switzer & JL Smith (1998) Adaptation of an enzyme to regulatory function: structure of *Bacillus subtilis* PyrR, a pyr RNA-binding attenuation protein and uracil phosphoribosyltransferase. *Structure* 6:337–350.
80. G Wistow (1993) Lens crystallins: gene recruitment and evolutionary dynamism. *Trends Biochem. Sci.* 18:301–306.
81. CM Dobson (2006) Protein aggregation and its consequences for human disease. *Protein Peptide Lett.* 13:219–227.
82. J Piatigorsky & G Wistow (1991) The recruitment of crystallins: new functions precede gene duplication. *Science* 252:1078–1079.
83. AG Murzin (1993) Can homologous proteins evolve different enzymatic activities? *Trends Biochem. Sci.* 18:403–405.
84. RA Jensen (1976) Enzyme recruitment in evolution of new function. *Annu. Rev. Microbiol.* 30:409–425.
85. JA Gerlt & PC Babbitt (2001) Divergent evolution of enzymatic function: mechanistically diverse superfamilies and functionally distinct suprafamilies. *Annu. Rev. Biochem.* 70:209–246.
86. ME Glasner, JA Gerlt & PC Babbitt (2007) Mechanisms of protein evolution. In EJ Toone (ed), Advances in Enzymology and Related Areas of Molecular Biology. Hoboken, NJ: Wiley.
87. GA Petsko, GL Kenyon, JA Gerlt et al. (1993) On the origin of enzymatic species. *Trends Biochem. Sci.* 18:372–376.
88. CA Orengo & JM Thornton (2005) Protein families and their evolution: a structural perspective. *Annu. Rev. Biochem.* 74:867–900.
89. SCG Rison & JM Thornton (2002) Pathway evolution, structurally speaking. *Curr. Opin. Struct. Biol.* 12:374–382.
90. SA Teichmann, SCG Rison, JM Thornton et al. (2001) The evolution and structural anatomy of the small molecule metabolic pathways in *Escherichia coli*. *J. Mol. Biol.* 311:693–708.
91. CA Orengo, AE Todd & JM Thornton (1999) From protein structure to function. *Curr. Opin. Struct. Biol.* 9:374–382.
92. PF Gherardini, MN Wass, M Helmer-Citterich & MJE Sternberg (2007) Convergent evolution of enzyme active sites is not a rare phenomenon. *J. Mol. Biol.* 372:817–845.
93. N Nagano, CA Orengo & JM Thornton (2002) One fold with many functions: the evolutionary relationships between TIM barrel families based on their sequences, structures and functions. *J. Mol. Biol.* 321:741–765.
94. SS Krishna & NV Grishin (2004) Structurally analogous proteins do exist! *Structure* 12:1125–1127.
95. LC James & DS Tawfik (2001) Catalytic and binding poly-reactivities shared by two unrelated enzymes: the potential role of promiscuity in enzyme evolution. *Prot. Sci.* 10:2600–2607.
96. B Rasmussen, FC Wilberg, HJ Flodgaard & IK Larsen (1997) Structure of HBP, a multifunctional protein with a serine proteinase fold. *Nature Struct. Biol.* 4:265–268.
97. CJ Jeffery (1999) Moonlighting proteins. *Trends Biochem. Sci.* 24:8–11.
98. CJ Jeffery (2003) Moonlighting proteins: old proteins learning new tricks. *Trends Genet.* 19:415–417.
99. CJ Jeffery (2009) Moonlighting proteins—an update. *Mol. BioSyst.* 5:345–350.
100. JH Hurley, AM Dean, DE Koshland & RM Stroud (1991) Catalytic mechanism of $NADP^+$-dependent isocitrate dehydrogenase: implications from the structures of magnesium isocitrate and $NADP^+$ complexes. *Biochemistry* 30:8671–8678.
101. NH Horowitz (1945) On the evolution of biochemical syntheses. *Proc. Natl. Acad. Sci. USA* 31:153–157.
102. MA Huynen, T Gabaldón & B Snel (2005) Variation and evolution of biomolecular systems: searching for functional relevance. *FEBS Lett.* 579:1839–1845.
103. L Dijkhuizen (1996) Evolution of metabolic pathways. *Soc. Gen. Microbiol. Symp.* 54:243–265.
104. LE Orgel (1998) The origin of life—a review of facts and speculations. *Trends Biochem. Sci.* 23:491–495.
105. EA Gaucher, JM Thomson, MF Burgan & SA Benner (2003) Inferring the palaeoenvironment of ancient bacteria on the basis of resurrected proteins. *Nature* 425:285–288.
106. M Pagel (1999) Inferring the historical patterns of biological evolution. *Nature* 401:877–884.
107. DH Bamford, RJC Gilbert, JM Grimes & DI Stuart (2001) Macromolecular assemblies: greater than their parts. *Curr. Opin. Struct. Biol.* 11:107–113.
108. A Roth & RR Breaker (1998) An amino acid as a cofactor for a catalytic polynucleotide. *Proc. Natl. Acad. Sci. USA* 95:6027–6031.
109. M Halic & R Beckmann (2005) The signal recognition particle and its interactions during protein targeting. *Curr. Opin. Struct. Biol.* 15:116–125.
110. T Yomo, S Saito & M Sasai (1999) Gradual development of protein-like global structures through functional selection. *Nature Struct. Biol.* 6:743–746.

111. CR Woese (1990) Evolutionary questions: the "progenote." *Science* 247:789.

112. LC James & DS Tawfik (2003) Conformational diversity and protein evolution—a 60-year-old hypothesis revisited. *Trends Biochem. Sci.* 28:361–368.

113. JW Schopf (1992) Evolution of the proterozoic biosphere: benchmarks, tempo and mode. In JW Schopf & C Klein (eds), The Proterozoic Biosphere. New York: Cambridge University Press.

114. G D'Alessio (1999) The evolutionary transition from monomeric to oligomeric proteins: tools, the environment, hypotheses. *Prog. Biophys. Mol. Biol.* 72:271–298.

115. JAG Ranea, A Sillero, JM Thornton & CA Orengo (2006) Protein superfamily evolution and the last universal common ancestor (LUCA). *J. Mol. Evol.* 63:513–525.

116. MY Galperin & EV Koonin (1999) Functional genomics and enzyme evolution—homologous and analogous enzymes encoded in microbial genomes. *Genetica* 106:159–170.

117. G Caetano-Anollés, HS Kim & JE Mittenthal (2007) The origin of modern metabolic networks inferred from phylogenomic analysis of protein architecture. *Proc. Natl. Acad. Sci. USA* 104:9358–9363.

118. E Zuckerkandl (1975) The appearance of new structures and functions in proteins during evolution. *J. Mol. Evol.* 7:1–57.

119. B Alberts (1998) The cell as a collection of protein machines: preparing the next generation of molecular biologists. *Cell* 92:291–294.

120. IS Povolotskaya & FA Kondrashov (2010) Sequence space and the ongoing expansion of the protein universe. *Nature* 465:922–926.

121. S Freilich, RV Spriggs, RA George et al. (2005) The complement of enzymatic sets in different species. *J. Mol. Biol.* 349:745–763.

122. JM Berg, JL Tymoczko & L Stryer (2007) Biochemistry, 6th ed. New York: Freeman.

123. DJ Voet & JG Voet (2004) Biochemistry, 3rd ed. New York: Wiley.

124. AM Lesk (2010) Introduction to Protein Science, 2nd ed. Oxford: Oxford University Press.

125. TE Creighton (1993) Proteins: Structures and Molecular Properties, 2nd ed. New York: Freeman.

126. WH Li (1997) Molecular Evolution. Sunderland, MA: Sinauer Associates, Inc.

127. AG Murzin, SE Brenner, T Hubbard & C Chothia (1995) SCOP—a structural classification of proteins database for the investigation of sequences and structures. *J. Mol. Biol.* 247:536–540.

128. CA Orengo, AD Michie, S Jones et al. (1997) CATH—a hierarchic classification of protein domain structures. *Structure* 5:1093–1108.

129. L Holm & C Sander (1998) Touring protein fold space with Dali/FSSP. *Nucleic Acids Res.* 26:316–319.

130. L Holm & C Sander (1993) Protein structure comparison by alignment of distance matrices. *J. Mol. Biol.* 233:123–138.

131. S Feo et al. (2000) ENO1 gene product binds to the c-*myc* promoter and acts as a transcriptional repressor: relationship with Myc promoter-binding protein 1 (MBP-1). *FEBS Lett.* 473:47–52.

132. JG Pastorino & JB Hoek (2003) Hexokinase II: the integration of energy metabolism and control of apoptosis. *Curr. Med. Chem.* 10:1535–1551.

133. J Stetefeld & MA Ruegg (2005) Structural and functional diversity generated by alternative mRNA splicing. *Trends Biochem. Sci.* 30:515–521.

134. HS Frank & MW Evans (1945) Free volume and entropy in condensed systems III. *J. Chem. Phys.* 13:507–532.

135. S Meier, S Grzesiek & M Blackledge (2007) Mapping the conformational landscape of urea-denatured ubiquitin using residual dipolar couplings. *J. Am. Chem. Soc.* 129:9799–9807.

136. E Yamaguchi, K Yamauchi, T Gullian & T Asakura (2009) Structural analysis of the Gly-rich region in spider dragline silk using stable-isotope labeled sequential model peptides and solid-state NMR. *Chem. Commun.* 28:4176–4178.

137. MJ Bayley, G Jones, P Willett & MP Williamson(1998) GENFOLD: a genetic algorithm for folding protein structures using NMR restraints. *Protein Sci.* 7:491–499.

CHAPTER 2
Protein Domains

We derive much of our pleasure as biologists from the continuing realization of how economical, elegant and intelligent are the accidents of evolution that have been maintained by selection.

David Baltimore (1975), Nobel lecture [1]

It is a general principle of evolution that multiple use is made of given resources.

J.E. Baldwin and H.E. Krebs (1981), [2]

In Chapter 1, we saw that protein structure can be subdivided into primary, secondary, tertiary, and quaternary structures. However, evolution does not work in such terms: the fundamental evolutionary unit is the domain. Domains also form the fundamental functional units. This chapter considers why domains are so important. We shall see that they are a convenient size for evolution to make use of, and they provide an instant benefit, allowing evolution to create new proteins and new functions at a stroke.

In complex organisms, a major problem is how to evolve new systems that are specific in their actions and do not compromise the selectivity of existing systems. Domains are fundamental in the solution to this problem: the addition of an extra domain provides an instant and readily evolved increase in the selectivity of binding. Domains also allow easily evolved regulation of binding, because intramolecular binding is inherently stronger than intermolecular binding. A weak intramolecular binding interaction can therefore be readily tuned to regulate intermolecular binding. This leads naturally to the evolution of embellishment: the addition of extra domains to regulate binding, a topic discussed many times in this book.

2.1 DOMAINS: THE FUNDAMENTAL UNIT OF PROTEIN STRUCTURE

2.1.1 Domains can be defined in a variety of ways

The words **domain** and **module** are used differently by different authors. We therefore start with the definitions used in this book:

> A domain is a polypeptide chain (or part of one) that can independently fold into a stable compact tertiary structure or fold. [3, 4]

The domain is the fundamental structural and evolutionary unit of proteins [5]. This is a definition based on a physical property that can be determined experimentally, although not necessarily very easily. It is therefore not a very useful definition for analyzing proteins by using sequence data. The most common alternative definition is based on sequence alignments, which is much simpler to apply: *A domain is a distinct sequence that is conserved in evolution*. The two definitions are related, in that most sequences that get duplicated and used elsewhere have this property because they are capable of forming independent folded structures. Domains are relatively small: the mean size is around 17 kDa [6, 7] (but the median is significantly smaller, because the distribution is very skewed), and the largest is approximately 35 kDa, depending on the definition of a domain (compare Figure 2.39).

> A **module** is a highly conserved sequence that is observed in different contexts in multidomain proteins.

Thus, modules are a subcategory of domains, and their definition looks a lot like the alternative definition of domain given above. The difference between modules and domains is that modules are always continuous peptide sequences, are found surrounded by different sequences in different proteins, and have clear module boundaries: in fact, most modules behave almost like independent beads on a string, with the "string" being the intervening random-coil sequences. Some examples are given below. Proteins consisting of multiple modules are sometimes called chimeric or **mosaic** proteins.

> A **fold** is a three-dimensional arrangement or topology of secondary structure elements.

Every domain or module adopts some kind of fold.

There have been several attempts to classify protein structures into hierarchical organizations, of which the best-known are the SCOP classification of Murzin et al. [5] and the CATH classification of Orengo et al. [8]. The two classification systems are similar, possibly the biggest difference being that SCOP requires human intervention to decide whether a given structure is a new fold or fits into an existing one, whereas CATH does it largely automatically. There is no obvious fixed boundary between one fold and another, so at some point an arbitrary decision has to be made; it is therefore not surprising that the boundaries differ for about 30% of proteins [9]. Both systems have the top level as being the structural class: mainly α, mainly β, mixed α and β, and a fourth class of small proteins with little regular secondary structure. The mixed class can also be subdivided into proteins with segregated α and β regions (α + β, typically antiparallel β sheet) and proteins with integrated α and β regions (α/β, many containing the β-α-β supersecondary structure). The next level down is the fold. Chothia proposed in 1992 that there are probably a small number of folds, of the order of 1000 [10]. Since then there has been a good deal of discussion about this number, with estimates ranging up to tens of thousands. However, it is now looking likely that the true number is rather less than 2000 [11]. CATH subdivides the fold category further into Architecture (arrangement of regular secondary structure) and Topology (the way in which the secondary structures are linked together).

The next level below the fold is the superfamily. Members of a superfamily have sequence similarities and have related function and structural features, implying that they are evolutionarily related. In other words, they are **homologous**. Strictly speaking, *homologous* means that two proteins are related by evolution. It has more loosely taken on an additional, and more easily proven, meaning of "having a similar sequence." The two meanings are almost the same: proteins that have sequences identical by more than about 30% are almost guaranteed to be evolutionarily related (**Figure 2.1**). An evolutionary relationship can often be defined more exactly by using two useful terms. An **ortholog** is a homolog that was created by a speciation event: two orthologs fulfill the same role in evolutionarily related organisms because they were derived from the same ancestral protein. By comparison, a **paralog** is a homolog that was created by gene duplication.

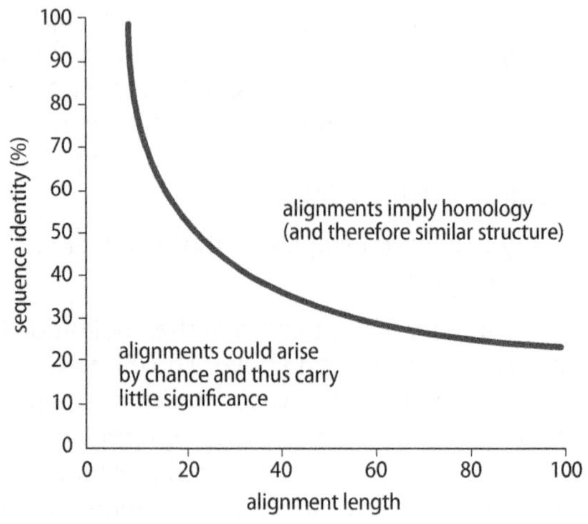

FIGURE 2.1
Proteins that have sequences identical by more than about 30% are almost certain to be homologous; that is, evolutionarily related. This also implies that they will have similar structure, and probably a similar type of function (compare with Figure 1.50). The degree of similarity needed depends on the length of the sequence—a shorter sequence has a higher probability of being similar just by chance and therefore needs to have greater similarity for a confident assignment as being homologous.

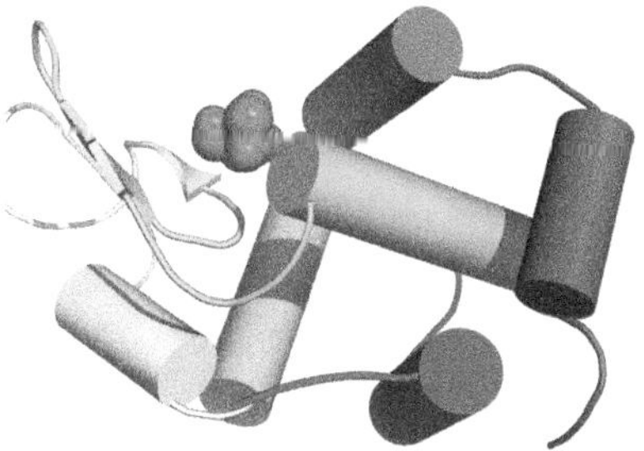

FIGURE 2.2
The structure of lysozyme (PDB file 2vb1) consists of an α-helical region and a mainly β-sheet region, with the active site being in a cleft between them (red balls). On some criteria the protein therefore contains two domains. However, on the criteria used here it is a single domain because there are no instances of one region occurring independently of the other, and they do not fold independently. It is likely that in some ancient protein these two regions consisted of "mini-domains" that were fused together.

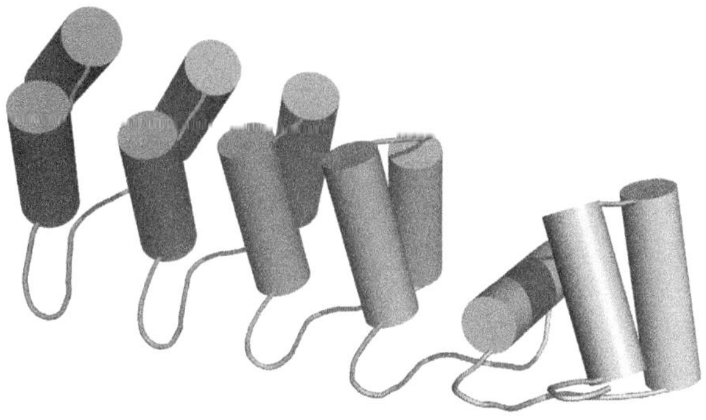

FIGURE 2.3
The ankyrin domain from the Notch receptor in *Drosophila* (PDB file 1ot8) consists of six ankyrin repeat sequences, which cannot fold independently.

The definition of domain used here is not universally agreed. Some authors have used the word to mean any structural unit that can be notionally obtained by slicing up a protein [12]. Thus, for example, lysozyme (**Figure 2.2**) is often described as having two domains. In many ways the two parts of lysozyme do indeed behave like domains: they are relatively independent; the active site is located between them; they move fairly separately; and the helical "domain" folds before the sheet "domain" [13]. However, they cannot fold independently into stable units and are never found separately in proteins. For this latter reason, SCOP and CATH both classify lysozyme as a single domain, the two separate parts being subdomains. Even more dramatically, several proteins are composed of multiple copies of small structural units; an example is ankyrin from the signaling protein Notch, discussed in Section 8.5.1 (**Figure 2.3**). Ankyrin can very clearly be dissected into multiple helix–turn units, individual units are unfolded, and the protein folds as one single cooperative event, but individual units have no structure or biological function [14]. The *domain* in this instance is thus the entire protein, not the individual units, which could be called subdomains.

2.1.2 Domains can usually be associated with specific functions

Domains often have clearly identifiable functions. A good example is the **Rossmann fold (*2.1)** (**Figure 2.4**), which functions to bind nucleotides such as $NADP^+$ or ATP. Many proteins that bind dinucleotides contain two Rossmann folds, each of which binds to one nucleotide. This domain structure is so closely linked to the nucleotide-binding function that observation of the structure is a very good predictor of the function.

By contrast, a common domain structure found in a wide variety of enzymes (approximately 10% of all enzymes) is the TIM barrel (so called because it was first characterized in the enzyme triosephosphate isomerase, abbreviated to TIM). This is a β barrel formed by a series of β-α-β **structure motifs** (**Figure 2.5**), in which the active site is always found in the same position, in the cleft defined by the loops at the C-terminal end of the barrel. Although the location of the active site is extremely predictable, the function is not: this is in fact the most versatile enzyme motif, with a very wide range of functions.

A very large number of bacterial proteins that bind to DNA, such as repressors, do so by a DNA-binding domain that contains a "helix–turn–helix" (HTH) motif, consisting of two α helices connected by a short loop. As a domain this is very small; indeed, one would normally expect such a short sequence not to

***2.1 Rossmann fold**
This fold was originally described by Michael Rossmann in 1970 and at that time was one of the very few examples of a protein **fold** that could clearly be identified with a function, namely nucleotide binding [99]. A typical structure is shown in Figure 2.4. It consists of a β-α-β-α-β structure motif and has several conserved sequence elements, notably a sequence GXGXXG (where X means any amino acid) that binds to the phosphate group of the nucleotide. The glycines are required for three reasons. First, the amino acid residues need to be small, because otherwise they would have a steric clash with the bound nucleotide. The consequent structural changes with a larger residue not only weaken the binding and destabilize the structure but also allow water access into the active site. Second, the loop needs to be conformationally flexible, which is only possible with glycines. Third, the first glycine is in the top right quadrant of the Ramachandran plot and is therefore in a conformation energetically unfavorable for all amino acids except glycine. This domain structure can be found in a very wide range of proteins that use nucleotides as cofactors.

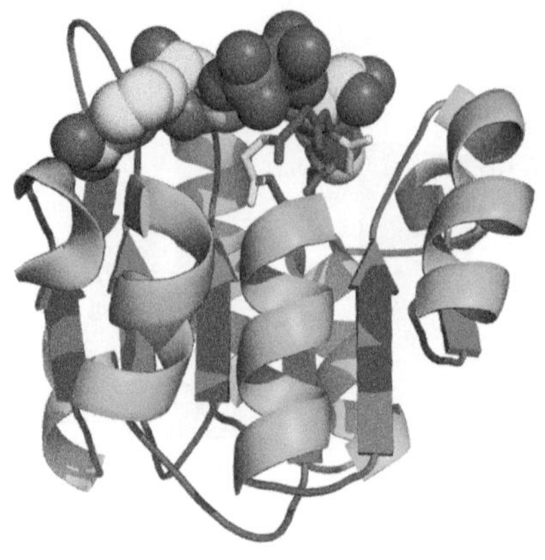

FIGURE 2.4
The Rossmann fold (PDB file 1u3w). This figure shows part of human alcohol dehydrogenase, which is bound to NAD^+ and Zn^{2+}. This is the enzyme that metabolizes alcohol, and it is therefore an enzyme of some importance. The figure shows NAD^+ in ball format and the GXGXXG sequence motif that is involved in binding NAD^+ as the loop in stick format near the center of the structure. Residues in blue are glycines, and Gly 3 of the motif is at top right of the loop, very close to a phosphate in NAD^+ (orange ball). It is just about possible to make out that the structure consists of a series of β-α-β supersecondary structures.

fold independently and therefore to constitute a structure motif rather than a domain as defined above. However, it has been shown that the HTH structure does in fact fold independently and is thus a genuine domain [15]. This domain (**Figure 2.6**, from *lac* repressor) inserts one of the two helices into the major groove of DNA, with the other helix running across the top almost at right angles. The recognition of specific DNA sequences is determined largely by the amino acid sequence of the "recognition helix," but there is no clear one-to-one correspondence between DNA and protein sequence. The intact repressor consists of a series of domains (**Figure 2.7**): from the N terminus, these are the HLH DNA-binding domain, a DNA-binding helix, two core domains, and a C-terminal helix that associates with the corresponding helix in other repressor molecules to form tetramers. Note that each domain has an independent function; this will be a recurring theme.

The two-hybrid screen, discussed in Chapter 11, provides an excellent example of the independent function of domains. The method is designed to look for domain:domain interactions, by attaching one of these domains to a DNA-binding domain and the other to a transcription activation domain. The screen only works if each of the domains involved works properly (that is, it has the same function as it does in its normal location) in its new context. The fact that this assay works well is a graphic illustration of the independence of domains.

In prokaryotes, roughly two-thirds of all proteins consist of two or more domains [16], and the proportion is even higher (roughly 80%) in eukaryotes [17, 18] (**Table 2.1**). Thus, the majority of proteins consist of multiple domains.

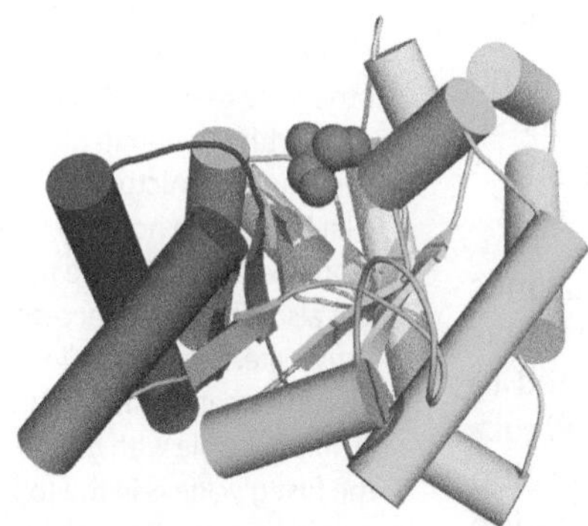

FIGURE 2.5
The structure of the β-barrel protein triosephosphate isomerase (PDB file 3gvg), color-coded from blue at the N terminus to yellow at the C terminus. The active site is shown (red spheres), sitting at the C-terminal end of the barrel (the C-terminal ends of the β strands). This is the same protein as that shown in Figure 1.34, but with a different view to emphasize the location of the active site.

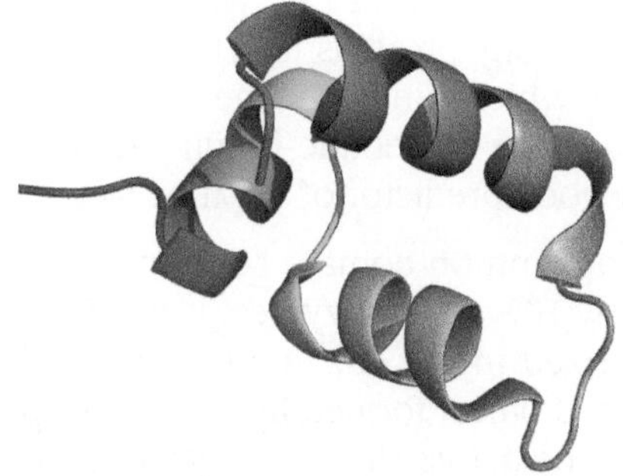

FIGURE 2.6
The helix–turn–helix protein, *lac* repressor (PDB file 2pe5). This figure shows the headpiece, the part of the protein that binds to DNA (compare Figure 1.33). The recognition helix, which binds to the major groove in DNA, is at the bottom.

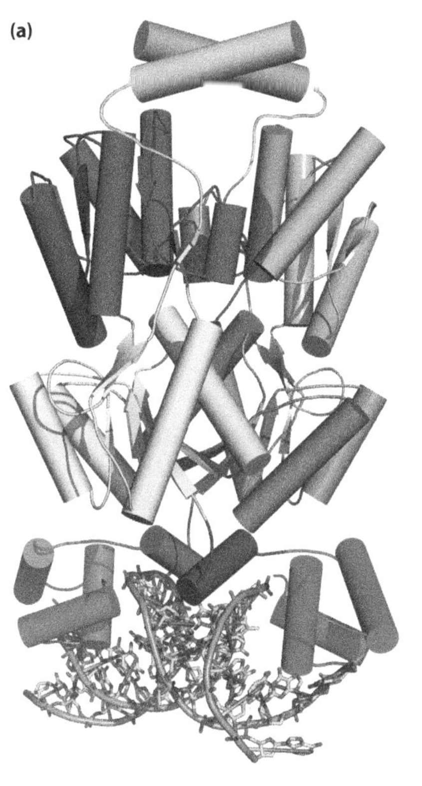

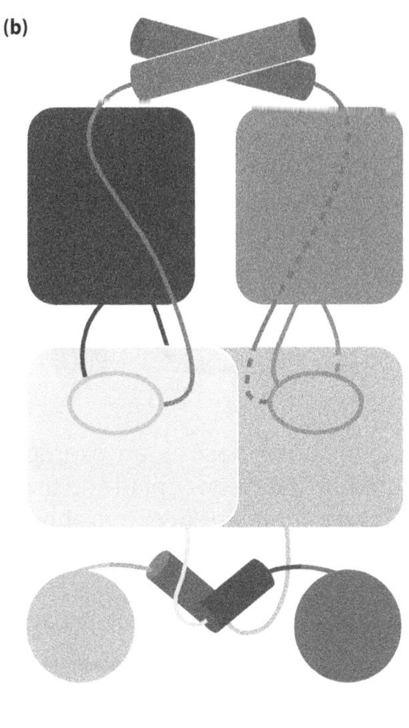

FIGURE 2.7
The domain structure of *lac* repressor (PDB file 2pe5). The figure shows a cartoon view of the structure (a) plus an almost equally helpful and much simpler diagrammatic view (b). The C-terminal tetramerization domain is absent from the crystal structure and is indicated in mustard yellow at the top. The intact protein is a dimer of dimers; shown here is the "core" dimer. One monomer is purple/blue/green/magenta/yellow, and the other is in similar but paler colors. Note that the protein therefore "crosses over" at the DNA-binding helix. The N-terminal core domain contains some sequence from the C-terminal end of the second core domain, corresponding to the long green helix at the front and the short β strands to the left of it (see panel (a)).

This is most obvious for enzymes, where it is extremely common to find the active site of the enzyme at the interface between two domains. An example is the enzyme adenylate kinase, which carries out the important transformation

$$\text{AMP} + \text{ATP} \rightleftharpoons 2\text{ADP}$$

This enzyme consists of three domains: a central domain flanked by two smaller nucleotide-binding regions. The binding sites for nucleotide binding are found at the interface between the central domain and each of the two smaller domains (**Figure 2.8**). The enzyme can be crystallized in both an open form and a closed form. In the closed form, the two smaller domains have rotated toward each other, forming a single globular structure that encloses the substrates.

TABLE 2.1 Percentage of single and multiple-domain proteins in different genomes

Genome group	Number of domains in protein					
	Exactly 1	Exactly 2	≥ 2	Exactly 3	≥ 3	≥ 4
Archaea	36	9	43	2	9	2
Bacteria	35	10	42	2	10	2
Yeast	22	5	57	1	5	3
Metazoa	23	4	52	1	4	7

These data are derived from fitting the sequences to domains as defined in SCOP. "Exactly 1" means that the whole sequence fits precisely to a single domain, whereas "≥ 2" means that only part of the sequence fits to a single domain, implying that there must be more than one domain present. Similar arguments apply to higher numbers. Thus, the percentage in the column "≥ 3", for example, is a minimum, because some of the proteins in the column "≥ 2" will in fact have three or more domains. (Data from G. Apic, J. Gough and S.A. Teichman, *J. Mol. Biol.* 310: 311–325, 2001. With permission from Elsevier.)

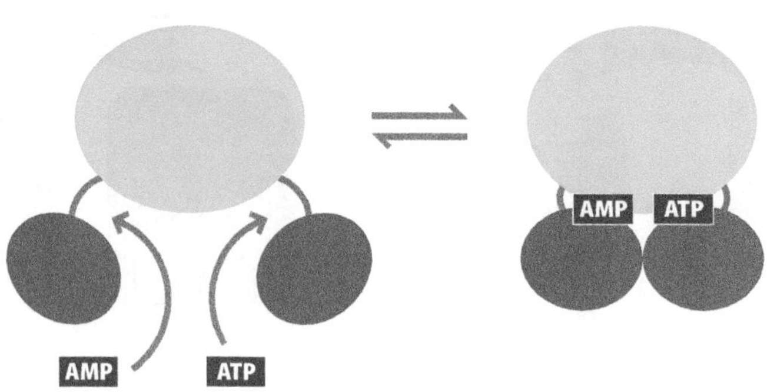

FIGURE 2.8
The structure of adenylate kinase. The two red domains are homologous. On binding AMP/ATP they close in on the central domain to make an enclosed active site.

Adenylate kinase is an example of a protein with **internal duplication**: a domain has been copied and used twice within the same protein. This is a rather common occurrence, possibly because genetically it is a fairly simple event. The protease chymotrypsin (discussed below) is another example. Sometimes this duplication is obvious from the sequence, but often the duplication happened so long ago that it can only be spotted from the three-dimensional structure, in which case the result is a protein that displays **pseudo-twofold symmetry**: the two halves have almost identical structures.

2.1.3 Domains are the basic building blocks of proteins

A domain can very often be considered as a single transferable functional entity. Transpose it to a different context and it does the same job; it can be used on its own or combined with other domains. It can often be treated as a single almost rigid entity, with flexibility coming from the linkers between domains. Thus, the vast majority of enzymes can be represented as two or more domains, with the active site(s) being at the interfaces between domains, implying that the backbone geometry of each domain does not need to move, only a few side chains in the binding site plus the linkers between domains. This book represents many domains as featureless shapes, because for most purposes the internal structure does not move much and does not matter much. Of course, in detail the structure and internal dynamics matter greatly, as discussed in Chapter 6.

It is worth noting that the living world is "modular" on many different scales [19]. Organisms consist of an assembly of parts, most obviously for insects such as centipedes, but it is true of all organisms, as the shortest consideration of comparative anatomy will show. Moreover, the same part is used in many different ways: fish gills have been turned into legs, wings, lungs, and so on [20]. The parts consist of separate organs with separate functions, the organs consist of separate tissues that interact in defined ways, and the tissues consist of different cell types that interact in different ways. The eukaryotic cell consists of several subcellular compartments, and each of these has its characteristic proteins. The domain is the next size down, being the fundamental building block of proteins. Below this come secondary and primary structure, amino acids, and atoms. This forms a fundamental tenet of the new science of systems biology. Or, to quote again from Jacob [19]: "Every object that biology considers represents a system of systems."

Many authors have suggested that, early in evolution, the basic unit was probably smaller than what we now call a domain, and consisted of a short piece of regular secondary structure such as a helix or β hairpin, perhaps 20 residues in length (see Figure 2.2). This is an enticing idea, not least because the construction of novel proteins by shuffling or mutation of sections is simpler when the sections are short. It is noted, for example, that a number of exons are roughly this length. An analysis of the structure of the bacterial ribonuclease

***2.2 Barnase**

Barnase is a ribonuclease from the bacterium *Bacillus amyloliquefaciens*. The structure is shown in **Figure 2.2.1**. It is an important protein: it has been much studied as a model system, mainly in protein folding studies, because it is small (110 residues) and contains a mixture of β-sheet and α-helix, with no disulfide bridges. It is also interesting because it is one of the smallest enzymes and because it binds extremely tightly to its natural inhibitor barstar, with very rapid association, as discussed in Chapter 4.

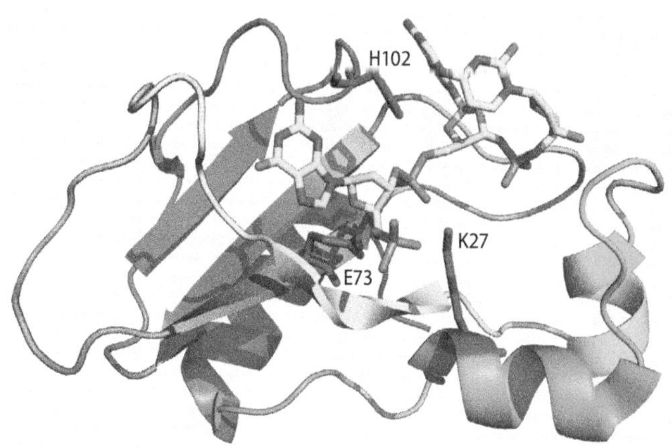

FIGURE 2.2.1
The structure of the ribonuclease barnase (PDB file 1brn). The two central bases of the deoxynucleotide inhibitor d(CGAC) are shown bound to the enzyme. The phosphodiester bond that is cleaved by the enzyme is in the center of the figure, roughly between H102 and K27. E73 is the catalytic base, and K27 is an electrophilic catalyst for attack on the phosphate.

barnase (*2.2) showed that it could be divided into six structural units, consisting of contiguous compact structures; of these, three were found to bind RNA and display weak RNAse activity [21]. It was therefore suggested that domains, and hence larger proteins, could have arisen by the shuffling and assembly of smaller primitive structural elements, roughly equivalent to structure motifs. This could well be the case; however, it still remains true that when evolution has shuffled segments of protein around in recent evolutionary history, the evidence is clearly that larger domain-sized pieces are the preferred unit.

2.1.4 Modules are transposable domains

Modules (as defined above) are found in both prokaryotes and eukaryotes, although they have a more significant role in eukaryotes [22]. Good examples can be found in proteins of the blood clotting cascade, several of which are composed entirely of modules strung together in different ways. The protein urokinase consists of three modules (**Figure 2.9**): an EGF domain (named after epidermal growth factor, which is the "founding father" of this domain and as its name suggests is a chemokine that stimulates the growth of epidermal cells), a kringle domain (named after a Danish pastry because the conventional sequence drawing looks like a kringle; **Figure 2.10**), and a chymotrypsin domain. Each domain has a function: the EGF domain stimulates EGF receptors, the kringle domain binds to lysines, and the chymotrypsin domain is a protease. Other proteins carry such module duplication to extremes: human apolipoprotein(a) has up to 38 copies of the kringle domain, whereas fibronectin contains 12 type I fibronectin modules interspersed by two type II modules and 15 type III modules [23].

Another other classic example of modules is the immunoglobulin superfamily, whose members are found not only in immunoglobulins but also as multiple repeats as components of cell surface receptors (**Figure 2.11**). The immunoglobulin fold and the fibronectin type III fold are very similar. This **superfold** is presumably particularly easily adapted to bind to different ligands.

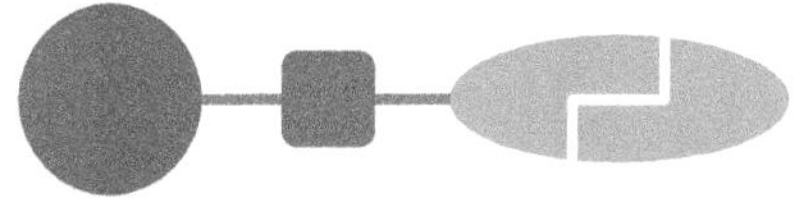

FIGURE 2.9
The domain structure of urokinase, which consists of an EGF domain (purple), a kringle domain (brown) and a chymotrypsin domain (cyan). Chymotrypsin consists of two subdomains and was originally formed by gene duplication. There are homologous proteins, such as the HIV protease, that still retain the two halves as identical separate polypeptides.

Most modules act to recognize and bind to specific ligands. They are a prominent feature of eukaryotic genomes in comparison with prokaryotic genomes, because they help to increase specificity, as discussed in Section 2.2. A small list of such modules is given in Table 8.1. It is significant that in many of these modules the N and C termini are close together and the ligand-binding site is on the opposite face. It is therefore possible to slot a module into an existing protein sequence, either at one end or in a surface loop, and thereby add

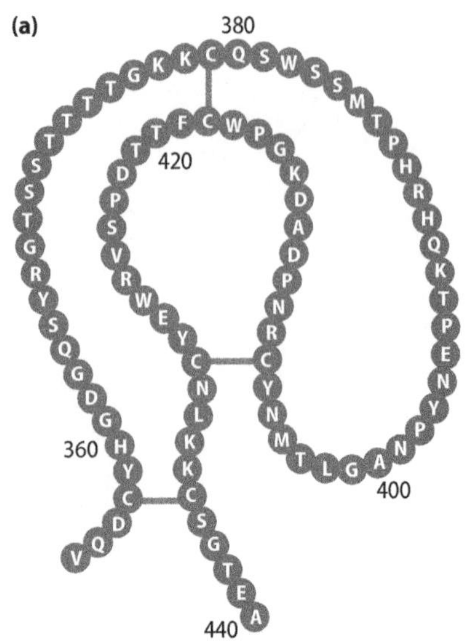

FIGURE 2.10
The kringle domain. (a) The sequence of the fourth kringle domain of plasminogen, which contains five kringle domains. It is drawn in the conventional representation, showing the location of the three disulfide bridges. (b) A kringle pastry. (With permission of SkagenHus, Wisconsin.)

a new ligand recognition site without perturbing the original function of the host protein.

These modules seem to act mainly individually, with little interaction between modules, and no "additional" function gained by the juxtaposition of two modules. Each kringle domain in a multimodule protein, for example, has a similar binding activity. It is, however, likely that multiple kringle domains can bind cooperatively, so that the cumulative affinity from a series of domains is much greater than the affinity of each one individually. Some modules seem to have lost their original functions entirely, and act purely as spacers. One can imagine that for proteins involved in blood clotting, a structure that is long and stringy would be of considerable advantage in forming interactions within a flowing medium and in permitting strong cross-links. Thus, a module with a function "purely as a spacer" can still be vital. Many more examples of modules are given in Li [24].

However, some modules do interact with their neighbors. Some of the modules in fibronectin function cooperatively [25, 26] (**Figure 2.12**), as do some modules in tissue plasminogen activator [27] (**Figure 2.13**). Some of the modules that recognize histone tail modifications (Section 4.4.3) do so as pairs, rather than as the more conventional one-module–one-modification, and the recognition sites are located in the interface between the two modules. The same is true for the human tumor suppressor protein BRCA1 [28]. However, these are the exception rather than the rule.

There are several websites that are designed to permit searches and visualization of modules and domains, of which two are Pfam (http://pfam.sanger.ac.uk/) and prodom (http://prodom.prabi.fr/), containing more than 10,000 different families. Such websites show that many modules have a rather limited range of neighbors in the protein sequence, with the same adjacent pairs occurring frequently, and also that module pairing tends to be organism-specific—another example of evolution taking whatever is most available at the time [29]. However, some are much more promiscuous. To take one example: the very versatile signaling module SH2 (Chapter 8) has been found in the

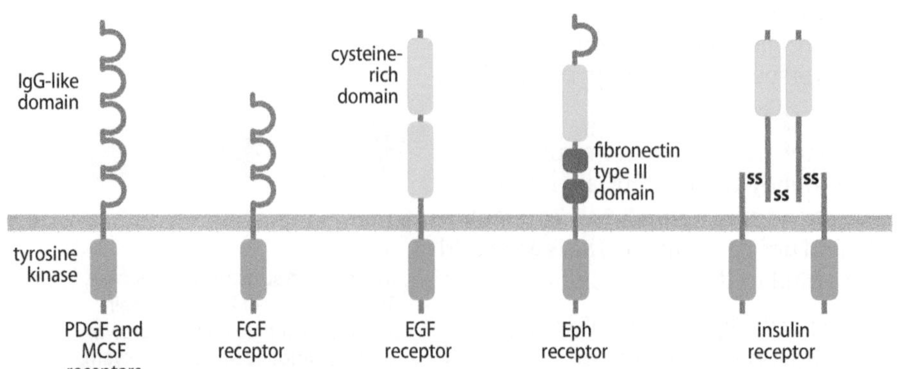

FIGURE 2.11
The modular structure of cell surface receptors. These often have multiple immunoglobulin folds on their extracellular surface and tyrosine kinases intracellularly, in a range of different architectures. Signaling pathways are a rich source of modular proteins, as discussed in Chapter 8.

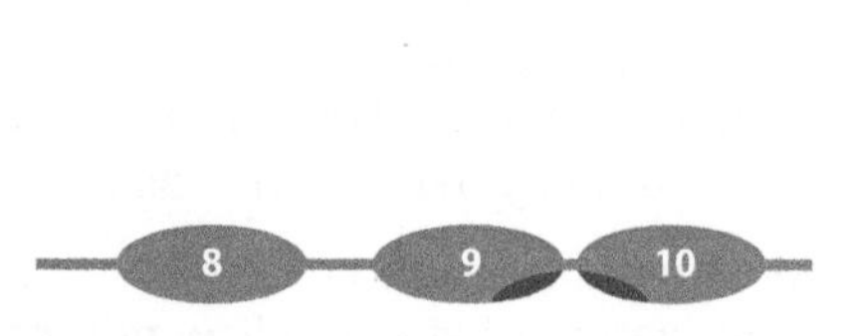

FIGURE 2.12
Fibronectin contains a large number of fibronectin type III modules. Most of these are probably independent, but modules 9 and 10 form a single binding site for integrins.

same polypeptide chain as 47 other modules. Most of these contain multiple modules, and almost all arrangements are possible. Thus, for example, SH2 is very commonly found associated with SH3 domains, and the organizations SH3-SH2, SH3-SH2-SH3, SH2-SH3, SH2-SH3-SH2, and SH3-X-SH2 (where X is a variety of other modules) are all found. It is abundantly clear that evolution has been able to "pick and mix" modules to achieve the desired combination of functions [30]. How it does this is considered below.

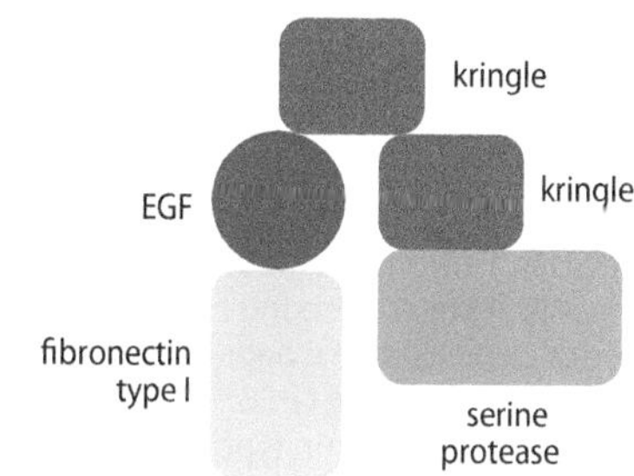

FIGURE 2.13
Tissue plasminogen activator consists of five modules, possibly in the fairly fixed hairpin geometry shown here.

2.2 THE KEY ROLE OF DOMAINS IN PROTEIN EVOLUTION

2.2.1 Multidomain proteins are produced by exon shuffling

If multidomain proteins are so common, how are they produced? That is, what biochemical mechanisms exist for taking a module from one place in the genome, copying it, and reinserting it elsewhere? There are several mechanisms, of which the most obvious is exon shuffling [18, 24, 31–33].

In eukaryotic genomes, but not prokaryotic ones, genes tend to be split up into several small sections called exons, separated by noncoding regions of DNA called introns (**Figure 2.14**). Prokaryotic cells have mechanisms for cutting out the introns and splicing the exons together. It was proposed many years ago that this provides an obvious mechanism for domain shuffling. One can easily see how modular proteins could have been put together in this way, as indicated in **Figure 2.15** for proteins in blood clotting [34].

The key feature required by models such as this is that an exon codes for a module. Is this in fact true? The answer is that sometimes they do, and sometimes they don't. Exons correspond more to structure motifs than to domains, because exon fragments are usually fairly short [32]. And in fact the typical size of a mammalian exon (180 bases or 60 amino acids) is a good size to constitute a motif. Exon boundaries can be identified straightforwardly, and they are often found either at domain boundaries or in loops, as expected from this theory. But there are many cases where they are also found in the middle of domains. Similarly, exon boundaries are conserved fairly well between species, but not always, and the phase of the exon boundaries (that is, at which base of the three-base codon it is spliced) is well conserved, as would be required by this model. A final argument is that "newer" proteins (that is, proteins that seem likely to have evolved more recently) have a greater tendency to have exon boundaries at domain boundaries. Thus, for example, the blood

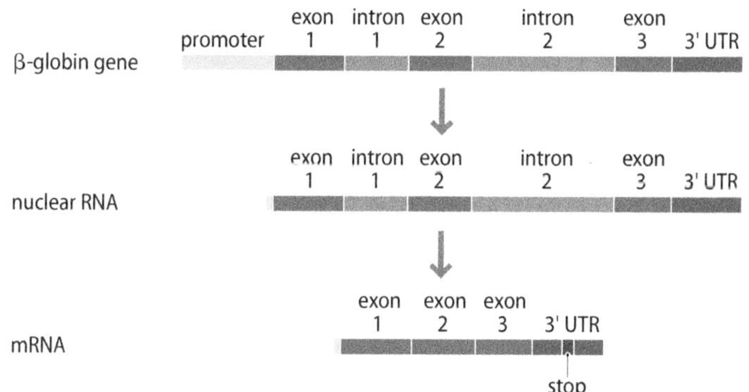

FIGURE 2.14
Eukaryotic DNA contains a large proportion of noncoding introns that are spliced out. This figure shows the structure of the gene for the β subunit of hemoglobin, and its processing into mature mRNA.

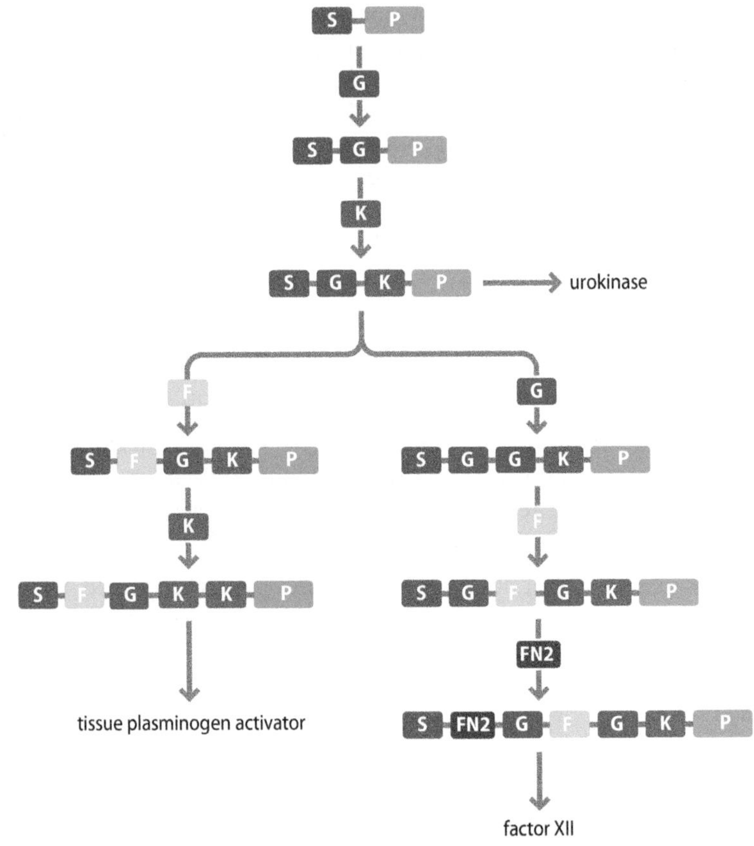

FIGURE 2.15
Hypothetical evolution of some of the proteins in the blood clotting cascade. It is suggested that all introductions of new modules arose by insertion of a new exon into existing introns. All module junctions in these proteins correspond to the same phase of intron. S, signal peptide; P, trypsin-like protease; G, EGF; K, kringle; F, finger; FN2, fibronectin type 2. (Redrawn from L. Patthy, Protein Evolution. Oxford, Blackwell, 1999. With permission from John Wiley & Sons Ltd.)

clotting proteins discussed earlier have domain boundaries that coincide with exon boundaries, in the same phase [24]. Therefore it seems likely that exons do indeed code for structural units, or at least did so at some time in the past.

It was proposed many years ago by Gilbert that introns were an ancient means of evolving new proteins, occurring before the split into prokaryotes and eukaryotes, and that prokaryotes subsequently discarded them [35]. This has become known as the intron-early theory. A competing theory, the intron-late theory, has suggested that introns arrived relatively late in evolution. A stronger version of this suggests that it was the arrival of introns that permitted a far more widespread shuffling of domains and provided the evolutionary mechanism to allow multicellular organisms to develop the large number of new genes to cope with multicellularity, or in other words that exon shuffling was the vital new development that made possible the metazoan radiation (the sudden development of multicellular organisms) [31, 36]. A related observation is that speciation (that is, the divergence of two populations to form a new species) seems to be related to novel combinations of domains [18].

There has been considerable discussion on this issue, and the current consensus is in broad favor of the intron-late theory. It is, however, clear that the history of introns is not as clearcut as either theory suggests, and that introns have been both partly gained and partly lost throughout evolution [37]. Whatever the outcome of this debate, it is clear that exon shuffling has been a major route for domain shuffling, in eukaryotes at least.

Further evidence for the impact of exon shuffling comes from triosephosphate isomerase, an enzyme that gets a mention not for the first or last time. It has an exon boundary that coincides with a domain boundary. In an attempt to "simulate" exon shuffling, the two separate parts were expressed separately. The two parts formed a complex that was active; more importantly, the two halves still had residual activity [38]. This observation provides a nice example of how exon shuffling could easily create enzymic activity from two weakly active precursor elements.

Evidence for fairly recent exon shuffling was also observed in trypanosomal phosphoglycerate kinase (PGK), in which the gene contains what is clearly part of an intron sequence [39]. Because the sequence of introns is essentially free of evolutionary constraints, this type of mechanism represents a very interesting way of introducing completely new amino acid sequences into a protein.

It is also worth noting that the well-known phenomenon of **alternative splicing** means that segments of proteins can be added, removed, or replaced relatively straightforwardly, not just in protein evolution but in modern-day proteins, again as long as the exon fragments correspond to structurally meaningful fragments [40]. This is an extremely common mechanism, and is, for example, responsible for much of the structural diversity in antigen recognition by antibodies (*5.6). Alternative splicing normally just adds extra domains, subdomains, or loops without changing the overall structure much. However, there are examples where alternative splicing leads to some rearrangement of the local secondary structure, and hence to more obvious changes in structure or function [41].

2.2.2 Multidomain proteins are also produced by other genetic mechanisms

Exon shuffling is not the only possible mechanism for domain shuffling. Of course, if the exon-late hypothesis is correct, it *cannot* be the only mechanism, because otherwise there would be no mechanism by which prokaryotes could shuffle genes. Prokaryotes are much better than eukaryotes at recombination [42]; they are also able to transfer genetic information to and from plasmids and move it about, both within an organism and from one organism to another. The most important of these routes, which is common in prokaryotes but not unknown in eukaryotes, is **horizontal transfer** from one species to another, followed by insertion of the new genetic information adjacent to an existing gene [43]. Bacteria can also shuffle modules between species by a process involving conjugation [44].

2.2.3 Evolution can proceed in jumps by three-dimensional domain swapping

We have seen that domain shuffling—the placing of one domain next to another in a new context—can be a very rapid way of generating a new biological function essentially in one genetic step. It is therefore a very powerful weapon in the arsenal of evolution. This section looks at another method for generating complexity in a very small number of genetic steps, which has probably been used widely throughout evolution, although it is certainly not the only way in which multidomain proteins have formed. It is called **three-dimensional domain swapping**, not to be confused with domain shuffling, which is a quite different thing. It has been discussed in several reviews [45–50].

Let us imagine that two domains have been fused together in a domain shuffling event [45]. Initially, they will have no fixed orientation with respect to each other, but one can envisage that a relatively small number of mutations could enable them to develop an interface, such that the two domains have a fixed orientation. Such a process must have occurred for every multidomain protein at some stage, and is thus ubiquitous.

Now let us place a second copy of the two-domain protein adjacent to the first. This gives us a structure with two well-defined intramolecular interfaces and two poor intermolecular interfaces, and therefore a weakly associating dimer.

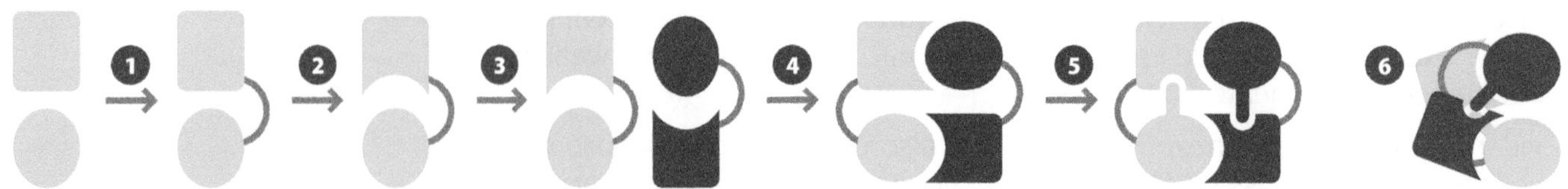

FIGURE 2.16
A hypothetical, but very likely, pathway for the evolution of multidomain proteins by three-dimensional domain swapping. We start with two protein domains. (1) Gene fusion. (2) Evolution of an interface. (3) Introduction of a second identical protein. (4) Rearrangement of the linkers to produce a *domain swap*. (5) Evolution of a second interface. (6) Usually the product is also twisted, which introduces much more potential for other stabilizing interactions between domains.

However, as shown in **Figure 2.16**, it only takes a rather small readjustment of the interdomain linker to produce a dimeric protein in which exactly the same interdomain interactions are present, except that this time the weak interactions are within one polypeptide chain, and the strong interactions are intermolecular: we have at a single stroke made a well-structured domain-swapped dimeric protein from a monomer. Because the interactions in the original and domain-swapped dimer are essentially identical, the energy required to form them is essentially identical [47]. Therefore, provided that the initial interdomain interface was energetically favorable, so will the new dimer interface be.

Once the domain swapping has occurred, it takes only a few mutations to stabilize it in this form [51], and a few more to stabilize the new intramolecular interface. Thus, in a few easy steps we have created a complex four-domain dimeric protein from monomers [52]. Of course, this structure can then evolve further to obscure its evolutionary origins: it is possible that several proteins with internal duplication were stabilized on the way by three-dimensional domain swapping [53]. In Section 2.3 we shall see the tremendous biochemical potential of interdomain interfaces: they can be used to generate catalytic sites and allosteric sites, for example. Thus, domain swapping is a potential method by which nature can readily evolve new ligand-binding [54], catalytic [55], and allosteric sites [56]. There is considerable evidence that this has indeed occurred, as detailed in the numerous reviews on this topic, for example [46, 47].

The "domains" can be as large as entire domains, but more frequently they are partial domains, for example an N- or C-terminal helix or a strand of a β sheet. It is also possible (although rare) for domain swapping to create a heterodimer [46].

2.2.4 Three-dimensional domain swapping still occurs

In Section 2.2.3, three-dimensional domain swapping was presented as an ancient evolutionary event. It can also be induced to happen *in vitro*, and there are many such examples. The only difference between the original and domain-swapped forms is the structure of the linkers. In many real examples, the linkers are bent by approximately 180° in the monomeric form but are extended in the domain-swapped form. Thus, by replacing a hinge-forming residue by a residue that prefers to be extended (for example, a glycine by a proline) or by altering the length of the linker it has been shown to be possible to favor the domain-swapped form [50, 57, 58].

Domain swapping effectively tangles the two monomers together: the equilibrium between the original and swapped forms is therefore usually very slow. Moreover, domain swapping can occur in two ways: it can either create a closed symmetric dimer or it can create an open-ended dimer with two "ends" ready to accept new monomers (**Figure 2.17**). It is possible that domain

FIGURE 2.17
Three-dimensional domain swapping can create open-ended dimers that have the potential to form amyloid.

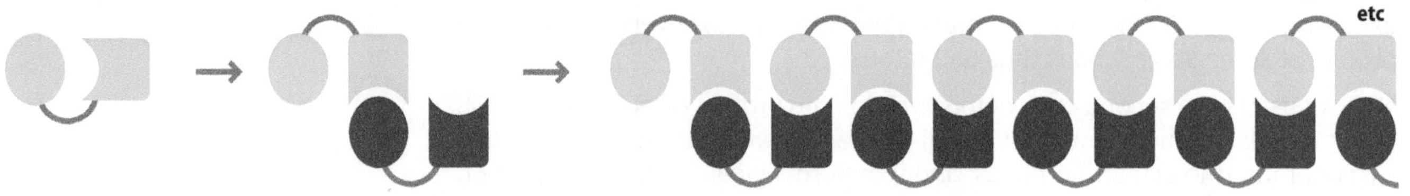

swapping is one of the mechanisms by which amyloid fibrils could form [48, 59]; certainly, fibrils formed in this way would be very slow and difficult to dissociate, which is one of the characteristics of amyloid fibrils.

2.2.5 Increased binding specificity is conferred by additional domain interactions

The smallest known genome is that of *Mycoplasma genitalium*, a parasite that lives inside the genitourinary tract. It has 521 genes, of which only 482 are coding regions [60]. However, only 382 of these are essential, as determined by single-gene knockouts [61], and it has been estimated that the minimal genome size needed for a living bacterial cell in a metabolically rich and stable environment (such as inside an animal host) is about 270 genes [62]. Most of these genes code for enzymes, and about half are involved in DNA synthesis and repair, and in transcription and translation. The rest have various tasks including membrane biosynthesis and transport; relatively few code for metabolism and very few for biosynthesis (no TCA cycle or electron transport genes!). A more typical bacterium has about 3000 genes. By contrast, a single-celled eukaryote such as yeast has 5000 genes, whereas humans have approximately 21,000 genes, which is not a lot more considering how much more complex we are. Where, then, does the complexity come from?

From comparisons of the sequences of different genomes, it has been shown that there are very few new enzyme functions or new structures in eukaryotes compared with prokaryotes, although in eukaryotes enzymes tend to be more specialized for a single function [63]. Chapter 1 described how the Last Universal Common Ancestor already had many of the protein folds and catalytic functions observed now. The most striking difference between eukaryotes and prokaryotes is that eukaryotic proteins are more complicated: they have a greater number of *additional domains* tacked on. This is strikingly evident when comparing combinations of domains, which show much greater variation in eukaryotes, and when going from lower eukaryotes (such as the worm) to higher ones (such as humans) [64]. The greater number of domains in eukaryotic proteins has also been suggested to be the reason why eukaryotic proteins are on average approximately 50% longer than the corresponding prokaryotic proteins [65]. Why is this so?

Consider the difficulty faced by a multicellular eukaryote when it needs to evolve a new biochemical function. This is of course not a pre-planned process: in fact, precisely the difficulty faced by the eukaryote is how to evolve such a function by tinkering with the "spare parts" available without messing up the existing functionality. Much the same difficulty is faced by a computer programmer in modifying existing code: it is all too easy to introduce bugs that affect the working of some other part of the program. A good way for programmers to avoid bugs is to organize the code into freestanding modules that are much less likely to interfere with each other. Exactly the same is true for biological systems, which are simpler to create and modify (that is, to evolve) if they are divided into independent subsystems, an analysis being pursued by Systems Biology.

The same modularity holds for the evolution of new function at the protein level. Let us imagine that an organism needs to develop a new signaling system to respond to a new signal. It already has several signaling systems that recognize similar molecules and do similar things. Inside the cell, there are already signaling pathways that recognize activated receptors and initiate responses (discussed in more detail in Chapter 8). The easiest way to evolve a new pathway is to duplicate an existing pathway (receptor plus downstream elements) and modify it in some way so as to be specific for the new stimulus. But how specific does it need to be? That is, if a new protein needs to recognize

***2.3 Off-rate, on-rate, and affinity**

For a binding event

$$A + B \underset{k_{off}}{\overset{k_{on}}{\rightleftharpoons}} AB$$

the *on-rate* is k_{on}[A][B] and is governed by the second-order rate constant for the forward reaction, and the *off-rate* is k_{off}[AB] and is governed by the first-order rate constant for the reverse reaction. The overall association constant K_a is given by $K_a = k_{on}/k_{off}$, and conversely the dissociation constant K_d is given by $K_d = k_{off}/k_{on}$. Thus, a tightly binding complex has a small K_d, a large K_a and typically a slow off-rate. Note the convention that a rate constant is written with a lowercase *k*, whereas an equilibrium constant is written with an uppercase *K*.

a new activated receptor, does it matter whether the cross-talk between the original pathway and the new pathway is 1%, 10%, or 50%?

It is always possible to evolve more specific binding, but at a cost. The cost is that the number of amino acids involved in the interface needs to be larger, and also the **off-rate (*2.3)** and on-rate will be slower. Imagine visually comparing two objects to see whether they are (in some particular respect) identical or not. The more selective you want to be, the longer the comparison takes. Biology has the same problem. So, effectively, the question in the previous paragraph comes down to a comparison of the *cost* of producing specific binding compared with the *benefit*. And the cost rises very sharply as the requirement for specificity gets tighter.

To get more specific, suppose that the recognition is between an SH2 domain and a phosphotyrosine (**Figure 2.18**). Typically, an SH2 domain recognizes phosphotyrosine plus the three residues C-terminal to the tyrosine. But there is usually considerable variation allowed in the recognition (**Table 2.2**), and indeed from looking at Table 2.2 one would conclude that it would be amazing if there were not already considerable cross-talk.Certainly there are not many combinations of amino acids left to choose from, to avoid a sequence that is not already recognized by something else. It would of course be possible to evolve a recognition site that recognized more amino acids. And to some extent this has already happened: some SH2 domains also recognize the amino acid N-terminal to the phosphotyrosine. But the *cost* of this would in general be large, because it would require retro-modification of all the existing systems too. Evolution does not work this way.

Instead, the solution is to reduce the *cost* of the recognition by adding extra parts. For example, one can add another domain that recognizes some other part of the receptor, or a domain that causes the SH2 to dimerize and recognize only the dimeric receptor, or a domain that keeps the SH2 domain in a particular orientation with respect to the membrane and therefore permits only certain binding modes, or a domain attached to the receptor that

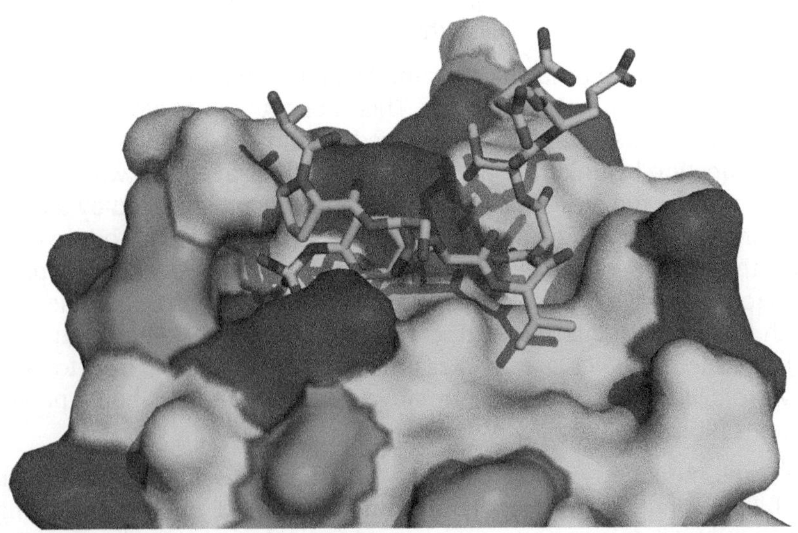

FIGURE 2.18
The interaction between an SH2 domain and a phosphopeptide (PDB file 1jyr). Positively charged residues on the SH2 domain are marked in red, and negatively charged ones in blue. The phosphotyrosine (phosphorus in orange) is recognized by a deep positively charged pocket, whereas hydrophobic residues just C-terminal to it (especially the valine seen at the bottom right of the peptide) are recognized by a shallower hydrophobic pocket (green). An alternative view of this complex can be found in Figure 8.8a.

TABLE 2.2 Consensus recognition sequences for SH2 domains

Domain	Sequence			
Abl	pY	E/T/M	N/E/D	P/V/L
Crk	pY	D/K/N	H/F/R	P/V/L
Fes	pY	E	X	V/I
Fgr	pY	E/Y/D	E/N/D	I/V
Fyn	pY	E/T	E/D/Q	I/V/M
Grb2	pY	I/V	N	I/L/V
Lck	pY	E/T/Q	E/D	I/V/M
Nck	pY	D	E	P/D/V
Rasa-N	pY	I/L/V	X	Φ
Rasa-C	pY	X	X	P
SH3BP2	pY	E/M/V	N/V/I	X
SHB	pY	T/V/I	X	L
Src	pY	E/D/T	E/N/Y	I/M/L
STAT1	pY	D/E	P/R	R/P/Q
STAT3	pY	X	X	Q
Syk-C	pY	Q/T/E	E/Q	L/I
Tns	pY	E	N	F/I/V
Vav1	pY	M/L/E	E	P

X represents any amino acid and Φ any hydrophobic amino acid. (Data from B.A. Liu, K. Jablonowski, M. Raina, M. Arcé, T. Pawson and P.D. Nash, *Mol. Cell* 22: 851-868, 2006. With permission from Elsevier.)

disfavors the original SH2 domain binding. The list is almost endless; given enough time and pressure, evolution will try them all and use anything that works. Most of these solutions have been adopted at some point by some organism. In Chapter 8 we shall come across numerous examples of cases where specificity is obtained by the addition of extra interacting domain pairs: an excellent review can be found in [66]. Exactly the same is true of RNA–protein interactions [67].

A particularly neat solution is to use a **scaffold** protein. This is an additional protein that for example binds to the receptor and will then allow only certain types of SH2 domain to bind subsequently. Admittedly, this requires the acquisition of an additional protein and so makes the system more complicated, but as a "quick fix" it is easy. And this is how evolution works: it does not work out the ideal "best" solution and apply it—rather, it takes whatever is there and tinkers with it to make it work. And if the new system is still not specific enough, then one can always add on more bits until it is. So the answer to the question posed four paragraphs above is that very often the ligand–receptor recognition is very nonspecific (possibly 20% cross-talk with other pathways, although this is very difficult to quantify) but that it is made much more specific by adding additional elements until the pathway is specific enough: 1% cross-talk, for example. Achieving specificity by adding extra proteins is easier to evolve, and leads to more rapid on- and off-rates, than the more direct approach of increasing the specificity of the recognition.

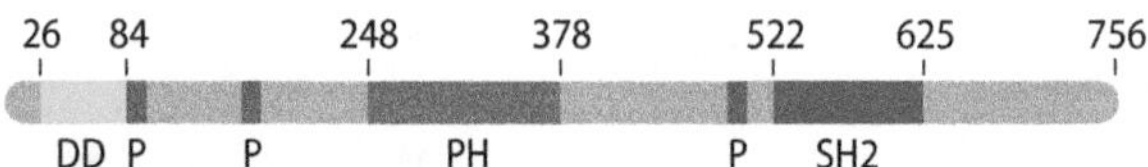

FIGURE 2.19
The sequence of SH2B1, an adaptor protein that connects domains together in several signaling pathways, particularly those related to insulin. DD, dimerization domain; P, proline-rich sequence; PH, pleckstrin homology domain; SH2, Src homology domain 2. Of the 756 residues in the mature protein, only 294, or less than 40%, are in a recognizable domain; the rest are presumably "random coil." This is not an atypical ratio, particularly in eukaryotic signaling proteins.

The consequence is that many signaling pathways are remarkably baroque: there are bits tacked on everywhere to embellish them: to modify binding, make it more specific, enable it to recognize or respond to other signals, link it to other pathways, or enhance or reduce the response. That's evolution for you.

2.2.6 Intramolecular binding is strong because the effective concentration is high

The addition of extra domains such as scaffolds to increase the specificity of binding is not the end of the baroque complexities that evolution can create. A particularly common and useful device is to use internal peptide sequences to regulate protein activity—another way in which the domain structure of proteins is used to biological advantage. A great many proteins, particularly in eukaryotes, consist not just of domains (plus linkers) but also of significant lengths of nonglobular and **natively unstructured** peptide (**Figure 2.19**). Some of these are used to bind to domains, both internally and externally, which turns out to be an extremely effective device, as discussed further in Chapters 4 and 8. To understand this effect, it is helpful to introduce a little maths. This relies on the concept of **effective concentration** and is based on the discussion of this topic in Creighton [68]. A similar idea was proposed in an interesting review [69], in which the effective concentration is given the name β.

If two groups A and B are covalently attached by being part of the same peptide chain, this limits their possible relative positions (**Figure 2.20**). It therefore makes them more likely to interact than if they were not connected. This effect can be described using the concept of effective concentration, which I write as [A/B] using a vaguely statistical nomenclature, implying the effective concentration of A for B. (Creighton uses the notation $[A/B]_i$.) Note that [A/B] = [B/A]. The effective concentration is defined as being the ratio between the affinity of A for B when they are connected and their affinity when they are not:

$$[A/B] = K_{\text{intermolecular}}/K_{\text{intramolecular}}$$

Note that these K are dissociation constants not association constants, so this definition looks different from Creighton's because he used association constants. The intramolecular dissociation constant is therefore given by

$$K_{\text{intramolecular}} = [A - B\ \text{free}]/[A - B\ \text{bound}]$$

It is thus a simple ratio between the amount of free and bound peptide. A $K_{\text{intramolecular}}$ of 1 implies that the peptide is 50% free and 50% bound, and a $K_{\text{intramolecular}}$ of 10 implies that the ratio of free to bound is 10, or in other words weaker binding. The fraction bound is [A – B bound]/[total A – B], or $1/(1 + K_{\text{intramolecular}})$.

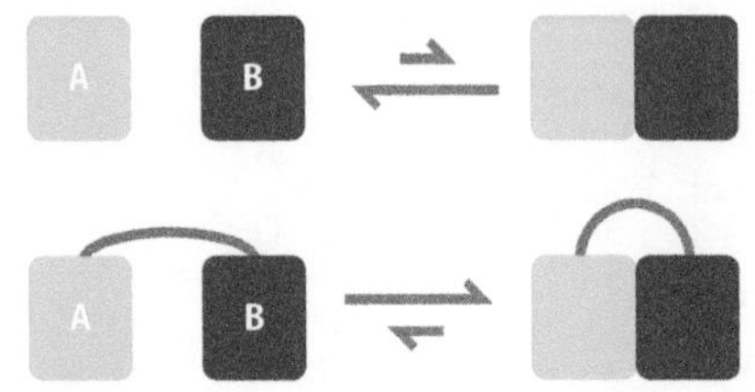

FIGURE 2.20
The interaction of two domains A and B is much more likely (that is, stronger) if they are tethered together. The degree of preference can be calculated as the *effective concentration* of A for B.

The effective concentration [A/B] is the concentration of unconnected (free) A that would be needed to get the same amount of binding to B as there is in the intramolecular A – B case. It is an important number because it allows us to calculate the benefit to binding of having the two binding groups within the same molecule. Theory suggests that [A/B] should depend on the number of rotatable bonds between A and B. It should also be greater when the interaction is highly directional (such as a covalent bond or a hydrogen bond, which is much more directional than a hydrophobic interaction). And it is zero if the molecule has a structure that does not allow them to interact. Experimental values are roughly in line with theory and can be surprisingly high. At the upper limit, when the structure of the molecule forces A and B to interact, [A/B] can be as large as 10^5 M; that is, there is an enormous increase in the binding affinity of two molecules if they are joined together in the appropriate way.

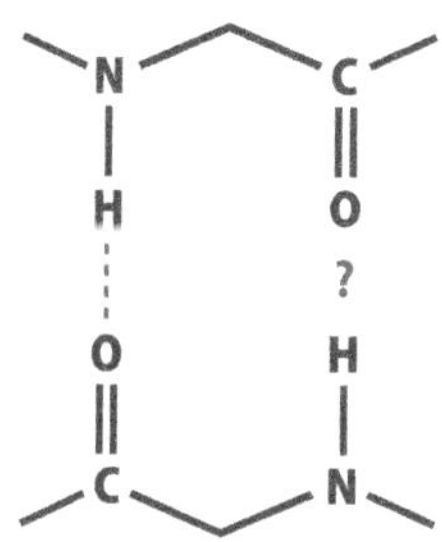

FIGURE 2.21
In a β sheet, once one hydrogen bond has formed, the neighboring one is already close to the correct orientation and is therefore much more likely to form than without the hydrogen bond.

2.2.7 Intramolecular interactions lead to cooperative hydrogen bonding

By how much does intramolecular binding strengthen hydrogen bonds? Let us consider the case where A and B are two hydrogen-bonding groups in a protein, for example two residues in a β sheet. What is [A/B] for two hydrogen-bonding groups when the adjacent residues are already hydrogen bonded; that is, how much easier is it to form a hydrogen bond in a β sheet when the sheet is already partly formed? For this interaction, the hydrogen-bonding groups are already held in an almost ideal geometry. The effective concentration is therefore large, in fact about 100 M (**Figure 2.21**). We can readily calculate how much this helps. From rearranging the equation above, we get

$$K_{\text{intramolecular}} = K_{\text{intermolecular}}/[A/B]$$

That is, if we know how strongly A and B interact when they are in separate molecules, and we know the effective concentration, we can easily work out how tightly they will interact when they are within the same molecule. The dissociation constant for forming a hydrogen bond between C=O and NH in water can be obtained by measuring the affinity of two small amides for each other. It turns out to be very weak. For example, the dimerization of urea has a K_d of 25 M, whereas the dimerization of *N*-methylacetamide has a K_d of 200 M [70]. It is very weak because it competes with the formation of hydrogen bonds to water. When you add in the loss in entropy due to bringing two molecules together, the overall affinity is so weak that it does not normally happen.

From the analysis above, we have $K_{\text{intermolecular}}$ as around 100 M, and [A/B] as also around 100 M. $K_{\text{intramolecular}}$ is therefore about 1. This means that if one hydrogen bond in a β sheet is formed, the addition of the next one has an equilibrium constant of about 1, meaning that it is neither favorable nor unfavorable.

However, as we saw in Chapter 1, hydrogen bonds are cooperative: each strengthens its neighbor. In particular, the formation of the second hydrogen bond strengthens the first. Although $K_{\text{intramolecular}}$ is only about 1, the formation of the second hydrogen bond is favorable overall. Each additional hydrogen bond is therefore increasingly favorable, leading eventually to a cooperative zipping up of the sheet.

2.2.8 Intramolecular domain:peptide binding facilitates autoinhibition

The other case of interest is the one that began Section 2.2.6: intramolecular binding of a peptide. The binding of two protein *domains* to each other can have a wide range of affinities, but the dissociation constant is often around nanomolar: it is strongly favorable because of the large number of interactions that can form at the interface. By contrast, the binding of a *peptide* to a domain is much weaker because the interface is smaller, and also because the peptide is usually unstructured when free and has to become ordered when bound: it therefore loses entropy (*3.1) when it binds, which further weakens the interaction (see below for a more detailed discussion). The affinity ($K_{\text{intermolecular}}$ as defined earlier) is more typically around micromolar. The effective concentration [A/B] for two groups within the same peptide chain depends on the number of intervening residues and is about 1 mM for a 10–20-residue separation [71], decreasing further for longer separations. Therefore for a peptide

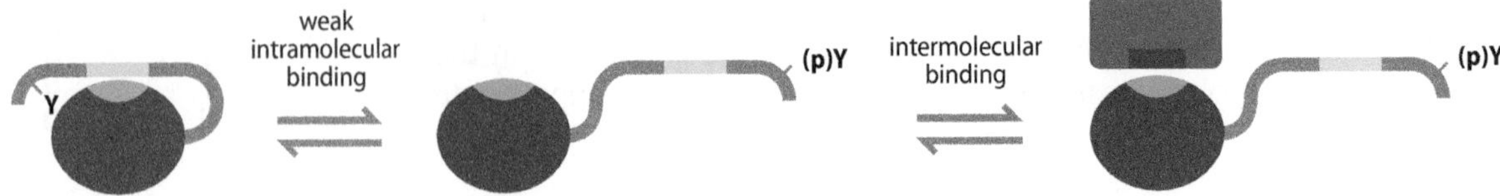

FIGURE 2.22
A typical example of autoinhibition. The binding site on a protein is covered by the relatively weak intramolecular binding of a peptide. This autoinhibition can be relieved for example by phosphorylation of a tyrosine on the peptide, allowing the protein to bind to its ligand.

close in sequence to its cognate binding domain, the intramolecular dissociation constant is again given by $K_{\text{intermolecular}}$/[A/B], and is about $10^{-6}/10^{-3}$ or 10^{-3}. In other words, only 1 in 1000 peptides is dissociated from the domain: it is tightly bound. A longer intervening sequence makes this weaker, but still leaves a very strong association.

Very commonly, the intramolecular binding is regulatory: the peptide binds to its target domain to shield that domain from further interactions, until the protein is somehow "switched on," either by some upstream activating event such as a signaling pathway or by (de-)phosphorylation, allowing the domain to be released from its intramolecular interaction to bind intermolecularly (**Figure 2.22**). This is a very important and frequent situation, often described as **autoinhibition**, because the intramolecular peptide inhibits the protein it is part of. Examples can be found very widely in biology and include the regulation of the calcium-binding protein calmodulin [72], the inhibition of a tandem PDZ domain in X11/Mnt [73], and the down-regulation of talin, which links the transmembrane integrin cell adhesion molecules to the actin cytoskeleton [74]. It is also seen in regulation of the chaperone proteins Hsp70 [75] and protein disulfide isomerase [76]: in both cases the hydrophobic binding site for unfolded ligands is protected by weak intramolecular binding of a peptide. Autoinhibition is particularly common in signaling and is discussed further in Chapter 8. In this situation, we need an intramolecular interaction that is strong enough to shield the domain and keep it inactive, but weak enough to allow the peptide to be released easily and rapidly. The required value of $K_{\text{intermolecular}}$ is therefore between approximately 1 and 0.1 (implying between 50% and 10% of the inactive domain being free). From the arguments above, this would imply a $K_{\text{intermolecular}}$ in the range 0.1–1 mM; that is, a weak interaction. This is of course easily arranged, by making the peptide a rather poor match to the intramolecular binding domain.

Such an arrangement is so common, and so useful, that it is worth restating the argument. A protein with a domain that recognizes a peptide sequence (for example the SH2-phosphopeptide system described in the previous section) can be prevented from binding, or in other words "turned off," by binding intramolecularly to a similar peptide sequence to its consensus phosphopeptide—but because the binding is intramolecular the peptide can differ significantly from the consensus and still bind tightly enough. Indeed, the inherent affinity $K_{\text{intermolecular}}$ is so weak that the peptide sequence can be very short—so short that such sequences can easily be present "by chance." The rather poor SH3-binding sequence motif RxPxxP, for example, can be found in 1 out of 20 randomly selected proteins [77], and is probably not functional in most of these. That is, it is too weak to bind significantly in an intermolecular interaction. However, within an intramolecular context, such a sequence binds tightly enough to be a useful autoinhibitory mechanism. Single point mutations can easily create or destroy an intramolecular binding interaction. This possibility has been used extensively in the evolution of signaling pathways as a way of modulating interactions: intramolecular interactions can be readily turned on or off, as described in the next section.

A real example is discussed by Zhou [78]. The binding of an SH3 domain to its nonconsensus intramolecular proline-rich peptide sequence connected by a rather short linker has a very weak $K_{\text{intermolecular}}$ of only 6 mM, implying that in a separate protein there would be no significant binding *in vivo*. However,

FIGURE 2.23
A possible mechanism for the autoinhibitory action of the acidic C-terminal tail of HMGB1. The two HMGB1 boxes (A and B) bind DNA on their inside faces. In the autoinhibited state, they may bind instead to the C-terminal tail. Binding to DNA would then liberate the tail to bind to histones, for example; alternatively, binding of the tail to a specific partner protein would liberate the HMGB1 boxes to bind DNA.

because they are fairly close in sequence there is an effective concentration estimated to be 13.5 mM, giving a calculated $K_{intramolecular}$ of about 0.5. The experimental value is 2—that is, they bind together with roughly one-third bound and two-thirds free at any point. Shortening the linker further makes it more difficult to bind intramolecularly, and the calculated $K_{intramolecular}$ rises to 2; the experimental value is 10. A related approach has been used to analyze the cooperative binding of domains to RNA: see [79], where $K_{intramolecular}$ values were expressed differently but were calculated to lie in a range from 1 to 10^{-3}.

A final example comes from the high-mobility group protein B1 (HMGB1), an abundant nonhistone chromatin-binding protein that binds nonspecifically to DNA via two homologous boxes. HMGB1 bends DNA when it binds, and is thought to function in aiding the bending of DNA during the formation of nucleoprotein complexes. HMGB1 has a 30-residue C-terminal tail composed entirely of acidic glutamate and aspartate residues, which down-regulates the activity of HMGB1. This is suggested to function in an autoinhibitory manner by binding to the DNA-binding surfaces of HMGB1 (**Figure 2.23**) and making them inaccessible [80]. Here, the intramolecular binding is inherently weak, being entirely electrostatic.

2.2.9 Intramolecular domain:peptide binding facilitates evolutionary change

When comparing **orthologous** proteins from different species, one usually finds that the domain architecture is well conserved. Any change in regulatory mechanism as a consequence of domain:domain interactions would therefore require a reasonably large-scale modification, as discussed above. However, because intramolecularly bound peptide sequences are so easily introduced or destroyed, they are much less well conserved. One example is shown in **Figure 2.24** [77].

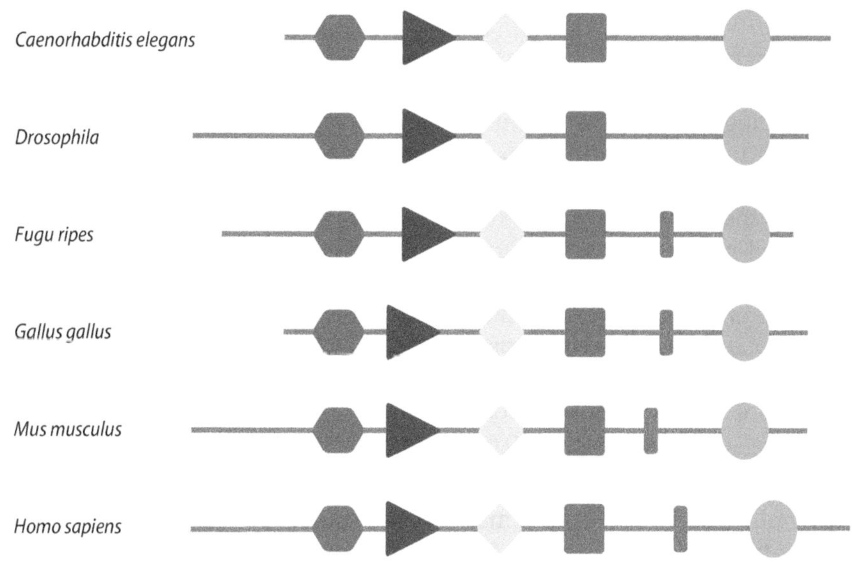

FIGURE 2.24
Intramolecular peptide binding motifs can be added or removed much more easily than folded domains. Domain architectures for SNA2α-like proteins are shown. The domains are highly conserved, but the retinoblastoma recognition sequence motif (mustard yellow) is only present in the higher eukaryotes.

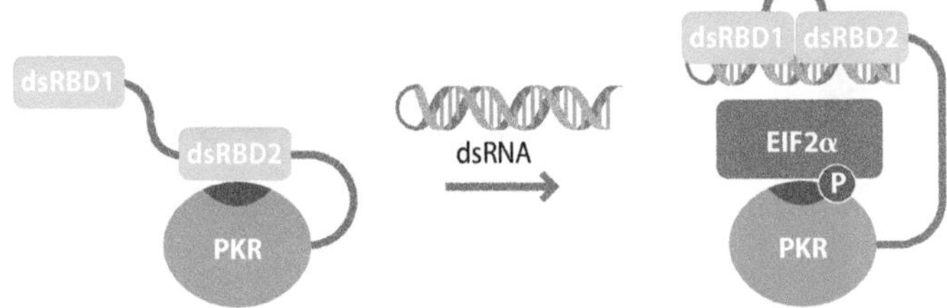

FIGURE 2.25
Protein kinase R (which protects the cell against viral infection by recognizing viral double-stranded RNA) is autoinhibited by its double-stranded RNA binding domain-2 (dsRBD2). When this domain binds to double-stranded RNA, the kinase becomes activated and can inhibit translation by phosphorylating eukaryotic initiation factor-2α.

Examples of intramolecular regulation can be found all over biology, and several are given elsewhere in this book, particularly in Chapter 8. Chapter 9 discusses RNA polymerase II TFIID, which is turned off by intramolecular interactions until bound to DNA. RNA binding presents other examples, for example protein kinase R (**Figure 2.25**) [81]. As a final example, the binding of transcriptional regulators to DNA can be switched off by intramolecular interactions, which block the DNA-binding site until they are activated. Thus, ETS proteins are a large family of DNA-binding regulatory proteins consisting of about 85 amino acid residues. Unwanted interactions with DNA are normally blocked by interactions with partner proteins. However, the ETS protein ETS-1 has a C-terminal helix that packs against the rest of the protein and inhibits DNA binding [82].

The recognition of short peptide sequences that bind to (and regulate the function of) protein domains is becoming a research area of its own. Such peptide sequences are often referred to in this context as Short Linear Motifs or SLiMs.

2.2.10 Binding specificity is increased by scaffold proteins

In Section 2.2.8, we saw that linking a binding domain to its cognate peptide sequence increases their mutual effective concentration and leads to stronger binding. This can be restated in another way: linking two recognition elements together on the same polypeptide chain or in the same complex increases their binding on-rate with little effect on their off-rate (*2.3). Because of the relationship $K_d = k_{off}/k_{on}$, this provides an equivalent explanation of the increase in affinity.

This effect is used extensively by **scaffold proteins**, which are discussed in more detail in Chapter 8. Such proteins contain multiple recognition domains that hold binding partners together, and they are very common in signaling pathways. The name scaffold is somewhat misleading, because it implies that the proteins have a rigid structure, like the scaffolding that is used in construction. However, in most cases the scaffolds are needed only to increase the effective concentration of a protein, by bringing two proteins together in a complex, and not to position them exactly. They do not need to be rigid, and in fact usually function better by being flexible and thereby allowing multiple binding interactions within the same complex. Their main function is to increase the specificity of signal transmission. Holding the partners together in one complex ensures that they interact with each other much more frequently than they interact with other proteins. Therefore, for example, if an activated protein A goes on to phosphorylate B, then the specificity of this event is improved if A finds the binding site on B much more frequently than it finds similar but "wrong" binding sites on other potential substrates (**Figure 2.26**). Clearly, the specificity of the transmission is increased by a factor $1/K_{intramolecular}$ compared with the intermolecular signal if they are held together in a complex by means of attachment to a common scaffold. Because $1/K_{intramolecular}$ can easily be 100 or greater, the effect is very large.

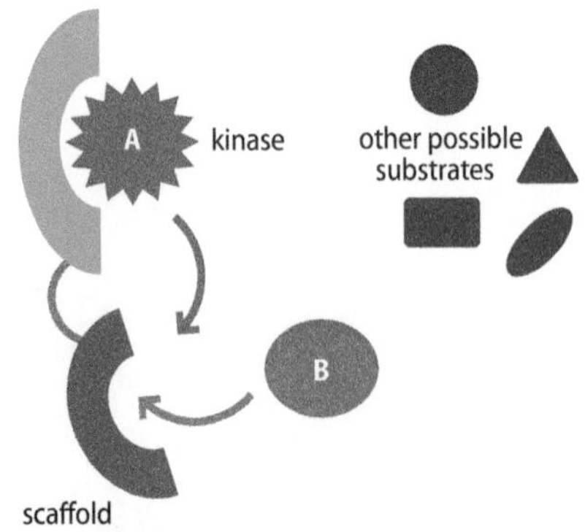

FIGURE 2.26
Scaffold proteins can increase the rate of phosphorylation of the desired substrate (B) at the expense of all other possible substrates merely by bringing *B* close to the kinase A so that the probability that the kinase will bind B is much greater than the probability of its binding any other substrate.

Scaffold proteins are normally discussed in the context of signaling pathways, but similar functions occur in many other systems. A nice example is provided by the blood clotting cascade [83–85]. The end product of the blood clotting

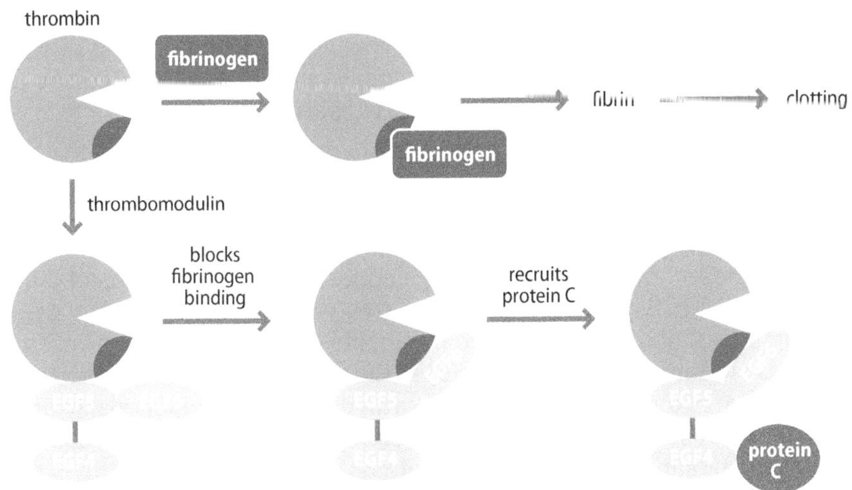

FIGURE 2.27
The scaffold protein thrombomodulin (which contains the domains EGF4, EGF5, and EGF6) deactivates thrombin once the blood clot has formed, to prevent excess clotting. It does this both by blocking the binding of the normal substrate, fibrinogen, and also by recruiting protein C and altering its conformation so that it gets cleaved by thrombin.

cascade is thrombin, a protease that cleaves fibrinogen into fibrin to cause clotting. However, once the damage has been repaired, thrombin is deactivated by the scaffold protein thrombomodulin (**Figure 2.27**). This protein has three functions. First, it binds to thrombin, via the fifth EGF domain, EGF5 (see Figure 2.27). Second, it blocks the binding site of thrombin for fibrinogen, via EGF6 binding, and thus slows down the cleavage of fibrinogen, effectively by reducing the availability of substrate. Third, it binds to protein C via the EGF4 domain. Protein C is normally a poor substrate for thrombin, but the rate of protein C cleavage is increased by its binding to thrombomodulin, in two ways. The first is a simple proximity effect as presented above: because the scaffold protein thrombomodulin is holding it close to the active site, it gets cleaved more often; the second is that the EGF3 domain probably also binds to protein C and activates it by causing a small rearrangement of the structure. Significantly, the scaffold protein has not affected the *proteolytic function* of thrombin at all: it merely alters the availability of different substrates.

It seems likely that a significant fraction of the modules found on blood clotting proteins (described earlier in this chapter) have been added on for similar reasons, namely to increase the substrate specificity of the attached protease. For example, the kringle domains of plasmin and plasminogen function in attaching the protease to fibrin.

This book is concerned with the principles of protein function rather than any practical applications. Nevertheless, it is worth noting in passing that the poor specificity of most kinases and phosphatases implies that they are unlikely to form good drug targets. Rather, specificity is provided by scaffolds. The clear implication is that scaffolds are good places to look for drug targets, particularly because scaffolds have little effect on the function of their constituent proteins but merely increase their local concentrations [86]. Blocking scaffold function is therefore less likely to lead to side effects. Scaffold function is known to be regulated in some cases by modifications such as phosphorylation or just protonation: these could be interesting systems to look at in modifying system functions.

2.2.11 Intramolecular binding is strong because it has less unfavorable entropy

The increased affinity arising from intramolecular binding is hardly a new observation. A helpful approach was originally proposed by Jencks in the 1970s [87–89] and developed by Williams [90], to explain the binding affinity of a ligand that interacts at two sites with its receptor. This approach sounds rather different from the analysis above, but it is actually similar, as I shall show.

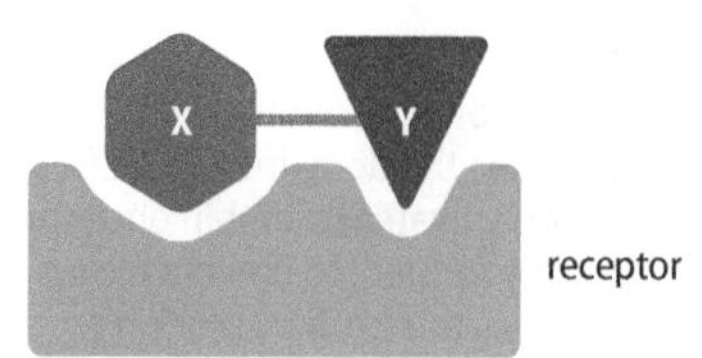

FIGURE 2.28
A ligand that interacts with its receptor in two sites, X and Y.

Consider a ligand that binds to its receptor at two sites X and Y (**Figure 2.28**). We will compare the binding affinity of this ligand to the binding of the two constituent parts and will find that the affinity of the intact ligand can be much larger than the sum of the affinity of the two parts.

The change in free energy for the binding of a ligand to a protein can be divided up as

$$\Delta G = \Sigma \Delta G_i + \Delta G_{t+r} + \Delta G_{conf}$$

In this equation, the ΔG_i are the free energies for the interactions of different binding groups, which here have only two components: ΔG_X and ΔG_Y, the binding interaction energies of the fragments X and Y. It is important to note that these terms are only the energies of the interactions themselves. They are just the sum of all the hydrogen bonds, van der Waals interactions, hydrophobic interactions, and so on, and are therefore not the same as we would measure for the binding of X and Y separately, as we shall see. This term is always favorable. It is opposed by the other two terms, which are unfavorable. ΔG_{t+r} is made up of all the mainly entropic factors that are unfavorable when two molecules are brought together to form one complex; it contains the translational and rotational entropy that is lost (**Figure 2.29**). In addition, ΔG_{conf} represents the conformational restriction that is placed on the full-length ligand when it binds, representing mainly the loss in rotational entropy of any bonds between X and Y that get restricted on binding. ΔG_{t+r} is approximately 50 kJ mol^{-1}, although it depends somewhat on the size of the ligand and the amount of motional freedom that is still available to the ligand when bound. ΔG_{conf} is approximately 5 kJ mol^{-1} per rotor frozen in the complex, although again it depends on how frozen each rotor actually is.

We are now in a position to calculate some numbers. Williams estimates that the total binding interaction energy ΔG_i for a hydrogen bond is approximately 20 kJ mol^{-1}. We may therefore use the equation above to estimate the total change in free energy for the binding of X to its receptor, if this interaction involves the formation of two hydrogen bonds. It is given by

$$\Delta G = \Sigma \Delta G_i + \Delta G_{t+r} + \Delta G_{conf} = -40 + 50 + 0 = +10 \text{ kJ mol}^{-1}$$

or, in other words, the interaction is so weak that it does not occur in solution because the favorable binding interaction is outweighed by the unfavorable loss in entropy on binding. The binding of Y alone has the same energy, if it also involves two hydrogen bonds. We now attach Y to X, which is connected to X by two rotatable bonds that become fixed in the complex. We now have

$$\Delta G = \Sigma \Delta G_i + \Delta G_{t+r} + \Delta G_{conf} = -80 + 50 + 10 = -20 \text{ kJ mol}^{-1}$$

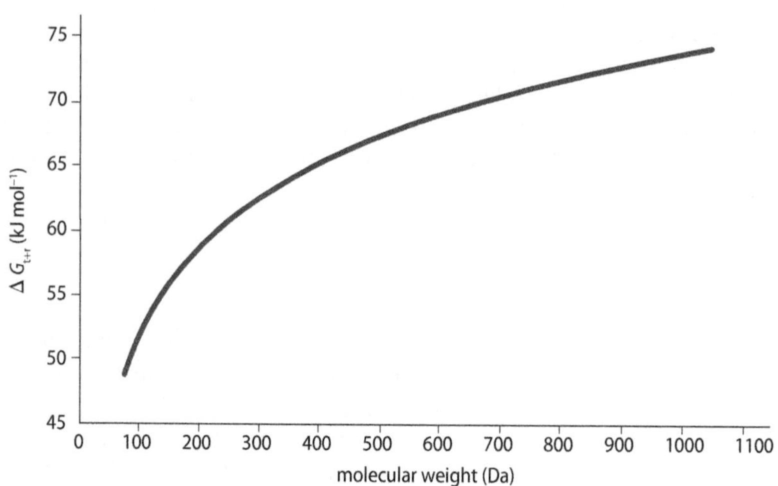

FIGURE 2.29
The dependence of ΔG_{t+r} (the unfavorable change in free energy on binding a ligand to a receptor arising from loss of translational and rotational freedom) on the molecular weight of the ligand. (Redrawn from D.H. Williams et al., *J. Am. Chem. Soc.* 113: 7020–7030, 1991. With permission from the American Chemical Society.)

This is a favorable interaction. This illustrates the statement made above, that the binding of the larger ligand X-Y is much more favorable than the sum of its parts, essentially because the unfavorable ΔG_{t+r} term is roughly the same no matter how big the ligand is, whereas the favorable bonding interactions continue to add up as the ligand gets bigger.

It is instructive to see how this approach is related to the effective concentration idea discussed in the previous section. The free energy of binding of X was calculated above to be +10 kJ mol^{-1}. Using the equation $\Delta G = -RT \ln K$ (see Binding and dissociation constants and free energy, *5.5), this translates to a dissociation constant of 60 M. In the terminology used in the previous section, this is $K_{\text{intermolecular}}$, where I use the terms intramolecular and intermolecular respectively to indicate the binding of X-Y and of X and Y separately. The binding of X-Y in the same way has $K_{\text{intramolecular}} = 3 \times 10^{-4}$. The effective concentration of X for Y in the complex is the ratio of these, namely 2×10^5 M. This is a large value for an effective concentration, but not unreasonable because X and Y are almost as fixed as it is possible to be. These two approaches are therefore related, providing complementary ways of viewing the same phenomenon.

2.3 POTENTIAL ADVANTAGES OF MULTIDOMAIN CONSTRUCTION

We saw above that more than two-thirds of proteins contain more than one domain. What selective advantage does this provide? Several possible reasons are listed below. For any given protein, not all will be true (hence the word "potential" in the heading), but each reason is certainly significant in some cases. This question has also been considered by others, among which is a nice analysis from **Cyrus Chothia (*2.4)** [91].

***2.4 Cyrus Chothia**
Cyrus Chothia (**Figure 2.4.1**) worked until recently at the Laboratory of Molecular Biology in Cambridge, UK, on protein structure, and is one of the founders of bioinformatics. In addition to his helix-packing work described in Chapter 1 and being the first to suggest a limited number of protein folds, he is one of the authors of the SCOP classification scheme for protein folds and has studied protein recognition and packing, particularly in antibodies, and conducted genome-wide analyses of proteins and protein evolution.

FIGURE 2.4.1
Cyrus Chothia. (Courtesy of C. Chothia and Medical Research Council, UK.)

2.3.1 Multidomain construction makes it simple to evolve a new function

To restate a comment made in Chapter 1: Nature is not teleological; it does not know where it is going. It is a tinker: it takes what is to hand and adapts it. As discussed in Chapter 1, mutations of existing proteins are a poor way of adapting something that is reasonably good at one function into something that is good at something else, because mutations can only search in regions of sequence space that are close to the starting point. They can therefore optimize a reasonably good solution, but do not provide a good mechanism for searching sequence space. They also search rather slowly, because it is likely to take a large number of mutations to create a new function.

By contrast, new functions can be created at a stroke by taking two existing functions and combining them, as we have seen. There is a nice analogy between evolution and the editing of documents, attributed to Stemmer [92]. Mutation is analogous to going through a document, deleting or adding letters and words one at a time. Domain shuffling is analogous to cutting and pasting. Just think how slow editing would be if one were restricted to changing one letter at a time. The advantage of some sort of domain shuffling is obvious. In fact, most proteins can be seen to have evolved this way. This is particularly clear for modular proteins, but it is equally true for more "normal" proteins. In particular, where an enzyme has to bind to two substrates it almost always binds one substrate on each domain. Thus, we can envisage that by far the easiest route for the evolution of a new enzyme whose function is to link together two molecules is to start from two separated domains that bind each of the molecules. This applies also to functions such as kinases, which again have two domains: one to bind the substrate and one the ATP. Importantly, the same is also often true for a lyase, which physiologically catalyzes the reverse reaction but equally well catalyzes the forward, linking, reaction. For example, adenylate kinase contains two Rossmann folds linked together, one for each nucleotide that it binds; the glycolytic enzyme phosphoglycerate kinase has two domains, one for each substrate that it binds; fatty acid desaturase has one domain that anchors it into the membrane and another that has a cytochrome *b*2 fold and performs the catalysis; and so on.

One implication of the present analysis is that most domains in any given protein should have only a single function: they should, for example, bind a ligand, locate the enzyme next to a membrane, or regulate activity, but should not carry out more than one of these functions. This is indeed generally true, and reinforces the concept of the domain as the fundamental evolutionary unit. There is nothing to prevent a domain from having different functions in different contexts, or to prevent evolution from developing a moonlighting function and using some domains for two genuinely different functions at the same time. This is, however, rare.

This simple route to the creation of new function also permits the creation of a bifunctional enzyme, simply by taking the genes for two enzymes and fusing them together. In the analysis of multidomain proteins mentioned above, the creation of a bifunctional enzyme from fusion was the most common functional change caused by the fusion [91]. The two enzymes concerned are usually sequential enzymes in a pathway, suggesting the possibility of channeling, as discussed further in Chapter 10.

2.3.2 Multidomain construction makes it simple to introduce control and regulation

Every biochemical system needs regulating, so that its activity fits in with that of surrounding systems. This does not mean that every reaction needs to be regulated: many reactions just go as fast as possible and are limited only by the amount of enzyme and substrates. However, many reactions *are* regulated. What mechanisms exist to do this? The simplest is to control the amount of enzyme present, by changing the rate of synthesis, degradation, or transport, or by moving the enzyme from one place to another, and this is indeed used extensively. However, more flexibly, the cell can change the activity of an enzyme reversibly. There are two main ways of doing this: by allosteric regulation or by covalent modification such as phosphorylation. We will consider these in more detail in Chapters 3 and 4, but it is helpful to make some comments here specifically on the role of domains.

Most allosteric enzymes have several domains, and many are oligomers. This is not essential, as discussed in Chapters 3 and 6, but it is certainly much the simplest way to achieve allostery. Allostery requires that a ligand binding at some distant position affects the activity: that is, it affects the shape or dynamics at the active site. How can this occur, and, more importantly, how can it evolve? By far the simplest is if (as is virtually always the case) the active site of the enzyme is at the interface between two domains, while the allosteric effector binds elsewhere, but still at or close to the interface, so that binding of the effector causes an alteration in the structure of the interface and therefore a change at the active site (**Figure 2.30**). Thus, a multidomain structure provides a simple way to generate conformationally adaptable binding sites for several ligands.

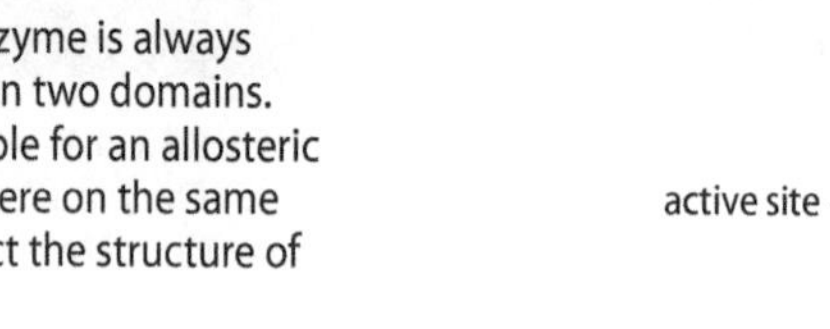

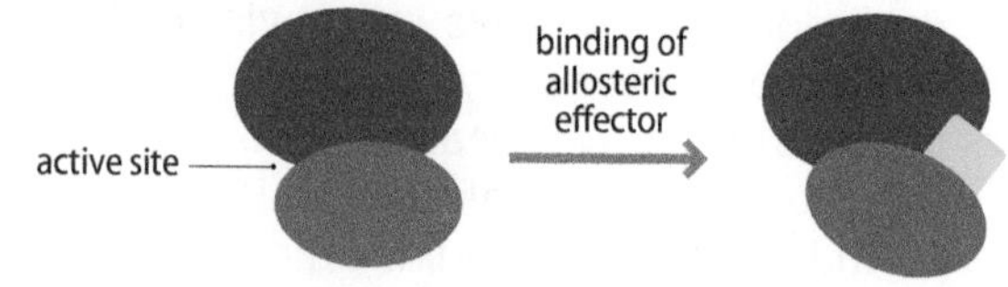

FIGURE 2.30
The active site of an enzyme is always in the interface between two domains. This makes it fairly simple for an allosteric effector to bind elsewhere on the same interface and thus affect the structure of the active site.

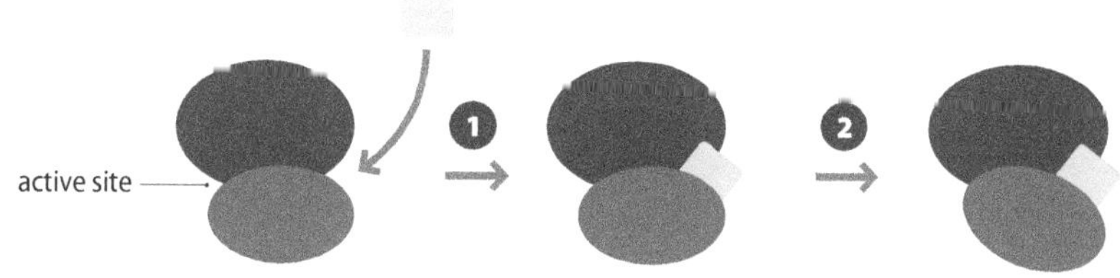

FIGURE 2.31
Possible route to evolution of an allosteric enzyme. (1) A relevant ligand binds weakly to the domain interface. (2) Gradually the enzyme evolves, both to bind the ligand more tightly (that is, at the physiologically relevant concentration) and also to use the binding to cause a change in the domain interface that leads to an alteration in the active site.

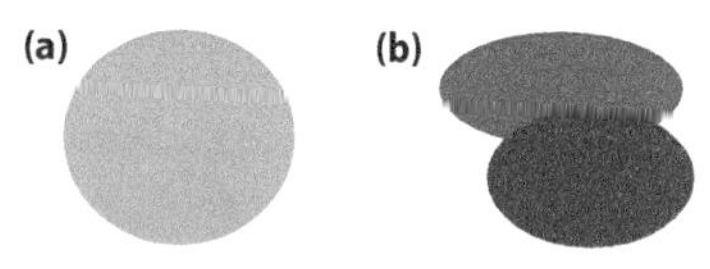

FIGURE 2.32
A single domain (a) is a basically convex surface, not well suited to form a binding pocket, whereas a two-domain protein (b) has ready-made concave pockets.

We can thus envisage the evolution of allostery in two simple stages (**Figure 2.31**). In the first stage, an enzyme is assembled from two domains, which of necessity have a flexible interface because alteration of this interface gives rise to the hinge bending that is essential to allow substrate and products in and out. In the second stage, a poor binding site elsewhere on the enzyme for a relevant allosteric effector is modified such that the affinity is in the correct physiological range, and that binding of the effector leads to a change in the domain interface and thus an appropriate change in the enzyme's activity. Real cases can be much more complicated than this, of which the most common example is that the two domains are in fact identical: that is, that the enzyme is an oligomer. This is in many ways a particularly elegant solution, because only one polypeptide chain has to be mutated to result in allosteric effects. In agreement with the embellishments presented in Chapter 1, evolution can come up with endless elaborations and decorations on this basic theme.

To a first approximation, a single-domain protein is a spherical object with entirely convex surfaces, whereas a two-domain or dimeric protein has a much more interesting shape with concave surfaces or pockets (**Figure 2.32**). One implication of this is that it is much easier to develop a binding site for a small molecule or for a protein partner from a two-domain protein.

The most common *covalent modification* that alters enzyme activity is phosphorylation. How does phosphorylation lead to a change in enzyme activity? Again, there are as many ways as random evolutionary processes can come up with. But one common way is to take an enzyme and add on an additional regulatory domain (**Figure 2.33**). This regulatory domain controls the activity of the catalytic domain, most obviously by decreasing it, for example by blocking the active site, although there are many examples of regulatory domains (not necessarily thought of in this way) that enhance activity by restructuring the active site. Phosphorylation of a suitable group on the regulatory domain then leads to dissociation or restructuring of the domain interface such that the enzyme's activity is altered.

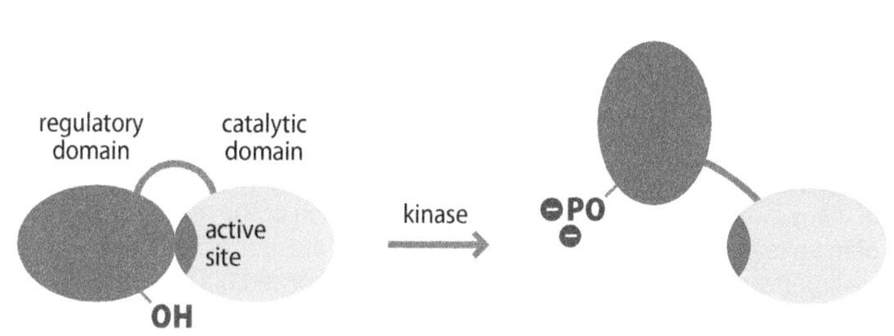

FIGURE 2.33
A simple mechanism for enzyme regulation by phosphorylation. Phosphorylation of a regulatory domain leads to dissociation of the domain from the catalytic domain, and therefore the uncovering of an active site. Note that this regulatory mechanism is much simpler to evolve than any functional change that might arise within a single domain, because all it requires is weak intramolecular binding and a flexible hinge.

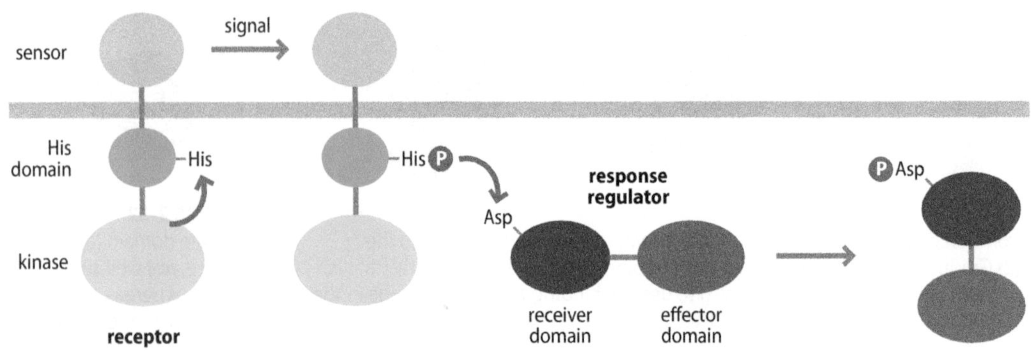

FIGURE 2.34
The bacterial two-component signaling system. This is a simpler version of Figure 8.34. An extracellular signal causes an intracellular kinase to phosphorylate a histidine residue on the receptor. Often this occurs via dimerization of the receptor. The phosphate is then transferred to an aspartate on a cognate response regulator. Most response regulators are two-domain proteins, containing a receiver domain (the domain that gets phosphorylated, which is conserved in structure and function in all two-component signaling systems) and an effector domain, which varies widely.

An example is the very widespread bacterial two-component signaling system, discussed in Chapter 8. An external stimulus leads to phosphorylation of a histidine residue on a membrane protein (a histidine kinase) (**Figure 2.34**). The phosphorylated histidine kinase then transfers the phosphate to an aspartate residue on an intracellular protein, termed a response regulator. Phosphorylation of the response regulator then leads to an intracellular response. Often this is done by a second domain on the response regulator that binds to DNA. Thus, phosphorylation of the response regulator gives rise to DNA binding, and therefore to altered transcription. The way this is done is again subject to many variations. In most cases, phosphorylation of the receiver domain (see Figure 2.34) leads to conformational changes in that domain that lead to dissociation of the effector domain (note again the clear distinction in function between the two domains) (**Figure 2.35**). The unphosphorylated form is inactive for two reasons: first, because the DNA recognition helix is sterically prevented from binding because of the presence of the receiver domain; and second, because the domain binds much more strongly to DNA as a dimer, and it can only dimerize once it has been freed from the receiver domain (**Figure 2.36**). (This is an example of cooperative binding, discussed in more detail in Chapter 3.) In some systems, the effector domain is an enzyme, one whose active site is buried in the interface to the receiver domain

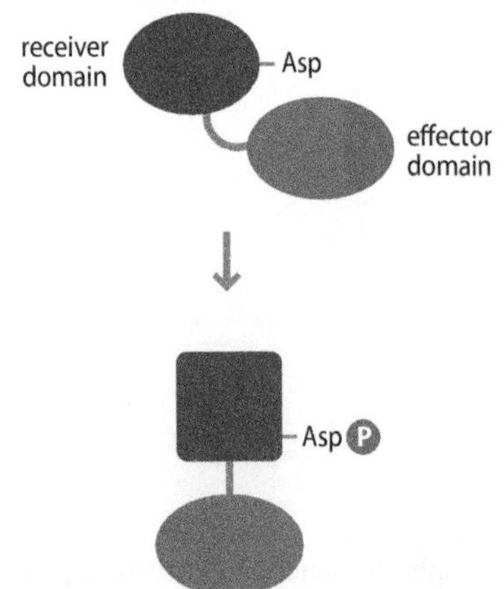

FIGURE 2.35
A common mechanism for phosphorylation to cause a downstream effect is that it leads to dissociation of the receiver domain from the effector domain. A simple example is the chemotaxis system CheB, in which the effector domain is an enzyme whose active site is blocked in the unphosphorylated complex.

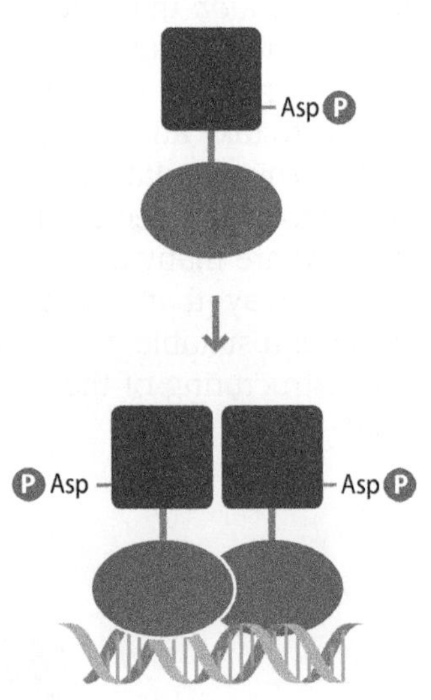

FIGURE 2.36
In many two-component systems, phosphorylation of the receiver domain leads to dissociation of the effector domain: but it also leads to dimerization of the RR, and hence to much enhanced binding of the effector domain to DNA.

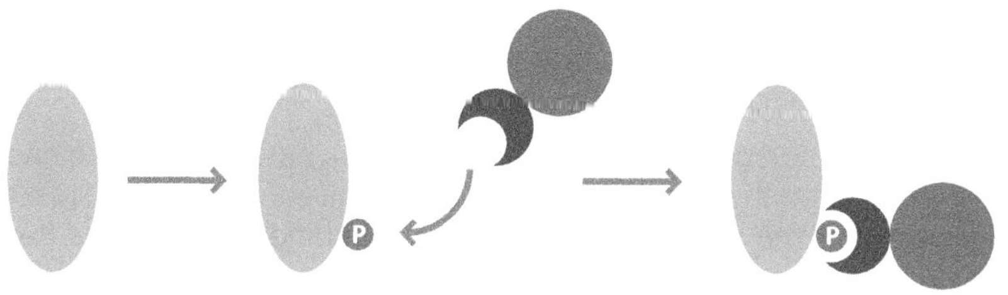

FIGURE 2.37
Phosphorylation of a protein enables the phosphorylated site to be recognized by a domain specialized in binding to such sites (SH2, PTB, 14-3-3, and so on), and hence enables the specific attachment of a second attached domain, for example an enzyme.

in the inactive unphosphorylated form and only becomes exposed after phosphorylation (see Figure 2.35).

An equally common mechanism for phosphorylation leading to a change in activity is one in which the phosphorylated residue is then recognized by a domain in a second protein: the phosphorylation functions mainly to bring two proteins close together (**Figure 2.37**). We shall see many times in this book that specificity is most easily evolved not by producing perfect and unique recognition but by the addition of multiple layers of less perfect recognition. In particular, bringing two proteins close together into a loose complex is an excellent way of improving the chances of a reaction between them. This type of regulation is how most signaling events achieve the desired specificity, as we shall see in Chapter 8.

There are other mechanisms for the regulation of activity by phosphorylation. The simplest is that the additional phosphate blocks the active site directly. This happens for example in the inactivation of the TCA cycle enzyme isocitrate dehydrogenase by phosphorylation. The regulation of kinases by phosphorylation is a rather special type of regulatory role, and this is discussed further in Chapters 4 and 8.

2.3.3 Multidomain construction makes an effective enzyme

We can illustrate the free energies of the different stages in a simple single-substrate enzyme reaction as in **Figure 2.38**. We start with separated enzyme and substrate, and the first step is for them to bind. One might naively suppose that this needs to be a favorable step, to encourage binding. However, in most enzymes under physiological conditions, binding is if anything slightly unfavorable (see Chapter 5). This is because the function of an enzyme is to make the reaction happen faster, or in other words to lower the transition-state energy $\Delta G^{\ddagger}$. This is the energy difference between the lowest and highest points on the pathway from free substrate and enzyme to products: therefore if the binding energy is favorable, the free energy for ES will be lower than that for E + S and the reaction will go more slowly.

Before the enzyme is ready to catalyze the reaction, it is usually necessary for some conformational change to occur, to bring the functional groups in the active site into the correct geometry. This conformational change is the one classically described as "induced fit," and it takes the enzyme from its resting state E to some activated state E*. As discussed in more detail in Chapter 6, this is not a completely accurate model for enzyme function, and there is certainly

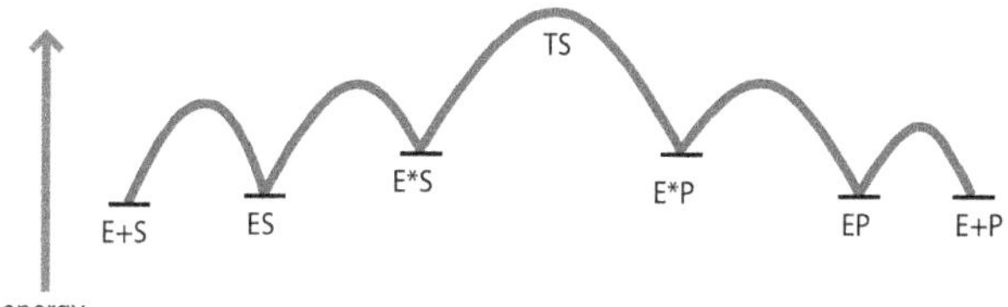

FIGURE 2.38
Representative energy diagram for an enzyme-catalyzed reaction.
E + S, separated enzyme and substrate;
ES, enzyme–substrate complex;
E*S, activated enzyme–substrate complex;
TS, transition state; P, product.

some rearrangement of the enzyme before the substrate binds. However, it remains useful to think of the conformational change to the activated enzyme and the binding step as separate events.

After this, the reaction can occur. The state at the highest point on the energy profile is by definition the transition state, and it is the energy required to reach there that controls the rate of the reaction by the standard Arrhenius transition-state equation

$$k = A\ \exp(-\Delta G^{\ddagger}/RT)$$

where the pre-exponential factor A is discussed in *6.4.

The enzyme lowers the activation energy $\Delta G^{\ddagger}$ by being complementary to the transition state, or in other words by having E* complementary to the transition state. The relevance to this discussion is that it must take care not to *raise* the activation energy by making either of the steps E + S → ES or ES → E*S more favorable than they need to be. I have already discussed the first of these steps. The second E → E* step is essentially a distortion of the enzyme from its resting or lowest-energy state. This does not necessarily mean that the step has to be energetically unfavorable, because some of the substrate binding energy can be (and is) used to reduce the energy cost of this step, as elegantly discussed by Jencks and others [87]. However, for both kinetic and thermodynamic reasons it is good to minimize the structural change required.

Structural rearrangement within a domain is certainly possible, but it is normally slow (because it requires the concerted movement of many different atoms, which is inherently unlikely) and is not an easy feature to evolve, because protein function normally requires a stable fold that is not easily altered. A much faster and simpler process is to have the active site at the interface between two domains, so that the E → E* rearrangement is a simple readjustment of a few residues in a hinge rather than a restructuring of the domain. Thus, location of the active site of an enzyme between two domains makes sense not only mechanistically but also kinetically and energetically.

2.3.4 Multidomain construction simplifies folding and assembly and stabilizes the protein

Proteins are synthesized as unfolded chains and subsequently have to fold up. *In vivo*, the nascent peptide chain is coated with chaperone proteins that prevent unwanted associations, but at some point these fall off and leave the protein chain to fold up. This needs to be a fairly rapid process, otherwise there are too many opportunities for the chain to interact with other cellular components and end up not folding properly. Small proteins can fold rapidly and without the need for further assistance, but larger proteins often require chaperones (*4.12) (such as the protein GroEL) that act as secluded spaces for them to fold up in without the risk of being interrupted by other cellular components. It is clear that one of the evolutionary constraints (although certainly not a strong one) on proteins is that they need to be able to fold efficiently to their native structure. This is much easier to organize in small domains than in large ones.

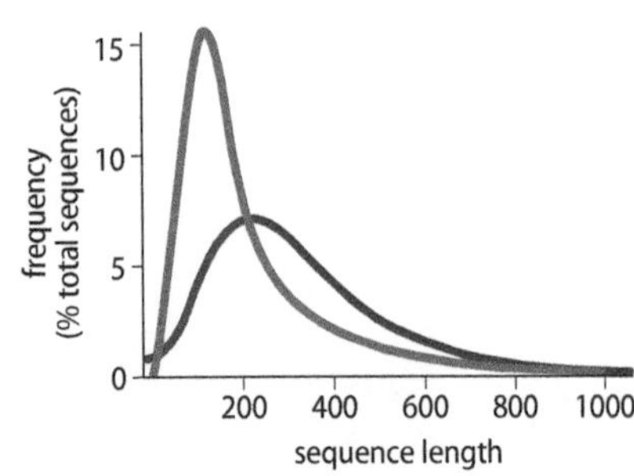

FIGURE 2.39
Approximate distribution of sequence lengths of domains in the PDB database (red) and of protein sequences in genomes (blue). (Redrawn from M. Gerstein, *Fold. Des.* 3: 497–512, 1998. With permission from Elsevier.)

Most modular proteins fold each module independently. That is, each bead on the string folds up by itself. Folding complexity, and thus the reciprocal of the folding rate, is very approximately proportional to n^3, where n is the number of amino acid residues. A large protein would therefore be expected to fold up much more slowly than a small protein. In particular, two domains of size n should fold up roughly eight times faster than a single domain of size $2n$. There is thus good reason to keep domains small, which indeed they are: the median domain size is approximately 120 residues long, whereas the median protein size is more like 200 residues long (**Figure 2.39**).

It may also be true, although this is far from clear as yet, that the need for a large domain to be able to fit within the GroEL cavity places limits on its physical size.

The overall stability of proteins (that is, the difference in free energy between folded and unfolded states under physiological conditions) is very small: only 20–60 kJ mol^{-1}. In comparison with the energy for a single hydrogen bond (a matter of much dispute, but probably 2 or 3 kJ mol^{-1}) or $RT/2$ (the average thermal energy per degree of freedom), which is 1.2 kJ mol^{-1}, this is not a great deal. Like everything else, the stability of proteins is under evolutionary control, and it is clearly advantageous for proteins to be of only limited stability. The stability of a protein determines the relative populations of folded and unfolded states. A protein with a free energy of unfolding of 30 kJ mol^{-1} has a probability of about 1 in 200,000 of being unfolded. This may not sound much, but it means that proteins spend a significant amount of time unfolded. An unfolded protein is much more likely to be degraded or removed, and one reason for the limited stability of proteins is presumably that it allows the cell to remove them when they are no longer wanted: a completely rigid and stable protein would be difficult to degrade. Conversely, a protein that is too unstable will bind less strongly and be degraded too rapidly (but see the discussion on natively unfolded proteins in Chapter 4).

In general, the stability of a protein is determined by the buried hydrophobic surface, which in turn depends on the volume of the protein. By contrast, the main way in which a protein is destabilized or unfolded is by interactions with the solvent; that is, it depends on the surface area of the protein. For a spherical molecule, the surface area is given by $4\pi r^2$ and the volume is given by $4\pi r^3/3$. The ratio of volume to surface area is thus $r/3$. In other words, as a very general rule, the stability of a protein increases in proportion to its radius. Thus, if there is an optimum stability of proteins, there must also be an optimum size. There is certainly a minimum size: proteins of less than about 40 amino acids without disulfide bridges are generally unfolded. Experimentally, the largest domain is approximately 35 kDa, although the size of folded proteins can go into the megadaltons by constructing them from multiple domains. Stability is likely to be only a small factor in controlling protein size, but it is nonetheless relevant. It should also be added that because most of the mechanisms for protein degradation require access to the protein surface, multiple domains help to stabilize proteins by reducing their accessible surface area.

2.4 PROTEINS AS TOOLS

It should already be clear that where possible this book pictures proteins as rigid units that perform largely mechanical operations; the detailed chemistry is treated as almost irrelevant. And it is worth pointing out that in prokaryotes only approximately 30–40% of proteins are in fact enzymes, and in eukaryotes the fraction is even lower, at approximately 20–30% [63]. That is, less than 40% of proteins actually change the chemical nature of their ligands. Many of the rest basically just bind to, and in so doing often change the conformation of, their partners. This analogy can be extended much further, and it is a useful exercise to consider to what extent proteins can be regarded simply as tools, performing essentially mechanical functions [93].

The properties of human tools can be described very generally (**Figure 2.40**):

- They recognize their substrate (e.g., spanner, screwdriver).
- They change in response to the substrate and its environment (for example pliers, adjustable spanner).
- They act on the substrate to change it (most tools).

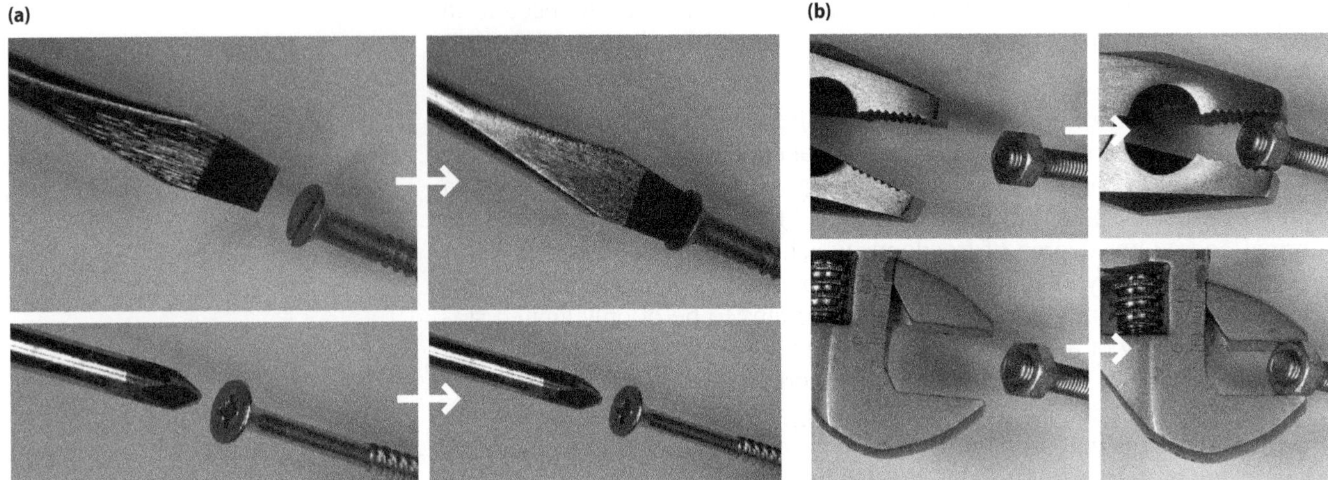

FIGURE 2.40
Properties of tools. (a) A tool needs to recognize its substrate. Different substrates need different, but related, tools. (b) A tool needs to change in response to the substrate. The change is often small, but very significant in terms of function.

Clearly, proteins do all of these. Tools work on the basis of several design principles (**Figure 2.41**):

- Two parts are made to move independently (for example pliers, adjustable spanner).
- There are a range of sizes but a common design (screwdriver and spanner sets).
- They share common parts (common handles, exchangeable ends).
- Where possible they are symmetric, or nearly so (scissors, pliers).
- And then there are specialist tools that do none of these.

To what extent do proteins do this?

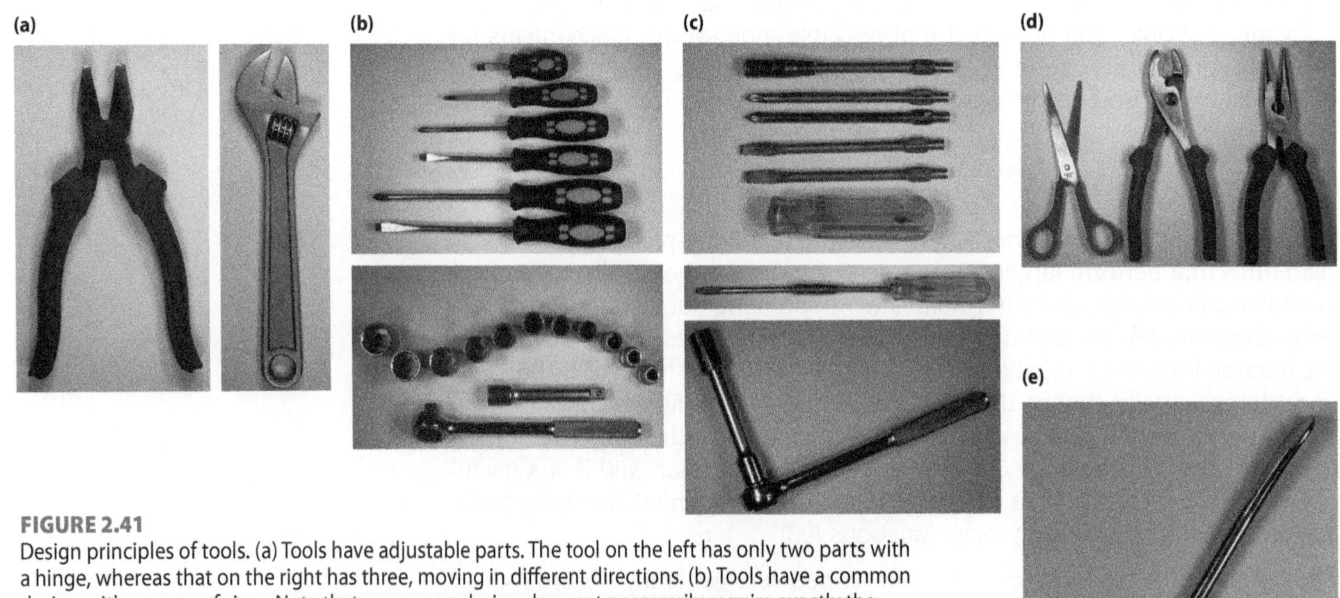

FIGURE 2.41
Design principles of tools. (a) Tools have adjustable parts. The tool on the left has only two parts with a hinge, whereas that on the right has three, moving in different directions. (b) Tools have a common design with a range of sizes. Note that a common design does not necessarily require exactly the same shape, but there are strong family resemblances. (c) Tools have common parts (note that the screwdriver set and the spanner set shown at the bottom of this figure and Figure 2.41b have a very useful "extra" part, namely an extension unit—commonly found in proteins too). (d) Tools have symmetry (but note that the pliers on the right are not in fact symmetric at all, although they are close to being symmetric). (e) A tool can be a specialist, with only a single use.

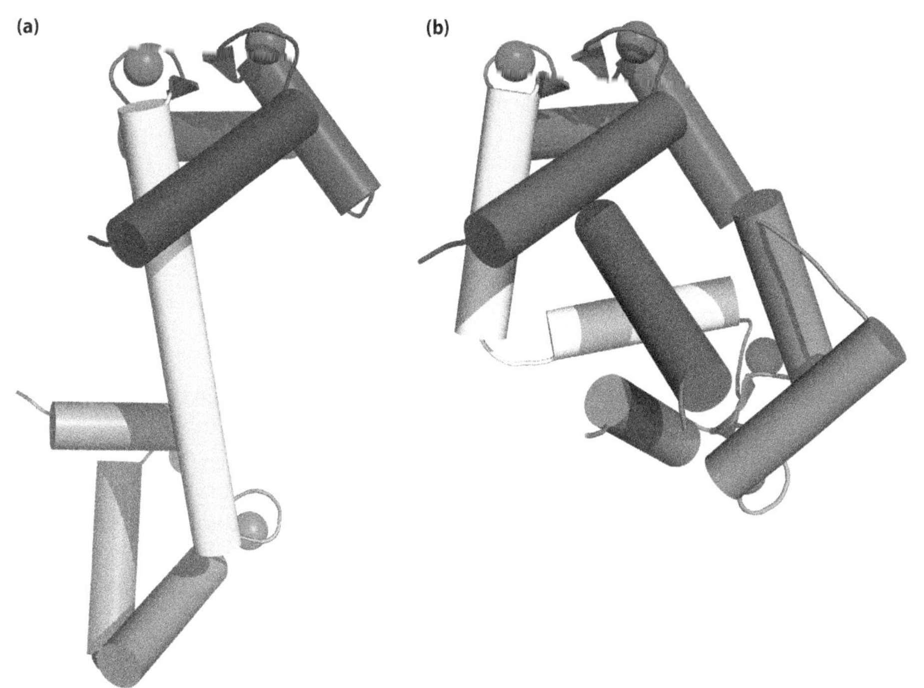

FIGURE 2.42
Calmodulin structures. (a) Free calmodulin (PDB file 1up5). (b) Calmodulin bound to the IQ motif from the cardiac $Ca_V1.2$ calcium channel (red) (PDB file 2f3y). Calmodulin has two domains, each consisting of two helix–loop–helix ("EF hand") motifs, with a calcium bound within the loop. The calcium ions are indicated by purple spheres. The three N-terminal helices are in blue, and the three C-terminal helices are in orange. In free calmodulin there is a long central helix, which makes up the fourth helix of the N-terminal domain plus the first helix of the C-terminal domain (green). In solution, this helix is very flexible in the middle. In complexes with ligands, the central helix folds up in the middle and encloses the ligand in a hydrophobic cavity made up from both termini, which constitute a single domain.

2.4.1 Tools have independently moving parts

This is one of the most notable features of proteins: many parts can move independently. But the movement is not a free-for-all wobbling, it is much more a movement of fixed units with hinges. We have already discussed the widespread use of two-domain structures in which substrate is grasped between the two domains as in pliers or a nutcracker; a good example is provided by calmodulin (**Figure 2.42**). Also very common is the use of a lid or flap that closes over the substrate (**Figure 2.43**). This is particularly common in proteases and nucleases, very probably because it is important in such cases to exclude water from the active site. Chapter 10 discusses several enzymes with swinging arms, another extremely useful device.

2.4.2 Tools have a common design but different sizes

At first glance, this seems to be an impossibility for proteins. How can related proteins have different sizes? And indeed this is rare. However, it is not unknown. Although proteins cannot just extend in three dimensions, they can certainly extend in one. Consider zinc fingers, for example. The Cys_2His_2 family is the largest class of eukaryotic transcription factors, and they are commonly used to recognize DNA, in which each finger fits into the major groove. They have been found with any number of fingers from 1 to 28 (except so far for 25) and higher: they therefore seem to be able to "grow" as much as necessary, to recognize longer sequences. Similarly, RNA is recognized by the Pumilio protein, which consists of tandem repeats, each repeat recognizing a single nucleotide. Sequence recognition can be made very specific simply by increasing the number of repeats (**Figure 2.44**) [81].

FIGURE 2.43
It is very common to find that enzymes contain flaps or lids that are usually disordered in the absence of substrate but close over the bound substrate.

Similarly, several proteins consist of repeated units, such as the β helix, the very common tetratricopeptide repeat (a 34-residue sequence that is found repeated up to at least 16 times), the ankyrin repeat, the leucine-rich repeat, and the armadillo repeat. Presumably the multiple repeats assist both specificity and affinity, although this is a somewhat crude way to achieve it! One might speculate that some of these proteins are evolutionarily recent: for example, two of the β helices known are antifreeze proteins and so could be

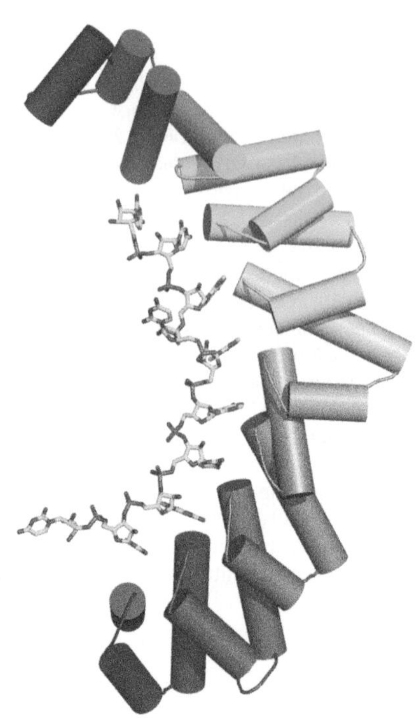

FIGURE 2.44
Human Pumilio 1 in complex with Puf5 RNA (PDB file 3bsx). Each pair of helices (except for the first three) recognizes a single base. The 5′ end of the RNA is at the bottom: the protein therefore recognizes the RNA sequence "backwards."

supposed to be a "recent" attempt to come up with a new protein. However, zinc fingers and tetratricopeptide repeats are widely found throughout nature and are presumably rather ancient structures.

2.4.3 Tools have common parts with variable "ends"

In human tools, this kind of design is very useful because it saves on materials and therefore on cost and weight. It also makes for very compact and adaptable tools. Biological systems do not care so much about cost and weight, but compactness and especially adaptability are of considerable importance. Therefore such designs are extremely common, and effectively constitute the case for modular proteins. We shall meet many such examples.

2.4.4 Some tools are symmetric

Proteins with mirror symmetry are of course never found because this would require D-amino acids. Human tools with mirror symmetry are also relatively rare. Scissors, for example, are not symmetric in this sense: ask any left-handed person, or try to cut your nails with your left hand. Scissors and pliers are (almost) rotationally symmetric, and so are oligomeric proteins. Of particular relevance are DNA-binding proteins, in which the symmetry of the protein is reflected in the symmetry of the palindromic DNA substrate. We shall discuss these more in Chapter 3.

2.4.5 Some tools have a specialist use

By "specialist" I mean tools that really only have a single function, such as bicycle tire levers. How often is it that a protein fold has only a single function? The answer depends to a large extent on how one defines a fold and a function. However, because the outstanding feature of evolution is tinkering, the basic answer is that any reasonably common fold has at some stage been picked up and used for something else. However, it is also true that the generation of new enzymes is a rather rare event. Thus, a survey in 2001 found that only 25% of superfamilies (roughly equivalent to folds, but rather more specific, as discussed at the start of this chapter) have members of different enzyme types, as defined by **EC classification number (*2.5)**. If two sequences are at least 40% identical the enzymes will be unlikely to have different functions, and above 30% identity the first three digits of the EC classification number can be predicted with at least 90% accuracy [94]. A recent review provides some examples of functional diversity within superfamilies, particularly those caused by structural embellishments [95].

***2.5 EC classification number**
The Enzyme Commission (EC) number is a hierarchical classification of the type of reaction catalyzed by an enzyme. The number has four parts, for example triosephosphate isomerase (TIM) has EC number 5.3.1.1. The first part describes the general type of reaction (1, oxidoreductase; 2, transferase; 3, hydrolase; 4, lyase; 5, isomerase; 6, ligase). The second part describes different versions of that basic reaction. For isomerases, these are 5.1, racemases and epimerases; 5.2, *cis/trans* isomerases; 5.3, intramolecular oxidoreductases; 5.4, intramolecular transferases; 5.5, intramolecular lyases; and 5.6, other isomerases. The third and fourth numbers describe yet further subdivisions of the main categories

2.5 SUMMARY

A domain is an independent structural and evolutionary unit. A module is similar, the only difference being that it is more versatile in its ability to be relocated. The CATH and SCOP classifications are useful ways of identifying domains. Domains usually have particular functions, in particular recognition and binding. They can therefore be copied and placed in a new context, and at a stroke they confer new binding properties. Almost always the new functions are simply additive: there are rather few cooperative functions that require two domains to act together, except for the common situation in which an enzyme's binding site is located at the interface between two domains. In eukaryotes the relocation of domains is usually achieved by exon shuffling, and in prokaryotes by recombination. Domain shuffling is a prominent feature of horizontal transfer. Domain shuffling should not be confused with three-dimensional domain swapping, which can be observed in several proteins and is likely to have been an important evolutionary mechanism by which new multidomain structures have arisen.

The function of many domains is to bind to other domains or peptide sequences. A weak or nonspecific domain–domain interaction can be made much stronger and more specific by tacking interacting domains onto each partner; such a solution can evolve with low cost and does not interfere with existing interactions. This embellishment has been a dominant pattern, particularly in mammalian evolution. An almost equivalent mechanism is to introduce scaffold proteins that bring together interacting partners. In both cases, the effect is to turn an intermolecular interaction into something much more like an intramolecular interaction. Because intramolecular interactions have a high effective concentration (that is, a less unfavorable entropy), they are stronger. An alternative way of stating this is that an inherently weak intramolecular binding can compete with a strong intermolecular interaction and hence forms an excellent regulatory or autoinhibitory mechanism, which is easily modulated.

This means that the large majority of proteins have multiple domains, especially in eukaryotes. Multidomain proteins have flexible interfaces between domains, which makes the development of allostery and covalent regulation (for example phosphorylation) much simpler. It also allows enzymes to recognize substrate and product, and makes folding simpler. The analogy with tools is helpful.

2.6 FURTHER READING

Molecular Evolution, by Li [24], is a clear and readable presentation of evolutionary mechanisms at a molecular level.

Protein Structure and Function, by Petsko and Ringe [96], is a beautifully illustrated and up-to-date book, containing many examples of domains and their functions.

The article by Doolittle [97] is an old but very readable account of the creation of modular proteins as a result of exon shuffling.

Doolittle's review [98] of domains, modules and domain shuffling is excellent. Doolittle is one of the "fathers" of domain shuffling.

A discussion of effective concentration from a slightly different angle can be found in Proteins: Structures and Molecular Properties, by Creighton [70].

Matrix Biology, volume 15, part 5, pp. 295–367, is devoted to the function and evolution of modular proteins and is well worth reading, especially [23] and [33].

2.7 WEBSITES

http://pfam.sanger.ac.uk/ *Pfam.*

http://prodom.prabi.fr/prodom/current/html/home.php *Prodom.*

http://supfam.mrc-lmb.cam.ac.uk/SUPERFAMILY/ *A database of structural and functional annotation for all proteins and genomes.*

http://www.bioinf.manchester.ac.uk/dbbrowser/PRINTS/index.php *PRINTS (protein motifs or fingerprints).*

http://www.bork.embl-heidelberg.de/Modules/ *Modules.*

http://www.cellsignal.com/reference/ *Intracellular protein interaction domains.*

http://www.ebi.ac.uk/interpro/ *Interpro, an excellent collection of databases, including CATH, SCOP and MODBASE.*

http://www.ebi.ac.uk/thornton-srv/databases/ProFunc/ *ProFunc (predicts function from structure).*

http://www.expasy.org/databases.html *Expasy proteomics server top page.*

http://www.expasy.org/prosite/ *ProSite.*

http://www.sanger.ac.uk/Software/ *Many other resources from the Sanger Institute.*

2.8 PROBLEMS

Hints for some of these problems can be found on the book's website.

1. Pick a protein that you are interested in, and find it in CATH and SCOP. Are the categorizations the same? If not, why not? Now look it up in databases that describe the functions of domains such as Pfam or Prodom. Do the annotations agree?

2. Find the sequence of *Rhodobacter sphaeroides* cytochrome c_2. Conduct a BLAST search for related sequences. Which are the most highly conserved residues? Find its structure in the Protein Data Bank. Does the structure explain why these residues are conserved?

3. The exon structure of human triosephosphate isomerase can be found in many places, for example the *ensembl* website http://www.ensembl.org (search for triosephosphate + isomerase + human, choose the TPI1 gene, and navigate to sequence → protein to get a colored view of the exon structure). Download the structure of triosephosphate isomerase, and look at it on a graphics system. Does the exon structure correspond to the three-dimensional structure as described here?

4. An interesting example of proteolytic processing is provided by the pro-apoptotic proteins caspases, which contain a pro-sequence that is cleaved off in the mature protein. By finding an appropriate website or review, describe the functional consequences of proteolytic processing of the pro-caspase.

5. Most eukaryotic modular proteins are (a) extracellular and (b) stabilized by disulfides. Suggest reasons why this should be so.

6. A common form of protein fold evolution is circular permutation, in which the N- and C-terminal ends of the protein are joined together and the gene is cut in a different location. Suggest how this might arise. How could you test this experimentally?

7. A class of tools not discussed in this chapter as analogies for proteins is power tools. There are two "classes" of modern power tools: those that require plugging in to a power source, and portable power tools with rechargeable batteries. What are the protein equivalents?

2.9 NUMERICAL PROBLEMS

N1. Protein A binds peptide p1 with a dissociation constant K_d of 1 μM. It binds to a different peptide, p2, with a K_d of 5 μM. What is the relative specificity of binding to p1 rather than p2? In other words, given low and equal concentrations of p1 and p2, how much more p1 is bound than p2?

N2. Adding extra amino acids to p1 can increase the specificity of binding. If each amino acid improves the affinity by a factor of 3, how much longer would the recognition site need to be to achieve a relative specificity of 1000?

N3. Alternatively, additional interacting domains could be added to A and p1. We can work out how large an effect this has. (a) If the affinity of A for p1 is 1 μM, what is the free energy of binding? (R = 8.31 J K^{-1} mol^{-1}; assume that this is done at 25°C). (b) Using the correction described in Section 2.2.11, calculate roughly the *intrinsic* affinity of A for p1; that is, after correcting for losses in entropy on binding A to p1, what is the free energy of interaction? (c) If the additional domain in isolation also has an affinity of 1 μM, and the attachment to A in the complex results in the loss of five freely rotatable bonds, use the method presented in Section 2.2.11 to calculate the overall affinity. Compare this with the result of N2. Which looks like a better way of increasing affinity?

N4. Prove the statement in Section 2.2.6 that the fraction bound = $1/(1 + K_{intramolecular})$.

N5. Zhou discusses a second interaction between an SH3 domain and a proline-rich peptide [78]. The affinity of the Rlk SH3 domain for the peptide QPSKRKPLPPLP when added as a separate peptide was 830 μM. However, when the peptide is attached to the SH3 domain with a 14-residue linker, the peptide is bound with a $K_{intramolecular}$ of 3. (a) What does this mean—for what proportion of the time is the peptide bound? Is this significant binding? (b) What is the corresponding effective concentration? Is this a low or high value for an effective concentration?

2.10 REFERENCES

1. D Baltimore (1976) Viruses, polymerases and cancer. *Science* 192:632–636.
2. JE Baldwin & H Krebs (1981) The evolution of metabolic cycles. *Nature* 291:381–382.
3. EV Koonin, YI Wolf & GP Karev (2002) The structure of the protein universe and genome evolution. *Nature* 420:218–223.
4. C Branden & J Tooze (1999) Introduction to Protein Structure, 2nd ed. New York: Garland Science.
5. AG Murzin, SE Brenner, T Hubbard & C Chothia (1995) SCOP - a structural classification of proteins database for the investigation of sequences and structures. *J. Mol. Biol.* 247:536–540.
6. M Gerstein (1997) A structural census of genomes: comparing bacterial, eukaryotic and archaeal genomes in terms of protein structure. *J. Mol. Biol.* 274:562–576.
7. SK Burley (2000) An overview of structural genomics. *Nature Struct. Biol.* 7:932–934.
8. CA Orengo, AD Michie, S Jones et al. (1997) CATH - a hierarchic classification of protein domain structures. *Structure* 5:1093–1108.
9. G Csaba, F Birzele & R Zimmer (2009) Systematic comparison of SCOP and CATH: a new gold standard for protein structure analysis. *BMC Struct. Biol.* 9:23.

10. C Chothia (1992) Proteins—1000 families for the molecular biologist. *Nature* 357:543–544.
11. M Levitt (2007) Growth of novel protein structural data. *Proc. Natl. Acad. Sci. USA* 104:3183–3188.
12. J Janin & SJ Wodak (1983) Structural domains in proteins and their role in the dynamics of protein function. *Prog. Biophys. Mol. Biol.* 42:21–78.
13. SE Radford, CM Dobson & PA Evans (1992) The folding of hen lysozyme involves partially structured intermediates and multiple pathways. *Nature* 358:302–307.
14. E Kloss, N Courtemanche & D Barrick (2008) Repeat-protein folding: new insights into origins of cooperativity, stability and topology. *Arch. Biochem. Biophys.* 469:83–99.
15. TL Religa, CM Johnson, DM Vu et al. (2007) The helix–turn–helix motif as an ultrafast independently folding domain: the pathway of folding of Engrailed homeodomain. *Proc. Natl. Acad. Sci. USA* 104:9272–9277.
16. M Gerstein (1998) How representative are the known structures of the proteins in a complete genome? A comprehensive structural census. *Folding Design* 3:497–512.
17. G Apic, J Gough & SA Teichmann (2001) Domain combinations in archaeal, eubacterial and eukaryotic proteomes. *J. Mol. Biol.* 310:311–325.
18. C Vogel, M Bashton, ND Kerrison et al. (2004) Structure, function and evolution of multidomain proteins. *Curr. Opin. Struct. Biol.* 14:208–216.
19. F Jacob (1970) The Logic of Life: A History of Heredity. Paris: Gallinard.
20. SB Carroll (2005) Endless Forms Most Beautiful. London: Weidenfeld & Nicolson.
21. H Yanagawa, K Yoshida, C Torigoe et al. (1993) Protein anatomy: functional roles of barnase module. *J. Biol. Chem.* 268:5861–5865.
22. P Bork, AK Downing, B Kieffer & ID Campbell (1996) Structure and distribution of modules in extracellular proteins. *Q. Rev. Biophys.* 29:119–167.
23. JR Potts & ID Campbell (1996) Structure and function of fibronectin modules. *Matrix Biol.* 15:313–320.
24. WH Li (1997) Molecular Evolution. Sunderland, MA: Sinauer Associates, Inc.
25. C Spitzfaden, RP Grant, HJ Mardon & ID Campbell (1997) Module–module interactions in the cell binding region of fibronectin: stability, flexibility and specificity. *J. Mol. Biol.* 265:565–579.
26. AR Pickford, SP Smith, D Staunton et al. (2001) The hairpin structure of the ^{6}F1^{1}F2^{2}F2 fragment from human fibronectin enhances gelatin binding. *EMBO J.* 20:1519–1529.
27. ID Campbell & AK Downing (1994) Building protein structure and function from modular units. *Trends Biotechnol.* 12:168–172.
28. JA Clapperton, IA Manke, DM Lowery et al. (2004) Structure and mechanism of BRCA1 BRCT domain recognition of phosphorylated BACH1 with implications for cancer. *Nature Struct. Biol.* 11:512–518.
29. CA Orengo & JM Thornton (2005) Protein families and their evolution: a structural perspective. *Annu. Rev. Biochem.* 74:867–900.
30. YI Wolf, SE Brenner, PA Bash & EV Koonin (1999) Distribution of protein folds in the three superkingdoms of life. *Genome Res.* 9:17–26.
31. L Patthy (1999) Genome evolution and the evolution of exon-shuffling: a review. *Gene* 238:103–114.
32. TW Traut (1988) Do exons code for structural or functional units in proteins? *Proc. Natl. Acad. Sci. USA* 85:2944–2948.
33. L Patthy (1996) Exon shuffling and other ways of module exchange. *Matrix Biol.* 15:301–310.
34. L Patthy (1999) Protein Evolution. Oxford: Blackwell.
35. W Gilbert (1987) The exon theory of genes. *Cold Spring Harb. Lab. Symp. Quant. Biol.* 52:901–905.
36. A Müller, RM MacCallum & MJE Sternberg (2002) Structural characterization of the human proteome. *Genome Res.* 12:1625–1641.
37. S Gudlaugsdottir, DR Boswell, GR Wood & J Ma (2007) Exon size distribution and the origin of introns. *Genetica* 131:299–306.
38. BL Bertolaet & JR Knowles (1995) Complementation of fragments of triosephosphate isomerase defined by exon boundaries. *Biochemistry* 34:5736–5743.
39. GB Golding, N Tsao & RE Pearlman (1994) Evidence for intron capture: an unusual path for the evolution of proteins. *Proc. Natl. Acad. Sci. USA* 91:7506–7509.
40. J Stetefeld, AT Alexandrescu, MW Maciejewski et al. (2004) Modulation of agrin function by alternative splicing and Ca^{2+} binding. *Structure* 12:503–515.
41. J Garcia, SH Gerber, S Sugita et al. (2003) A conformational switch in the Piccolo C_2A domain regulated by alternative splicing. *Nature Struct. Mol. Biol.* 11:45–53.
42. M de Château & L Björck (1996) Identification of interdomain sequences promoting the intronless evolution of a bacterial protein family. *Proc. Natl. Acad. Sci. USA* 93:8490–8495.
43. JP Gogarten & JP Townsend (2005) Horizontal gene transfer, genome innovation and evolution. *Nature Rev. Microbiol.* 3:679–687.
44. M de Château & L Björck (1994) Protein PAB, a mosaic albumin-binding bacterial protein representing the first contemporary example of module shuffling. *J. Biol. Chem.* 269:12147–12151.
45. MJ Bennett, MP Schlunegger & D Eisenberg (1995) 3D domain swapping: a mechanism for oligomer assembly. *Protein Sci.* 4:2455–2468.
46. Y Liu & D Eisenberg (2002) 3D domain swapping: as domains continue to swap. *Protein Sci.* 11:1285–1299.
47. MP Schlunegger, MJ Bennett & D Eisenberg (1997) Oligomer formation by 3D domain swapping: a model for protein assembly and misassembly. *Adv. Protein Chem.* 50:61–122.
48. M Jaskólski (2001) 3D domain swapping, protein oligomerization, and amyloid formation. *Acta Biochim. Pol.* 48:807–827.
49. J Heringa & WR Taylor (1997) Three-dimensional domain duplication, swapping and stealing. *Curr. Opin. Struct. Biol.* 7:416–421.
50. F Rousseau, JWH Schymkowitz & LS Itzhaki (2003) The unfolding story of three-dimensional domain swapping. *Structure* 11:243–251.
51. A Canals, J Pous, A Guasch et al. (2001) The structure of an engineered domain-swapped ribonuclease dimer and its implications for the evolution of proteins toward oligomerization. *Structure* 9:967–976.
52. G D'Alessio (1995) Oligomer evolution in action? *Nature Struct. Biol.* 2:11–13.
53. AJ Murray, SJ Lewis, AN Barclay & RL Brady (1995) One sequence, two folds: a metastable structure of CD2. *Proc. Natl. Acad. Sci. USA* 92:7337–7341.
54. AD Cameron, B Olin, M Ridderström et al. (1997) Crystal structure of human glyoxalase 1. Evidence for gene duplication and 3D domain swapping. *EMBO J.* 16:3386–3395.
55. ME Newcomer (2002) Protein folding and three-dimensional domain swapping: a strained relationship? *Curr. Opin. Struct. Biol.* 12:48–53.
56. L Vitagliano, S Adinolfi, F Sica et al. (1999) A potential allosteric subsite generated by domain swapping in bovine seminal ribonuclease. *J. Mol. Biol.* 293:569–577.
57. SM Green, AG Gittis, AK Meeker & EE Lattman (1995) One-step evolution of a dimer from a monomeric protein. *Nature Struct. Biol.* 2:746–751.
58. M Bergdoll, MH Remy, C Cagnon et al. (1997) Proline-dependent oligomerization with arm exchange. *Structure* 5:391–401.
59. KJ Knaus, M Morillas, W Swietnicki et al. (2001) Crystal structure of the human prion protein reveals a mechanism for oligomerization. *Nature Struct. Biol.* 8:770–774.
60. CM Fraser, JD Gocayne, O White et al. (1995) The minimal gene complement of *Mycoplasma genitalium*. *Science* 270:397–403.

61. JI Glass, N Assad-Garcia, N Alperovich et al. (2006) Essential genes of a minimal bacterium. *Proc. Natl. Acad. Sci. USA* 103:425–430.

62. K Kobayashi, SD Ehrlich, A Albertini et al. (2003) Essential *Bacillus subtilis* genes. *Proc. Natl. Acad. Sci. USA* 100:4678–4683.

63. S Freilich, RV Spriggs, RA George et al. (2005) The complement of enzymatic sets in different species. *J. Mol. Biol.* 349:745–763.

64. RR Copley, J Schultz, CP Ponting & P Bork (1999) Protein families in multicellular organisms. *Curr. Opin. Struct. Biol.* 9:408–415.

65. L Brocchieri & S Karlin (2005) Protein length in eukaryotic and prokaryotic proteomes. *Nucleic Acids Res.* 33:3390–3400.

66. T Pawson & P Nash (2003) Assembly of cell regulatory systems through protein interaction domains. *Science* 300:445–452.

67. R Singh & J Valcárcel (2005) Building specificity with nonspecific RNA-binding proteins. *Nature Struct. Mol. Biol.* 12:645–653.

68. TE Creighton (1984) Proteins: Structures and Molecular Principles. New York, WH Freeman.

69. M Mammen, SK Choi & GM Whitesides (1998) Polyvalent interactions in biological systems: implications for design and use of multivalent ligands and inhibitors. *Angew. Chem. Int. Ed.* 37:2755–2794.

70. TE Creighton (1993) Proteins: Structures and Molecular Properties, 2nd ed. New York: Freeman.

71. M Mutter (1977) Macrocylization equilibriums of polypeptides. *J. Am. Chem. Soc.* 99:8307–8314.

72. KP Hoeflich & M Ikura (2002) Calmodulin in action: diversity in target recognition and activation mechanisms. *Cell* 108:739–742.

73. JF Long, W Feng, R Wang et al. (2005) Autoinhibition of X11/Mint scaffold proteins revealed by the closed conformation of the PDZ tandem. *Nature Struct. Mol. Biol.* 12:722–728.

74. BT Goult, N Bate, NJ Anthis et al. (2009) The structure of an interdomain complex that regulates talin activity. *J. Biol. Chem.* 284:15097–15106.

75. J Jiang, K Prasad, EM Lafer & R Sousa (2005) Structural basis of interdomain communication in the Hsc70 chaperone. *Cell* 20:513–524.

76. LJ Byrne, A Sidhu, AK Wallis et al. (2009) Mapping of the ligand-specific site on the b′ domain of human PDI: interaction with peptide ligands and the x-linker region. *Biochem J* 423:209–217.

77. V Neduva & RB Russell (2005) Linear motifs: evolutionary interaction switches. *FEBS Lett.* 579:3342–3345.

78. HX Zhou (2006) Quantitative relation between intermolecular and intramolecular binding of pro-rich peptides to SH3 domains. *Biophys. J.* 91:3170–3181.

79. Y Shamoo, N Abdul-Manan & KR Williams (1995) Multiple RNA binding domains (RBDs) just don't add up. *Nucleic Acids Res.* 23:725–728.

80. M Watson, K Stott & JO Thomas (2007) Mapping intramolecular interactions between domains in HMGB1 using a tail-truncation approach. *J. Mol. Biol.* 374:1286–1297.

81. BM Lunde, C Moore & G Varani (2007) RNA-binding proteins: modular design for efficient function. *Nature Rev Mol. Cell. Biol.* 8:479–490.

82. KR Ely & R Kodandapani (1998) Ankyrin(g) ETS domains to DNA. *Nature Struct. Biol.* 5:255–259.

83. M Overduin & T de Beer (2000) The plot thickens: how thrombin modulates blood clotting. *Nature Struct. Biol.* 7:267–269.

84. MJ Wood, BAS Benitez & EA Komives (2000) Solution structure of the smallest cofactor-active fragment of thrombomodulin. *Nature Struct. Biol.* 7:200–204.

85. P Fuentes-Prior, Y Iwanaga, R Huber et al. (2000) Structural basis for the anticoagulant activity of the thrombin–thrombomodulin complex. *Nature* 404:518–525.

86. RB Russell & P Aloy (2008) Targeting and tinkering with interaction networks. *Nature Chem. Biol.* 4:666–673.

87. WP Jencks (1975) Binding energy, specificity, and enzymic catalysis: Circe effect. *Adv. Enzymol.* 43:219–410.

88. MI Page & WP Jencks (1971) Entropic contributions to rate accelerations in enzymic and intramolecular reactions and the chelate effect. *Proc. Natl. Acad. Sci. USA* 68:1678–1683.

89. WP Jencks (1981) On the attribution and additivity of binding energies. *Proc. Natl. Acad. Sci. USA* 78:4046–4050.

90. DH Williams, E Stephens, DP O'Brien & M Zhou (2004) Understanding noncovalent interactions: ligand binding energy and catalytic efficiency from ligand-induced reductions in motion within receptors and enzymes. *Angew. Chem. Int. Ed.* 43:6596–6616.

91. M Bashton & C Chothia (2007) The generation of new protein functions by the combination of domains. *Structure* 15:85–99.

92. MA Fuchs & C Buta (1997) The role of peptide modules in protein evolution. *Biophys. Chem.* 66:203–210.

93. RJP Williams (1993) Are enzymes mechanical devices? *Trends Biochem. Sci.* 18:115–117.

94. AE Todd, CA Orengo & JM Thornton (2001) Evolution of function in protein superfamilies, from a structural perspective. *J. Mol. Biol.* 307:1113–1143.

95. BH Dessailly, OC Redfern, A Cuff & CA Orengo (2009) Exploiting structural classifications for function prediction: towards a domain grammar for protein function. *Curr. Opin. Struct. Biol.* 19:349–356.

96. GA Petsko & D Ringe (2004) Protein Structure and Function. New York: Sinauer.

97. RF Doolittle (1985) The genealogy of some recently evolved vertebrate proteins. *Trends Biochem. Sci.* 10:233–237.

98. RF Doolittle (1995) The multiplicity of domains in proteins. *Annu. Rev. Biochem.* 64:287–314.

99. MG Rossmann, D Moras & KW Olsen (1974) Chemical and biological evolution of a nucleotide-binding protein. *Nature* 250:194–199.

CHAPTER 3
Oligomers

As we have seen in Chapter 2, a multidomain structure has a powerful evolutionary advantage: it allows proteins to acquire new functions, and it gives proteins suitable concave binding sites for binding ligands or allosteric effectors. The same is almost equally true of oligomeric proteins, to which we now turn. Most proteins are in fact oligomers: there must therefore be a powerful advantage to being oligomeric. As we shall see, there are actually several quite different advantages, of which the two main are that it allows proteins to be allosteric (Section 3.2) and that it allows proteins to bind cooperatively (Section 3.3). It is also presumably the case that constraints within a symmetric homo-oligomeric interaction make it simple for a homo-oligomeric interaction to evolve [2]: that is, oligomers can evolve rapidly and easily (**Figure 3.1**). Evolution never ignores something that can provide a quick gain with little effort!

Proteins can form homo-oligomers by self-association, or hetero-oligomers by binding to a different protein. This chapter is mainly concerned with homo-oligomers. Protein complexes made up of different components are discussed in Chapter 9.

We may ... ask why molecular evolution should have so frequently favoured the appearance and maintenance of oligomeric globular proteins. That it should be so must mean that there are functional advantages of some kind, inherent in the oligomeric state, and absent or difficult to achieve in the monomeric state.

Monod, Wyman & Changeux (1965), [1]

The truth is rarely pure and never simple.

Oscar Wilde (1854–1900), The Importance of Being Earnest, Act 1

3.1 WHY DO PROTEINS OLIGOMERIZE?

In *Escherichia coli*, the average oligomerization state of proteins is 4. Thus, a protein that exists as a monomer is a relatively uncommon event. A survey of human enzymes shows that approximately two-thirds are oligomers [3]. An oligomeric protein creates some problems in assembly, in that at least two molecules of protein need to be present in the same place at the same time. It also makes the protein larger, and therefore less mobile, so there must be good reasons why proteins have evolved to exist as oligomers [4]. In this section, we explore what these are.

3.1.1 Oligomerization shelters and regulates the active site

We have already seen in Chapter 2 that proteins with multiple domains have big advantages over single-domain proteins. Oligomeric proteins are of course a simple and effective way of creating proteins with several domains. The evolution of a stable oligomeric protein from a monomer is rather simple, in that mutations affect both monomers of a dimer and have "twice the effect" of mutations in a heterodimer. It therefore takes rather few mutations to create (or of course destroy) a stable protein dimer (see Figure 3.1).

In passing, it is of interest to look at oligomeric interfaces. Are they different from the rest of the protein surface? Yes [5, 6]. The interface surface is much more like the interior of a protein than its surface [7], unlike for example the

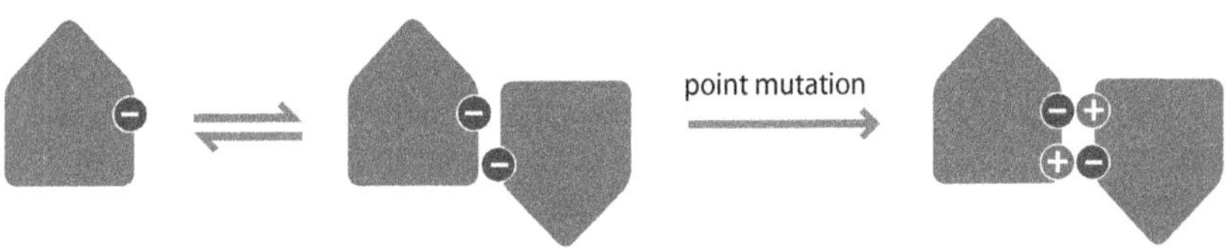

FIGURE 3.1
A monomeric protein with a negative charge on the surface forms a weak dimer. A single point mutation, creating a complementary positive charge, generates two symmetrical stabilizing interactions and therefore has a strong influence on the dimerization equilibrium. If it is advantageous for the protein to be dimeric, the mutation will be quickly established.

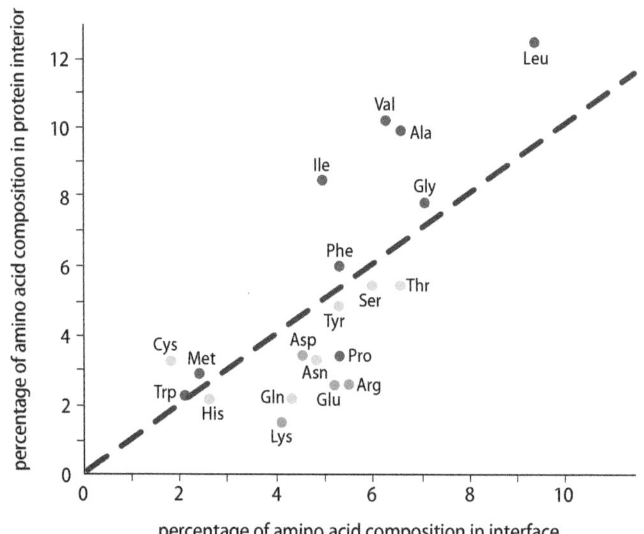

FIGURE 3.2
Comparison of amino acid frequencies in oligomeric interfaces with frequencies in the protein interior. The interface is relatively abundant in charged amino acids (cyan) and deficient in hydrophobic amino acids (red). Note the anomalous position of proline, which has both hydrophobic and hydrophilic properties, as described in Chapter 1. The dashed line is $y = x$. (Adapted from C. J. Tsai et al., *Protein Sci.* 6:53–64, 1997. With permission from John Wiley and Sons.)

interaction surface in antigen/antibody or protease/inhibitor contacts, which are normal protein surfaces [8]. They are, as one might expect, more hydrophobic, particularly if one excludes a "rim" of more solvent-exposed amino acids around the edge of the interface [7]. However, they also tend to be rich in arginines, as well as to some extent the other charged amino acids, which form buried salt bridges (**Figure 3.2**) [5]. A salt bridge exposed on a protein surface is not particularly energetically favorable, because the favorable **enthalpy (*3.1)** of binding is almost exactly balanced by the unfavorable

*3.1 Entropy and enthalpy

The following description is not my original idea (though I wish it were); it comes from [25].

The Second Law of Thermodynamics says that a reaction will go spontaneously if the overall entropy of the universe increases as a result. The *entropy* of a system is determined by the number of different ways in which the system can be arranged. Boltzmann's (*5.3) famous equation is

$$S = k \ln W$$

where W is the number of ways, and k is Boltzmann's constant, equal to the gas constant R divided by Avogadro's number. Therefore, for example, two molecules have more entropy than one, because there are more ways in which they can be arranged; a random mixture has more entropy than an ordered one; every degree of freedom in a molecule such as a bond rotation or vibration, or translational movement, has an associated entropy. This is why entropy is often described as the amount of *disorder* in a system. Thus, the Second Law is essentially saying the obvious: that you end up with the most probable outcome.

Therefore, if we want to know whether a reaction will go in the forward direction or not, the question is whether it leads to an increase in entropy. This entropy comes in two parts: the entropy change in the reacting molecules themselves, and the entropy change in the surroundings. The first of these is written ΔS, the Δ symbol just meaning 'observable change in', and is calculated from Boltzmann's equation by considering changes in translational motion, bond vibrations, and so on. The second, the change in the entropy of the surroundings, is essentially the same as the *heat* given off (or taken up) during the reaction. The heat produced by a reaction is called the *enthalpy* and has the symbol ΔH: it depends on things such as changes in bond energy. A reaction that gives off heat is called exothermic, and one that takes up heat is called endothermic. An exothermic reaction has a negative value of ΔH because the total heat energy of the molecules decreases. If the reaction gives off heat, this heats up the surroundings, which gives them more random thermal motion, and therefore they have more entropy. The change in entropy also depends on the temperature: if the surroundings are very cold, a small amount of heat will cause a big increase in disorder, whereas if it is already hot, the same amount of heat will cause a smaller increase in disorder, because it is disordered already. Therefore the change of entropy of the surroundings can be described as $\Delta H/T$, where T is the absolute temperature in kelvins. Because an exothermic reaction (negative ΔH) corresponds to an increase in entropy of the surroundings, this change is actually $-\Delta H/T$.

Thus, the total entropy change is given by the sum of these two, namely $-\Delta H/T + \Delta S$. This is what determines whether a reaction goes, and **Gibbs (*3.2)** called this the *free energy*. It now has the symbol $-\Delta G$. To put it into units of energy it needs to be divided by the temperature: $-\Delta G/T = -\Delta H/T + \Delta S$, which of course is the well-known expression

$$\Delta G = \Delta H - T\Delta S$$

Thus, this equation is really saying that to analyze whether a reaction will go we need to consider two things: whether the change in chemical energy contained in the bonds and other interactions is favorable; and whether the result is more *probable*, in a statistical sense. When applied to a binding event, it is essentially saying that binding will occur if there is a favorable binding energy (that is, enthalpy), but not if this restricts the conformational freedom of the molecules involved too much.

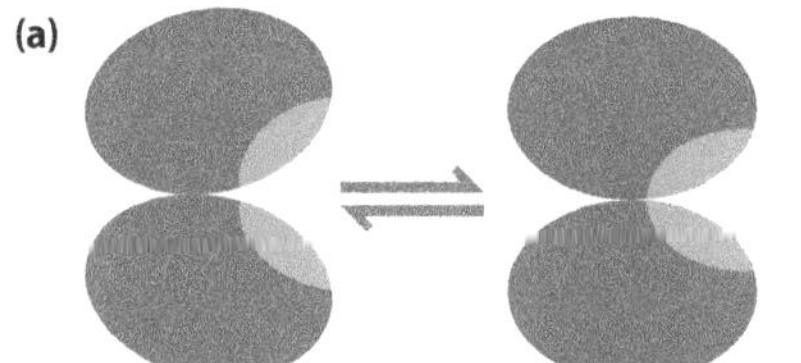

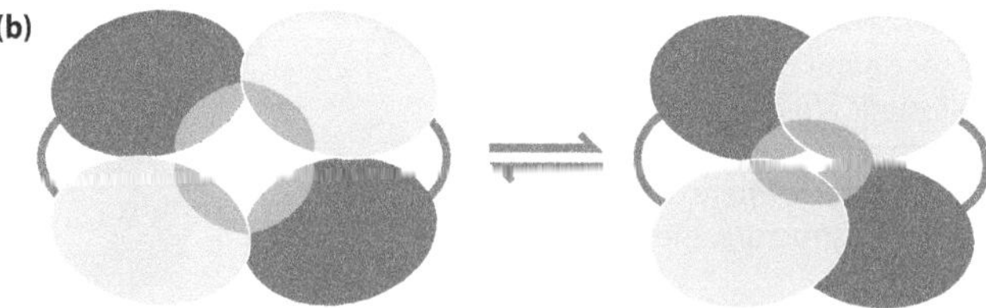

FIGURE 3.3
An oligomeric enzyme can regulate the accessibility of the active site relatively easily just by changing the dimer interface. This is true for a dimeric single-domain enzyme (a), and even more true for a dimeric two-domain enzyme (b), which is a much more common situation.

entropy of keeping the two charges together. The favorable enthalpy is not large, because the charges are screened (*4.4) by water and ions. Indeed, a recent study suggests that, in solution, exposed salt bridges are slightly unfavorable [9]. However, inside a protein interface, the dielectric constant (*4.2) is much smaller and therefore the enthalpy of binding is greater: salt bridges here are therefore much more favorable. They not only stabilize the oligomer interface but also provide a high degree of specificity and complementarity to the structure, because an unsatisfied or poorly structured salt bridge has much less favorable energy.

The advantages arising from oligomerization are closely related to those described in Chapter 2 for multiple domains (including the argument for stability, not discussed further here). Let us start with the reason given in Section 2.3.3: oligomers make a better enzyme. If the active site of an enzyme can be located close to the dimerization interface (which it usually is), this creates many advantages. Flexibility within the domain interface (or in an extreme case, reversible dissociation of the dimer) allows the active site to open out to permit more rapid substrate binding and product release, but also allows the two monomers to close over the active site during the reaction itself and thus shield the active site from solvent (Section 4.1.3) (**Figure 3.3**). It also decreases the energy input needed to change and shield the active site, and therefore, for the reasons given in Chapter 2, allows the enzyme to catalyze reactions faster.

A more restrictive access to the active site is also of benefit in that it allows the enzyme greater opportunity to be selective about what substrates can enter. At a very crude level, an exposed active site can allow very large substrates to enter, whereas a more restrictive active site imposes a size limit on substrates (**Figure 3.4**).

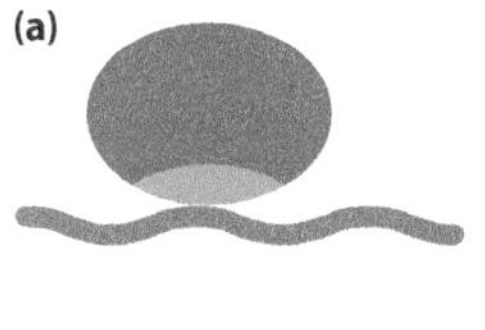

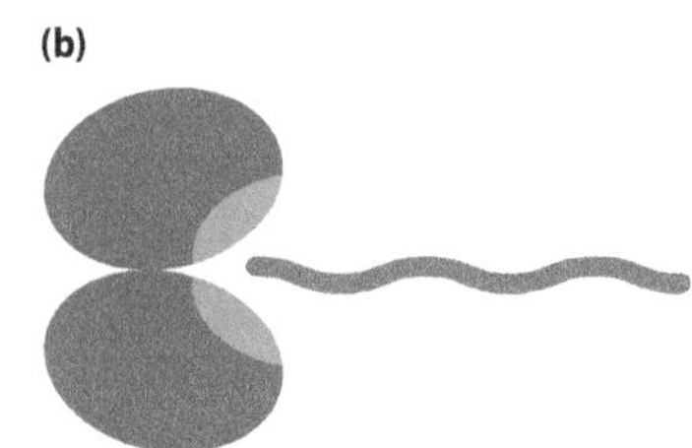

FIGURE 3.4
An exposed active site can accommodate a large substrate, whereas a more restricted active site, as in a dimer, can only accommodate a smaller substrate. Here, the monomer (a) can act equally as an *exolytic* or *endolytic* enzyme, whereas the dimer (b) can only act as an exolytic enzyme.

***3.2 Josiah Willard Gibbs**

Josiah Willard Gibbs (**Figure 3.2.1**) is one of the greatest American scientists of all time, having originated much of chemical thermodynamics (particularly of course in the area of free or usable energy) and vector analysis, and developed statistical mechanics. He was a very private man, whose writing is remarkably dense and difficult to read, who once wrote (according to Wikipedia), "A mathematician may say anything he pleases, but a physicist must be at least partially sane."

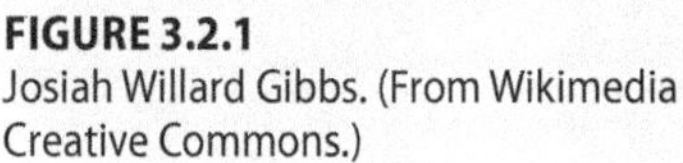

FIGURE 3.2.1
Josiah Willard Gibbs. (From Wikimedia Creative Commons.)

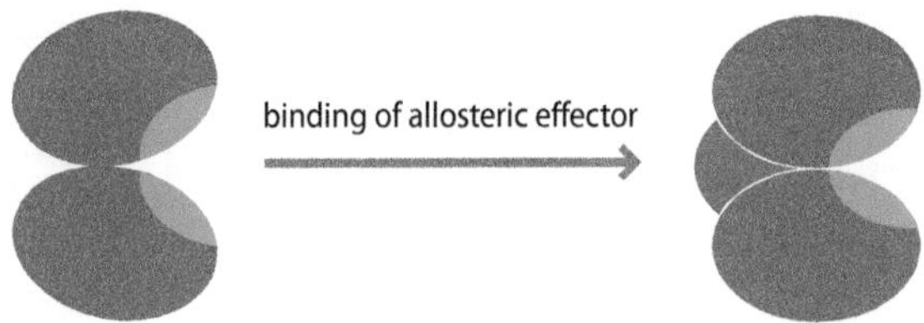

FIGURE 3.5
Dimeric enzymes permit the simple evolution of allosteric effectors. Provided that the active site is at the dimer interface, such ligands can bind elsewhere on the interface and alter its structure, thereby affecting the active site.

As noted in Chapter 2, placement of the active site at the interface between two domains also creates the possibility of **allostery**: that is, by keeping the domains rigid but altering the interface between them, it is rather straightforward to create a wide range of binding sites and connections between the domains (**Figure 3.5**). We shall investigate this in more detail shortly. In general, oligomerization provides an easy route to the evolution of additional binding sites.

One example is provided by our old friend triosephosphate isomerase, which is a dimer. When a monomer was engineered by shortening a loop at the interface, the activity of the monomer was found to be less than that of the dimer, because some of the loops that normally bind substrate were more disordered in the monomer than they are in the dimer [10]. This example illustrates that the detailed structure and behavior of the active site can be regulated by the dimer interface. Thus, minor changes in the interface are likely to affect function and can be seized upon and tinkered with by evolution to create allosteric function and other regulatory variations.

Finally, the active site of enzymes is usually hydrophobic, and thus inherently prone to unwanted aggregation and general binding of undesirable hydrophobic molecules. This can be decreased if the active site is at least partly shielded from solvent (**Figure 3.6**).

3.1.2 Oligomerization provides improved enzyme functionality

An enzyme requires specific binding (that is, a reasonably low K_m) together with catalysis against specific substrates (that is, a reasonably high k_{cat}), as described in more detail in Chapter 5. Both the binding and the catalysis are normally supplied by amino acids located at the interface between two domains, and often between two monomers of an oligomer. Mutation of these residues can affect just the residues themselves, and therefore just the binding and catalysis, or it can affect the oligomer interface, and therefore have additional effects on binding and allostery. Conversely, a change to the oligomer interface can have effects that affect only binding, only catalysis, only allostery, or any combination of these. Oligomerization thus provides a protein with a greatly increased potential for further evolution.

An interesting example is provided by the amino acid dehydrogenases [11]. Leucine dehydrogenase (LDH) and glutamate dehydrogenase (GDH) perform similar reactions, namely the oxidative deamination of an amino acid to the corresponding 2-oxo acid. The two enzymes have similar sequences and structures. Among the differences are mutations to residues in the active site, which alter the binding site, and therefore the specificity for which amino acid is bound, but not the catalytic residues. However, one other difference is that LDH is an octamer, whereas GDH is a hexamer. This difference leads to a small

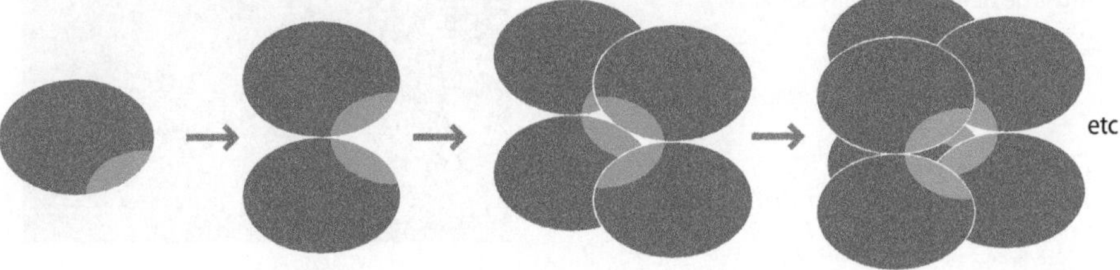

FIGURE 3.6
The active site of an enzyme is usually hydrophobic (cyan patch). This can lead to random dimerization and then on to greater degrees of uncontrolled aggregation. A dimer limits the amount of association that can occur.

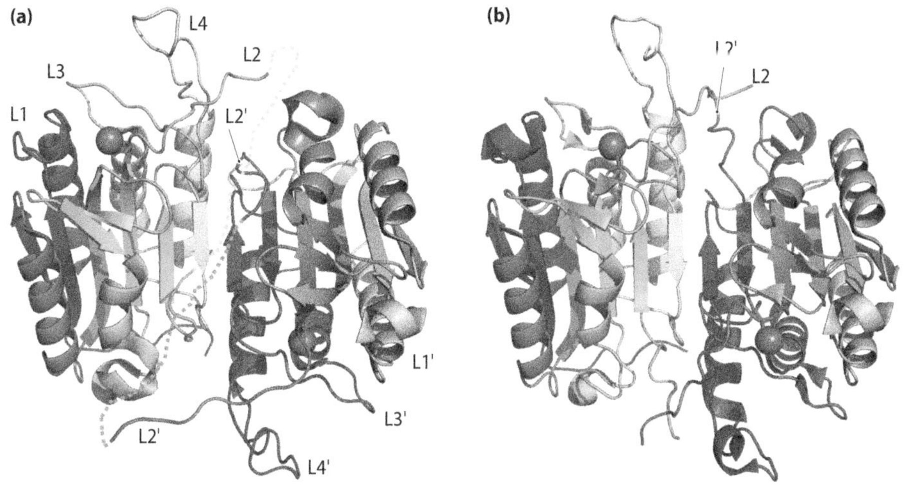

FIGURE 3.7
Activation of caspase-7 by proteolytic cleavage. (a) Structure of the uncleaved zymogen (PDB file 1k86). The dimer interface runs approximately vertically down the center; one monomer is color-coded from blue to green, and the other runs from yellow to red. The active site is Cys 186, indicated by magenta sticks and an orange ball, and sits at the start of loop 2 (L2). The substrate is held in a cleft whose bottom is formed by L3 and the walls by L1 and L4. L2 determines the position of the cysteine, and also interacts with L4. It is a very long loop, running right across the face of the enzyme, and is disordered in the crystal, meaning that no electron density is visible: the approximate position of the chain is indicated by a dashed line. Note that the start of the L2′ loop (that is, the L2 loop on the second monomer) is folded back tightly on itself, meaning that L2′ does not interact with L2. (b) Structure of the cleaved active enzyme (PDB file 3h1p). The cleavage is within L2, meaning that the two cleaved ends are now free to diverge further. In particular, the start of the loop, indicated here by L2′, now lies in a different orientation in which it can interact with L2 and L4 of the other monomer and stabilize them, thus producing an active site that binds productively to substrate. Thus, the dimeric nature of the protein is the key to its proteolytic activation. Other caspases are thought to be activated in a similar way.

change in the packing of the interface between monomers, which causes a change to the shape of the binding pocket, and therefore also affects the binding specificity. Thus, the binding specificity is determined not just by the amino acids in the binding site but also by the quaternary structure. A similar observation was made for tyrosyl-tRNA and tryptophanyl-tRNA synthetases [12].

There are many examples of enzymes that are essentially inactive as monomers and only become active as oligomers. An important example is provided by **caspases**, which are the central components of the apoptotic machinery and initiate irreversible proteolysis leading to cell death. Inappropriate activation of caspases leads to unwanted apoptosis, so the process is tightly regulated. One way in which this is done is that caspases are only active as dimers. The active site is composed of several loops and is only formed on proteolytic activation of the zymogen. The activated dimer is asymmetric, with the cleaved loop of one monomer forming the active site but bolstered by the cleaved loop of the other monomer, which has to move extensively to support the active conformation (**Figure 3.7**) [13].

3.1.3 Oligomerization makes symmetric dimers

The dimer formed by the association of two monomers is almost always symmetric. This is of no direct benefit, unless the substrate or ligand for the protein is also symmetric. A large number of the proteins that bind to DNA have taken advantage of this property by binding to palindromic DNA sequences, as discussed in more detail later in this chapter.

It is of relevance to note that different types of symmetry are possible. In particular, protein monomers may assemble in a head-to-head manner in which the interactions between monomers are symmetric, or in a cyclic head-to-tail manner in which monomers assemble into a ring. The former leads to even numbers of monomers in an oligomer, whereas the latter may lead to even or odd numbers. The former is roughly 10 times more common, not least because it is easier to evolve, explaining the observation that oligomers containing even numbers of monomers are considerably more common than oligomers containing odd numbers [14].

3.1.4 Coding errors, coding efficiency and linkers are not convincing reasons

The advantages described above apply equally well to a dimeric protein and to a single polypeptide chain containing two domains connected by a linker. As we have seen in Chapter 2, modular proteins such as a two-domain protein with a linker are common, particularly in eukaryotes. Sometimes modular proteins form well-structured oligomers, but more often in modular proteins the modules are completely or partly independent and do not form a well-defined oligomeric structure. Thus, most "true" oligomeric proteins in fact consist of different peptide chains rather than a single chain.

It is not immediately obvious why this should be. Indeed, there are advantages in constructing a protein from a single chain. An oligomer has to assemble from its constituent monomers. In the crowded environment of a cell, even if two protein transcripts are created in quick succession from the same mRNA message, it is not a trivial matter for the two monomers to find each other. Because eukaryotic cells are much larger than prokaryotic cells, this is likely to be a bigger problem for eukaryotes; this is probably one reason why it is more common in eukaryotes to find oligomeric proteins formed from a single polypeptide chain rather than two separate chains.

One argument in favor of two separate chains concerns errors in transcription or translation. A two-domain protein is twice as long as a single-domain protein, and therefore the chance of making an error is twice as great. The error rate in translation is not that different between eukaryotes and prokaryotes, and is roughly one in 4000 amino acids. Thus in a single-domain protein of 200 amino acids, one in 20 proteins contain an error, whereas in a two-domain protein of 400 amino acids, one in 10 contains an error. Many of these errors will not affect the function, but some will. If the error disrupts the function of the domain it is in, it will disrupt the function of the entire two-domain protein, but it will only disrupt one domain of the two single-domain proteins. Thus, the effect of an error is rather less severe in single-domain proteins. And the expenditure of energy in correcting the error is also less severe in single-domain proteins, because the cell only wasted half the energy in making the incorrect single domain. There is again a slight advantage for eukaryotes in having longer chains because of their better transcriptional proof-reading and post-translational control.

Another, and probably more powerful, argument concerns the linkers between the domains. As noted above, the big advantage of an oligomeric protein comes when the interface between the two domains is flexible. A linker necessarily constrains this flexibility. This is not an insuperable problem: most obviously the cell could create a long linker. However, a long linker has its own problems in that it is more likely to get recognized as an unfolded peptide and become degraded, or be involved in unwanted interactions with other proteins (**Figure 3.8**).

A protein constructed from two identical single domains has greater *coding efficiency* than a two-domain protein: that is, the genome can be shorter because it only needs to code for a protein half the length. For most organisms this is not a significant issue. In eukaryotes, most of the DNA does not

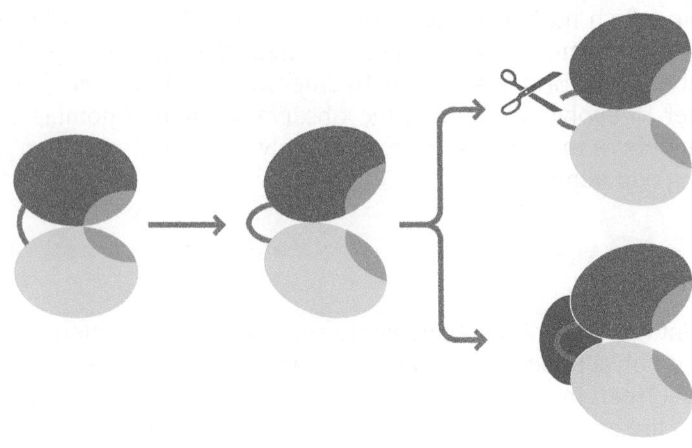

FIGURE 3.8
A two-domain protein with a linker between the domains has the problem that conformational change of the domain interface leads to loosening of the linker and therefore a greater potential for degradation or unwanted association of the linker.

code for protein anyway: less than 2% of our DNA codes for protein or RNA or has a clear regulatory role. Prokaryotes have typically less than 15% of their genome as "junk" DNA, but even so have very little evolutionary pressure to keep the genome size small. However, some prokaryotes have clearly had considerable pressure to reduce their genome size. The obvious example is the smallest known bacterial genome, *Mycoplasma genitalium*, which lives in primate genitourinary tracts and is streamlined to reproduce its genome very rapidly: it has reduced its genome size to 580 kilobases, or 482 coding genes. However, even for this organism, roughly 20% of the genes are nonessential, and 12% of the genome is noncoding, though of course not necessarily "junk" [15]. Thus, for most organisms, coding efficiency is not a relevant factor in protein architecture.

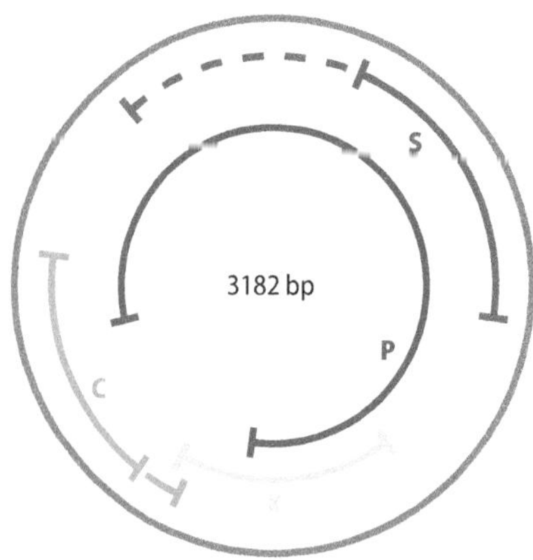

FIGURE 3.9
The genome structure of the hepatitis B virus. The virus genome codes for only four proteins, two of which have pre-regions (dashed). The overlapping sequences are in different reading frames.

The exception is viruses, which have extreme pressure to be efficient in their use of DNA. For example, a reasonably typical virus, hepatitis B, has 3182 nucleotides [16]. The genome codes for four polyproteins (**Figure 3.9**). *All* nucleotides are used for coding proteins, but in addition 1583 bases (50%) of the genome are used twice, with different codon phases! And naturally viral coats are composed of very large numbers of identical proteins assembled into oligomers. Interestingly, the association between viral coat proteins is often mediated by three-dimensional domain swapping, as discussed in Chapter 2. It has been suggested that this provides a strong interaction without increasing the size of the protein, and also allows the flexibility needed to accommodate quasi-equivalent contacts in the assembly of the coat [17].

3.2 ALLOSTERY

3.2.1 Most enzymes are not allosteric

Although it is not necessary for a protein to be oligomeric to have **allosteric** properties, it certainly makes its evolution easier, and the large majority of allosteric proteins are indeed oligomers. This is because allostery requires the binding of a ligand at one location to affect the structure of the active site. By far the simplest way to do this is if both sites are at domain interfaces (see Figure 3.5). A recent review gives many examples of allosteric enzymes, and in virtually all cases the allosteric effects work in this way [18]. (As the exception that proves the rule, hexokinase is allosteric yet monomeric; however, the monomer is actually formed by internal duplication of an ancestral protein and is therefore effectively a dimer. An appealing model for this allosteric mechanism is presented by Aleshin et al. [19].)

Allostery is relatively uncommon. There is no need for every enzyme to be allosterically regulated. Many just need to go as fast as possible, and for a series of enzymes in a pathway it is only necessary to regulate one of them because this will act as a volume control to alter the **flux (*3.3)** throughout the

***3.3 Flux**
A detailed mathematical description of metabolic pathways arose in the late 1960s from Kacser and others, generally known as metabolic control analysis, which is described in A. Cornish-Bowden, Fundamentals of Enzyme Kinetics, 3rd ed. London: Portland Press, 2004. It shows for example that the importance of an enzyme in determining the flux (flow) through a pathway can be characterized by its control coefficient, which must be between 0 and 1. All enzymes have non-zero control coefficients, implying that there is, strictly speaking, no such thing as a 'rate-determining step' because all enzymes play some part in the flux through a pathway, and for example increasing the rate of an enzyme that has a large control coefficient alters the control coefficients of other enzymes in the system. However, some enzymes are much more important than others, and in many real cases there may well be one enzyme that is effectively rate-limiting.

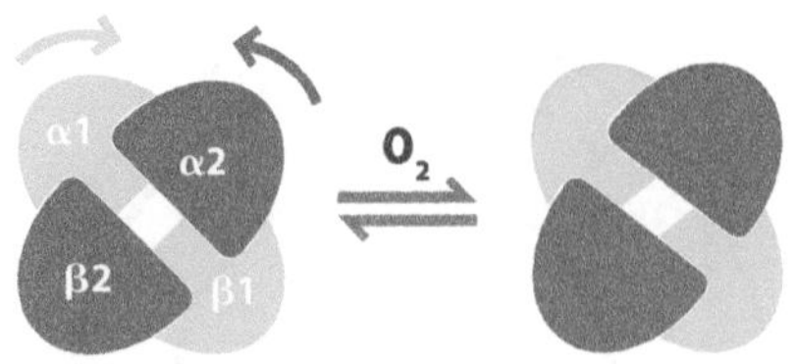

FIGURE 3.10
Hemoglobin is an α_2–β_2 tetramer, organized as two αβ pairs. Binding of oxygen leads to the rotation of one αβ pair relative to the other.

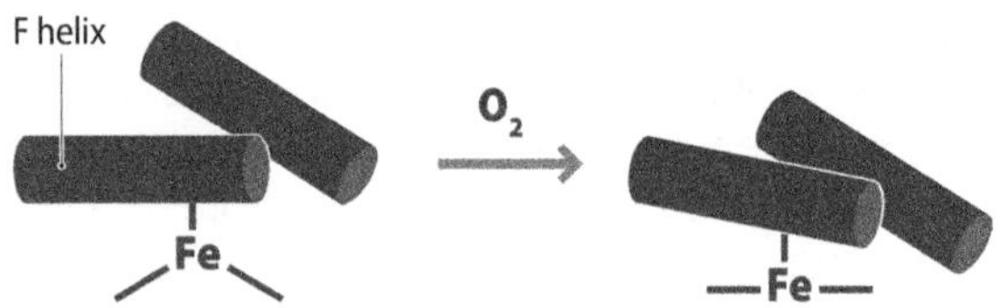

FIGURE 3.11
In reduced hemoglobin, the Fe^{2+} ion at the center of the heme is slightly too large to fit inside the heme ring, and therefore sits just above it, causing a bending of the heme plane. In oxidized hemoglobin, the Fe–N bonds are slightly smaller and the Fe^{3+} ion is able to fit within a flat heme. Thus, oxidation leads to a movement of the iron atom by about 0.6 Å. Because the heme is ligated to a histidine in helix F, this pulls helix F down by a similar amount (1.0 Å). The helix is a relatively rigid rod, and thus the movement is transmitted by a lever-like action to larger changes elsewhere, affecting the packing of helix F against its neighbors, particularly helix C.

whole pathway. I am not aware of any systematic survey of which enzymes are allosterically controlled, but Laskowski et al. [18] note that many allosteric enzymes come at the start of biosynthetic pathways, particularly for amino acids, which presumably are important because of the high throughput of these pathways. This makes a lot of sense, because biosynthetic pathways represent considerable metabolic effort and it is therefore useful for the cell to regulate them carefully. We shall see in Chapter 10 that the large and complicated machines called multienzyme complexes also tend to occur only in the most expensive biosynthetic pathways, presumably for the same reason.

3.2.2 Hemoglobin is the classic example of allostery

The function of hemoglobin is to bind to oxygen in the lung, where oxygen pressure is high, and transport it to muscle, where oxygen pressure is low. The structure of hemoglobin, determined by Max Perutz (*11.13) in 1960, and the subsequent structure-based explanation of allostery, remain the best model for allostery. Hemoglobin is roughly an oligomer of four single-domain myosins assembled into a tetrahedron. It is actually composed of two α chains and two homologous β chains assembled in two αβ pairs; that is, the interactions between an α and its β partner (α_1–β_1 and α_2–β_2) are stronger and more numerous than any others (**Figure 3.10**). The binding of oxygen causes a conformational change in the hemoglobin tetramer, with one αβ pair rotating with respect to the other by about 15° (see Figure 3.10). This is often described using detailed molecular structures, but actually the change is more easily (and hopefully more instructively) considered using a very simplified model.

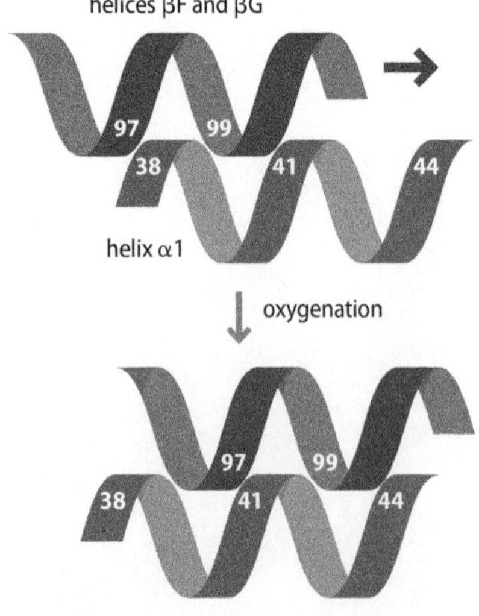

FIGURE 3.12
The interface between helix C on chain α1 and the helix F/G turn on chain β2 fits together as knobs into holes in one of two ways. The shifting of the F helix moves it out of one arrangement into the other. Each arrangement is stabilized by hydrogen bonds.

The binding of oxygen to the iron atom in the center of a heme ring leads to a very small change in the geometry of the heme ring, basically because an oxygen-bound Fe is slightly smaller than an oxygen-free Fe and fits more neatly into the center of the heme, allowing it to be flat (**Figure 3.11**). Because this change is then transmitted to the F helix adjacent to the heme, and helices move as rather rigid units (Section 1.3.2), the helix can act as a "lever" to magnify the conformational change, which becomes 2.5 Å at the end of helix F. The tight packing of the interior of the protein means that there are only two stable conformations for the interface. These are governed by the "ridges into grooves" principles described in Chapter 1: once the lever formed by helix F has moved sufficiently far, it tips the helix–helix interface between helix F on subunit α_1 and helix C on subunit β_2 (and the corresponding interface between helix C on subunit α_1 and helix F on subunit β_2) such that it slides into the other allowed position (**Figure 3.12**). In each of these conformations, the structure is

***3.4 Thermodynamic cycle**

Free energy is a so-called state function, meaning that the change in free energy from going from one state to another is independent of the path taken. It can therefore be a very useful device to draw a reaction going by two possible paths. The change in free energy is the same for both paths, which allows useful deductions to be drawn about the changes in one free energy relative to another (see Figure 3.13). Many thermodynamic variables are state functions (for example enthalpy and entropy), meaning that thermodynamic cycles can be constructed for these variables also.

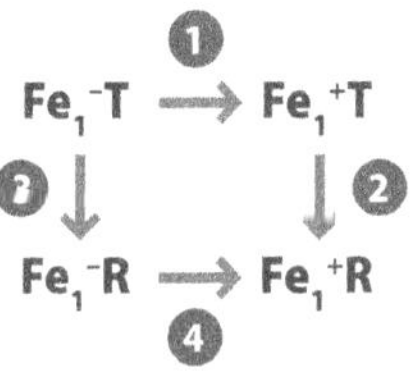

FIGURE 3.13
A thermodynamic cycle to illustrate the energetic link between oxygen binding and conformational change.

stabilized by a set of hydrogen bonds between monomers, which are of course different in the oxygen-free and oxygen-bound states.

There is thus an energetic link between the binding of oxygen to an iron atom in one subunit and the position of the interface between that subunit and its neighbor in the opposite αβ pair (for example α_1 and β_2, which because of the symmetry of the complex is the same for the other interfaces): once oxygen has bound, then the alternative interface geometry is stabilized. We may picture this by using a **thermodynamic cycle (*3.4)**, as shown in **Figure 3.13**. Subunit 1 can either have oxygen bound (Fe_1^+) or not (Fe_1^-) (where the subscript indicates domain 1, and the + and – signs indicate oxygen bound or not), and it can have its interface in one of two conformations, normally described as T (tense) or R (relaxed). It can therefore exist in one of four states. What we have seen is that binding of oxygen (Fe_1^- to Fe_1^+, step 1) makes the change from T to R (step 2) more favorable. However, the interface geometry represented by T and R is of course the interface between two domains. In the *other* domain there has been a change from T to R, but in the absence of oxygen (step 3). Because the total energy change has to be the same whether we go by steps 1 and 2 or steps 3 and 4, this necessarily implies that binding of oxygen to the other domain (equivalent to step 4, but now for domain 2; that is, Fe_2^- to Fe_2^+) is now more favorable. And by symmetry, the same goes for the other oxygens.

Thus, binding of oxygen to one domain makes the binding of oxygen to the other domains more favorable. And this change is cumulative, such that the binding of the third oxygen is even more favorable, and the fourth more again (the fourth oxygen binds between 100 and 1000 times more tightly than the first). This alters the oxygen binding curve from the normal saturation binding behavior seen for myoglobin to a sigmoidal curve (**Figure 3.14b**). This of course has a tremendous physiological advantage, because it means that at the oxygen pressures typically found in the lung, most of the hemoglobin is bound to oxygen, whereas at the oxygen pressures typically found in muscle, most is not bound: therefore hemoglobin can deliver almost its full load of four oxygens. This would not be true for myoglobin, which would deliver far fewer oxygens (Figure 3.14a).

(a)

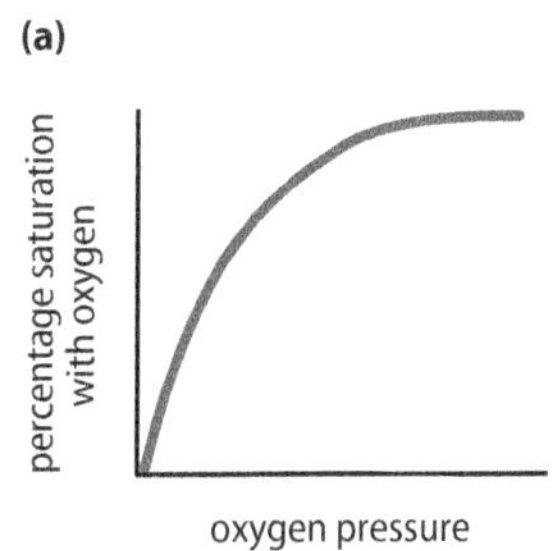

(b)

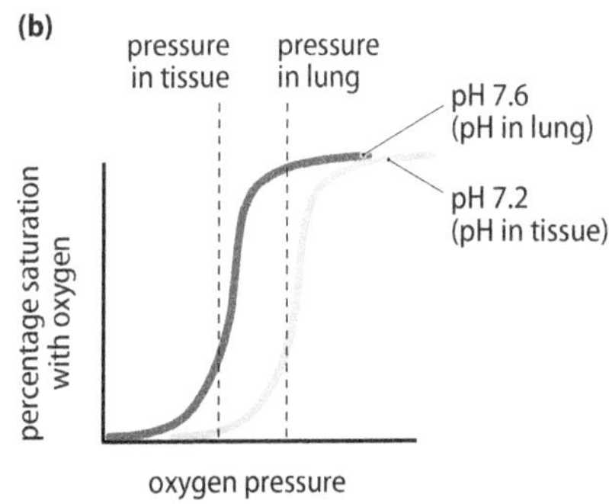

FIGURE 3.14
Oxygen saturation curves for myoglobin and hemoglobin. (a) Myoglobin has a typical noncooperative saturation curve. (b) Hemoglobin has a cooperative, sigmoidal, curve. This means that a relatively small change in oxygen pressure between lung and muscle can lead to a large change in the amount of oxygen bound. Thus, at the higher oxygen pressure in the lung, hemoglobin is almost completely saturated with oxygen (dashed line on right, blue curve), while at the lower oxygen pressure in tissue, almost all the oxygen is discharged (dashed line on left, green curve). The effect of pH is discussed in Section 3.2.3.

3.2.3 Oxygen affinity in hemoglobin is fine-tuned by other effectors

The oxygen binding curve is further modified by two allosteric effectors. It should be no surprise to find that both effectors [bisphosphoglycerate (BPG) and H^+] bind at the interface between domains. BPG binds in the central cavity between all four domains. Protons bind at several sites, some of which are clearly at domain interfaces, although others are less so. Binding of H^+ (that is, a change to a lower pH) causes a shift of the binding curve to the right, known as the Bohr effect (named after its discoverer, Christian Bohr, the father of the atomic physicist Niels Bohr) (see Figure 3.14b). Actively respiring muscle tissue generates CO_2, which dissolves in the blood as bicarbonate, generating H^+:

$$CO_2 + H_2O \rightleftharpoons H^+ + HCO_3^-$$

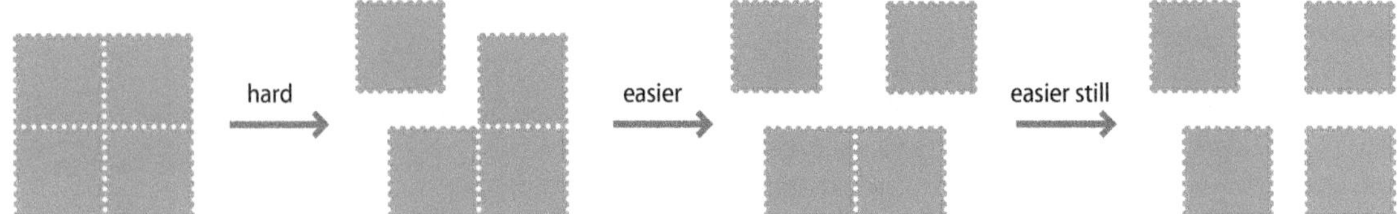

FIGURE 3.15
When removing stamps from a block of four, each successive stamp gets progressively less difficult (by a factor of 2 for each stamp). This forms an analogy with allosteric behavior in oligomeric proteins.

Thus, the pH close to respiring tissues is lower than in the lung. Because of the Bohr effect, this permits a greater amount of oxygen to be delivered to the tissue: almost one more oxygen per hemoglobin tetramer.

BPG decreases the affinity of hemoglobin for oxygen (by binding preferentially to the nonoxygenated form) and is a useful physiological regulator, which among other things is responsible for high-altitude adaptation. At high altitude, the concentration of BPG is increased, inducing lower oxygen affinity, so that more oxygen can be delivered to tissues. BPG is useful for another vital physiological reason as well. A fetus needs to be able to extract oxygen from its mother's blood. Fetal hemoglobin therefore has a higher affinity than maternal hemoglobin for oxygen. This is accomplished because fetal hemoglobin replaces the β subunit by a γ one and has an $\alpha_2\gamma_2$ structure. This form is not regulated to the same extent by BPG and so binds oxygen more tightly.

The allosteric behavior described here, in which binding gets more favorable as more oxygen molecules are bound, was elegantly described in earlier editions of the Stryer Biochemistry textbook (regrettably it has been removed in later editions) as being analogous to removing postage stamps from a block of four stamps (**Figure 3.15**). Removal of the first stamp requires the breaking of two intersubunit interactions and is difficult. Removal of the second stamp requires the breaking of only one interaction and is easier. And breaking the third interaction is again twice as easy because it yields two stamps for one break. Although the analogy is somewhat strained, it is an excellent illustration of the value of oligomeric interactions to produce allostery.

3.2.4 There are two main models for allostery

Models are an attempt to simplify the truth and thereby reach a better understanding by being able to stand back and look at the overall picture without needing to worry about all the details. They are therefore inevitably inaccurate in some details. However, once we appreciate the limitations of the model, they are very powerful aids to understanding. For example, we now know that Newtonian mechanics is a "model," which fails under certain circumstances, most importantly at high speeds and energies. Nevertheless, for almost all purposes it is just as accurate as relativistic calculations, and much easier to understand, which makes it a very useful model.

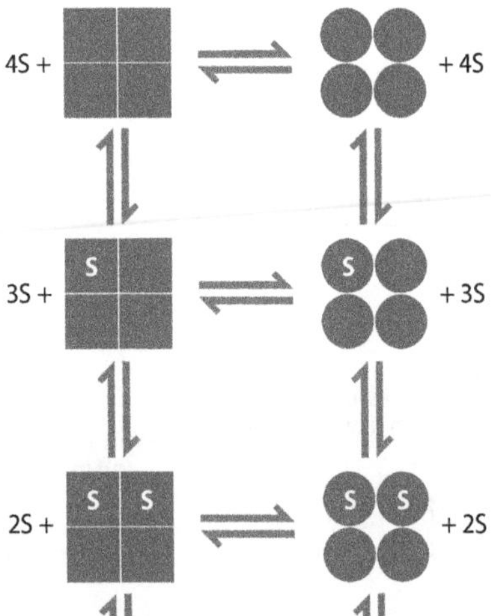

The most important models for understanding allostery were based on hemoglobin, which has a large amount of both structural and kinetic data. The simplest and most elegant was proposed by Monod, Wyman, and Changeux in 1965 [1], in a paper still worth reading. In their model, often called the MWC or symmetric model, hemoglobin is a tetramer of identical subunits, each of which can exist in one of two states: a T (tense) state, which has low oxygen affinity, and an R (relaxed) state, with higher oxygen affinity (**Figure 3.16**). The affinity of the R state is actually similar to that of myoglobin: the allostery arises from the **weaker affinity (*3.5)** of the T state than that of the nonoligomeric myoglobin, because the interactions between the monomers in the T state hinder the binding of oxygen. Hemoglobin is always symmetric, meaning that it contains either four T or four R monomers. In the T state there are more interactions between monomers: this is why it is called Tense, and why it is the preferred conformation in the absence of oxygen. Hemoglobin is in

4S + ⇌ + 4S
3S + ⇌ + 3S
2S + ⇌ + 2S
S + ⇌ + S
T form R form

FIGURE 3.16
The MWC model for the binding of a ligand to a tetrameric protein. S = substrate; T and R represent tense and relaxed, respectively. T has more interactions than R between subunits. It is the major form present in the absence of S. R has a higher affinity than T for S, and therefore in the presence of S the equilibrium shifts to favor the R form.

***3.5 Weaker affinity**
The text notes that the allosteric modification on going from myoglobin to hemoglobin corresponds to a decrease in oxygen affinity in the T state. It is much easier to evolve a decrease in affinity or activity than an increase, and virtually all regulatory interactions correspond to losses of activity rather than increases. Compare for example with Section 9.2, in which the changes in the activity of the RNA hydrolase resulting from its being present within the exosome all result in a decrease in activity.

equilibrium between the T_4 and R_4 states. Binding of oxygen stabilizes the R state, which implies that as each oxygen is added, the overall equilibrium is pushed increasingly toward the fully bound R_4 state.

This model is very simple: it requires only three parameters (the affinity for each state plus the relative stabilities of the two states), it explains almost all the experimental data, and it explains in a simple way how activators and inhibitors work (by binding and stabilizing the R or T state respectively).

However, the model is not a universal explanation. It presupposes that intermediate or asymmetric states cannot exist, and such states have clearly been seen. (And hemoglobin is not in fact made of four identical subunits, so in detail the MWC model *must* be "wrong.") And it is unable to explain *negative* cooperativity, the situation in which the addition of each substrate molecule decreases the affinity for further substrates. Negative cooperativity is not seen in hemoglobin, but it is in other proteins.

Soon afterward, Koshland et al. [20] proposed an alternative model, often called the *sequential* or KNF model. This model does not require the quaternary structure to be symmetric but instead models the binding as a sequential process (**Figure 3.17**) in which the binding of oxygen to one monomer changes it from T to R in an **induced fit** process. In so doing it alters the affinity of each monomer for its neighbors, and therefore also for the substrate. Thus, again, it

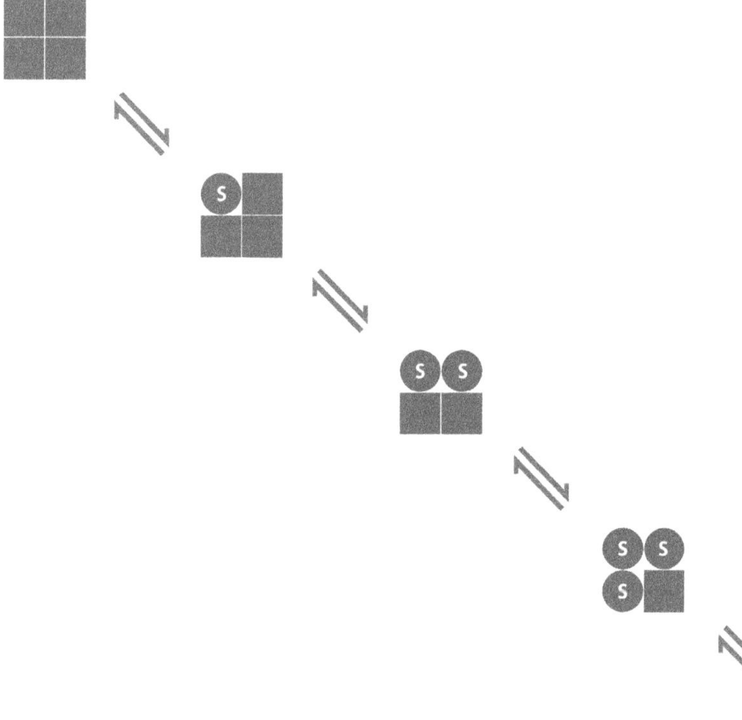

FIGURE 3.17
The sequential or KNF model for ligand binding to a tetrameric protein.

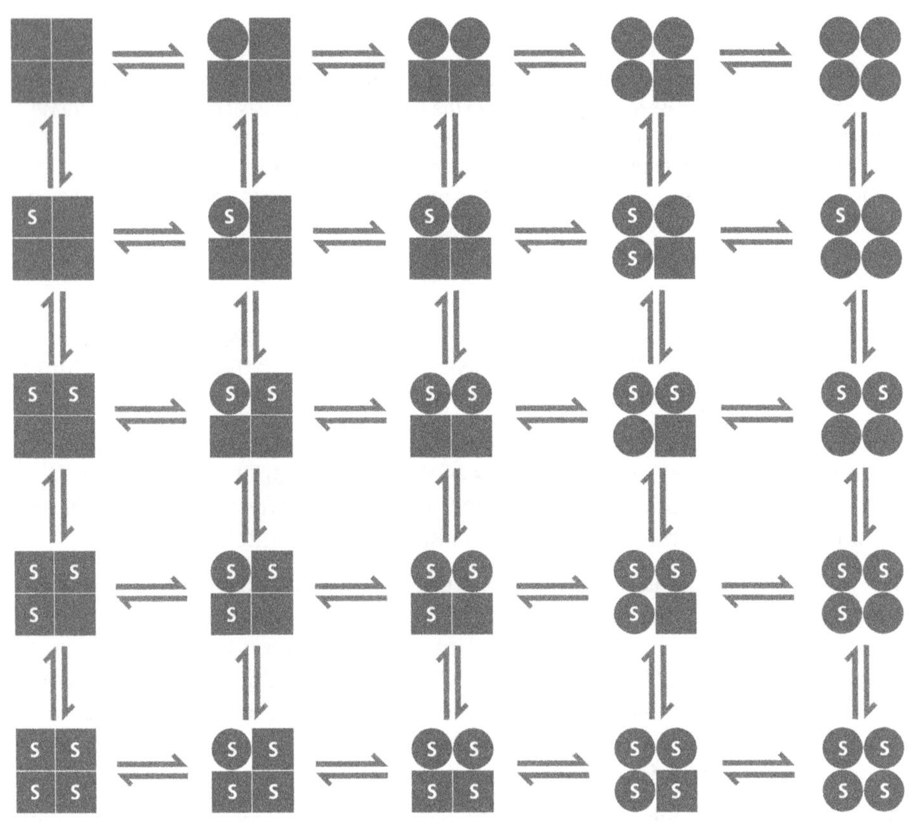

FIGURE 3.18
A general scheme for ligand binding to a tetrameric protein, which includes both the MWC and the sequential models as special cases. The MWC model has only the left and right columns, and the sequential model has only the diagonal.

accounts for the increased affinity as more oxygen-binding sites are occupied. The model is not quite as simple and elegant as the MWC model and requires one more parameter. It can, however, account for negative cooperativity, and in such systems it works much better than the MWC model. It also explains the numerical results for hemoglobin just as well as the MWC model.

It was soon pointed out, by Eigen on the award of his Nobel Prize in 1967 [21], that both these models are simplifications of a general scheme (**Figure 3.18**). The MWC model uses only the left and right edges of this scheme, and the KNF model uses only the diagonal.

Does this mean that the more complete model (see Figure 3.18) is better? There are certainly situations in which a more complete model is necessary. However, for most purposes the MWC model is the easiest to understand and explains the data perfectly well; it is therefore for most purposes the "best," even though we need to be aware that it has failings.

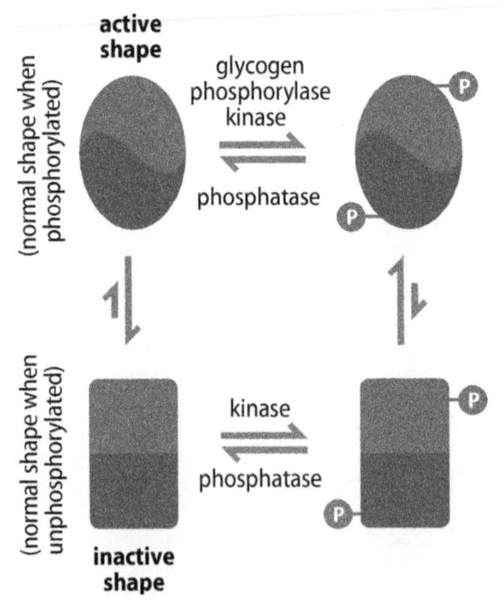

FIGURE 3.19
Phosphorylation of glycogen phosphorylase alters the equilibrium between two possible conformations and favors the active conformation, thereby markedly increasing its activity. Glycogen phosphorylase is phosphorylated by phosphorylase kinase, which in turn is phosphorylated and activated by cAMP-dependent protein kinase. This kinase is activated by an increase in cAMP concentrations, as a consequence of the activation of a receptor triggered by epinephrine (adrenaline). Thus, epinephrine leads to an increase in intracellular glucose, in readiness for the "fight or flight" response.

3.2.5 Glycogen phosphorylase is another good example of allostery

The other classic example of allostery is glycogen phosphorylase, usually known just as phosphorylase. Glycogen forms the major storage of glucose in the body; it is degraded and reformed continuously to keep the level of glucose in the blood constant. Phosphorylase removes glucose units from glycogen and is thus vital for the regulation of blood glucose levels [22]:

$$\text{glycogen}_n + \text{P}_\text{i} \rightleftharpoons \text{glycogen}_{n-1} + \text{glucose-1 phosphate}$$

The enzyme is regulated by phosphorylation: the enzymes glycogen phosphorylase kinase and phosphatase control the phosphorylation state. In the unphosphorylated form it has low activity, whereas in the phosphorylated form it is active. This occurs because of a conformational change as a result of phosphorylation, which can be represented in outline as in **Figure 3.19**. We

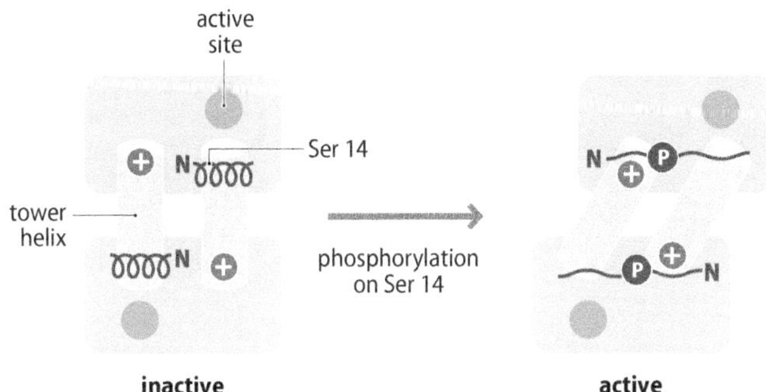

FIGURE 3.20
The mechanism for activation of glycogen phosphorylase by phosphorylation on Ser 14. The phosphorylation creates a negative charge, which is attracted to a positive charge on the tower helix of the other monomer. This leads to unwinding of the N-terminal helix and a reaching out of the N terminus toward the tower helix, and hence to a shearing of the tower helices. This uncovers the active site and activates the enzyme.

can, however, understand the role of homo-oligomerization in mediating this change by looking at the structure in slightly more detail, as shown in **Figure 3.20**.

Phosphorylase is a dimer. It has a helix, usually called the "tower helix," protruding from one monomer and contacting the other, this being the main contact between the two monomers [23]. The residue that is phosphorylated is serine 14, which is in a somewhat disordered region in the unphosphorylated state. The unphosphorylated form is inactive for what is essentially a geometric reason: entry to the active site is blocked by the tower helix. When serine 14 is phosphorylated, it gains a negative charge, and the N terminus of the protein becomes ordered such that the negatively charged phosphate shifts by some 36 Å and binds to a positively charged region of the protein surface, which is a positively charged patch on the tower helix of the opposite monomer (see Figure 3.20). The result is that the tower helices swing over and cause the monomers to slide. There is thus a major change in the quaternary structure, which also requires some changes to the tertiary structure, notably the interface between the two monomers. The tower helices are now tilted, which allows access to the active site and produces an active enzyme.

Phosphorylase is in reality more complicated than this simple picture, because its activity is also regulated by several allosteric effectors (Problem 4).

When viewed at this level, the mechanism is a rather simple geometrical change. This is something of an oversimplification, but it does provide the essential details. A similar mechanism *could* occur in a monomeric protein, as long as such a protein had two independently moving parts or domains (**Figure 3.21**). And indeed this is essentially what does happen in many protein systems that are regulated by phosphorylation, such as the two-component signaling systems discussed in Chapter 8. So what is the advantage of being dimeric? Or, to put it in a more provocative but slightly anthropomorphic way, why did nature bother going to the trouble of evolving a complicated dimeric system when a monomer would work? The answer of course is that although a monomeric system works, it is not as effective. In particular, by being a monomer it is less *cooperative*, and it therefore has a less sharp transition between being "on" and "off." In the next section, I discuss the reason behind this important behavior.

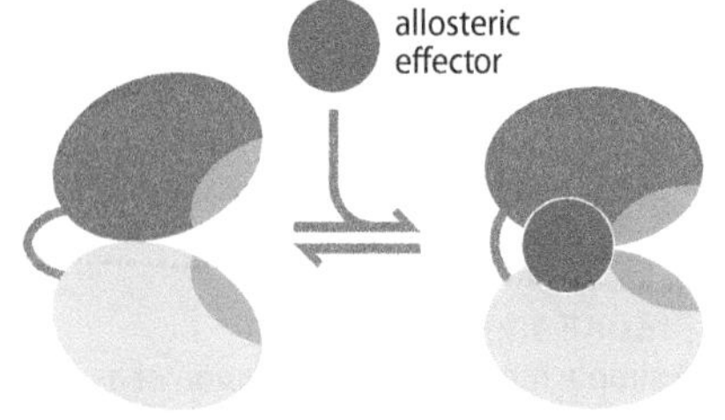

FIGURE 3.21
An allosteric mechanism in a two-domain (but monomeric) enzyme.

The allosteric change in the classic example aspartate transcarbamoylase is similar to the one described here for phosphorylase, but it is more complicated in that it involves a lever-like rotation of two symmetric halves, each composed of trimers of catalytic and regulatory domains [24].

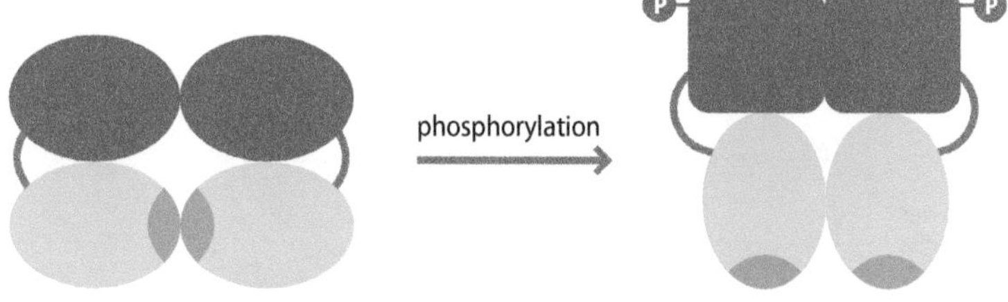

FIGURE 3.22
A typical bacterial two-component signaling system, as discussed in Chapter 8. Phosphorylation of the top domain (the receiver domain) leads to a change in conformation, which leads to dissociation of the bottom domain (the effector domain). This domain is then able to dimerize and bind to DNA.

3.3 COOPERATIVE BINDING OF DIMERS TO DNA

3.3.1 Cooperativity can be understood by using thermodynamics

Why is a dimeric system potentially more cooperative, as stated above? Let us consider a simple dimeric system, such as the typical bacterial two-component system (Section 8.2.8). The essential parts of the downstream end of this signaling system are shown in **Figure 3.22**. In the inactive state, the effector domain is unable to bind to DNA because it is sterically prevented from doing so. Phosphorylation of the receiver domain leads to dissociation of the effector domain, which is then able to bind to DNA. This system is a dimer; why does this make it more cooperative, and what exactly does cooperative mean in this context?

Let us simplify this model even further to the essential elements, as in **Figure 3.23**. There is an inactive conformation (square) and an active one (circle): the protein can either be a monomer or a dimer. To a first approximation, the energy required for the conformational change (steps 1 and 3) is the same for two monomers as it is for one dimer, because the main interface affected by the phosphorylation is between the two domains within one monomer. The total change in free energy from top left to bottom right must be the same whichever way round the loop we go, or in other words

$$\Delta G_1 + \Delta G_4 = \Delta G_2 + \Delta G_3$$

If $\Delta G_1 = \Delta G_3$, as just suggested, this must also mean that $\Delta G_4 = \Delta G_2$. In other words, there is little advantage to being dimeric so far.

The big advantage comes in the next step, the binding to DNA. Here we have two possibilities, as in **Figure 3.24**. The difference between the two binding events is that in the dimer the two monomeric domains are held together in a more or less rigid manner, whereas in the monomers they are not. The favorable binding energy between protein and DNA is the same for both association events (ignoring any binding between what is drawn in Figure 3.24 as a red linker between the domains). There is, however, a large *unfavorable* energy associated with the binding, which is the loss of translational and rotational entropy of the protein as it binds (Section 2.2.11). This energy penalty is present for the binding of both monomer and dimer, but it is much larger in the monomer because two molecules that were previously independently mobile both lose their mobility and therefore their entropy, whereas the two molecules in the dimer have already lost most of their relative mobility and therefore have less entropy to lose. In the language used in the section Entropy and enthalpy (*3.1), the number of ways of arranging the two protein molecules in the dimer is already much lower than it is for the two monomers, because the two molecules in the dimer are constrained to be adjacent in the same orientation. The difference in the entropy loss in these two cases, when translated into an energy, can be as large as 50 kJ mol^{-1} [25], although it is in fact usually much smaller.

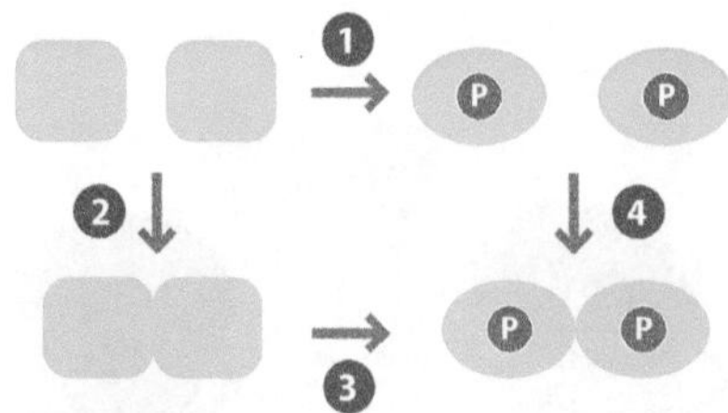

FIGURE 3.23
A simple model for the activation of a dimeric protein by phosphorylation. This is shown as a thermodynamic cycle (*3.4). The dimeric phosphorylated protein can then go on to bind to DNA.

How does this help explain cooperativity? The phosphorylation step acts essentially as a switch: the signal should be off in the unphosphorylated state, and on in the phosphorylated state. That is, we need to have the monomer not

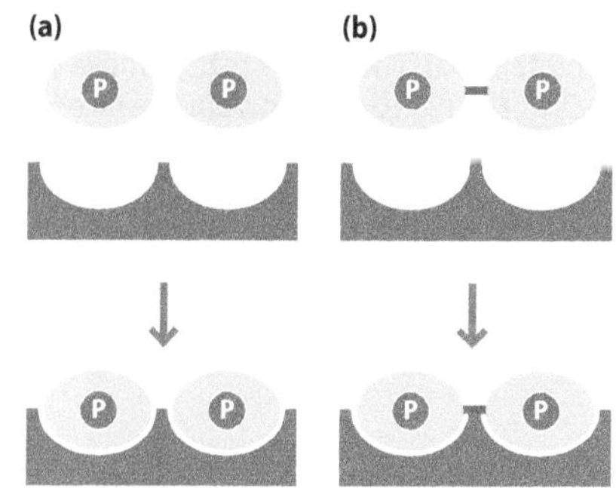

FIGURE 3.24
The binding of a protein to a dimer-binding site on DNA. In (a), the protein binds as two independent monomers; in (b) the protein is already associated together as a dimer before it binds.

binding, but the dimer binding. If sometimes the monomer binds, or the dimer dissociates, the switch is less effective. The extent of binding is measured by the equilibrium constant between free and bound, which is given by

$$K = \exp(-\Delta G/RT)$$

where ΔG is here the difference in free energy between free and bound (see Binding and dissociation constants, *5.5). A difference in the free energy between monomer and dimer of 50 kJ mol^{-1} equates to a massive factor of 10^9 in the binding constant. In other words, binding as a dimer can make the switch better (that is, it can decrease the fraction of monomer that is bound or the fraction of dimer that is not bound) by up to 10^9. This is what is meant by *cooperative*: binding together as a dimer makes the switch more effective. Actually, real systems do not need to be nearly as perfect as this (and they are not), but the potential is there.

3.3.2 Sequence-specific binding to DNA is a problem

As noted in Section 3.1.3, one advantage of being oligomeric is observed in proteins that bind to DNA. The main problem faced by proteins that bind to DNA is how to recognize their specific binding site. The normal way in which DNA-binding proteins find their binding site is to bind nonspecifically and then search in one dimension along the DNA (Section 4.2.3). A typical bacterial genome is about 10^7 base pairs long, and some proteins bind to only a single site within the genome. This implies that for it to spend as much time bound to the single specific site as it does to all the nonspecific sites put together, it needs to be able to discriminate between specific and nonspecific sites with a discrimination on the order of 10^7; and if it is required to spend most of its time bound to the specific site, it needs to do even better. The protein therefore has to be able to bind to DNA strongly enough for it not to dissociate often but weakly enough for it to be able to move rapidly along the DNA, but then once it has recognized its specific site it needs to be able to bind tightly and stay there sometimes for extended periods (tens of minutes, for example): that is, it needs to have a slow off-rate. A typical DNA-binding protein steps from one base to the next at about 10^6 base pairs per second, and is attached mainly by electrostatic interactions. It binds very weakly to nonspecific sequences, with a K_d of about 1–2 mM, whereas binding to the specific site has an effective K_d of about 1 pM, or 10^9 times stronger, in agreement with the analysis above [26]. Thus, there is a major problem in how to recognize and bind to a *specific* DNA sequence much more strongly than a weak one. In this, the protein is greatly helped by the cooperativity arising from dimerization.

The specificity of a protein–DNA interaction can be increased by making the recognition interface longer. If we assume that DNA sequences have the four bases occurring at random, a given DNA sequence of length n occurs with a probability of 4^{-n}. Thus a 3-mer occurs on average with $p = 0.0156$ or 1.6×10^5 times in a genome of 10^7 bases; a 4-mer with $p = 0.0039$ or 4×10^4 times; a 6-mer 2500 times; an 8-mer 150 times; a 10-mer 10 times; and a 12 mer less than once. In other words, to recognize a single site within the genome we need to be able to recognize a sequence of at least 12 bases. In reality it is not quite so simple; nevertheless, this gives a good feel for roughly the size of interface needed. Recall that most proteins recognize DNA within the major groove, and that the major groove has a periodicity of 10 base pairs in B-DNA. Thus, for a protein to bind to a 12-mer it needs to wrap itself around the DNA

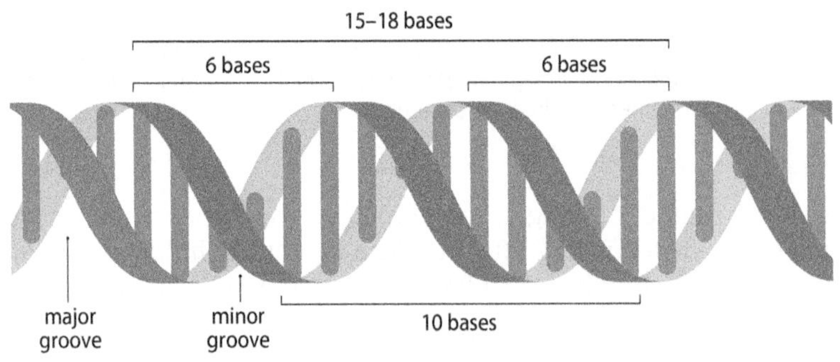

FIGURE 3.25
A protein that binds to the major groove of B-DNA and has to come in from one face and bind at least 12 bases in total must bind two groups of 6 bases, and therefore has to span a total of 15–18 bases.

for rather more than one turn. Although this is of course possible, and many proteins do indeed do this, such a geometry requires the protein to have a very twisted binding site, which is difficult to evolve.

Most DNA-binding proteins therefore adopt a different strategy: they recognize the DNA from a single direction (**Figure 3.25**). This removes the difficulty of having to wrap around the DNA, but it creates a different problem: how to make a protein that can span what now needs to be a longer DNA sequence. This is roughly 12 base pairs, but split into two six-base sequences with an intervening (minor groove) sequence, so roughly 16 bases long altogether. Of course, the solution is to use a dimeric protein. This solves two problems in one. The first problem is the one just discussed: how to cover a contiguous sequence of about 16 bases with one protein. The second is how to maximize the speed of the sequence search. We shall see in Chapter 4 that a large rigid interface requires a significant amount of dehydration in the transition state, which makes binding a very slow process. It is much faster if it is possible to have a flexible interface that is able to form in a series of smaller steps, effectively zippering onto the DNA. If a DNA-binding protein binds as a dimer (particularly when in fact it binds as one monomer first followed by a second monomer, or at least as a dimer with a flexible joint, so that it does not need to bind as a single rigid block), then the interface for a dehydrated transition state is half as large, and so the binding is much more than twice as fast. In other words, rapid scanning and tight specific binding are greatly facilitated by having dimeric proteins with a flexible interface (**Figure 3.26**).

If even greater specificity is required, the protein adds extra domains that interact with adjoining DNA sequences, in a manner reminiscent of the discussion of binding specificity in Section 2.2.5. For example, homeodomains are transcription factors that control development, and there are many that have similar sequences. Yeast mating type is determined by the transcription factors Mcm1 or Mat **a**1, but alone they bind DNA very weakly. Strength and specificity of binding is achieved by their both being bound as a heterodimer (in different geometries) to the homeodomain Mat α2 [27], which also binds DNA weakly as a monomer but recognizes specific DNA sequences and binds tightly as a heterodimer, in two quite different geometries (**Figure 3.27**).

FIGURE 3.26
A protein that binds to DNA as a dimer. Nonspecific binding is weak and therefore the protein often dissociates on one side or the other. This weakens the attraction between protein and DNA further, and allows the protein to scan rapidly along the DNA. Binding to a specific sequence is stronger and therefore both monomers within the dimer bind to DNA simultaneously.

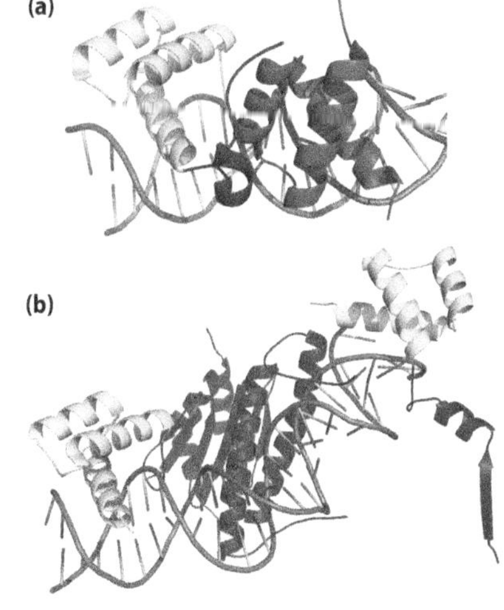

FIGURE 3.27
Binding of yeast mating factor MAT α2 to DNA. In both examples, MAT α2 (green) binds in the same way, with its C terminal helix binding into the major groove. (a) MAT α2 binds as a heterodimer with MAT **a**1 (purple) upstream of haploid-specific genes and represses them. A C-terminal extension of MAT α2 (red), which is unstructured in the free protein, forms a helix and docks against a hydrophobic surface of MAT a1 (PDB file 1le8). (b) MAT α2 can also bind as a heterotetramer with MCM1 (red) upstream of a-specific genes and repress them. In this complex, an N-terminal extension of the left-hand MAT α2 (blue), which again is unstructured in the free protein, forms a two-stranded sheet that binds to MCM1 by b-strand extension (*6.2). MCM1 binds as a dimer, and the other monomer of MCM1 also binds to MAT α2, although this time the N-terminal extensionof MAT α2 forms a different conformation and binds to an adjacent protein in the crystal (PDB file 1mnm). In both examples, the heterodimeric complex bends the DNA significantly; monomeric complexes are not bent.

Binding as a dimer creates another feature, already described in Chapter 2: because in most cases the protein dimer has rotational symmetry, so does the DNA that it recognizes. In other words, the DNA sequence is palindromic (**Figure 3.28**). In this context, it is worth noting that most DNA *in vivo* is methylated, on specific adenines and cytosines. DNA methyltransferases fall into two main categories: Type I and Type II. The Type I enzymes all recognize symmetric sequences and are themselves symmetric [28]. By contrast, Type II methyltransferase can recognize asymmetric hemimethylated DNA, and are not necessarily symmetric.

3.3.3 The *trp* repressor recognizes DNA by hinge bending

The *trp* repressor represents a classic example of a DNA-binding protein. It is also a classic example of transcriptional regulation. It controls the operon for the synthesis of tryptophan in bacteria such as *E. coli*. In the absence of tryptophan it does not bind DNA; it therefore allows expression of the genes needed to make tryptophan. However, in the presence of tryptophan there is a conformational change in the protein, allowing it to bind to DNA and halt the synthesis of tryptophan.

The structure of *trp* repressor is a helix–turn–helix (HTH). This is a well-known DNA-binding fold, in which the first helix is called the recognition helix and binds into the major groove of DNA (**Figure 3.29**). For the dimer to be able to bind to DNA, the recognition helices in the two monomers need to be approximately parallel and 34 Å apart, because this is one complete turn of the DNA helix. In the absence of tryptophan, the two helices are twisted toward one another, and the protein does not bind to DNA.

When tryptophan is present, it binds in a cavity between the recognition helix and helix 3. This causes a conformational change, such that the two recognition helices in the two monomers are now in the correct orientation to bind (**Figure 3.30**). In other words, the dimer interface stays the same, but what has altered is the angle between two of the helices.

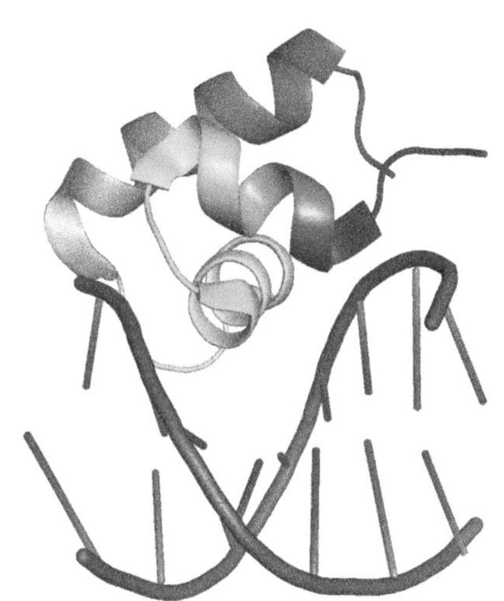

FIGURE 3.29
A helix–turn–helix repressor, actually the *lac* repressor headpiece (PDB file 2pe5). The recognition helix (the first helix of this motif) is in green. There is a recognition "code" for sequence-specific recognition of DNA by this motif [49], but it is still not robust enough for us to be able to predict the DNA sequence from the protein sequence.

5' A T G A **C G** T C A T 3'
3' T A C T **G C** A G T A 5'

FIGURE 3.28
The DNA binding site for the yeast transcription factor GCN4 is palindromic: the four-base sequence on one strand (bold) is the same as the four-base sequence on the other strand read in the opposite direction.

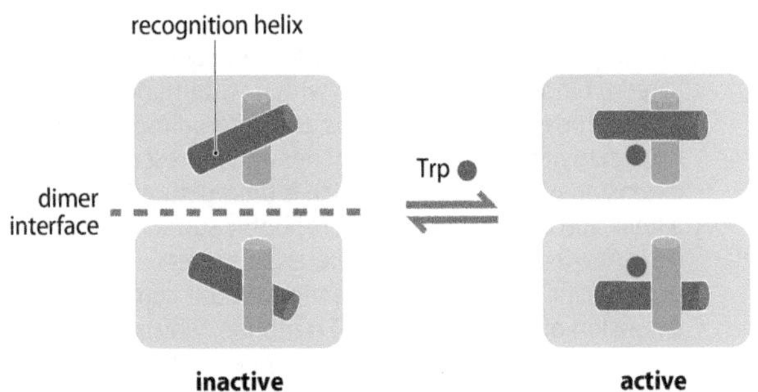

FIGURE 3.30
The DNA-binding domain of *trp* repressor has its recognition helices at an incorrect angle to bind to DNA in the absence of tryptophan. Addition of tryptophan causes the interhelical angle to alter, making the recognition helices parallel and approximately 34 Å apart, and therefore able to slot into two successive major grooves.

3.3.4 CAP recognizes DNA by rotation around the dimer interface

Catabolite gene activating protein, or CAP, is familiar to most biochemistry and genetics students because of its involvement in the regulation of the *lac* operon, which prevents the expression of genes that are needed in the transport and metabolism of lactose when the preferred energy source, glucose, is present (**Figure 3.31**). Expression of the *lac* operon is directly controlled by the binding of the *lac* repressor, which in turn is regulated by the binding of lactose, and also by the synthetic chemical isopropyl β-D-thiogalactoside (IPTG). However, this "switch" is not particularly efficient: the amount of expression differs by only a factor of 20 as a result. Additional selectivity mechanisms are therefore needed. One of these is the creation of two additional operator sites, one within the *lacZ* gene and one overlapping the CAP-binding site (not shown here). The other is the creation of a second regulatory mechanism, CAP. It is possible that the additional mechanisms are embellishments: elaborations evolved to provide additional specificity. However, there are many DNA-binding proteins with much less leaky switches, and it seems more reasonable to suggest that the weak "primary" switch provides a means of permitting some background activity together with a flexible response to different cellular environments.

In the presence of glucose, the protein CAP (also known as CRP) is unable to bind to DNA. However, when the concentration of glucose is decreased, the concentration of cAMP in the cell increases; the cAMP binds to CAP, alters its conformation, and causes it to bind to DNA at a site upstream of the operator. This complex forms a binding site for RNA polymerase, which is thus able to bind to the *lac* operon promoter and start transcription of the operon (see Figure 3.31d).

The structural change in this case is again to allow the two recognition helices to become parallel and correctly spaced. However, the mechanism that allows this to happen is different from that in *trp* repressor. CAP has two domains:

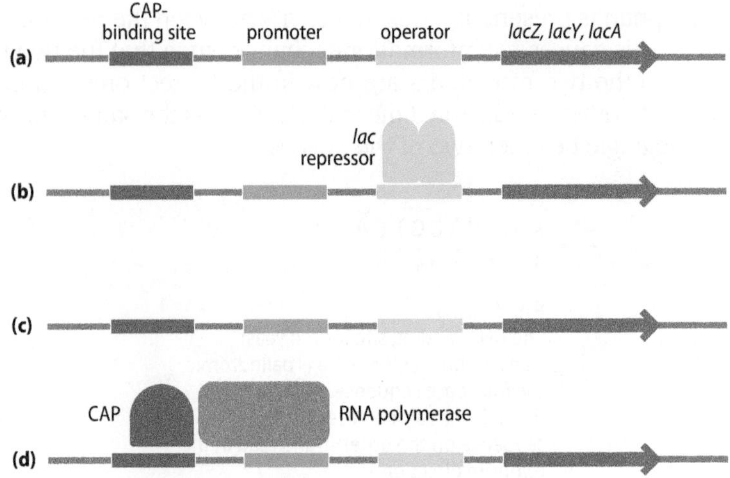

FIGURE 3.31
Operation of the *lac* operon. This operon expresses the three genes required for the metabolism of lactose in *E. coli*; it is only required when lactose is present *and* when glucose (the preferred substrate) is not. (a) The operon consists of a CAP-binding site, followed by a promoter (to which RNA polymerase binds), followed by the operator site and then the three *lac* genes *lacZ*, *lacY*, and *lacA*. (b) In the absence of lactose, *lac* repressor binds to the operator site and prevents expression of the genes. (c) In the presence of lactose but a high concentration of glucose, the repressor does not bind. However, expression of the *lac* genes is weak because the RNA polymerase does not bind well to the promoter. (d) In the presence of lactose and the absence of glucose, the concentration of cAMP is high, so cAMP binds to the CAP protein. This enables CAP to bind to DNA. It increases the affinity of RNA polymerase for the promoter, greatly enhancing the transcriptional rate.

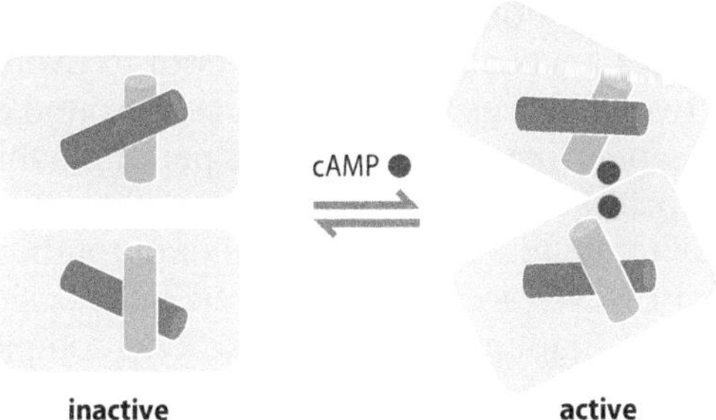

FIGURE 3.32
Inactive CAP protein is similar to inactive *trp* repressor, in that the DNA recognition helices are at an incorrect orientation to bind. CAP consists of two domains: a dimerization domain (green), and a DNA-binding domain, represented by the two helices. Binding of cAMP leads to a change in the dimer interface, which puts the recognition helices parallel.

the HTH DNA-binding domain, and a second domain that forms a homodimer interface and binds to cAMP. (In passing, note again that each domain has a clear function, and the overall function of the protein is made up simply from a combination of the separate functions of its domains.) On binding of cAMP, the structure of the interface is altered, so as to bring the recognition helices into the correct orientation (**Figure 3.32**).

These two DNA-binding proteins—*trp* repressor and CAP—thus achieve the same overall effect: a mechanical rotation of a rigid helix to create a parallel dimeric arrangement. However, they do it by different mechanisms: *trp* repressor has a rotation of the helix within a fixed framework, whereas CAP has a fixed framework but a rotation of the dimer interface. There is a useful analogy with spectacles, which also form a symmetric dimer (actually one with a plane of symmetry rather than an axis of symmetry). A pair of spectacles has to be arranged to fit onto a face. Normally, the way this is done is analogous to what happens in *trp* repressor: the angle within each symmetric half is altered to give the correct geometry (**Figure 3.33a**). It would of course be possible to do something analogous to CAP, and keep the angle between lens and earpiece fixed but vary the angle of the dimer interface (Figure 3.33b). Note that a "monomer" of spectacles binds extremely poorly, as do monomers of *trp* repressor or CAP: the dimer binds much more strongly than two separate monomers. This is a nice example of the importance of cooperativity in binding.

CAP binding causes a marked bending of DNA, by a total of almost 90°. This achieves several results, one of which is an extra degree of specificity, in that the bending requires a sequence (TG) capable of being bent in this way. The bending probably also assists in recognition of the complex and helps to bring other protein components together.

3.3.5 DNA recognition by a symmetric leucine zipper

The *trp* repressor and CAP are both HTH proteins. Another major class of protein that recognizes DNA is the leucine zipper, represented here by GCN4. Like most DNA-binding proteins, GCN4 is a classic example of the "one domain represents one function" paradigm: it has a leucine zipper domain that binds DNA, and a separate domain that leads to translational activation.

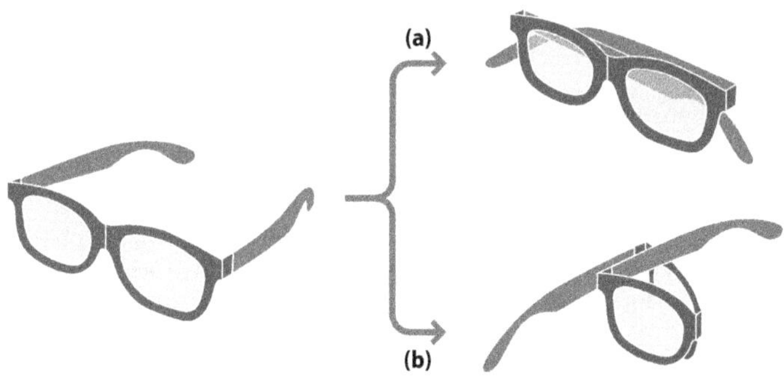

FIGURE 3.33
Two ways of folding up a pair of spectacles. (a) The normal way, by bending at the interface between lens and earpiece. (b) An alternative way, by bending at the bridge.

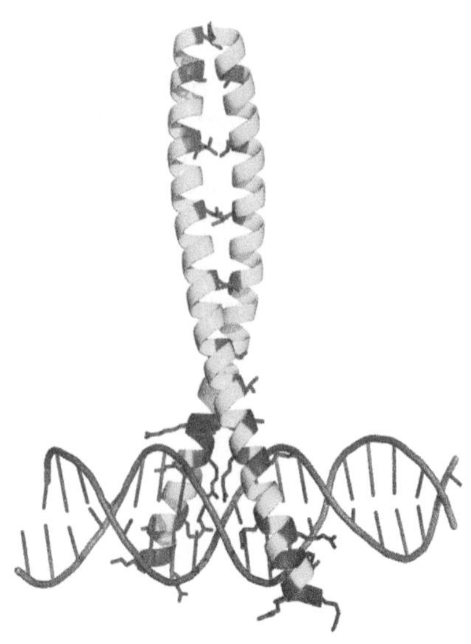

FIGURE 3.34
A coiled coil: the Fos–Jun AP1 dimer (PDB file 1fos). Leucine residues are highlighted in magenta, and basic residues in blue. The sequences of Fos and Jun are similar, so this coiled coil is almost symmetrical, and looks very similar to the symmetrical GCN4 dimer.

The leucine zipper forms a **coiled-coil** structure, an extremely common way of producing oligomers because it is so easy both to evolve and to regulate [29]. It consists of two α helices, which are wound round each other (**Figure 3.34**). The standard α helix has 3.6 amino acids per turn, which means that the rotation per amino acid is 360/3.6 = 100°. It therefore takes 18 amino acids (18 × 100° = 5 × 360) to make a complete number of rotations and get back to where you started. However, in a coiled coil, the helix is slightly twisted, and consequently it takes 3.5 amino acids to make one complete turn, or seven amino acids to make two turns. This explains why coiled-coil structures typically have seven-residue amino acid repeats (a "heptad repeat") (**Figure 3.35**), as suggested originally by Francis Crick (*1.13). In particular there is a leucine every seven residues, conventionally at position *d* in the sequence.

The leucines are all on the same face of the helix, and two helices can wind around each other such that the leucines from the two helices sit side by side and form a hydrophobic line up the interface (see Figure 3.34). Usually the residues at position *a* in the heptad are also hydrophobic and fill the intervening contacts.

GCN4 is a transcription factor found in yeast; it regulates the expression of a large number of genes concerned with the biosynthesis of amino acids in response to starvation. The sequence of the DNA-binding region is often described as bZip, meaning basic zipper. There is a basic region of about 20 amino acids that contains seven arginines and one lysine, which is followed by the leucine zipper region. When free, the basic region is unstructured; however, when it is bound to DNA it forms a continuous helix in which the basic regions from the two monomers bind to DNA in the major groove, gripping the DNA in what is known in wrestling as a scissors grip (see Figure 3.34) [30].

Because GCN4 is a symmetric dimer it binds to symmetric DNA, with the sequence **GATGA**CG**TCATC**. The sequence that is recognized is the outer five base pairs (in bold), with the inner two base pairs acting as a spacer between them. It also binds to a pseudopalindromic sequence shorter by one base pair: d(**GATGA**C**TCATC**).(**GATGA**G**TCATC**). This is symmetric in the outer sequences but not in the center. GCN4 is therefore able to bind to DNA structures of different lengths. In the two structures the coiled coil has a slightly different orientation where it connects into the basic region, and the DNA is slightly bent. As a consequence, the interactions between protein and DNA are identical.

(a) *a b c d e f g*
ϕ L

(b)

```
Fos  KRRIRRERNKMAAAKSRNRRRE
Jun  KAERKRMRNRIAASKSRKRKLE

Fos  LTDTLQAETDQLEDEKSALQTEIANLLKEKEKLEFILAAH
      abcdefgabcdefgabcdefgabcdefgabcdefga
Jun  RIARLEEKVKTLKAQNSELASTANMLREQVAQLKQKVMNH
```

FIGURE 3.35
Dimerization of the Fos–Jun coiled coil. (a) Coiled-coil sequences typically consist of heptad repeats: residue *a* is hydrophobic (represented by ϕ), whereas residue *d* is leucine. (b) The sequences of Fos and Jun. The N-terminal sequences are very basic (magenta), whereas the C-terminal sequences contain leucines at every *d* position (brown) and hydrophobic residues at most *a* positions (cyan). Also highlighted are two occurrences of oppositely charged pairs of residues at the spatially close *e* and *g* positions on the two strands (green), discussed below (see also the locations of *e* and *g* in Figure 1.37 and the helical wheel in Figure 1.38).

GCN4 represents one of a large class of DNA-binding proteins. The three leucine zippers GAL4, PUT3, and PPR1 are bZip proteins, which contain a short linker between the basic and zipper regions (respectively the blue and magenta regions of Figure 3.34). They all bind to the pseudopalindromic sequence CGG-N_x-CCG, but with different numbers of bases in the center. For GAL4, PUT3, and PPR1, x is respectively 11, 10, and 6 bases long. This difference is very simply related to the length and structure of the linkers. In GAL4 the linker is in an extended conformation, in PUT3 it has a β-sheet conformation (which is slightly less extended), and in PPR1 it forms a β-turn and is therefore much shorter [31]. The bZip structure is thus very suitable as a DNA-binding structural motif because it is readily adapted to different biological needs.

3.3.6 DNA recognition by a heterodimeric leucine zipper

The dimers that we have looked at so far have all been strictly homodimeric, namely having two copies of the same protein. The next level up in complexity, and also in regulatory potential, is heterodimers, but in this section the two proteins are different but related. We shall see that the additional complexity allows the system to have more complex functions, and in particular to be open to more complex regulation. This means that homodimers tend to be observed in prokaryotes, whereas hetero-oligomers tend to be observed in eukaryotes, and particularly the higher eukaryotes. Thus, for example, leucine zippers in yeast tend to be homodimers, whereas in higher eukaryotes they tend to be heterodimers.

In particular, heterodimers often form the key elements of eukaryotic transcription factors. Because eukaryotic transcription factors regulate the expression of genes in eukaryotes, misfunctions are often linked to cancers and developmental problems, and hence they are key targets for biomedical research. This tends to make them interesting proteins, and they have received a good deal of attention.

Fos and Jun are eukaryotic transcription factors; they bind to DNA, regulating a wide variety of functions. They are both activated by MAP kinase cascades, as described in Chapter 8. Malfunctions in such cascades mean that signals are either switched on or switched off for much longer than they should be. The activity of Fos is normally switched off by rapid degradation of Fos mRNA. However, Fos can also be expressed from retroviral DNA (a form known as v-Fos), in which form it does not contain the 3′-end degradation signal. Therefore, v-Fos leads to a longer-lasting signal and ultimately to cancer. The normal cellular form of Fos, c-Fos, is therefore described as a proto-oncogene, because it is closely related to a true oncogene (*8.8).

Similarly, the normal cellular form of Jun, c-Jun, is efficiently ubiquitylated and is therefore proteolytically digested and removed, whereas a form of Jun encoded by a virus, v-Jun, is not. Jun is thus also a proto-oncogene.

Fos and Jun form the two halves of a heterodimeric transcription factor known as AP1 [32]. These proteins contain a bZip sequence and a transactivation domain (**Figure 3.36**). Thus, in common with almost every other protein discussed in this book, their activities can be dissected into discrete domains—one domain to bind to DNA, one domain to dimerize, and a third domain to recognize the transcription machinery—which merely need to be plugged together appropriately to work correctly. (This feature forms the rationale behind the two-hybrid screening method, discussed in Chapter 11.) The basic regions of Fos and Jun have similar sequences (see Figure 3.35), which means that the DNA sequences that they recognize are similar and that the sequence recognized by the Fos–Jun dimer is approximately palindromic. Indeed, the dimer is so symmetric that in the crystal structure of AP1 bound to DNA there

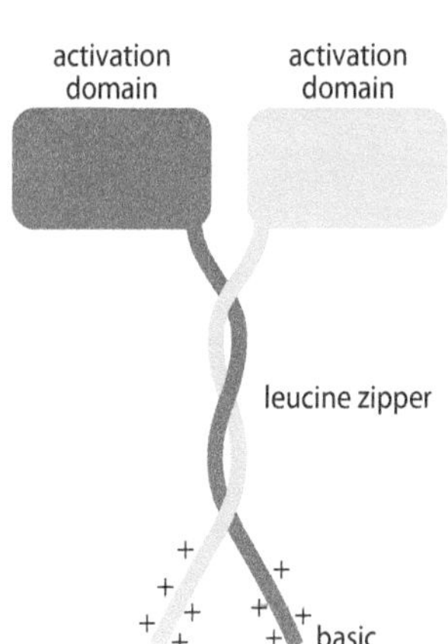

FIGURE 3.36
The structure of the Fos–Jun AP1 heterodimer consists of a basic-leucine zipper motif (bZip) followed by an activation domain.

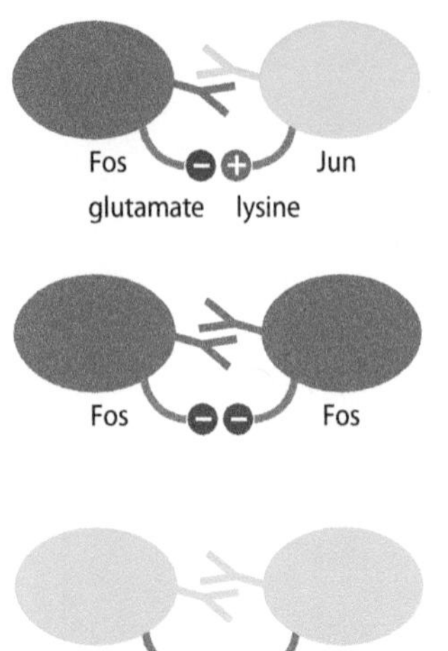

FIGURE 3.37
The Fos–Jun heterodimer is stabilized by favorable electrostatic interactions between side chains (see Figure 3.34, residues highlighted in green), whereas Fos–Fos and Jun–Jun homodimers are destabilized.

are two molecules of each, and Fos–Jun recognizes DNA in both orientations [33]. The basic regions of Fos and Jun are virtually identical to that of GCN4, meaning that they all interact with DNA in exactly the same way.

Fos and Jun both contain leucine zipper sequences, which means they can in principle form homodimers as well as heterodimers. However, they also have charged residues within the sequence, as shown in **Figure 3.37**: Fos has a negatively charged glutamate, which is located opposite a positively charged lysine in Jun. This makes the Fos–Jun heterodimer stable. The Fos–Fos homodimer has two glutamates facing each other and is therefore unstable; it does not dimerize. The Jun–Jun homodimer has two lysines. Because the lysine side chain is longer and more flexible than that of glutamate, the side chains can move apart, so that they do not destabilize the homodimer. The Jun–Jun homodimer is therefore reasonably stable, although it is only one-tenth as stable as the heterodimer [34].

Full activity of AP1 requires the presence of both transactivation domains; it is therefore only obtained for the heterodimer. Both Fos and Jun can be phosphorylated, by different kinases, which regulates the assembly of the dimer. Having a heterodimer rather than a homodimer permits a greater degree of control over the activity of the dimer than would be possible for a homodimer.

3.3.7 Max and Myc form a heterodimeric zipper with alternative partners

A slightly higher degree of complexity, and therefore also of regulatory control, is demonstrated by the DNA-binding proteins Max and Myc [35]. These are also bZip proteins, although they contain an additional HTH domain between the basic and zipper regions, and are described as b-HTH-Zip proteins. As mentioned in Chapter 1, the HTH motif in these proteins is not used for DNA binding but for dimerization, by which two HTH domains dimerize to form a single four-helix bundle. It is a tempting speculation that evolution picked on this mechanism because it already "knew" about HTH motifs, because of their DNA-binding ability, but tinkered with them to turn them into something with a quite different function. The proteins bind to the palindromic DNA sequence CACGTG.

The first component of this system to be identified was Myc, which is a proto-oncogene like Fos and Jun. It was hard to identify what its function might be, because overexpression of Myc had very little phenotypic effect. The reason for this behavior became clear once it was shown that Myc functions only as a heterodimer with the protein Max [36]. (This is the protein's name in yeast: in mice the orthologous protein is attractively known as Myn.) The system is more complex than Fos–Jun, because Max is also able to form an alternative heterodimer as well as a homodimer (**Figure 3.38**).

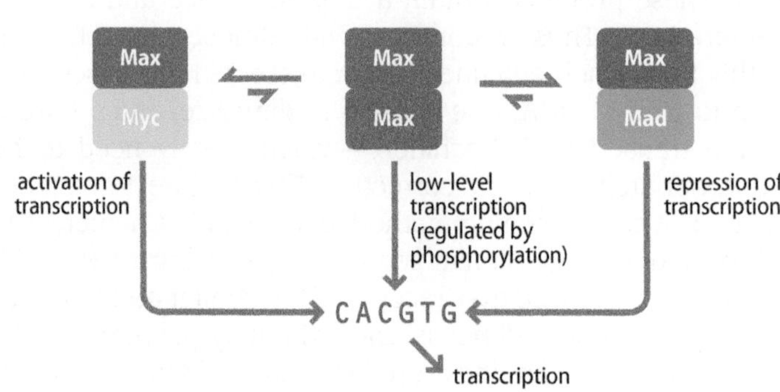

FIGURE 3.38
The Max–Myc–Mad system. According to the authors of the original paper, the naming of Mad (the second protein in the Mad–Max heterodimer) has nothing to do with a movie of the same name, but comes from "Max dimerization."

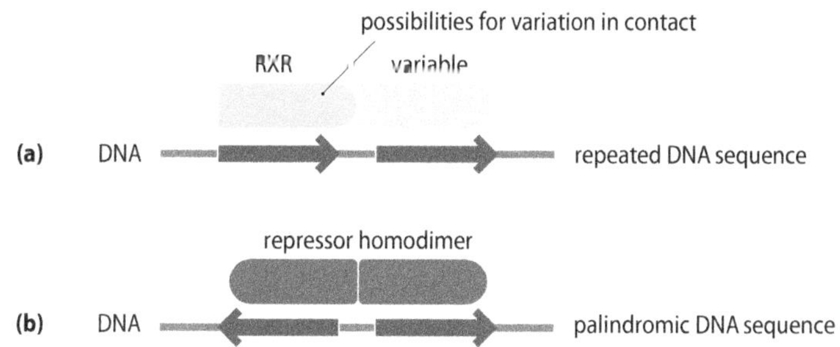

FIGURE 3.39
Comparison between lipophilic hormones and bacterial repressors. (a) The lipophilic hormone receptors bind to DNA in a tandem orientation. The upstream partner is always the 9-*cis* retinoic acid receptor, RXR. The spacing between the two binding sites is variable (see Table 3.1), allowing variation in the contact between the two receptors. (b) By contrast, bacterial repressors (and coiled-coil transcription factors) bind a palindromic sequence.

The Max homodimer binds to CACGTG and leads to low-level expression, in what can be considered a basal background level. Max itself has no activation domain, and significant effects on transcription therefore occur only when it forms a heterodimer with Myc or Mad, which do have activation domains. In a similar way to Fos and Jun, the heterodimer (with either Myc or Mad) is more stable than the homodimer, implying that low concentrations of Myc or Mad will produce a heterodimer. Also similarly to Fos and Jun, dimerization affinity is determined by ionic interactions within the zipper region [37–39]. Formation of the heterodimer with Myc leads to an activation of transcription. However, Mad binds more tightly than Myc to Max, meaning that in the presence of both Myc and Mad, the Mad heterodimer is formed in preference (explaining why the overexpression of Myc has little effect) [40]. The Mad–Max heterodimer leads to repression of transcription.

Thus, suitable changes in the concentrations of Myc and, particularly, Mad lead to large changes in gene transcription and provide a greater degree of regulation than in the simpler Fos–Jun system. The levels of Max tend to be fairly constant, but the concentrations of Myc and Mad are closely regulated throughout the cell cycle; other interactions also effectively remove them and thus further regulate activity [41]. Further regulation is provided, as in Fos and Jun, by phosphorylation of the different components.

3.3.8 DNA recognition by a tandem dimer

In most signaling systems, binding of a signal to a cell surface receptor leads to an intracellular effect such as a phosphorylation, which is then converted via a signaling pathway to an effect on a DNA-binding protein and thus ultimately to an alteration in gene expression. The lipophilic hormones are different (and arguably more primitive). They comprise steroid hormones, retinoic acid, vitamin D, thyroid hormone, and others, which are all lipophilic enough to be able to diffuse across the cell membrane and therefore interact directly with intracellular components. They also tend to have more sustained long-term developmental effects on the cell than most of the more typical hormones that bind to cell surface receptors. Typically, these lipophilic hormones bind to an intracellular receptor and convert it from an inactive conformation to an active one. The active conformation then binds to DNA [42].

The receptors bind DNA in a head-to-tail geometry (**Figure 3.39a**). This means that the DNA sequences they bind to have tandem repeats, in contrast with the typical bacterial repressor-binding sequences that bind to *trp* repressor and CAP, which are palindromic because they are bound by dimers with rotational symmetry (Figure 3.39b). In common with many (but by no means all) DNA-binding proteins, the dimeric receptor is not stable in solution: it dimerizes only when bound to DNA.

The receptors bind such that the upstream element is always the 9-*cis*-retinoic acid receptor RXR, which binds to the sequence AGGTCA. The downstream component also binds to the sequence AGGTCA, but it is determined by the spacing between the two repeated sequences, as shown in **Table 3.1**. This use

TABLE 3.1 Lipophilic hormone receptor targets

RXR.RXR	AGGTCA*n*AGGTCA
RXR.RAR	AGGTCA*nn*AGGTCA
RXR.VDR	AGGTCA*nnn*AGGTCA
RXR.TR	AGGTCA*nnnn*AGGTCA
RXR.RAR	AGGTCA*nnnnn*AGGTCA

RXR = 9-*cis* retinoic acid receptor; RAR = all-*trans* retinoic acid receptor; VDR = vitamin D1 receptor; TR = thyroid hormone receptor. (Based on a Table in F. Rastinejad, *Curr. Opin. Struct. Biol.* 11:33–38, 2001. With permission from Elsevier.)

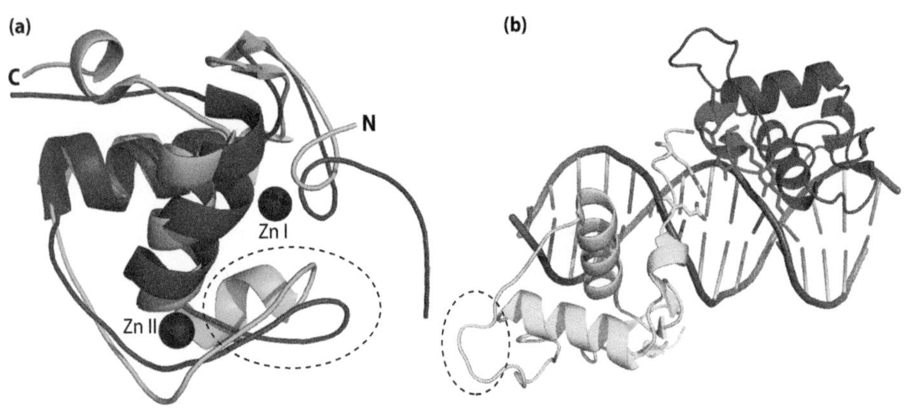

FIGURE 3.40
Cooperative binding of RAR and RXR to DNA. (a) Comparison of the structures of the retinoic acid receptor (RAR), free (red; PDB file 1hra) and bound to DNA and to the RXR receptor (cyan; PDB file 1dsz). The two zinc atoms are shown for the free receptor. There is a significant structural change in the Zn II region, indicated by the dashed line. (b) The structure of the complex of RXR (left, green) and RAR (right, red). Residues Arg 75 and Gln 72 of RXR are indicated by sticks in the center of the diagram pointing toward the DNA. The Zn II region of the RAR receptor is shown in magenta. Many of the residues in this region make contacts with DNA, and others make contact with the RXR receptor. Note how the restructuring of the Zn II region not only permits specific recognition of DNA but also forms the interface with RXR. In other complexes, RXR contacts its downstream partner using quite different residues; for example, in the TR complex it binds in the opposite orientation and uses mainly Arg 38, Arg 48, and Arg 52 (indicated by the dashed line). (a, Based on a figure in F. Rastinejad, *Curr. Opin. Struct. Biol.* 11:33–38, 2001. With permission from Elsevier.)

of one fixed domain and one variable one is reminiscent of the "tools with common parts" idea discussed in Chapter 2. DNA is of course helical, which means that an increased spacing between binding sites not only puts them further apart, it also rotates them around the helix. Therefore the interface between the two receptors has a quite different geometry depending on the spacing.

Earlier in this chapter, I discussed why it is that binding of a dimer leads to cooperativity. One aspect not discussed there is that on binding of the first monomer (here the upstream RXR receptor), the binding interactions stabilize the conformation of the "head" (downstream) end of the receptor, as shown in **Figure 3.40**. The head interacts with the bases between the two AGGTCA sequences, implying that the structure of the head depends on the spacing of the two sequences. This means that the binding interactions that determine which downstream element binds to the second AGGTCA sequence include interactions with the RXR head, which help to increase the specificity for the correct receptor. In other words, binding of the two receptors to DNA is cooperative with their binding to each other [43]. This of course provides a strong rationale for why the two receptors do not dimerize in solution. Such a mechanism no doubt has a role in the binding of most dimers, but it is clearer here because the binding site is more obviously specific to the downstream element.

3.4 ISOZYMES

In this chapter I have been emphasizing that heterodimers have an advantage over homodimers because they offer an additional level of regulation. This final section presents a rather different example of hetero-oligomers, and one that is so classical and "metabolic" that some years ago it used to form a core part of every biochemistry course; however, now it tends to get crowded out by other, no doubt more worthy, topics, namely isozymes. Isozymes are pairs of enzymes that have different sequences but catalyze the same reaction in the same host. They are often found in different tissues. An interesting class of isozymes comprises oligomeric enzymes composed of two different monomers that can be present in varying ratios, which give the resultant enzyme different properties. They therefore have some functional commonality with systems such as Myc–Max, in that different combinations of related monomers have different functional consequences. The strong association of different monomers to form an isozyme is selected by evolution and functionally useful [4].

Lactate dehydrogenase is an important enzyme in animals, because it forms part of a vital shuttle system. In muscle, energy is obtained by glycolysis.

However, it often happens that complete oxidation of glucose through the TCA cycle and the electron transport chain is not necessary, and all that happens is oxidation as far as pyruvate. This produces rather little energy in comparison with electron transport, but it has the major advantage that the carbon skeleton of glucose is not used up and can therefore be recycled: once pyruvate has been converted to acetyl-CoA the carbons cannot be converted back into glucose and end up either being oxidized completely or (regrettably, for many humans at least) being converted into fatty acids. Therefore, as long as the muscle is not required to work hard, it makes more sense to obtain the required energy by oxidation of glucose to pyruvate and then return the pyruvate in the blood to the liver, where it is reduced back to glucose again. This is of course wasteful of energy, but it keeps all the serious metabolism within the liver and gives the body greater flexibility for central metabolism.

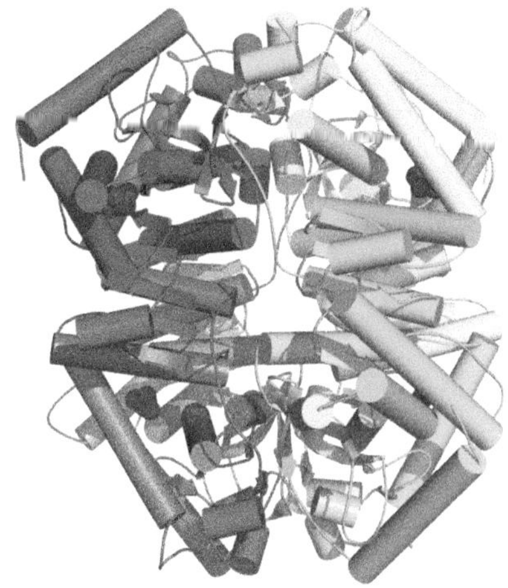

FIGURE 3.41
The structure of human lactate dehydrogenase (PDB file 1i10). This is the M_4 form: the L_4 form is very similar. The structure includes four oxamates (red balls), which are substrate-like inhibitors.

Pyruvate is not a very stable compound. So instead the body performs one further reaction on pyruvate, by reducing it to lactate:

$$\text{pyruvate} + \text{NADH} \rightleftharpoons \text{lactate} + \text{NAD}^+$$

It is lactate that is actually recycled in the blood back to the liver, and in the liver the first step is to oxidize it back to pyruvate in the reverse reaction.

Both these reactions are performed by lactate dehydrogenase. The reaction as drawn above is a reduction, and it runs in the forward direction in reducing environments such as skeletal muscle and liver, but in the reverse direction in more oxidizing environments such as heart muscle, which has a greater supply of oxygenated blood. As is well known, an enzyme can only make a reaction faster, it cannot change the position of equilibrium. However, in the heart there is a different form of the enzyme, which both has a higher affinity for NAD^+ and undergoes inhibition by pyruvate. Both of these reasons tend to make the heart enzyme better at catalyzing the aerobic conversion of lactate to pyruvate, in contrast with the enzyme found in liver and skeletal muscle, which is better at catalyzing the anaerobic conversion of pyruvate to lactate. Other organs in the body require gradations of activity somewhere between these two extremes. The body has evolved an elegant and economical system for achieving graded enzyme activities, by constructing the enzyme as a tetramer (**Figure 3.41**). The four subunits can be one of two forms, known as H (heart) and M (muscle), implying that five possible tetramers can occur, which are known as *isozymes*. The different isozymes are found in different ratios in different tissues, as shown in **Figure 3.42**. They are also found in different ratios during development of the fetus, with increasingly higher proportions of the aerobic H form and less of the anaerobic M form in the heart as the fetus moves from the anaerobic womb to the aerobic world outside [44].

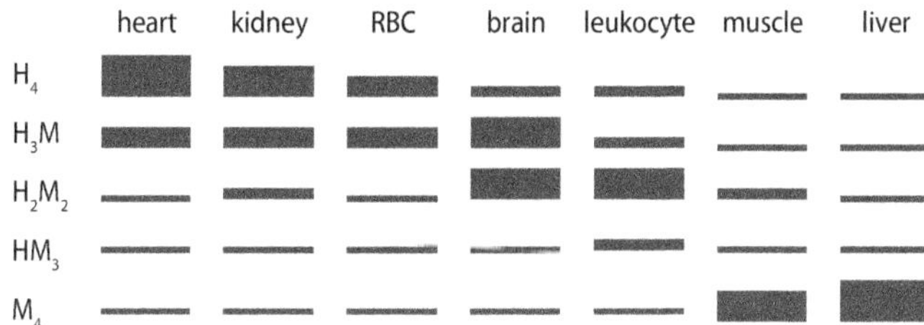

FIGURE 3.42
Relative concentrations of different isozymes of lactate dehydrogenase in human tissue. RBC, red blood cells. (Redrawn from K. Urich, Comparative Animal Biochemistry Berlin: Springer Verlag, p. 542, 1990. With kind permission of Springer Science + Business Media.)

3.5 SUMMARY

Approximately two-thirds of human proteins are oligomers. There are several reasons, of which the two chief ones are:

1. In an oligomer, the active site is almost always in a cleft between two domains, which protects it, allows access to the active site to be opened and closed easily, and makes allosteric effects easier.

2. Oligomerization provides a much larger binding surface and therefore provides increased cooperativity, especially in binding to DNA.

Almost all allosteric enzymes are oligomers, mainly because an allosteric effector can bind close to the oligomer interface, and therefore allosteric effects are relatively simple to evolve in an oligomer. The classic example of allostery is hemoglobin, in which the structural change is almost entirely at the domain interfaces. Two main models have been proposed that both match the data well. Of these, the MWC symmetry model is intuitively simpler and is a good description most of the time, but it does not fit all cases, for example negative allostery. The "true" situation must be described by a much more complicated model; however, this model offers little understanding of what is going on.

Many proteins only bind to DNA as dimers, and most bind as symmetric head-to-head dimers. The increased affinity from cooperativity can be explained as a smaller unfavorable entropy of binding. The *trp* repressor and CAP bind to DNA in very similar ways using a helix–turn–helix, but use different structural changes to achieve it. Many other proteins bind using a basic region followed by a leucine zipper coiled coil, which provides considerable flexibility in the spacing of the two halves of the DNA-binding site. In eukaryotes these proteins tend to be heterodimers, which provides further opportunities for regulation. The lipophilic hormone receptors bind differently again, using a head-to-tail arrangement that allows a range of downstream receptors to bind, also depending on the spacing between the two binding sites.

Isozymes form a set of oligomeric proteins with different subunit compositions, and thereby allow a gradation of activity that can be used differently in different places in the body.

3.6 FURTHER READING

An elegantly written review by Goodsell and Olson [45] discusses symmetry in oligomeric structures and approaches many of the topics discussed here but from a different point of view: it therefore forms a good contrast.

Introduction to Protein Structure, by Branden and Tooze [46], is particularly good on DNA-binding proteins. There are good accounts of hemoglobin in most biochemistry textbooks.

The paper by Monod, Wyman, and Changeux [1] remains a classic that is well worth studying. For more detail on the thermodynamic arguments presented here, see Williams et al. [25]: this is not an easy read but is very clearly written.

Isozymes are discussed in all major biochemistry textbooks, especially older ones.

3.7 WEBSITES

http://3did.irbbarcelona.org/ *Interacting domains.*

http://3dcomplex.org/ *Three-dimensional complexes.*

http://cluspro.bu.edu/login.php *ClusPro (docking).*

http://consurftest.tau.ac.il/ *ConSurf (functional regions in proteins).*

http://dunbrack.fccc.edu/ProtBuD/ *ProtBuD (identifies asymmetric units).*

http://nic.ucsf.edu/asedb/ *Alanine scanning energetics protein hotspots.*

http://prism.ccbb.ku.edu.tr/prism/ *PRISM (protein interactions).*

http://viperdb.scripps.edu/ *VIPERdb (virus structures).*

http://www.bioinformatics.sussex.ac.uk/protorp/ *PROTORP (protein interfaces).*

http://www.boseinst.ernet.in/resources/bioinfo/stag.html *Proface (several programs).*

http://www.ebi.ac.uk/thornton-srv/databases/cgi-bin/valdar/scorecons_server.pl *Scorecons (scores residue conservation).*

http://www.piqsi.org/ *PiQSi (quaternary structures database).*

3.8 PROBLEMS

Hints for some of these problems can be found on the book's website.

1. Phosphofructokinase is an early enzyme in glycolysis, and one of the classic examples of allosteric control, because it is the major regulatory enzyme in glycolysis. It is activated by ADP and GDP and inhibited by phosphoenolpyruvate (PEP). The structural mechanism that controls this allostery is described by Schirmer and Evans [47]. Study that paper, and describe (at the sort of level used here for hemoglobin and glycogen phosphorylase) how the allosteric change works. In particular, you should discuss (a) the quaternary structure of the enzyme, (b) how the quaternary structure changes on activation from the T state to the ADP-activated R state, (c) the role of Glu 161 and Arg 162 in this transition, explaining why the substrate fructose 6-phosphate is bound more tightly in the R state.

2. It is often suggested that one advantage of homo-oligomerization is that it keeps the genome size small. What kind of evidence would support this hypothesis? Is there any evidence to show that it is in fact true?

3. Chapter 1 shows that structure is conserved much more than sequence, and that function is conserved somewhere in between; in other words, two proteins with 20% sequence similarity are very likely to have the same fold and fairly likely to have the same function. Where does oligomerization state (quaternary structure) fit into this scheme? That is, if a protein is a hexamer, will it be a hexamer in homologs?

4. What are the regulators of glycogen phosphorylase activity? Rationalize their effects.

5. What are the sequences of Fos and Jun? Show the basic regions and the charge–charge interaction described here.

6. Creatine kinase is another example of an isozyme that forms different compositions of oligomers in different tissues. Describe how this affects its function.

3.9 NUMERICAL PROBLEMS

N1. One test of the specificity of DNA binding discussed in Section 3.3.2 is the number of cutting sites for restriction enzymes. A survey of the number of cut sites produced in the *E. coli* genome, using several restriction enzymes, produced between 470 and 1567 cuts per enzyme; the actual numbers for *Bam*HI, *Bgl*I, *Eco*RI, *Eco*RV, *Hin*dIII, *Kpn*I, *Pst*I, and *Pvu*II were 470, 1567, 610, 158, 517, 497, 846, and 1431, respectively. Is this consistent with the probabilities described in Section 3.3.2? Each enzyme recognizes a six-base sequence, and the *E. coli* genome is about 4.72 megabases long. (Data from [48].)

N2. Section 3.2.4 says that the MWC model has only three parameters, whereas the sequential model has four. Check the original paper on the sequential model [20]: is this really true? What did Koshland et al. conclude about the way in which hemoglobin binds oxygen?

N3. Section 3.3.1 says that a difference in the free energy between monomer and dimer of $50\,\mathrm{kJ\,mol^{-1}}$ equates to a factor of 10^9 in the binding constant. Justify this statement. ($R = 8.31\,\mathrm{J K^{-1} mol^{-1}}$; assume that this is done at 25 °C.)

3.10 REFERENCES

1. J Monod, J Wyman & J-P Changeux (1965) On the nature of allosteric transitions: a plausible model. *J. Mol. Biol.* 12:88–118.
2. J Park, M Lappe & SA Teichmann (2001) Mapping protein family interactions: intramolecular and intermolecular protein family interaction repertoires in the PDB and yeast. *J. Mol. Biol.* 307:929–938.
3. NJ Marianayagam, M Sunde & JM Matthews (2004) The power of two: protein dimerization in biology. *Trends Biochem. Sci.* 29:618–625.
4. RN Perham (1975) Self-assembly of biological macromolecules. *Phil. Trans. R. Soc. Lond. B* 272:123–136.
5. CJ Tsai, SL Lin, HJ Wolfson & R Nussinov (1997) Studies of protein–protein interfaces: a statistical analysis of the hydrophobic effect. *Prot. Sci.* 6:53–64.
6. J Janin, S Miller & C Chothia (1988) Surface, subunit interfaces and interior of oligomeric proteins. *J. Mol. Biol.* 204:155–164.
7. J Janin, RP Bahadur & P Chakrabarti (2008) Protein–protein interaction and quaternary structure. *Quart. Rev. Biophys.* 41:133–180.
8. L Lo Conte, C Chothia & J Janin (1999) The atomic structure of protein–protein recognition sites. *J. Mol. Biol.* 285:2177–2198.
9. JH Tomlinson, S Ullah, PE Hansen & MP Williamson (2009) Characterization of salt bridges to lysines in the protein G B1 domain. *J. Am. Chem. Soc.* 131:4674–4684.
10. TV Borchert, KVR Kishan, JP Zeelen et al. (1995) Three new crystal structures of point mutation variants of monoTIM: conformational flexibility of loop-1, loop-4 and loop-8. *Structure* 3:669–679.
11. PJ Baker, AP Turnbull, SE Sedelnikova et al. (1995) A role for quaternary structure in the substrate specificity of leucine dehydrogenase. *Structure* 3:693–705.
12. S Doublié, G Bricogne, C Gilmore & CW Carter (1995) Tryptophanyl-tRNA synthetase crystal structure reveals an unexpected homology to tyrosyl-tRNA synthetase. *Structure* 3:17–31.
13. SJ Riedl & Y Shi (2004) Molecular mechanisms of caspase regulation during apoptosis. *Nature Rev. Mol. Cell Biol.* 5:897–907.
14. AJ Venkatakrishnan, ED Levy & SA Teichmann (2010) Homomeric protein complexes: evolution and assembly. *Biochem. Soc. Trans.* 38:879–882.
15. CM Fraser, JD Gocayne, O White et al. (1995) The minimal gene complement of *Mycoplasma genitalium*. *Science* 270:397–403.
16. V Bichko, P Pushko, D Dreilina et al. (1985) Subtype ayw variant of Hepatitis B virus: DNA primary structure analysis. *FEBS Lett.* 185:208–212.
17. M Bergdoll, MH Remy, C Cagnon et al. (1997) Proline-dependent oligomerization with arm exchange. *Structure* 5:391–401.
18. RA Laskowski, F Gerick & JM Thornton (2009) The structural basis of allosteric regulation in proteins. *FEBS Lett.* 583:1692–1698.
19. AE Aleshin, C Kirby, XF Liu et al. (2000) Crystal structures of mutant monomeric hexokinase I reveal multiple ADP binding sites and conformational changes relevant to allosteric regulation. *J. Mol. Biol.* 296:1001–1015.
20. DE Koshland, G Némethy & D Filmer (1965) Comparison of experimental binding data and theoretical models in proteins containing subunits. *Biochemistry* 5:365–385.
21. M Eigen (1967) Kinetics of reaction control and information transfer in enzymes and nucleic acids. *Nobel Symp.* 5:333–369.
22. LN Johnson & M O'Reilly (1996) Control by phosphorylation. *Curr. Opin. Struct. Biol.* 6:762–769.
23. LN Johnson (1992) Glycogen phosphorylase: control by phosphorylation and allosteric effectors. *FASEB J.* 6:2274–2282.
24. KL Krause, KW Volz & WN Lipscomb (1985) Structure at 2.9-Å resolution of aspartate transcarbamoylase complexed with the bisubstrate analog *N*-(phosphonoacetyl)-L-aspartate. *Proc. Natl. Acad. Sci. USA* 82:1643–1647.
25. DH Williams, E Stephens, DP O'Brien & M Zhou (2004) Understanding noncovalent interactions: ligand binding energy and catalytic efficiency from ligand-induced reductions in motion within receptors and enzymes. *Angew. Chem. Int. Ed.* 43:6596–6616.
26. RB Winter, OG Berg & PH von Hippel (1981) Diffusion-driven mechanisms of protein translocation on nucleic acids. 3. The *Escherichia coli lac* repressor–operator interaction: kinetic measurements and conclusions. *Biochemistry* 20:6961–6977.
27. C Wolberger (1999) Multiprotein-DNA complexes in transcriptional regulation. *Annu. Rev. Biophys. Biomol. Struct.* 28:29–56.
28. GG Kneale (1994) A symmetrical model for the domain structure of type I DNA methyltransferases. *J. Mol. Biol.* 243:1–5.
29. RA Kammerer (1997) α-Helical coiled-coil oligomerization domains in extracellular proteins. *Matrix Biol.* 15:555–565.
30. D Pathak & PB Sigler (1992) Updating structure-function relationships in the bZip family of transcription factors. *Curr. Opin. Struct. Biol.* 2:116–123.
31. JWR Schwabe & D Rhodes (1997) Linkers made to measure. *Nature Struct. Biol.* 4:680–683.
32. T Curran & BR Franza (1988) Fos and Jun—the AP-1 connection. *Cell* 55:395–397.
33. JNM Glover & SC Harrison (1995) Crystal structure of the heterodimeric bZIP transcription factor c-Fos–c-Jun bound to DNA. *Nature* 373:257–261.

34. EK O'Shea, R Rutkowski & PS Kim (1992) Mechanism of specificity in the Fos-Jun oncoprotein heterodimer. *Cell* 68:699–708.

35. EM Blackwood & RN Eisenman (1991) Max: a helix-loop-helix zipper protein that forms a sequence-specific DNA-binding complex with Myc. *Science* 251:1211–1217.

36. B Amati, MW Brooks, N Levy et al. (1993) Oncogenic activity of the c-Myc protein requires dimerization with Max. *Cell* 72:233–245.

37. P Lavigne, LH Kondejewski, ME Houston et al. (1995) Preferential heterodimeric parallel coiled-coil formation by synthetic Max and c-Myc leucine zippers: a description of putative electrostatic interactions responsible for the specificity of heterodimerization. *J. Mol. Biol.* 254:505–520.

38. P Lavigne, MP Crump, SM Gagné et al. (1998) Insights into the mechanism of heterodimerization from the ^{1}H-NMR solution structure of the c-Myc-Max heterodimeric leucine zipper. *J. Mol. Biol.* 281:165–181.

39. SK Nair & SK Burley (2003) X-ray structures of Myc-Max and Mad-Max recognizing DNA: molecular bases of regulation by proto-oncogenic transcription factors. *Cell* 112:193–205.

40. DE Ayer, L Kretzner & RN Eisenman (1993) Mad—a heterodimeric partner for Max that antagonizes Myc transcriptional activity. *Cell* 72:211–222.

41. AS Zervos, J Gyuris & R Brent (1993) Mxi1, a protein that specifically interacts with Max to bind Myc-Max recognition sites. *Cell* 72:223–232.

42. F Rastinejad (2001) Retinoid X receptor and its partners in the nuclear receptor family. *Curr. Opin. Struct. Biol.* 11:33–38.

43. F Rastinejad, T Perlmann, RM Evans & PB Sigler (1995) Structural determinants of nuclear receptor assembly on DNA direct repeats. *Nature* 375:203–211.

44. WH Li (1997) Molecular Evolution. Sunderland, MA: Sinauer Associates, Inc.

45. DS Goodsell & AJ Olson (2000) Structural symmetry and protein function. *Annu. Rev. Biophys. Biomol. Struct.* 29:105–153.

46. C Branden & J Tooze (1999) Introduction to Protein Structure, 2nd ed. New York: Garland.

47. T Schirmer & PR Evans (1990) Structural basis of the allosteric behaviour of phosphofructokinase. *Nature* 343:140–145.

48. GA Churchill, DL Daniels & MS Waterman (1990) The distribution of restriction enzyme sites in *Escherichia coli*. *Nucleic Acids Res.* 18:589–597.

49. M Suzuki & N Yagi (1994) DNA recognition code of transcription factors in the helix-turn–helix, probe helix, hormone receptor, and zinc finger families. *Proc. Natl. Acad. Sci. USA* 91:12357–12361.

CHAPTER 4

Protein Interactions *in vivo*

It ain't so much the things we don't know that get us into trouble. It's the things we do know that just ain't so.

Artemus Ward (1834–1867), humorist

A young scientist who hopes to make discoveries is badly advised if his teacher tells him, "Go round and observe," and he is well advised if his teacher tells him, "Try to learn what people are discussing nowadays in science. Find out where difficulties arise, and take an interest in disagreements. These are the questions which you should take up."

Karl Popper (1972), [1]

Proteins are normally studied in dilute aqueous solution. However, this is not their normal working environment. The inside of a cell is very different from the typical conditions used to study proteins, and it imposes some major constraints on what proteins can do, and how they can do it [2]. In this chapter, we explore the consequences of the protein environment, and the inherent physical limitations on proteins, that have greatly influenced the way in which proteins work. We will see that the physical shape of the cell and of the proteins that fill it have a major effect on association constants and on-rates. And we will see how proteins get round these problems: by using membranes for their associative processes, by being processive, and by having arms of various types that can reach out from the main protein body to stick to binding partners and reel in the protein. In the final part of this chapter, we see how proteins are covalently modified, and how their limited stability against misfolding has both good and bad consequences.

4.1 FACTORS INFLUENCING COLLISION RATES

4.1.1 On a small scale, random processes have much more significant effects

Many of the equations and laws that we use to describe fundamental events such as binding and free energy are essentially probabilistic laws. When for example we measure the affinity of binding of A to B, we are saying that *on average* the rate of association is $k_{on}[A][B]$ and the rate of dissociation is $k_{off}[AB]$, and the affinity is the ratio of one to the other. On a molecular scale, these numbers are defining the probability of two molecules sticking together or dissociating. Such probabilities are good descriptions of most systems that we can measure, because the numbers of molecules involved are so large. 10 ml of a 1 μM solution contains 6×10^{15} molecules, a number so large that the effect of random deviation is too small to measure. However, the same is not true in the cell. For example, the volume of a solid macroscopic object is essentially completely invariant with time, but a protein has a root mean square volume fluctuation about 0.2% of the volume of the protein, as a result of bond and angle fluctuations, whereas a small molecule (one-hundredth of the volume) has volume fluctuations about 2% of its volume. An *Escherichia coli* cell contains only about 2 million protein molecules, and many proteins are present at only a single copy per cell, or a "concentration" of 2 nM. At pH 7, the number of protons in an *E. coli* cell is about 50. When we consider that many of these are nominally bound to proteins, there cannot be enough free protons to go round. Therefore a protonation event on a molecular scale must be quantized: the proton must come off somewhere else before it can attach in a new site. In fact any binding event is stochastic: on average, binding and dissociation will happen in the ratio described by the affinity constant, but for any particular molecule at any time there is always a finite and reasonably large probability of an "on" switch being "off," for example.

The consequence is that individual molecules and individual cells can do unpredictable things, and random fluctuations at a molecular scale are very large. The only reason that anything works at all is that the system shows an

***4.1 Rate constant**
There is an important distinction between a *rate constant* and a *rate*. For a reaction such as

$$A \rightarrow B$$

the *rate constant* for the forward reaction is typically written with a small k, for example k_f, and has units of $M^{-1} s^{-1}$. As one would expect from the name, this is a constant (for a given set of experimental conditions such as temperature, pH, and ionic strength). The *rate* is then given by the product of the rate constant and the concentration(s) of the substrate(s). In this case the rate is just $k_f[A]$, and thus has units of s^{-1}. Thus, the rate depends on the concentration of A present, whereas the rate constant does not.

average behavior, either over a large number of molecules or cells or over a long period. (A "long period" is a period long compared with molecular interactions, which is a timescale of perhaps nanoseconds. One second would therefore be a "long" period.) Thus, when we talk about switches being on or off, or a binding event producing a change, or an enzyme catalyzing a reaction, we need to remember that on a molecular scale all this looks much more chaotic: order only emerges when we average over many thousands or millions of events.

4.1.2 Diffusion occurs by a random walk

The ultimate limit to how fast an enzyme-catalyzed reaction can go is determined by physics, not by chemistry. Once the chemical reaction is fast enough, an enzyme rate depends on how fast substrate can diffuse into the active site, and how fast the product can diffuse out [3]. This rate is proportional to the diffusional collision rate, which in turn depends on the size of the molecules and on the viscosity of the medium: we shall discuss these later.

The maximum diffusional collision **rate constant (*4.1)** for small molecules in water hitting a macromolecular target is about $10^9\ M^{-1}\ s^{-1}$. This enables us to calculate how often a substrate is likely to hit an enzyme, purely by random diffusive processes. If the substrate is present at a concentration of 100 μM, for example, the collision rate to a single enzyme molecule is simply $k_{on}[A]$:

$$\text{Rate} = 100\,\mu\text{M} \times 10^9\,\text{M}^{-1}\,\text{s}^{-1}$$

In other words, it will collide with the enzyme 10^5 times per second. Diffusion is thus a fairly rapid effect.

We can also calculate how fast molecules move. Einstein showed in 1905 that the average kinetic energy of a particle in one direction is $kT/2$, where k is Boltmann's constant and T is the absolute temperature. The average kinetic energy is simply $mv^2/2$, where m is the mass and v the average velocity, implying that $mv^2/2 = kT/2$, or $v = (kT/m)^{1/2}$. For a small protein at room temperature this gives an average velocity of about $15\,\text{m}\,\text{s}^{-1}$ ($50\,\text{km}\,\text{h}^{-1}$, or 35 miles per hour), and for a small molecule it is considerably larger. For water molecules, for example, the mean velocity is about $500\,\text{m}\,\text{s}^{-1}$ or $1800\,\text{km}\,\text{h}^{-1}$ (1100 miles per hour).

However, on a molecular scale diffusion occurs by a "random walk" process, similar to Brownian motion: molecules move in some direction for a very short distance (typically less than the diameter of a solvent molecule) until they collide with a solvent or solute molecule, at which point they either bind or "bounce off" in a different direction. In a random walk, molecules often come back close to where they started, so that effectively they spread out in a spherical distribution from their starting point. The net distance traveled from the starting point is proportional to the square root of the time taken, implying that diffusion by random walk over short distances is rapid, but that over long distances it is much slower. This sounds like a remarkably inefficient way of getting from one place to another, and indeed it is. We often speak loosely of substrates "looking for" the active site or proteins "looking for" their binding partner. In reality the diffusive process is random and therefore much less efficient.

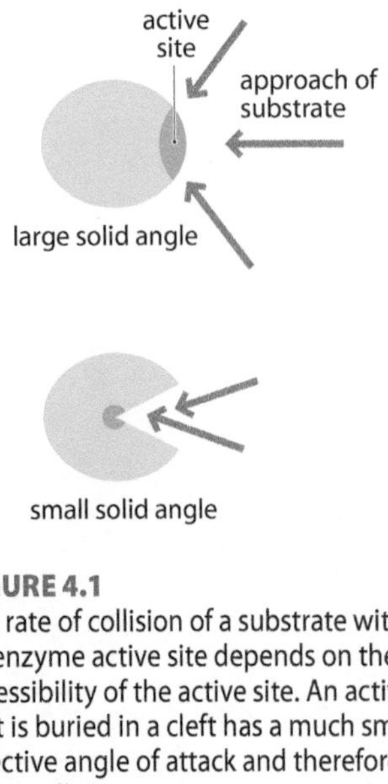

FIGURE 4.1
The rate of collision of a substrate with an enzyme active site depends on the accessibility of the active site. An active site that is buried in a cleft has a much smaller effective angle of attack and therefore a lower collision rate.

4.1.3 The collision rate is limited by geometrical factors

The rate of successful collisions also depends on a shape factor. A substrate has to find its way into the active site of an enzyme; the rate of this process depends on how large and accessible the active site is (**Figure 4.1**). If the active site is well exposed on the surface of the enzyme, it presents a large solid angle (the three-dimensional equivalent of an angle) to the substrate, and diffusion is rapid. If we imagine the substrate being fired toward the enzyme from different directions, then the rate at which it hits the active site is proportional to the fraction of the surface area of the enzyme that is occupied by the binding site. It is therefore only a few percent of the maximum possible rate, particularly as the substrate also has to be in the correct orientation to bind [4].

However, if the active site is not exposed but is buried in a cleft, then the successful collision rate is decreased, because the solid angle needed is much smaller (see Figure 4.1). And most active sites are indeed buried. They need to be buried, because the enzyme needs to be able to control the orientation of the substrate very precisely; it also needs to control the electrostatic environment, which normally means that it needs to limit access to water. Thus, for any real enzyme, the requirement for a buried active site greatly decreases the successful collision rate and therefore the maximum possible turnover rate of the enzyme. This effect has been estimated to lead to a decrease in rate of about 10-fold [5].

This decrease is only important if enzymes are so efficient that their rates are close to being limited by diffusion. **Table 4.1** lists the value of the specificity constant (*5.18) k_{cat}/K_m for several very fast enzymes: k_{cat}/K_m is the apparent second-order rate constant for an enzyme and thus corresponds to the apparent maximum turnover rate, as described in Chapter 5. It is clear that these enzymes are indeed going at the diffusion-controlled limit or in fact significantly faster, bearing in mind that the limited area of the active site limits the "theoretical" maximum rate to about 1% of the diffusion limit; that is, to about $10^7\ M^{-1}\ s^{-1}$. (We return shortly to the question of how it is that rates can go faster than is theoretically possible!) It is therefore clear that our simple picture of "firing the substrate into the active site" is too simple. Some of the complexities are discussed further in the following few sections. Thus, in general it *does* matter that the active site has a restricted access: at least some enzymes have clearly become so efficient that diffusion of substrate into the active site is close to the diffusion-controlled limit.

We have already seen that most active sites are at the interface between two domains. In Chapter 2, a possible reason for this location was presented in evolutionary terms: it is easiest to create an active site here, and this location also makes it easier to create a new function, add allosteric effects (Section 3.2),

TABLE 4.1 Values of k_{cat}/K_m for some enzymes

Enzyme	Substrate	k_{cat}/K_m ($M^{-1}\ s^{-1}$)
Acetylcholinesterase	Acetylcholine	1.5×10^8
Carbonic anhydrase	Carbon dioxide	8.3×10^7
Catalase	Hydrogen peroxide	4.0×10^8
Fumarase	Fumarate	1.6×10^8
Fumarase	Malate	3.6×10^7
Superoxide dismutase	Superoxide	2.8×10^9
Triosephosphate isomerase	Dihydroxyacetone phosphate	7.5×10^5
Triosephosphate isomerase	Glyceraldehyde 3-phosphate	2.4×10^8
Lysozyme	$(NAG\text{-}NAM)_3$	83
Glucose isomerase	Glucose	7.4

Abbreviation: NAG-NAM, *N*-acetylglucosamine–*N*-acetylmuramic acid disaccharide.

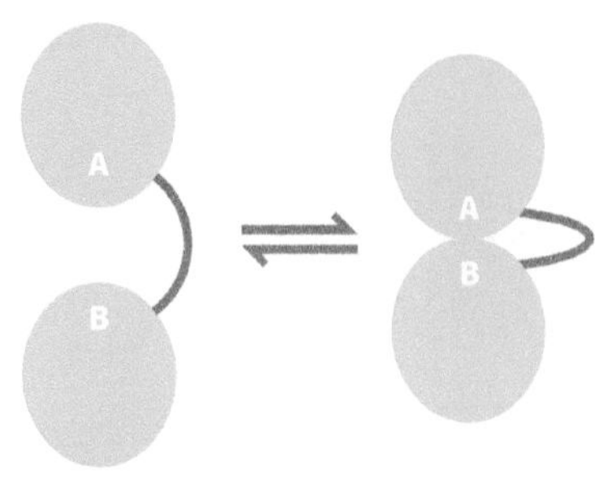

FIGURE 4.2
The rate of diffusion of substrate into the active site of an enzyme is maximized when the substrate binds at the interface between two domains that have a flexible linker such that they can open out to allow the substrate to bind and then close to catalyze the reaction.

and enhance control and regulation. It was also justified in terms of the energetics of enzyme reactions: this location minimizes the energy input needed to produce the activated enzyme conformation E*, and therefore increases the rate of reaction. Here, we see yet another reason: maximization of collision rate.

The arguments outlined above create a dilemma for enzymes: how can they simultaneously have a buried active site, so as to increase control and selectivity, and an exposed active site, to maximize the rate of the reaction? The solution is to have the active site in the interface between two domains, such that the domains can open out to allow the substrate in and products out, but then close for the duration of the reaction to provide the necessary enclosed environment (**Figure 4.2**). An essentially equivalent solution, used by triose-phosphate isomerase and many proteases, is to have a flap or lid that closes once the substrate is bound.

This solution to the problem does potentially lead to some slowing down of the reaction, in that it means that the reaction rate is potentially limited by the rate at which the two domains open and close, or the flap opens. Table 4.1 shows that several enzymes have k_{cat}/K_m values of around $10^8\,M^{-1}\,s^{-1}$. Does this mean that flaps need to open and close at this rate? No, because as always we need to be careful about our terms and units, and distinguish between a *rate* and a *rate constant*. The figures in Table 4.1 are effectively *rate constants*. To get the *rate*, we need to multiply the rate constant by the substrate concentration. This is made clear by considering the binding of enzyme to substrate:

$$\text{Enzyme} + \text{Substrate} \xrightarrow{k} \text{Products}$$

For this reaction, the second-order *rate constant* is k, and the actual *rate* is

$$\text{Rate} = k[\text{Enzyme}][\text{Substrate}]$$

If we consider this from the point of view of the enzyme, then per molecule of enzyme, the rate is given by

$$\text{Rate}/[\text{Enzyme}] = k[\text{Substrate}]$$

The physiological concentration of enzyme substrates can vary enormously, but for example for acetylcholine in the synapse after stimulation of the nerve it can rise well above 0.1 mM or 10^{-4} M. Therefore although the rate constant for the hydrolysis of acetylcholine is extremely fast, the *rate* of the reaction catalyzed by acetylcholinesterase is significantly slower, at about $10^4\,s^{-1}$. The point of this calculation is that if the hydrolysis of acetylcholine is not to be slowed down by domain movements, the domain movement needs to be at least this fast (namely $10^4\,s^{-1}$ rather than $10^8\,M^{-1}\,s^{-1}$). In fact, a rate of $10^4\,s^{-1}$ for a domain movement is readily achievable. Thus, for many enzymes, domain movement does not cause a significant decrease in turnover rate, and certainly nothing like the decrease that would be caused by an inaccessible active site. Therefore what has evolved is the fastest solution (governed of course by the limitations of what evolution can achieve), namely hinged domain movement.

4.1.4 Collision rates can be increased by electrostatic attraction

The typical bimolecular rate constant for the interaction of two proteins is about $10^6\,M^{-1}\,s^{-1}$, which reflects the diffusion-controlled rate after considering the steric factors considered in the previous section [6]. However, in the previous section we saw that a number of enzymes have collision rates much faster than this, and in some cases a maximum turnover rate that is faster than physically possible, at least using the simple arguments used there. How can this be explained? In this section and the next, we see that it is due in large part to electrostatic effects, which do two different things: they bring opposite charges together and, more importantly, they act to *steer* molecules as they

*4.2 Dielectric constant

The force between two charges q_1 and q_2 separated by a distance r is roughly $F = q_1q_2/4\pi\epsilon r^2$, where ϵ is the dielectric constant. In a vacuum ϵ has a value of 1. Organic solvents such as hexane have values of about 2, whereas water has a value of 80. Therefore water greatly decreases the force between two charges, and stabilizes them by solvating the charges and screening them from each other. The inside of a protein has an ϵ of about 4 and is therefore much less good at stabilizing charges.

*4.3 Ionic strength

Adding extra ions to the solution increases the ionic strength of the solution. The ionic strength μ is given by the formula

$$\mu = \tfrac{1}{2}\Sigma z_i^2 C_i$$

where z_i is the charge on each ion i and C is its concentration, and the sum is over all the charged species present. In blood, for example, there is roughly 130 mM Na^+, 110 mM Cl^- and a few other ions, so very roughly we could say that there is 150 mM singly charged positive ions and 150 mM singly charged negative ions, giving an ionic strength of 0.5 × (0.15 + 0.15) M, or 150 mM.

*4.4 Electrostatic screening

The effect of increasing ionic strength on electrostatic interactions was considered by Debye and Hückel in 1923. Both scientists made enormous contributions to physical chemistry: Debye studied dipole moments (for which the unit is named the debye), specific heat at low temperatures, atomic structure, and the effect of temperature on diffraction patterns (the Debye–Waller factor, B); his assistant, Hückel, is equally if not more famous for describing the π-electron density in aromatic systems. Their 'extended' model says that the concentration of a charged species should be converted to an *activity* by multiplying the concentration by an activity coefficient γ:

$$a = \gamma C$$

The values of γ can be calculated at 25°C by

$$\log \gamma_i = \frac{-0.509\, z_i^2 \sqrt{\mu}}{1 + (3.29\, \alpha_i \sqrt{\mu})}$$

where α is an empirically determined effective diameter of the hydrated ion in nanometers, and has values of about 0.25 for ammonium ions (for example lysine side chains), 0.3 for K^+ and Cl^-, and 0.4 for Na^+ and carboxylate ions (glutamate and aspartate side chains).

Applying this formula gives γ values of 0.7 for lysines and 0.74 for glutamate and aspartate. The overall attractive force is therefore given by $F = q_1q_2/4\pi\epsilon r^2$, where the charges q are decreased by the factor γ, and so the overall force is decreased by a factor $\gamma_{Lys}\gamma_{Glu} = 0.7 \times 0.74$ or about one-half.

An alternative but related approach is to calculate the *Debye length*, which is the characteristic distance over which electrostatic forces operate. This is given by

$$r = \left(\frac{\epsilon KT}{2N_0 e^2 \mu}\right)^{1/2}$$

where N_0 is Avogadro's number. For a physiological ionic strength, r is about 8 Å.

approach each other, so that they approach from the right direction and in the right orientation to bind. These two effects are known as translational and orientational steering, respectively.

It seems self-evident that if two interacting molecules have opposite charges they will experience some electrostatic attraction and therefore come together faster, whereas if they have the same charge they will experience some electrostatic repulsion and come together more slowly. This naive expectation is indeed true, although the effect is not very large. It has been estimated that the rate of collision of enzyme and substrate should be roughly twice as fast with charges of +1 and −1 in comparison with no charge, and four times as fast for charges of +1 and −2 [7]. More importantly, electrostatic repulsion has a bigger effect than electrostatic attraction, although again it is not a large factor [8]. Translational steering is thus a significant effect but not an enormous one.

The effect is decreased even further by electrostatic screening. The electrostatic interaction falls off rather slowly with distance. We saw in Chapter 1 that it actually falls off as r^2. However, the attractive force is also decreased by a factor given by the **dielectric constant (*4.2)** of the medium. Water has a dielectric constant of 80, implying that it screens charges rather well: it makes the attractive force much weaker than it would be in a nonpolar solvent, or conversely it means that the electrostatic attraction is only significant once the two charges get rather close together.

The interaction is also weakened if there are other charged species in solution, in other words if the **ionic strength (*4.3)** is high. There are indeed other ions: in particular, ions such as Na^+, K^+, and Cl^-. These charges congregate round oppositely charged ions and solvate them, thereby making them less "charged" and decreasing the attractive force between oppositely charged groups. At physiological concentrations, they decrease the effective charge by very roughly a factor of 2, so again the favorable interaction is weaker. (The calculations are described in **electrostatic screening, *4.4**.) All of this means that charge–charge attractions and repulsions only extend out for about 10 Å [2, 9]. In the naive example quoted above, the enzyme–substrate translational rate enhancement almost disappears at physiological ionic strength (roughly 150 mM NaCl).

4.1.5 Collision rates are also increased by electrostatic steering

A particularly interesting and relevant case concerns what happens when a charged substrate gets very close to an enzyme surface, but not at the active

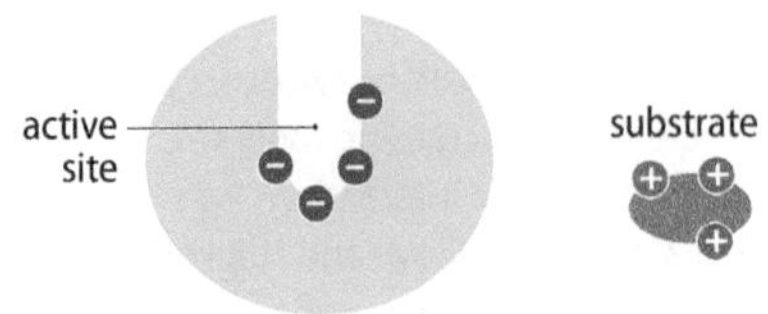

FIGURE 4.3
When the substrate has an opposite charge to the active site, they attract each other.

site (**Figure 4.3**). Here, the electrostatic attraction is going not through water but through the protein. A protein interior is generally reckoned to have a dielectric constant of around 4: it is much less polar than water. Therefore the electrostatic interaction is much stronger than it would be in water. But if the protein is in the way, how does this help? It helps because the arrangement of charges within a protein is often asymmetric, constituting a **dipole moment**. Such an arrangement gives the protein charge a directionality and provides a force that helps to move the substrate toward the oppositely charged end of the dipole. For example, in acetylcholinesterase (discussed above as an example of "unnaturally fast enzymes") there is a strong dipole aligned with the active site cleft, which acts to orient the substrate and pull it into the cleft, much as iron filings are attracted to a magnet (**Figure 4.4**) [10]. The substrate (acetylcholine) is positively charged, and the arrangement of charges on the enzyme provides a strong orientation for the approaching substrate, ensuring that there is a high probability that it will approach from the correct direction. This effect is electrostatic orientational steering, often simply called *electrostatic steering*. It is also noteworthy that in acetylcholinesterase the negative charges that line the active site cleft are almost entirely covered by a "lining" of aromatic residues that make the cleft hydrophobic or "oily," and presumably decrease the direct electrostatic attraction so that the substrate does not stick to the sides of the cleft as it moves down it [10]. An oily lining is found in other examples, such as in both the central spindle and the central channel in F_0F_1 ATPase discussed in Chapter 7, and in the membrane water channel aquaporin, which has an oily face on one side to allow water molecules to slide, and a hydrophilic face on the other to keep them aligned properly as they do so. A similar effect was observed for superoxide dismutase, an enzyme with a very high rate constant of about 10^9 M^{-1} s^{-1}. It reacts with the negatively charged superoxide ion, and the charge distribution on the enzyme directs the substrate into the active site, enhancing the collision rate about 30-fold [11]. It does this even though the overall charge on the protein is negative, roughly –4: in other words, the electrostatic steering depends more on the distribution of charges than on the net charge, which is actually unfavorable.

Electrostatic steering was used above to describe the steering of a substrate toward the active site. It also operates to rotate the substrate as it approaches the active site, so that it is pointing in the correct orientation when it arrives. The combined effect of both kinds of steering in the action of triosephosphate isomerase has been calculated to be in the range 10–100-fold at physiological ionic strength [12]. The translational steering depends on protein net charge, but the orientational steering depends much more importantly on the three-dimensional arrangements of the charges, particularly those close to the binding site [13]. Significantly, the orientational steering seems in general to be more important than translational steering [4].

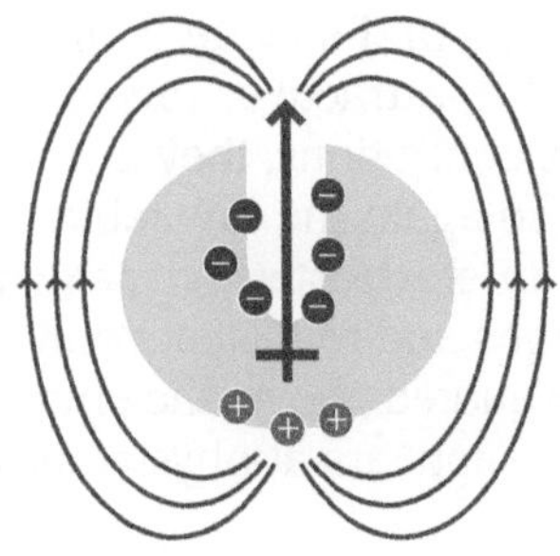

FIGURE 4.4
The enzyme acetylcholinesterase has an electric dipole aligned with the active site cleft, which acts to direct a positively charged substrate into the cleft. The lines of force (electric field lines) on a positively charged substrate are indicated.

4.5 Cytochrome *c

Cytochrome *c* functions as an electron carrier, mainly within the electron transport chain. It is therefore small (Section 4.1.9), because it has to shuttle rapidly between electron donor and receptor. The electron is carried on a heme group (**Figure 4.5.1**) and is located mainly on the iron at the center, which alternates between Fe^{II} and Fe^{III} (oxidation states 2 and 3). The iron atom has six ligands. Four are provided by the heme, one is a histidine, and the sixth is a methionine sulfur (Met 80) in mitochondrial cytochrome *c*. On oxidation, there is increased mobility in a hydrogen-bonded channel leading from Met 80 to the surface, which may help partner proteins in identifying the cytochrome as being oxidized. Cytochromes are in general soluble proteins, although some have lipid anchors to keep them close to the membrane.

Cytochromes are ubiquitous and have slow mutation rates, presumably because of the large number of proteins with which they are required to interact. Thus, human and chimpanzee mitochondrial cytochrome *c* are identical, and the rhesus monkey protein differs at only one position.

Studies of cytochrome *c* have undergone a resurgence in recent years, with the discovery in 1996 that this very well-studied protein **moonlights** as a signal for apoptosis.

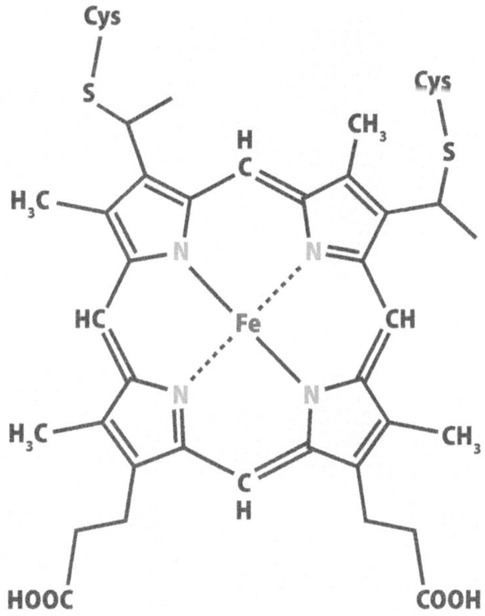

FIGURE 4.5.1
The structure of heme, as found in cytochrome *c*. The prosthetic group is attached to the protein through two cystines.

Electrostatic steering is also important in protein:protein interactions. **Cytochrome *c* (*4.5)** is a particularly interesting example. Its function is to carry electrons from an electron source to an electron acceptor, usually as part of an electron transport chain. It is found in several such chains, and it therefore needs to recognize a large number of different proteins. To transfer the electron as fast as possible, it needs to be oriented such that the electron (which is located on the iron–heme system toward one edge of the enzyme) is presented to its redox partners. One mechanism to enable this is that there is a ring of positively charged residues around the heme that match rings of negatively charged residues on its redox partners, and thereby help to orient it correctly (**Figure 4.5**) [14].

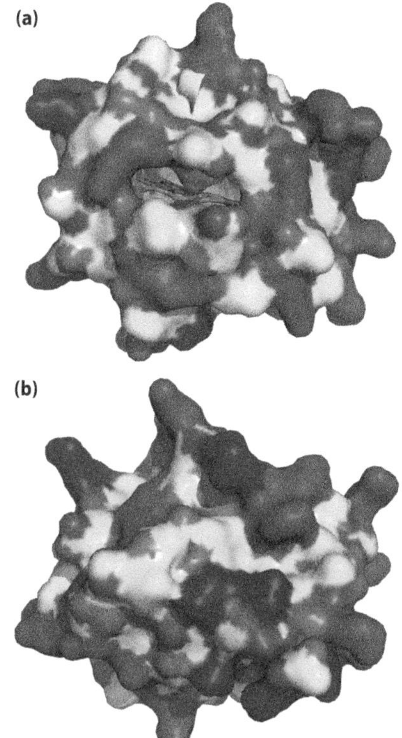

FIGURE 4.5
The distribution of electric charge on the surface of cytochrome *c*. (a) Front: the heme group is shown in magenta. This is therefore the face presented to partner proteins, because the electron transfer is to or from the heme. Note the ring of red (basic) residues around the heme. (b) Back. Note the preponderance of blue (acidic) groups in the center of the face, giving cytochrome *c* a dipole.

4.1.6 Protein binding takes place via an encounter complex

Such matching of charged regions is common, particularly for proteins that bind to nucleic acids, which almost always have a positively charged region at the interface. However, this solution to long-range steering is for many proteins not a practical solution, because they cannot manipulate the charge around the active site in this way. Therefore it is commonly observed that proteins have charged regions elsewhere on the protein, which serve to orient the protein correctly with respect to its binding partner (**Figure 4.6**) [15]. As two proteins approach, they form an **encounter complex** with opposite charges attracting, but not fully in the correct orientation. Molecular dynamics calculations suggest that the key requirement for the formation of an encounter complex is that at least two polar contacts be made, to anchor the two proteins in a suitable orientation; the rest can zip up later [16, 17]. As long as this charge–charge interaction holds the proteins together for longer than the time it takes for the proteins to rearrange, it will significantly speed up association in the correct orientation (at least 1000-fold) [9]. And typically molecules do stay close together for significant times, therefore permitting this rearrangement. This is because of the nature of Brownian diffusion, which typically takes many small excursions from its initial position before moving a significant distance. In a molecular dynamics simulation of an encounter between two proteins, each collision between the two proteins only held them together for 0.4 ns [6]. This is not long enough for one protein to reorient with respect

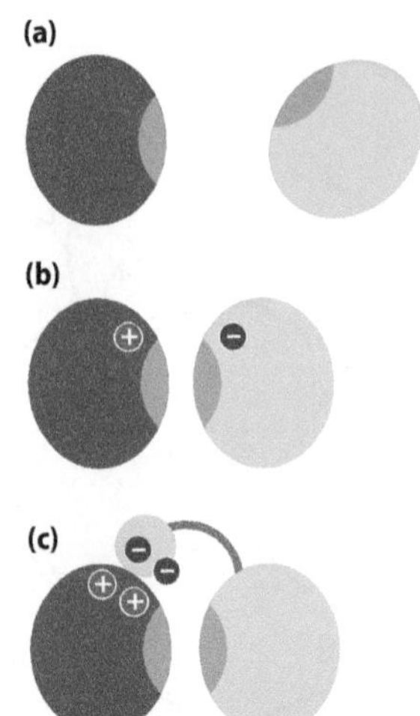

FIGURE 4.6
The correct orientation of protein encounters is aided by electrostatic interactions. (a) The association rate of two proteins is limited by the requirement for them to come together in the correct relative orientation. (b) The association rate is increased by electrostatic steering. (c) Electrostatic steering can also use unstructured arms or domains to provide interactions over a longer distance. Such an arrangement also gives the protein greater flexibility in the locations of charged residues elsewhere on the surface.

to the other, which takes a time of the order of 5 ns. However, the encounter complex lasts for about 6 ns before Brownian motion finally pushes it apart, giving the molecules time to search a substantial conformational space before they finally dissociate.

Calculations suggest that for the complex of β-lactamase with its inhibitor, the encounter complex more often dissociates than leads to productive binding: that is, more often than not the charge–charge interaction is *not* strong enough to hold the proteins together every time there is a collision. Nevertheless, it increases the overall rate of binding by increasing the probability of a correctly formed complex. It is suggested that in this case, stronger electrostatic binding, sufficient to guarantee a correctly formed complex, would be counterproductive because it would require excessive desolvation of the protein pair during formation of the complex, therefore leading to overall slower binding [9]. Most researchers in this field agree that the encounter complex is not desolvated—it has not yet proceeded to the stage at which the peeling off of water molecules is necessary. It is thus an "arm's length" complex rather than a close embrace.

The preceding paragraph illustrates a somewhat tricky concept, namely the difference between the encounter complex (which we may consider an intermediate state corresponding to the first significant contact between two proteins) and the *transition state* for binding, which is classically defined as the state from which half the molecules will go on to bind and the other half will separate again [16]. The transition state can either precede or follow the encounter complex, depending on the proteins involved and in particular on how rapidly they bind once they meet. In the case of β-lactamase discussed above, the encounter complex precedes the transition state. For molecules binding close to the diffusion-controlled rate, the transition state must precede the encounter complex, because by the definition of diffusion-controlled binding, every encounter complex goes on to bind. A corollary of this point is that encounter complexes should be relatively highly populated in weak and rapidly interacting complexes, as indeed seems to be true [18].

Such arguments are supported by a wide range of experimental observations. Thus, for example, the rate at which the nuclease barnase (*2.2) binds to its physiological inhibitor barstar is very high (as it needs to be, to prevent barnase from digesting the RNA in its host organism): in water it is 10^9 M^{-1} s^{-1} at an ionic strength of 10 mM, and a remarkable 10^{10} M^{-1} s^{-1} at zero ionic strength (that is, faster than the unaided diffusion limit). Addition of salt to the medium decreases the association rate up to 10^5-fold, as a result of the screening of the electrostatic interaction [4]. Thus, barnase uses electrostatics to increase the diffusional collision rate with barstar, and also in electrostatic steering so that they tend to approach each other in the correct orientation [19].

A nice example is provided by the binding of trypsin to its inhibitor BPTI (bovine pancreatic trypsin inhibitor), discussed in [4]. The on-rate is 10^6 M^{-1} s^{-1}, not an unreasonably fast rate but certainly fast considering the very precise geometric fit between protease and inhibitor. Trypsin and BPTI are both positively charged, so there is unfavorable electrostatic translational steering. Mutation of Lys 15 in the active site of BPTI to alanine decreases the overall

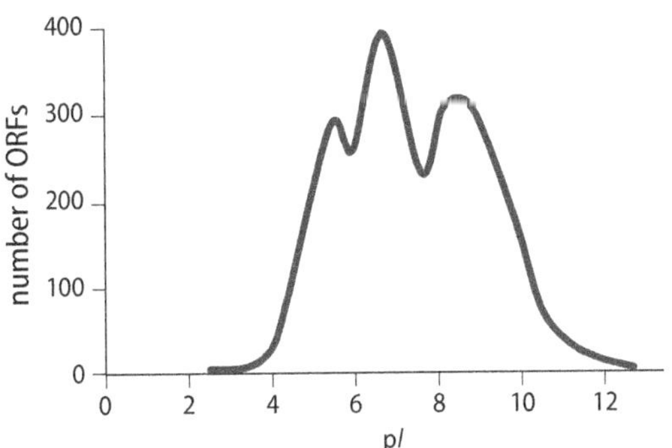

FIGURE 4.7
The approximate distribution of p*I* values predicted for proteins in *Drosophila melanogaster*, based on the genome. Similar plots are obtained for other eukaryotic organisms. ORF, open reading frame. (Redrawn from R. Schwartz, C.S. Ting and J. King, *Genome Res.* 11:703–709, 2001. With permission from Cold Spring Harbor Press.)

positive charge on BPTI and therefore must decrease the electrostatic repulsion between them. Nevertheless, it actually decreases the on-rate by a factor of 250, an effect attributed to the role of Lys 15 in electrostatic orientational steering: this is another example of the observation that orientational steering is usually more important than translational steering.

4.1.7 Electrostatic repulsion is also important for limiting interactions

We shall see shortly that a big constraint on proteins in cells is that cells are very crowded. This implies that electrostatic interactions between proteins must be very significant. And indeed, considering the number of charged residues on the outside of proteins and their high concentration, why do proteins not just stick together in one horrible sticky mess?

There seem to be two main answers to this question. The first is that electrostatic interactions are generally weak, because of the ionic strength inside cells. Electrostatic interactions can (as observed above) increase the affinity of proteins for each other by orders of magnitude, and this will certainly make them stickier, so that as proteins diffuse past each other they will form transient interactions and therefore ultimately diffuse more slowly, but this is still not sufficient to make them "stick together in one horrible sticky mess": it just makes them stick together more as they pass. The second answer is much more interesting: they tend all to have the same charge. Therefore they will all repel each other.

A plot of the predicted **p*I*** values of cellular proteins is shown in **Figure 4.7**. This plot shows that proteins have p*I* values above and below 7, and thus have a range of both positive and negative charges at pH 7; this does not seem to be consistent with the statement above. However, if we plot the p*I* values for membrane and soluble proteins separately (**Figure 4.8**), we can see that in fact most integral membrane proteins have a high p*I* (and are therefore positively charged at neutral pH), whereas most soluble proteins have a low p*I* (and are therefore negatively charged at neutral pH). At neutral pH, this will tend

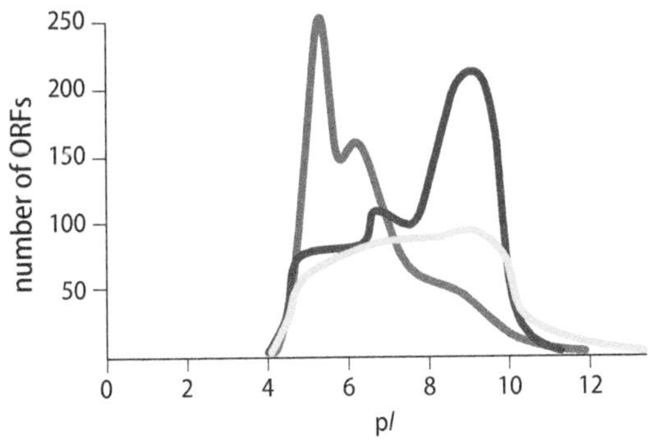

FIGURE 4.8
Distribution of p*I* values for eukaryotic proteins by location in the cell, based on SWISS-PROT annotations. Cytoplasmic, integral membrane, and nuclear proteins are shown by red, blue, and green lines, respectively. (Redrawn from R. Schwartz, C.S. Ting and J. King, *Genome Res.* 11:703–709, 2001. With permission from Cold Spring Harbor Press.)

to make the soluble proteins repel each other but attach to the membrane proteins. Membranes themselves tend to be negatively charged, which is presumably why membrane proteins have an opposite charge. Therefore soluble proteins are in general repelled by membranes but attracted by the membrane proteins. An analysis of the distribution of residue types in proteins from different cellular locations also showed the interesting result that extracellular proteins have fewer charged residues but more polar residues than intracellular proteins, a result that has been suggested to relate to the much lower ionic strength outside the cell [20]. Thus protein charge, like everything else, is seen to be tightly regulated in evolution.

The plot of Figure 4.8 shows that there are many proteins for which this vast generalization is not true. And the figure does not take account of perturbed pK_a values, nor of the very many phosphorylation events that make proteins even more negatively charged. Nevertheless, it shows the power of simple electrostatic charge to modify molecular interactions.

4.1.8 Macromolecular crowding increases the amount of protein association but slows its rate

The inside of a cell is remarkably highly crowded. This is demonstrated most forcefully by the very elegant drawings of David Goodsell, as found in his book The Machinery of Life and in the always interesting 'Molecule of the month' articles on the Protein Data Bank website (http://www.rcsb.org), and also in [21]. **Figure 4.9**, for example, depicts the interior of a bacterial cell and shows that the cell is very crowded, and there is remarkably little free water present in cells (that is, water that is not adjacent to a solute), with most proteins

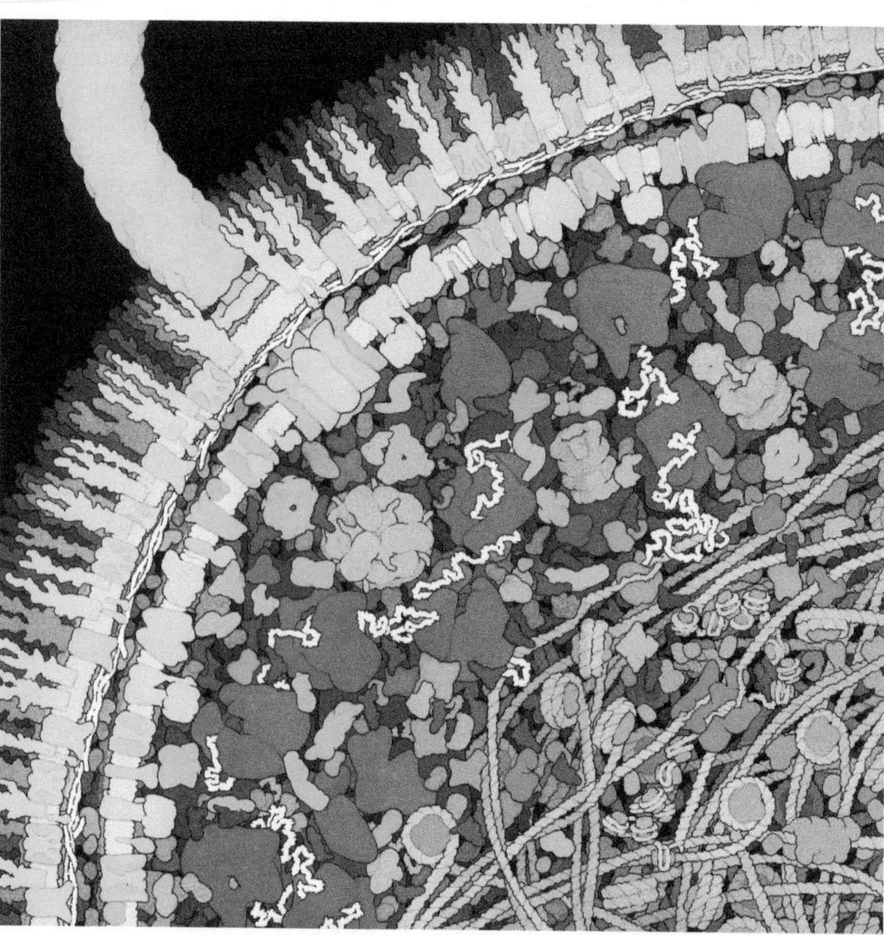

FIGURE 4.9
An impression of the inside of an *E. coli* cell. The cell wall is in green, packed with transmembrane proteins. The large green object at the top left is the bacterial flagellum. The cytoplasm is in blue and purple, and includes ribosomes (large purple objects), tRNA (L-shaped maroon objects) and mRNA (white). The nucleoid region is in orange and brown, with DNA wrapped around HU protein (bacterial nucleosomes). (Courtesy of David S. Goodsell, The Scripps Research Institute.)

being solvated by only one or two layers of water. A remarkably similar overall picture was obtained by direct experimental observation, using cryoelectron tomography (**Figure 4.10**) [22]. In addition, the very high concentration of proteins means that a protein molecule is unable to diffuse freely in water because it will very soon run up against another protein. The mean distance between proteins in the cell has been estimated to be less than 100 Å [23] and in many cases less than 50 Å [24]. Because proteins have a diameter of about 35 Å, crowding is a very real phenomenon. Or, to put it another way, proteins occupy between 10 and 40% of the total fluid volume in cells [25, 26]. In eukaryotic cells the situation is even worse, because eukaryotic cells are criss-crossed by large numbers of fibers such as microfilaments and microtubules. These are close enough together for them to act as a cage for larger proteins, and they severely restrict mobility for smaller ones.

This feature of a high protein concentration inside cells is generally referred to as *cellular crowding* or *macromolecular crowding* [27]. This name emphasizes that the problem is not the high concentration of each individual protein (which is in fact low for most proteins): it is the fact that there is a very high macromolecule concentration overall. The macromolecules take up much of the available space in the cell and thus produce a very large *excluded volume*, this being the space taken up by the macromolecules, implying that the space left for other molecules to function in is both small and oddly shaped—effectively, other solutes are only free to function and diffuse in the spaces left by all the other macromolecules that are present.

One implication of this concept of excluded volume is that the effect of macromolecular crowding depends strongly on the size of the molecule studied in comparison with the size of the crowding agent. In particular, if the molecule studied is *larger* than the crowding agent, its available space (total volume minus excluded volume) becomes significantly decreased as the concentration of the crowding agent increases, whereas if the molecule studied is *smaller*, it is almost unaffected. Because the main crowding agent in cells is proteins, this means that proteins are strongly affected by crowding (in ways discussed below), whereas small molecules are not.

This tricky idea of excluded volume is neatly explained by an analogy (**Figure 4.11**). Consider a beaker filled with ball bearings. For randomly packed balls, about 65% of the total volume is occupied, but even though the remaining 35% is "empty," it is impossible to get any more ball bearings into the beaker. This volume is excluded to ball bearings and to anything bigger. However, it is available to smaller particles, such as sand. If we pour sand into the beaker, it will fill up the spaces: in fact, apart from a decreased total volume, the mobility of the sand is not greatly affected by the ball bearings. In reality, the sand again only fills up about 65% of the available space, the remainder again being "empty," and again available to yet smaller particles such as water.

(a)

(b)

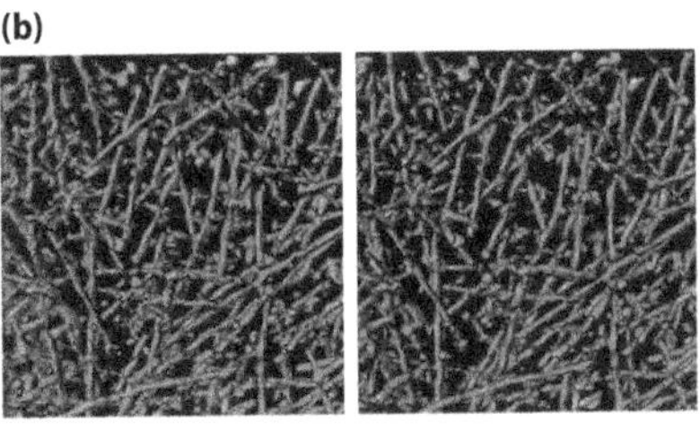

FIGURE 4.10
Image of a *Dictyostelium* (slime mold) cell, obtained by cryoelectron tomography. The figure shows mainly the actin network. (a) A volume 815 nm × 870 nm × 97 nm, with actin filaments in red, ribosomes in green, and membranes in blue. (b) Stereo image (crossed-eye fusion) of an actin layer from the upper left part of (a), showing a crowded network of branched and crosslinked filaments. (From O. Medalia et al., *Science* 298:1209–1213, 2002. With permission from AAAS.)

(a)

(b)

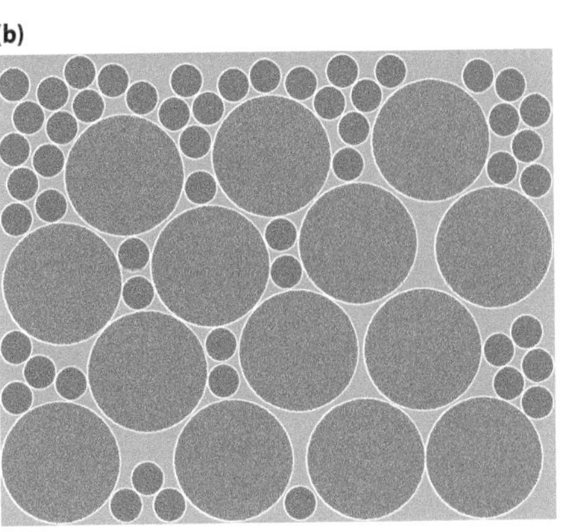

FIGURE 4.11
Excluded volume depends on the size of the crowding agent compared with the molecule being measured. (a) In this example, the container is filled with ball bearings: it cannot contain any more. The green area is therefore excluded to ball bearings. (b) However, it is accessible to smaller particles (red). And after the red particles have filled the space available to them, there is still considerably more space available to yet smaller particles, such as water (blue). (After D. Hall and A.P. Minton, *Biochim. Biophys. Acta* 1649:127–139, 2003. With permission from Elsevier.)

TABLE 4.2 Some examples of effects of macromolecular crowding

Observation	Magnitude; comment	Reference
Self-association of apomyoglobin	Monomer in water; mostly dimer in >200 g l^{-1} added protein	[27]
Stabilization of thrombin against thermal denaturation	Probably due to formation of heat-stable oligomers	[27]
Self-association of pyruvate dehydrogenase	50–90% of 22S species converted to 55S in 30 g l^{-1} PEG	[27]
Association of T4 gene 45 protein and gene 44/62 protein complex	50-fold increase at 75 g l^{-1} PEG	[27]
Optimum temperature for DNA ligase from *Thermus thermophilus*	37 °C to 60 °C in 20% PEG	[27]
Decrease in rate of glyceraldehyde-3-phosphate dehydrogenase	30-fold in 300 g l^{-1} protein; due to tetramer formation?	[27]
Acceleration of actin polymerization	3-fold at 80 g l^{-1} polymer	[27]
Acceleration of T4 polynucleotide kinase	Orders of magnitude; due to stabilization of oligomeric enzyme	[27]
Enhancement of spectrin self-assembly	10-fold increase in 20% dextran	[33]
Acceleration of amyloid formation	8-fold in 150 g l^{-1} dextran	[25]
Stabilization of pH 2 molten globule state of cytochrome c relative to unfolded	2.5 kJ mol^{-1} (150-fold change in K) at 370 g l^{-1} dextran	[25]
Folding–unfolding transition of a WW domain	Folded state 4.5 kJ mol^{-1} more stable in 25% crowder (simulation)	[34]
Refolding of reduced lysozyme	Abolished at high dextran because of self-association of the unfolded form	[120]
Enhancement of rate and extent of fibrous or amyloid assemblies	Range of proteins including tubulin and α-synuclein	[29]

Abbreviation: PEG, poly(ethylene glycol).

Macromolecular crowding has several very important consequences, some examples of which are presented in **Table 4.2**. First, it means that the degree of association of proteins is much higher than it would be in water by up to several orders of magnitude. **Figure 4.12** demonstrates a typical effect [25]. The association of larger oligomers is favored, so that for example a tetramer is more stabilized than a dimer. Larger oligomers, such as the GroEL heptamer (*4.12) or the FtsZ high-molecular-weight aggregate [28] (Section 7.1.1), are even more strongly favored. Crowding increases both the extent and rate of formation of fibrous or rod-shaped assemblies, including sickle hemoglobin, tubulin, FtsZ, and the Parkinson's disease-related protein α-synuclein [29–31].

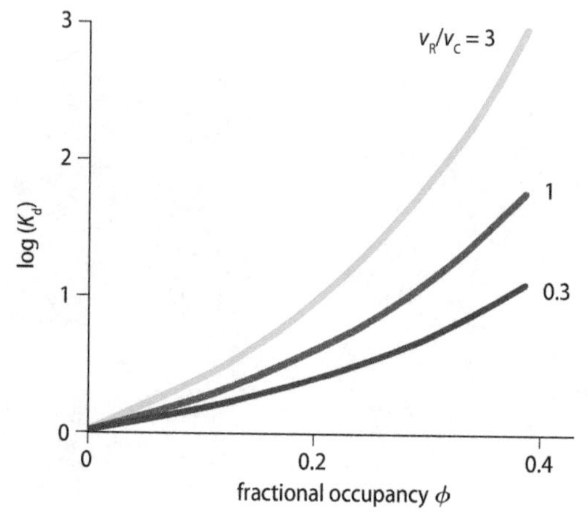

FIGURE 4.12
The effect of macromolecular crowding on dimerization. The horizontal axis shows the fractional occupancy of the volume: in cells, the most severe crowding corresponds approximately to $\phi = 0.4$. The vertical axis is the logarithm of the dimerization constant. Values are shown for three different ratios of the volume of the monomer R and the crowder C. The graph shows that crowding at $\phi = 0.4$ increases the dimerization constant about 30-fold for proteins of the same size, or 400-fold for a monomer three times larger in volume than the crowder. (After D. Hall and A.P. Minton, *Biochim. Biophys. Acta* 1649:127–139, 2003. With permission from Elsevier.)

As we might expect from the discussion above, the effect is larger for larger monomers. The effect also depends on the shape of the oligomer, so that roughly spherical oligomers are more favored than elongated ones [32].

Second, the high degree of association effectively makes proteins more sticky, and means that the *rate* at which protein complexes can assemble is very severely retarded. This can be illustrated by a simple analogy. Imagine that you enter a room at a party and want to meet a particular type of person (I leave the choice to your imagination). As the number of people in the room increases, initially the rate of such a meeting is increased, because there are more people there, and therefore very probably more of the desired kind of person. However, once the room gets really crowded, the rate goes down again because it gets so difficult to diffuse randomly around the room that you get stuck in one place or can only move along the few channels available (an illustration of excluded volume).

An alternative analogy is that of a car journey. The time taken to get to the goal depends on the speed of the car (that is, the viscosity of the medium and the size of the solute), the time spent at stop lights (binding), and the route taken (macromolecular crowding). Macromolecular crowding affects all three of these, although it has most effect on the third.

This decrease in the rate of association applies most obviously to reaction rates, which also initially increase but then decrease with macromolecule concentration [27], as illustrated in **Figure 4.13**. The point at which this change from increase to decrease occurs varies between systems, meaning that sometimes the observation is merely a steady increase in rate, whereas sometimes more complicated effects occur [33]. In general, crowding decreases the rate of fast associations and increases the rate of slow associations [34].

Unfolded proteins have a bigger radius of gyration and occupy a larger volume than folded proteins. Because crowding favors states that require less volume, it stabilizes the folded state against the unfolded state. However, the effect is not large. Moreover, because the **molten globule** state is only slightly more expanded than the native state, crowding is expected to have rather little effect on the equilibrium between molten globule and native states, and will have its largest effect in converting unfolded protein to a molten globule [35]. It is therefore possible that **natively unstructured proteins** adopt a more folded conformation inside cells than they do in dilute aqueous solution, but the effect is probably rather small, as discussed later in this chapter.

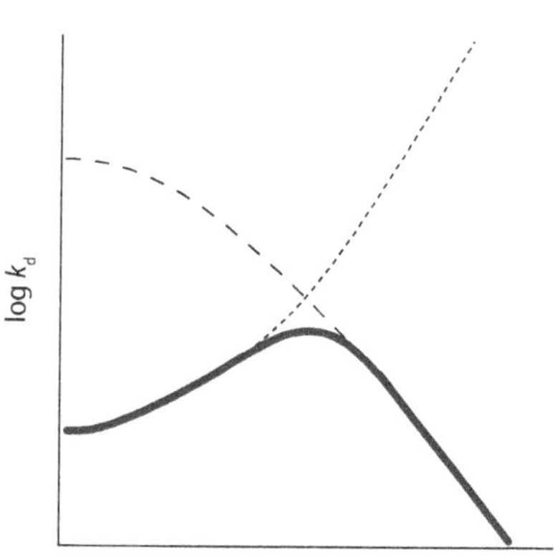

FIGURE 4.13
Schematic dependence of reaction rate on volume occupancy. The rate initially increases because there is a higher concentration of molecules (short dashed curve), but then decreases because the diffusion rate decreases (long dashed curve). The overall result is shown by the solid curve. (Adapted from A.P. Minton, *Int. J. Biochem.* 22:1063–1067, 1990. With permission from Elsevier.)

One of the most crowded cellular compartments is the mitochondrion. However, the diffusion rate measured within the mitochondrion is actually very similar to that in the cytoplasm [36]. It has been suggested that this is because the distribution of proteins in the mitochondrion is far from uniform, with most proteins being attached to membranes, leaving the interior freer for diffusion. This observation supports the view that many or even most proteins are not freely floating around in the cell but are localized in metabolic complexes or metabolons, probably attached to membranes (Section 9.4).

It is reasonable to expect that evolution has optimized proteins for the degree of macromolecular crowding to which they are normally subjected—that is, that the degree of self-association of proteins has been optimized by natural selection within their operating milieu. Interestingly, this means that when we study proteins in dilute aqueous solution we could expect the amount of association to be decreased and suboptimal, and possibly that the stability and degree of foldedness are also suboptimal [37]. This has particularly important implications when we come to think about weak associations between proteins to form protein complexes in Chapter 9. It also of course means that extrapolation from the properties we can measure in dilute aqueous systems to real cells is far from straightforward.

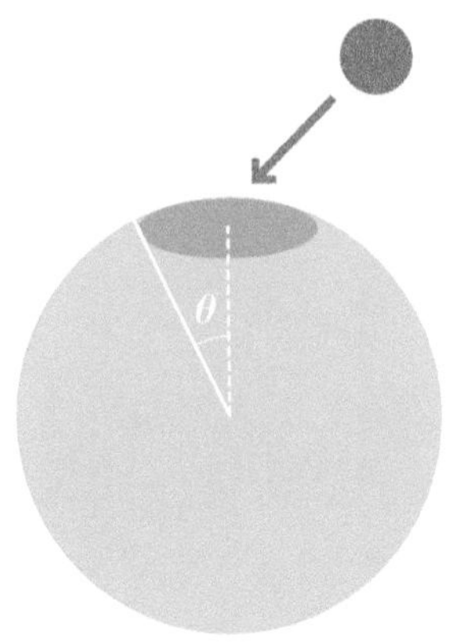

FIGURE 4.14
The active site of the target protein is only a fraction of the total area, implying that the collision rate with the active site is reduced. Calculations show that the collision rate is decreased in proportion not to the area of the active site [$\sin^2(\theta/2)$] but to its linear extent [$\sin(\theta/2)$]. The decrease in collision rate is therefore much less severe than one might expect.

4.1.9 Larger proteins diffuse more slowly

The high concentration of molecules in the cell means that the viscosity inside cells is very high: the cytoplasm is more like thick soup than water. There have been several measurements of the viscosity inside cells, which have shown that the viscosity in eukaryotic cells is about 5 times that in water, whereas in prokaryotic cells it is even worse, about 10 times [38–40]. Interestingly, the periplasm was even more viscous, being about 30 times more viscous than water [39], whereas the nucleus has a similar viscosity to the cytoplasm [41].

The inhomogeneity inside cells has another important consequence, namely that the effective viscosity is greater for big molecules such as proteins, which therefore diffuse even more slowly than you might expect, and smaller for water, which experiences a viscosity very similar to that of pure water [36, 42]. In particular, very large proteins and assemblies such as the ribosome are almost immobilized because they are trapped by the "cages" formed by microfilaments and other fibers: the "soup" description has therefore been suitably modified as "bouillon with vermicelli" [43].

The diffusional collision rate of two spherical molecules A and B in solution is given by the Smoluchowski formula [3]:

$$k_a = 4\pi DR$$

where D is the diffusion coefficient for the two molecules to come together and is the sum of the two individual diffusion coefficients ($D = D_A + D_B$), and R is the radius of the target, which again is the sum of the radii of the targets on the two individual molecules, namely $R = r_A + r_B$. [In the more common case where the active site is a small fraction of the surface of the protein target, then the rate is proportionally slower, as discussed in the first part of this chapter. Nonintuitively, the collision rate is decreased not by the fractional *area* of the active site but by the fractional *radius*, so that the Smoluchowski formula is modified to $k_a = 4\pi DR\sin(\theta/2)$: **Figure 4.14**.] In turn, D is given by the Stokes–Einstein relation:

$$D = kT/6\pi\xi r$$

where k is Boltzmann's constant, T is the absolute temperature, ξ is the viscosity of the solution, and r is the radius of the diffusing particle. In the case where the ligand A is much smaller than the enzyme B, which has an active site that is small compared with the overall area of radius r_S, then $r_B >> r_A$ and $D_B << D_A$, leading to a combined formula approximately

$$k_a = kTr_S/3\xi r_A$$

The collision rate thus depends on the viscosity of the medium. As discussed above, the viscosity inside cells is greater than that of water by a considerable amount. However, because of macromolecular crowding, the effective viscosity depends on the size of the molecule: the viscosity is much greater for proteins than it is for small molecules. We can therefore more or less ignore the question of viscosity if we are only concerned with the diffusion of small molecules.

The collision rate thus depends most critically on the size r_A of the diffusing molecule: small molecules diffuse faster than large ones. If we consider the binding of a ligand to an enzyme, it is therefore clear that in general we should expect the ligand to diffuse to the enzyme rather than the other way round.

This concept is reflected by the sizes of actual proteins, as shown elegantly again by Goodsell [44]. **Figure 4.15** shows the sizes of several proteins, given to scale, along with membranes, tRNA, and DNA. The proteins at the top left are toxins. The function of toxins is to diffuse rapidly to their site of action, so as expected these are very small. The next two rows are electron transport proteins such as cytochrome *c*; their function is to carry electrons. Normally

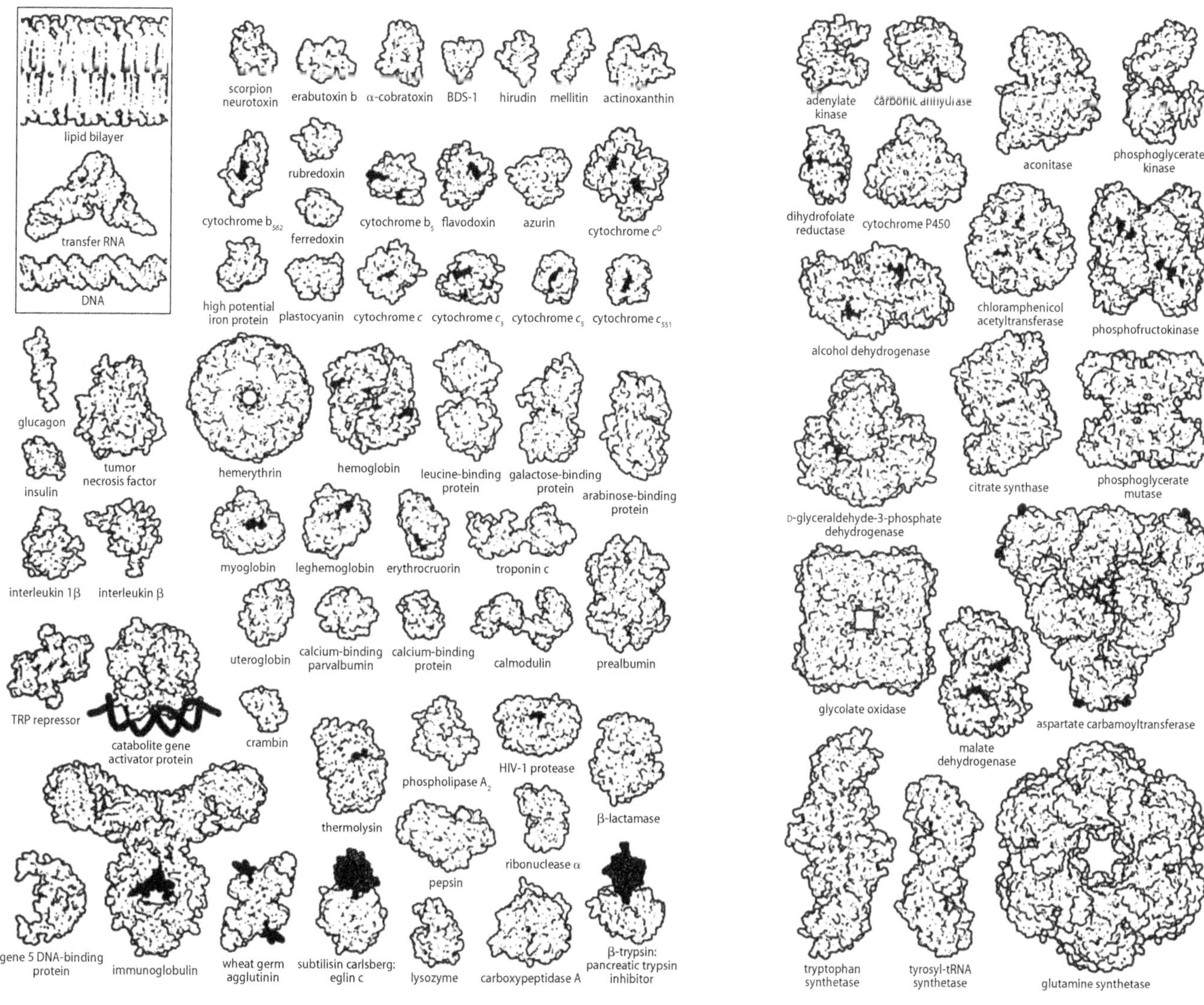

they need to do this very rapidly, so again they are small. The selection of proteins at the far left are hormones; like the toxins, they need to diffuse rapidly and are small. And in the center of the left side are carrier proteins, which again in general are small for the same reason that electron transport proteins are small.

FIGURE 4.15
The relative sizes of a range of proteins, plus membranes, tRNA, and DNA. (From D.S. Goodsell and A.J. Olson, *Trends Biochem. Sci.* 18:65–68, 1993. With permission from Elsevier.)

It is striking that the enzymes (on the right of Figure 4.15) are in general much larger than these carriers, hormones, and toxins, not least because they are usually oligomers (for reasons discussed in Chapter 2). This is because it doesn't really matter how slowly these molecules diffuse, as long as their substrates and products can diffuse rapidly to them. And the most striking result is that enzymes such as proteases, lipases, and nucleases (bottom left), whose substrates are macromolecules and therefore cannot diffuse rapidly, are again small. We can also see that proteins that bind to the macromolecular ligand DNA (bottom left) are small too.

It has been argued [45, 46] that thc concentration of free water in cells is so low that the cytoplasm should be considered more like a gel than a solution, or even a liquid crystal [47]. This argument leads to some fascinating speculations about phase transitions as being partly responsible for movement or compartmentation inside the cell [48], and the (decreased) role of diffusion. However, the idea has yet to be proven. One of the main arguments in its support is the high concentration of protein inside cells. This is undoubtedly the

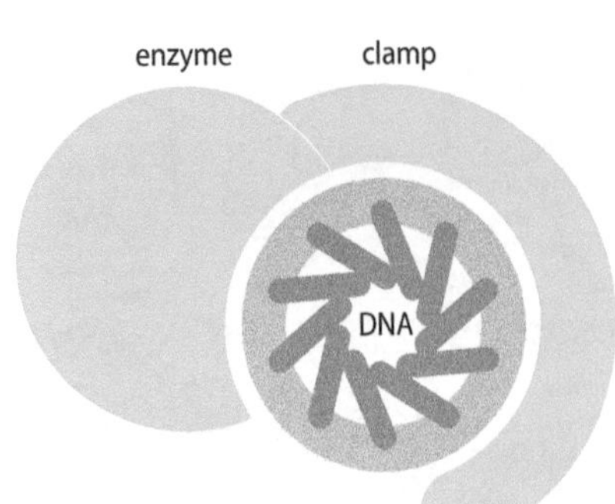

FIGURE 4.16
Many processive enzymes that operate on DNA or RNA have a clamp that holds them onto the substrate and prevents them from dissociating.

case but, as suggested in the previous section, proteins are probably distributed very anisotropically inside cells, for example being concentrated close to membranes. The argument is therefore possible but (on current evidence) unlikely.

On the molecular scale of proteins, *viscosity* is in fact the wrong word, because molecules are held back not by viscous drag but by Brownian dynamics. In fact, on the molecular scale many of our everyday concepts such as viscosity, gravity, and concentration have much less meaning, as expounded in a fascinating and thought-provoking paper [2].

4.2 HOW PROTEINS CAN FIND THEIR PARTNERS FASTER

4.2.1 Processivity decreases the off-rate from polymeric substrates

We have seen that enzymes diffuse slowly and are large. The exception is the enzymes that act on large and immobile substrates. Many of these catalyze multiple reactions on the same substrate: DNA polymerase adds not just a single nucleotide but many; cellulases digest not just one glycosidic bond but many. That is, they are **processive** enzymes. It is therefore a very clear advantage if they are somehow restrained to be close to their substrate. It would be very inefficient if DNA polymerase dissociated after the addition of each nucleotide and then had to find a suitable binding site all over again. DNA polymerase and reverse transcriptase achieve this processivity by having a clamp that locks them onto the DNA helix and prevents them from leaving (**Figure 4.16**).

Many cellulases use a different mechanism: they have one or more carbohydrate-binding modules in addition to the catalytic domain. The function of the modules is to attach the enzyme to the polymeric substrate and hold it there long enough for it to be able to perform several digestions (**Figure 4.17**). They need to bind weakly enough that the enzyme can eventually dissociate and move elsewhere to find another patch of substrate. The affinity of the binding domains is typically relatively weak (in the micromolar range), presumably being the optimum balance between binding too strongly and not allowing the enzyme to dissociate, and binding too weakly and not holding the enzyme adjacent to its substrate for long enough. Many enzymes (and **lectins, *4.6**) use several weakly binding modules rather than one very tightly binding one [49]; as argued many times in this book, a combination of multiple weak binding is not only easier to evolve and change but binds and releases faster, and offers more potential for regulation. Lectins have an interesting strategy: many are divalent, which allows them to bind cooperatively to large substrates and cross-link them together. Not only does this increase the affinity, it is also a good way of forming large macromolecular aggregates—a method also used by many immunoglobulins and the basis for the form of **immunoprecipitation (*4.7)** known as **hemagglutination (*4.8)**. The binding domain from glucoamylase (which digests starch) is of interest because it has evolved a hinge region to allow the enzyme to scout around the surface next to the binding domain, as well as two binding sites that function to twist the starch strands apart and therefore make the substrate more available to the enzyme (**Figure 4.18**) [50].

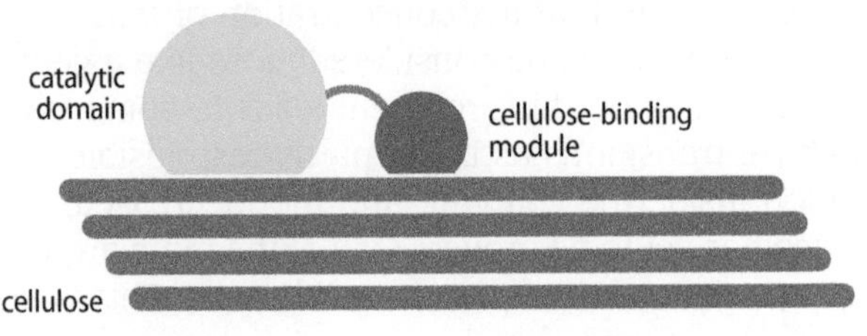

FIGURE 4.17
Cellulases (and other hydrolases that act on polysaccharides) generally have a catalytic domain plus one or more carbohydrate-binding domains that prevent the enzyme from dissociating from their substrate.

***4.6 Lectins**

Lectins are proteins whose function is to bind to saccharides. They generally recognize the ends of saccharide chains that are attached to proteins, for example integral proteins in the cell wall (**Figure 4.6.1**). They are used by eukaryotes for the recognition of specific proteins or cells, typically in cell–cell recognition. They are used, for example, in the recruitment of white blood cells to sites of inflammation, in the recognition of pathogens, in sperm–egg interaction, and by liver cells to bind and remove particular glycoproteins from the circulation. Many lectins bind to more than one saccharide and therefore cause clumping or agglutination of animal cells, which can be used for example in the clinical diagnosis of blood types. They can also be used to characterize and purify glycoproteins.

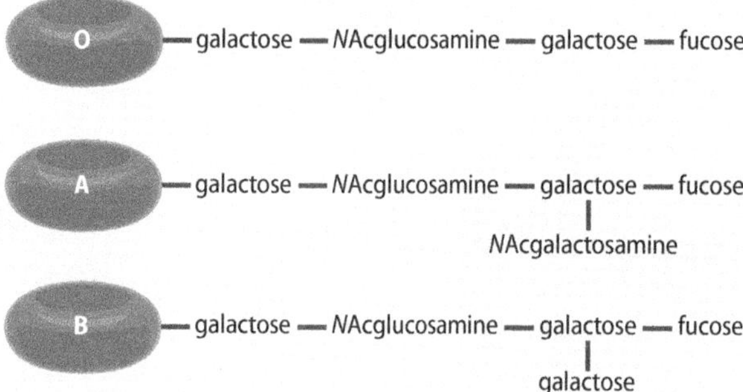

FIGURE 4.6.1
The difference between the A, B, and O blood groups lies in glycoconjugates attached to red blood cells. The normal test for blood type uses antibodies, but it is also possible to use lectins that specifically recognize the terminal galactose or *N*-acetyl galactosamine. Lectins have two or more binding sites and therefore lead to the aggregation of cells and 'clumping', in a similar way to that caused by bivalent antibodies.

Cellobiohydrolase is a processive enzyme that chops off disaccharide units from cellulose chains. It does this by having a tunnel that seems to perform several functions: helping to peel off a cellulose strand from a fiber, keeping the enzyme attached to its substrate in a similar way to the clamp on DNA polymerase, twisting the cellulose strand as it goes through the tunnel and therefore making it easier to hydrolyze, and acting as a ratchet to prevent the enzyme from moving backward on the polysaccharide chain and falling off (**Figure 4.19**)—an impressive range of functions!

4.2.2 Searching is much faster in two dimensions

It seems very reasonable, and can be shown fairly simply, that a search in two dimensions is much faster than a search in three dimensions (**Figure 4.20**). It is easy to imagine why this might be so. When we enter a dark room and want to switch on the light, we normally feel for the wall, and then feel along the wall until we hit the light switch. If the light switch is not located on the two-dimensional wall but is instead hanging on a cord in a three-dimensional volume, the search is usually significantly slower. Thus, what we do to find a light switch is to undertake a three-dimensional search (to find the wall) followed by a two-dimensional search along the wall to find the switch. Proteins very probably do the same, typically using a membrane as the two-dimensional surface: that

***4.7 Immunoprecipitation**

In immunoprecipitation, an antibody is attached to a bead. Therefore when it binds to its protein antigen, this attaches the antigen to the bead, which can be separated and the protein eluted to characterize it. It is particularly useful for examining protein complexes, because proteins will co-immunoprecipitate with their binding partners. One can for example attach a tag to a protein of interest such as glutathione S-transferase (GST), and immunoprecipitate with an antibody against GST. This will also precipitate any proteins bound to the protein of interest. In this form it is often known as a *pull-down assay*. Any bound proteins can then be identified by using SDS-PAGE. 'Simple' immunoprecipitation can also be used to identify and characterize proteins, for example to see whether the protein is present in a particular preparation or cell type. In this method, typically a soluble antibody is added, followed by an immobilized antibody-binding protein such as Protein A attached to beads: the target protein is then absorbed onto the beads.

*4.8 Hemagglutination

Hemagglutination is a form of immunoprecipitation, in which bivalent IgG or multivalent IgM antibodies are mixed with red blood cells. If the antibodies recognize an antigen on the surface of the red blood cell, they bind. This results in multiple weak interactions that cross-link red blood cells together (**Figure 4.8.1**). The resulting clumping of red blood cells provides a simple diagnostic marker for the presence of the antigen, for example in blood typing.

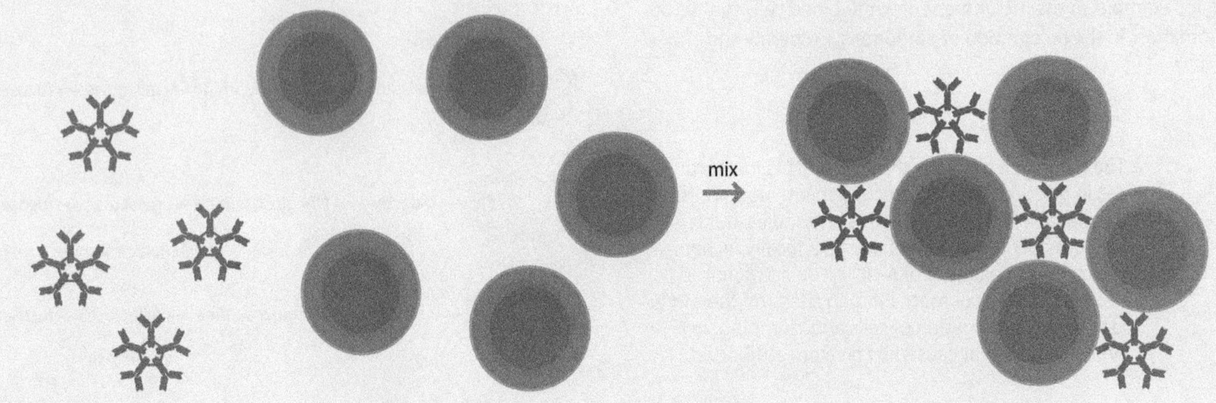

FIGURE 4.8.1
IgM antibodies have five bivalent antibodies in one assembly and therefore contain ten recognition sites. Incubation of IgM with red blood cells carrying a suitable epitope (for example, specific for a given blood type) results in clumping of the cells.

is, the quickest search strategy is to locate the target on a membrane, so that a searcher has only to find the membrane and then search along it.

Mathematically, the mean time for random diffusion to a target is proportional to R/r in three dimensions (where R is the radius of the search space, and r is the radius of the target; R/r therefore describes the size of the target relative

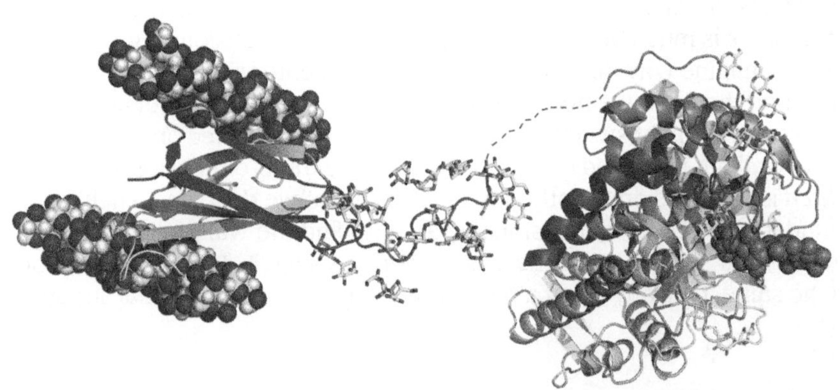

FIGURE 4.18
A model of the structure of fungal glucoamylase. It consists of an N-terminal catalytic domain, shown on the right, bound to an inhibitor, acarbose (brown spheres). At the C-terminal end of this domain is a long linker that winds around the domain: it starts at the left center of the domain, winds around the back and emerges at the top right, where it is glycosylated. This then forms an extended and heavily glycosylated linker (purple dashed line), which acts as a semi-rigid rod to hold the two domains apart [121]. At the C terminus is a starch-binding domain. This domain has two binding sites for starch helices, which are shown as spheres on the left. It binds the two helices almost at 90° to each other, and thereby pulls the (largely parallel) bundles of starch apart, thus making hydrolysis easier [122]. The SBD attaches the protein to starch, and the linker allows the catalytic domain to move around on the surface of the starch granule. The end of the SBD where the linker is attached is dynamic, giving the linker further flexibility [50]. This figure was constructed using the crystal structure of the catalytic domain from *Aspergillus awamori* (PDB file 1agm) and the NMR structure of the SBD from *Aspergillus niger* (PDB file 1ac0), together with a model of starch helices and part of the linker.

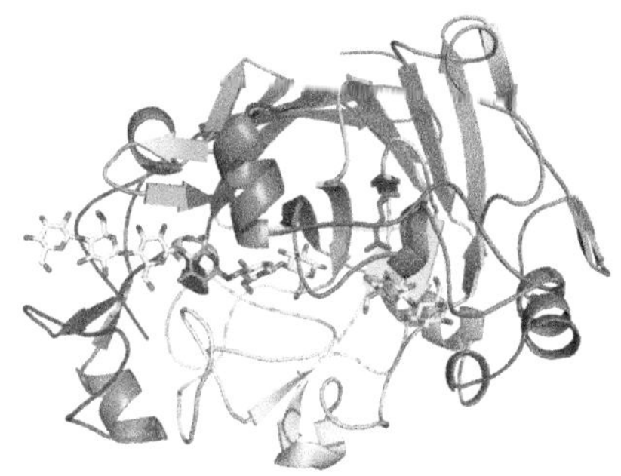

FIGURE 4.19
Cellobiohydrolase (PDB file 7cel). The active site's general base is Glu 217 (magenta), and the protein is bound to cellohexaose and cellobiose, representing a cleaved cellooctaose. Note how the saccharide passes through a tunnel in the enzyme and how the saccharide substrate is twisted as it passes through the tunnel.

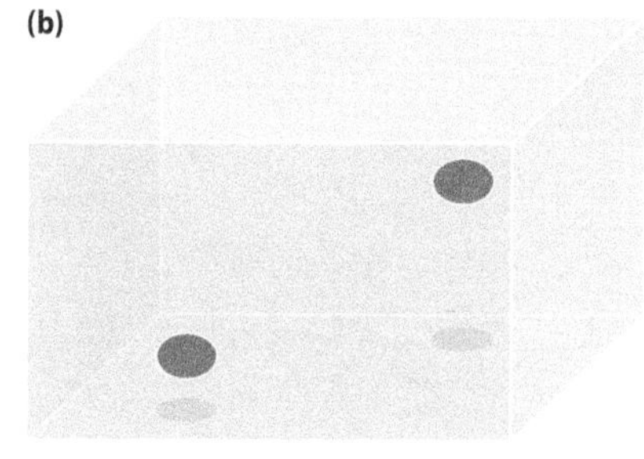

FIGURE 4.20
Searching is faster in two dimensions (a) than in three (b).

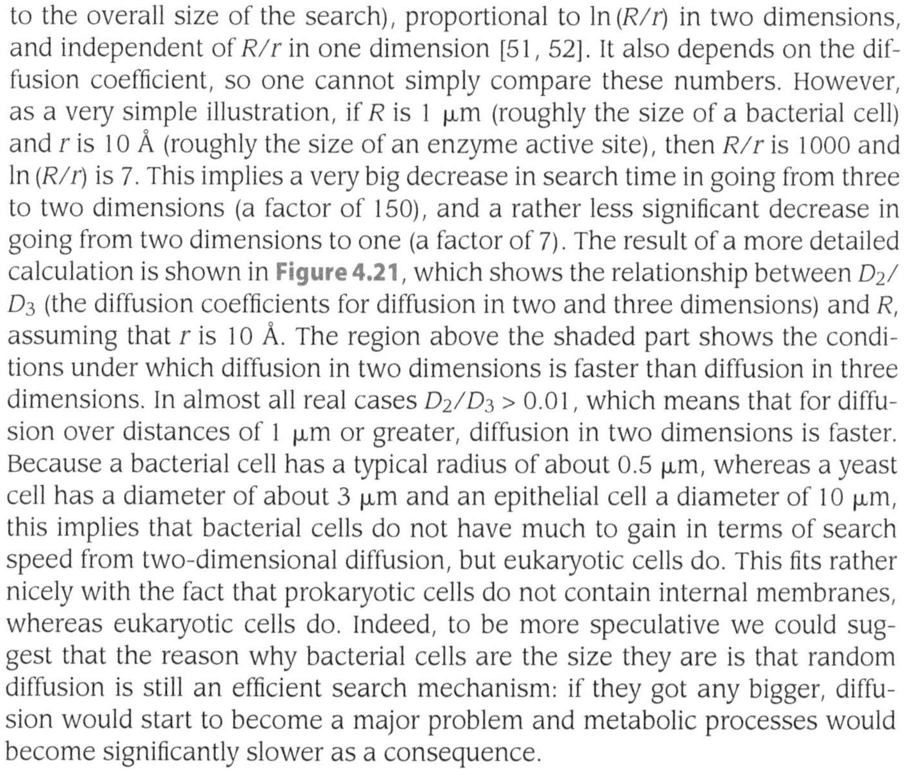

to the overall size of the search), proportional to ln (R/r) in two dimensions, and independent of R/r in one dimension [51, 52]. It also depends on the diffusion coefficient, so one cannot simply compare these numbers. However, as a very simple illustration, if R is 1 μm (roughly the size of a bacterial cell) and r is 10 Å (roughly the size of an enzyme active site), then R/r is 1000 and ln (R/r) is 7. This implies a very big decrease in search time in going from three to two dimensions (a factor of 150), and a rather less significant decrease in going from two dimensions to one (a factor of 7). The result of a more detailed calculation is shown in **Figure 4.21**, which shows the relationship between D_2/D_3 (the diffusion coefficients for diffusion in two and three dimensions) and R, assuming that r is 10 Å. The region above the shaded part shows the conditions under which diffusion in two dimensions is faster than diffusion in three dimensions. In almost all real cases $D_2/D_3 > 0.01$, which means that for diffusion over distances of 1 μm or greater, diffusion in two dimensions is faster. Because a bacterial cell has a typical radius of about 0.5 μm, whereas a yeast cell has a diameter of about 3 μm and an epithelial cell a diameter of 10 μm, this implies that bacterial cells do not have much to gain in terms of search speed from two-dimensional diffusion, but eukaryotic cells do. This fits rather nicely with the fact that prokaryotic cells do not contain internal membranes, whereas eukaryotic cells do. Indeed, to be more speculative we could suggest that the reason why bacterial cells are the size they are is that random diffusion is still an efficient search mechanism: if they got any bigger, diffusion would start to become a major problem and metabolic processes would become significantly slower as a consequence.

Two-dimensional searching on membrane surfaces is indeed a very widely used method in biological systems, as we shall consider further when we look at signaling systems in Chapter 8. It is particularly important for the formation of protein:protein complexes, perhaps because protein diffusion in three dimensions is relatively slow. It is estimated that about 20–30% of microbial proteins are located at the membrane [53], either by having transmembrane sequences or covalent modifications that attach them to a membrane surface. Many of these have no obvious need to be at the membrane surface (that is, they are not channels or transporters, for example) and presumably are attached to the membrane merely because this makes it faster to locate their binding partners. Eukaryotic cells have a very large amount of internal membrane: in hepatocytes, for example, it is estimated as 50 times the area of the cell membrane (quoted in [42]). It is therefore not difficult for proteins to find a membrane to bind to.

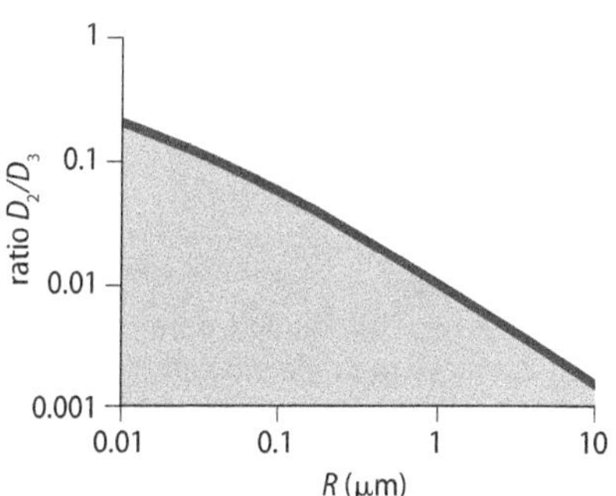

FIGURE 4.21
The ratio between the diffusion coefficients in two versus three dimensions, as a function of R, the radius of the search space, calculated for a target with a radius of 10 Å. Within the white region, diffusion in two dimensions is faster than in three dimensions. (Redrawn from G. Adam and M. Delbrück, in Structural Chemistry and Molecular Biology (A. Rich and N. Davidson, eds), pp. 198–215, 1968. San Francisco: W.H. Freeman.)

Eukaryotic cells are about 1000 times larger than prokaryotic cells, implying that the random walk search process in eukaryotes is much slower than in prokaryotes. It is also striking that a higher proportion of eukaryotic proteins are associated with membranes. Presumably these two facts are related by the increased ease of searching in two dimensions rather than three (see Figure 4.20). Moreover, it is becoming clear that the membranes on different eukaryotic organelles have different lipids and different surface charges. Therefore the search problem can be significantly decreased not only by searching in two dimensions but also by searching only on specific organelles.

4.2.3 Searching is slightly faster again in one dimension

As discussed above, searching in one dimension is even faster than in two, although not enormously so. One-dimensional searching certainly occurs. The most obvious example is in DNA-binding proteins. There has been considerable research on how DNA-binding proteins find their specific target. The general answer is that they bind randomly somewhere on DNA and then diffuse along it until they find their specific site [54], even in eukaryotes [55]. This provides an increase in rate of a factor of up to 10^4 in comparison with a simple three-dimensional search [56]. There is good evidence that proteins also sometimes come off and jump to other sites. Considering both the length of eukaryotic DNA and the fact that it is covered by proteins and is usually supercoiled (thus bringing strands close together), for eukaryotes this is likely to be overall a faster search method than merely linear searching [57].

The actual search method used probably varies from protein to protein. An instructive example is provided by the human Oct-1 transcription factor, which is a member of the POU family and consists of two helix–turn–helix DNA-binding domains connected by a flexible linker. Oct-1 is attached to DNA at two almost equivalent sites. Under physiological conditions, this protein shuffles along DNA in a one-dimensional search. However, it also "jumps" from one segment of DNA to another. This jump can involve complete dissociation of the protein from the DNA. However, at the high concentrations of DNA typically found in the nucleus, it can also jump by a stepping movement (**Figure 4.22**) [58]. The two jumping processes probably occur at comparable frequencies. This provides a nice example of the benefit of binding via two domains.

For low-dimensional searching to be fast compared with searching in a higher dimension, the diffusion rate must be fast. (In reference to Figure 4.21, D_2/D_3 cannot be too small; otherwise it becomes more efficient to remain in three dimensions.) In other words, it is important that the searcher does not get "stuck" on the one- or two-dimensional surface. The surface of DNA, for example, is locally bumpy: does this not slow down diffusion too much? The answer seems to be no, because the proteins that search along DNA in this way interact almost exclusively by electrostatic interactions, therefore allowing them to maintain a lubricating solvent layer between the protein and the DNA and thus slide more smoothly. (Interestingly, this also implies that one-dimensional searching along DNA should involve a helical rotation of the protein to track along the phosphate backbone, an implication borne out by most experiments including single-molecule visualization [59].) The same is probably true for positively charged proteins searching on negatively charged membrane surfaces [3].

FIGURE 4.22
Oct-1 has two DNA-binding domains and can move to a different part of DNA either by sliding along the DNA or else by jumping, either in a completely dissociative mechanism (a) or by stepping from one strand to the next (b). (After M. Doucleff and G.M. Clore, *Proc. Natl. Acad. Sci. USA* 105:13871–13876, 2008. With permission from the National Academy of Sciences.)

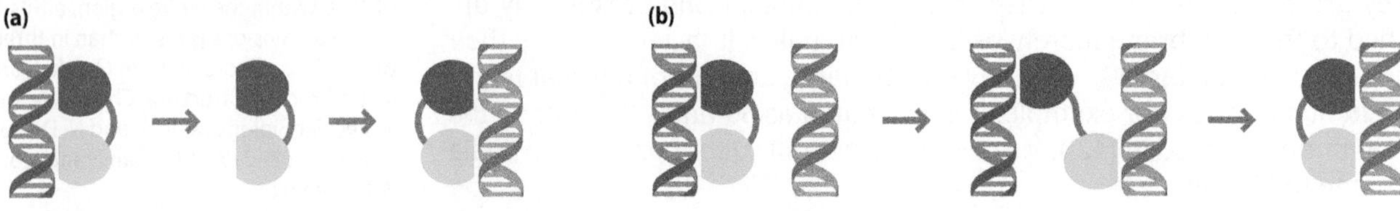

The other obvious example of a one-dimensional search is proteins that bind to microtubules. Much of the transport in eukaryotic cells involves organelles attached to motor proteins such as dynein and kinesin, which track along microtubules toward the minus or plus end, respectively. These proteins are usually processive (Chapter 7): they make many steps before dissociating.

At this point it may be helpful to introduce an analogy between cells and airports [21]. A cell has to direct incoming and outgoing traffic to the correct places without them interfering with each other, and it has to target cargo to the correct location; so does an airport. The airport does this with a variety of mechanisms. Cargo (such as baggage) to be targeted to the correct destination is usually tagged and then sent through a transport system that recognizes the tags and sorts the cargo appropriately. The cell does something similar. Most airports have the departures and arrivals physically separated, and streams them both so that they rarely cross. In particular, arriving passengers who want to leave the airport are usually directed so that they meet up with their baggage at a late stage and then are rapidly directed to an exit: the same thing happens in cells. In cells the streaming is still poorly understood but clearly involves one-dimensional movement in packages: the airport bus serves a similar function. And so on; see Problem 1.

4.2.4 Searching is faster in smaller compartments

The previous section argued that membranes are important in providing surfaces on which proteins can recognize each other. They are also of course important in separating one part of the eukaryotic cell from another. Once the cell has evolved a mechanism for tagging proteins and directing them to the appropriate compartment, then organelles are tremendously useful for keeping different processes separate, and allow the concentrations of metabolites and proteins to be very different in different parts of the cell. In general, the different organelles within the cell do indeed have well-defined biological functions such as energy generation, waste disposal, secretion, and post-translational processing.

One benefit of having different functions segregated is that when the cell needs to mount a response it can localize the metabolic changes to a limited number of parts of the cell and therefore does not need to perturb the metabolism elsewhere too much.

4.2.5 Sticky arms are useful for short-range searching

We have seen above that molecular association can be accelerated considerably by electrostatics. Opposite charges attract, and the attraction is a reasonably long-range interaction. This is the reason why electrostatic forces have been emphasized so far. The other kinds of molecular interaction that can occur in water are, by contrast, much shorter in range. In particular, the hydrophobic interaction is essentially a contact interaction: the two surfaces have to be close enough together that they can expel the water molecules in the interface between them, which is where the energy for the interaction comes from. Similarly, hydrogen bonding requires water to be expelled and a very precise placing of the two interacting groups before the interaction becomes favorable. Both interactions are therefore not useful as a way of attracting ligands from a distance. However, once the ligand is close enough, then both are very useful interactions. The next few sections discuss how these short-range interactions can be used to speed searches.

By way of illustration, I offer two analogies. One is the chameleon, which eats flies by sitting still and flicking its tongue out when a fly gets close enough. The tongue sticks to the fly and allows the chameleon to drag in the fly without needing to move its body. Binding is therefore achieved by having only a small

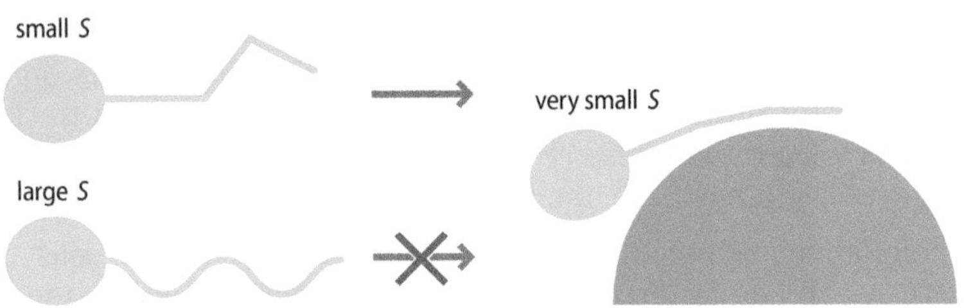

FIGURE 4.23
When a peptide binds to a target, it loses almost all of its entropy (*S*). Therefore for effective binding, it should not have too much entropy to start with; that is, it should be relatively rigid. This is often accomplished by having a high proportion of prolines in the peptide, which impart rigidity.

part of the chameleon that moves; this is a very effective mechanism. Another is closer to what proteins actually do, and goes back to the party analogy introduced in Section 4.1.8. If you are in a very crowded room, and cannot readily diffuse around the room but want to interact with a particular person who is close but not close enough to talk to, one way to do this would be to reach out an arm, grab hold of the person, and then effectively wait until random diffusion moves the intervening people out of the way and allows you to pull in the target. I would not recommend this as a technique to be used in parties, but it is widely used by proteins, which have fewer concerns over social niceties.

The sticky arm used in this way by proteins needs to have some particular properties. First, it needs to be hydrophobic. This is something of a problem, because hydrophobic peptides tend to coil up in water. It therefore also needs to be fairly stiff and extended. It is a big advantage for the arm to be stiff, because this decreases the inevitable loss of entropy (*3.1) when the arm binds to its target. This needs explaining. The free energy of binding is the sum of the enthalpy change and the entropy change, corresponding respectively to the bond energies formed and the amount of disorder lost in the arm:

$$\Delta G = \Delta H - T\Delta S$$

A sticky arm is usually a single peptide chain with a relatively small area. It therefore cannot form many bonding interactions with its target and thus is limited to a relatively small favorable enthalpy of binding. This means that it cannot afford to have too unfavorable an entropy of binding: it cannot afford to lose too much flexibility, otherwise it will prefer to be free, loose, and floppy rather than stick to its target but be more rigid (the favorable enthalpy will not be large enough to overcome the unfavorable entropy). The bound peptide is of necessity fairly rigid and low in entropy, so this means that the free peptide must already be reasonably rigid, so that it does not lose too much entropy on binding (**Figure 4.23**). The solution to a peptide that is hydrophobic, stiff, and extended is to have it rich in proline residues.

4.2.6 Proline-rich sequences make good sticky arms

Proline has a hydrophobic side chain and is thus genuinely hydrophobic. However, it also has a more electron-rich carbonyl group than other amino acids and is thus a good hydrogen bond acceptor: in other words it is well hydrated and forms strong hydrogen bonds and is therefore also hydrophilic. Polyproline sequences are in fact very soluble in water, much more so than most poly-amino acids. Proline has a limited backbone conformational mobility, and polymers of proline tend to form highly extended structures known as **polyproline II helices** in water, which have a 120° rotation from one amino acid to the next and thus form a threefold helically twisted structure (**Figure 4.24**). This also implies that polyproline does not have any tendency to coil up into a hydrophobic cluster, which makes the surface readily available for binding, and (like the chameleon's tongue) gives its host protein a good "reach." Proline-rich regions in proteins are typically near either the N terminus or the C terminus and therefore do form genuine "arms" that project out from the main body of the protein [60]. In addition, because proline has a

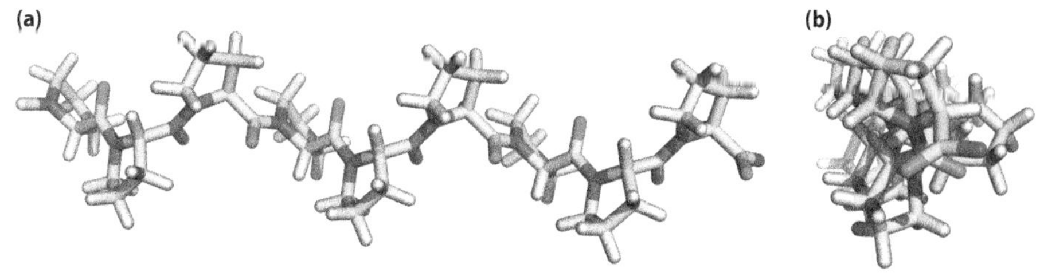

FIGURE 4.24
The structure of an ideal polyproline II helix. (a) From the side. (b) From the end. There is a 120° rotation from one residue to the next. This means that every third residue is in an identical orientation.

limited conformational mobility, polyproline sequences are already rather rigid and therefore the loss in entropy on binding to a protein target is small. All these reasons make polyproline sequences ideal as sticky arms. As discussed in the next section, peptide sequences also have rapid on- and off-rates. This makes sticky arms, and proline-rich sticky arms in particular, highly suitable as devices for protein binding within signaling systems, which often require rapid on- and off-rates. Some examples are presented in Chapter 8.

A sticky arm should not be completely rigid: imagine how difficult it would be for a protein to have a rigid arm permanently protruding from the surface (**Figure 4.25a**). In fact, the ideal sticky arm has rigid sections with flexible linkers, not unlike a real arm with elbows (Figure 4.25b). This provides the optimum combination of rigidity (and therefore minimal loss in entropy) with flexibility. An interesting example comes from pyruvate dehydrogenase (Chapter 10), which uses proline-rich linkers to produce jointed motion of domains. These linkers contain alternating alanine and proline residues, with occasional glycines. The glycines act as hinges, by providing sites of greater flexibility. The $(AP)_n$ sequences form somewhat flexible polyproline II helices. Replacing this sequence by only prolines produced a more rigid polyproline II helix, whereas replacing it by only alanines gave an α-helical peptide [61]. In both cases the resultant enzyme was less active, presumably because the linker was too inflexible. The function of a proline-rich arm can be altered in a graduated way by substituting other amino acids for prolines. Replacement by glycine decreases affinity by about 4 kJ mol^{-1} per glycine, by a combination of entropy and enthalpy, whereas alanine has a smaller effect [62]. Because kT is 2.5 kJ mol^{-1} at room temperature, this represents a very significant drop in affinity.

4.2.7 Sticky-arm interactions have fast on- and off-rates

Proline-rich regions provide a unique combination of rapid on- and off-rates with reasonably strong binding (up to approximately micromolar) and are therefore used extensively in situations where these characteristics are important: typically this means signaling systems. As discussed in Chapter 8, signaling relies on the recognition of one domain or peptide motif by another domain. There are no fewer than three well-characterized and common signaling domains (SH3, WW, and EVH1, the last of these being a subset of the Wasp homology 1 or WH1 domains) that recognize proline-rich sequences [63]. Proline-rich sequences typically form polyproline II helices, which have a 120° rotation per amino acid and therefore a three-residue repeat. The standard proline-rich recognition sequence has a PXXP motif, in which the two prolines are in equivalent positions in the helix but one turn apart: they are therefore well placed to interact with a hydrophobic protein surface, which is typically made from aromatic residues in the binding protein (**Figure 4.26**). The

FIGURE 4.25
The importance of a hinged sticky arm. A protein that had a rigid protruding arm (a) would be unwieldy and get tangled up with other parts of the cell. A hinged arm (b) has almost the same entropy and is much more maneuverable.

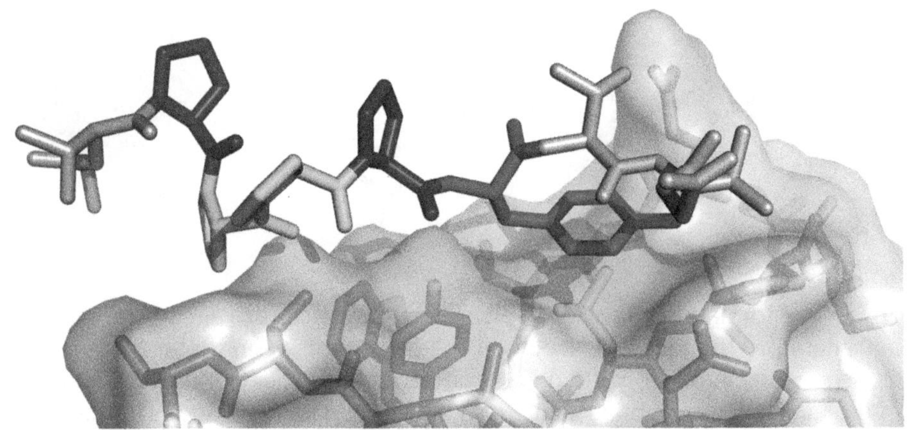

FIGURE 4.26
The structure of the complex of the WW domain of YAP65 with a proline-rich peptide, GTPPPPYTVG (PDB file 1jmq). The four prolines form a polyproline II helix, with prolines 2 and 3 (yellow) contacting the protein surface, and the other two (blue) keeping the peptide in the polyproline II helix. The tyrosine ring (magenta) also contacts the protein surface, which has a complementary hydrophobic surface composed of Trp 39, Tyr 28, and Leu 30 (green).

recognition surface can be fairly rigid and exposed, producing rapid on- and off-rates and strong binding, and requiring only six or seven residues in the proline-rich core sequence.

Interestingly, because of the threefold symmetry of the polyproline II sequence, it looks almost the same from either direction, and it even has its carbonyl groups in such similar places that it can hydrogen bond to the same groups in both orientations. This means that SH3 domains have the ability to bind proline-rich sequences in both orientations (**Figure 4.27**). The orientation is determined by the position of a positively charged residue that interacts with a negatively charged region on the SH3 domain: peptides with sequence +XXPXXP (where + is a positively charged residue) bind in a "class I" orientation, whereas those with sequence PXXPX+ bind in the opposite "class II" orientation. Some SH3 domains, including that in the important Src signaling system, can bind peptides in both orientations, which may endow them with the ability to switch binding partners to generate radically different geometries.

An example related to signaling is that several proteins involved in cytoskeletal remodeling have proline-rich regions, including activators of the Arp2/3 complex which regulates actin polymerization downstream of several receptors [64]. Profilin has a polyproline-binding sequence, but it also binds actin and regulates assembly of the actin filament. Presumably again the value of proline-rich sequences is that they can bind and release rapidly. Some of these

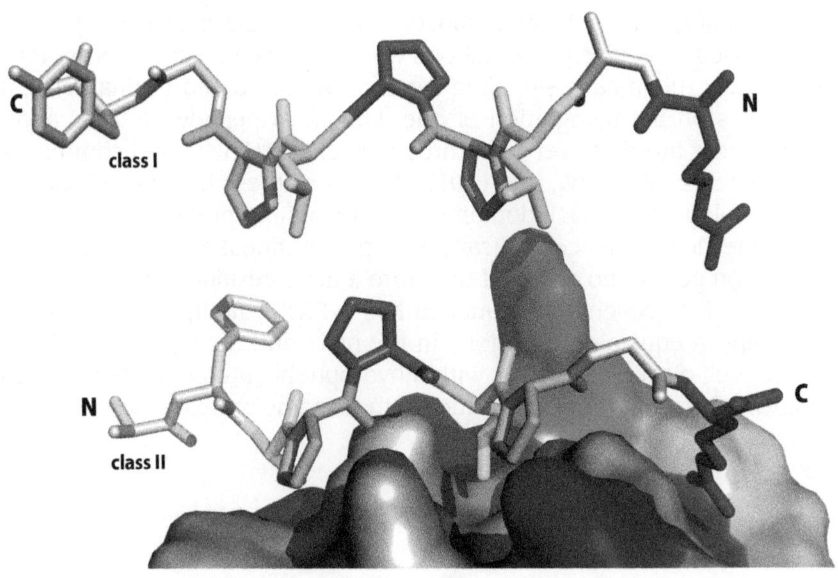

FIGURE 4.27
Interactions of a Src SH3 domain with proline-rich peptides (PDB files 1prm and 1rlq). The protein can bind to peptides in both orientations. It has a surface containing two hydrophobic pockets (left and center) separated by ridges, and a positively charged pocket (right, blue). Peptides in a polyproline II conformation can bind with a proline (yellow) plus an adjacent hydrophobic residue (green) in the hydrophobic pockets, and an arginine (red) in the positive pocket. The orientation of the peptide is determined by whether the arginine is at the N-terminal end (as in the class I peptide RALPPLPRY) or the C-terminal end (as in the class II peptide AFAPPLPRR).

polyproline sequences also bind to SH3 and WW domains, which normally belong to signaling pathways, thus suggesting a link between receptor signaling and actin polymerization (and hence between signaling and movement along the cytoskeleton). A further link between signaling and the cytoskeleton is the observation that some "unconventional" myosins (namely myosins other than myosin II, the standard actin-binding muscle protein) contain SH3 domains that seem to have a function in localizing them to polarized actin patches, sites for cell growth [65, 66]. It is likely that evolution adapted the polyproline-rich sequence from one function to the other because of its favorable properties but was then faced with the problem of cross-talk between the old and new functions. One potential fate would have been that the new function was selected against and removed. However, in this case, the old and new functions were sufficiently related that evolution tinkered with the new function, adding extra embellishments to improve selectivity, with the result that similar proteins can be successfully used in both systems, without cross-talk except where such a function is desirable.

Sticky-arm interactions involving proline are regulated in several different ways. A particularly common method is to include serines or threonines within the proline-rich sequence and phosphorylate them reversibly [63]. This mechanism is used in RNA polymerase II, whose C-terminal domain contains a long tandem repeat section containing the sequence YSPTSPS. Phosphorylation of the two serines and subsequent dephosphorylation serve to regulate the binding of a large number of additional enzymes to the transcriptional machine (Section 9.3.3).

Sticky arms are not restricted to proline. Nor do they need to be extended peptides, although this is the normal situation because most "random coil" peptides are inherently extended (Chapter 1). Peptides that are not proline-rich have a weaker inherent binding affinity (they are less sticky), partly because of their greater mobility and therefore greater loss in entropy on binding, and partly because they usually form weaker interactions, for reasons discussed above. Proteins that bind to non-proline-rich peptides therefore tend to need longer binding sites than the roughly six or seven residues used by SH3 or WW domains to bind proline-rich regions [63]. An example of such a binding site is the protein β-catenin (known as Armadillo in *Drosophila*), which has the remarkable structure shown in **Figure 4.28**. This presents a long and extended binding site, well suited to binding an unstructured peptide arm and giving the possibility of binding in a graded manner, more like a volume control than an on/off switch, and therefore integrating information from different signals [67].

A large number of other sticky arms are known. Some of these are terminal peptides, and some are internal and are more appropriately called linear motifs [68]. They have a wide range of functions including signaling, markers for destruction or post-translational modification, and transport, and some

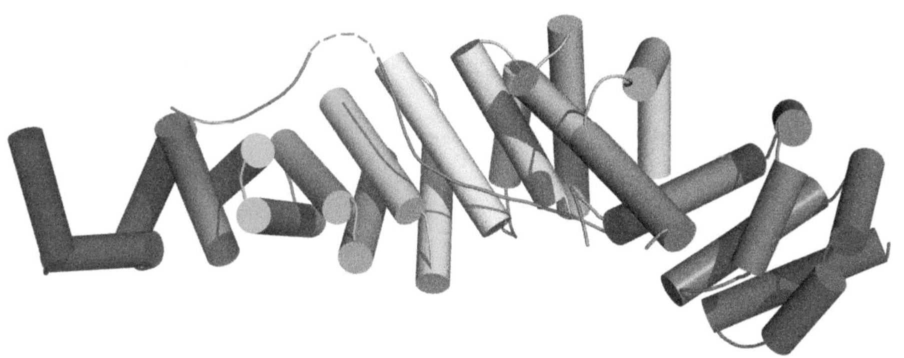

FIGURE 4.28
The β-catenin Armadillo repeat bound to a peptide from adenomatous polyposis coli protein. The peptide (mauve) runs along the repeat; on average about four residues are recognized by each repeat. The dashed region is disordered in the crystal (PDB file 1v18). The Armadillo repeat is a rare example of a tool with a variable size (Section 2.4.2).

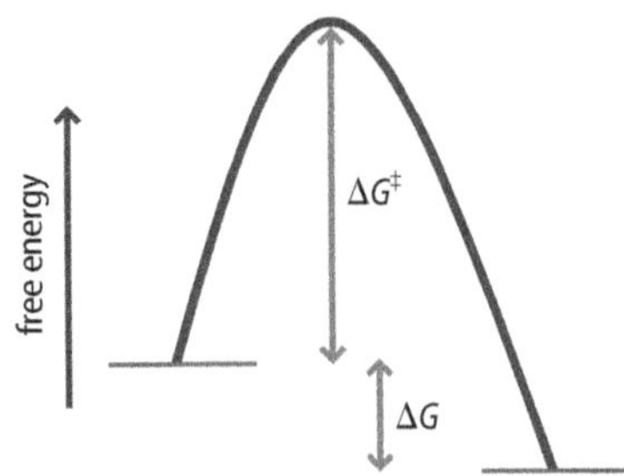

FIGURE 4.29
Free-energy diagram for the binding of one molecule to another.

websites are listed at the end of this chapter. One set of terminal motifs is the signal sequences that target the host protein to different cellular locations (Section 7.5.1). In contrast to the identification of domains, which can be done fairly straightforwardly with sequence alignments, the identification of linear motifs has proved a difficult problem. This is because they are shorter (no more than 10 residues) and more variable, both in their sequence and in their phylogenetic distribution [69].

Where stronger and slower interactions are required, it makes sense to use entire modules as sticky elements, thus scaling up the interaction. Several extracellular modular eukaryotic proteins seem to do this (see, for example, Section 2.1.4). It seems likely that in some cases these modules act as multiple weak sticky binding sites, for example in thrombomodulin [70].

4.2.8 Sticky arms are fast because they zip up rather than lock on

So far we have considered mainly the *energetics* of binding. How about the *rate*? As noted already in Chapter 2, a large interface usually means a slow on- and off-rate, whereas a small interface (such as that presented by a peptide sticky arm) tends to mean a fast on- and off-rate. Why is this? **Figure 4.29** shows the standard free-energy diagram for a reaction, which applies equally well to binding. The rate of this reaction is given in the normal way by

$$\text{rate} = A\exp(-\Delta G^{\ddagger}/RT)$$

where $\Delta G^{\ddagger}$ is the activation energy, which is by definition the difference in energy between the starting structure and the highest point on the energy profile. For a binding reaction, what determines the highest energy point? It is almost always the removal of hydrating water molecules. Water forms hydrogen bonds to the surfaces of the two interacting molecules in their free state and is not present in the bound state (**Figure 4.30**). However, when A and B are bound, these water molecules are now in bulk solvent and free to bind to each other. Therefore roughly the same number of hydrogen bonds are formed in both free and bound states, and the overall change in hydrogen-bonding enthalpy (*3.1) is small. However, at some point during the binding, we have to have a situation where the solvating waters have been removed but are not yet liberated to bulk solvent. At this point we have lost several hydrogen bonding contributions. At roughly 15–20 kJ mol^{-1} per hydrogen bond, this represents a large energy barrier and is thus the major contributor to the rate.

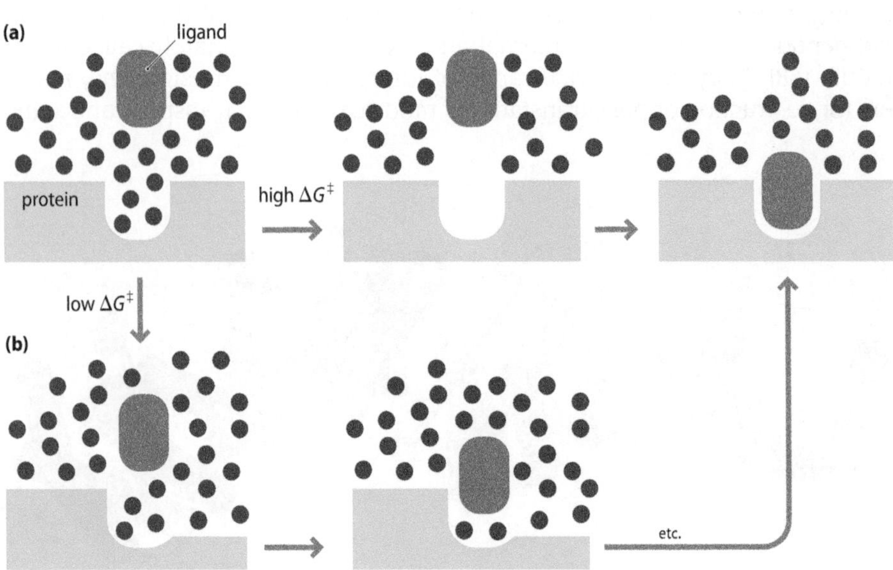

FIGURE 4.30
The binding of a ligand to a protein involves desolvation of their binding surfaces: removal of the hydrating waters (brown). (a) If this has to be done completely before binding occurs, it represents a major energy barrier. (b) The barrier is much smaller if the water molecules can be removed one by one in a zippering process. This diagram is grossly oversimplified; in reality it is unlikely that all the waters within the interface need to be removed.

(Astute readers will be comparing this figure of 15–20 kJ mol^{-1} for the free energy of a hydrogen bond with the figure of 2–3 kJ mol^{-1} given in Chapter 2. The difference is very instructive. Chapter 2 was discussing the average energy for a hydrogen bond as a contribution to the free energy of protein folding, and was therefore comparing the free energy of peptide–peptide hydrogen bonds with that of peptide–water hydrogen bonds. This *difference* in free energy is small. Here, what matters is the *total* free energy of a hydrogen bond, which is of course much greater.)

In fact, this large activation energy for desolvation means that, where possible, the desolvation will not occur in one step but will take place as a series of smaller steps: in each, a solvent molecule is removed and is allowed to diffuse into bulk water. That is, desolvation, and the consequent binding of the two molecules, normally occurs more like a zipper than a rigid docking (see Figure 4.30b). Like many good ideas in biochemistry, this one is not exactly new [71] (see also Section 5.2.8).

To return to the theme of this chapter, a small contact surface area means that the number of solvent waters is relatively small, and therefore the activation energy for desolvation is small and the binding is rapid. For a peptide ligand, the rate is enhanced by the flexibility of the peptide, which makes it easier for it to "zip" onto its target and for the solvent waters to squeeze out at the sides.

4.3 NATIVELY UNSTRUCTURED PROTEINS

Sections 4.2.5 through 4.2.8 demonstrated the importance of sticky arms for rapid protein:protein interaction. We now move on to consider a different kind of sticky arm: natively unstructured proteins.

4.3.1 Natively unstructured proteins are common

It is now clear that a significant fraction of proteins, estimated as about 4% in bacteria but 30% in eukaryotes [72], contain natively unfolded regions in the cell *in vivo*. Remarkably, about 25% of mammalian proteins are predicted to be entirely natively unfolded [73]. We normally think of an unfolded protein as lacking the specific shape that allows it to function correctly, and therefore as non-functional. What, then, is the significance of so much natively unfolded sequence, particularly when one considers that unfolded protein chains are much more likely to get degraded or aggregated than folded chains?

It should first be mentioned that it has been suggested that, *in vivo*, supposedly "natively unfolded" proteins may be more structured than they seem to be *in vitro*. Macromolecular crowding, as discussed earlier in this chapter, stabilizes folded proteins against unfolded ones, and it is therefore likely that proteins inside cells are more folded than they are in dilute aqueous solution. For example, it has been shown that the intrinsically disordered protein FlgM becomes partly structured at 400 g l^{-1} glucose, with the C-terminal end becoming folded but the N-terminal end remaining unstructured [74]. However, the effect depends on the difference in radius of gyration between unfolded and folded and is therefore usually fairly small, implying that macromolecular crowding may change the equilibrium but still leaves "natively unfolded" proteins mostly genuinely unfolded.

This topic is rather new and unfamiliar, and the picture is still far from clear. There have been a number of reviews [72, 73, 75–80], which generally agree on the main points. There is not a single class or function of natively unfolded proteins. In fact, they fall into two main groups. First, there are proteins that are permanently unfolded, which have a range of different functions for which a nonglobular state is preferred and which therefore "need" to be disordered. It is for example quite likely that parts of many scaffold and signaling proteins are permanently natively unfolded, because their function is mainly to regulate the degree of association between two proteins, as discussed in Chapter 2, and this depends largely on the (random coil) distance between them. The nucleoporin proteins that line the nuclear pore (Chapter 7) are probably natively unfolded, because their function is to act as a curtain-like barrier across the pore. The protein resilin is an elastic protein found in several insects that allows them to jump distances many times their body length, and it is natively unfolded (although cross-linked). The major protein constituents in saliva, the proline-rich proteins, are randomly coiled because their function is either to act as lubricants for food swallowing or else to bind and remove polyphenols (tannins) from food [81]. In both cases they need to be extended with a large reach to perform this function. Many of these proteins are also of **low complexity**; that is, their sequences consist largely of multiple repeats of simple sequences [82]. It is possible that this is because they are recent products of evolution and arose by **repeat expansion**, which is a simple way for evolution to generate new sequence when the exact sequence does not matter.

Second, there is a more interesting group of proteins that are unfolded but become ordered when bound to their target protein. There are probably two main benefits to be gained from such behavior. The first is that it allows the protein to retain *specificity* in its binding, without having to sacrifice so much in binding *rate*; the second, which is equally important if not more so, is that it permits specific binding without having very *strong* binding. Two other benefits have also been proposed: that because the protein changes its conformation as it binds, it can adapt its conformation to fit different target proteins; and that the unfolded conformation allows the protein to be degraded more rapidly and thus assists in the regulation of interactions that need to be turned off rapidly. We consider these in turn.

4.3.2 Natively unstructured proteins permit specific binding with fast on-rate

Recall the arguments presented a little earlier in this chapter: a selective interaction requires many amino acids in the interface, which must be positioned exactly correctly to favor the correct interaction and disfavor the incorrect ones. This therefore requires a reasonably large interface. However, a large interface also tends to mean a slower interaction, because the requirement for an exact matching of the orientation is more stringent; there are more water molecules in the interface to be removed, and there are therefore fewer options for zipping up the surface in a flexible way. This would therefore seem to imply that a specific interaction is necessarily slow. However, this is where protein disorder can be an advantage. If the protein only folds up to its native structure as it binds to its partner, then it can zipper up flexibly, and the water molecules in the interface can be liberated one by one. Therefore the binding can be fast. At the same time, it can still be strong (if this is needed) because an unfolded protein can fold up as it binds to its partner to make a large surface area at the interface.

FIGURE 4.31
Fly fishing.

This idea—that a protein can interact initially with a binding partner that is unfolded, and that the binding interface develops as the binding partner folds—has been called *fly casting* [83]. The reference is to fly fishing (**Figure 4.31**), in which the angler throws or "casts" a "fly" (the bait for the fish) over a considerable distance, giving a better chance of catching the fish. Because the partner is unfolded, it has a large "capture radius" or reach, and therefore the probability of binding, and in turn the rate of binding, is increased. There is evidence that in the initial interaction (the encounter complex), a natively unfolded protein remains largely unfolded; it only folds subsequently, as required by this theory [84, 85]. Furthermore, it seems that at least in some cases the bound protein retains significant mobility.

Natively unfolded proteins often have DNA as their ligand. When the rate acceleration arising from fly casting was tested with a simple model on a protein–DNA complex, it turned out to be only a factor of 1.6 [83]. However, by including more realistic assumptions about electrostatics into the calculation, an increased acceleration was calculated [86].

This all sounds good, but there must be a drawback somewhere. One has already been mentioned: that a natively unfolded protein is inherently more likely to be degraded or aggregated. However, this may be a small price to pay if the cell needs rapid interactions. The other drawback is that although one gains speed because of the flexibility of the disordered protein, one also loses speed because it takes time for the protein to fold. However, folding can be very fast, so although this does represent a real effect, it is not as severe as the alternative of very slow binding of a globular chain.

4.3.3 Natively unstructured proteins provide specific binding without strong binding

A specific interaction (that is, one in which the correct binding partner is bound, but similar although incorrect partners are not) is one that necessarily has many interactions between the binding partners, and therefore almost automatically a specific interaction is a tight interaction: that is, there is a highly favorable enthalpy (*3.1) of binding. This means that there are many interactions to be broken in detaching the complex, and thus a tight interaction is normally characterized by a slow off-rate (*2.3). In many biological contexts this is a bad thing, because it slows everything down. For example, in the context of a signaling pathway it leaves the signal either switched on or switched off until the proteins dissociate again. The interaction can be made weaker, and the off-rate increased, by making the entropy (*3.1) of binding less favorable: that is, by having the free protein natively unfolded so that the binding energy is now the sum of the favorable binding interaction plus the unfavorable folding energy (**Figure 4.32**). In this way, the affinity and the off-rate can be adjusted to almost any desired value by varying the free energy of folding of the unbound protein. It has been observed that some natively unfolded proteins are "almost" folded whereas others are not; this behavior is a natural consequence of this adjustment of the free energy [87].

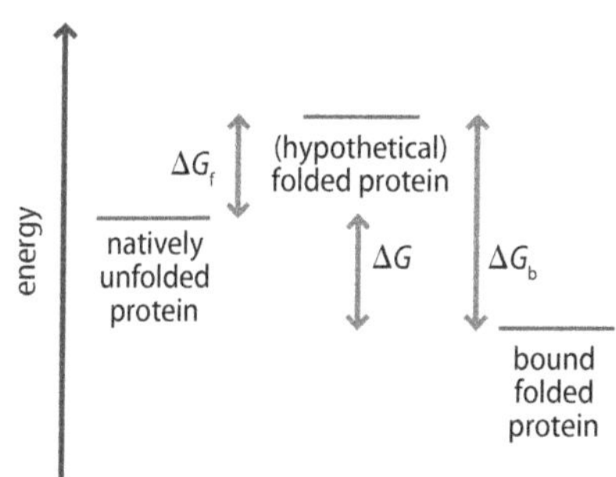

FIGURE 4.32
The change in free energy ΔG for the binding of an unfolded protein to a target can be considered as the sum of two terms: a hypothetical unfavorable term for the folding of the free protein (ΔG_f) plus a favorable binding energy of the folded protein (ΔG_b). The folding energy ΔG_f is hypothetical because binding of the unfolded protein does not in fact involve folding *before* binding: the two occur in a stepwise manner as it binds. Thus, the overall change in free energy ΔG can be adjusted to any desired value by altering the folding energy, or in other words by adjusting the stability of the unfolded protein relative to the bound protein. This of course applies equally well to the reverse step: the free energy for detachment of protein from its target and simultaneous unfolding is likewise adjustable. Thus, the free energy for detachment, and therefore the energy barrier and accordingly the rate, can be adjusted to whatever value is desired.

One such example is the intrinsically disordered protein FlgM mentioned above. In a "crowded" medium the N-terminal end remains unfolded but the C-terminal end becomes structured. It is of relevance that the C-terminal end is a fly-caster; that is, its function is to bind to a target protein, the transcription factor σ^{28}, whereupon it becomes structured [74]. Therefore the C-terminal domain is poised in an "almost" folded state, whereas the N-terminal domain has no need to try to fold.

4.3.4 Natively unstructured proteins may provide other benefits

A folded protein has to bind partners that have an approximate structural complementarity, and it therefore has a limited range of possible interactions. It has been argued that natively unfolded proteins may have a greater degree of flexibility in their folded structure and therefore may be able to bind a greater diversity of partners than their natively folded counterparts [75, 88]. This is an interesting idea, but so far there is little actual evidence to support it, for example in the form of structures of complexes [80].

It has also been argued that because disordered proteins are subject to rapid turnover in the cell, there may be an advantage to being unfolded if very tight regulation of the protein function is required, because a natively unfolded protein could presumably be removed rapidly when no longer wanted [72, 78].

In support of this argument, it has been noted that many natively unfolded proteins operate in signaling and transcriptional regulation, which typically require rapid switching of the signal. Again, there is so far little evidence for this idea.

4.4 POST-TRANSLATIONAL MODIFICATION OF PROTEINS

This chapter is about the behavior of "real" proteins in cells, in contrast with the more ideal proteins in dilute aqueous solution that we normally consider. So far we have concentrated on interactions between proteins. The last two sections of this chapter discuss rather different aspects of real proteins in cells: their covalent modifications, and their folding and misfolding.

4.4.1 Covalent modifications modify protein function

Throughout most of this book (and indeed in most textbooks) one could easily get the impression that once proteins have been made they just float around in the cytoplasm and get on with their job. This is very far from the case. Both the birth and the death of proteins are tightly regulated. Their location is regulated, and for many proteins this changes throughout their life; their assembly into complexes is regulated; and they frequently undergo covalent modifications that affect their function. Many of these aspects are still rather poorly understood. In this section I discuss covalent modifications, a key part of the regulation of protein function. Location of proteins forms the contents of Chapter 7.

Post-translational modifications have two main functions. One is to regulate the activity of the protein. Such modifications are usually reversible; they include phosphorylation, methylation, and acetylation. Some are an irreversible turning on, such as the conversion of protease zymogens to active enzymes, and the creation of short peptide hormones from longer precursors. And some serve to give the protein some special specific property, such as proline hydroxylation in collagen, heme attachment in cytochromes, and creation of the fluorophore in green fluorescent protein. The other main function is to control where the protein goes; this includes lipidation and glycosylation. Evolution being a tinker, many modifications have multiple roles depending on the context, of which the most dramatic example is **ubiquitylation (*4.9)**. Post-translational modification is very common, particularly in eukaryotes. For example, it is estimated that as a result of a combination of post-translational modification and **alternative splicing**, the human proteome contains approximately 10 times more proteins than there are genes.

The most familiar, and the most common, covalent modification is phosphorylation. But there are many others, some of which are listed in **Table 4.3**. Evolution has seized on anything that could potentially be useful and given it a try. One of the most drastic modifications is the complete removal of an internal domain: a self-splicing protein domain known as an **intein (*4.10)**.

4.4.2 Phosphorylation

Phosphorylation has several advantages that make it a useful covalent modification. It is a simple reaction, requiring an activated phosphate such as ATP, and a rather simple mechanism involving an in-line nucleophilic attack (**Figure 4.33**). This makes it easy to evolve new **kinases** as the need arises. Proteins have many potentially phosphorylatable side chains. This is good because it gives plenty of opportunity for evolving new phosphorylation sites, but bad if it leads to indiscriminate phosphorylation. It is worth noting that because kinases use an RNA-like cofactor (ATP), they are probably a very early evolutionary

*4.9 Ubiquitin

The name ubiquitin derives from the word *ubiquitous*, because the protein is found very widely in cells. It is a fascinating protein, which (like many other examples in the book) has been adapted by evolution for a range of functions. It is a small and very stable protein whose main function is as a label for protein degradation (**Figure 4.9.1**). The main details were worked out by Ciechanover, Hershko, and Rose, who were awarded the Nobel Prize in Chemistry in 2004. Ubiquitin is activated by attachment to an E1 activating enzyme (through a thioester link to the C terminus), and thence is transferred to one of a range of E2 conjugating enzymes (**Figure 4.9.2**). Each E2 is associated with several E3 ubiquitin ligase enzymes, which recognize different substrate proteins. Proteins are targeted for degradation mainly because they are recognized as being denatured, misfolded, oxidized, or otherwise damaged, whereupon they are bound by the E2–E3–ubiquitin complex, and ubiquitin is attached by means of its C terminus to a lysine side chain on the targeted protein. After this, further ubiquitin molecules are linked together through lysine 48. This polyubiquitylated complex is recognized by the proteasome (*4.13), which removes the ubiquitins and recycles them, and chops the target protein up into small peptides.

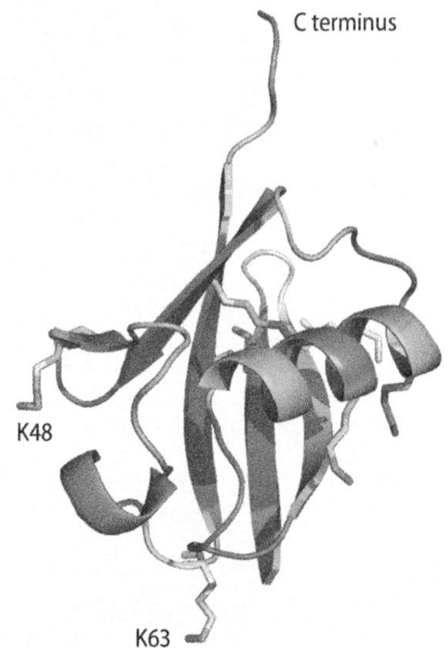

FIGURE 4.9.1
The structure of ubiquitin (PDB file 1ubq). This is colored by crystallographic *B* factor (Section 11.4.5): regions in yellow and red are the most mobile. This shows that the C terminus of the protein is the most mobile part. The structure also shows the six lysines, which (as expected) are mobile at the ends of their side chains. The two most widely used lysines are K48 (typically used for polyubiquitinylation targeting for proteasomal degradation) and K63 (used for a range of other functions, including endocytosis).

Ubiquitylation can result in a remarkable number of other cellular consequences, including the regulation of DNA repair, the cell cycle, cellular location, and immune function (related to the degradative function), as well as a range of apparently nonproteolytic functions such as vesicular trafficking and histone modification. Most of these cellular consequences are, however, a result of ubiquitin's targeting particular proteins for degradation. The cell cycle, for example, is regulated in part by the presence of a protein called cyclin B, whose concentration rises markedly at the start of mitosis but which needs to be removed again at the end of mitosis. This latter is achieved by ubiquitylation (**Figure 4.9.3**). The different functions of ubiquitin are distinguished by the mode of attachment of ubiquitin (that is, which lysine residue) and by the number of ubiquitins attached. All seven lysines of ubiquitin can be used for attachment, the frequencies at K48, K11, K63, K6, K27, K29, and K33 being respectively 29%, 28%, 17%, 11%, 9%, 3%, and 3% [124]. Some of these pathways involve specific deubiquitylating enzymes. In the human genome there are two E1 enzymes, about 40 E2, and about 300 E3, plus about 90 deubiquitylation enzymes.

There are a range of proteins with similar sequences to that of ubiquitin, termed ubiquitin-like modifiers, which have their own specific E1, E2, and E3 enzymes as well as deubiquitylation homologs. These have different functions again; one of them (SUMO) acts as an antagonist to ubiquitin.

The various functions of ubiquitin have been described in a *Nature* Insight supplement [125].

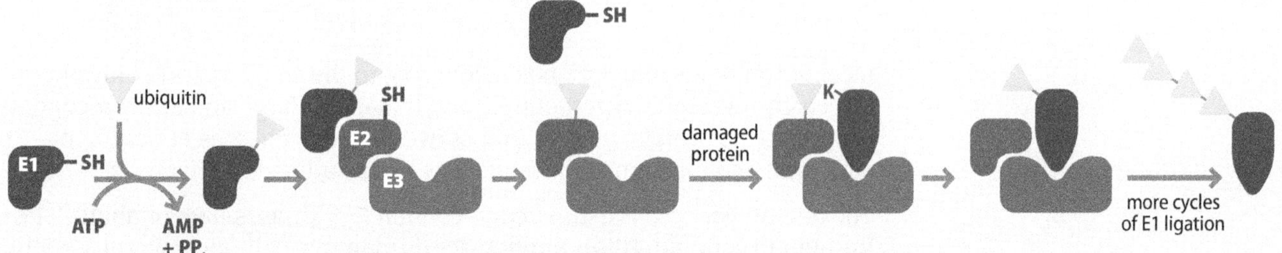

FIGURE 4.9.2 (above)
Polyubiquitinylation targets proteins for destruction by the proteasome. E1 (ubiquitin-activating enzyme) activates the C terminus of ubiquitin by making a thioester. It then binds to an E2 (ubiquitin-conjugating enzyme) complexed to one of a range of E2-specific E3 (ubiquitin ligase) enzymes. The ubiquitin is transferred to a thiol group on E2, and from there to a lysine side chain on a target protein, which is recognized by E3. (In the terminology of Chapters 2 and 8, E3 is acting as a scaffold, bringing together enzyme and substrate.) Further ubiquitins are then added by means of Lys 48 (compare with Figure 4.9.1).

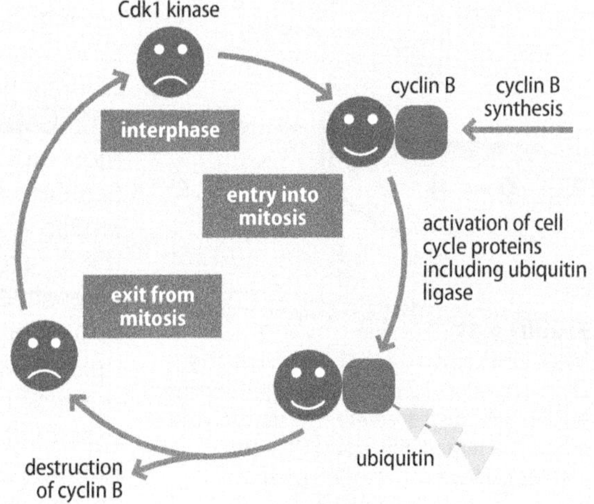

FIGURE 4.9.3 (right)
Within the cell cycle, entry into mitosis is stimulated largely by activation of the Cdk1 kinase. This kinase is activated by binding to the protein cyclin B, as described in Section 4.4.2. Thus, entry into mitosis is controlled by a rapid increase in the concentration of cyclin B. For cells to exit from mitosis, Cdk1 has to be turned off again, which is done by ubiquitylation of cyclin B and subsequent digestion by proteasomes. The ubiquitin ligase that attaches ubiquitin is activated by Cdk1 itself.

TABLE 4.3 A few covalent modifications of proteins

Modification	Site	Comments
Phosphorylation	Ser, Thr, Tyr	Regulates activity. Regulates assembly
Acetylation	Lys	Creates part of histone code in chromatin
Methylation	Lys	Creates part of histone code in chromatin
Methylation	Arg	
Lipid attachment	Cys, C terminus	Attaches protein to membrane
SUMOylation	Lys	Role in transport, transcriptional regulation, apoptosis
Ubiquitylation	Lys	Regulates transport and degradation, plus histone readout
Limited proteolysis		Activates proteases (zymogens) in extracellular location (e.g. chymotrypsin); activates hormones (e.g. insulin)
Attachment of *N*-acetylglucosamine	Ser, Thr	Regulates activity in enzymes involved in glucose metabolism
Glycosylation	Asn, Ser/Thr	Eukaryotes. Recognition, membrane protein folding
Hydroxylation	Pro	Collagen: to facilitate triple helix formation. Irreversible
ADP ribosylation	Arg, Glu, Asp	As part of signaling, DNA repair and apoptosis
Sulfation	Tyr	Irreversible and probably required for activity
Carboxylation	Glu	Creates γ-carboxyglutamate (Gla), a calcium ligand

mechanism. Phosphorylation is of low cost to the cell, requiring only one ATP. It is easily reversed when required, and it makes a large and easily recognized change to the protein, adding almost two negative charges at neutral pH. The structural and functional consequences were outlined in Chapter 2.

The human genome is estimated to contain 518 kinases [89], or about 1.7% of the human genome. This is about twice as many as in the fly or worm, underlining their important roles in development and regulation. Most of these kinases belong to the same superfamily, implying that they have diverged from a common ancestor. For many of them, a biological role can be predicted because of their homology to orthologs with known function. However, their exact targets are known in very few cases. And, indeed, it is becoming clear that most kinases are not that specific. As discussed in Chapters 2 and 8, evolution achieves specificity not by engineering a highly precise recognition between kinase and substrate but by adding extra domains, both to the kinase and to the substrate, to increase or inhibit the binding of kinase to substrate, and to colocalize the kinase and substrate. In signaling pathways, kinases typically phosphorylate multiple sites. Naturally, evolution has tinkered with these multiple phosphorylations and turned them to advantage by using them to create new binding sites and regulatory features, as discussed in Chapter 8; but essentially they are a reflection of the rather poor substrate specificity of most kinases. The correct functioning of many proteins requires the appropriate combination of phosphorylations at different sites. A good example is provided

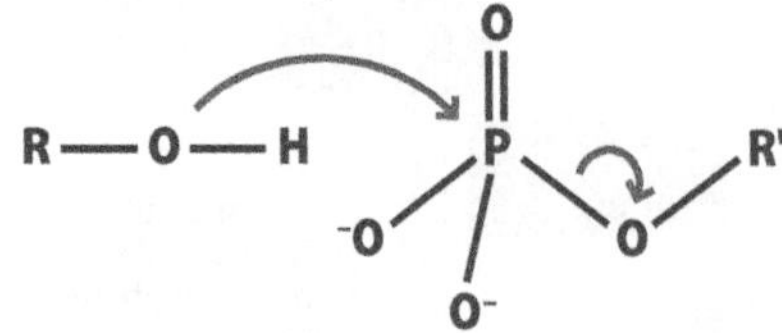

FIGURE 4.33
Mechanistically, phosphorylation is a very simple reaction, involving a straightforward nucleophilic displacement. The uncatalyzed reaction is remarkably slow [123], and kinases catalyze the reaction by an amazing factor of up to 10^{20}.

*4.10 Inteins

Inteins are self-splicing proteins, and require no other companion molecules to splice [126]. They are essentially a post-translational modification, though a rather unusual one because the modification consists of the removal of a central intein domain, fusing together the two extein domains on either side, in a similar way (although by a completely different mechanism) to the splicing out of intron RNA from mRNA. The roughly 135-residue intein domain holds the two ends of the splice close together, and typically starts with Ser or Cys (drawn as an XH side chain to the left of the intein domain in **Figure 4.10.1**), and ends with Asn. In addition, the C-terminal extein starts with Ser, Thr, or Cys. The mechanism shown in Figure 4.10.1 consists of four nucleophilic displacements, and the final result is a normal polypeptide chain, containing a Ser, Thr, or Cys at the splice site.

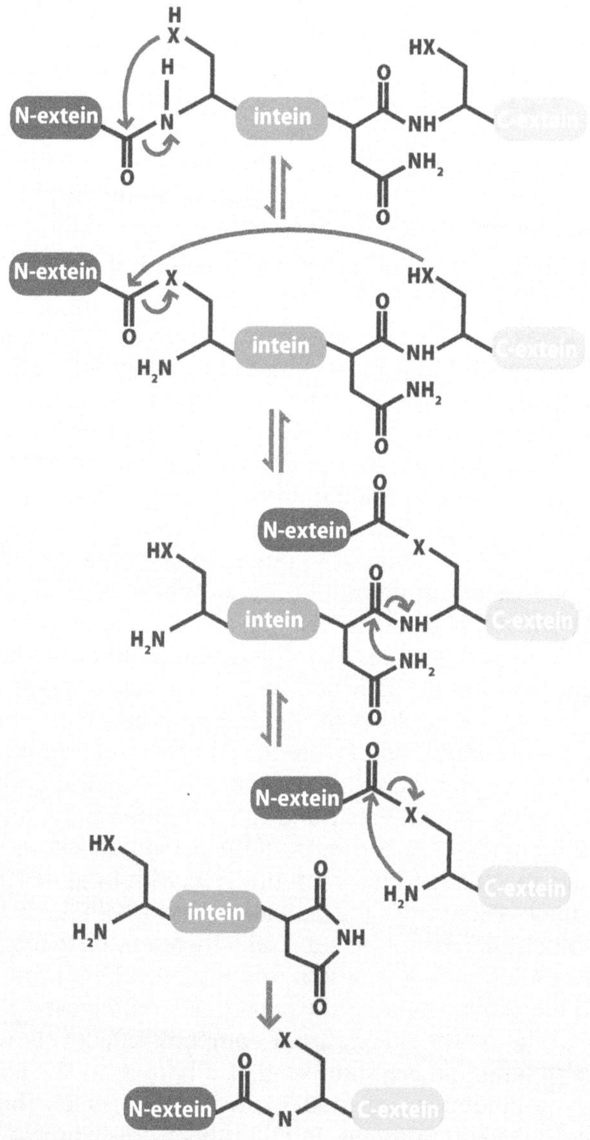

FIGURE 4.10.1
The mechanism of intein splicing involves four nucleophilic displacements. In the first step, the side chain of a nucleophilic Ser or Cys (the first amino acid of the intein domain) attacks the preceding amide bond, leading to an acyl rearrangement to create a thioester (considerably more reactive than an amide). This is then attacked by the first residue of the C-terminal extein domain, to produce a branched ester or thioester. The side chain of the Asn that forms the last residue of the intein then attacks its own backbone carbonyl to remove the intein. Finally, the backbone amine from the C-extein attacks the preceding ester to make a standard polypeptide.

Inteins are found in prokaryotes, archaea, and unicellular eukaryotes, but only sporadically: there are only a few hundred such sequences identified so far [in September 2009, the Intein Registry (http://tools.neb.com/inbase/list.php) contained 100 eukaryotic, 232 prokaryotic, and 157 archaeal genes]. It is thought that they are probably 'parasitic' sequences, originally inserted into the DNA by a 'homing endonuclease' that is usually found as an additional domain inserted within the intein domain sequence and which functions to insert the DNA coding for the intein into the genome.

There has been much interest in inteins since their original discovery in 1987, mainly because of their potential in biotechnology for linking together protein domains, for example for tagging proteins, studying protein interactions, and segmental isotope labeling for NMR studies (an example is given in **Figure 4.10.2**). There have been several successes, but inteins are certainly not a simple solution. This is mainly because successful ligation together of the two exteins probably requires some degree of association of the exteins before the splicing reaction; in other words, one cannot simply take two domains linked by an intein and fuse them together.

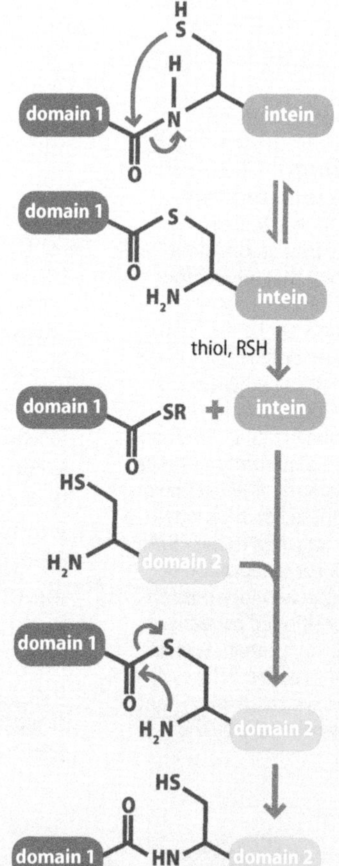

FIGURE 4.10.2
A biotechnological application of inteins. Here, the fusion of domain 1 with an intein (created genetically by fusing the genes) is induced to self-cleave by addition of an external thiol, RSH. The thioester thus formed is subsequently cleaved by a second domain starting in a cysteine, to give a two-domain protein.

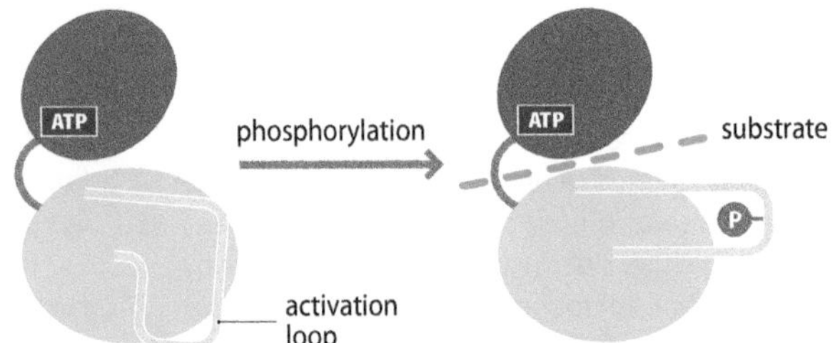

FIGURE 4.34
In the active kinase (right), the peptide substrate is held in a cleft between the two domains, and is positioned by hydrogen bonds to the activation loop, which is in the lower domain. The ATP is held in the other domain. The activation loop is locked into the correct position by phosphorylation of a tyrosine residue.

by the C-terminal domain (CTD) of RNA polymerase II, discussed in Chapter 9, which undergoes a series of phosphorylations and dephosphorylations during transcription: not only does it require phosphorylation and dephosphorylation in the right order, but many of these are also rather nonspecific. For many proteins, many of the phosphorylation sites probably have little or no biological significance. It is estimated that about one-third of the roughly 25,000 human proteins are phosphorylated, many of them at several locations by more than one kinase, from which it is clear that phosphorylation must be a rather nonspecific reaction: there are just not enough kinases to go round (see [90] for a readable overview, and Problem N2 in Chapter 8).

As noted above, protein kinases arose by divergent evolution from the same basic pattern. This is not so much because the reaction is inherently difficult (although actually it is: the uncatalyzed rate of phosphate hydrolysis is remarkably slow, meaning that kinases increase the rate by factors of up to 10^{20}, making this one of the greatest catalytic enhancements known) but because regulation of kinase activity is crucial to cellular function: the structure of protein kinases lends itself well to regulation [91]. Protein kinases have a two-domain structure, in which the ATP is bound to one domain (via a GXGXXG sequence motif) and the peptide substrate is bound in a cleft between the two domains and is oriented correctly by interactions with an *activation loop* in the other domain (**Figure 4.34**). Residues needed for the reaction are contributed by both domains. Protein kinases are activated by themselves being phosphorylated, which fixes the activation loop in the correct position to orient the substrate and active-site residues (see Figure 4.34). This structure therefore achieves two important aims: it permits recognition of only the correct substrate (by the residues in the binding cleft and the activation loop) and it ensures that the kinase is virtually 100% turned off if the activation loop is not phosphorylated. This basic mechanism has been tinkered with in all sorts of ways. In some kinases, the unphosphorylated activation loop sits in the active site cleft and physically blocks access to the substrate, therefore further turning off the enzyme (**Figure 4.35a**). In others, this role is played by a separate part of the protein called the pseudosubstrate sequence or loop (Figure4.35b). This loop binds weakly to the active site cleft but competes effectively with the substrate because the interaction is intramolecular (Chapter 2). The pseudosubstrate is removed by binding to an additional protein, which therefore acts as an additional activating protein. In still others (for example the

FIGURE 4.35
Mechanisms for deactivation and activation of kinases. (a) In some kinases, the activation loop in its unphosphorylated form physically blocks the substrate-binding site. (b) In others, there is a different part of the protein called the pseudosubstrate loop that mimics the substrate and binds in the substrate-binding site. This loop can be moved out of the way by binding of the appropriate partner protein. (c) In still others, phosphorylation of the activation loop also repositions a dimerization interface, therefore leading to dimerization. This can have several effects, such as activation or relocation of the kinase. (d) In yet others, a glutamate residue attached to the PSTAIRE helix is required for correct location of the ATP relative to the peptide substrate. The glutamate is positioned correctly by binding of a partner protein. Such a mechanism is used to activate the cyclin-dependent kinases, which are important components of the cell cycle, by the binding of cyclins.

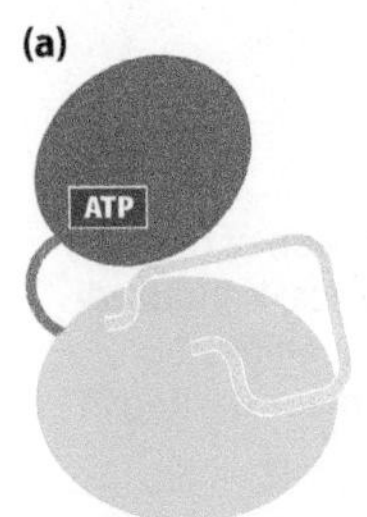

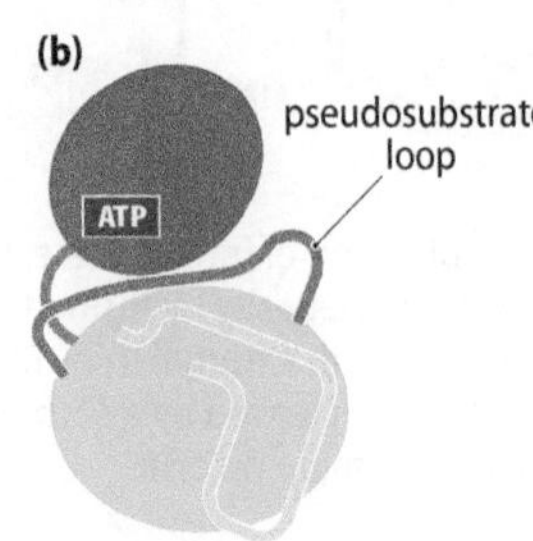

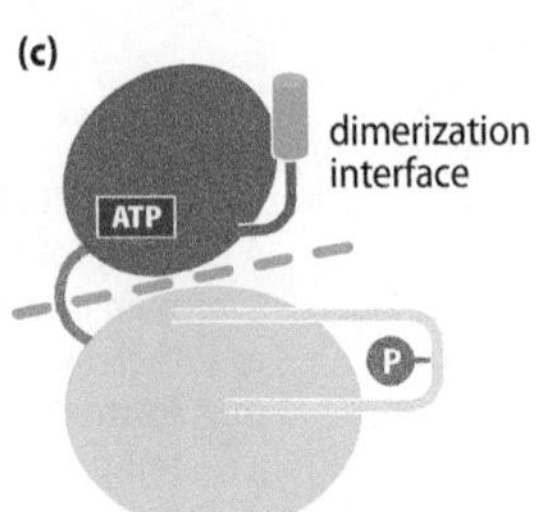

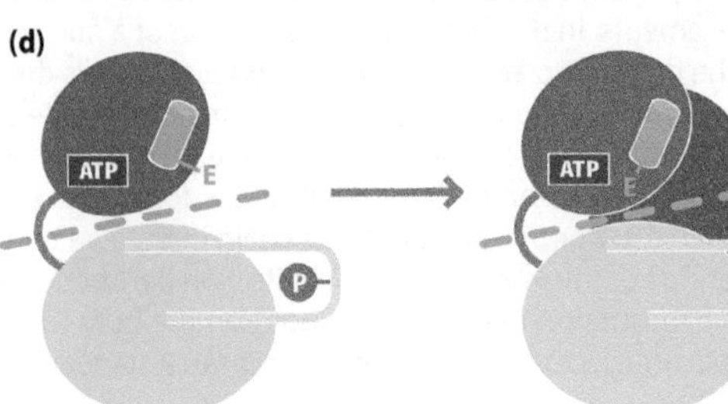

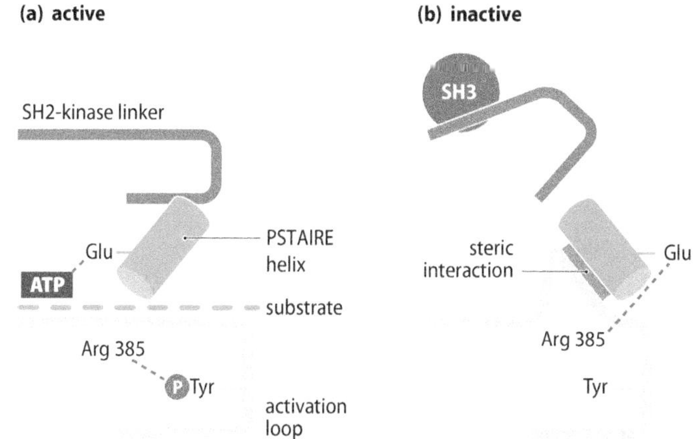

FIGURE 4.36
The activation of Src kinases such as Hck or Lck. (a) In the active form, the glutamate from the αC or PSTAIRE helix orients ATP for phosphorylation of the substrate. This helix is held in place by interactions with the linker between the kinase domain and its preceding SH2 domain. The substrate is held in place by interactions with the activation loop, which in turn is held in place by the phosphorylated tyrosine, which makes a hydrogen bond to Arg 385. (b) The enzyme can be inactivated by one or both of dephosphorylation of the phosphotyrosine or reorientation of the αC helix. Dephosphorylation of the phosphotyrosine leads to loss of the hydrogen bond to Arg 385, which can now form an alternative hydrogen bond to the glutamate. The consequent reorientation of the αC helix is stabilized by a steric contact with the now reorganized activation loop. The helix can be reoriented by a quite different mechanism, namely restructuring of the linker to the SH2 domain, produced by binding of an SH3 domain to it. (Redrawn from F. Sicheri, I. Moarefi and J. Kuriyan, *Nature* 385: 602–609, 1997. With permission from Macmillan Publishers Ltd.)

mitogen-activated protein kinase ERK2), phosphorylation of the activation loop also leads to repositioning of a dimerization interface, thereby leading to dimerization of the kinase: this alters the properties of the kinase, for example creating a nuclear localization sequence that allows the kinase to enter the nucleus (Figure 4.35c). And in yet others, such as the cell cycle regulatory kinases (cyclin-dependent kinases or CDKs), a second regulatory site has been developed in which a vital glutamate (E in Figure 4.35d), which is part of the "PSTAIRE" helix and orients the ATP in the active conformation, is incorrectly positioned and so turns off the kinase unless the helix is rotated to the correct position by binding to an additional protein, a cyclin. The position of the PSTAIRE helix can also be controlled by the orientation of the linker leading up to it (**Figure 4.36**) [92]. This is for example how the kinase is turned off in Src, as discussed in Section 8.1.8.

It is worth restating the mechanism described above for kinase activation. In the *inactive* kinase, the activation loop is disordered and therefore cannot bind to the substrate. Once the activation loop has been phosphorylated, it is fixed in an active conformation and is able to bind substrate. Different kinases have different inactive, disordered, conformations, but all kinases have rather similar active conformations. Exactly the same is true for the small GTPases, discussed in Chapter 7: the *inactive* form (GDP-bound) differs between different proteins, but the active (GTP-bound) form is "tense" and similar in all GTPases. It would be reasonable to speculate that at least one reason for this behavior is the same as that proposed for natively unstructured proteins: that the loss of entropy in rigidifying the disordered state partly offsets the increased enthalpy in the new bonds created in the active form, and helps balance the energetics of the switch. It is also true that it is much easier to evolve a switch in which only one of the two states needs to be highly ordered.

Phosphatases are even less specific. There are only about one-third as many phosphatases as kinases, implying that each one must on average dephosphorylate three times as many proteins than each kinase phosphorylates (that is, on average 60 proteins, in contrast with 20 for kinases). As with kinases, the activity of phosphatases is made more specific by linking them to modules that locate them to particular regions of the cell, by a variety of module types including SH2 domains [93].

The specificity of phosphatases is actually much lower even than implied by the previous paragraph. In the human genome there are about 130 tyrosine kinases but 100 tyrosine phosphatases, implying that tyrosine dephosphorylation is only slightly less specific than the phosphorylation [94]. (One could argue that this is reasonable because the main role of tyrosine phosphorylation is the initiation of signaling pathways, which strongly require specificity.)

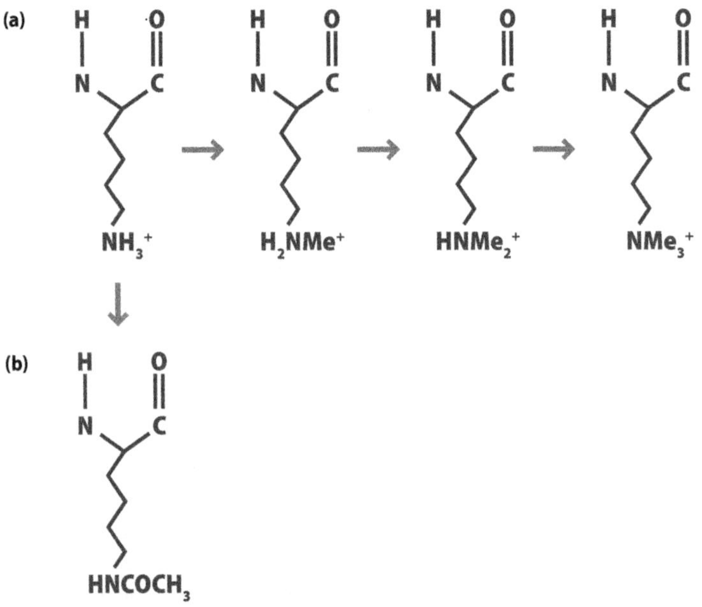

FIGURE 4.37
Modification of lysines. (a) By methylation, which has no effect on the charge but makes the side chain more hydrophobic and larger. There are three distinct methyllysines possible. All are physiologically relevant, but the most important is trimethyllysine. (b) By acetylation, which removes the charge.

However, there are about 350 Ser/Thr kinases but only 35 Ser/Thr phosphatases, plus 50 dual-specificity phosphatases, making the dephosphorylation of serines and threonines a very non-specific process (see Problem N2 in Chapter 8).

4.4.3 Methylation and acetylation

Proteins can also be modified by the addition of one, two, or three methyl groups to lysine side chains, or by lysine acetylation (**Figure 4.37**). These are mutually exclusive modifications. Like phosphorylation, they are also biochemically quite simple, low-energy conversions that can be reversed (although not as readily as phosphorylation). Methylation occurs through the action of S-adenosyl methionine (another RNA-like cofactor), which means that one cycle of addition and regeneration requires the loss of all three phosphates from an ATP as well as the involvement of methyltetrahydrofolate, and so is significantly more costly than phosphorylation. Lysine acetylation (via coenzyme A, another RNA-like cofactor) changes the net charge on the side chain by one. It is therefore not quite such an obvious effect as phosphorylation, and it is also less well localized, because lysine side chains are longer and more flexible than serine or threonine. Lysine methylation has no effect on charge and is thus a significantly more subtle modification.

Such modifications are very common at the N termini of **histones**, and they have been suggested to form a "code" for the functional state of the histone [95]. Like phosphorylation, these changes are reversible. However, they are frequently rather long-lived changes and can last through several cell divisions. They often persist in the fertilized egg and therefore regulate the way in which the DNA of the egg is read. Such effects are described as *epigenetic*, meaning that they are not coded for by DNA.

DNA is wrapped around histones to form nucleosomes (**Figure 4.38**). The nucleosome core consists of an octamer containing two copies each of the histones H2A, H2B, H3, and H4 arranged roughly in a sphere. The N-terminal tails of these proteins protrude from the nucleosome. The tails of H2B and H3 have a region that interacts strongly with DNA, and another that interacts with the tail of an H4 histone from another nucleosome. However, much of the tail is thought to interact with additional proteins, which regulate the expression of DNA and the unwinding of nucleosomes. A general theme of this book is that evolution works by tinkering. It would therefore be unlikely that anything

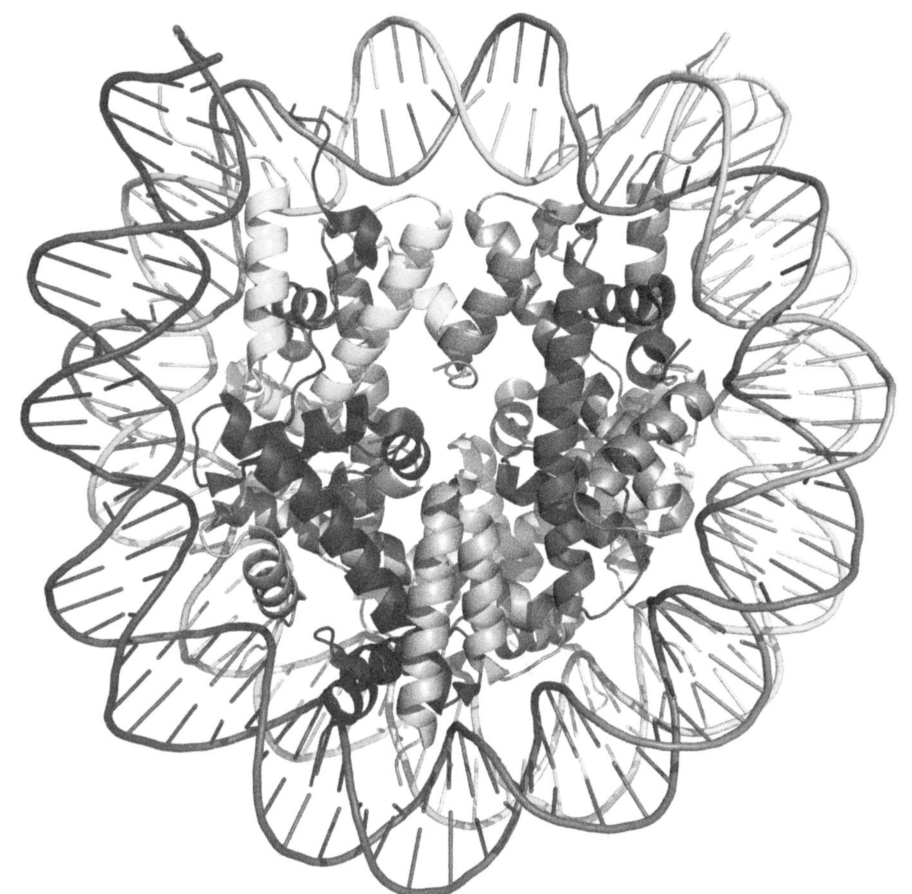

FIGURE 4.38
The *Drosophila* nucleosome core is formed from two copies each of histones H2A, H2B, H3, and H4 (PDB file 2pyo). The two H3 histones are colored green, H4 is colored red/salmon, H2A is colored blue, and H2B is colored olive green. These histones are homologous, and share the 'histone fold'. A total of 147 base pairs of DNA are wrapped round these proteins to make the nucleosome core. The core is capped by histone H1, which sits over the DNA exit and entry points. H1 is not shown in this figure, but it sits at the top of the structure. The nucleosome proper is completed by about 80 base pairs connecting one core particle to the next. The interactions between DNA and protein are rather nonspecific, as one might expect given that virtually all DNA has to be packaged into nucleosomes regardless of its sequence. The DNA is not uniformly bent: it is kinked at particular sites. The N-terminal tails of the histones protrude through the DNA and are accessible to the surface: it is these sequences where all the 'histone code' modifications occur. They are in general disordered and therefore do not appear in crystal structures, which limits our understanding of their mode of action.

as precise as a universal "histone code" had appeared as a result of evolution. However, there are general rules, and tinkering implies that things that work will have been picked up, adapted, and improved on, suggesting that both acetylation and methylation are useful devices, even if not used consistently. Acetylation decreases the positive charge on the histone and therefore tends to make it bind less tightly to DNA. It therefore generally encourages gene expression. By contrast, trimethylation of H3 at Lys 4 causes gene expression, whereas methylation at Lys 9 or Lys 27 leads to gene silencing. Confusingly, dimethylation at Lys 4 can cause gene silencing. As is also true for phosphorylation, modification at one site often requires prior modification at others, and reading of the code often requires the modification of several sites simultaneously. Similarly (as we see repeatedly throughout this book), increased specificity in the code is provided by combinations of markers that are read by combinations of recognition domains (**Figure 4.39**) [96]. Naturally, acetylation and methylation of lysines is by no means restricted to histones [97] and is for example found in signaling pathways. Acetylation is more common in

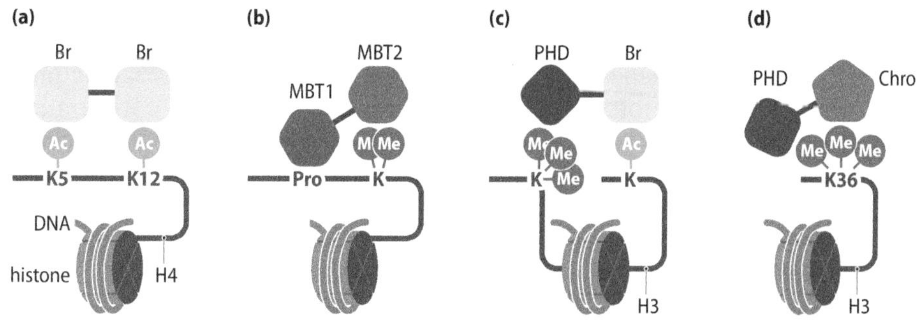

FIGURE 4.39
Some examples of increased specificity of binding to histone proteins, using recognition by multiple domains. (a) Two adjacent acetylated lysines on histone H3 or H4 are recognized by a pair of bromodomains (Br) in hTAF1. Lysine methylation is typically recognized by several aromatic residues, which form a hydrophobic cage that binds preferentially to the hydrophobic but charged methyl groups. It is likely that the methyl protons form 'aliphatic hydrogen bonds' to the centers of the aromatic rings. (b) Modules MBT1 and MBT2 from L3MBTL1 possibly recognize proline and dimethyllysine, respectively. (c) Adjacent PHD (plant homeo domain) and bromodomains on BPTF recognize a trimethylated lysine and an acetylated lysine, respectively, on histone H3. (d) The PHD finger of Rco1p increases the affinity of the Eaf3p chromobarrel (Chro) to a trimethylated lysine on histone H3. (Redrawn from S. D. Taverna, H. Li, A. J. Ruthenburg, C. D. Allis and D. J. Patel, *Nat. Struct. Mol. Biol.* 14: 1025–1040, 2007. With permission from Macmillan Publishers Ltd.)

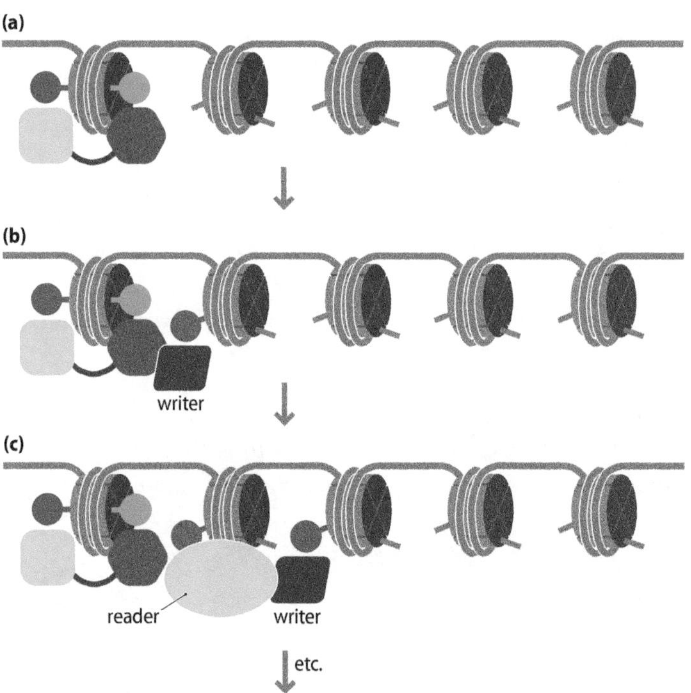

FIGURE 4.40
Histone reader–writer complexes enable the propagation of a signal along chromatin. (a) A regulatory protein recognizes a particular code combination (compare Figure 4.39). (b) A histone-modifying enzyme ('writer') binds to the regulatory protein and modifies the neighboring histone. (c) The modification is recognized by a reader protein, which recruits another writer, and is the correct length to span the distance from one nucleosome to the next, so that the writer can then modify the next nucleosome, and so on.

mammals than other organisms, but it is not yet clear whether this reflects a lack of information about other organisms or a real novel adaptation by mammals in the face of evolutionary pressure (presumably not least because phosphorylation has developed almost as far as it can).

The distance between one nucleosome particle and the next is fixed, which has permitted the evolution of a very neat mechanism for propagating covalent modification along a chain of nucleosomes, using *reader–writer complexes* (**Figure 4.40**). Modification of one nucleosome by a "writer" (histone acetylase or methylase) then permits the modification to be read by a reader–writer complex, which has the correct dimensions and charge distribution to allow the writer to extend to the next nucleosome and modify that. A chain of such complexes thus allows modification to be passed along a chain of nucleosomes, and therefore allows the reading for example of a whole gene at once.

Probably all of these histone-modifying enzymes are held by some kind of scaffold, most probably a rather unstructured protein that holds the nucleosome and enzymes in close proximity to encourage correct interactions and discourage incorrect ones (Chapter 2).

4.4.4 Glycosylation

Glycosylation is the addition of glucose or other sugars to amino acid side chains, most commonly asparagine, serine, or threonine. It is estimated that more than 50% of human proteins are likely to become glycosylated at some point in their existence, making this a very significant modification. The sugar is usually transferred from a nucleotide such as UDP. Glycosylation occurs in prokaryotes and archaea, where it is found mainly on externally facing proteins and membrane proteins as well as in secreted proteins [98]. Its primary function there seems to be to stabilize the protein and protect it against digestion and degradation [20]. It is found much more widely in eukaryotes, where it forms another good example of evolution tinkering with an existing function and using it for completely different purposes. In eukaryotes it also clearly has its original role in stabilizing proteins and directing the folding of membrane proteins. However, glycosylation in eukaryotes also has a major role in monitoring and regulating the folding and secretion of proteins (as discussed below), and also in cell–cell recognition and therefore in the development of multicellular organisms.

4.5 PROTEIN FOLDING AND MISFOLDING

4.5.1 Protein folding is often rapid and thermodynamically controlled

The folded form of a protein is in general the thermodynamically most stable state. Therefore for a protein to fold, it merely needs to be synthesized in an unfolded form and allowed to fold by itself. There is good evidence that protein sequences have had modest evolutionary pressure that has forced them to fold rapidly and reliably and to have essentially only a single stable fold at the global minimum, thus preventing the formation of alternative structures [99]. (This is not true for RNA, which can and does form multiple secondary structures.) The route to the folded state is sometimes described as *minimally frustrated*, meaning that there are no significant energetic barriers to folding under native conditions. The word "modest" above is intended to imply that correct folding is usually not the main factor controlling the sequence, and that in consequence many proteins need a little help with folding in the intracellular environment, which is very crowded and therefore far from an ideal folding environment. However, the limited evidence available suggests that the *pathway* of folding is not nearly as highly conserved as the folded structure. The bacterial proteins Protein G and Protein L have a very similar and simple fold, yet fold by different pathways [100, 101], as shown by Φ **value analysis (*4.11)**; mutations to proteins have in several instances been shown to produce a change in the folding pathway (but a very similar final fold). It may therefore be that for many proteins the folding pathway is not important.

There has been a large amount of work done on how proteins fold. The most obvious conclusion from this work is that different proteins fold differently: there is no uniform "mechanism" for folding in that sense. The most common route seems to be that secondary structures, particularly α-helices, flicker in and out of existence in "unfolded" proteins and become stabilized by interactions between themselves and between other parts of the protein chain. In particular, hydrophobic clusters formed by spatially close side chains stabilize the chain into more folded conformations. This initial collapse can be very rapid (faster than milliseconds). The protein then goes through some kind of searching process, in which the side chains explore different geometries to find a close-packed arrangement. At some point, designated the transition state, the protein interior becomes sufficiently organized that it can "zip up," expelling any remaining water molecules from the protein interior as it does so. However, it is clear that the relative rates and importance of these different stages are very different in different proteins: in some proteins, for example, the hydrophobic collapse comes very early.

Unfolded proteins are not functional and are a danger to the cell (except probably for natively unstructured proteins; see Section 4.3). They have exposed hydrophobic surfaces, which are likely to interact with each other and with other cellular components, leading to inappropriate and unwanted interactions, and possibly to aggregation or else to proteolytic digestion. The cell is thus usually concerned to ensure that proteins get folded up as quickly as possible after translation. Folding can often be cotranslational: the proteins start to fold as soon as they emerge from the ribosome. Indeed, there is fairly good evidence [102] that multidomain proteins fold in this way, domain by domain, in some cases apparently with translational pauses between domains to allow the folding of one domain before translation of the next [103].

*4.11 Φ value analysis

This is a method for determining the amino acids important for the transition state in the folding reaction, introduced by Alan Fersht [127]; it analyses the effect of mutations, such as replacement of the native amino acid by alanine. In general, mutation of an amino acid might be expected to destabilize the folded state. If this amino acid is still unfolded in the folding transition state, the mutation will not affect the stability of the transition state, and hence the energy of the transition state is not altered (**Figure 4.11.1a**). This energy can be measured from the rate of folding, k_f, which is given by $k_f = \exp[-(\Delta G_{\ddagger-U}/RT)]$, where $\Delta G_{\ddagger-U}$ is the change in free energy between the transition state and the unfolded state. The effect of the mutation is then the difference in free energies of the mutant and the wild type, and is proportional to the ratio of the folding rates: $\Delta\Delta G_{\ddagger-U} = -RT \ln k_f^{wt}/k_f^{mut}$. This is most usefully expressed as a ratio of the change in free energy between unfolded and folded, and is given the symbol Φ: $\Phi = \Delta\Delta G_{\ddagger-U}/\Delta\Delta G_{F-U}$. In principle the value of Φ should range between 0 (see Figure 4.11.1a), in which the mutated residue is not involved in the folding transition state, and 1 (Figure 1b), in which the mutated residue has the same effect on the stability of the transition state as it does on the folded protein. And in many cases this is indeed true, with many residues having Φ close to 0 and only a few having large values, as might be expected.

Protein G and Protein L have very similar folds, consisting of a helix packed against a four-stranded sheet formed by two adjacent hairpins (**Figure 4.11.2**). They are probably homologous (that is, a product of divergent evolution). For Protein L, Φ value analysis shows that residues with $\Phi > 0.3$ are located mainly in the first β hairpin (strands 1 and 2) (see Figure 4.11.2a). In other words, the first hairpin forms the nucleus of the folding transition state [100]. By contrast, for Protein G, the folding transition state is the second hairpin (see Figure 4.11.2b) [101]. Interestingly, when residues in the first hairpin of Protein G were replaced by residues designed to produce a preferred hairpin, the overall folding rate increased 100-fold, and only the first hairpin was present in the folding transition state [128]. Therefore the folding pathway can be altered relatively simply without changing the final folded conformation.

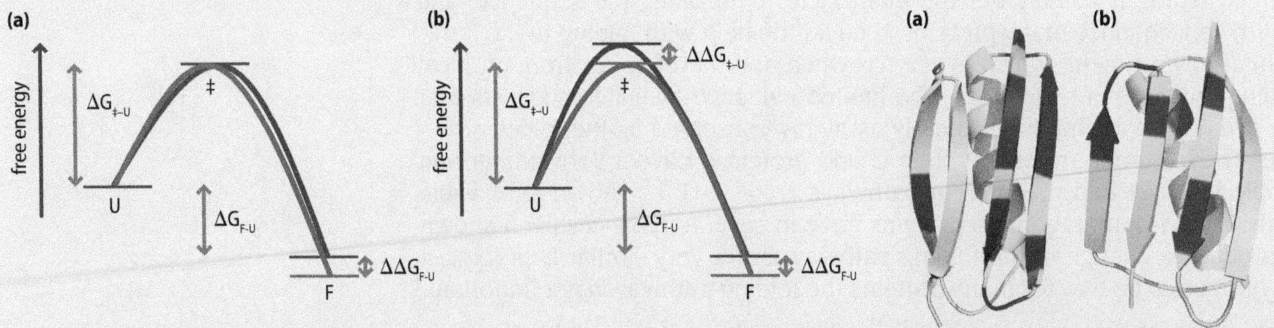

FIGURE 4.11.1
The rationale for Φ analysis. (a) If a residue is not involved in the folding transition state, mutation of that residue will have no effect on the stability of the folding transition state, and thus on the folding rate. This means that $\Phi = 0$. (b) In contrast, if the conformation and interactions of the residue are the same in the transition state as in the folded protein, then $\Phi = 1$.

FIGURE 4.11.2
The structures of Protein L (a) and Protein G (b). Residues involved in the folding transition state (that is, with Φ values greater than 0.3) are colored red. In Protein L these residues are mainly in the first β hairpin (right), whereas in Protein G they are mainly in the second hairpin (left).

Proteins are produced starting with the N terminus. There are some examples where the presence of a folded N-terminal domain helps the folding of a second domain that is C-terminal to it [104, 105], but in general it seems that the domains in multidomain proteins fold independently. Interestingly, there is some evidence that in proteins containing tandemly repeated domains, evolution has suitably modified adjacent repeats so that they do not form interdomain associations and thus do not interfere with each other's folding [104].

In most eukaryotic proteins, the nascent proteins are first coated with **chaperones (*4.12)** to prevent them from misfolding. Subsequent folding steps are frequently regulated in a controlled environment (such as inside a chaperonin), to ensure that they happen properly. The cell can also use additional proteins such as **peptidylprolyl isomerases** to make sure that folding takes place correctly. Many proteins require some form of post-translational modification such as glycosylation. Such proteins are translocated off the ribosome into the **endoplasmic reticulum** (ER), where they fold and are subsequently modified before being released [106] (**Figure 4.41**).

*4.12 Chaperones

In the nineteenth century, a single woman in public was often accompanied by a chaperone, an older or married woman who acted to prevent her friend from engaging in inappropriate contacts with the opposite sex. By analogy, a protein chaperone functions to prevent unfolded proteins from engaging in unwanted interactions, by binding to exposed hydrophobic regions in a rather nonspecific manner. Some chaperones, such as Hsp70 (which includes DnaK in *E. coli*), bind to unfolded proteins and protect them from aggregation and unwanted binding. Subsequent exchange of bound ADP for ATP leads to dissociation. Such proteins are used for example in the transport of newly synthesized unfolded chains to the mitochondrial membrane (Section 7.5.3). The name Hsp stands for heat shock protein, indicating that such proteins are up-regulated after heat shock or other stressful conditions. These are very abundant proteins in the cell, especially in stressed cells.

Other chaperones function to aid protein folding; these are the chaperonins. The most common is GroEL (Hsp60 in bacteria), which forms a hollow barrel-like structure made up of two rings each of seven subunits (**Figure 4.12.1**). At each end there is a cap made of seven copies of GroES (Hsp10 in bacteria). Unfolded protein binds to a hydrophobic region inside the GroEL barrel. ATP hydrolysis leads to a conformational change that buries the hydrophobic surface of GroEL, releases the bound protein, and allows it to fold. Entry of more unfolded protein into the other end of the barrel releases the protein into the cytoplasm. If by this stage it is folded, the job of the chaperone is done; if it is not, the protein can re-enter the barrel as many times as needed for it to fold. The chaperonin does not accelerate protein folding; in fact it slows it down. However, it does greatly increase the fraction of protein that ends up folded, by preventing the formation of aggregates.

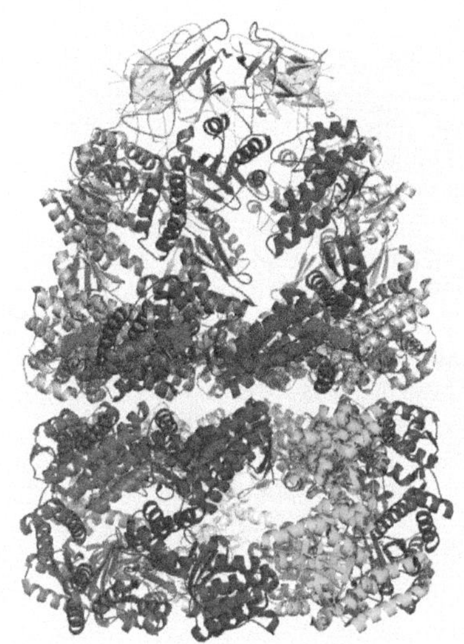

FIGURE 4.12.1
The structure of the chaperonin GroEL–GroES complex (PDB file 1pf9). The figure shows two heptameric GroEL rings, one in shades of red and one in shades of blue: the structure is clearly asymmetrical. The red ring is capped by a ring of seven GroES proteins (shades of green). The position of ADP is shown by the magenta spheres. Unfolded proteins bind within the cylindrical cavity in the upper GroEL ring.

4.5.2 All proteins have a limited lifespan, especially unfolded ones

In the ER, there is a detailed quality control system for identifying wrongly folded proteins and degrading them. This remarkable system looks for the signatures of misfolded proteins, such as exposed hydrophobic patches, exposed cysteine residues, incomplete glycosylation patterns, and aggregation. In particular, glycosylated proteins within the ER undergo a complicated series of glycosylations and deglycosylations that act as checks to distinguish between correctly folded and misfolded proteins (see Figure 4.41) [107]. Proteins failing these controls, plus proteins expressed cytoplasmically and detected as being misfolded, are tagged with multiple ubiquitin (*4.9) chains and fed into cytoplasmic proteolytic machines called **proteasomes (*4.13)** [108]. Experimental

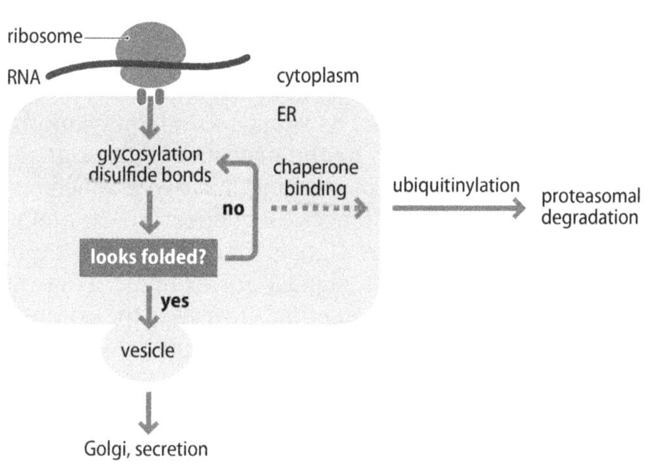

FIGURE 4.41
The quality control mechanism in the endoplasmic reticulum. Proteins for secretion or membrane insertion are translated directly off the ribosome into the ER. Here they are glycosylated and can be disulfide bonded. The glycosylation is monitored carefully by the quality control mechanism, and seems to be the main marker of correct folding. Folded proteins are sent off to the Golgi for secretion, whereas incorrectly folded proteins are put through several cycles of glycosylation and deglycosylation. Proteins that persistently fail the quality control are picked up by chaperones and exported from the ER, linked to ubiquitin (*4.9), and degraded in proteasomes in the cytoplasm.

*4.13 Proteasome

The proteasome is a cytoplasmic molecular machine whose function is to degrade proteins that have been tagged as unwanted or wrongly folded by the attachment of multiple ubiquitin molecules. It forms a large barrel (**Figure 4.13.1**) composed of $(\alpha_7\beta_7)_2$ rings, capped by a regulatory particle that recognizes the polyubiquitin tag, unfolds the protein, and feeds it into the proteasome, a process that requires ATP. The protease active site is located in the center of the barrel on the β subunits: the $\alpha_7\beta_7$ interface forms an 'antechamber' that presumably functions to feed unfolded protein into the active site in a regulated manner. In archaea, all the β subunits are identical, but in eukaryotes they are different, with different substrate specificities. The peptides generated are typically seven to nine residues long. Proteasomes are processive, in that they do not release their targets until they have been fully degraded.

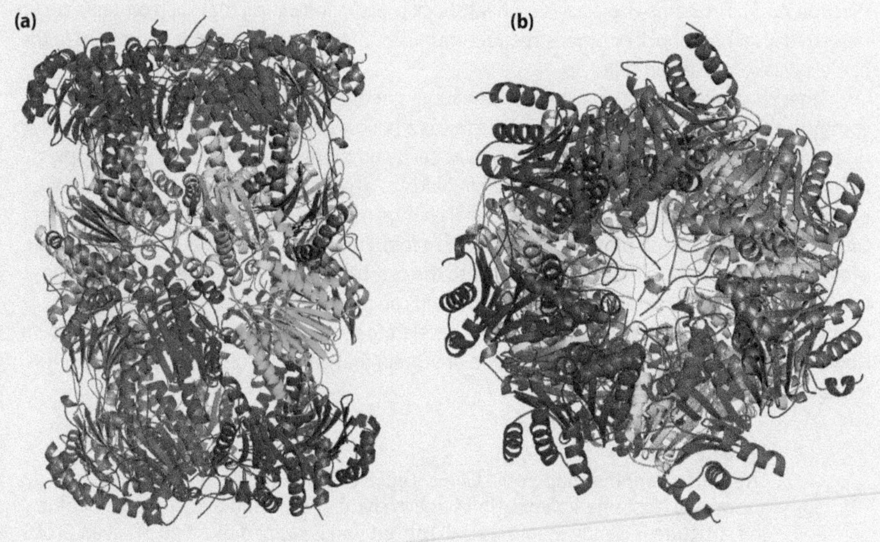

FIGURE 4.13.1
The structure of the proteasome. (a) A side view of the proteasome: this is the yeast (*S. cerevisiae*) 20S proteasome (PDB file 1ryp). The seven α subunits are shown in different shades of red and brown, and the seven β subunits are shown in greens for one ring and blues for the other. There is some interpenetration of the different rings. The active sites are within the β subunits. In the intact proteasome there is a cap on each end that regulates entry into the barrel-like interior. (b) Top view, with the same color scheme, showing the narrow entrance.

results suggest that up to 30% of proteins may fail these quality control tests and therefore get degraded immediately [109]. Some proteins are poor folders, and more than half of such proteins get rejected as misfolded before they emerge from the secretory system. (So maybe we should not be so disappointed if our own efforts at expressing soluble proteins sometimes fail!)

So far, we have considered only the degradation of misfolded protein, which happens as quickly as possible because such proteins are potentially dangerous. However, almost all proteins have evolved to have a limited lifespan, which like every other feature is under evolutionary control. There is therefore a mechanism for labeling proteins with their lifespan, and destroying them when they reach this point. As we might expect, there are actually several such mechanisms, which work together in complicated ways.

The basic rule is that **housekeeping proteins**, which are needed at roughly constant levels all the time, have very long lifetimes, whereas proteins involved in short-timescale functions, such as signaling proteins and those that regulate the cell cycle, have very short lifetimes. These latter are degraded in the same way as discussed above, by being labeled with polyubiquitin. They have short sequence motifs that act as signals for ubiquitylation: presumably this happens at a rate determined by the exposure of the signal and the frequency with which it is picked up by the ubiquitin ligase machinery. Similarly, it has been shown that PEST sequences (sequences 10–50 amino acids long enriched in Pro, Glu, Ser, or Thr) are signals for degradation, as are disordered regions; and that exposure of such signals, for example by phosphorylation, markedly enhances the rate of degradation. However, the commonest label is the N-terminal residue (the "N-end rule"). Hydrophobic residues (Phe, Leu, Trp, Ile, and Tyr), and in prokaryotes basic residues (Arg, Lys, and His) act as degradation signals, whereas small residues (Gly, Ser, Ala, and Cys) are stabilization signals (**Table 4.4**) [110]. Therefore N-terminal processing of proteins, by

TABLE 4.4 Dependence of half-life of cytoplasmic proteins on their N-terminal residue

N-terminal residue	Half-life
Met, Gly, Ala, Ser, Thr, Val	> 20 h
Ile, Glu	30 min
Tyr, Gln	10 min
Pro	7 min
Leu, Phe, Asp, Lys	3 min
Arg	2 min

either the removal or the addition of residues, will also alter their lifetime. In eukaryotes the N-terminal residue is recognized by E3 ubiquitin ligases, which direct the protein to the proteasome, whereas in prokaryotes it is recognized by an adaptor protein, ClpS, that directs them to the ClpAP protease [111]. (Prokaryotes do not possess the ubiquitin system; however, there are similarities, both in the chaperone-like proteins and in the overall structure of the ClpAP protease, which looks very similar to the proteasome.) All these signals interact with each other and no one signal is dominant [112], which means that the mechanism is still far from being understood in detail.

The stability of a folded protein *in vitro* is usually surprisingly low. (*In vivo*, the stability may be somewhat greater as a result of cellular crowding, although recent *in vivo* experiments using NMR suggest that for the possibly special case of ubiquitin itself, interactions with other proteins actually make it less stable [113].) A typical protein has a stability of about 20–60 kJ mol^{-1} compared with the unfolded state; remarkably, the stability shows no correlation with the size of the protein. This is equivalent to unfolded populations of between 3×10^{-4} and 3×10^{-11}. Thus, although some proteins may effectively never unfold *in vivo*, others certainly do, having a significant unfolded population, and many proteins will unfold and refold at least once during their functional lifetime. *Locally* unfolded structures are even more common. Indeed it has been suggested that a significant fraction of the proteins that are recognized as unfolded and degraded by the proteasome are not really unfolded at all, and are just tagged by an overzealous policing system. If true, this would show how important it is to the cell to remove any hint of misfolded protein before it gets a chance to cause aggregation. An important example is the protein responsible for cystic fibrosis, which is a transmembrane chloride channel. The most common mutation leading to cystic fibrosis is a three-nucleotide deletion that deletes F508. This mutant is recognized as misfolded and is degraded, causing a deficiency in chloride channels. However, this mutant would in fact function if it ever got a chance to arrive at the membrane.

4.5.3 Amyloid is a consequence of protein misfolding

Most proteins function in what is essentially a single folded conformation; any other conformation is to some extent unfolded and nonfunctional and is removed by the cell. Some exceptions to this have been discussed earlier in this book and include natively unfolded proteins and metastable proteins such as three-dimensional domain-swapped dimers. However, a surprisingly large number of proteins have been found to exist in two states: the normal folded state and a pathological misfolded state called amyloid. As noted in Chapter 1, this does not violate the "Anfinsen hypothesis," that a protein adopts the most stable conformation, because this misfolded state is only stable at high protein concentration. In other words, provided that the protein concentration remains low, the lowest-energy state is the unaggregated form; however, if the concentration of the protein gets too high, the most stable form is an aggregate. This state is a polymeric fibrous structure known as **amyloid**, composed entirely of β sheet. This type of sheet is called a cross-β sheet, because the β strands are perpendicular to the direction of the fiber (**Figure 4.42**). A β sheet is a stable

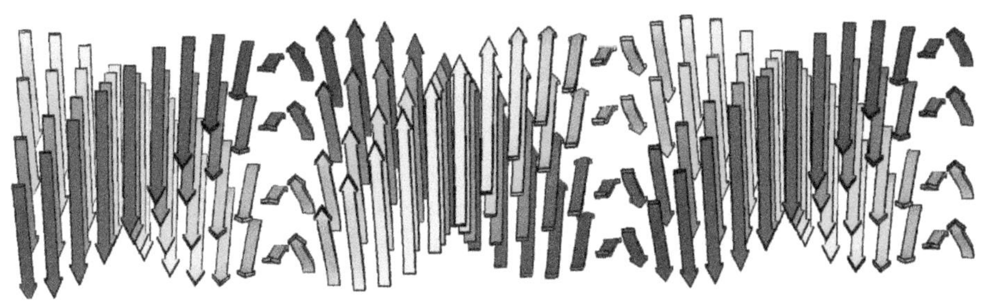

FIGURE 4.42
The generally accepted model of amyloid fibers, called a cross-β sheet, because the β strands are perpendicular to the direction of the fiber. This is a parallel in-register stack, which is commonly assumed to be the predominant form, although it is clear that some proteins form antiparallel stacks. There is a pronounced twist to the fiber, which can often be seen in electron micrographs.

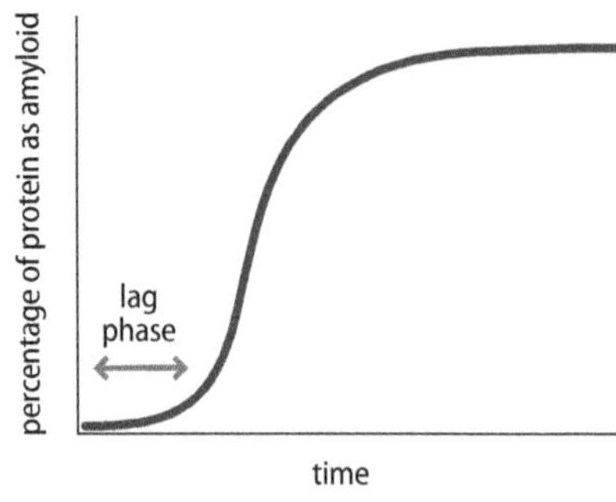

FIGURE 4.43
The formation of amyloid typically has a lag phase, followed by a faster polymerization step. *In vitro*, the lag phase can often be manipulated to an experimentally convenient range (hours or days); *in vivo*, the lag phase is often many years.

structure with no strong requirements for the nature of the sequence, other than some rather general requirements for alternating size and polarity in the center of the sheet. A wide range of proteins can therefore fold up into such a structure. Nevertheless, for a single protein molecule, the β-sheet form is higher in energy than the native state, and so it has a very low population. However, once a nucleus of β sheet has been made, it can act as a platform on which a stable fibrous structure can be built, because for each monomer added there is an additional stabilization energy from hydrogen bonding to the existing nucleus. The formation of amyloid therefore typically has a long lag phase during which the nucleus or seed is made, followed by a more rapid polymerization step (**Figure 4.43**). (This is the same principle that leads to the formation of very stable tubulin fibers from monomeric tubulin, discussed in Chapter 7).

In detail, it is likely that the cross-β structures formed by different proteins are different: the strands have different lengths, and the number of strands within each layer (perpendicular to the axis of the fiber) is different. In fact, even a single polypeptide chain can adopt more than one cross-β structure. The Alzheimer's disease peptide Aβ forms a β hairpin, which then polymerizes in parallel layers. However, depending on the conditions under which the fiber is formed, the resulting amyloid fiber can contain either two dimers per layer arranged in an antiparallel manner, or three in an equilateral triangle [114]. Similar polymorphism probably occurs in other amyloids also. It is possible that one mechanism for the formation of stable amyloid fibers arises from three-dimensional domain swapping (Chapter 2), in which the domain swapping stabilizes a β sheet formed by two adjacent monomers.

Amyloid formation usually requires a high concentration of the protein, plus some mechanism for allowing the normal native state to unfold enough to provide a significant population of partly unfolded protein. This mechanism can be a mutation (implying that many amyloid diseases result from destabilizing mutations of normal cellular proteins); it can just happen randomly ("sporadic" disease); it can occur as a result of a breakdown in the body's normal mechanisms for removing misfolded proteins (so amyloidosis tends to occur much more often in older people); or it can occur as a result of medical procedures. Thus, hemodialysis leads to a partial unfolding of the immune system protein β_2-microglobulin, which leads to its accumulation as amyloid deposits all over the body but particularly in joints. The Alzheimer's peptide Aβ is natively unfolded and is a small fragment cleaved off a much larger precursor protein, although it is still not clear what the "normal" biological function of either the peptide or the precursor is. By contrast, the diseases caused by **prion (*4.14)** proteins, such as bovine spongiform encephalopathy (BSE, or "mad cow disease"), result from misfolding of a normal cellular protein, although again the function of the correctly folded protein is still not clear. In all these cases, the amyloidosis is probably simply a result of amyloid's being formed faster than the body is able to remove it.

In most cases, the cross-β structure is formed by only a fraction of the protein molecule. In prions there seems to be a substantial fraction of the protein that has low structural content in the cellular form PrP^C and that remains relatively unstructured in the fibrillar amyloid form PrP^{Sc}.

Amyloid is an unwanted and dangerous consequence of protein misfolding. There is, however, some evidence that evolution has found several biological functions for amyloid [115]. Bacteria and fungi use it as an extracellular structural component; some egg-laying animals apparently use it to provide a protective coat on the egg; and it has been shown that, in humans, amyloid provides a fibrous location for facilitating the production and storage of melanin [116].

*4.14 Prion

A prion is an infectious agent composed entirely of protein, and it is responsible for scrapie in sheep, 'mad cow' disease (bovine spongiform encephalopathy, BSE), and Creutzfeldt–Jakob disease (CJD) in humans. The prion protein PrP can exist in two forms: the normal cellular form, PrP^C, and the disease form, PrP^{Sc} (with Sc standing for scrapie). In many cases, the function of the cellular form is still unknown. PrP^{Sc} accumulates in the brain, leading to neurological damage, although the exact mechanism is still unclear. PrP^{Sc} is an amyloid protein, and acts as a seed to propagate the conversion of PrP^C into PrP^{Sc} (**Figure 4.14.1**). It is very likely that the disease can be passed on by eating PrP^{Sc}: a small proportion of the amyloid escapes the digestive and immune systems and passes into the animal to produce prion disease. In cows, BSE is thought to have been transmitted by feeding them with infected cerebrospinal material, and the human disease Kuru is thought to have been caused by ritual cannibalism in which the brain of a dead relative was eaten. The rate of transmission is much faster within a species than across species barriers; there is strong evidence that 'new variant CJD' in the UK was caused by eating beef from cows with BSE, and consequently has a slower transmission rate than BSE from cow to cow. Amyloid protein is not easily destroyed, and hence some cases of prion infection in humans are thought to have been caused by incomplete sterilization of surgical tools.

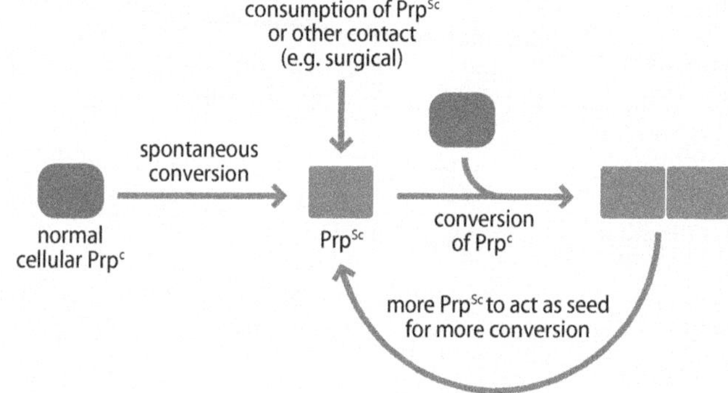

FIGURE 4.14.1
The currently accepted mechanism for prion disease.

4.6 SUMMARY

Molecules in the cell move mainly by diffusion. This is a random walk, driven by Brownian motion, and is an inefficient way of moving around. The diffusion rate of a substrate into an enzyme active site is governed mainly by the area of the active site, which is small: hence most enzymes have flaps or lids over the active site, or a flexible domain interface, thus opening up the active site to allow faster access of substrate; the lid is then closed for the reaction itself.

The collision rate is increased a little by electrostatic attraction, and it is increased significantly by electrostatic steering, which ensures that the substrate often hits the enzyme at the active site, in the right orientation. Very often, protein:protein interactions form an initial encounter complex, which lasts long enough to allow the proteins to alter their conformations and reorient, and therefore to form productive complexes. Proteins in the same cellular compartment have a tendency to have the same overall charge, thus deterring them from forming unwanted interactions.

The high intracellular concentration of proteins leads to macromolecular crowding, which produces a higher degree of protein association and slows down association rates, particularly for large proteins. Larger molecules diffuse more slowly, which has clearly been an important constraint on enzyme size, in that enzymes that act on polymeric substrates need to diffuse to their substrate (by contrast with most enzymes, which can sit still and wait for substrate to diffuse to them), and hence are small. Such enzymes also tend to be processive: they stay attached to their substrate over many catalytic cycles.

Another way of increasing the collision rate, particularly for protein:protein interactions, is to limit the search to two dimensions by attaching both proteins to a membrane surface. Membranes also help by limiting the size of the compartment to be searched. Even faster association occurs in one dimension, such as along a DNA strand.

Searching is speeded up by sticky arms, which allow a protein to reach out and grab its neighbors, forming associations that are rapidly formed and rapidly broken. Proline-rich sequences are very good at this and are therefore often used in signaling. Natively unfolded proteins may also work in the same way, by fly casting. This permits rapid binding, and also gives specificity without sacrificing speed.

Proteins are often covalently modified, for example by phosphorylation, methylation, acetylation, glycosylation, and proteolytic cleavage. Apart from proteolysis, this provides reversible regulatory mechanisms.

Proteins fold rapidly, although in the cell they are often assisted by chaperones. Misfolded proteins are usually removed rapidly by specific degradatory mechanisms; all proteins have a limited lifetime, defined mainly by their N-terminal residue. Proteins sometimes misfold into amyloid, which has a β-sheet structure. This is harmful when the rate of amyloid formation exceeds its degradation rate.

4.7 FURTHER READING

The Machinery of Life, by Goodsell [117], is a captivating book filled with thought-provoking pictures of what biological systems actually look like at a molecular level.

Lesk, in Introduction to Protein Science [118], provides a good discussion of protein folding.

The classic paper by Berg and von Hippel on collision rates [3] is a closely argued review well worth reading carefully.

Random Walks in Biology, by Berg [119], discusses diffusion clearly and elegantly.

4.8 WEBSITES

http://elm.eu.org/ and http://scansite.mit.edu/ *Identification of linear motifs.*

4.9 PROBLEMS

Hints for some of these problems can be found on the book's website.

1. Make a list of the ways in which the design of airports facilitates the movement of people and baggage to their correct destinations. Do these have analogies with the way that cells direct movement, and if so how?

2. Many peptide hormones are synthesized as inactive longer precursors called pre-pro-peptides, and undergo at least two proteolytic cleavage reactions (to produce pre-peptide and hormone, respectively). Suggest why such a complicated system may have evolved. Your answer should include a discussion of insulin and pro-opiomelanocortin.

3. Would you expect a housekeeping protein to have intrinsically disordered regions? Why?

4. What is the fastest possible diffusion-limited turnover rate for an enzyme, and why? How is it that some enzymes go faster than this?

5. Cytochrome *c* can be denatured by lowering the pH to 2. On addition of dextran (molecular weight about 35,000 Da) to a solution of cytochrome *c* at pH2, the circular dichroism (CD) spectrum was observed to change, as shown in **Figure 4.44** [35]. The CD spectrum at high concentration of dextran was similar to that for the molten globule conformation. Explain what is happening here. What concentration of dextran is necessary for half the protein present to be folded? Could dextran be used to fold the protein to the native state? Would you expect a different behavior if a dextran polymer of lower molecular weight were used, and if so, what?

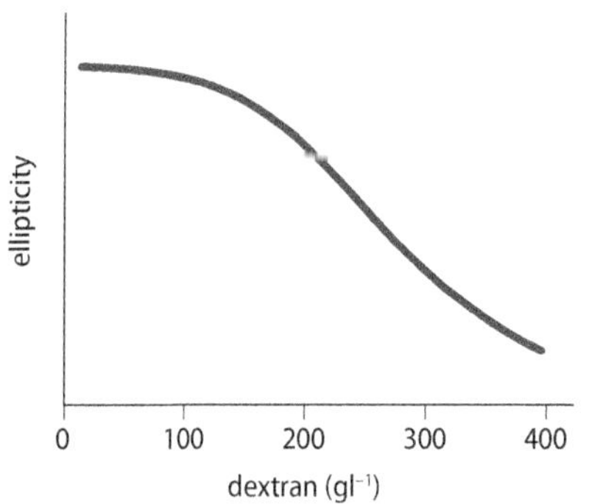

FIGURE 4.44
The change in ellipticity of cytochrome *c* with dextran.

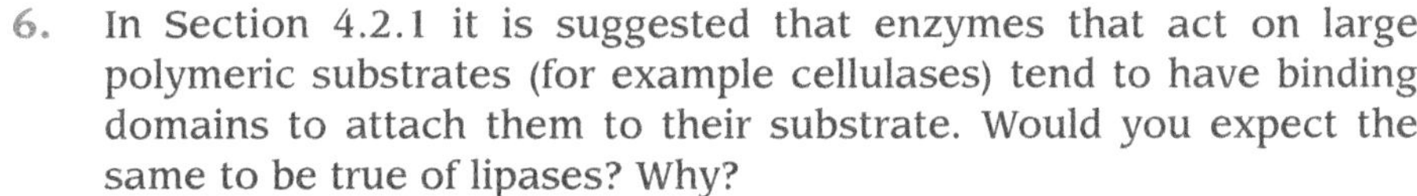

6. In Section 4.2.1 it is suggested that enzymes that act on large polymeric substrates (for example cellulases) tend to have binding domains to attach them to their substrate. Would you expect the same to be true of lipases? Why?

7. If proline-rich arms are as useful as implied in the text, they should be found in a very wide range of biological contexts. Is this actually true?

8. What is the significance of electrostatics in the discussion of fly casting in the context of protein–DNA binding in Section 4.3.2?

9. C.M. Dobson has suggested that *all* proteins may be capable of forming amyloid. Comment on this suggestion.

10. For discussion:

(a) We typically calculate reaction, collision or diffusion rates on the basis of "average" concentrations. If in fact many molecules are present in very small numbers in the cell, and are therefore not in any sense "average," is this a correct or meaningful calculation?

(b) Over what timescale is averaging acceptable or meaningful?

(c) If the number of free protons in the cell is so low (see, for example, question N2), is it meaningful to talk about intracellular pH at all?

(d) If a random walk is so inefficient, how does anything get to where it is meant to be within a reasonable time?

(e) How should we modify experiments to approximate the true cellular conditions, for example macromolecular crowding?

4.10 NUMERICAL PROBLEMS

N1. An *E. coli* cell is approximately a cylinder with a length of 2 μm and a diameter of 0.65 μm. What is its volume in μm^3? And in litres? The volume of a cylinder is $\pi r^2 h$, where r is the radius and h is the length. One litre is 1 dm^3.

N2. How many protons need to enter an *E. coli* cell to change the pH from 7.0 to 6.5? Avogadro's number (the number of molecules in a mole) is about 6×10^{23}.

N3. If a channel in *E. coli* has a conductance of 10^6 protons per second, how long will it take to change the pH inside the cell from 7 to 6.5? Does this seem a reasonable number?

N4. If a substrate is present at a concentration of 1 μM and its motion through the cell is unhindered (that is, it has a typical small-molecule diffusion coefficient), how often will it hit its target enzyme per second?

N5. A typical plant cell has a diameter of about 50 μm. What is the approximate ratio of the rate of search for an enzyme active site in three, two and one dimensions in a plant cell? What does this imply?

N6. If the free energy of a protein for folding is 30 kJ mol^{-1}, for what fraction of the time is it unfolded?

4.11 REFERENCES

1. K Popper (1972) Conjectures and Refutations: the Growth of Scientific Knowledge, 4th ed. London: Routledge & Kegan Paul.
2. G Albrecht-Buehler (1990) In defense of "nonmolecular" cell biology. *Int. Rev. Cytol.* 120:191–241.
3. OG Berg & PH von Hippel (1985) Diffusion-controlled macromolecular interactions. *Annu. Rev. Biophys. Biophys. Chem.* 14:131–160.
4. J Janin (1997) The kinetics of protein–protein recognition. *Proteins Struct. Funct. Genet.* 28:153–161.
5. HV Westerhoff & GR Welch (1992) Enzyme organization and the direction of metabolic flow: physicochemical considerations. *Curr. Top. Cell. Regul.* 33:361–390.
6. SH Northrup & HP Erickson (1992) Kinetics of protein–protein association explained by Brownian dynamics computer simulation. *Proc. Natl. Acad. Sci. USA* 89:3338–3342.
7. RA Alberty & GG Hammes (1958) Application of the theory of diffusion-controlled reactions to enzyme kinetics. *J. Phys. Chem.* 62:154–159.
8. GG Hammes & RA Alberty (1959) The influence of the net protein charge on the rate of formation of enzyme–substrate complexes. *J. Phys. Chem.* 63:274–279.
9. T Selzer & G Schreiber (2001) New insights into the mechanism of protein–protein association. *Proteins Struct. Funct. Genet.* 45:190–198.
10. DR Ripoll, CH Faerman, PH Axelsen et al. (1993) An electrostatic mechanism for substrate guidance down the aromatic gorge of acetylcholinesterase. *Proc. Natl. Acad. Sci. USA* 90:5128–5132.
11. K Sharp, R Fine & B Honig (1987) Computer simulations of the diffusion of a substrate to an active site of an enzyme. *Science* 236:1460–1463.
12. RC Wade, RR Gabdoulline & BA Luty (1998) Species dependence of enzyme-substrate encounter rates for triose phosphate isomerase. *Proteins Struct. Funct. Genet.* 31:406–416.
13. RC Wade, RR Gabdoulline, SK Lüdemann & V Lounnas (1998) Electrostatic steering and ionic tethering in enzyme–ligand binding: insights from simulations. *Proc. Natl. Acad. Sci. USA* 95:5942–5949.
14. E Margoliash & HR Bosshard (1983) Guided by electrostatics, a textbook protein comes of age. *Trends Biochem. Sci.* 8:316–320.
15. BW Pontius (1993) Close encounters: why unstructured, polymeric domains can increase rates of specific macromolecular association. *Trends Biochem. Sci.* 18:181–186.
16. RR Gabdoulline & RC Wade (2002) Biomolecular diffusional association. *Curr. Opin. Struct. Biol.* 12:204–213.
17. G Schreiber (2002) Kinetic studies of protein–protein interactions. *Curr. Opin. Struct. Biol.* 12:41–47.
18. X Xu, W Reinle, F Hannemann et al. (2008) Dynamics in a pure encounter complex of two proteins studied by solution scattering and paramagnetic NMR spectroscopy. *J. Am. Chem. Soc.* 130:6395–6403.
19. C Frisch, AR Fersht & G Schreiber (2001) Experimental assignment of the structure of the transition state for the association of barnase and barstar. *J. Mol. Biol.* 308:69–77.
20. MA Andrade, SI O'Donoghue & B Rost (1998) Adaptation of protein surfaces to subcellular location. *J. Mol. Biol.* 276:517–525.
21. DS Goodsell (1991) Inside a living cell. *Trends Biochem. Sci.* 16:203–206.
22. O Medalia, I Weber, AS Frangakis et al. (2002) Macromolecular architecture in eukaryotic cells visualized by cryoelectron tomography. *Science* 298:1209–1213.
23. R Phillips, J Kondev & J Theriot (2009) Physical Biology of the Cell. New York: Garland Science.
24. P Mentré (1995) L'eau Dans la Cellule [Water in the Cell]. Paris: Masson.
25. D Hall & AP Minton (2003) Macromolecular crowding: qualitative and semiquantitative successes, quantitative challenges. *Biochim. Biophys. Acta* 1649:127–139.
26. SB Zimmerman & SO Trach (1991) Estimation of macromolecule concentrations and excluded volume effects for the cytoplasm of *Escherichia coli*. *J. Mol. Biol.* 222:599–620.
27. SB Zimmerman & AP Minton (1993) Macromolecular crowding: biochemical, biophysical, and physiological consequences. *Annu. Rev. Biophys. Biomol. Struct.* 22:27–65.
28. G Rivas, JA Fernandez & AP Minton (2001) Direct observation of the enhancement of noncooperative protein self-assembly by macromolecular crowding: indefinite linear self-association of bacterial cell division protein FtsZ. *Proc. Natl. Acad. Sci. USA* 98:3150–3155.
29. G Rivas, F Ferrone & J Herzfeld (2004) Life in a crowded world. *EMBO Rep.* 5:23–27.
30. VN Uversky, EM Cooper, KS Bower, J Li & AL Fink (2002) Accelerated α-synuclein fibrillation in crowded milieu. *FEBS Lett.* 515:99–103.
31. MD Shtilerman, TT Ding & PT Lansbury (2002) Molecular crowding accelerates fibrillization of α-synuclein: could an increase in the cytoplasmic protein concentration induce Parkinson's disease? *Biochemistry* 41:3855–3860.
32. AP Minton (1981) Excluded volume as a determinant of macromolecular structure and reactivity. *Biopolymers* 20:2093–2120.
33. AP Minton (2001) The influence of macromolecular crowding and macromolecular confinement on biochemical reactions in physiological media. *J. Biol. Chem.* 276:10577–10580.
34. HX Zhou, GN Rivas & AP Minton (2008) Macromolecular crowding and confinement: biochemical, biophysical, and potential physiological consequences. *Annu. Rev. Biophys.* 37:375–397.
35. K Sasahara, P McPhie & AP Minton (2003) Effect of dextran on protein stability and conformation attributed to macromolecular crowding. *J. Mol. Biol.* 326:1227–1237.

36. AS Verkman (2002) Solute and macromolecule diffusion in cellular aqueous compartments. *Trends Biochem. Sci.* 27:27–33.

37. RJ Ellis (2001) Macromolecular crowding: an important but neglected aspect of the intracellular environment. *Curr. Opin. Struct. Biol.* 11:114–119.

38. MB Elowitz, MG Surette, PE Wolf, JB Stock & S Leibler (1999) Protein mobility in the cytoplasm of *Escherichia coli*. *J. Bact.* 181:197–203.

39. CW Mullineaux, A Nenninger, N Ray & C Robinson (2006) Diffusion of green fluorescent protein in three cell environments in *Escherichia coli*. *J. Bacteriol.* 188:3442–3448.

40. HP Kao, JR Abney & AS Verkman (1993) Determinants of the translational mobility of a small solute in cell cytoplasm. *J. Cell Biol.* 120:175–184.

41. T Misteli (2001) Protein dynamics: implications for nuclear architecture and gene expression. *Science* 291:843–847.

42. K Luby-Phelps (2000) Cytoarchitecture and physical properties of cytoplasm: volume, viscosity, diffusion, intracellular surface area. *Int. Rev. Cytol.* 192:189–221.

43. HV Westerhoff (1985) Organization in the cell soup. *Nature* 318:106.

44. DS Goodsell & AJ Olson (1993) Soluble proteins: size, shape and function. *Trends Biochem. Sci.* 18:65–68.

45. GH Pollack (2001) Is the cell a gel—and why does it matter? *Jpn. J. Physiol.* 51:649–660.

46. GH Pollack (2003) The role of aqueous interfaces in the cell. *Adv. Colloid Interf. Sci.* 103:173–196.

47. HJ Morowitz (1984) The completeness of molecular biology. *Israel J. Med. Sci.* 20:750–753.

48. H Walter & DE Brooks (1995) Phase separation in cytoplasm, due to macromolecular crowding, is the basis for microcompartmentation. *FEBS Lett.* 361:135–139.

49. U Kishore, P Eggleton & KBM Reid (1997) Modular organization of carbohydrate recognition domains in animal lectins. *Matrix Biol.* 15:583–592.

50. K Sorimachi, MF Le Gal-Coëffet, G Williamson, DB Archer & MP Williamson (1997) Solution structure of the granular starch binding domain of *Aspergillus niger* glucoamylase bound to β-cyclodextrin. *Structure* 5:647–661.

51. TE Creighton (1993) Proteins: Structures and Molecular Properties, 2nd ed. New York: Freeman.

52. G Adam & M Delbrück (1968) Reduction of dimensionality in biological diffusion processes. In A Rich and N Davidson (eds) Structural Chemistry and Molecular Biology. San Francisco: W.H. Freeman, 198–215.

53. SA Teichmann, C Chothia & M Gerstein (1999) Advances in structural genomics. *Curr. Opin. Struct. Biol.* 9:390–399.

54. H Kabata, O Kurosawa, I Arai et al. (1993) Visualization of single molecules of RNA polymerase sliding along DNA. *Science* 262:1561–1563.

55. J Gorman, A Chowdhury, JA Surtees et al. (2007) Dynamic basis for one-dimensional DNA scanning by the mismatch repair complex Msh2-Msh6. *Mol. Cell* 28:359–370.

56. RB Winter, OG Berg & PH von Hippel (1981) Diffusion-driven mechanisms of protein translocation on nucleic acids. 3. The *Escherichia coli lac* repressor–operator interaction: kinetic measurements and conclusions. *Biochemistry* 20:6961–6977.

57. DM Gowers & SE Halford (2003) Protein motion from non-specific to specific DNA by three-dimensional routes aided by supercoiling. *EMBO J.* 22:1410–1418.

58. M Doucleff & GM Clore (2008) Global jumping and domain-specific intersegment transfer between DNA cognate sites of the multidomain transcription factor Oct-1. *Proc. Natl. Acad. Sci. USA* 105:13871–13876.

59. J Gorman & EC Greene (2008) Visualizing one-dimensional diffusion of proteins along DNA. *Nature Struct. Mol. Biol.* 15:768–774.

60. MP Williamson (1994) The structure and function of proline-rich regions in proteins. *Biochem. J.* 297:249–260.

61. SL Turner, GC Russell, MP Williamson & JR Guest (1993) Restructuring an interdomain linker in the dihydrolipoamide acetyltransferase component of the pyruvate dehydrogenase complex of *Escherichia coli*. *Protein Eng.* 6:101–108.

62. EC Petrella, LM Machesky, DA Kaiser & TD Pollard (1996) Structural requirements and thermodynamics of the interaction of proline peptides with profilin. *Biochemistry* 35:16535–16543.

63. BK Kay, MP Williamson & P Sudol (2000) The importance of being proline: the interaction of proline-rich motifs in signaling proteins with their cognate domains. *FASEB J.* 14:231–241.

64. IM Olazabal & LM Machesky (2001) Abp1p and cortactin, new "handholds" for actin. *J. Cell Biol.* 154:679–682.

65. K Tanaka & Y Matsui (2001) Functions of unconventional myosins in the yeast *Saccharomyces cerevisiae*. *Cell Struct. Funct.* 26:671–675.

66. BL Anderson, I Boldogh, M Evangelista et al. (1998) The Src homology domain 3 (SH3) of a yeast type I myosin, Myo5p, binds to verprolin and is required for targeting to sites of actin polarization. *J. Cell Biol.* 141:1357–1370.

67. L Shapiro (2001) β-Catenin and its multiple partners: promiscuity explained. *Nature Struct. Biol.* 8:484–487.

68. F Diella, N Haslam, C Chica et al. (2008) Understanding eukaryotic linear motifs and their role in cell signaling and regulation. *Front. Biosci.* 13:6580–6603.

69. V Neduva & RB Russell (2005) Linear motifs: evolutionary interaction switches. *FEBS Lett.* 579:3342–3345.

70. M Overduin & T de Beer (2000) The plot thickens: how thrombin modulates blood clotting. *Nature Struct. Biol.* 7:267–269.

71. ASV Burgen, GCK Roberts & J Feeney (1975) Binding of flexible ligands to macromolecules. *Nature* 253:753–755.

72. AL Fink (2005) Natively unfolded proteins. *Curr. Opin. Struct. Biol.* 15:35–41.

73. AK Dunker, I Silman, VN Uversky & JL Sussman (2008) Function and structure of inherently disordered proteins. *Curr. Opin. Struct. Biol.* 18:756–764.

74. MM Dedmon, CN Patel, GB Young & GJ Pielak (2002) FlgM gains structure in living cells. *Proc. Natl. Acad. Sci. USA* 99:12681–12684.

75. PE Wright & HJ Dyson (1999) Intrinsically unstructured proteins: re-assessing the protein structure–function paradigm. *J. Mol. Biol.* 293:321–331.

76. AK Dunker, JD Lawson, CJ Brown et al. (2001) Intrinsically disordered protein. *J. Mol. Graphics. Model.* 19:26–59.

77. P Tompa (2002) Intrinsically unstructured proteins. *Trends Biochem. Sci.* 27:527–533.

78. HJ Dyson & PE Wright (2005) Intrinsically unstructured proteins and their functions. *Nature Rev. Mol. Cell Biol.* 6:197–208.

79. VN Uversky (2002) Natively unfolded proteins: a point where biology waits for physics. *Protein Sci.* 11:739–756.

80. PE Wright & HJ Dyson (2009) Linking folding and binding. *Curr. Opin. Struct. Biol.* 19:31–38.

81. G Luck, H Liao, NJ Murray et al. (1994) Polyphenols, astringency and proline-rich proteins. *Phytochemistry* 37:357–371.

82. P Tompa (2003) Intrinsically unstructured proteins evolve by repeat expansion. *BioEssays* 25:847–855.

83. BA Shoemaker, JJ Portman & PG Wolynes (2000) Speeding molecular recognition by using the folding funnel: the fly-casting mechanism. *Proc. Natl. Acad. Sci. USA* 97:8868–8873.

84. K Sugase, HJ Dyson & PE Wright (2007) Mechanism of coupled folding and binding of an intrinsically disordered protein. *Nature* 447:1021–1025.

85 Y Huang & Z Liu (2009) Kinetic advantage of intrinsically disordered proteins in coupled folding-binding process: A critical assessment of the "fly-casting" mechanism. *J. Mol. Biol.* 393:1143–1159.

86. Y Levy, JN Onuchic & PG Wolynes (2007) Fly-casting in protein–DNA binding: frustration between protein folding and electrostatics facilitates target recognition. *J. Am. Chem. Soc.* 129:738–739.

87. M Fuxreiter, I Simon, P Friedrich & P Tompa (2004) Preformed structural elements feature in partner recognition by intrinsically unstructured proteins. *J. Mol. Biol.* 338:1015–1026.

88. RW Kriwacki, L Hengst, L Tennant, SI Reed & PE Wright (1996) Structural studies of $p21^{Waf1/Cip1/Sdi1}$ in the free and Cdk2-bound state: conformational disorder mediates binding diversity. *Proc. Natl. Acad. Sci. USA* 93:11504–11509.

89. G Manning, DB Whyte, R Martinez, T Hunter & S Sudarsanam (2002) The protein kinase complement of the human genome. *Science* 298:1912–1934.

90. P Cohen (2002) The origins of protein phosphorylation. *Nature Cell Biol.* 4:E127–E130.

91. M Huse & J Kuriyan (2002) The conformational plasticity of protein kinases. *Cell* 109:275–282.

92. F Sicheri, I Moarefi & J Kuriyan (1997) Crystal structure of the Src family tyrosine kinase, Hck. *Nature* 385:602–609.

93. LJ Mauro & JE Dixon (1994) 'Zip codes' direct intracellular protein tyrosine phosphatases to the correct cellular 'address'. *Trends Biochem. Sci.* 19:151–155.

94. S Arena, S Benvenuti & A Bardelli (2005) Genetic analysis of the *kinome* and *phosphatome* in cancer. *Cell Mol. Life Sci.* 62:2092–2099.

95. A Munshi, G Shafi, N Aliya & A Jyothy (2009) Histone modifications dictate specific biological readouts. *J. Genet. Genom.* 36:75–88.

96. SD Taverna, H Li, AJ Ruthenburg, CD Allis & DJ Patel (2007) How chromatin-binding modules interpret histone modifications: lessons from professional pocket pickers. *Nature Struct. Mol. Biol.* 14:1025–1040.

97. XJ Yang & E Seto (2008) Lysine acetylation: codified crosstalk with other posttranslational modifications. *Mol. Cell* 31:449–461.

98. C Schäffer, M Graninger & P Messner (2001) Prokaryotic glycosylation. *Proteomics* 1:248–261.

99. AL Watters, P Deka, C Corrent et al. (2007) The highly cooperative folding of small naturally occurring proteins is likely the result of natural selection. *Cell* 128:613–624.

100. DE Kim, C Fisher & D Baker (2000) A breakdown of symmetry in the folding transition state of protein L. *J. Mol. Biol.* 298:971–984.

101. EL McCallister, E Alm & D Baker (2000) Critical role of β-hairpin formation in protein G folding. *Nature Struct. Biol.* 7:669–673.

102. S Batey, AA Nickson & J Clarke (2008) Studying the folding of multidomain proteins. *HFSP J.* 2:365–377.

103. TA Thanaraj & P Argos (1996) Ribosome-mediated translational pause and protein domain organization. *Prot. Sci.* 5:1594–1612.

104. JH Han, S Batey, AA Nickson, SA Teichmann & J Clarke (2007) The folding and evolution of multidomain proteins. *Nature Rev. Mol. Cell Biol.* 8:319–330.

105. S Batey & J Clarke (2008) The folding pathway of a single domain in a multidomain protein is not affected by its neighbouring domain. *J. Mol. Biol.* 378:297–301.

106. CM Dobson (2003) Protein folding and misfolding. *Nature* 426:884–890.

107. C Hammond & A Helenius (1995) Quality control in the secretory pathway. *Curr. Opin. Cell Biol.* 7:523–529.

108. J Roth, GHF Yam, JY Fan et al. (2008) Protein quality control: the who's who, the where's and therapeutic escapes. *Histochem. Cell Biol.* 129:163–177.

109. U Schubert, LC Antón, J Gibbs et al. (2000) Rapid degradation of a large fraction of newly synthesized proteins by proteasomes. *Nature* 404:770–774.

110. A Bachmair, D Finley & A Varshavsky (1986) *In vivo* half-life of a protein is a function of its amino-terminal residue. *Science* 234:179–186.

111. A Mogk, R Schmidt & B Bukau (2007) The N-end rule pathway for regulated proteolysis: prokaryotic and eukaryotic strategies. *Trends Cell Biol.* 17:165–172.

112. P Tompa, J Prilusky, I Silman & JL Sussman (2008) Structural disorder serves as a weak signal for intracellular protein degradation. *Proteins Struct. Funct. Bioinf.* 71:903–909.

113. K Inomata, A Ohno, H Tochio et al. (2009) High-resolution multidimensional NMR spectroscopy of proteins in human cells. *Nature* 458:106–109.

114. AK Paravastu, RD Leapman, WM Yau & R Tycko (2008) Molecular structural basis for polymorphism in Alzheimer's β-amyloid fibrils. *Proc. Natl. Acad. Sci. USA* 105:18349–18354.

115. DM Fowler, AV Koulov, WE Balch & JW Kelly (2007) Functional amyloid—from bacteria to humans. *Trends Biochem. Sci.* 32:217–224.

116. SK Maji, MH Perrin, MR Sawaya et al. (2009) Functional amyloids as natural storage of peptide hormones in pituitary secretory granules. *Science* 325:328–332.

117. DS Goodsell (1998) The Machinery of Life. New York: Springer-Verlag.

118. AM Lesk (2010) Introduction to Protein Science, 2nd ed. Oxford: Oxford University Press.

119. HC Berg (1993) Random Walks in Biology, expanded ed. Princeton: Princeton University Press.

120. B van den Berg, RJ Ellis & CM Dobson (1999) Effects of macromolecular crowding on protein folding and aggregation. *EMBO J.* 18:6927–6933.

121. G Williamson, NJ Belshaw & MP Williamson (1992) *O*-glycosylation in *Aspergillus* glucoamylase. Conformation and role in binding. *Biochem. J.* 282:423–428.

122. SM Southall, PJ Simpson, HJ Gilbert, G Williamson & MP Williamson (1999): The starch binding domain from glucoamylase disrupts the structure of starch. *FEBS Lett.* 447:58–60.

123. C Lad, NH Williams & R Wolfenden (2003) The rate of hydrolysis of phosphomonoester dianions and the exceptional catalytic proficiencies of protein and inositol phosphatases. *Proc. Natl. Acad. Sci. USA* 100:5607–5610.

124. P Xu, DM Duong, NT Seyfried et al. (2009) Quantitative proteomics reveals the function of unconventional ubiquitin chains in proteasomal degradation. *Cell* 137:133–145.

125. D Nath & S Shadan (eds) (2009) Insight: the ubiquitin system. *Nature* 458:421–467.

126. FB Perler (2005) Protein splicing mechanisms and applications. *IUBMB Life* 57:469–476.

127. A Matouschek, JT Kellis, L Serrano & AR Fersht (1989) Mapping the transition state and pathway of protein folding by protein engineering. *Nature* 340:122–126.

128. S Nauli, B Kuhlman & D Baker (2001) Computer-based redesign of a protein folding pathway. *Nature Struct. Biol.* 8:602–605.

CHAPTER 5
How Enzymes Work

There are many excellent books on enzymes, some of which are listed at the end of this chapter. Many of these deal either with kinetics or with mechanisms, or with both. However, my concern in this chapter is to emphasize those factors of most relevance to the themes of this book, which are mainly about binding and evolution rather than catalytic mechanism. We shall see that a good proportion of the catalytic power demonstrated by enzymes arises from their ability to bind substrates and to change in response to substrate binding, topics that are fundamental to the function of all proteins, not merely enzymes. Therefore, I hope to demonstrate that enzymes are not "special" in that sense. The material in this chapter will prove challenging for some readers, because this chapter is more chemical than any of the others and so the chemistry is explained in particular detail.

Never a dull enzyme.

Arthur Kornberg, Nobel Prize for Medicine (1959), title of autobiographical article [1]

It is not generally appreciated how little is understood about the mechanism by which enzymes bring about their extraordinary and specific rate accelerations.

W. P. Jencks (1969), first sentence of Catalysis in Chemistry and Enzymology [2]

The specific effect [of invertin and emulsin] on glucosides might thus be explained by the assumption that the contact between molecules necessary to produce a chemical reaction can only take place if they have complementary geometrical shape. To give an illustration I could say that enzyme and glucoside must fit together like lock and key. ... The finding that the efficacy of enzymes is limited to such a high degree by molecular geometry could also be of considerable use to biochemical research.

Emil Fischer (1894), original in German, translation by author [3]

5.1 ENZYMES LOWER THE ENERGY OF THE TRANSITION STATE

5.1.1 What is the transition state?

Essential to the account given here is the energy-level diagram for a chemical reaction, shown in **Figure 5.1**, representing the **transition state** model of catalysis. In such diagrams, the vertical axis usually represents **free energy (*5.1)**, and the horizontal axis is usually called the reaction coordinate. Often this axis has no particular meaning, but basically it means anything that changes during the reaction. For example, if the reaction involves the breaking of a bond, this axis could represent the length of the bond. The substrate is on the left and the product on the right, and in between these the molecule needs to cross an energy barrier. The top of the energy barrier is defined as being the transition state (TS). Because it is at the top of the energy profile, it is highly unstable, so it has a very short existence, of the order of a bond vibration (10^{-12} s). Thus, one cannot isolate or directly observe transition states. Nevertheless they are very important, because by definition they are at the top of the energy barrier.

This diagram is a major simplification of what really happens, in that it ignores everything else in the molecule except that one bond. In real life, the energy required to break a bond depends not just on the length of that bond but also on many other things, such as the angle between the bond and a neighboring bond. In that case, we should draw the diagram not as a one-dimensional surface but as a two-dimensional surface, as in **Figure 5.2**. This diagram makes it clear that there is in general not a single pathway for the reaction, although there is always one that involves the lowest energy barrier and therefore is the most likely. There are in fact infinitely many, although most of these involve crossing an energy barrier so high that they will never happen and we can ignore them. However, it is clear that by making changes elsewhere in the molecule we could potentially affect the shape of this surface and allow the reaction to proceed by a different pathway. In a real molecule, the surface has not just two dimensions but many more, and is thus rather hard to draw or imagine.

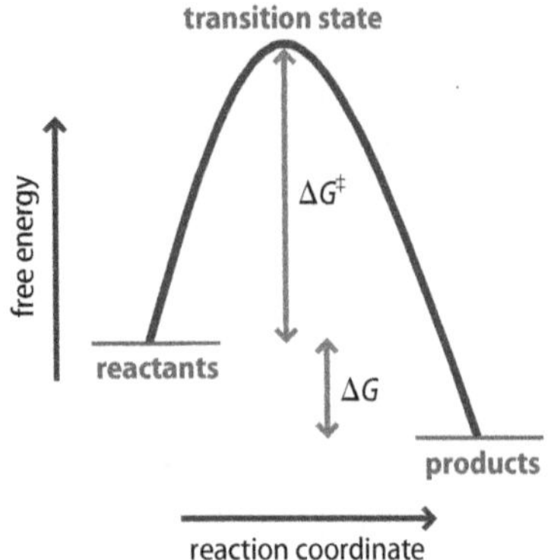

FIGURE 5.1
The transition state model for chemical reactions. ΔG is the change in free energy for the reaction, whereas $\Delta G^\ddagger$ is the activation energy for the forward reaction. The activation energy for the reverse reaction is ($\Delta G + \Delta G^\ddagger$) and is therefore larger, and hence the reverse reaction is slower than the forward reaction. This is implicit in the drawing, which shows the products as being of lower energy (more stable) than the reactants, and hence the equilibrium lies on the side of products.

The thermal energy of a molecule is generally quoted as being $kT/2$ per degree of freedom, where a degree of freedom means for example a bond rotation. It follows a **Boltzmann distribution (*5.2)**, which means that a small number of molecules have much more energy; or, equivalently, each degree of freedom (for example the vibrational energy in a bond) will very occasionally have a much larger energy than the average. If this energy happens to lie in the bond to be broken, the molecule will have enough energy to break the bond. Standard transition state theory says that once the molecule gets to the top of the curve in Figure 5.1, it can either cross over and become product, or it can go back and return to substrate, with a certain probability called the transmission coefficient (*6.4) that varies between 0 and 1, depending on how rough the top of the energy curve is and thus how strongly the substrate is hindered from going over the top. In the simplest description, which we will use, the transmission coefficient has a value of 1, so that once it gets to the top then it crosses over and becomes product. Crucially, the rate of the reaction (k) depends only on the height at the top compared with the starting position (and for example is independent of the shape of the energy curve or how many other states there might be on the way up or down), and it is given by an exponential function, often called the **Arrhenius (*5.4)** equation:

$$k = Ae^{-E_a/RT}$$

where R is the gas constant, T is the temperature, and E_a is an important concept, the **activation energy**, which is also written as $\Delta G^\ddagger$, read as "delta G double dagger." Thus, very simply, if we want a reaction to go faster, we need to decrease the size of $\Delta G^\ddagger$ (or increase the temperature, an option that is usually not open to cells). This means that we need to either decrease the energy of the TS or increase the energy of the substrate. As we shall see, enzymes do both. We should note that the rate is *exponentially* dependent on $\Delta G^\ddagger$, which means that a change in $\Delta G^\ddagger$ causes a proportionately larger change in the rate.

***5.1 Free energy**
The free energy is the amount of energy available to do work. We always calculate and measure changes in free energy rather than the total amount. A favorable reaction is by definition one that proceeds with the release of energy, and thus there is less energy available at the end than there was at the start: in other words, the change in *free energy* is negative. In biological systems, the term free energy always means the Gibbs free energy, which is the free energy at constant temperature and pressure. It is given the symbol G, and changes in free energy are given the symbol ΔG, the Δ being the conventional way of indicating a change. For more on this, see *3.1.

***5.2 Boltzmann distribution**
A distribution of atomic or molecular velocities that fits the distribution described by **Boltzmann (*5.3)** (or more properly, by Maxwell and Boltzmann, the Maxwell–Boltzmann distribution). It is given by

$$\frac{N_i}{N} = \frac{g_i e^{-E_i/kT}}{\Sigma g_i e^{-E_i/kT}}$$

where N_i is the number of particles in state i with energy E_i, N is the total number of particles, g_i is the number of states with energy E_i, and k is the Boltzmann constant. It looks like **Figure 5.2.1**.

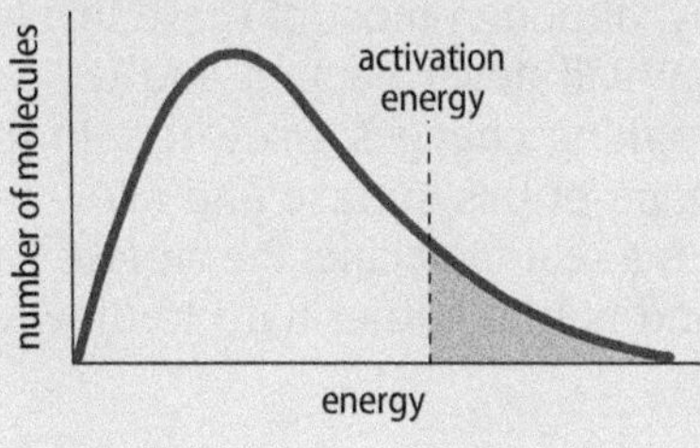

FIGURE 5.2.1
A Boltzmann distribution, showing the proportion of molecules with a given energy. As the temperature rises, the distribution moves to the right and flattens out. If a molecule has an energy greater than some threshold value called the activation energy, then it has sufficient energy to undergo a reaction.

*5.3 Ludwig Boltzmann

Ludwig Boltzmann (**Figure 5.3.1**) was born in Austria in 1844; he was an advocate of the atomic theory, which was opposed by most of his contemporaries. His students included Svante Arrhenius and Walther Nernst. He developed a kinetic theory of gases, deriving the Boltzmann distribution that describes the velocity of molecules in a gas; he also gave statistical descriptions of matter that formed the foundation of classical statistical mechanics and helped describe what temperature means, and he characterized the nature of entropy. His theories relied on the existence of atoms, so he spent much of his career arguing the case for atoms, becoming embroiled in philosophy. This no doubt contributed to bouts of depression. During one such bout he committed suicide. His tombstone in Vienna bears the inscription $S = k\ln W$, the best definition of entropy, where k is the Boltzmann constant.

FIGURE 5.3.1
Ludwig Boltzmann. (Courtesy of Wikimedia Commons.)

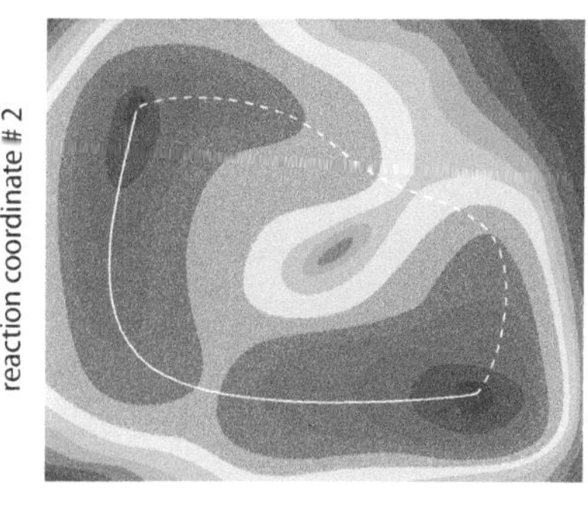

FIGURE 5.2
A transition state model drawn using two reaction coordinates. In this example, the energy levels are color-coded from red (lowest energy) through yellow and green to blue and violet. The lowest-energy transition state, and thus the preferred reaction pathway, is shown by the solid line. However, the reaction pathway shown by the dashed line is only slightly higher in energy, and therefore reactions could proceed by this route almost as fast. In reality many reactions (particularly enzyme-catalyzed reactions) will have a very large number of reaction coordinates and can therefore travel on a very large number of (fairly similar) pathways of similar energy; however, one would need to draw a diagram in multi-dimensional space to show it.

An enzyme-catalyzed reaction more typically has an energy profile that looks like **Figure 5.3**. This is different from Figure 5.1 because the enzyme has allowed the reaction to proceed along a different pathway (we shall see how later). This pathway has an intermediate in an energy well. In general the intermediate is a stable compound that can be isolated, and often characterized (crystallized for example) and studied. Because the intermediate is similar in energy to the TS, it will have a structure fairly similar to the TS. Thus one of the best ways of seeing what the TS might look like is to look at intermediates.

The product is drawn with a lower energy than the substrate, implying that the product is more stable than the reactant; that is, that the reaction equilibrium will lie on the product side. (See **Binding and dissociation constants and free energy, *5.5**.) This implies that the activation energy going from right to left is larger than the activation energy going from left to right, or in other words that the back reaction is slower than the forward reaction. And this statement in turn is equivalent to saying that the equilibrium constant favors the forward reaction, because the equilibrium constant $K = k_{\text{forward}}/k_{\text{back}}$.

*5.4 Svante Arrhenius

Svante Arrhenius (1859–1927) (**Figure 5.4.1**) was Swedish and is most famous for his idea that the conductivity of salt solutions depends on the dissolved ions; he also developed the idea of activation energy. The first of these was the subject of his PhD thesis, which he almost failed because the committee did not believe him. He wrote several books to bring science to the wider public (Destiny of the Stars, Smallpox and its Combating, and others) and had a broad range of interests including immunology (toxins), geology (as part of which he proposed that CO_2 might act as a greenhouse gas), and astronomy (propounding the idea that life may have been carried to Earth as spores by radiation pressure, an idea now known as panspermia)

FIGURE 5.4.1
Svente Arrhenius. (Courtesy of Wikimedia Commons.)

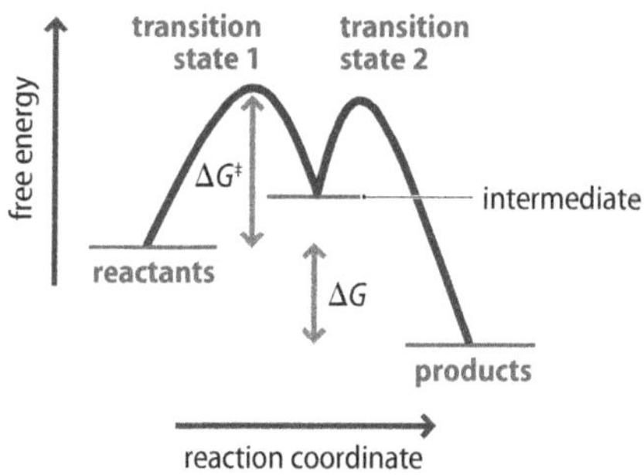

FIGURE 5.3
An energy profile for an enzyme-catalyzed reaction. The reaction pathway is different from that shown in Figure 5.1 because there is now an intermediate. There are also two transition states. The rate is determined by the highest energy barrier, $\Delta G^{\ddagger}$.

It is important to be clear about the difference between a TS and an intermediate: this is illustrated by **Figure 5.4**.

5.1.2 Enzymes lower both enthalpy and entropy barriers in the transition state

The well-known equation for the Gibbs free energy (*5.1) is

$$\Delta G = \Delta H - T\Delta S$$

where H and S are the enthalpy and entropy (*3.1), respectively. The free energy says whether a reaction goes forward or backward: a negative ΔG means it goes forward, whereas a positive ΔG means it goes backward. More precisely, a negative ΔG means that at equilibrium the products are at a higher concentration than the reactants. Enthalpy is the heat energy in the system, and in the context of a chemical reaction this typically means bond energy. Entropy is essentially the amount of disorder in the system. So in other words, this equation is telling us that the free energy of a reaction depends on a balance between bond energies (so that in general a product with more and/or stronger bonds is more stable) and disorder (a more disordered product, for example one with more molecules, is preferred).

The same is also true for the activation energy:

$$\Delta G^{\ddagger} = \Delta H^{\ddagger} - T\Delta S^{\ddagger}$$

This equation tells us something similar to the previous one. It says that if you want to decrease the activation energy, you need to make the activation enthalpy more negative or the activation entropy more positive. Let us take a

***5.5 Binding and dissociation constants and free energy**

When two molecules bind together, they do so with a certain affinity or binding constant. Consider the equilibrium

$$A + B \rightleftharpoons AB$$

As in any equation, the equilibrium constant for this reaction is given by the concentration of the things on the right divided by the concentrations of things on the left:

$$K = [AB]/[A][B]$$

Because this is an association reaction, the binding constant is an association constant, often written K_a. The units of K_a are reciprocal concentration such as M^{-1}. Alternatively, we could look at the *dissociation*

$$AB \rightleftharpoons A + B$$

whose dissociation constant K_d is given by $K_d = [A][B]/[AB]$. K_d has units of concentration. Traditionally, chemists describe binding equilibria by using association constants, whereas biochemists use dissociation constants. The dissociation constant has the advantage that K_d is directly a concentration: it is roughly the concentration at which AB is half dissociated and half free, so it is the midpoint concentration. Therefore a dissociation constant of 1 mM is weak binding, whereas 1 nM is strong binding. Obviously the two constants are related by $K_d = 1/K_a$.

The equilibrium can also be represented by a *free energy*: the energy input needed to get from one side to the other, or conversely the amount of energy that can be extracted. Biochemists always use the Gibbs free energy ΔG, which is the free energy at constant pressure, the only conditions that normally matter. ΔG is given by

$$\Delta G = -RT\ln K$$

where R is a constant (the gas constant), with a value of $8.31\ \text{J}\ \text{K}^{-1}\ \text{mol}^{-1}$, and T is the temperature in kelvins. [In this book I use only joules (J), not calories (cal). It was agreed in 1948 that the standard unit of energy should be the joule, not the calorie; unfortunately, biochemists have still not entirely caught up with this concept. I have to confess, however, to still using the non-SI unit the ångström (Å), because it seems far more intuitive than nanometers (nm), even though they are only a factor of 10 different: 1 nm is 10 Å.] This equation can be rewritten as

$$K = \exp(-\Delta G/RT)$$

a form that is sometimes more useful.

Therefore, for the dissociation direction,

$$\Delta G = -RT\ln K_d$$

A dissociation equilibrium that lies far to the right has a large K_d and therefore ΔG is large and negative; a dissociation equilibrium that lies to the left has a positive ΔG, and one that is exactly poised in the middle has $K = 1$ and a zero ΔG. ΔG therefore tells us whether the reaction will proceed or not: if it is negative it will, and if it is positive it will not. For many purposes it is easier to work in energies than equilibrium constants. Equilibrium constants can often be very large (or small) numbers, whereas free energies, being logarithms of the equilibrium constants, are much more manageable numbers. Further, for linked reactions, free energies add, whereas equilibrium constants multiply. For convenience, the relationship between K and ΔG at the standard temperature of 298 K (25 °C) is given in the following table:

K	ΔG (kJ mol^{-1})
1	0
10	-5.9
100	-11.9
1000	-17.8
10,000	-23.7

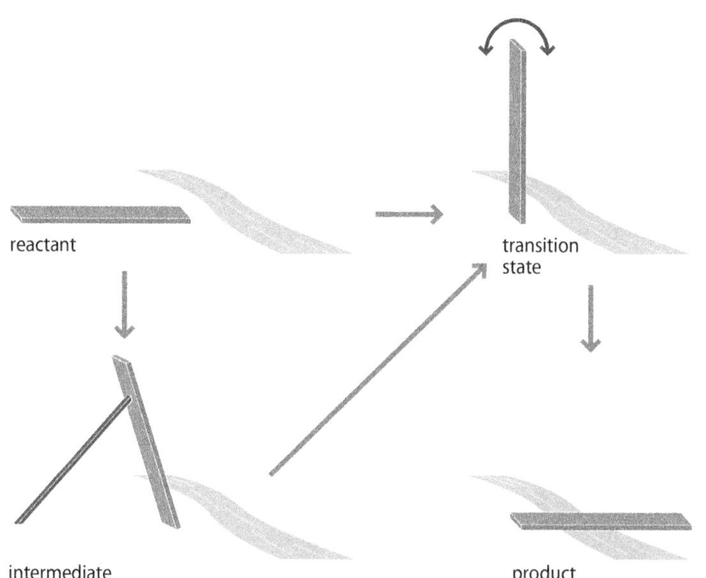

FIGURE 5.4
An illustration of the difference between an intermediate and a transition state. The illustration depicts flipping a plank end-over-end to get it to cross a stream. The *transition state* is a very unstable state, which the plank spends virtually no time in, and it has the highest energy in the pathway. From here, the plank must either continue over the energy barrier to the other side or fall back to roughly where it came from. By contrast, the *intermediate* is a (meta)stable state, which looks rather similar to the transition state but in which the system can rest, as long as it is not buffeted by external energy.

concrete example: the TS for the hydrolysis of an ester by hydroxide (**Figure 5.5**). In this reaction, the TS must look like something between the starting molecules and the products. It therefore clearly must have the C=O double bond partly broken and the new O–C single bond partly formed. In this case, how can the enzyme help to decrease the activation energy?

First, it can make the activation *enthalpy* more favorable (negative). It can do this most obviously by weakening the C=O bond in the starting molecule. Enzymes frequently do this, by electrophilic catalysis and general acid catalysis, as we shall see shortly. This concept is often expressed as "enzymes are complementary to the TS, not to the substrate," an idea dating back to Linus Pauling (*1.8) in 1948 [4]. For this specific example, we can see that the TS differs from the substrate in that there is more negative charge on the oxygen, and the shape of the molecule is different: in the substrate the C=O carbon is flat (trigonal), whereas in the TS it must be becoming more tetrahedral. Therefore the enzyme can stabilize the TS by putting a positively charged group close to the carbonyl oxygen (which is unfavorable for the substrate but favorable for the TS) and by having a structure that matches better a tetrahedral carbon than a trigonal one. In other words, the enzyme can *stress* the substrate: effectively this is the implication of its being complementary to the TS and not to the substrate.

Second (and often more importantly) it can make the activation *entropy* more positive. The size of $\Delta S^{\ddagger}$ depends on the disorder in the substrates compared with the disorder in the TS. In the TS, everything needs to be very well positioned: the formation of a new bond requires very exact positioning of the atoms, because bonds are very highly directional. By contrast, there is no such order required in the substrates: for the reaction in Figure 5.5 to occur in water, there is nothing to specify the relative positions of the two reactants before the reaction starts, and therefore they lose a large amount of entropy in forming the TS. Thus *in general it is the unfavorable activation entropy that makes the overall activation energy large.* This implies that an enzyme can lower the activation entropy either by making the TS less ordered or the substrates more

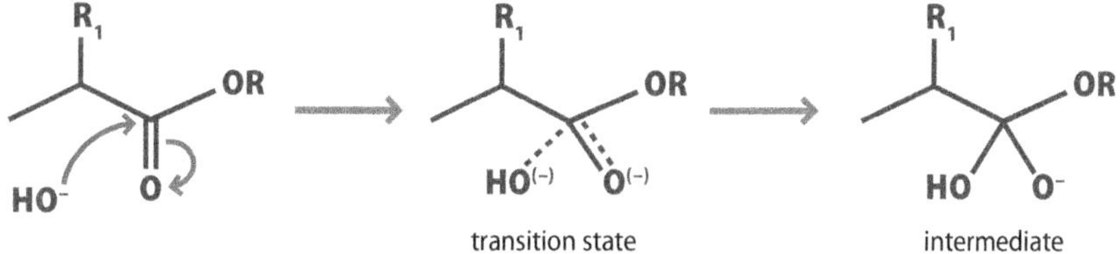

FIGURE 5.5
The hydrolysis of an ester by a hydroxyl ion. This figure only shows the first part of the reaction, up to the intermediate. The rest of the reaction can be found in Figure 5.10.

ordered. There is little scope for making the TS less ordered, but there is a lot of scope for making the substrates more ordered. Thus, a major proportion of the rate enhancement produced by enzymes comes from the fact that in the enzyme–substrate complex, the reactants are *already* highly ordered. That is, they are bound into the desired positions: they have already lost most of the entropy that they are going to lose by the time they reach the TS.

This idea is at its most obvious in a reaction that involves the linkage together of two chemicals, or the transfer of a chemical group from one molecule to another. The entropy requirement in the TS is large, because the two molecules have to be held exactly in the right position for the bond to be formed. But if the two molecules are already bound to the enzyme, in the correct positions, then this entropy has already been lost before the reaction starts. This is what is meant by the enzyme's using part of the binding energy to increase the energy of the substrates.

We can express this in another way. There is a limit to the amount of energy that can be gained by a substrate binding to an enzyme. Evolution could so arrange it that nearly all of this energy is obtained, and the substrate binds tightly. Actually this would in any case be a very bad thing, as we shall see. But it is preferable to bind the substrates so that the bound state is destabilized by comparison with the free substrate, for example by binding the substrate adjacent to suitably charged groups so that the C=O bond is weakened. This makes the binding weaker: in effect it uses some of the potential binding energy to destabilize or stress the bound substrate (**Figure 5.6**). In addition, the binding energy is used to bind the substrates so that they are fixed in an orientation appropriate for the reaction that is about to happen. Thus, enzymes decrease the activation energy both by decreasing the enthalpy of the TS and also by decreasing the entropy of the substrates. Of these, the largest factor is usually the decrease in entropy by binding substrates in a correctly fixed orientation: this factor alone can increase the rate of a reaction by a factor of about 10^8, though it is usually much less than this [5, 6]. This effect has been given many different names, including *propinquity* or *proximity*, *orientational catalysis*, binding in a *near-attack conformation*, and more poetically as the *Circe effect*, after the goddess Circe in Homer's Odyssey who lured men to her and then transformed them into pigs [7]. A list of 25 almost equivalent terms is given by M.I. Page [5].

In summary, an enzyme achieves a high proportion of its catalytic effect by having a binding site that matches the TS of the reaction, particularly in terms of the charge distribution. (In other words, what matters is that the enzyme should be complementary in electrostatics rather than in shape to the TS.) This has two main effects: it markedly decreases the entropic requirements for the reaction, and it stabilizes the TS relative to the free substrate. This was described more than 30 years ago, and despite enormous advances in molecular biology and in computation it remains the key to enzyme function [8, 9].

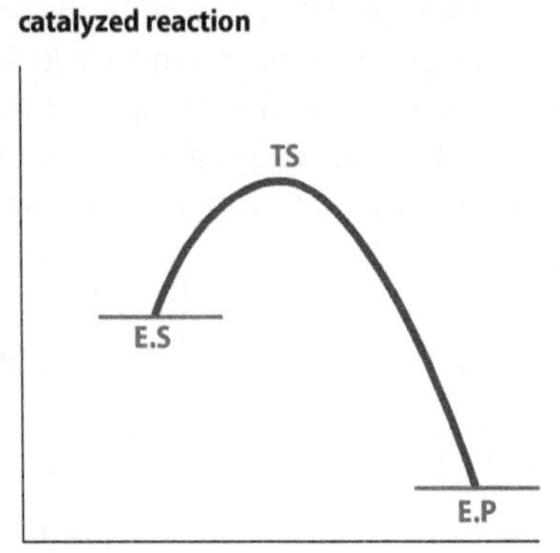

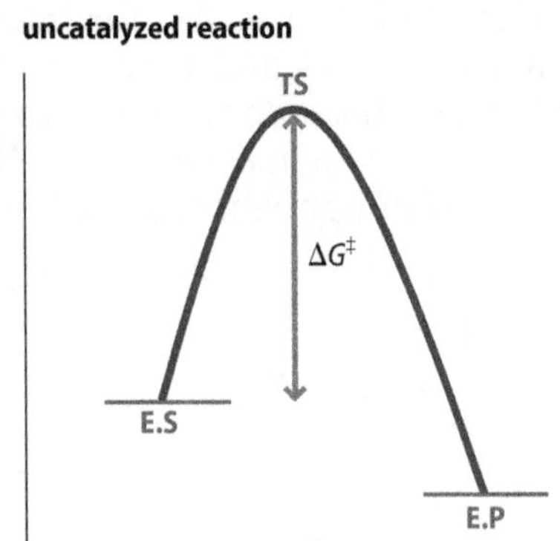

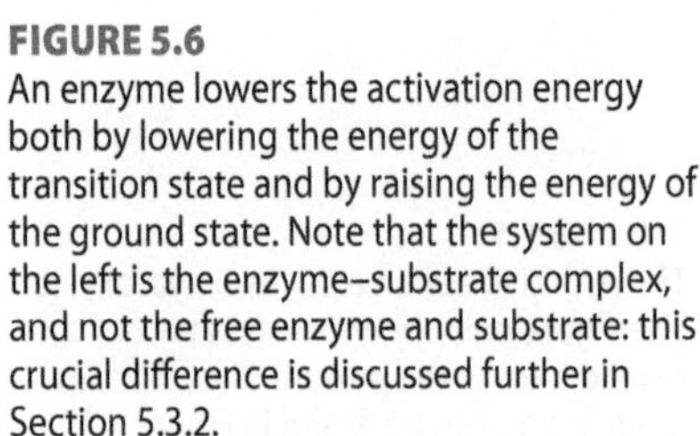

FIGURE 5.6
An enzyme lowers the activation energy both by lowering the energy of the transition state and by raising the energy of the ground state. Note that the system on the left is the enzyme–substrate complex, and not the free enzyme and substrate: this crucial difference is discussed further in Section 5.3.2.

5.1.3 Catalytic antibodies demonstrate the strong entropic contribution

The previous section showed that an enzyme can achieve a very large enhancement in rate merely by binding substrates in the correct orientation, without any chemical catalysis. This observation formed the basis for the idea of catalytic antibodies, also known as abzymes [10]. Peter Schultz suggested that if one could prepare an **antibody (*5.6)** whose antigen was something resembling the TS, then it should be a good catalyst, because (1) it will bind the substrates(s) in the correct orientation to react, and (2) it will be complementary to the TS. It will therefore decrease both the activation entropy and the activation enthalpy. By further rounds of mutation one might hope to be able to introduce suitable catalytic groups next to the substrate and thus increase the catalytic effect. Some support for this idea comes for example from the observation that if you remove any of the catalytic triad (Ser, His, or Asp) from the active site of the serine protease subtilisin you completely remove its ability to perform chemical catalysis of peptide bond cleavage, and yet it still catalyses hydrolysis 1000-fold.

This does indeed work. A large number of such catalytic antibodies have been generated, and they do turn out to be able to catalyze reactions by factors between 10^3 and 10^6, these factors being mainly due to a decrease in the entropy of the reactants. It has proved remarkably difficult to improve on this, though. As discussed below, no doubt some of this is due to the importance of creating an active site in which water is removed, which is not easy to achieve in a catalytic antibody.

5.2 CHEMICAL CATALYSIS

5.2.1 Chemical reactions involve movement of electrons

The next few sections require some discussion of chemistry. Remarkably, this is almost the only place in this book where "real" chemistry is mentioned: an important theme of this book is that most protein function can be understood by considering proteins as featureless blobs, in which the actual chemistry is not important for understanding the function.

A chemical reaction normally consists of the making or breaking of a bond. A bond is formed by two electrons: to make a C–H bond, for example, you would normally use the electron that surrounds the proton nucleus and one of the free (unpaired) electrons from the carbon, and put them together into a bonding orbital (**Figure 5.7**). In the bonding orbital, the electrons are mainly located between the two nuclei. Because electrons are negatively charged and nuclei are positively charged, this is a favorable interaction and the bond is stable.

Although the previous paragraph is perfectly correct, it is not in fact the way that bonds are normally made, because hydrogen is not normally found as a single atom containing one electron, and carbon does not normally have unpaired electrons either. Free or unpaired electrons are very reactive and unstable, and are called free radicals. So hydrogen is normally found either

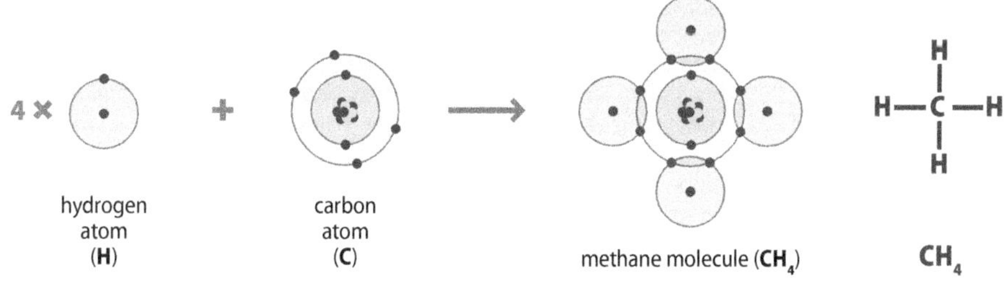

FIGURE 5.7
A molecule of methane (CH_4) is made up from four hydrogen atoms, each carrying one electron, and a carbon carrying four electrons in the outer shell. The electrons pair up to form bonds, such that each hydrogen is surrounded by a closed shell of two electrons, and the carbon is surrounded by a closed shell of eight electrons.

*5.6 Antibody

Antibodies are unique to vertebrates and protect them against infection by recognizing and binding to foreign material such as bacteria and viruses. They are also called immunoglobulins (Ig).

Structure of antibodies

All antibodies are constructed from immunoglobulin modules (**Figure 5.6.1**), which are about 110 amino acid residues long. Antibodies all have the same basic structure, consisting of two identical light chains and two identical heavy chains held together by disulfide bonds (**Figure 5.6.2**). The light chain is made of two domains, whereas the heavy chain is made of four or five. The heavy chains are glycosylated. Antibodies can be split by the protease papain at the hinge region to produce Fab fragments, each of which binds one antigen, and Fc regions. The site of recognition of the *antigen* is at the tip of the arms and is derived partly from the light chain and partly from the heavy chain—in fact, from the N-terminal domain in each chain, also called the variable domain. Each of these domains contains three hypervariable loops, also known as complementarity-determining regions, which are the main sources of variation: changes to these loops allow the production of antibodies to roughly 10^8 different antigens (see Figure 5.6.1). These loops are each only 5–10 amino acids long, implying that the binding site of an antibody is relatively small.

There is some change in structure of the antibody when it binds to its antigen, but the change is rather small. Antibodies that bind small ligands usually have quite deep holes to fit the ligand, whereas antibodies that bind proteins have a much larger and flatter surface. It would be fair to say that prediction of the structure of an antibody on the basis of its sequence is reasonably accurate (there are several websites for this purpose, including WAM, PIGS, and Rosetta Antibody), but prediction of the ligand or of the binding affinity is still much less accurate.

Because of their two Fab domains, antibodies are bivalent: they bind antigen at two different and independent binding sites. Furthermore, the distance and angle between the two binding sites are variable because of the hinge region. This has two main advantages. First, as discussed in Chapter 2, binding at two sites can markedly increase the affinity. Second, binding to antigens with more than one binding site (for example, proteins on a cell surface) means that antibody binding can lead to aggregation of the antigen. If the antigen is attached to a cell surface, it therefore leads to cell clumping (**Figure 5.6.3**).

There are five different classes of human antibody: IgA, IgD, IgE, IgG, and IgM; IgG forms the main class of circulating antibodies. The various classes differ in the structure and function of their Fc part, which determines what is done with the antibody once it has bound to an antigen. The heavy chain (including all of the Fc region) is known by the corresponding Greek letter, namely α, δ, ϵ, γ, and μ, respectively. In IgG, the Fc part activates the complement system, which destroys pathogens by punching holes in their cell membrane, and it also binds to specific receptors on white blood cells such as macrophages, which destroy foreign bodies by engulfing them. The IgG antibody thus acts rather like the adaptors discussed in Chapter 8: it links a specific antigen to a common cellular response. The heavy chain has four Ig modules (namely one variable and three constant) for IgG, IgA, and IgD (as drawn in Figure 5.6.2), but five for IgM and IgE.

IgM and IgD are early-stage antibodies, expressed on the surface of naive B cells (see below). IgM is also produced as a secreted pentameric form, in which five Fc tails are held together in a ring by a J or joining protein. (Presumably the pentameric structure gives it greater affinity to multivalent antigens, as discussed above.) Similarly, IgA is produced as a dimer, also with the Fc tails held together by a J protein, and is found in secretions such as tears and saliva. Finally, IgE is involved in the degranulation of mast cells and basophils and forms the trigger for allergic reactions. Each IgE Fc region (that is, ϵ chain) binds to an Fc receptor on the cell surface. Binding of an antigen with more than one binding site leads to the cross-linking together of two receptors (in a mechanism reminiscent of tyrosine kinase-linked receptors) and stimulates the cells to release histamine.

FIGURE 5.6.1
The immunoglobulin (Ig) fold, which is two parallel β sheets (a β sandwich) or a squashed β barrel. The three hypervariable loops, which confer the binding specificity, are on the right (cyan, green, and brown), of which the third loop (brown) is the one that contains most variation. All the hypervariable loops, except the third loop in the light chain, have only a small set of accessible conformations, known as canonical structures. Variability in these loops comes from the amino acids that provide hydrophobic, charged, etc., binding sites for the antigen. Each Ig module in the constant region of antibody heavy chains is coded for by a different exon.

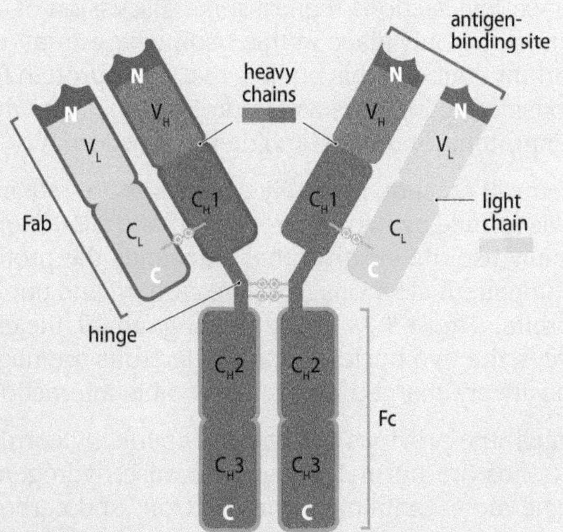

FIGURE 5.6.2
The structure of an immunoglobulin G (IgG) antibody consists of two light chains [green, each consisting of two Ig modules: one variable (V_L) and one constant (C_L)] and two heavy chains [brown, each consisting of four Ig modules: one variable (V_H) and three constant (C_H1, C_H2, and C_H3)]. The C_H2 module is glycosylated. The chains are held together by disulfides (pale brown lines). The structure can be divided up into the Fc region plus two Fab regions. The Fab regions contain the antigen-binding sites at their ends, and the Fc region binds effectors, cell membrane, or other antibodies. There is a flexible hinge between Fc and Fab, allowing the antigen recognition sites to adopt variable geometries.

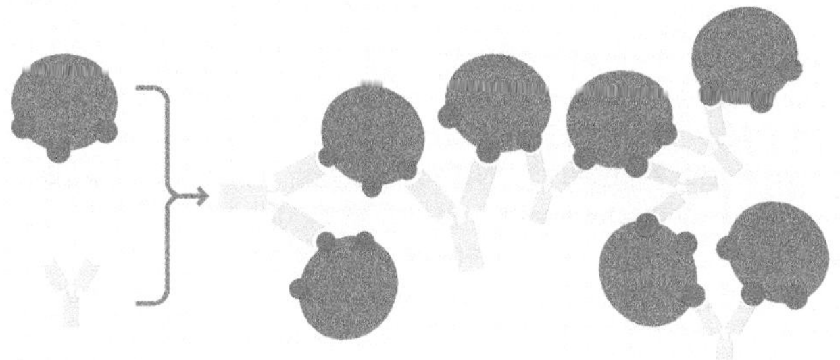

FIGURE 5.6.3
Binding of a divalent antibody to a cell expressing multiple copies of the antigen leads to cross-linking, and therefore to clumping and precipitation of the cells.

The generation of antibody diversity

Antibodies are produced by B cells. Each B cell makes only one genetically unique antibody. Most of these are secreted, but some are attached to the cell membrane, where they act not as antibodies but as antigen receptors: they signal to the B cell when an antigen is bound. This is vital, because most B cells exist in a quiescent state until stimulated by the binding of a suitable antigen, at which point they proliferate (*clonal expansion*) and start to secrete large numbers of antibodies.

Humans probably produce between 10^{10} and 10^{12} different antibodies. We clearly do not have that many antibody genes. The production of the large range of antibodies therefore uses a method called V(D)J recombination. The gene locus for the heavy chain in germ cells (that is, cells not actively making antibodies) contains a set of 40 genes coding for the V segment, 25 coding for D, and 6 coding for J. After this comes the gene clusters coding for each of the constant regions. The intact heavy chain is produced by the selection of a D segment, which is joined to a J segment. This DJ combined DNA is then placed next to a V segment (**Figure 5.6.4**). This rearranged DNA is unique to each B cell, and each B cell therefore produces just one of the $40 \times 25 \times 6 = 6000$ possible proteins. The figure shows a combination of V3, D2, and J4. The RNA produced by transcription from this DNA is then spliced to remove all the DNA between the chosen J segment and the C segment appropriate for the immunoglobulin class: here a Cμ cluster is selected because this B cell will produce an IgM antibody.

Light chains are produced in the same way, except that there is rather less diversity because there is no D segment and there are only five J segments. There are, however, two different types of light chain, namely κ and λ. This means that in total humans can produce 200 κ (40×5) and 120 λ chains, making 320 light chains, thus yielding $320 \times 6000 = 2 \times 10^6$ different antibodies. This falls far short of the 10^{12} mentioned above.

The additional very large degree of variation comes partly from variation in gene sequence at the joins between V, D, and J segments. When the DNA is cut and rejoined as described above, different numbers of nucleotides can be lost or added. This is particularly significant in the DJ join, which encodes the third hypervariable region. Loss or gain of nucleotides is likely to lead to frameshift mutations, which are incapable of producing the correct protein sequence downstream; such mutations are discarded during maturation of the B cell, leading to the death of the B cell.

A further large degree of variation comes from the *secondary repertoire*. The process described above produces 'naive' B cells. Each of these has copies of its specific antibody displayed on the cell surface acting as an antigen receptor, and it exists as only a single cell or a very small number of cells. After stimulation of the B cell by binding of an antigen to the antigen receptor, the cell proliferates and differentiates. As the cells proliferate, mutations are introduced into the V segment of the antigen-binding site at a very high rate, approximately 10^6 times more than the normal mutation rate (*somatic hypermutation*). This high rate generates roughly one mutation per V region per cell division. The rare occurrence of a mutation that leads to a significant increase in binding affinity to the antigen leads to such cells binding most of the available antigen, and therefore these cells continue to proliferate while the others die by apoptosis. This forms a very effective selection mechanism for successful mutations.

Finally, differentiating B cells are able to switch the class of antibody they are making; that is, they can change the constant chain region with which the VDJ segment is associated (see Figure 5.6.4). Naive B cells produce mainly IgM and IgD, but at this point they can irreversibly change to producing different classes.

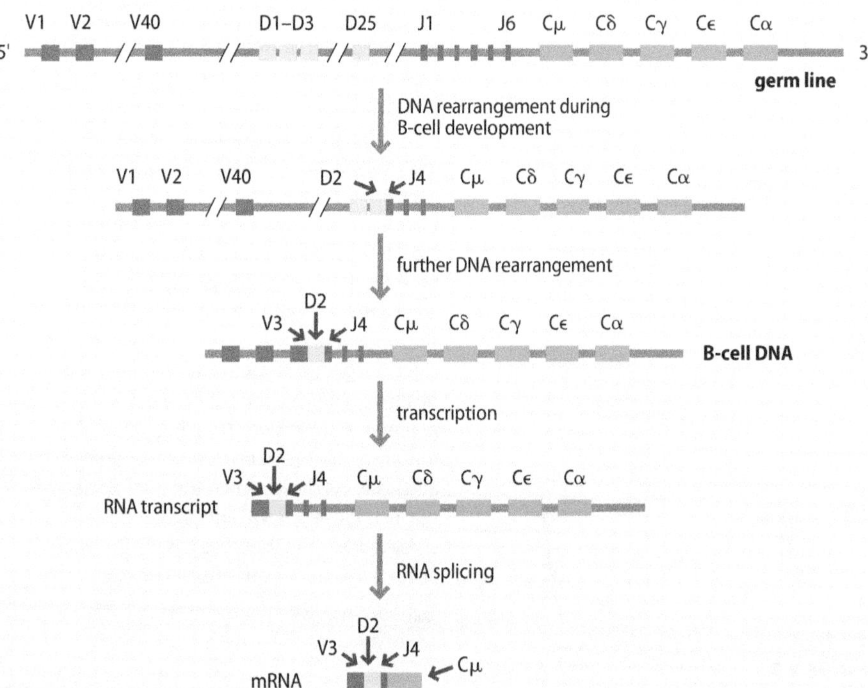

FIGURE 5.6.4
The human immunoglobulin heavy-chain locus contains 40 V segments, 25 D segments, and 6 J segments (note that this is a completely different J from the proteins used to join IgM and IgA monomers together), followed by the genes for the different constant regions. These are in fact encoded by a series of exons and are therefore much longer than implied by this diagram. The D and part of the J segments encode the third hypervariable loop. The intact heavy chain is produced by selecting one of each segment, first by joining D and J and then by joining this to V. The RNA produced from this is then further truncated by splicing out the sequence between the desired J and C segments, to produce the mature mRNA that codes for the antibody.

H^+	$H^\cdot$	H^-
one proton	one proton one electron	one proton two electrons

FIGURE 5.8
Three very different states of hydrogen. The H^+ ion is just a proton, and is very stable (provided it can be solvated or attached to a basic atom). The H^- (hydride) ion is moderately stable, but reactive: it is a strong reducing agent. The $H^\cdot$ radical has an unpaired electron and is very reactive. The addition of each electron increases the negative charge by one.

*5.7 Nucleophile

The word nucleophile means 'loving the nucleus'. The atomic nucleus is of course positively charged, so nucleophiles are generally negatively charged or electron-rich groups, which are able to attack positively charged electrophiles (*5.8) and form bonds with them. The best nucleophiles are those that have a nicely exposed and available electron density. Examples of good nucleophiles are OH^- and RS^- (where R represents any organic group), which both have an easily available electron. Cl^- is also a fairly good nucleophile, although not as good as OH^- because the electron density in Cl^- is more spread out and less localized. A phosphate group is even less good because the negative charge is spread around several oxygen atoms, and therefore the net charge on each is less than 1. There are also many neutral molecules that are still nucleophilic because they have electrons in **lone pair (*5.9)** orbitals. For example, water and ammonia are nucleophilic for this reason, although they are nowhere near as good as OH^- because they 'hold on' to their electrons more tightly. In reactions described here I use Nu to describe a nucleophile.

bonded to another atom, in which case the bond contains a pair of electrons shared by both nuclei, or else it has no electrons at all (the familiar proton), or it has two: the hydride ion, which is rather energetic but still relatively common (**Figure 5.8**). A hydrogen with just one electron on it is a hydrogen radical $H^\bullet$ and is so reactive that it is almost never observed. Bonds are in fact normally formed by taking *both* the bonding electrons from the same partner, which avoids having to have unpaired electrons at any point. We normally view the reaction as the molecule with the electrons (the **nucleophile, *5.7**) *attacking* the other molecule (the **electrophile, *5.8**). Conventionally, this movement of

*5.8 Electrophile

The word electrophile means 'loving electrons', and electrophiles are positively charged or electron-deficient groups. A typical electrophile is therefore an atom that is attached to a strongly electronegative atom (that is, an atom with an excess of electrons, typically in groups 5, 6, and 7 of the periodic table, such as N, O, or Cl), which pulls electron density away. It therefore forms a partly exposed nucleus, which can then be attacked by a **nucleophile**. The best common electrophile is a carbonyl (C=O) carbon. Not only is the carbon attached to an electronegative oxygen, and therefore partly positively charged because the oxygen is pulling electrons away from it, but it is also forming a double bond. A single bond has the electrons mainly between the two connected nuclei (**Figure 5.8.1a**) and is therefore stable and strong, because the electrons are attracted to the positive nuclei on each side. However, in a double bond there is very significant electron density to the side of the bond (Figure 5.8.1b), in π orbitals. This makes the bond weaker because the electrons are not shielding the two nuclei from each other so well. Therefore the π-orbital bond in a double bond is significantly weaker than a normal single bond.

For similar reasons the phosphorus atom in a phosphate group is an electrophile: it is surrounded by electronegative oxygens and forms a double bond. It is not as good an electrophile as a carbonyl carbon, because phosphorus is more electronegative than carbon; that is, it has more electrons around it. And carbons that form single bonds to electronegative atoms, such as $R_3C–OH$ or $R_3C–Br$, are also electrophilic, but they are less good electrophiles because the electronegative atom is attached by only a single bond.

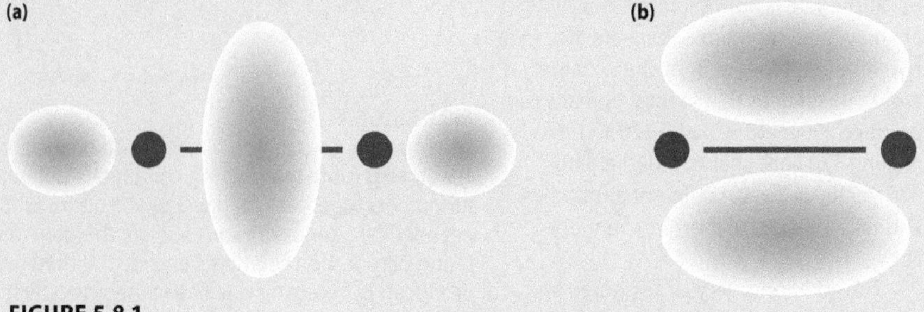

FIGURE 5.8.1
The approximate electron density in (a) a single bond (2p σ_g bonding orbital) and (b) a double bond (2p π_u bonding orbital).

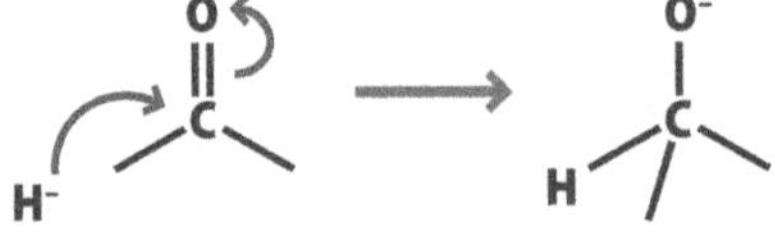

FIGURE 5.9
The reactant hydride ion consists of a proton plus two electrons. In the reduction reaction shown, these two electrons are both used to make a bond to a carbon atom. In the process, the two electrons forming the π (double) bond from C to O migrate from their position shared by both atoms onto the oxygen, therefore leaving the oxygen with a single negative charge. The curly arrows (red) each show the movement of a *pair* of electrons.

***5.9 Lone pair**

This is the term used to describe a pair of electrons that do not form a bond but are in a non-spherical orbital and are therefore available to act as **nucleophiles**. For example, the oxygen atom has eight electrons (atomic number 8). Two of these are in a spherical s orbital close to the nucleus and never do anything. In the formation of water, two of the remaining six electrons from oxygen combine with an electron each from hydrogen atoms to make bonds, each of which then contains one electron from the oxygen and one from the hydrogen. The oxygen is then surrounded by the inner shell of two electrons described above and an outer shell of eight electrons. Each of these is a full shell; this is thus a stable arrangement. However, these eight outer electrons are not random. They are in four pairs, two of which are the bonds to the hydrogens, plus two more called *lone pairs*. The four pairs of electrons repel each other and thus point in a roughly tetrahedral direction (**Figure 5.9.1**). The lone pairs are often written as two dots. Similarly, nitrogen in ammonia forms a closed eight-electron shell using three bonds and one lone pair.

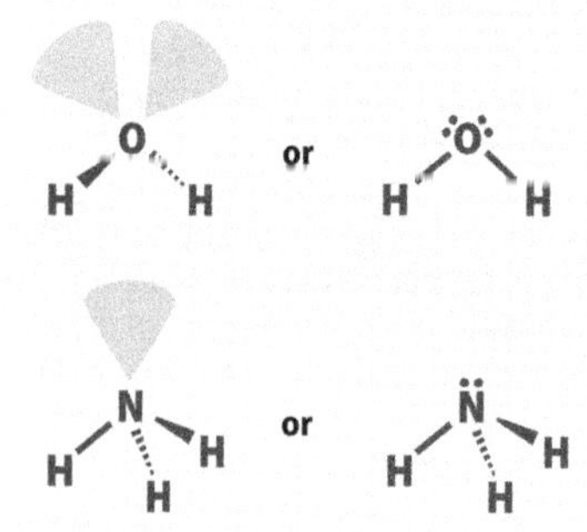

FIGURE 5.9.1
Lone pairs in water and ammonia. The electrons all try to get as far apart from each other as possible, so both molecules are roughly tetrahedral.

electrons from one molecule to another to form the bond between them is drawn with **curly arrows (*5.10)**, as shown in **Figure 5.9**.

5.2.2 A good leaving group is important

Many reactions involve the breaking of a bond. As noted above, this requires both the electrons from the bond to move onto an atom. The part of the molecule containing this atom is called the *leaving group* (**Figure 5.10**). For the leaving group to be able to leave successfully and complete the reaction, it has to be stable. If it is not stable, it will simply recombine again to regenerate the original molecule (see Figure 5.10a). In general, the better the leaving group, the faster the reaction. The easiest way to measure whether something is a good leaving group is to measure its pK_a (Section 1.1.5). Because the pK_a is the pH at which the group is half protonated and half unprotonated, it describes

***5.10 Curly arrows**

Curly arrows are the normal way for drawing the mechanism of a chemical reaction. Because a bond consists of two electrons, the curly arrow starts on the source of these electrons, which is a negative charge, a lone pair (*5.9), or another bond, and it finishes in the middle of where the new bond will go (**Figure 5.10.1**). If the reaction is the breaking of a bond, the curly arrow points to where the electrons will go, for example to make a charged atom.

When drawing a reaction using curly arrows, one has to make sure that all the electrons start and end in sensible places and that the total charge on the molecule is conserved. Thus, for example, drawing the reaction shown in **Figure 5.10.2** is incorrect because it implies that the carbonyl carbon ends up with an excess of electrons: the incoming electrons generally need to displace electrons elsewhere. Reaction mechanisms therefore often contain a series of curly arrows that shuffle electrons around (**Figure 5.10.3**). A drawing like this implies that all the electrons move at essentially the same time, and therefore all the bonds involved break or form simultaneously.

This explanation is slightly confusing, because of course a negative charge of −1 consists of only one electron. Why therefore do I describe it as though it had two? Consider an OH^- ion. What do we get if we remove the electron that carries the negative charge? We get a neutral OH, in which the oxygen atom is surrounded by a total of seven electrons: two in the bond to the hydrogen, four more in two lone pairs, and one unpaired electron. In other words, we have the highly reactive hydroxyl radical $OH^•$. When we add back the electron to make the hydroxyl ion OH^-, it pairs up with the existing unpaired electron to make a pair of electrons, which are the electrons that move with the curly arrow. The total charge on the hydroxyl is, however, only −1.

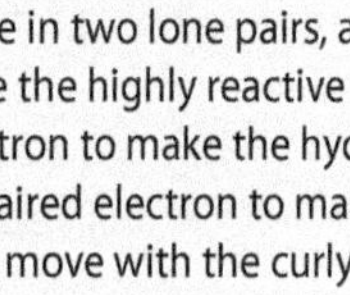

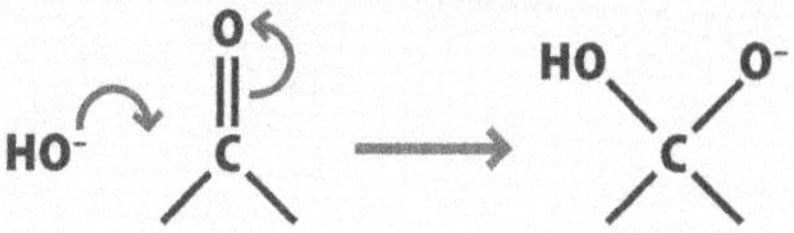

FIGURE 5.10.1
The formation of a new bond between O and C, and the breaking of one bond in the C=O double bond.

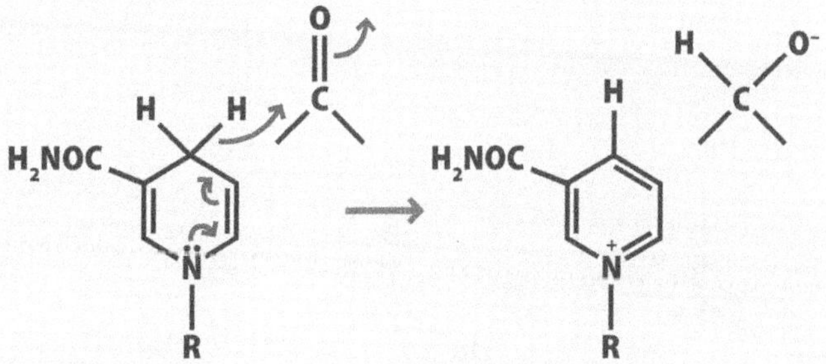

FIGURE 5.10.2
An incorrect curly arrow diagram.

FIGURE 5.10.3
The reduction of a ketone to an alcohol by NADH.

FIGURE 5.10
The second part of the reaction for hydrolysis of an ester by hydroxide (see Figure 5.5 for the first part). The intermediate can break down (via a transition state, not shown) by one of two routes. In (a) the bond to OH is broken, and the reactants are re-formed. In (b) the bond to OR is broken, and the ester is hydrolysed. The OH^- and OR^- are called the *leaving groups* in these two reactions. Which of these two routes is followed depends on the relative stabilities of the products, and thus on their pK_a.

how tightly the group binds to a proton, or in other words how "unhappy" it is in an unprotonated state. **Table 5.1** lists the pK_a values for some typical leaving groups. Thus, for example, the pK_a for HPO_4^{2-} is just over 7, implying that it is a good leaving group because it is happy to exist as HPO_4^{2-} at pH 7. By contrast, the pK_a for PO_4^{3-} is 12.4, which means it is a worse leaving group. This emphasizes the importance of protonating a group to make it a better leaving group. In particular, note that the pK_a for $-NH^-$ is 25, meaning that this is a very bad leaving group. The pK_a is so high that in water it will never leave. However, if protonated, it becomes at least a moderately good leaving group. And indeed it is a general rule that leaving groups leave better, and the reaction goes faster, if the leaving group is protonated. This implies that an enzyme that catalyzes the hydrolysis of peptide bonds (such as a protease) needs to be able to protonate the amine leaving group, otherwise the reaction will be so slow that it will be unobservable. We consider this in the next section.

5.2.3 General acid and general base catalysis are ubiquitous

In the previous section, we saw that for many leaving groups it is essential that they should be protonated before they can leave. Thus, an important catalytic mechanism used by almost all enzymes at some point is protonation. Protonation not only makes leaving groups better; it also makes electrophiles better. Consider again the attack of a nucleophile on a carbonyl group (**Figure 5.11**). The carbonyl carbon is here acting as an electrophile (*5.8) and is being

TABLE 5.1 pK_a values for some typical leaving groups

Protonated (pH below pK_a)	pK_a	Unprotonated (pH above pK_a)
$C{=}OH^+$	−7	C=O
−COOH (Glu, Asp)	4	$-COO^-$
ImH^+ (His)	6–7	Im
$H_2PO_4^-$	7.2	HPO_4^{2-}
$-NH_3^+$ (Lys)	10.5	$-NH_2$
ArOH (Tyr)	10.5	$-ArO^-$
−SH (Cys)	12	$-S^-$
HPO_4^{2-}	12.4	PO_4^{3-}
H_2O	14	OH^-
−OH (Ser, Thr)	18	$-O^-$
$-NH_2$	25	$-NH^-$

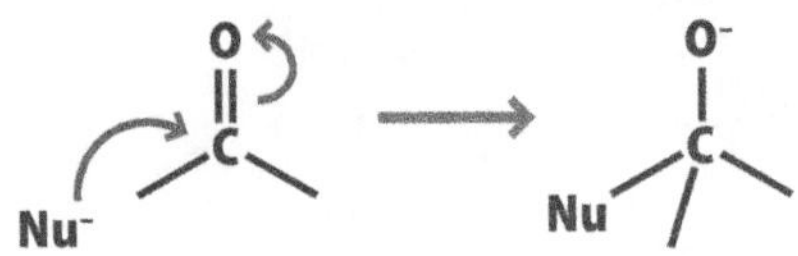

FIGURE 5.11
The attack of a nucleophile (Nu) on a carbonyl group.

attacked by an electron-rich nucleophile (*5.7). Clearly the approach of the nucleophile, and the attack from its electrons, will be easier the more positively charged the carbon. This is most easily achieved by protonating the oxygen (**Figure 5.12**), which shifts the electrons in the C–O bond more toward the oxygen and makes the carbon more positive. Thus, any reaction will go faster if the atom or group receiving the electrons can be protonated first. This mechanism is called *general acid catalysis* and is extremely common. The word "general" implies that any acid will do, as compared to *specific* acid catalysis, where the rate of the reaction depends on the hydrogen ion concentration (the pH) but not on the concentrations of particular acids.

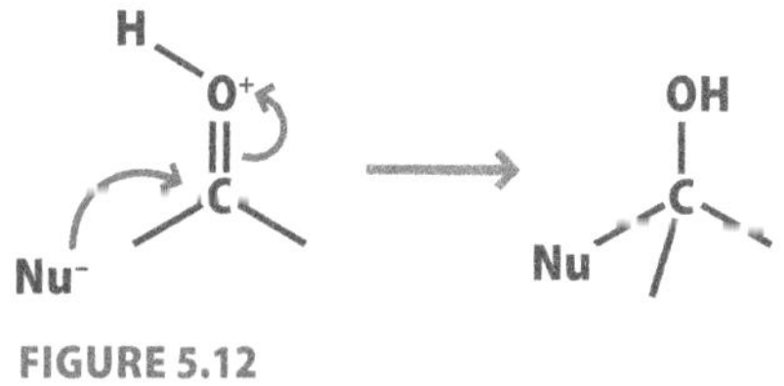

FIGURE 5.12
The attack of a nucleophile on a protonated carbonyl group is much easier, because the product is much more stable in water.

What kind of acids are available to proteins? In the absence of cofactors (see below), the only acidic groups present are protein side chains (see Table 5.1). The carboxyl group of Glu or Asp has a pK_a of about 4, which means that at pH 7 it is mainly deprotonated, with only one residue in 10^3 being protonated at any time. They are therefore not ideal as general acids. The pK_a of His is close to neutral, making it a much better general acid, and indeed for this reason histidine residues are very commonly found to be involved in catalytic mechanisms. A transfer of a proton from $HisH^+$ to $-NH^-$, to make His and $-NH_2$, is energetically favorable because the pK_a of His is lower than the pK_a of $-NH^-$.

This would imply that a protein more or less has to use His as its general acid catalyst. However, it is possible to change the pK_a of a group a long way by suitably arranging the local environment (see Table 1.1). The pK_a of a carboxylate is the pH at which it is half in the form $-COO^-$ and half in the form $-CO_2H$. If there is a negatively charged group close by, this will destabilize the $-COO^-$ form and therefore increase the pK_a; conversely, a neighboring positive charge will lower the pK_a. The energy of a charged $-COO^-$ also depends on the dielectric constant (*4.2): a high dielectric constant, such as in water, stabilizes a charge, whereas a low dielectric constant, such as in the inside of a protein, destabilizes it. Therefore putting a negatively charged carboxylate inside a protein destabilizes it and raises its pK_a. Similarly, putting a protonated group such as $-NH_3^+$ inside a protein destabilizes it and lowers its pK_a. In this way, pK_a values can easily be altered by 1–2 pH units or more: in acetoacetate decarboxylase the pK_a of lysine is lowered by 4.5 pH units as a result of being held in a hydrophobic environment with an adjacent lysine to destabilize a positive charge [11]. It is in fact common to find glutamates in the active sites of proteins and acting as general acids, with pK_a values close to 7.

General base catalysis is essentially the opposite of general acid catalysis. Consider again the nucleophilic attack on a carbonyl shown in Figure 5.11. A good nucleophile (*5.7) is negatively charged, because this makes its electrons more available. Just as an electrophile is better if protonated, so a nucleophile is better if deprotonated. In particular, OH^- is a much better nucleophile than H_2O. Therefore a good way for an enzyme to catalyze proteolysis is to remove a proton from the attacking water molecule, to make a hydroxyl ion (**Figure 5.13**). This is general base catalysis: again the "general" means that it does not matter what base is used to remove the proton. The obvious base to use is again His. Glu and Asp have a lower pK_a, which means that they will not readily accept a proton unless their pK_a is perturbed to be close to neutral. As described above, this is perfectly possible, and many Glu and Asp residues do indeed act as general bases. RS^- and RO^- (Cys, Tyr, Ser, and Thr) are even better bases, but for them to be happy in an initially unprotonated state requires their pK_a values to be perturbed by at least 5 pH units, and so this is rare. The exception is Tyr, with a pK_a of 10.5, which only needs perturbing by 3–4 pH units. It is, however, actually very rare to find Tyr acting as a general base. Alternatively, Lys (pK_a 10.5) could also act as a general base if its pK_a could be perturbed by 3–4 pH units. This does occur, and there are numerous examples.

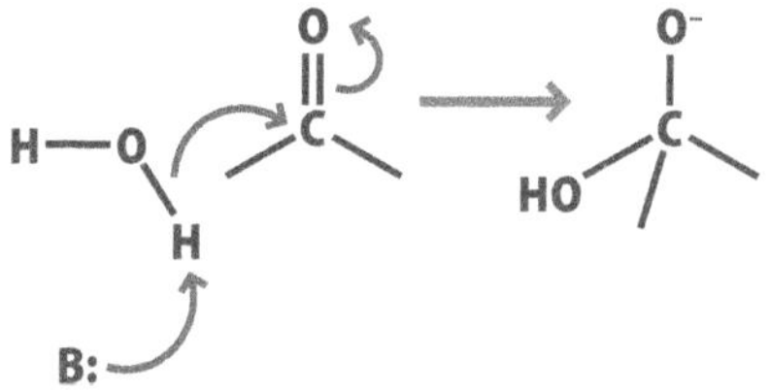

FIGURE 5.13
An example of general base catalysis. A base (B, shown with a lone pair of electrons suitable for removing a proton) removes a proton from water, making a much better nucleophile, the hydroxyl ion.

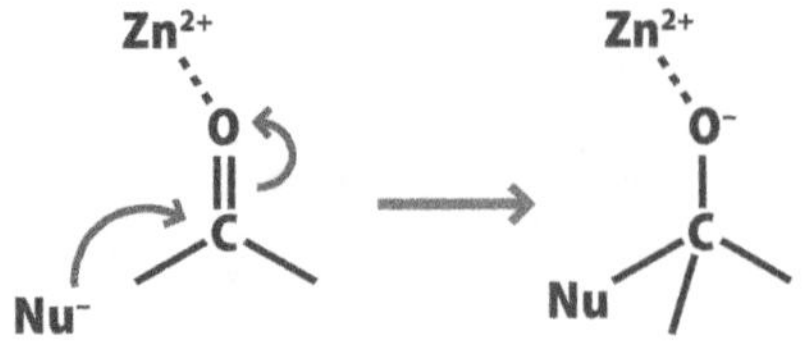

FIGURE 5.14
Activation of an electrophilic center by an electrophilic catalyst. In this (common) example, a zinc ion is the electrophilic catalyst: it is binding to the carbonyl group and activating it for attack by a nucleophile. Other metals can do similar things. A particularly common case is Mg^{2+} activating phosphates.

The best (in other words, strongest) bases are those with the highest pK_a, implying that the best general base catalysts will have a high pK_a. However, as we have just seen, such a catalyst is only useful as a base if it is in its deprotonated state. For an enzyme in water at neutral pH, this means that the pK_a of the base should be at or below 7. Combining these two requirements implies that the best general base catalyst (for a typical enzyme operating at close to neutral pH) has a pK_a near 7. Similarly, the best general acid catalyst also has a pK_a near 7.

A further implication of this is that any side chains in a protein that are identified to have pK_a values a long way from their standard values are probably involved in the catalytic step, because evolving a local environment that stabilizes an appropriately altered pK_a requires considerable evolutionary pressure, which is only likely to arise when an improved catalytic step can be generated. This can be a useful way to identify catalytically important residues.

5.2.4 Electrophilic catalysis is also common

Electrophilic catalysis is closely related to general acid catalysis. For the activation of an electrophile, it is not necessary that this be done by a proton: anything that pulls electrons away from the electrophile would do. Most commonly, a positively charged metal ion can be used (**Figure 5.14**). The metal ion acts as the electrophile, and activates the carbon. As discussed in Section 5.2.7, a Schiff base is a very good electrophile, and therefore several cofactors, in particular pyridoxyl phosphate, form Schiff bases with protein side chains to catalyze reactions.

5.2.5 Thermolysin uses all these mechanisms

We are now in a position to look at a real enzyme and understand how it works. Thermolysin is a bacterial protease. It is actually an example of an enzyme that uses a cofactor, as discussed below; in this case the cofactor is a zinc ion: thermolysin is a metalloprotease.

Thermolysin acts as a general acid catalyst, using a histidine to assist the leaving group (**Figure 5.15**). It also acts as a general base, using a glutamate to remove a proton from water to make a better nucleophile (**Figure 5.16**). In addition, the zinc ion is positioned so that it is close to the carbonyl oxygen; it acts as an electrophilic catalyst by polarizing the carbonyl group (**Figure 5.17**).

It is also noteworthy that at the same time the zinc ion also destabilizes the substrate, because it weakens the C=O double bond: in other words it stresses the substrate, and so raises its free energy, making the activation energy smaller. In addition it stabilizes the TS by stabilizing the charged oxygen. Thus, it both destabilizes the substrate and stabilizes the TS, helping to lower the activation energy "at both ends." The zinc ion is one of several electropositive groups in the vicinity of the negative charge that develops on the carbonyl oxygen during the reaction and becomes a full charge in the tetrahedral intermediate. These positive groups stabilize the negative oxygen in the intermediate and thus also the TS, within a hydrophobic pocket that is found also in serine proteases, where it is known as the "oxyanion hole."

FIGURE 5.15
Thermolysin uses His 231 as a general acid to protonate the leaving group amine. This converts it from NH^- (pK_a 25) to NH_2 (pK_a 10.5). The histidine itself has a pK_a of about 6, making it ideal as a leaving group (see Section 5.2.3). The double-headed arrow is a shorthand, implying a two-stage reaction in which the electrons move onto the oxygen and then back again.

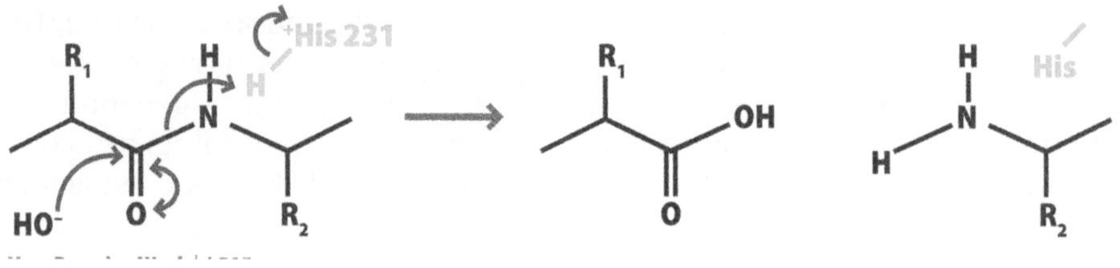

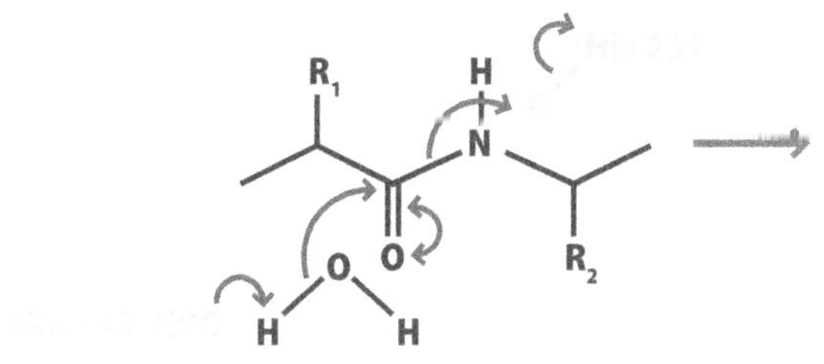

FIGURE 5.16
Thermolysin uses Glu 143 as a general base to deprotonate an attacking water, converting it to the much more nucleophilic hydroxide ion.

The zinc ion also binds to the oxygen atom of the attacking water molecule. This lowers the pK_a of the attacking water molecule from 16 to about 5, a remarkable change. It is thus acting independently as an electrophilic catalyst in two different ways—a very efficient use of a metal by the enzyme!

The general base catalyst, Glu 143, is at the base of a deep pocket in the enzyme. The attacking water molecule is bound to it; when the substrate binds, it buries the glutamate carboxylate and the water within the hydrophobic (low dielectric) protein interior. This has a number of consequences. First, it raises the pK_a of the glutamate to 5.3 so that it is basic enough to deprotonate the water. Second (and in a similar way to the zinc ion), it destabilizes the bound substrate with respect to the TS, by having a negative charge buried deep within the protein: in the TS this charge has moved and can be stabilized. And third, this charged system pre-organizes the oxygen atoms in the glutamate into the correct position so that they are correctly oriented to conduct the subsequent attack on the carbonyl: in other words, it lowers the activation energy by decreasing the loss of entropy in forming the TS.

In all these ways, the initial location of the glutamate acts to destabilize the substrate–enzyme complex and thereby lower the energy barrier for the reaction. The free energy for this effectively comes from the free energy for folding of the protein, which is less favorable than it would have been if the glutamate were not buried.

The overall thermolysin-catalyzed reaction (see Figure 5.17) has two steps with an intermediate state and two transition states, as shown in Figure 5.3, and the rate of the catalyzed reaction depends simply on the height of the highest peak. If the first peak is higher, the "rate-determining step" is the formation of the tetrahedral intermediate, whereas if the second peak is higher, the rate-determining step is the breakdown. Enzymes evolve under the influence of evolutionary pressure. It may of course be that the evolutionary pressure is not to increase the rate of the reaction. A classic example is the enzyme ribulose 1,5-bisphosphate carboxylase/oxidase (RuBisCo), the enzyme that fixes CO_2 in plants. This is a notoriously slow enzyme, with a turnover of roughly 3 s^{-1} compared with a more typical rate of thousands per second for many enzymes. In this case a stronger evolutionary pressure is to make sure that the enzyme does not react with oxygen rather than CO_2. However, for most enzymes there is at least some evolutionary pressure to increase the rate. In a reaction such as that catalyzed by thermolysin there are two steps, and the evolutionary pressure acts only on the slower one, because this is the step that determines the overall rate of the reaction. In fact in thermolysin the slower step is the

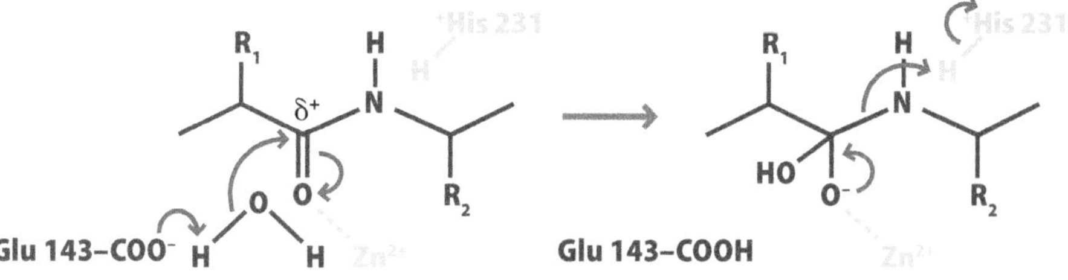

FIGURE 5.17
Thermolysin uses the zinc ion to polarize the electrophilic carbonyl and hence act as an electrophilic catalyst.

second one [12]. This is no doubt partly because the first step is aided by the presence of the zinc ion, which has several roles and helps to increase the rate markedly, as discussed above.

It is worth noting that because the evolutionary pressure is always on the slowest step, we might expect that, over the course of evolution, the activation energy barriers for all the steps will end up roughly similar. This is indeed true, as we shall see for triosephosphate isomerase.

In summary, thermolysin uses a variety of methods to catalyze the reaction, and some active site groups, such as the zinc ion, are used in several different ways. In this sense it is a sophisticated catalyst. However, each individual part is a standard well-understood catalytic mechanism: there is nothing particularly "clever" about what it does. The enzyme has taken a lot of different small rate enhancements and added them together constructively to get a big rate enhancement. As we shall see many times, that is basically how evolution works. A detailed analysis of enzymes has shown that most catalyze the "easy" reactions and involve general acid–base and nucleophilic catalysis [13]: why would an enzyme catalyze difficult chemistry when it can achieve its goals by catalyzing something easier?

5.2.6 Nucleophilic catalysis implies a change in mechanism

Most proteolytic reactions are catalyzed not by metalloenzymes such as thermolysin but by serine proteases such as trypsin and chymotrypsin. We can speculate that, because these enzymes do not have the benefit of the zinc ion, it is the first step that has had the evolutionary pressure. And these enzymes have replaced the first step by a different reaction altogether, which is faster than attack by hydroxyl because the nucleophile used is better: it is a serine hydroxylate (RO^-) rather than a hydroxyl (HO^-) (**Figure 5.18**). Thus, serine proteases require an extra two steps in comparison with thermolysin, namely the initial attack and the breakdown of the enzyme-bound intermediate. Nevertheless, because it is (presumably) the initial nucleophilic attack that is the slower step, the overall reaction is still faster with this mechanism. (The second nucleophilic attack, on the covalently bound acyl enzyme, is faster because the substrate is now an ester not an amide, which is inherently more reactive.) To repeat earlier comments, it makes no difference to the rate how many steps there are; what matters is only the highest activation energy. Thus, we can compare the energetics of thermolysin with that of serine proteases

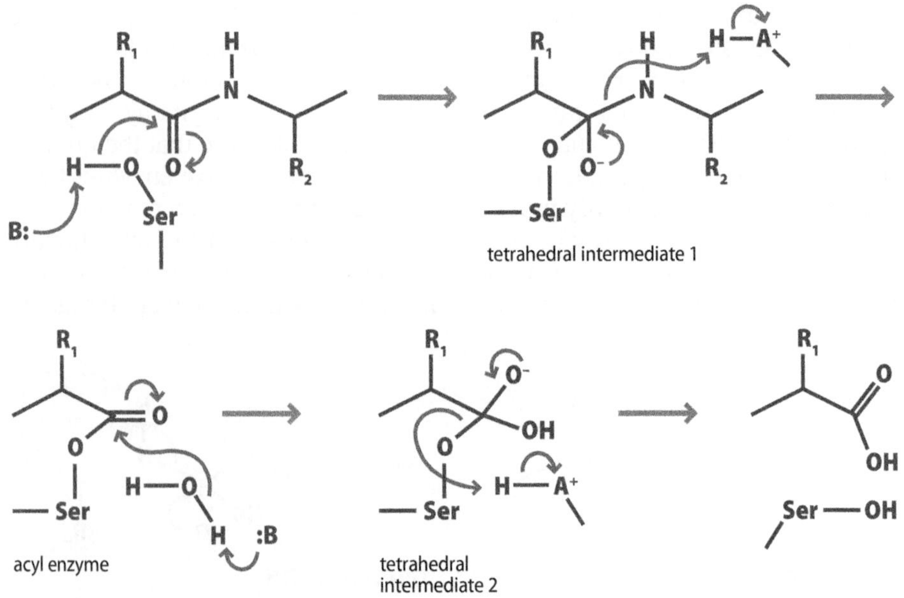

FIGURE 5.18
The mechanism of serine proteases. B and A represent general base and general acid catalysts, respectively; in many cases these are the same amino acid side chain taking on a dual function. This is an example of nucleophilic catalysis because the serine hydroxyl of the enzyme is directly involved in the catalytic mechanism, acting as a nucleophile to the amide carbonyl. It is a catalyst because at the end of the reaction it is again free.

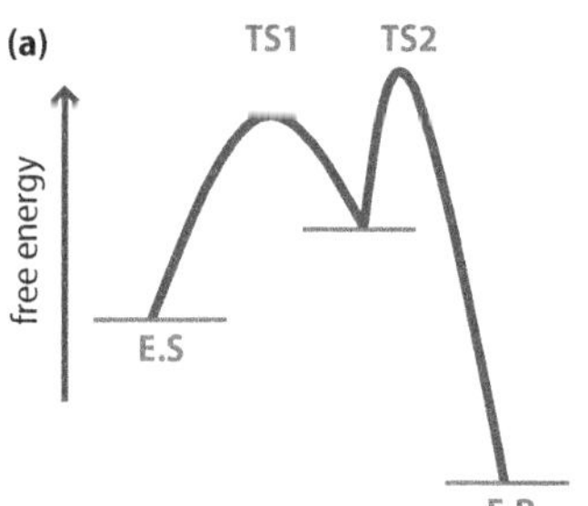

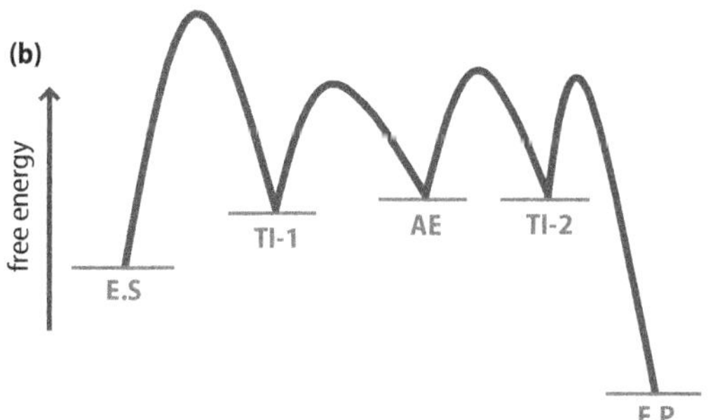

FIGURE 5.19
Schematic free-energy diagrams for proteolysis (a) by thermolysin and (b) by serine proteases. The energetics of the final step is the almost same in both reactions. Note that the rate-determining step (the step with the highest overall activation energy) is different in the two reactions. TI and AE refer to tetrahedral intermediate and acyl enzyme (compare Figure 5.18).

(**Figure 5.19**), to see that the inclusion of the zinc ion has led to a change in the rate-determining step. Because this mechanism uses a nucleophilic group within the enzyme to catalyze the reaction, it is often called *nucleophilic catalysis*. The most nucleophilic side chain in the standard 20 amino acids is the cysteine anion (RS^-), which explains why cysteine is the second most common side chain (after histidine; see Chapter 1) found as a catalytic residue [13]. (The hydroxyl ion, as found in serine and threonine, is almost as good but is much less readily produced because its pK_a is higher; see Table 1.1.) It is very common for the enzyme-catalyzed reaction to follow a different pathway (that is, to use a different mechanism) from the uncatalyzed reaction.

5.2.7 Enzymes often use cofactors and coenzymes

Proteins have only a limited range of functional groups, which limits the chemistry that they are able to do. In Chapter 1 we noted that the earliest form of life is thought to have been an RNA world, in which RNA molecules carried the genetic information and also performed all the catalytic steps. Where proteins existed, they were initially merely scaffolds to strengthen and protect the RNA. However, proteins have taken over almost all the catalytic functions once performed by RNA because they have a much wider range of chemistry available to them. But it is still not enough. An obvious lack is that, with the exception of the sulfhydryl group of cysteine, amino acids do not contain any group that can catalyze oxidation and reduction (redox) reactions. Therefore proteins have appropriated anything available that is able to be oxidized or reduced. This includes a range of RNA-related molecules, which presumably were available in the RNA world, such as NADH, FAD, and FMN. They also use coenzyme A, *S*-adenosyl methionine, UDP-glucose, CDP-diacylglycerol, and GTP and CTP as used in the molybdenum cofactor (**Figure 5.20**); they also use transition metal ions, particularly iron. In the prebiotic world iron was a good choice because it was easily available and can equilibrate between Fe^{2+} and Fe^{3+}, but in an oxygen-rich atmosphere iron is more of a problem because Fe^{2+} is easily oxidized to Fe^{3+}, in which form it is much less available. Iron is, however, used extensively in redox proteins, both as heme and as iron–sulfur clusters. This is further evidence that proteins "crystallized" in something like their current form rather early on in evolution and have not changed in any major way since, apart from the extensive use of extra domains, as discussed in Chapters 2 and 8.

In fact, metals form part of the structure of a remarkable 47% of enzymes, and are part of the active site in 41% [14]. The metals are used in approximate proportion to their availability, either now or in a prebiotic environment. Interestingly, zinc is used to a noticeably higher extent in eukaryotes than in prokaryotes. This is likely to be a consequence of the large number of zinc and RING finger domains added in prokaryotes. Zinc is a relatively recent element to be used extensively, because in the prebiotic world zinc was probably mainly locked up as zinc sulfides.

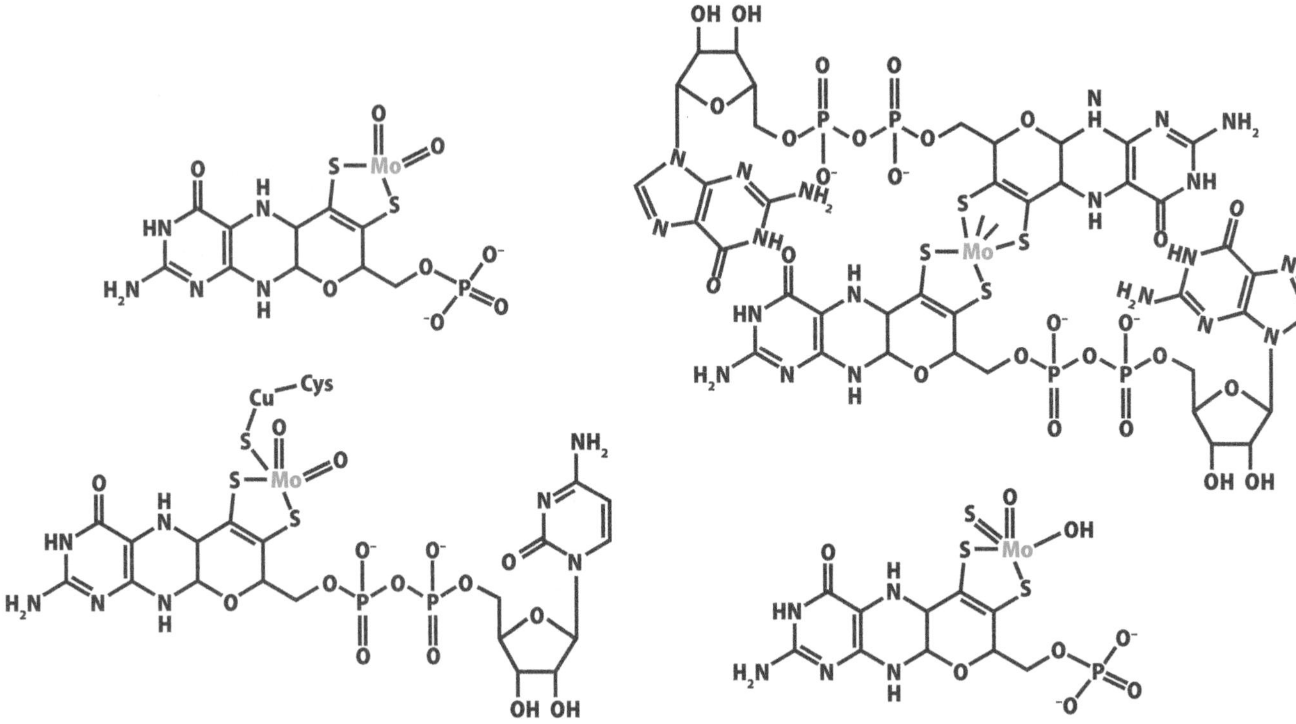

FIGURE 5.20
A selection of cofactors containing molybdenum, used in a range of redox reactions. These cofactors are synthesized from ATP, GTP, and CTP.

The other function that amino acids are not very good at is group transfer reactions. If we look at the functions of the common cofactors we see that they are basically required for either redox or group transfer (**Table 5.2**). Indeed, if we classify redox reactions as electron transfer and light capture as photon transfer, they are *all* involved in transfer reactions.

Proteins use as cofactors anything that is useful to them. They use most of the common metals in one way or another, plus a large range of organic compounds. Sometimes they even convert their own amino acids to something useful. Thus, the fluorescent group in green fluorescent protein is made from three consecutive amino acids (serine, tyrosine, and glycine) (**Figure 5.21**) (see Section 11.2.4).

As an example of the versatility of cofactors, **Figure 5.22** presents some of the reactions that are catalyzed by pyridoxyl phosphate, the biologically active form of vitamin B6. This cofactor reacts with free amino groups, forming a C=N group, which is known as a Schiff base. This group is easily protonated, making it a good electrophile (*5.8). It is able to catalyze a remarkable range of reactions, including the removal of a proton from a reactive carbon (racemization of α-amino acids), decarboxylation, and transamination (the replacement of CH–NH_2 by C=O, as found in the biochemically important reactions that exchange α-amino acids with α-keto acids). It is therefore a very common and essential cofactor.

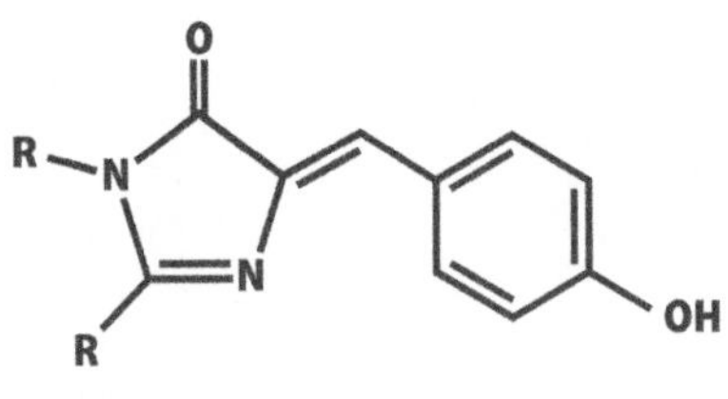

FIGURE 5.21
The fluorophore from green fluorescent protein.

To return to the first point in this section, it is interesting to note that RNA enzymes ("ribozymes") use nothing more sophisticated than metal ions in their active sites. Considering that RNA is actually very good at binding to a wide range of other molecules, this is a little surprising. Possibly it is because proteins are so much better as catalysts that they have replaced any attempts by RNA, except for the remaining RNA enzymes such as the ribosome, which presumably is so essential (and so primitive) that it cannot be replaced.

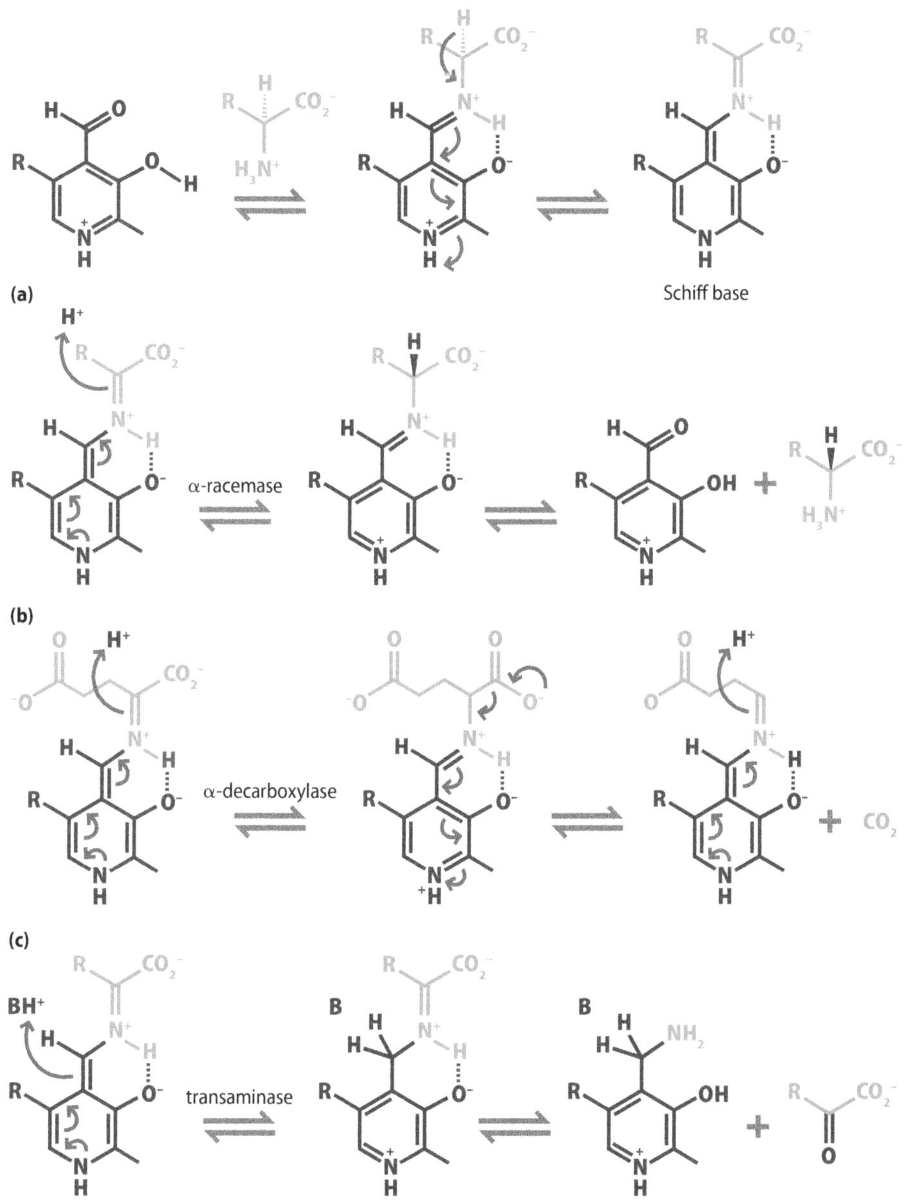

FIGURE 5.22
Some of the reactions catalyzed by enzymes using the cofactor pyridoxyl phosphate. The R group at the left is $CH_2\text{-}OPO_3^{2-}$. The aldehyde group of pyridoxyl phosphate forms a Schiff base with amines (top right), creating a versatile electrophilic sink that can be used for a range of reactions. In the transaminase reaction shown here, an α-amino acid is converted to an α-keto acid. The original aldehyde is re-formed by a reverse reaction with a different α-keto acid to regenerate an α-amino acid.

TABLE 5.2 Common coenzymes

Coenzyme/ cofactor	Function
NADH/NADPH	Redox
FMN, FAD	Redox
Quinones	Redox
Iron–sulfur clusters	Redox
Nicotinamides	Redox
Heme	Redox
Chlorophyll	Light capture
Retinal	Light capture
Biotin	Transfer of carboxyl
CDP-diacylglycerol	Transfer of phosphatidate
Cobalamin (vitamin B_{12})	Transfer of alkyl
Coenzyme A (pantotheine)	Transfer of acyl
Lipoic acid	Transfer of acyl
Pyridoxyl phosphate	Transfer of amine (and other functions)
S-adenosyl methionine	Transfer of methyl
Tetrahydrofolate	Transfer of single carbon
Thiamine pyrophosphate	Transfer of aldehyde
UDP-glucose	Transfer of glucose

5.2.8 Enzymes control water in the active site

Nearly all active sites are in a cleft or pocket, shielded from water. Very often (as we have already seen in Chapter 2) an enzyme will bind its substrate and then close a flap or domain over the top of the active site to isolate it completely from solvent. In this way, active sites tend to be completely cut off from solvent, and water is excluded except for any water molecules that are required as part of the mechanism. There are three main reasons why this is so.

The first reason is that water is fairly reactive and could perform side reactions. The classic and best example of this is in kinases. These catalyze the addition of a phosphate group, using nucleophilic attack of the substrate onto an activated phosphate (**Figure 5.23**). They use a variety of means to make the phosphorus more electrophilic and the P–O bond weaker. In that case, any water in the active site should also be able to react in the same way (**Figure 5.24**). The result is hydrolysis of the phosphate—for example the conversion of ATP to ADP—which is an undesirable side reaction. Water is not as good a nucleophile as serine, but there is a lot more water than serine, so this is a perfectly feasible reaction. It is avoided by excluding water from the active site; all kinases have their active sites buried in the interface between two domains and bind the two substrates one on each domain. The active complex is only set up by domain closure once both substrates have bound; this is an example of the induced-fit mechanism described below. In this way the enzyme ensures that there is no water present in the catalytically active complex.

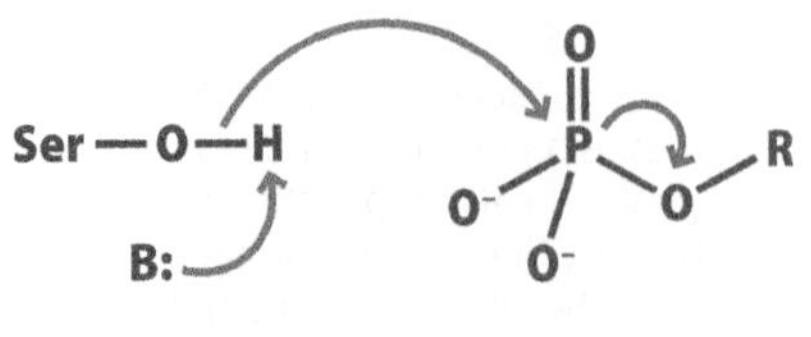

FIGURE 5.23
The reaction catalyzed by a serine kinase. There has been considerable debate about this mechanism, but the general consensus is that it is a direct in-line nucleophilic displacement, as shown here. Kinases also make use of electrophilic catalysis from Mg^{2+} ions coordinated to the oxygens, as well as additional electrophilic side chains.

The second reason for the exclusion of water is more general and therefore more important: water has a very high dielectric constant ϵ (*4.2). The energy of a charge is proportional to q/ϵ, where q is the size of the charge. This means that a charge inside a protein ($\epsilon \approx 4$) is about 20 times less stable than a charge in water ($\epsilon \approx 80$). Less stable means more energetic and more reactive. Thus, any kind of interactions that depend on charge, such as hydrogen bonds or electrophilic catalysis, will be stronger and more effective in the absence of water. It is no coincidence that most synthetic organic chemistry reactions are performed in very hydrophobic solvents. The manipulation of dielectric constant by enzymes is a crucial part of their catalytic power. Look for example at the catalysis by thermolysin of proteolytic cleavage (see Figure 5.17). The zinc ion has a crucial role, which is immensely strengthened by the fact that the environment is generally hydrophobic (low ϵ). The enzyme disfavors buried charges, such as the glutamate carboxylate and the zinc ion, and favors (stabilizes) the less highly charged TS; in other words, it lowers the activation energy by burying both the glutamate and the zinc ion in a hydrophobic environment.

It is worth noting that this implies that active sites of enzymes are chemically unusual, in that they are hydrophobic but often contain suitable placed charges. This means that the amino acids in the active site are often more reactive than normal: for example, the serine in the active site of serine proteases reacts with a range of chemicals that do not react with any other serines in the protein, an observation that has greatly helped to identify the active site in many proteins.

The third reason is almost equally important: water has a strong **dipole**, which orients itself so that it solvates charges. We are familiar with this in that water molecules cluster round a charged ion, with the electronegative end of water pointing toward a positively charged metal, for example (**Figure 5.25**). As we have seen, any chemical reaction involves the movement of negatively charged electrons: this is what the curly arrows depict. If there are water molecules close by, they try to reorient themselves so that they solvate the charges. Thus, during a reaction, the water molecules try to "follow" the reaction. This acts as a drag on the reaction and slows it down considerably [15]. Electrons can travel very fast, whereas the reorientation of water molecules is much slower. Calculations suggest that this factor alone can slow reactions down by a factor of 10^{10}, although the effect is not usually so extreme. Therefore enzymes desolvate the TS, not just to lower the activation energy but also to avoid slowing the reaction [15]. Of course, they still need to have dipoles present to stabilize charges in the active site, but because these are derived from fixed atoms within the enzyme they do not move during the reaction and therefore do not slow it down.

FIGURE 5.24
Hydrolysis of a phosphate (such as ATP), using the catalytic center of a kinase.

Water molecules normally solvate a protein surface very effectively. As discussed in more detail in Chapter 4, removing a water molecule can be a very energetically demanding process. So, for example, if kinases had to remove all the water molecules from the active site before they could close, this would represent a very large energy barrier and would be a very slow process. However, this is not the way that it happens. Water molecules can usually be removed one at a time, more like a zipper closing than a rigid lid closing, so that the height of the activation energy barrier never gets very large.

5.3 ENZYMES RECOGNIZE THE TRANSITION STATE, NOT THE SUBSTRATE

5.3.1 The lock and key and induced-fit models

When **Emil Fischer (*5.11)** proposed his **lock and key model** in 1894, it was a revolutionary idea: that nature could make molecules that could hold chemicals like a lock. It represented a big step forward in our understanding of enzymes. Indeed, this idea expresses much of what this chapter has been about so far. It is often said that the lock and key model does not explain how the enzyme lowers the energy of the TS, and this is certainly true if we use the original description in which the lock matches the substrate. However, if the lock has a structure that matches the *transition state* more than the substrate, it will hold the reactants into the correct positions, destabilize the substrate, and stabilize the TS; it is thus a good description of much of what enzymes do. Catalytic antibodies work basically by being a lock that fits the TS.

The lock and key model has been replaced by the **induced-fit model**. This model suggests that the enzyme is not rigid but can adapt its structure to fit the substrate. The enzyme has one conformation in the free state, but when the substrate binds, the energy of binding causes the enzyme to change its conformation, to form a catalytically competent state. Essentially this is what is implied by the typical sequence

$$E + S \rightarrow ES \rightarrow E^{*}S \rightarrow EP$$

where the conversion of ES to E*S represents the transition to an activated or catalytically competent form.

The original proposal [16] was used to explain how it is that an enzyme can work more slowly with a smaller substrate than with the "correct" one. The lock and key model explains how an enzyme can be selective against a larger substrate: the lock just does not fit the key. But it cannot explain why an enzyme should not work against a smaller substrate. The idea that the substrate forces the enzyme to change shape can explain this very nicely (**Figure 5.26**). This idea has since been taken up to suggest that the enzyme in the resting state may be complementary to the substrate, but on binding to the substrate it undergoes a change to an activated form; however, this is not necessary for the model, and as we have seen it can equally well form part of the lock and key model.

The induced-fit model is indeed a better explanation of enzyme mechanism than the lock and key model. Partly this is because it accepts that enzymes are not rigid and that the conformation of the enzyme in the TS (or correspondingly in intermediates, where they exist) is different from that in the free enzyme. It would be better to modify the original theory to recognize that the substrate is also not rigid, so that when they bind there is a mutual structural and dynamic change: effectively that the enzyme not only *stresses* the substrate (that is, puts it into an environment where it is unstable) but also *strains*

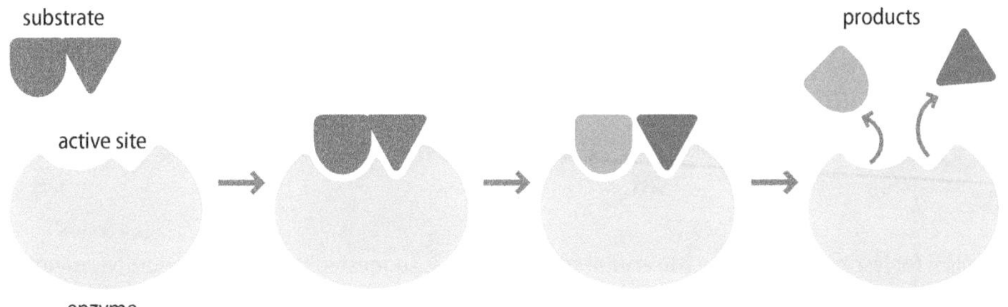

FIGURE 5.26
The induced-fit model.

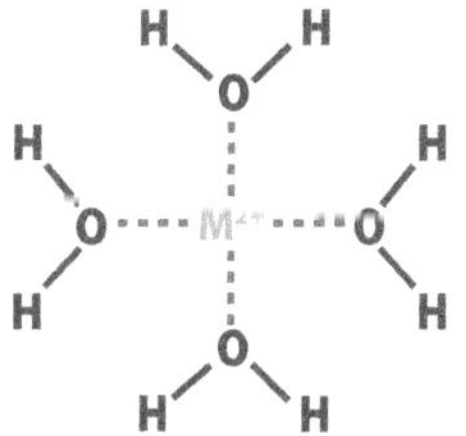

FIGURE 5.25
Solvation of a metal ion by water.

***5.11 Emil Fischer**

Emil Fischer (**Figure 5.11.1**) is best known for his work on sugars—in which he was one of the first chemists to appreciate the importance of stereochemistry—and purines. It was this work that earned him the Nobel Prize in Chemistry in 1902. It also gave rise to his **lock and key model** to explain the stereospecificity of enzyme reactions; this was a revolutionary idea, given that the concept of an enzyme as a molecule of defined composition or structure was still much debated. Subsequently, however, he turned to work on amino acids and proteins, in which he characterized several amino acids and was the first to describe the peptide bond and synthesize a dipeptide.

FIGURE 5.11.1
Emil Fischer. (Copyright Science Museum/SSPL.)

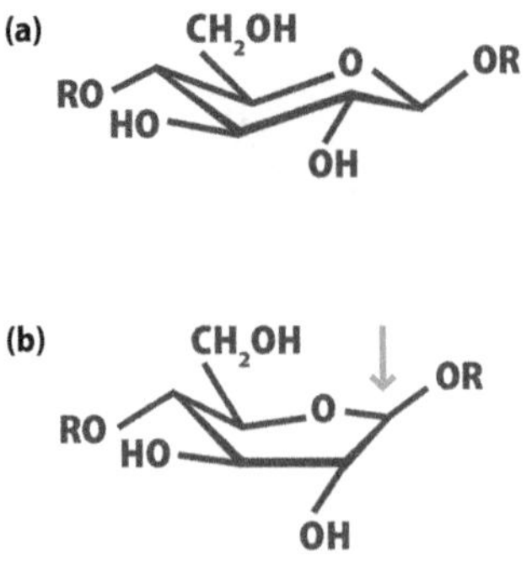

(a)

(b)

FIGURE 5.27
Two conformations of the hexose sugar ring. (a) The chair. (b) The half-chair. The chair form is more stable in solution. In the lysozyme crystal, ring D (at the cleavage site) is forced into a half-chair conformation. This is much more favorable for cleavage because the glycosidic carbon (arrowed) is already almost planar, which is the conformation it needs in the transition state.

it (that is, physically deforms it into a conformation from which it is easier to react).

The argument for substrate strain put forward in the previous paragraph is usually illustrated by lysozyme. Lysozyme catalyzes the digestion of polysaccharides and is found for example in tears, where it helps to kill off bacteria. It was the first enzyme to have its crystal structure determined (in 1965), and the crystal structure of a complex with its substrate immediately suggested how the enzyme worked. In the complex, it was only possible to fit the substrate into the active site if the sugar ring that gets hydrolyzed was distorted from its normal chair structure to a half-chair (**Figure 5.27**). This is a very attractive idea, because the half-chair structure is much better at stabilizing a positive charge on the glycosidic carbon, implying that lysozyme should catalyze sugar hydrolysis by a combination of general acid catalysis (to protonate the leaving group sugar) and destabilization of the substrate together with stabilization of the positively charged TS (**Figure 5.28**). Lysozyme was therefore suggested to work by a chemically somewhat unusual SN_1 reaction rather than the much more common SN_2 reaction: that is, it was suggested that the leaving carbohydrate leaves first, to give a positively charged half-chair carbonium ion, after which water is added. This mechanism is found in almost every biochemistry textbook up to 2001, when it was shown that actually this mechanism is probably incorrect (a good example to show that even textbooks get it wrong sometimes): in fact, lysozyme acts in a much more conventional way, by nucleophilic catalysis forming a covalent intermediate, which breaks down in a subsequent step (**Figure 5.29**) [17]. This is another example of a really great idea that is so good that it has to be true, but unfortunately isn't.

Despite the revision needed to the lysozyme mechanism, substrate strain certainly does happen. It takes a lot of energy (enthalpy) to distort small molecules significantly, so the magnitude of the strain is often small [18], and it almost always makes more sense to think of the distortion as stabilizing the TS in preference to the substrate, rather than any direct catalytic effect of strain [19]. It is also relevant to add that there is no evidence in general for substrates being bound in unusual or high-energy conformations [20].

The classic example of an enzyme that proceeds by induced fit is adenylate kinase, already mentioned in Chapter 2. The essential point about adenylate kinase is that it is a kinase, and in common with all other kinases, as noted above, it has to exclude water from the active site. The enzyme has two domains (both Rossmann folds): ATP binds to one domain and AMP to the other. Only when both substrates have bound does the enzyme close to form the catalytically competent form, with water excluded. Here the rationale for the conformational change is not so much to alter the structure of the enzyme to match the TS, but much more to exclude water.

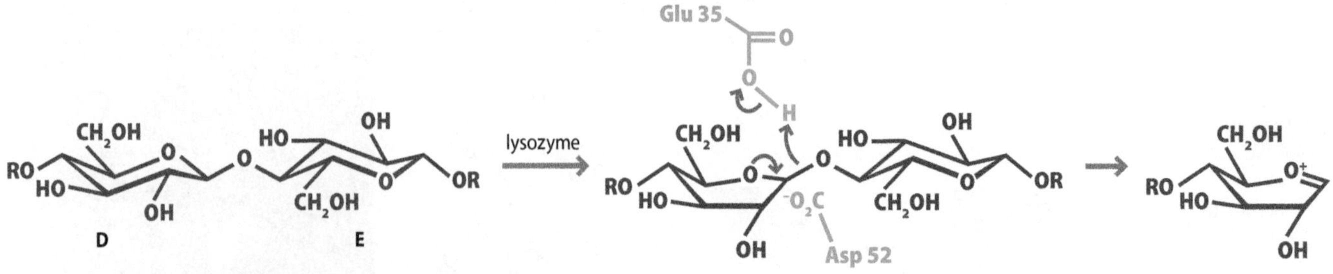

FIGURE 5.28
The original suggested mechanism for lysozyme, based on the crystal structure. It was suggested that ring D undergoes general acid-catalyzed cleavage, the acid being Glu 35; this was made possible by the destabilization of the chair form of ring D and its twisting into the half-chair, thus preparing it for the formation of the carbonium ion shown on the right. Asp 52 acts to stabilize the positive charge (and destabilize the substrate). The carbonium ion is rapidly attacked by water to produce a hexose sugar. The naming of the sugar rings originates from the crystal structure.

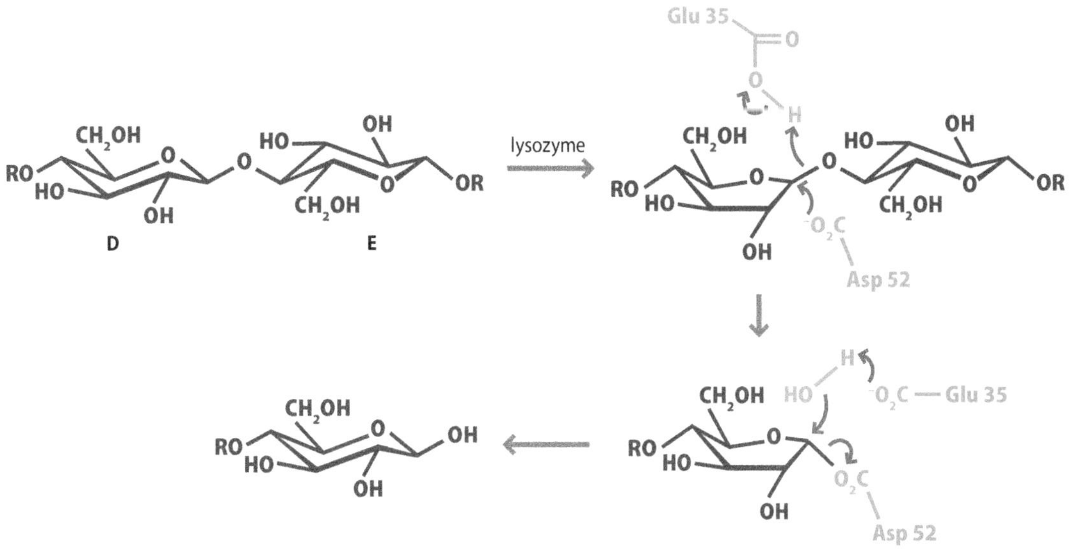

FIGURE 5.29
The more probable catalytic mechanism for lysozyme. This is a much more typical mechanism, found in a large number of glycosyl hydrolases and transferases. Asp 52 acts as a nucleophilic catalyst, aided by Glu 35 acting as a general acid. This generates an enzyme-bound intermediate. In a second step, a water molecule replaces the hydrolysed sugar, and Glu 35 now acts as a general base, deprotonating the water so that it can attack the sugar, liberating a free sugar.

The concept of an enzyme as remodeling itself to fit the substrate also very nicely explains most aspects of allostery, as discussed in Chapter 3.

The induced-fit model is a reasonably good description of how enzymes work. However, it is now clear that the model is not quite right. Rather than the binding of the substrate "forcing" or inducing a change to the enzyme, it is actually much more accurate to think of the free enzyme as being in a dynamic state, in which most of the time it looks much like the "free state" (as seen in crystal structures of free enzyme, for example) but it also undergoes motional processes in which it can adopt other conformations, some of which look much more like the active conformation. The substrate then binds probably not to the ground state but to one of the more active conformations, and in so doing it alters the conformational ensemble of the enzyme so that on average it now looks like the bound conformation. There is thus an alteration in a pre-existing conformational equilibrium. This model has been called the *conformational selection* or *pre-existing equilibrium* model and is widely agreed to be in general a more accurate model than induced fit. It is discussed in more detail in Section 6.2.

5.3.2 An enzyme should not bind strongly to its substrate

We have now considered most of the factors that enable enzymes to catalyze reactions. There remains one more, which is important but often overlooked. How tightly should an enzyme bind to the substrate? We have already seen that usually the enzyme sacrifices some of the potential binding energy so as to be complementary to the TS rather than the substrate; in other words, it *could* bind more tightly to the substrate but to achieve catalysis it is better not to. In this section we see that the optimal affinity is actually rather weak.

The affinity of an enzyme for its substrate is normally expressed by the K_m or **Michaelis (*5.12)** constant. The famous **Michaelis–Menten equation (*5.13)**

$$V = V_{max} \frac{[S]}{[S] + K_m}$$

shows that the rate V of a reaction is half maximal when the substrate concentration [S] is equal to K_m (**Figure 5.30**). (The square brackets around [S] just mean "the concentration of S," or more exactly the **activity** of S.) Thus, K_m can be considered to be the same as the dissociation constant (*5.5) for the ES complex under most circumstances: it is the concentration of substrate at

*5.12 Leonor Michaelis

Leonor Michaelis (**Figure 5.12.1**) was a German physiologist and physician who was forced to leave Germany in 1922 because he had disputed the correctness of a pregnancy test proposed by a leading German physiologist named Emil Abderhalden, which was meant to be able to detect madness. He went first to Japan and then to America, where he not only studied enzyme kinetics but also discovered that keratin could be dissolved in thioglycolic acid, a phenomenon that is now used widely to create the permanent wave or perm.

FIGURE 5.12.1
Leonor Michaelis.

*5.13 Michaelis–Menten equation

The Michaelis–Menten equation is

$$V = V_{max} \frac{[S]}{[S] + K_m}$$

and shows the hyperbolic relationship between the reaction rate V, the maximum reaction rate V_{max}, the substrate concentration [S], and the 'Michaelis constant' K_m, which is the substrate concentration at which the reaction is proceeding at half its maximum rate (see Figure 5.30). At low substrate concentration, where [S] is much lower than K_m, the rate is linearly proportional to [S], which seems intuitively reasonable. The maximum rate is achieved only when the enzyme is saturated with substrate, at which point the substrate concentration makes no difference.

The equation can be derived in several ways, depending on what assumptions are made. As shown in many textbooks, the original assumptions of Michaelis and **Menten (*5.14)** are overly restrictive. A helpful analogy is to think of the enzyme as a shuttle bus, and the substrate as people going from one place to another. With small numbers of people, the rate of movement of people on the shuttle bus just increases linearly with the number of people, but as the number of people wanting to travel increases, the rate of movement eventually reaches a constant when all the shuttle buses are full.

which half of it is bound to the enzyme. Values of K_m for different enzymes vary widely, from about 0.1 μM to 0.1 M. What determines its value?

If we compare the value of K_m for a particular enzyme with the physiological concentration of the substrate, it is usually found that K_m is between 1 and 10 times the value of [S]. In other words, for most enzyme reactions *in vivo*, the substrate affinity is weak enough for more than half of the enzyme not to be bound. This sounds counter-intuitive: why is it so?

The explanation is based on the energetics of an enzyme-catalyzed reaction. We first need to look a little at the standard energy diagram. We know that the equilibrium position for an association reaction depends on the concentrations involved. Thus, for the equilibrium

$$A + B \rightleftharpoons AB$$

it is easily shown (and it makes good sense) that if we increase the concentration of A or B in the solution, then more of the complex AB is formed: this essentially is **Le Chatelier's principle**. In other words, the difference in energy

*5.14 Maud Menten

Maud Menten (**Figure 5.14.1**) was one of the first Canadian women to obtain a medical doctorate, in 1911. At that time women were not allowed to do research in Canada, so she went to Germany, where she worked on enzyme kinetics with Leonor Michaelis, and then to America. She also was the first person to separate proteins by using electrophoresis, in 1944; this is now of course something done by every biochemistry undergraduate. In addition, she painted, climbed mountains, and went on several expeditions to the Arctic. She was finally promoted to full professor at the University of Pittsburgh in 1948 at the age of 70, one year before her death. Considering she co-authored what must surely be the most famous equation in biochemistry, this seems hardly fair.

FIGURE 5.14.1
Maud Menten.

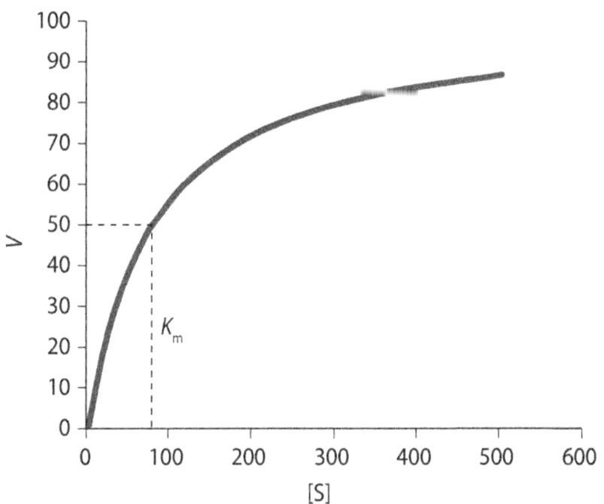

FIGURE 5.30
Enzyme kinetics obeying the Michaelis–Menten equation. The graph is calculated for a V_{max} of 100, and it is worth noting that at any reasonable substrate concentration, the rate still does not get close to V_{max}. The dashed lines indicate $V_{max}/2$, giving a K_m of 80.

between A + B and AB depends on their concentrations. Mathematically this is expressed in the equation

$$\Delta G = \Delta G^\circ + RT\ln([\text{products}]/[\text{reactants}])$$

stating that the free energy ΔG differs from the standard free energy ΔG° depending on the concentrations of products and reactants.

If we translate this into an energy diagram, it means that when we draw the step from A + B to AB, the difference in energy varies depending on their concentrations. In the context of an enzyme reaction, the obvious concentrations to choose are those found *in vivo*.

This means that in the normal situation in which K_m is greater than [S], the free energy of ES is higher than that of E + S; in other words, it is energetically unfavorable to form the ES complex (**Figure 5.31a**). In the other scenario, in which K_m is less than [S], formation of the complex is energetically favorable (Figure 5.31b). The figure makes it clear why this occurs. The activation energy for the reaction is the difference in free energy from the TS down to the lowest point on the pathway. If the binding of E to S is energetically favorable (see Figure 5.31b), ES is at a lower energy than E + S and the activation energy is larger; thus the reaction is slower. This is often expressed by saying that tight binding of enzyme to substrate creates a *thermodynamic pit*, which requires energy to get out of.

In the scenario of Figure 5.31a, in which binding is unfavorable, the catalytic rate is faster because there is no thermodynamic pit. In contrast, the concentration of ES is necessarily less than half what it could be, so the reaction is going more slowly than it would if all the substrate were bound. However, this effect is outweighed by the energetic factor, because the rate depends exponentially on the activation energy:

$$k = Ae^{-E_a/RT}$$

This means that it is much more important to ensure that the activation energy is low than to ensure that all the substrate is fully bound.

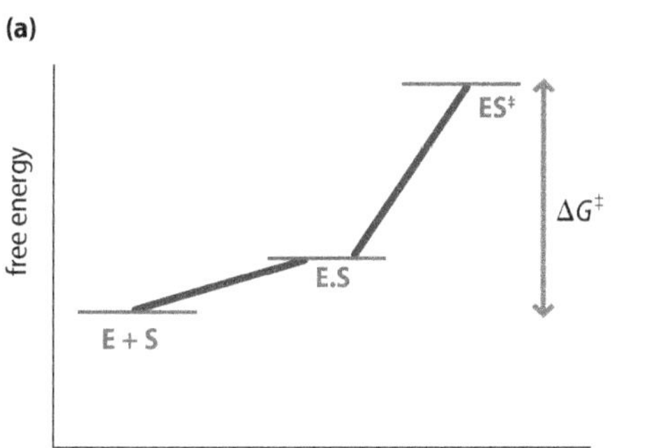

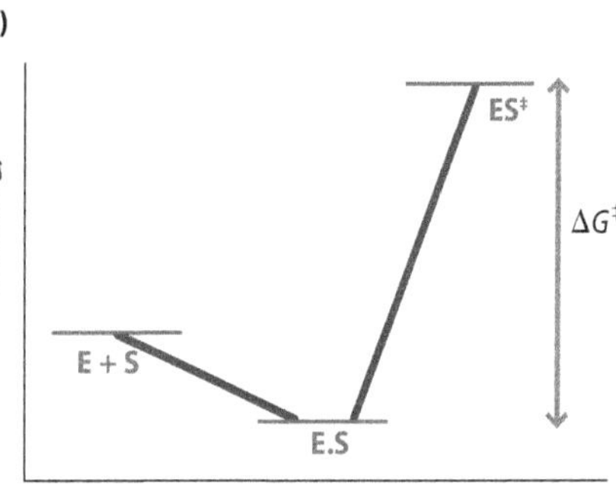

FIGURE 5.31
The thermodynamic pit that occurs when the substrate binds tightly. (a) Weak binding of the substrate (K_m > [S]). The activation energy Is determined by the height of the transition state. (b) Strong binding of the substrate (K_m < [S]). The activation energy is larger, because the complex has to dig itself out of the thermodynamic pit before it can start to react.

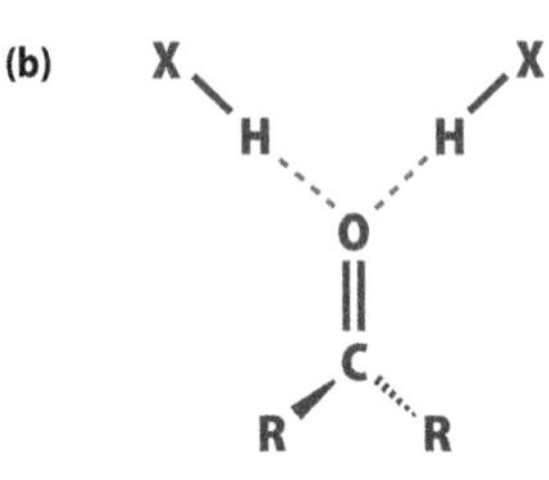

FIGURE 5.32
Binding of a substrate is usually organized such that the substrate is selectively destabilized but the transition state is stabilized. Most carbonyl groups that are attacked by nucleophiles in an enzyme-catalyzed reaction are bound with hydrogen bond donors 90° to the plane of the carbonyl (b), despite the fact that the planar geometry (a) provides more stabilization energy for the transition state. The reason is that this planar geometry also provides good stabilization to the substrate, whereas the out-of-plane geometry (b) does not.

An interesting consequence of this general rule is that the binding constant (or K_m) of an enzyme for its specific substrate is often no stronger than the affinity for a non-specific substrate: again, this is because some of the binding energy for the specific substrate is being used to lower the TS energy.

A good example of this concept is an exception that proves the rule. One would expect that enzymes that catalyze nucleophilic attacks on carbonyl groups (such as proteases) would stabilize the TS by having hydrogen-bonding groups in the optimal geometry for the TS; that is, in the same plane as the carbonyl group (**Figure 5.32a**). However, in general enzymes have their hydrogen-bonding groups at 90° to this (Figure 5.32b). In this geometry, the stabilization of the TS is almost as good, but the stabilization of the ground state is much weaker; hence the binding of the substrate remains weak, and the free energy of activation is decreased maximally [21].

The analysis above is relevant only if the constraints determining enzyme activity require the rate to be optimized. This is not always the main role of an enzyme. It is easily shown [6] that the rate of an enzyme-catalyzed reaction is most sensitive to substrate concentration when K_m = [S]. In other words, if the function of the enzyme is to keep metabolite levels roughly constant (which tends to be true for many intracellular enzymes), the enzyme will do this best when K_m = [S]; it therefore has a similar optimum position to that described above. However, if the role of the enzyme is to produce a steady flow of product whatever the substrate concentration (which is often true for extracellular enzymes), the enzyme should be working at saturation conditions ([S] >> K_m), where the enzyme rate is independent of the substrate concentration (see Figure 5.30). There is therefore no universal rule: the ideal relationship between K_m and [S] is different for different enzymes.

5.3.3 Binding and catalytic rate are closely interrelated

Enzymes are highly specific: they only catalyze the reaction of the correct substrate. This means that they need to bind the correct substrate but reject (that is, not bind) incorrect ones. The simplest way to recognize only the correct substrate is to have functional groups on the enzyme positioned so that they are in appropriate places to bind only to the substrate. But this creates a problem, because it would give rise to binding that is too strong. An enzyme has therefore to decrease its affinity for the correct substrate while retaining its specificity. It does this by two related mechanisms. First, it undergoes an **induced-fit** conformational change when the substrate binds: this change requires energy and in effect serves to decrease the affinity for the substrate [7]. Second, the resultant binding site can often be conceptually divided into a *specificity subsite*, which recognizes key parts of the substrate, and a *reaction subsite*. Binding of the substrate to the reaction subsite is effectively unfavorable, because the catalytic power of the enzyme derives from stabilizing (in other words binding) the TS rather than the substrate; the binding energy all derives from the specificity subsite. This is probably the reason why enzymes tend to attach substrates to large cofactors: because the binding affinity of the substrate is very weak, the enzyme needs to have something else it can bind to, typically such as a nucleotide (for example CoA, nicotinamide or flavin nucleotide) [18].

We should bear in mind that the division of enzyme binding into two subsites would not work if it were actually done with a real enzyme, because binding and catalysis are inextricably linked. This is nicely illustrated by kinetic data on substrates for the protease elastase given in **Table 5.3**. For both pairs of substrates, the longer substrate does not (as we might expect) have tighter binding (a lower K_m); it actually has a faster rate (a higher k_{cat}), despite the fact that the additional residue is not in the active site. That is, the additional

TABLE 5.3 Kinetic data for elastase, showing the interdependence of K_m and k_{cat}

Substrate	k_{cat} (s^{-1})	K_m (mM)	k_{cat}/K_m (s^{-1} M^{-1})
Ac-Ala-Pro-Ala-NH_2	0.09	4.2	21
Ac-Pro-Ala-Pro-Ala-NH_2	8.5	3.9	2200
Ac-Gly-Pro-Ala-NH_2	0.02	33	0.5
Ac-Pro-Gly-Pro-Ala-NH_2	2.8	43	64

(Data from R. C. Thompson and E. R. Blout, *Biochemistry* 12: 51–57, 1973. With permission from the American Chemical Society.)

binding has been used to stabilize the binding to the TS, rather than to bind the substrate more tightly. For the same reason, mutation of a key binding residue can for example affect the catalytic rate rather than (as expected) the substrate binding affinity, thus confusing any detailed analysis of the role of individual amino acids [22].

We have already seen that enzymes bind to the TS rather than the substrate, and they sacrifice some of the potential binding energy to stress the substrate. Here we have a second reason why enzymes should not bind tightly to their substrates: to improve their specificity of binding to the correct substrate. Enzymes make use of much less binding energy than is potentially available; they do not bind as tightly as they could. We note that if the enzyme has to restrict itself to weak binding, one of the few ways of increasing the rate of the reaction is to increase the concentration of enzyme. And several enzymes in the cell do indeed have very high concentrations, considerably higher than their substrates.

5.3.4 Transition-state analogs make good enzyme inhibitors

We have just seen that an enzyme should bind tightly to the TS but not to the substrate. Therefore if you want to design a drug that should bind tightly to an enzyme and **inhibit (*5.15)** its function, the drug should look like the TS and not like the substrate or product: that is, it should be a *transition-state analog*. This is in general difficult to achieve, because the structure of the TS is usually not known in detail and it is probably difficult to synthesize a compound with the same properties as the TS. For example, in a bond-cleaving or bond-making reaction, the TS will probably contain a partial bond, which is longer than a normal bond. Making a compound with a "long" bond requires some ingenuity.

As an example, the enzyme purine nucleoside phosphorylase (PNP) uses phosphate to break the bond between ribose and an attached purine base, as part of nucleoside scavenging. In particular, it clears deoxyguanosine from the body, an essential process in dividing T cells. Experiments and theoretical calculations have shown that the TS for the reaction looks something like **Figure 5.33b**: that is, the ribose ring oxygen has a partial positive charge, the phosphate–ribose bond is partly formed, the ribose–inosine bond is partly broken (although not by much, as the bond is still almost intact), and the inosine ring nitrogen is protonated. On this basis, the analog shown in Figure 5.33c was synthesized, which has a much more positive charge where the ribose oxygen was, a slightly longer and less polarized ribose–inosine bond, and a protonated inosine nitrogen. This compound binds to the enzyme at least 10^5 times more strongly than the substrate (picomolar affinity), and has proved a useful lead for drug design [23]. By inhibiting PNP it targets dividing T cells and is thus effective in the treatment of T-cell leukemia.

*5.15 Enzyme inhibitors

The activities of enzymes can be slowed down by inhibition. Inhibitors work in different ways. Some inhibitors react with the enzyme in the active site and are called irreversible inhibitors, suicide inhibitors, or catalytic poisons. Some of the components of protease inhibitor cocktails (used to inhibit cellular proteases while purifying proteins from cell extracts) such as PMSF (phenylmethylsulfonyl fluoride) work in this way. Irreversible inhibitors are uncommon as therapeutic drugs.

Many inhibitors resemble the substrate, product, or transition state (the most strongly binding inhibitors resemble the transition state). Such inhibitors bind in the active site and are called competitive inhibitors. They compete with substrate for binding and therefore increase the K_m of the enzyme. Effects of competitive inhibitors can be reversed by increasing the substrate concentration. In the presence of a competitive inhibitor, the Michaelis–Menten equation is modified to

$$V = V_{max} \frac{[S]}{[S] + K^{app}_m}$$

where K^{app}_m is related to K_m by $K^{app}_m = K_m \times (1 + [I]/K_i)$, where [I] is the concentration of inhibitor and K_i is the inhibition constant. All enzymes are inhibited by the product of the reaction, which generally acts purely by competitive inhibition.

There are two other standard classes of inhibitor, called non-competitive and uncompetitive. Non-competitive inhibition is rare: see Question 4 at the end of this chapter. Uncompetitive inhibition arises when the inhibitor binds only to the enzyme–substrate complex and not to the free enzyme. Examples are relatively rare, but for example the effect of Li^+ on *myo*-inositol monophosphatase is of this type; it is important for the effect of Li^+ in treating manic depression. Uncompetitive inhibition modifies the Michaelis–Menten equation to

$$V = V^{app}_{max} \frac{[S]}{[S] + K^{app}_m}$$

where K^{app}_m is as above and $V^{app}_{max} = V_{max}/(1 + [I]/K_i)$. Many inhibitors have mixed effects, combining both competitive and uncompetitive inhibition.

Enzymes can also be inhibited allosterically. Allosteric effects are discussed in detail in Chapter 3. It is worth noting that a feature of allosteric effects is that they are almost always *cooperative*. In other words, the effect of an inhibitor on rate is more marked than for a simple inhibitor. The effect is very similar to the difference in oxygen affinity between myoglobin and hemoglobin (see Figure 3.14).

5.4 TRIOSEPHOSPHATE ISOMERASE

5.4.1 Triosephosphate isomerase uses many catalytic mechanisms

In this final section, we study one of the classic examples of enzyme catalysis, and still one of the best-understood enzymes: triosephosphate isomerase, or TIM. TIM catalyzes one of the reactions in glycolysis, namely the interconversion of dihydroxyacetone phosphate (DHAP) and glyceraldehyde 3-phosphate (G3P) (**Figure 5.34**). *In vivo* (the forward direction of glycolysis, and the direction I shall call *forward* here), the reaction normally goes from left to right. This is not a regulated or allosterically controlled reaction: it just needs to run as fast as possible. And indeed it is a very fast reaction, running roughly 10^{10} times faster than the equivalent simple base-catalyzed reaction in water.

In essence, the reaction moves an H from C^1 to C^2. However (as always) it is important to be precise: the reaction moves a proton plus two electrons; that is, it moves a hydride ion, H^-. It is in fact an internal redox reaction: C^1 is oxidized while C^2 is reduced. Moving protons is easy: moving hydride is not. Therefore the reaction does not proceed by the movement of hydride: it moves

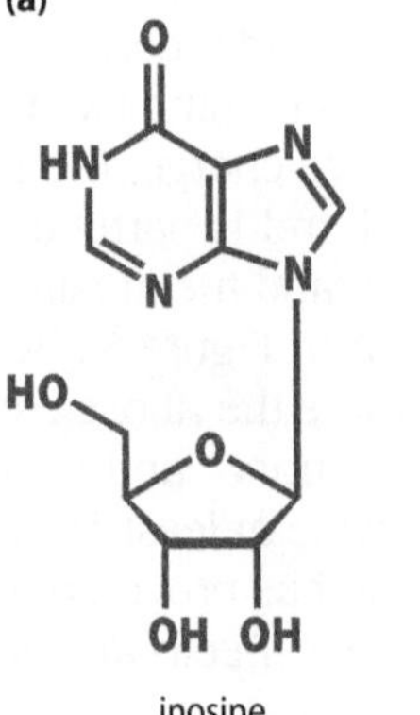

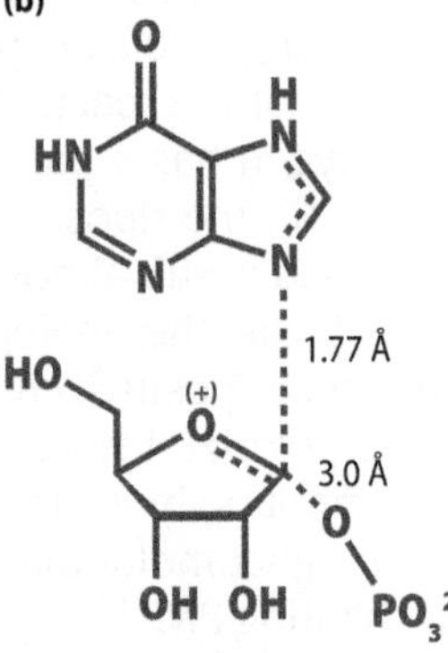

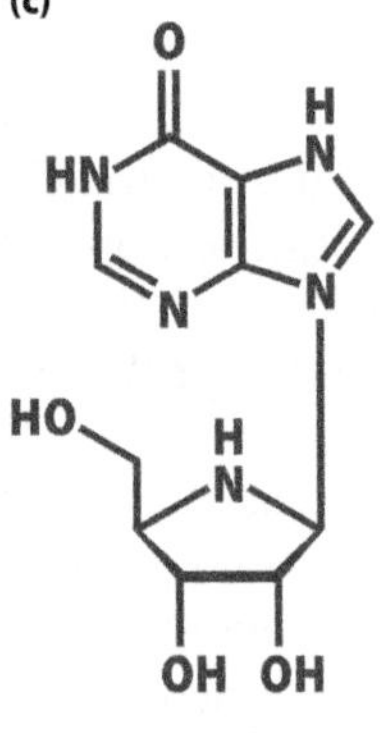

FIGURE 5.33
A transition state analog. (a) Inosine, a substrate for purine nucleoside phosphorylase. (b) The transition state. (c) An enzyme inhibitor that resembles the transition state and binds more than 10^5 times more strongly than the substrate. The pK_a of N^7 (the nitrogen in the five-membered ring) is similar to that predicted for the transition state.

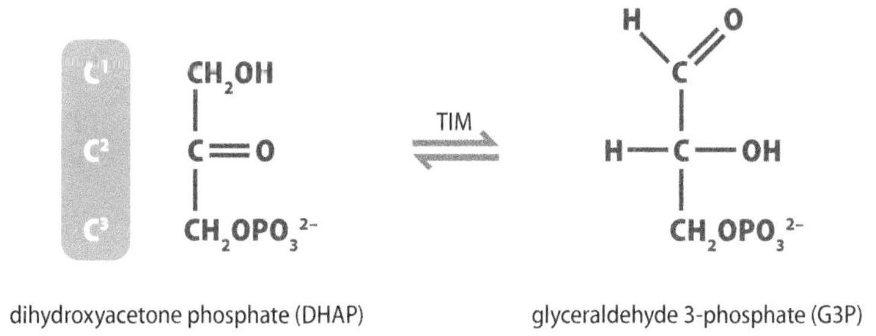

dihydroxyacetone phosphate (DHAP) glyceraldehyde 3-phosphate (G3P)

FIGURE 5.34
The reaction catalyzed by triosephosphate isomerase. The nomenclature for the three carbon atoms is shown at the left.

protons and electrons, but not bonded together. Thus, the reaction (both the reaction in water and the reaction pathway catalyzed by TIM) is as shown in **Figure 5.35**. Both of these steps are the same well-known type of chemical reaction, known as keto–enol tautomerization. Step 1 produces an enediol (that is, two OH groups attached to a double bond), which can break down by keto–enol tautomerization either to produce G3P by removal of the OH proton on C^1, or to go back to the starting material DHAP by removal of the OH proton on C^2.

Step 1 would be expected to be catalyzed by a general base at one end and a general acid at the other end, as shown in Figure 5.35. Similarly, step 2 also requires a general base and general acid.

It is instructive to look at some of the reasoning that led to the proposed mechanism.

1. If the reaction is observed using as starting material DHAP with a radioactive tritium atom ^3H attached to C^1 (**Figure 5.36**), some of the tritium ends up on C^2. This is the hydrogen that is removed by the general base in step 1 and then added on by the general acid in step 2 (see Figure 5.35). One can imagine several ways in which this might happen, but by far the most obvious is that the amino acid that acts as the general base in step 1 is the same one as the general acid in step 2: B1 and A2 are the same amino acid. Structurally this makes sense because they are required to be in a similar location.
2. In the crystal structure, there is a residue that looks appropriate to act as this general acid/base, namely Glu 165. To check this, Glu 165 was mutated. As expected, mutation to an alanine decreases the rate more than 10^6-fold. Strikingly, even as small a mutation as Glu to Asp decreased the rate 10^3-fold. The position of the carboxylate is critical. It is thought that the carboxylate is able to juggle the proton in between steps 1 and 2, to allow it to position it in exactly the right place (**Figure 5.37**).

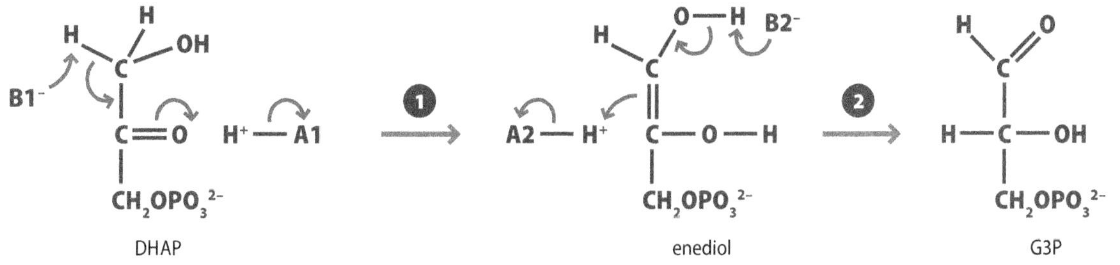

DHAP enediol G3P

FIGURE 5.35
Mechanism of the interconversion of DHAP and G3P. This is the pathway followed by the TIM-catalyzed reaction, but it is also close to the chemically preferred route. The intermediate has a C–C double bond with an alcohol (OH) group at each end; it is therefore called an enediol. The general acids and bases are called A1, A2 and B1, B2 for clarity.

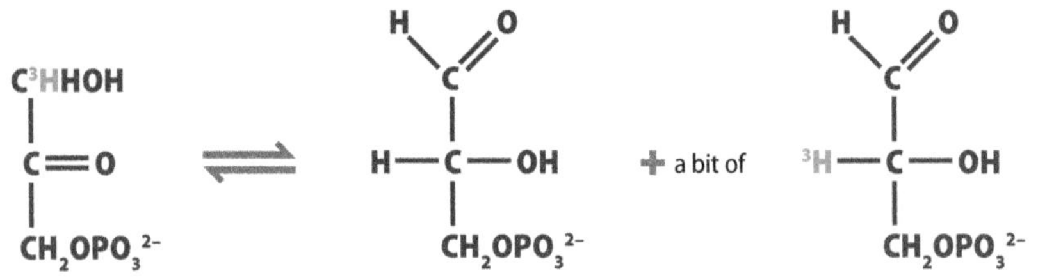

FIGURE 5.36
Some of the tritium from C^1 gets transferred to C^2.

3. Similarly, the general acid residue A1 was shown to be His 95. This is also probably the general base B2. [See **His 95 in trisoephosphate isomerase, *5.16**.]
4. There is a loop that can be seen by crystallography to fold over the substrate-binding site when the substrate is bound, thus shielding the active site from water. It moves by about 7 Å. Kinetic experiments using a mutant in which the loop has been removed show that the loop stabilizes the TS, removes water from the active site, and increases the rate about 10^5-fold. It is also important for holding the substrate in the correct orientation. In the wild-type enzyme, essentially 100% of substrate molecules are converted to product; in the loop deletion mutant, only one molecule in six is converted, the rest dissociating in mid-reaction and subsequently decomposing to give an unwanted side product [24]. This was graphically described as the enzyme "losing its grip" on the substrate [25].
5. NMR experiments show that the loop moves in and out even in the absence of substrate. Furthermore, the rate at which the loop opens is almost identical to the rate of product release. In other words, it seems that the enzyme has evolved the following mechanism: it binds and encloses the substrate within the active site; it holds the substrate there long enough to allow the reaction to occur, safely within an enclosed and dehydrated location; and it then releases the product. This is effectively an induced-fit process, or more exactly a conformational selection process (Chapter 6).
6. We can draw the TS for step 1 roughly as in **Figure 5.38**, in which all of the bonds are partly made or broken. The TS has an increased electron density on the C^2 carbonyl oxygen, providing an opportunity for the enzyme to use electrophilic catalysis and stabilize this charge. The electrophile is probably Lys 12, although His 95 (the same residue that acts as the general acid/base) also probably functions in this way.
7. Very good use was made of *kinetic isotope effects* in the analysis of this mechanism. If a reaction involves the breaking of a C–H bond (for example, step 1) or an O–H bond (for example, step 2), then the reaction is slower if the H is replaced by a heavier isotope such as ^2H or ^3H. This is because the C–^3H bond is stronger than the C–^1H bond. It is straightforward to calculate how large the effect should be: in fact, it should make the reaction about 20 times slower if the bond is fully broken in the TS. However, for this reaction there is *no* measurable kinetic isotope effect at all: it proceeds at the same

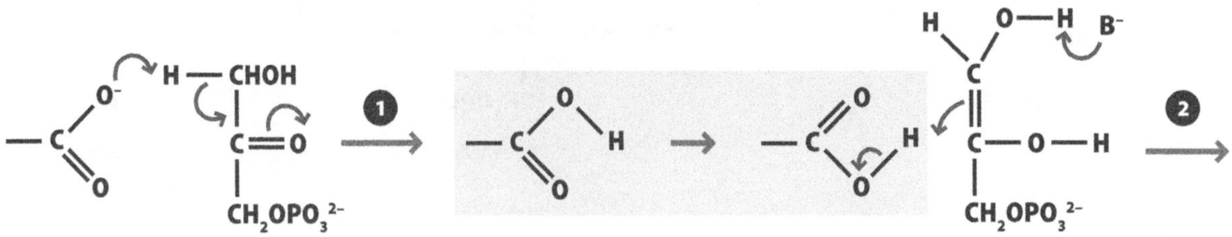

FIGURE 5.37
Glu 165 is thought to be able to rotate to pass the proton from one side to the other (shading), to facilitate its role as both general acid and general base.

***5.16 His 95 in triosephosphate isomerase**

His 95 has a more complicated role than implied simply by calling it a general acid/base. As a general acid one would expect it to be protonated, as are the vast majority of such histidines. However, His 95 has a pK_a of less than 4.5 and is therefore deprotonated at neutral pH [25]. Furthermore, it seems that His 95 also functions as an electrophilic catalyst, by making a hydrogen bond to the C^2 carbonyl and therefore partly destabilizing the double bond [32]. Again, it would be a much better electrophilic catalyst if it were positively charged. Knowles has offered some speculation as to why it might have such a low pK_a [24], of which the most convincing to me is that a low pK_a for imidazolium (protonated to neutral) may also imply a low pK_a for imidazole (neutral to deprotonated), which may bring the pK_a close to that of the enediol intermediate and therefore make proton transfer between O^1 and O^2 very rapid.

rate with ^1H or ^3H. This (along with numerous other proofs) shows that the slowest step in the reaction is actually neither step 1 nor step 2: it is the loss of the product from the enzyme, which obviously is not affected by isotopic substitution.

Detailed kinetic measurements such as those in point 6 above showed that the energy profile for TIM is as shown in **Figure 5.39**. This diagram is worth studying carefully. In the forward direction, the activation energy is the energy required to go from the lowest energy (E + S) to energy barrier TS4, loss of product, as noted above. In the reverse direction, the activation energy is the energy from E.G3P or E.enediol (both rather similar in energy) to energy barrier TS2 (which is marginally higher than the barrier from E+G3P to TS4). Thus for the reverse reaction there *is* a kinetic isotope effect, because the slowest step requires the breaking of an O–H bond.

5.4.2 Triosephosphate isomerase is an evolutionarily perfect enzyme

In Figure 5.39, the product E + G3P is drawn at a higher energy level than the starting material E + DHAP. This is partly because of the relative free energies of the two compounds, but largely because, physiologically, G3P is at a higher concentration than DHAP. It is striking that all the energy barriers and all the troughs have a similar energy.[1] The presumed reason for this effect was noted earlier in this chapter: evolutionary pressure will always act on the highest barriers and lowest troughs, to lower and raise them respectively. Therefore over the course of evolutionary history one might expect that different barriers and troughs will in turn become targets for improvement. Once a barrier is no longer the highest, there is no evolutionary pressure to make it any lower. This will tend to result in the appearance shown in Figure 5.39.

[1] Next time you listen to Handel's *Messiah*, consider the aria "Every valley shall be exalted, and every mountain and hill made low" (Isaiah 40:4): this is a good illustration of the principles described here.

In fact the highest barrier (for the forward reaction) is barrier 4. In the reverse direction, this step consists simply of the diffusion of G3P into the active site

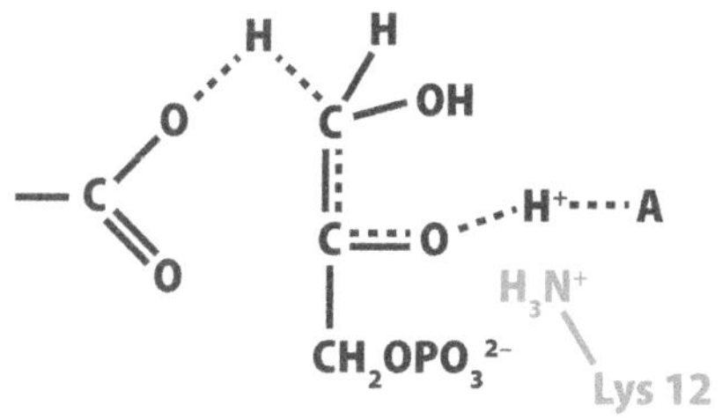

FIGURE 5.38
Electrostatic catalysis is provided from Lys 12.

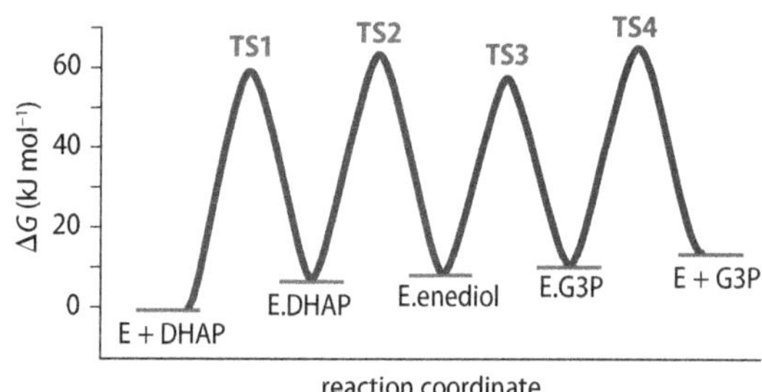

FIGURE 5.39
The energy profile for the reaction catalyzed by triosephosphate isomerase. The difference in free energy between (E + DHAP) and (E + G3P) is determined by the relative concentrations of DHAP and G3P *in vivo*, and the differences in energy for steps 1 and 4 also depend on the concentrations of DHAP and G3P *in vivo*.

***5.17 Second-order rate**

A first-order rate is one that depends only on the concentration of one reactant. For example, radioactive decay depends only on the amount of radioactive substance present, and the hydrolysis rate of a compound in water is dependent only on the concentration of that compound. In both cases we can write rate = k[A], where k is a *first-order* rate constant. (It is a rate *constant* not a rate because it is constant; the rate itself is variable because it depends on the concentration of the compound A.) By contrast, a second-order rate depends on the concentrations of two compounds and on a second-order rate constant. Thus, the rate of reaction of A and B is given by k[A][B], where here k is a second-order rate constant. One way to tell the order of the rate constant is from the units of the constant. The units on both sides of an equation must always match. If a rate is measured in $M\,s^{-1}$ (because it is measuring the rate of change of a concentration) and a concentration in M ($mol\,l^{-1}$), then a first-order rate constant must have units of s^{-1}, whereas a second-order rate constant must have units of $M\,s^{-1}$.

(plus any associated conformational change). It is not difficult to work out whether the back reaction is going at a diffusion-controlled rate, because we know that the diffusion-controlled rate is roughly $10^9\,M^{-1}\,s^{-1}$ (Chapter 4). It just requires a little mathematics, leading to the conclusion that the **second-order rate (*5.17)** constant for a bimolecular reaction is given by the **specificity constant (*5.18)** k_{cat}/K_m.

TIM has a specificity constant of about $2.4 \times 10^8\,s^{-1}\,M^{-1}$, which is essentially at the diffusion limit. This means that the height of the activation energy barrier for TIM is fixed by diffusion rates: it is limited by the physics of how fast molecules can diffuse, and not by how fast the chemical reactions happen. There is thus no way in which the enzyme can work any faster. Evolution has done as much as it can. This is the reason why **Jeremy Knowles (*5.19)** (the enzymologist who did much of this work) described TIM as an evolutionarily perfect enzyme.

***5.18 Specificity constant**

The specificity constant for an enzyme obeying Michaelis–Menten kinetics is given by the ratio k_{cat}/K_m, where k_{cat} is the turnover number; that is, $V_{max}/[E]_0$ (the maximum possible rate per mole of enzyme). It is a useful ratio because it defines how specific an enzyme is for a substrate. A poor substrate can still bind tightly to the enzyme (sometimes more tightly than the optimum substrate), or alternatively it can react rapidly when bound even though it binds very weakly; only by combining the two measurements do we get a good idea of whether it is really a good substrate or not [26].

The specificity constant has another important application, namely to define the apparent second-order rate constant (*5.17) for a bimolecular reaction (for example, enzyme plus a single substrate).

As shown in Section 5.3.2, the Michaelis–Menten equation is

$$V = V_{max}\frac{[S]}{[S]+K_m}$$

This is the most common way for it to be written. However, V_{max} clearly depends on the concentration of the enzyme, $[E]_0$. In fact it can be written as $k_{cat}[E]_0$, where k_{cat} is usually called the *turnover number*, which is the rate at which the active site turns over substrate into product. In this form, the equation is

$$V = \frac{k_{cat}[E]_0[S]}{[S]+K_m}$$

Very often the substrate concentration [S] is quite a lot less than K_m. In this case, [S] + K_m is almost the same as K_m, so we can simplify this to

$$V = \frac{k_{cat}}{K_m}[E]_0[S]$$

This looks very much like the normal rate equation for a second-order reaction: recall that for a reaction A + B → C, the forward rate is given by k[A][B]. Therefore the specificity constant k_{cat}/K_m can be treated like the effective rate constant for the reaction (as long as it was measured under conditions where [S] was significantly less than K_m).

There is a physical limitation to k_{cat}/K_m: it cannot be greater than the collision rate between enzyme and substrate (an enzyme cannot react until the reactant is actually bound to the enzyme). Therefore k_{cat}/K_m provides a measure of how close the enzyme is to being diffusion-controlled; that is, with a rate limited only by the rate at which substrate can bind. Because the bimolecular collision rate is about $5 \times 10^8\,M^{-1}\,s^{-1}$, a diffusion-controlled rate ('evolutionarily perfect', as discussed in Section 5.4.2) implies a k_{cat}/K_m of this order.

***5.19 Jeremy Knowles**

Jeremy Knowles (**Figure 5.19.1**) was a British chemist who spent most of his career at Harvard University, where he investigated catalysis by enzymes, most notably triosephosphate isomerase. He was Dean of the Faculty of Arts and Sciences from 1992 to 2002 and again from 2007 to 2008, and was recognized as an extremely effective administrator.

FIGURE 5.19.1
Jeremy Knowles. (Courtesy of President and Fellows of Harvard College.)

5.5 SUMMARY

For most enzymes, the largest factor leading to catalytic rate enhancements is when the enzyme binds its substrates in a geometry that already has them positioned correctly for reaction; that is, it increases the entropy of the transition state relative to the free substrate. Enzymes are also complementary to the transition state rather than the substrate, which means that they stabilize the transition state and destabilize the bound substrate; in other words, the enzyme also lowers the enthalpy of the transition state. The truth of this is demonstrated by the success of catalytic antibodies, which often use no other mechanism to catalyze reactions.

Enzymes use a variety of mechanisms to catalyze reactions. They very often act as both a general acid and a general base, to protonate a leaving group and activate a nucleophile or base. They also use electrophilic catalysis. Thermolysin is a good example of all these mechanisms. In addition, many enzymes use nucleophilic catalysis. This mechanism implies that the reaction mechanism of the enzyme-catalyzed reaction becomes different from the uncatalyzed reaction. They also use a range of cofactors and coenzymes, particularly to catalyze group transfer reactions, because amino acid side chains do not possess a suitable functionality. Finally, enzymes tightly control the access of water molecules to the active site, not least to prevent the reorientation of water molecules from slowing down the reaction.

The induced-fit model is a good general model to describe how enzymes respond to ligand binding. Thermodynamic considerations show that most enzymes should bind weakly to their substrates, with K_m being lower than [S]. This can be achieved by a separation between the specificity subsite and the reaction subsite.

Triosephosphate isomerase is a good illustration of many of these points. It is particularly interesting because it goes so fast that the limiting step is the diffusion of products off the enzyme: it is thus an evolutionarily perfect enzyme.

5.6 FURTHER READING

The best general book is Structure and Mechanism in Protein Science, by Alan Fersht [26], which is not an easy read but is well worth studying carefully. I also recommend the key paper by W.P. Jencks [7] as well as his book Catalysis in Chemistry and Enzymology (which has the paper reprinted at the end) [2], about which the same could be said. There are a range of books on enzyme mechanisms and catalysis. The major reference book is by C. Walsh, Enzymatic Reaction Mechanisms [27], but somewhat simpler reading is provided by many others, for example Fundamentals of Enzymology by N.C. Price and L. Stevens [28].

Also worth a look is A.J. Kirby and F. Hollfelder's From Enzyme Models to Model Enzymes [29], which is mainly about model enzyme systems but has some very insightful things to say about active sites and mechanisms.

5.7 PROBLEMS

Hints for some of these problems can be found on the book's website.

1. In Section 1.4.9 the comment is made that tyrosine and lysine are unusual amino acid residues to find as general acids for catalyzing a keto–enol tautomerization. Why are they unusual? What would be more normal and why?

2. Explain why most enzyme-catalyzed reactions have pH rate profiles looking something like **Figure 5.40**.

3. If you want to optimize an enzyme-catalyzed reaction, which is the critical number that you need to optimize: k_{cat}, K_m or k_{cat}/K_m?

4. Describe what is meant by *competitive* inhibition and *non-competitive* inhibition. Why is competitive inhibition more common? Non-competitive inhibition is actually rather rare, except by protons. Explain why this should be so.

5. In bacteria and archaea, enzymes comprise about 30–40% of the proteome. Yet in eukaryotes, they comprise only 18–29% of the proteome [30]. Why is this?

6. If an enzyme is going at a diffusion-limited rate in the forward direction, does this mean it will also be going at a diffusion-limited rate in the reverse direction? What would it depend on?

7. The three most common metals found in enzymes are (in order) magnesium, zinc, and iron. Suggest reasons why this might be so. Aluminum is the most abundant metal in the Earth's crust and is the third most common element after oxygen and silicon. Yet aluminum and silicon are very uncommon in proteins. Why?

8. The footnote in Section 1.4.12 introduces a method called *metabolic flux control analysis*, which characterizes how important an enzyme is for regulating the flux through a pathway. One of its main conclusions is that altering the turnover rate of one enzyme in a pathway often causes other rates to change in compensation, with the consequence that the overall flux through the pathway is less affected than you might expect; and that the often-described concept of a *rate-controlling step* is not often really true. (a) Describe metabolic flux control analysis. (b) Explain what a flux control coefficient is. (c) Explain why the flux control coefficient is close to zero for an enzyme that is very fast. (d) Thus, explain why there is no evolutionary pressure on such an enzyme; in particular, why increasing the rate of the reaction catalyzed by triosephosphate isomerase would not increase the rate of glycolysis. See, for example, Cornish-Bowden [31].

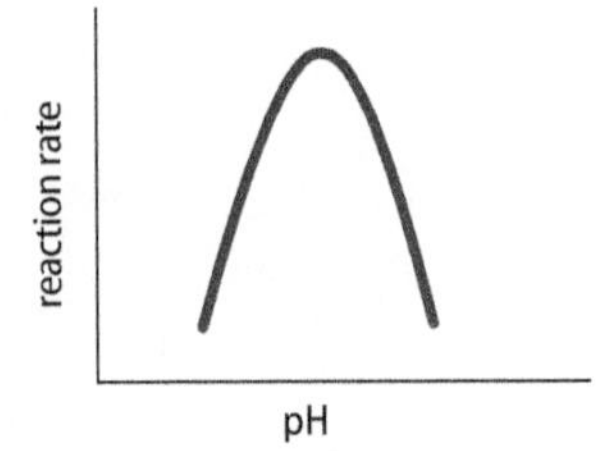

FIGURE 5.40
A typical enzyme pH profile.

5.8 NUMERICAL PROBLEMS

1. From the Arrhenius equation, what fractional rate increase would you expect from a 10° increase in temperature, if the activation energy is 50 kJ mol^{-1}?

2. An enzyme-catalyzed reaction is catalyzed with a general base and a general acid, with pK_a values of 6.5 and 7.5 respectively. Draw the pH rate profile and mark in the pK_a values.

3. Use the Michaelis–Menten equation (Section 5.3.2) to determine how much slower a reaction gets when the substrate concentration is decreased from saturation to (a) K_m, (b) $K_m/2$, (c) $K_m/10$.

4. In the first-order reaction A → B, the concentration of A is 1 mM at zero time, and 0.5 mM after 1 min. What will it be another 20 s later?

5. Enzymes accelerate the rate of reactions by lowering the activation energy. If the rate of reaction is given by the standard Arrhenius equation $k = Ae^{-E_a/RT}$, and E_a is decreased by 4 kJ mol^{-1} (roughly the energy of one hydrogen bond) in the presence of the enzyme, how much faster is the reaction at standard conditions (25 °C)? Assume that A is the same for the uncatalyzed and catalyzed reactions. $R = 8.31\,JK^{-1}\,mol^{-1}$.

5.9 REFERENCES

1. A Kornberg (1989) Never a dull enzyme. *Annu. Rev. Biochem.* 58:1–31.
2. WP Jencks (1987) Catalysis in Chemistry and Enzymology. New York: Dover Publications.
3. E Fischer (1894) Einfluss der Configuration auf die Wirkung der Enzyme. *Chem. Ber.* 27:2985–2993.
4. L Pauling (1948) Chemical achievement and hope for the future. *Am. Sci.* 36:51–58.
5. MI Page (1991) The energetics of intramolecular reactions and enzyme catalysis. *Phil. Trans. R. Soc. Lond. B* 332:149–156.
6. MI Page (1984) The energetics and specificity of enzyme-substrate interactions. In MI Page (ed.) The Chemistry of Enzyme Action. Amsterdam: Elsevier, 1–54.
7. WP Jencks (1975) Binding energy, specificity, and enzymic catalysis: Circe effect. *Adv. Enzymol.* 43:219–410.
8. S Marti, M Roca, J Andrés et al. (2004) Theoretical insights in enzyme catalysis. *Chem. Soc. Rev.* 33:98–107.
9. A Warshel, G Narayszabo, F Sussman & JK Hwang (1989) How do serine proteases really work? *Biochemistry* 28:3629–3637.
10. TS Scanlon & PG Schultz (1991) Recent advances in catalytic antibodies. *Phil. Trans. R. Soc. Lond. B* 332:157–164.
11. MC Ho, JF Menetret, H Tsuruta & KN Allen (2009) The origin of the electrostatic perturbation in acetoacetate decarboxylase. *Nature* 459:393–397.
12. RL Stein (1988) Transition-state structural features for the thermolysin-catalyzed hydrolysis of *N*-(3-[2-furyl]acryloyl)-Gly-LeuNH_2. *J. Am. Chem. Soc.* 110:7907–7908.
13. GL Holliday, DE Almonacid, JBO Mitchell & JM Thornton (2007) The chemistry of protein catalysis. *J. Mol. Biol.* 372:1261–1277.
14. KJ Waldron, JC Rutherford, D Ford & NJ Robinson (2009) Metalloproteins and metal sensing. *Nature* 460:823–830.
15. WR Cannon, SF Singleton & SJ Benkovic (1996) A perspective on biological catalysis. *Nature Struct. Biol.* 3:821–833.
16. DE Koshland (1958) Application of a theory of enzyme specificity to protein synthesis. *Proc. Natl. Acad. Sci. USA* 44:98–104.
17. DJ Vocadlo, GJ Davies, R Laine & SG Withers (2001) Catalysis by hen egg-white lysozyme proceeds via a covalent intermediate. *Nature* 412:835–838.
18. WP Jencks (1997) From chemistry to biochemistry to catalysis to movement. *Annu. Rev. Biochem* 66:1–18.
19. M Štrajbl, A Shurki, M Kato & A Warshel (2003) Apparent NAC effect in chorismate mutase reflects electrostatic transition state stabilization. *J. Am. Chem. Soc.* 125:10228–10237.
20. GR Stockwell & JM Thornton (2006) Conformational diversity of ligands bound to proteins. *J. Mol. Biol.* 356:928–944.
21. L Simón & J Goodman (2010) Enzyme catalysis by hydrogen bonds: the balance between transition state binding and substrate binding in oxyanion holes. *J. Org. Chem.* 75:1831–1840.
22. GJ Narlikar & D Herschlag (1998) Direct demonstration of the catalytic role of binding interactions in an enzymatic reaction. *Biochemistry* 37:9902–9911.
23. GA Kicska, L Long, H Hörig et al. (2001) Immucillin H, a powerful transition-state analog inhibitor of purine nucleoside phosphorylase, selectively inhibits human T lymphocytes. *Proc. Natl. Acad. Sci. USA* 98:4593–4598.
24. JR Knowles (1991) To build an enzyme.... *Phil. Trans. R. Soc. Lond. B* 332:115–121.
25. JR Knowles (1991) Enzyme catalysis: not different, just better. *Nature* 350:121–124.
26. AR Fersht (1999) Structure and Mechanism in Protein Science. New York: WH Freeman.
27. CT Walsh (1979) Enzymatic Reaction Mechanisms. New York: WH Freeman.
28. NC Price & L Stevens (1989) Fundamentals of Enzymology, 2nd ed. Oxford: Oxford University Press.

29. AJ Kirby & F Hollfelder (2009) From Enzyme Models to Model Enzymes. Cambridge, UK: Royal Society of Chemistry.

30. S Freilich, RV Spriggs, RA George et al. (2005) The complement of enzymatic sets in different species. *J. Mol. Biol.* 349:745–763.

31. A Cornish-Bowden (2004) Fundamentals of Enzyme Kinetics, 3rd ed. London: Portland Press, ch 12.

32. EA Komives, LC Chang, E Lolis et al. (1991) Electrophilic catalysis in triosephosphate isomerase: the role of histidine-95. *Biochemistry* 30:3011–3019.

CHAPTER 6

Protein Flexibility and Dynamics

Everything should be made as simple as possible, but no simpler.

Albert Einstein (10 June 1933), The Herbert Spencer Lecture (paraphrased)

From the time that Berzelius coined the term protein, scholars have enjoyed the implied secondary reference of Proteus, the god who went about his business by changing shape and form according to the circumstances.

Gurd & Rothgeb (1979), [1]

Throughout most of this book, protein domains are treated as being rigid and "tool-like." And indeed for many purposes, this treatment is perfectly adequate. However, in reality proteins are not rigid: they have internal motions on a wide range of timescales and distance scales (**Figure 6.1**). This has a number of consequences for the way in which proteins work. The biggest aspect that this affects is enzyme activity. We have already seen the induced-fit model, which requires internal mobility. In this chapter, we see that *conformational selection* (or some balance between the two) is an even better model. A related area is protein:protein interactions, which again are not rigid-body interactions; they require mutual readjustment of the two partners. In many cases, internal mobility is helped by the presence of buried waters, which confer additional flexibility. The theoretical underpinning to much of this is the so-called "new view" of proteins, which sees proteins as a population of conformers, some of which can have structures quite different from the average, and which sees change between conformations not as a simple pathway between two states but as a change in populations. The consequences of these ideas are still being worked through in current research, both experimentally and theoretically, making the subject of this chapter an actively developing field (more than any other in this book except perhaps Chapter 9).

6.1 TIMESCALES AND DISTANCE SCALES OF MOTIONS

6.1.1 Rapid motions are local and uncorrelated

The motions in proteins derive ultimately from thermal energy. A protein is constantly bombarded by water and solute molecules, which induces vibrations and librations (rocking motions) of individual atoms and groups, at rates of about 10^{11} to 10^{14} s^{-1}. The faster movements are bond vibrations; the slower ones are librations. These are very small *harmonic* oscillations, meaning that they vibrate symmetrically about the average position. They are also *uncorrelated*, meaning that vibration of one bond is not related to the motion of another. These motions lead to slightly larger-scale motions, typically rotations of single bonds such as those in exposed amino acid side chains, which occur on timescales of around 10^{-9} s. Motions of this sort occur all over the protein, although single-bond rotations are more restricted inside the protein because of greater steric clashes. Think of a protein rather like Jell-O® (English: jelly): a defined overall shape but wobbling internally.

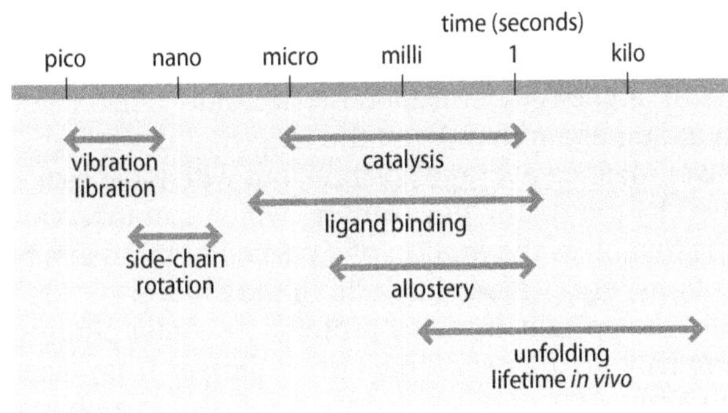

FIGURE 6.1
Protein motions span a very wide range of lifetimes.

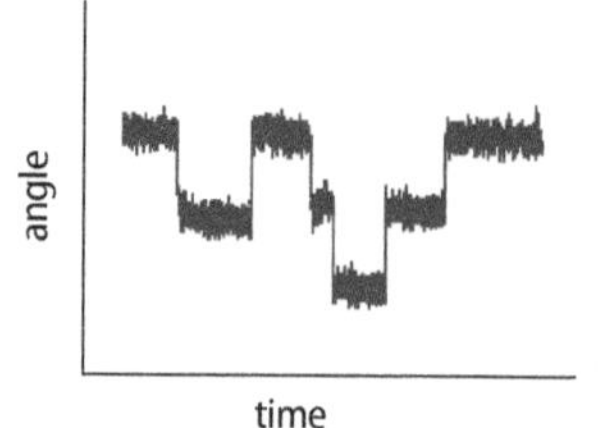

FIGURE 6.2
Typical motions in proteins involve rapid local motion with infrequent ("slow") changes to a different conformation. These "slow" changes involve rapid transitions; they are slow because they do not happen often.

At this point it is important to be clear exactly what this "timescale" statement means, because similar such statements crop up many times throughout this chapter. It means that you will observe a change in side-chain position roughly 10^9 times per second. However, each of these changes is in itself very rapid: the actual movement is perhaps 100 times faster. Thus if you could watch the side chain you would see it wobble around one position for about 10^{-9} seconds and then undergo a very rapid flip to a new position, typically of roughly 120° to a new staggered position (**Figure 6.2**). This is because the energy responsible for the flip comes from the thermal motions, which are on the $10^{11}\,s^{-1}$ timescale or faster. Side-chain rotation is less frequent than bond vibration because it requires a more **correlated** motion: it requires some solvent molecules to move out of the way and therefore to be moving at similar rate in the same direction. Correlated movements are less probable, because they require that several atoms happen to be moving in the same direction at similar rate. This is an example of an important general principle: the more correlated is the motion, the less likely it is, and the slower (less frequent) it is. This is hardly a new idea [2–5], but it gets rediscovered periodically, as indeed do a large number of ideas in science.

Proteins also undergo rigid-body rotations. Most obviously, the whole protein moves and tumbles as a rigid molecule. The **correlation time** for protein rotation is about 10 ns (10^{-8} s), depending on the size of the protein. The correlation time for rotation can be predicted well from the size and shape of the protein. Typically a protein rotates faster in some directions than in others ("anisotropic rotation"), because it is not completely spherical. However, many proteins are fairly close to being spherical and rotate no more than twice as fast in one direction as in another. In two-domain proteins, the two domains have some independence of motion. One can break down the motions into those in which both domains move together and those in which the two domains move in different directions (hinge bending).

These types of motion are well understood, because there are several established experimental methods that can be used to probe them. One of the main such methods is NMR (Section 11.3). A well-established NMR method uses measurements of relaxation rates to derive figures for rigid-body rotation as well as local fluctuations in conformation. The picture that emerges (**Figure 6.3**) is that the N- and C-terminal ends of many proteins have considerable mobility: they wave about essentially completely randomly, buffeted by the solvent. The motion is unrestrained by the rest of the protein if the linker connecting the chain segment to the more rigid part is more than about 10 residues long, with a graduated amount of motion in between. There may also be internal loops that have a higher degree of flexibility, some being completely disordered and others being more ordered. The amount of order depends in a moderately predictable way on the length of the loop and the number of contacts between the loop and the body of the protein. This also applies to the linkers between different domains: some have considerable flexibility whereas others are virtually rigid, depending again mainly on the length of the linker. The rest of the protein has a well-defined average position, but with local oscillations. In NMR this is described by the **order parameter**, which can take values between 0 and 1; 0 means that all of the motion of a particular bond can be described by rapid local motions, and 1 means that all of the motion comes from the rigid-body rotation of the protein. In many proteins the order parameter has values close to 0.9, implying that most of the motion derives from overall tumbling of the protein but about 10% of the motion is more rapid wobbling.

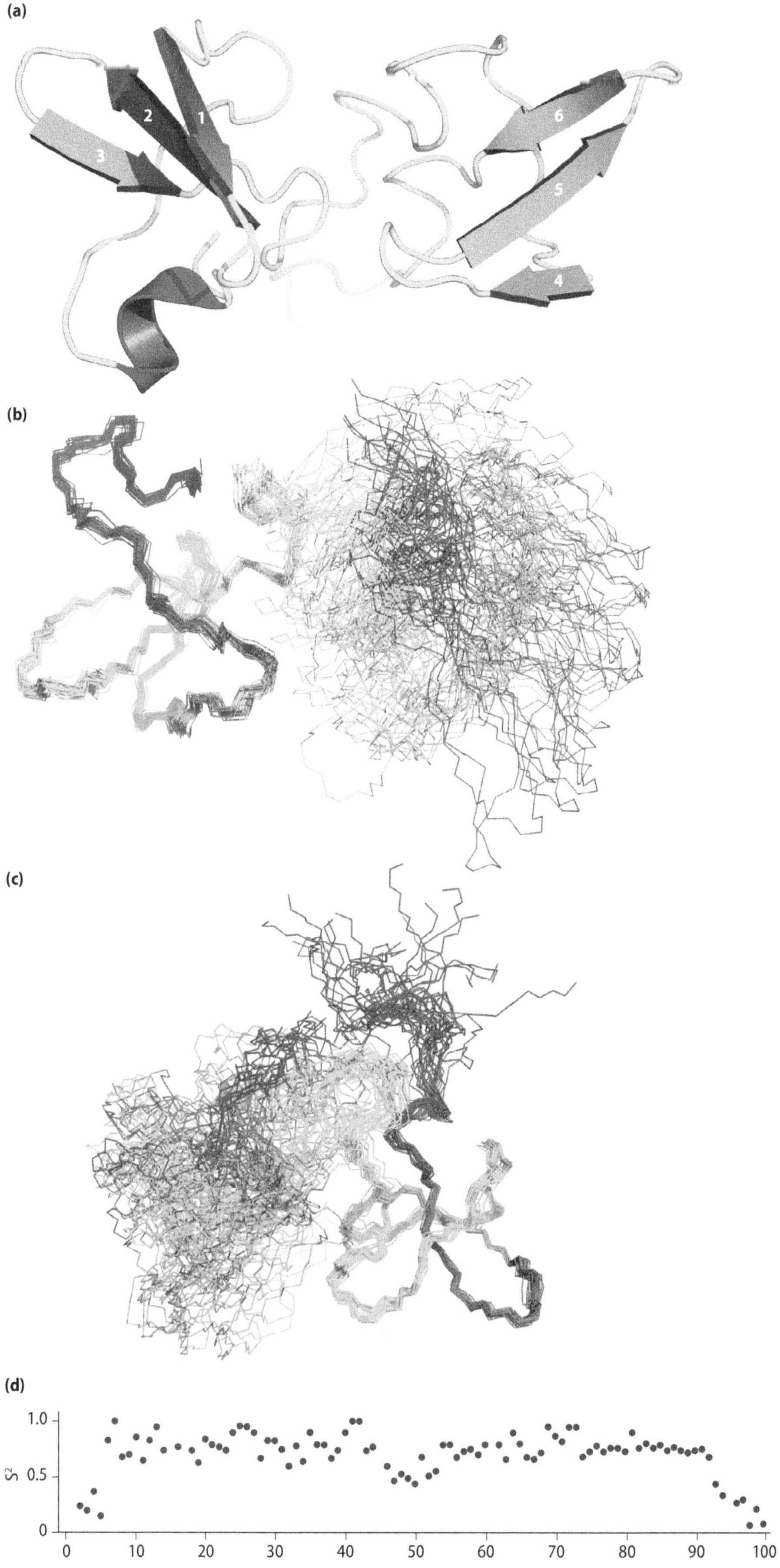

(a)

(b)

(c)

(d)

FIGURE 6.3
Typical NMR-derived results describing the internal dynamics of a protein. The protein is a two-domain docking protein used to link catalytic domains together to form a cooperative multienzyme cellulolytic enzyme called a cellulosome. (a) A cartoon of the structure. (b, c) Overlays of 30 low-energy NMR structures, overlaid on either the N-terminal domain (b) or the C-terminal domain (c). These overlays show that each domain is well defined, but the join between them is not.
(d) The order parameter S^2 is obtained from ^{15}N relaxation data; it characterizes the proportion of the motion of the NH bond vector that comes from slow molecular tumbling, the rest coming from more rapid local motion. It ranges from 0 (rapid unconstrained motion) to 1 (complete rigid body motion). In this protein, S^2 is close to 1 over most of the protein, although with significant variation, and dips below 0.5 at both termini, indicating random coil mobility at both ends. There is a region of lower S^2 in the center of the protein, around residue 50, showing that the two domains have some degree of independent motion, as shown by the NMR-calculated structures. (Adapted from T. Nagy et al., *J. Mol. Biol.* 373: 612–622, 2007. With permission from Elsevier.)

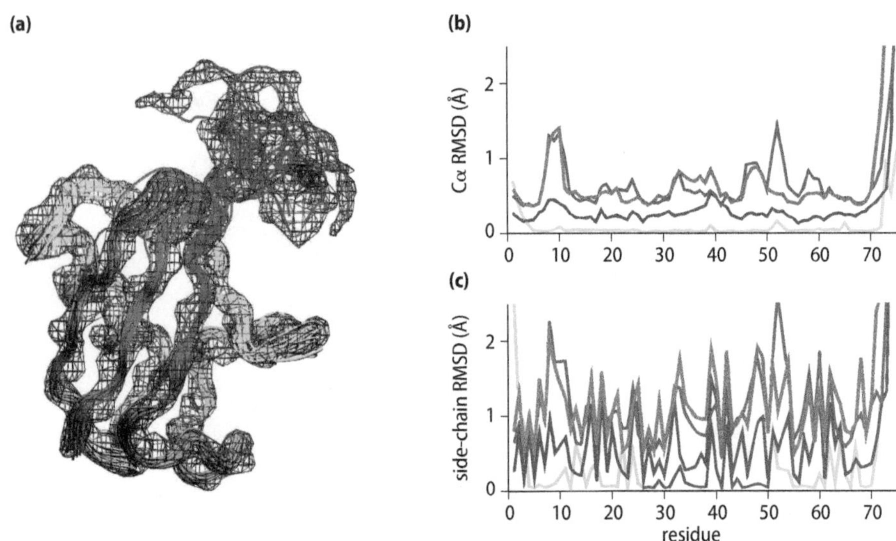

FIGURE 6.4
The structure of ubiquitin, as determined by a molecular dynamics calculation that uses the NMR-derived order parameters as restraints. (a) A backbone trace of 15 representative structures, obtained by using a clustering procedure from a large number of calculated structures. The structures are colored from red at the N terminus to blue at the C terminus and are enclosed within an atomic density map representing the 20% amplitude isosurface of the density of the atoms in the main chain; that is, 80% of the structures are contained within this surface. (b, c) The root mean squared deviation (RMSD) between Cα (b) and side-chain (c) atoms as a function of the sequence. The gray, blue, red, and green lines show, respectively, the RMSD differences calculated from the order-parameter–restrained calculation; an NMR ensemble; a standard molecular dynamics calculation; and a set of X-ray structures. (From M. Vendruscolo and K. Lindorff-Larsen, *Nature* 433:128, 2005. With permission from Macmillan Publishers Ltd.)

6.1.2 Local motions produce global disorder

The discussion of order parameters in the previous section makes it sound as though proteins are almost rigid, but in fact an order parameter of 0.9 still leaves the protein considerable flexibility. As an example, a molecular dynamics simulation was performed for the very stable protein ubiquitin, to see what this means for the instantaneous positions of the atoms. The result is shown in **Figure 6.4** [6]: each backbone atom wobbles by about 1 Å RMSD (root mean squared deviation) from its average position. As bond lengths are in the range 1–1.5 Å, this represents significant motion on a local scale. In this study, as in similar studies by others, the backbone was found to be less mobile than the side chains, which were suggested to be almost liquid-like in their mobility. In particular, this study found that some side chains adopted only a single rotamer throughout the entire simulation, whereas the majority occasionally adopted other rotamers. More recent studies with a larger amount of experimental data came to a similar conclusion, although with a rather smaller spread of backbone positions [7].

The mobility found here (see Figure 6.4) seems to be very large. However, one must remember that many of the motions are correlated. In other words, the motions are not random 1 Å movements: rather, units of structure may all move together. Thus, on a local scale and relatively short timescale, the protein is indeed fairly fixed in its geometry.

There have been several related studies that have considered side-chain rotamers in proteins. Typically, different high-resolution crystal structures of the same protein in different crystal environments show that side chains in the protein interior adopt the same rotamer position in each structure, whereas side chains on the surface are much more variable. NMR results agree well with crystallography, showing that most interior side chains adopt essentially a single rotamer that is the same as the one seen by crystallography, and that side chains that are observed to adopt multiple conformations in solution tend to adopt multiple conformations in different crystal structures [8, 9]. Moreover, as one might expect, rotation of a side-chain methyl group around its axis is very fast (because this rotation requires almost no other displacements nearby to accommodate it), whereas rotation around for example a valine χ_1 dihedral angle (thereby rotating the entire side chain and approximately interchanging the positions of the two methyl groups) is slower (**Figure 6.5**). Such studies imply that side-chain mobility follows a wide distribution of motions and timescales, with no obvious steps or hierarchies [10]. Side chains that are more mobile tend to be attached to more mobile backbones, although the

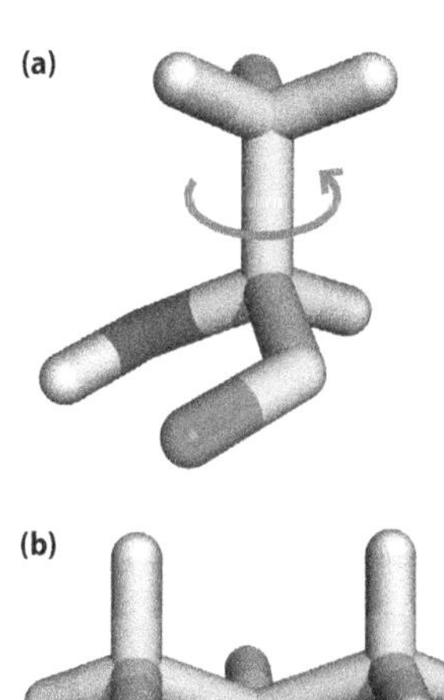

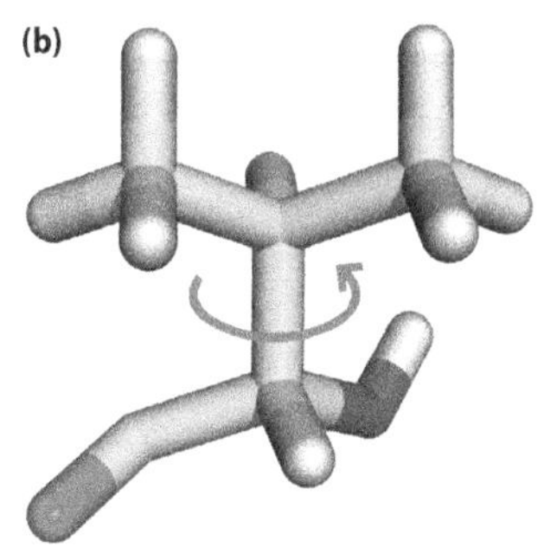

FIGURE 6.5
Side-chain rotation rate depends on steric effects. (a) Rotation around the χ_1 dihedral angle of an alanine residue is fast because it requires very little displacement of neighboring atoms. (b) Rotation around the χ_1 dihedral angle of a valine residue is slower because the side chain is much larger and less symmetrical, and its rotation requires the significant movement of surrounding atoms.

correlation is weak [11]. The slowest motions are those that require the most movement of surrounding groups. This wide distribution of motions is significant: an influential early paper on protein mobility proposed a hierarchical model of protein mobility, in which different magnitudes of motion happen essentially as subpopulations within different tiers, each characterized by different energy barriers and motional scales [12]. Such a model is attractive for theoretical modeling because it allows a protein to be treated like a glassy state. However, it now looks as though motions are less hierarchical and more "fractal": motions occur simultaneously on many size and time scales.

6.1.3 Larger-scale motions are more correlated and are therefore slower

The motions discussed so far are local and rapid, occurring on timescales up to about 10 ns (10^{-8} s). Such motions occur in every protein in a similar way. However, motions can also occur on a much larger scale. These motions are less general, and they occur to different extents and at different rates depending on the nature of the protein. For example, crystal structures of many enzymes indicate large-scale motions that can be characterized as motions of one domain relative to another. Most importantly, these include the kind of induced-fit motions discussed in Chapter 5 that enable an enzyme to close around its substrates and form a catalytically competent enzyme: these are discussed later. Crystal structures do not in general give any indication of the timescale of these motions (Chapter 11). However, kinetic and spectroscopic analyses of enzymes imply that these domain-hinging motions happen at rates of the order of the enzymatic turnover rate, these being often on the millisecond timescale or slightly faster. Many proteins have buried active sites and therefore need a large-scale motion if they are to function, to allow substrates in and out. A classic example is the binding of oxygen to myoglobin. In the crystal structure of myoglobin, there is no route by which oxygen can diffuse through the protein structure into its heme-binding site. There must necessarily be a movement of the protein to allow it to pass through, which at room temperature occurs at a rate of about $10^4\,s^{-1}$ [13]. These kinds of motion are essential to the function of the protein; they are therefore the most biologically important.

NMR has been used to study numerous examples of protein dynamics. The classic example is of the flipping of aromatic rings. Tyrosine and phenylalanine rings are symmetrical about the Cγ–Cζ axis (**Figure 6.6**). A 180° flip around the Cβ–Cγ bond therefore has no effect on the structure of the protein, but it can be observed by NMR because it leads to an averaging of the frequencies of the signals of the two symmetry-related aromatic protons (**Figure 6.7**). Analysis of NMR spectra of proteins shows that aromatic rings flip on a

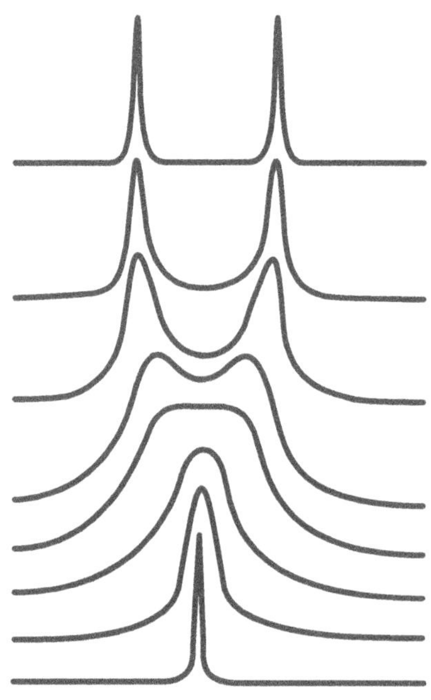

FIGURE 6.7
The appearance of the Hδ/Hδ′ or Hε/Hε′ protons of an aromatic ring depends on the exchange rate around the Cβ–Cγ bond. Slow rotation gives rise to two separate signals (top), whereas fast rotation gives rise to one single averaged signal (bottom). Intermediate rates give rise to various intermediate broadened lines.

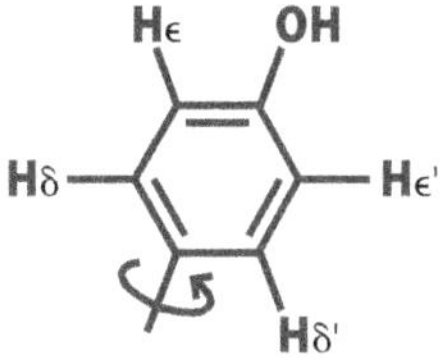

FIGURE 6.6
A 180° rotation of a tyrosine or phenylalanine ring around the Cβ–Cγ bond gives rise to a structure that is chemically identical. However, it can be detected easily with NMR because it gives rise to an averaging of the chemical shift [frequency] of the two symmetry-related protons Hδ and Hδ′ or Hε and Hε′.

wide range of timescales: many flip at rates of $10^6 s^{-1}$, although some flip at rates slower than $10^3 s^{-1}$, these being aromatics deeply buried within the structure [14]. The experimentally determined activation volume for an aromatic ring flip is about 50 Å^3 [15]. Rotation of an aromatic ring requires a concerted movement of the atoms packed against the two faces of the ring, which from calculations have an approximate volume of 50 Å^3, which is consistent with the rate of ring flip being dependent simply on the protein volume that has to be displaced to allow the rotation. The slowest ring flips have been seen in very tightly restricted structures: the slow flips mentioned above were seen in basic pancreatic trypsin inhibitor, a very rigid small protein held together by three disulfide bridges. Thus, although on a fast timescale all proteins behave rather similarly, on slower timescales they behave differently, depending on their packing arrangement.

***6.1 Blind men and the elephant**

A traditional story, probably originally from India, tells of several blind men who were examining an elephant. One felt its legs and declared that the elephant was like a pillar. One felt its tail and declared it was like a rope; one felt its trunk and declared it was like a tree branch; one felt its tusk and said it was like a spear; and so on. The moral of the story depends on what you wish to make of it. For our purposes the moral is closest to the Hindu interpretation: each experimental investigation is incomplete, and we cannot rely on one observation alone to present the complete picture.

6.1.4 Slower motions are more protein-specific than fast motions

It is becoming clear that motion in proteins happens on all timescales, although with very different magnitudes. The picture is not yet clear, but it seems likely that the internal structures of different proteins allow them to convert rapid thermal motions into slower and potentially functionally important motions in different ways. This can be considered as a channeling of many rapid small-scale uncorrelated motions into a small number of highly correlated "hinge-bending" motions. We now consider three cases, hopefully illustrative of the kind of channeling processes that may occur. The types of motion are different. So far, however, it is unclear whether this difference is "real" or merely apparent because different kinds of measurement technique were used—a situation similar to that of the **blind men and the elephant (*6.1)**.

The first example is that of barnase (*2.2), a small ribonuclease. It consists of a β sheet with loops at each end, which are used to bind the substrate and catalyze its hydrolysis. Molecular dynamics simulations show that the protein has large-scale flexibility: the sheet flexes in a hinge-like manner, bringing the loop regions at each end into the appropriate geometry for catalysis [16, 17]. Similar conformational changes are also seen by comparing crystal structures of free and inhibitor-bound protein. NMR studies show that a similar motion occurs in solution, on a timescale of about $1000 s^{-1}$ [18]. However, they also show that there is a faster motion, about 1000 times more frequent (that is, about $10^6 s^{-1}$), which involves the same region of the protein but is less concerted (**Figure 6.8**). Thus, in this case the channeling of rapid uncorrelated motions into infrequent correlated motions occurs by an increased localization of the

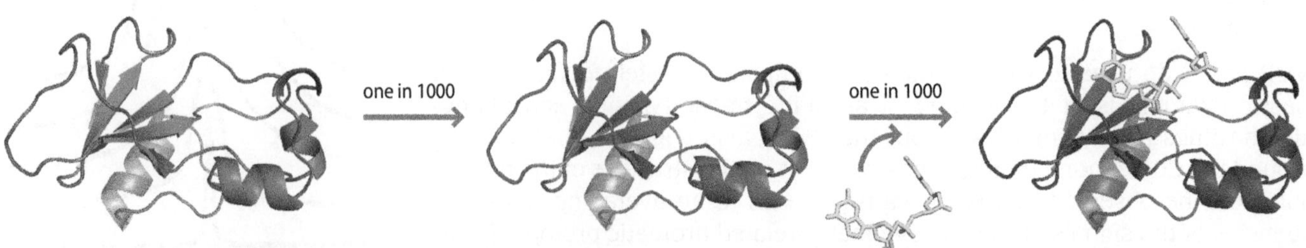

FIGURE 6.8
The mobility of barnase, as deduced from NMR experiments. The enzyme undergoes rapid uncorrelated thermal motions, on a timescale of nanoseconds, produced by the constant bombardment of the protein by solvent. These are colored pink to indicate a uniform small-scale motion. Roughly once in 1000 vibrations; that is, on a microsecond timescale, these random motions happen to occur in a more correlated direction, leading to a larger-scale mobility, corresponding to a flexing in the center of the protein (red, center panel). The red color is intended to indicate a concentration of the thermal motions into a small number of global modes. Once in 1000 of these, and thus on a millisecond timescale, these motions again happen to occur in a more concerted way, leading to a hinge bending motion. This motion is suggested to be the type of productive motion that enables substrate (green) to bind in an active conformation and get cleaved. There is thus a focusing or channeling of random motions into a few modes. (From D.J. Wilton et al., *Biophys. J.* 97:1482–1490, 2009. With permission from Elsevier.)

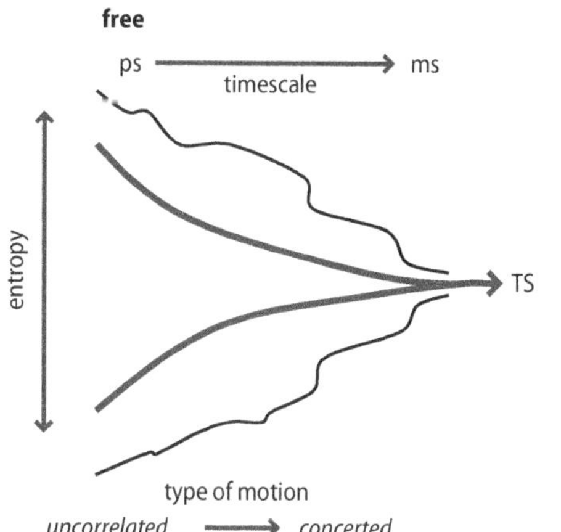

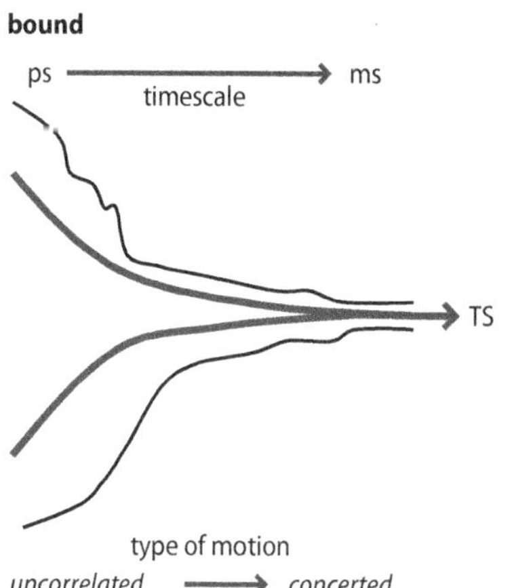

FIGURE 6.9
It is suggested that (in barnase at least, but probably in some form in all enzymes) a channeling of motion is required, in that random uncorrelated motions with high entropy very occasionally come together in an appropriate way to lead to a small number of correlated motions, which push the enzyme into a more catalytically active state. It seems likely (although it remains to be established more clearly) that this channeling occurs in a more focused way in the presence of a substrate or analogous ligand. Effectively this is a restatement of the conformational selection theory described in Section 6.2.

motions to one end of the sheet. On a fast timescale, motions in the protein are uniformly distributed and uncorrelated. Larger-scale motions are progressively less frequent because they require a more concerted movement within the protein. One can therefore draw a parallel between the degree of correlation in the motion, the timescale, and the loss of entropy required (**Figure 6.9**). It is clear that the faster-timescale motions provide the energy for the slower motions [19]. In other words, constant bombardment of the protein by solvent leads to rapid small-scale motions. It is these motions that provide the protein with the kinetic energy it needs to perform the conformational changes required for catalysis; the only real difference between the two types of motion is that the correlated motions are much less probable, explaining why such motions are so much slower.

The second example is another β-sheet protein, the B1 domain from streptococcal Protein G. This is a small protein containing a four-stranded sheet with a helix running diagonally across one face, and it has been studied extensively as a model system. A study of mobility of the peptide bonds in the protein showed that some peptide planes remained approximately static, whereas others displayed a rocking motion (**Figure 6.10**) [20]. This motion is occurring at a rate of about $10^6 s^{-1}$. In particular, it can be seen that the peptides with large rocking amplitudes occur in chains, linking together peptide groups from different β strands that are hydrogen bonded across the sheet. This flexibility was described as a standing wave across the sheet; the location of the standing wave is related to the location of hydrophobic side chains in the protein, such that amino acids with hydrophobic side chains are more firmly anchored, whereas exposed hydrophilic amino acids have greater flexibility. In other words, there is a large-scale flexing of the protein across the β sheet and its location is determined at least in part by hydrophobic packing. Interestingly, the largest amplitude of motion was observed to occur at one edge of the protein, which is the edge that interacts with immunoglobulin G to form a continuous sheet by **β-strand extension (*6.2)**. It was suggested that the motion may be allowing Protein G to flex so as to match up properly against its binding partner; in other words, the slow correlated motion is (as in barnase) likely to be functionally important.

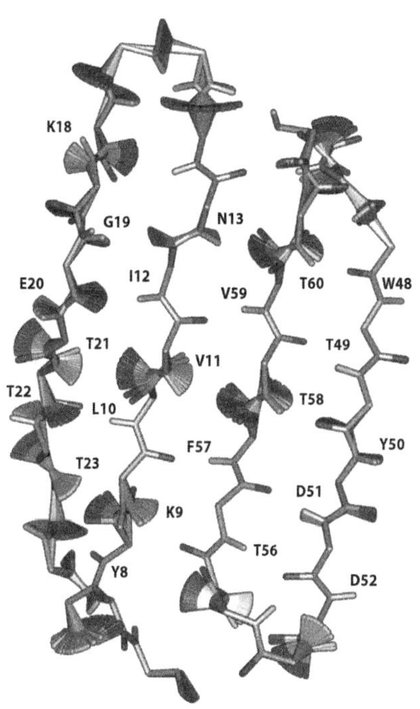

FIGURE 6.10
Correlated rocking motions in Protein G. These are "wobbling" motions of the peptide plane, hinged around the Cα-Cα axis, and describe "standing waves" of oscillation. The standing waves are much larger in some regions than others, and they occur in a correlated way across different strands in the sheet. The more rigid regions are suggested to be stabilized by steric interactions between hydrophobic side chains. (From G. Bouvignies et al. *Proc. Natl. Acad. Sci. USA* 102:13885, 2005. With permission from the National Academy of Sciences.)

*6.2 β-strand extension

A β-sheet protein often has one or both edges of the β sheet exposed. These edges seem to be good sites for interactions with other peptide chains, which can form additional antiparallel strands hydrogen bonded to the sheet [94]. Such interactions are seen in the binding of Protein G to IgG and in the formation of cross-β structure in amyloid. A nice example is found in a surface-associated protein from the pathogenic bacterium *Staphylococcus aureus* that binds to fibronectin: the interaction attaches the bacterium to human cells and is used to help it invade them. Fibronectin contains multiple tandem fibronectin type I modules (Chapter 2), which form a roughly continuous β strand on one face. The bacterial protein FnBPA forms an extra β strand that zips up along the β-strand face of fibronectin, allowing it to bind with high affinity (**Figure 6.2.1**).

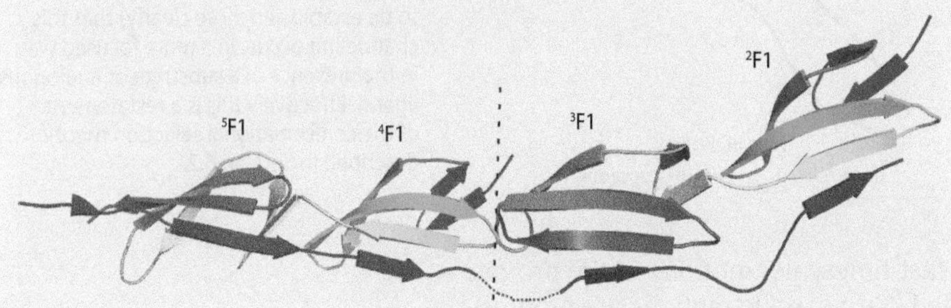

FIGURE 6.2.1
The structures of fibronectin type I domains 2, 3 and 4, 5 bound to *Staphylococcus aureus* FnBPA peptides (PDB files 2rkz and 2rky, respectively) [100]. The two pairs were crystallized separately and have been placed in a suitable orientation. The join is indicated by the dashed lines. The FnBPA peptide (red) forms an extension to the sheets of both domain pairs, presumably explaining its high affinity.

A subsequent study from the same group derived the solution ensemble shown in **Figure 6.11**, which aims at representing the distribution of structures that could be found if one were able to take snapshots of the protein over the course of several seconds of real time; that is, long enough for it to sample all accessible conformations. This ensemble has an RMSD spread of backbone coordinates of 0.7 Å, roughly similar to the ubiquitin calculation discussed above; however, because the Protein G calculation includes the slower-scale motions as well as rapid ones, one might expect it to have a larger RMSD than the ubiquitin calculations. Thus, it actually represents a significantly tighter spread of structures than in the ubiquitin study [21]. The study allowed the authors to propose a model for the energy landscape of Protein G, which contains a relatively large flattish area at the bottom with a "diameter" of about 1.3 Å, with steep and smooth sides (**Figure 6.12**). Other studies on different proteins have shown similar patterns, suggesting that the greatest large-scale flexibility is located at the active site [22–25]. This of course is what one might expect, given the requirement for enzymes to be complementary to the transition state rather than the substrate or product. That is, the structure of an enzyme keeps it rigid but allows segmental mobility (correlated motions in functionally relevant directions) around the active site—such as hinge bending.

The third example is ubiquitin (*4.9), which has a structure similar to that of Protein G, consisting of a β sheet covered on one face by a helix. In the cell, ubiquitin interacts with a large range of other proteins. A slow fluctuation (on a timescale between 10^9 and $10^6\,s^{-1}$) was characterized in ubiquitin, in which a loop between two strands and the C-terminal end of the protein can adopt a

FIGURE 6.11
Ensemble of structures of the B3 domain of Protein G, obtained with molecular dynamics simulations restrained by NMR data. Although this figure looks similar to many NMR protein structures, what it is showing is very significantly different. The spread of structures in the typical NMR ensemble (as discussed in Chapter 11) represents how similar different attempts at a "best" solution are to each other. By contrast, the spread of structures in this figure represents the authors' estimate of the true spread of positions during the course of the molecular dynamics trajectory. In other words, the spread of structures shown here represents the thermally accessible spread at ambient temperature, and not the (experimentally restrained) width of the minimum. (From P.R.L. Markwick, G. Bouvignies and M. Blackledge, *J. Am. Chem. Soc.* 129: 4724, 2007. With permission from the American Chemical Society.)

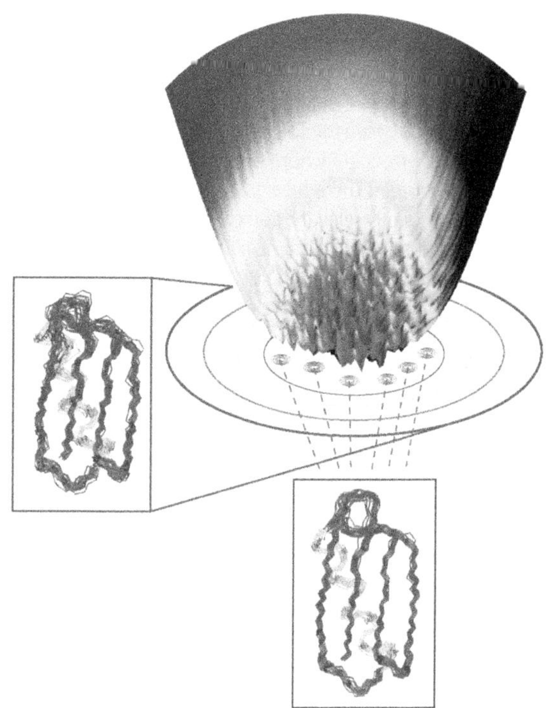

FIGURE 6.12
A schematic potential-energy landscape of the B3 domain of Protein G. This is suggested to be a steep-sided well, but with a fairly large and flat bottom that contains several low-energy conformations. Each of these conformations is a tightly grouped set of conformations, as indicated by the representative ensemble in the blue box. The entire thermally accessible conformational space is indicated by the left-hand structure in the red box, which is the same as that shown in Figure 6.11. (From P.R.L. Markwick, G. Bouvignies and M. Blackledge, *J. Am. Chem. Soc.* 129:4724, 2007. With permission from the American Chemical Society.)

range of structures, described as a pincer-like motion (**Figure 6.13**) [26]. This fluctuation allows the surface of ubiquitin to take up a range of conformations, appropriate for binding to the different complementary surfaces of its binding partners.

This survey of slower motions is far from complete, reflecting our current limited understanding. However, it is becoming clear that (by contrast with rapid motions) these motions are different in different proteins and are clearly related to their function, not only in their locations but also in their timescales. They are therefore an evolved property of the protein. In other words, it is not only the structure of a protein that is important for its function, it is also the slow dynamics. There is, for example, emerging evidence that different members of the PDZ family of signaling modules have evolved to have very different motional properties (relevant to their function) despite very similar sequences [27]. By contrast, the rapid dynamics common to all proteins apparently has rather little connection to function [28].

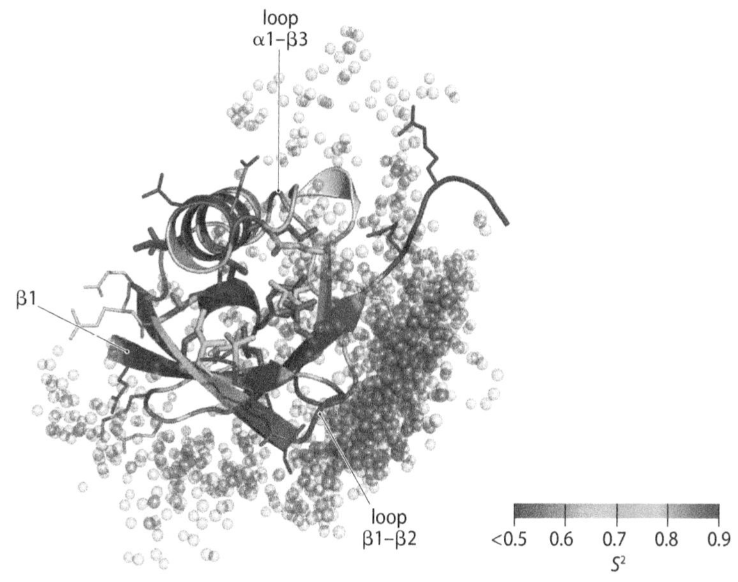

FIGURE 6.13
Slow dynamical fluctuation of ubiquitin. The flexibility of the protein backbone is colored from blue (least) to red (greatest). In particular, the β1–β2 loop is found to be the most mobile region, on a microsecond to millisecond timescale. The gray spheres indicate the locations of protein atoms that contact ubiquitin in a range of crystal structures. The main region of contact for ubiquitin is at the bottom right, coincident with the region of greatest mobility. In other words, ubiquitin contacts its partners mainly via a region of the protein that displays the largest flexibility on a slow (microsecond to millisecond) timescale. (From O.F. Lange et al., *Science* 320:1471–1475, 2008. With permission from AAAS.)

***6.3 Normal modes**

The oscillation of a weight on the end of an ideal spring can be described as *harmonic*: that is, the restoring force is proportional to the displacement, as a consequence of which there is a simple sinusoidal motion, in which the period of oscillation does not depend on the amplitude of motion but only on the *spring constant* (that is, the stiffness of the spring) and the mass of the weight. This means that given the value of the spring constant, one can easily calculate the oscillatory motion of the weight. One can treat a protein as a coupled oscillatory system, because the forces between all atoms can be calculated reasonably accurately. One has to assume that the motions are harmonic (parabolic), which is something of a big approximation; however, given this assumption, then a system containing n atoms will give rise to n calculated patterns of coupled harmonic motion, which are called normal modes. They can be calculated even for a large protein by fairly straightforward matrix methods. Many of these describe local rapid oscillations, but the lowest-frequency normal modes are collective motions of much of the protein. There have been numerous reports suggesting that the functionally relevant deformations of proteins are closely similar to one of the low-frequency normal modes. For example, the motions of ubiquitin discussed in Section 6.1.4 are similar to the normal modes. Normal modes can be calculated straightforwardly from a protein structure, and indeed a database is available, linked to the Chime molecular graphics system (http://cube.socs.waseda.ac.jp/pages/jsp/index.jsp) [95].

The motions discussed here are probably closely related to **normal modes (*6.3)**, a relatively straightforwardly calculated property. Thus, it may turn out that these functionally important motions may be fairly readily and accurately predicted.

6.1.5 Correlated motions can occur over several hydrogen bonds

It is of considerable interest and importance to ask how large the correlated units are in proteins: over what kind of distance might one reasonably expect correlated motion? This is important because the sites of correlated motion represent the rigid blocks of a protein that might be able to transmit conformational change. As yet there is no clear answer to the question, but there are some clues, from both experiment and simulations.

Helices in membrane proteins can transmit information from one side of the membrane to the other, roughly four helical turns long. In agreement with their more "mechanical" role (Chapter 1), helices are the strongest and most rigid structural units. However, correlated motions in hydrogen-bonded networks probably normally extend over no more than four or five hydrogen bonds. For example, a recent study on Protein G showed that removal of a side-chain carboxyl from the N-terminal end of a helix (by a decrease in pH) led to cooperative loosening only of the first turn of the helix, whereas protonation of another carboxylate led to weakening of a set of about eight hydrogen bonds across three strands of a β sheet [29]. The boundaries to the correlated effects were defined by the "standing wave" fluctuations described earlier in this chapter (see Figure 6.10) [20]. The conclusion to be drawn is that hydrophobic packing and hydrogen bonds act cooperatively to define structural units, which can extend over reasonably large areas of a protein, but usually much less than the typical size of a domain. The importance of hydrogen bonds is that they are highly directional and are associated with a large change in overall energy if broken. The well-structured units are closely related to the hydrophobic core of proteins. It would therefore be reasonably accurate to say that proteins are made of rather rigid and highly correlated hydrophobic cores, separated by more hydrophilic and mobile channels.

Computer simulations are more ambiguous. Some suggest reasonably well localized cooperative blocks, which increase in size as the protein becomes less stable [30], but others have implied that dynamic correlations can lead to distant regions becoming correlated: this is discussed further below. Hydrogen bonds are again seen to be important in defining the cooperative units, presumably because they are the most strongly directional of local interactions [31].

6.2 CONFORMATIONAL SELECTION

6.2.1 Proteins populate a conformational landscape

There is currently considerable interest in protein mobility, which one can ascribe to the fact that both computational and experimental methods have recently become developed enough to be applicable to real proteins, and thus for the first time we are able to provide answers to interesting questions. One outcome of this interest is the popularization of the view that proteins exist not as a single conformation but as a dynamic ensemble. In the language used in Section 1.3.4, one can represent the ensemble by a conformational energy diagram (**Figure 6.14**), this being a very simplified representation in two dimensions of the very high-dimensional conformational space accessible to a protein. As discussed above, the surface of this energy diagram is probably fractal: it is rough on every distance scale. Proteins will populate such a landscape in a Boltzmann distribution. Thus, low-energy barriers are crossed easily, and proteins can adopt many similar conformations, for example those in which exposed side chains have different orientations or in which there are minor changes in backbone angles. Larger conformational changes, such as the rotation of buried side chains or the movement of a hinged loop on the surface, have larger energy barriers that are crossed less frequently, depending again on a Boltzmann distribution. This is often described as an *energy landscape.*

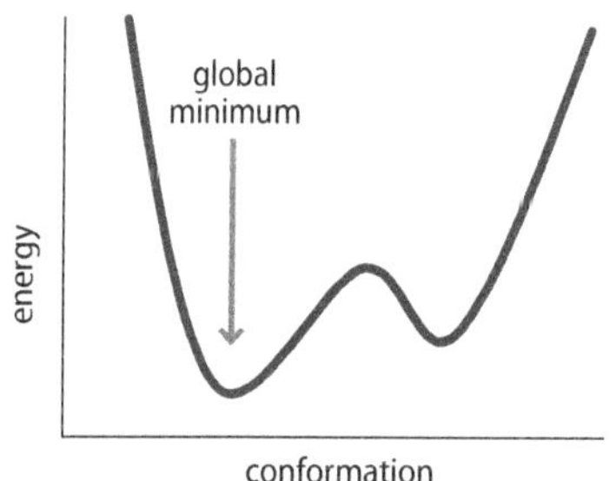

FIGURE 6.14
A conformational energy diagram of a protein, or energy landscape. The vertical axis is free energy, conformational energy, or potential energy, and the horizontal axis is a highly simplified representation of the true multidimensional conformational space. In this figure there is a global minimum, but there is also a second minimum with an energy barrier to the global minimum. The protein will populate the energy diagram in accordance with the Boltzmann distribution. Therefore, depending on the size of the energy barrier and the difference in energy between the two minima, it will populate both minima and will be in equilibrium between both states.

Such ideas have been important in the development of a conceptual framework for protein folding. Cyrus Levinthal noted in 1968 that the number of possible conformations of a protein is astronomically large [32]. This means that it is not possible for a protein to search every possible conformation to find the global energy minimum, because it would take far too long. He therefore proposed that proteins fold on a pathway, which places increasingly tight limits on the conformations allowable to the protein as it folds and so guides the folding. A large body of research since then has tried to characterize such pathways. This has not proved easy, very largely because folding is such a highly cooperative process. A protein can be observed in a folded state or an unfolded state, but observation of any intermediate states is difficult or impossible (Section 4.5.1).

Proteins can be denatured by chemicals such as urea or guanidine, or partly denatured by extremes of pH. The "denatured state" that can most easily be studied is therefore either chemically different from the folded protein, being either highly protonated or highly deprotonated, or else bound to urea or guanidine, which makes proteins more extended than is expected for a random-coil protein [33]. This means that the unfolded state that is normally studied is not in the same solution conditions as the folded state (which is studied at more neutral pH, in the absence of denaturants). Most studies of denatured proteins therefore require some extrapolation to be able to make conclusions about folding processes under normal conditions, adding further uncertainty to our understanding of the unfolding equilibrium.

The general picture emerging is that an unfolded protein is far from being random. One of the main driving forces for protein folding is hydrophobic packing. In an "unfolded" protein, hydrophobic groups tend to associate together, producing a protein that is much more compact than a truly random-coil chain and has already lost a significant fraction of the entropy that has to be lost in forming the folded protein. (Significantly, natively unstructured proteins have a lower proportion of hydrophobic side chains than natively folded ones. They are therefore very likely to be much more like true random coil in solution.) Protein folding is often studied experimentally by a sudden dilution of denaturants or a sudden change in pH. Under these conditions, the protein backbone very rapidly (faster than microseconds) undergoes a hydrophobic collapse, after which the secondary structure elements start to form. In general, helices form more rapidly than sheets, because their folding is entirely a local structural rearrangement, whereas sheet formation requires interactions between amino acids that are farther apart in the sequence.

A key insight arising from the landscape view is that proteins do not fold along a single pathway. When an unfolded protein is produced as described above, by rapid dilution of denaturants or pH change, different molecules will be in different conformations. Each molecule finds itself in a different place on the multidimensional "folding funnel" (**Figure 6.15**) and finds its own way down to the bottom, representing the folded protein. This naturally means that many proteins will pass through the same region of the folding funnel on their way to the bottom, and end up following similar routes; many proteins will therefore fold on a similar pathway, although in detail they will follow different routes and have different energy barriers on the way. However, there will be other molecules that follow different routes. It is thus too simple a picture to think of a folding pathway. The more sophisticated model described here has arisen largely as a consequence of computer simulations of folding and is often called the "new view" of protein folding [34].

This model has been discussed at some length because it applies not just to folding but also to the conformations of "folded" proteins. In the same way that different proteins fold up in different ways, so folded proteins can adopt a range of conformations. And just as proteins fold along different pathways, so the transition from one "conformation" to another, for example as a consequence of ligand binding, is not simply a matter of going from one energy well to another across a transition state; the process needs to be described by a set of pathways, many of which will pass across a similar transition state,

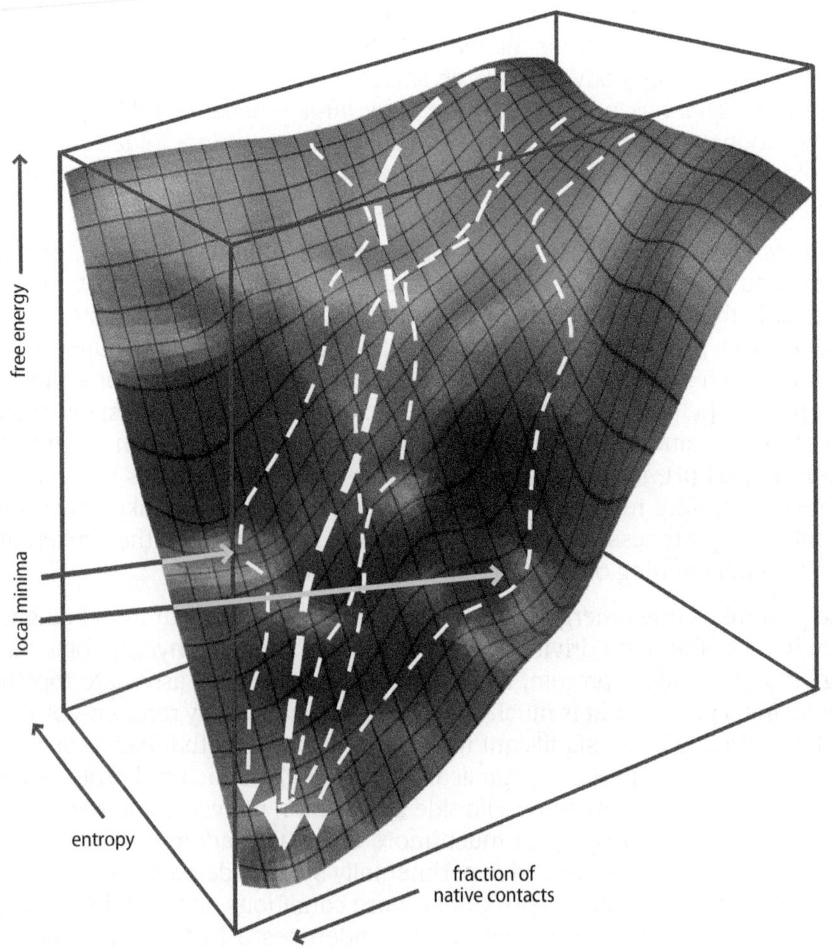

FIGURE 6.15
The folding funnel. This diagram represents only a fraction of the total funnel, which in reality is much more complicated. As the free energy of a protein conformation decreases, its entropy tends to decrease and the fraction of native contacts (that is, the contacts that are present in the native protein) increases. In the unfolded state, the protein can be anywhere on the conformational surface, but as it folds it tends to follow one or a relatively small number of pathways, indicated by the dashed lines. Thus, there is no single folding pathway, although there may be relatively few routes populated to any significant extent. Many of these go smoothly down to the global minimum (for example the thicker line), but a few fall into local minima as they pass, and some may stay there, leading at least to a slower route for some molecules to the global minimum. These data are for the B1 domain of Protein G [97], but many other single-domain proteins probably fold in a similar way, at least qualitatively.

but some of which may pass across different routes [35]. The situation is rather like the wave/particle duality found in physics. Rather than thinking of a protein as a single object that follows a defined path from one conformation to another, we should think of it as a wavelet: a bunch of conformations, which move collectively from one location to another. In physics, it is in practice more convenient to treat an electron as a particle for some applications, and as a wave in others. In the same way, most of the time (and throughout the rest of this book) we can think of a protein as a macromolecule with a single defined conformation, but we must always remember that in reality it exists as a set of conformations, a small proportion of which can be very different from the average. Most strikingly, as noted in Chapter 4, a small fraction (between about 10^{-4} and 10^{-11}) of proteins at any one time is unfolded.

6.2.2 Conformational selection is a better model than induced fit

In the previous chapter, we saw that **induced fit** is a better description of how enzymes work than is **lock and key**. The induced-fit model recognizes that proteins are not rigid, and they accommodate themselves to their bound ligand. An enzyme often has to change its structure during the reaction, for example to allow the substrate to bind or to become complementary to the transition state [36]. Numerous kinetic studies have implicated the existence of some sort of activated conformation, represented by E* in the typical reaction

$$E + S \rightleftharpoons ES \rightleftharpoons E^*S \rightarrow E^*P \rightleftharpoons EP \rightleftharpoons E + P$$

It is assumed that E represents the ground-state conformation, and E* represents some activated (induced-fit, closed, or, in the MWC allosteric model, R) conformation.

The landscape view of protein conformation encourages us to look again at this model. In the landscape view, a protein cannot be viewed simply as being "in the ground-state conformation": it is an ensemble of structures, many of which resemble the ground state (widely assumed to be roughly equivalent to the crystal structure) but some of which will be significantly different, including structures resembling the activated conformation. It may therefore be more appropriate to think of the change in enzyme conformation not as an induced fit, in which the substrate binds to free enzyme and forces it to change conformation, but as **conformational selection**, also known by a range of other names, including population shift and pre-existing equilibrium [37]. In this model, the free enzyme samples a range of conformations, including the activated conformation, and the binding of substrate *selects* conformations that provide a favorable free energy of binding from this ensemble. Binding to such a conformation leads to a redistribution of conformations around the conformational space available, creating more molecules in the activated conformation [38]. Therefore there is no binding-induced structural change as such: the structural change has already occurred before binding. This idea is in fact inherent in the MWC allosteric model, as discussed in Chapter 3. Recently, numerous examples have been described of such processes, including the case of ubiquitin, described above [26], and calmodulin [39]. It therefore has a good experimental basis. Conformational selection is also well supported by computer calculations. For example, a study looked at four protein protein complexes for which there are crystal structures of both free and bound protein [40]. In each case, the bound conformation differs from the free. Calculations on the free protein, however, showed that the motions exhibiting the largest conformational change are correlated with the structural change on binding; in other words, the free protein already includes the bound conformation within its structural ensemble. However, there is likely to be some subsequent rearrangement of the two partners once bound, as discussed in the next section.

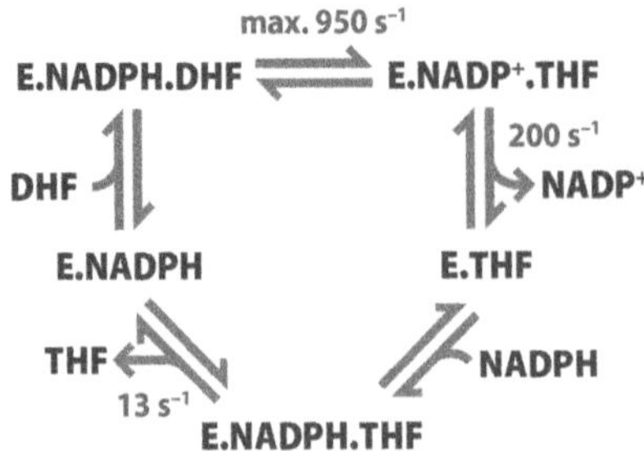

FIGURE 6.16
The kinetically most important catalytic cycle for dihydrofolate reductase (E). The main catalytic reaction, the reduction of dihydrofolate to tetrahydrofolate using NADPH as cofactor, is shown at the top; the species at top left is therefore the "Michaelis complex," and that at top right is the product. There is then a preferred route for recharging the enzyme with its two substrates, each step setting up the enzyme for the subsequent reaction. The rate constants for three of the steps, measured with pre-steady-state kinetics at 298 K and pH 6, are indicated in red.

There are some strong arguments in favor of such a model. One is kinetic [41]: if binding has to induce a conformational change, there is a limit to how fast reactions can occur. This is governed by the rate at which conformational change can occur and by the fraction of bound enzyme that will be in the activated conformation at equilibrium. Calculations suggest that for many enzymes these induced-fit changes are too slow to be consistent with observation.

Another argument is based on experimental observations of enzymes during their catalytic cycles. The clearest example is dihydrofolate reductase (DHFR), which catalyzes the reduction of dihydrofolate to tetrahydrofolate (THF) using NADPH as the reducing cofactor; it is an important metabolic enzyme because the THF is used as a source of methyl groups, for example in the biosynthesis of thymine. The main physiological catalytic cycle of DHFR is shown in **Figure 6.16** [42]. Several of the stages in the cycle can be characterized spectroscopically, meaning that it is possible to measure the rate at which they interconvert. These are indicated by the red numbers in Figure 6.16: they occur relatively slowly. This includes the catalytic step, the process of hydride transfer itself, a point to which we shall return shortly.

The intermediates in the cycle can all be studied in isolation and characterized in solution. Strikingly, for *each* intermediate, it was shown to adopt at least one alternative conformation, which in each case resembled the ground state of either the next stage in the cycle or the previous stage in the cycle or both. Even more remarkably, the rate of conformational exchange between the ground state and the alternative conformation matches well with the observed rate of movement around the cycle, where these are known. The very strong implication of this is that at each stage in the reaction cycle, conformational selection is operating, rather than induced fit. Thus, at each stage the enzyme is in equilibrium with the next stage, and the rate at which the next stage is reached is dependent on the rate of this conformational exchange [43]. In particular, the rate of catalytic turnover of DHFR is not the hydride transfer reaction but the release of THF, which in turn is reliant on the rate of protein opening [11].

DHFR is therefore a convincing example of conformational selection rather than induced fit. Importantly, this is true for the catalytic step just as much as the others. In particular, *the turnover rate for catalysis is determined not by the rate of the chemical reaction but by the rate at which the enzyme alters conformation around it* [44–46]. This is an important result, and is probably applicable to a wide range of enzymes, including cyclophilin A [47], ribonuclease [48], phosphoglucomutase, and others, reviewed in Boehr et al. [11]. A very clear example is protochlorophyllide reductase, an enzyme that is homologous to DHFR and has a closely related catalytic cycle. This enzyme is notable as one of only two enzymes known whose catalytic step is driven by light, appropriately for this enzyme because it is one of the enzymes on the biosynthetic pathway of chlorophyll. For this enzyme it was possible to characterize all the steps spectroscopically, and it was found that the initial hydride transfer occurs on a picosecond timescale, which is many orders of magnitude faster than the overall reaction rate [49]. The overall reaction rate is constrained by conformational change of the enzyme, as shown by low-temperature studies, which permitted a separation of the catalytic step from subsequent conformational changes [50].

In this context, it is relevant to note that a wide variety of enzymes have turnover rates within a rather limited range, from about 50 to 5000 s^{-1}. This is despite the fact that the uncatalyzed rates span an enormous range (**Figure 6.17**) [51]. It is tempting to speculate that, for most (if not all) enzymes, the limiting factor for the turnover rate is conformational change within the enzyme rather than the chemical reaction itself.

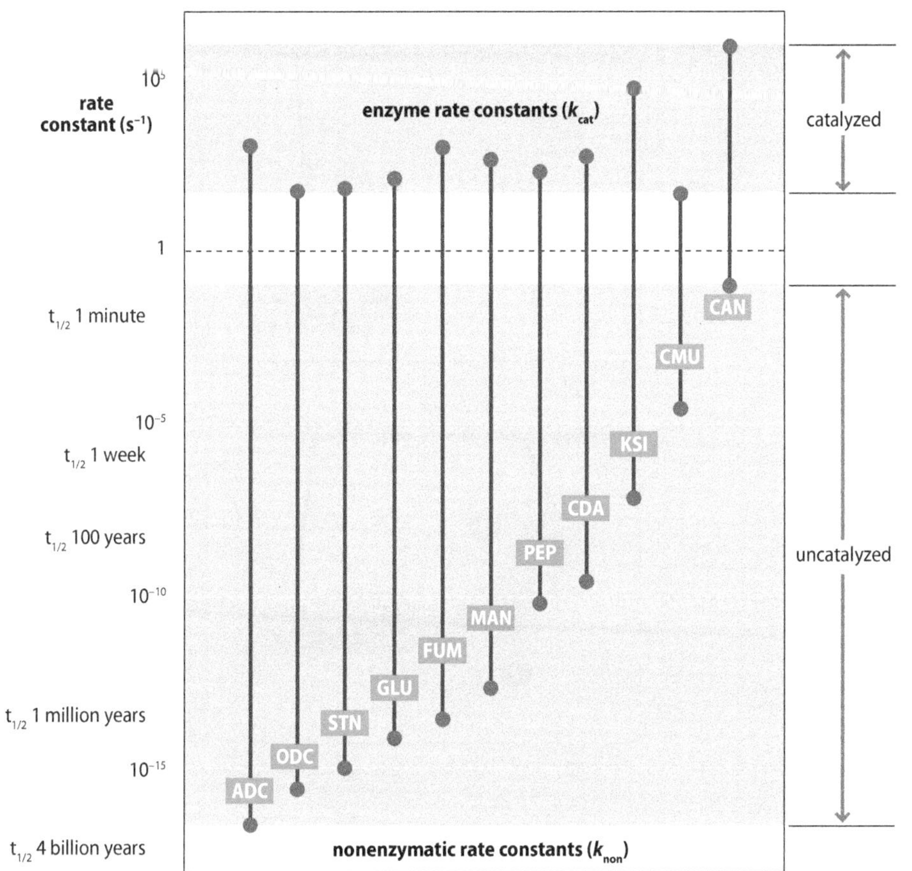

FIGURE 6.17
A comparison of the rate constants for reactions uncatalyzed and enzyme-catalyzed at 25 °C. The labels are: ADC, arginine decarboxylase/oxygenase; ODC, orotidine 5′-phosphate decarboxylase/oxygenase; STN, staphylococcal nuclease; GLU, sweet potato β-amylase; FUM, fumarase; MAN, mandelate racemase; PEP, carboxypeptidase B; CDA, *Escherichia coli* cytidine deaminase; KSI, ketosteroid isomerase; CMU, chorismate mutase; CAN, carbonic anhydrase. (From R. Wolfenden and M.J. Snider, *Acc. Chem. Res.* 34: 938–945, 2001. With permission from the American Chemical Society.)

6.2.3 Conformational selection and induced fit are two ends of a continuum

When conformational selection first emerged, it looked like a major improvement on induced fit, and advocates of the former presented it as an answer to all questions of protein binding [52]. However, it is now becoming clear that these two mechanisms are not alternatives; rather, they represent two ends of a continuum of possibilities.

Consider the possible stages by which a ligand may bind to a protein. This can be represented by a conformational energy diagram of the type drawn previously as Figure 6.14. This diagram is shown in **Figure 6.18**, which indicates that there are two routes by which free enzyme plus ligand can arrive at the bound state. One involves a change in conformation of the enzyme, or the creation of an ensemble of states of different conformations, and the selection of a suitable conformation that is complementary to the ligand, followed by binding of the ligand to this conformation (blue arrows). The other involves ligand binding followed by an induced-fit structural change of the enzyme (green arrows). However, this is something of a simplification, and it is more complete to draw

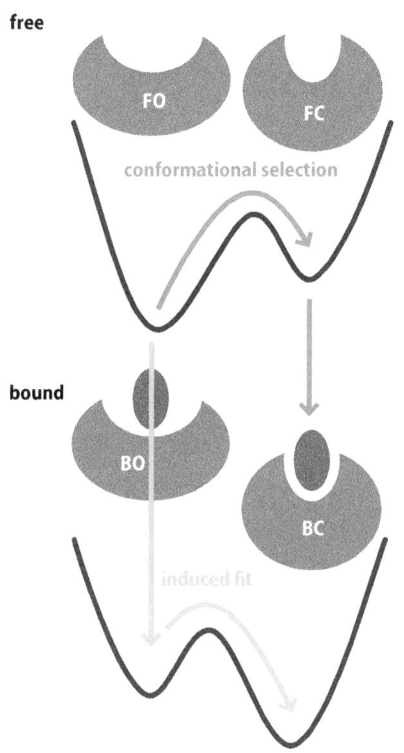

FIGURE 6.18
Energy landscape for the binding of a ligand to an enzyme. In this simple diagram, the enzyme has two conformations, O (open) and C (closed). The enzyme is in equilibrium between these two conformations, the position of the equilibrium being different depending on whether the enzyme is free (F) or bound (B) to ligand. There are two possible routes to get from the free open conformation at top left to the closed bound conformation at bottom right. Either the enzyme changes conformation first and then binds (*conformational selection*, blue arrows) or it binds first followed by a conformational change (*induced fit*, green arrows).

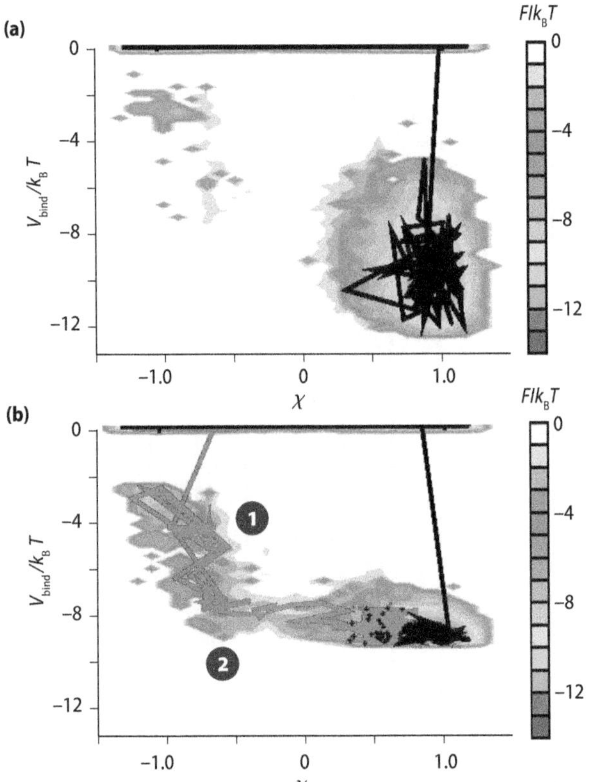

FIGURE 6.19
Two-dimensional energy surface for the binding of an enzyme to a ligand. The horizontal axis (χ) is conformation, and the vertical axis is the binding energy, ranging from 0 (free) to about $-9k_BT$. Thus, in this figure (by contrast with Figure 6.18), binding is not all or nothing but is a continuum. (a) A pure conformational selection, in which the conformation changes first, followed by binding. (b) A mainly induced-fit binding, in which the preferred route is for binding first followed by conformational change. These panels represent binding trajectories simulated for the binding of glutamine to glutamine-binding protein. Some trajectories in (b) followed a conformational selection route, shown by the black lines. The points labeled 1 and 2 are discussed in the text. (From K. Okazaki and S. Takada, *Proc. Natl. Acad. Sci. USA* 105:11182–11187, 2008. With permission from the National Academy of Sciences.)

a two-dimensional surface (**Figure 6.19**). One dimension is the standard rather vaguely defined "conformation," while the other describes how "bound" the ligand is. It could be argued that a ligand is either free or bound, and that there can be no intermediate positions. However, once we consider an ensemble of conformationally mobile proteins, this is no longer true. For example, a protein can (and does) "zip up" along a ligand, as described in Section 4.2.8, in the process of which it becomes progressively more bound.

The conformational selection mechanism is described by the route shown in Figure 6.19a (across and down): the conformation changes before the binding occurs. By contrast, the induced-fit mechanism is described by the route in Figure 6.19b (down and across): the ligand first binds, and then there is a conformational change. Thus, both routes are possible, depending on the shape of the energy surface. Or the preferred route may be somewhere in between, neither one nor the other.

Figure 6.19 is the result of a simulation [53]. The difference between the calculations shown in the two panels lay in the balance between short-range interactions such as van der Waals and hydrophobic interactions, and long-range interactions such as electrostatic: Figure 6.19b places a higher weight on long-range interactions. The clear implication of this is that ligands that bind primarily by electrostatic interactions will tend to follow an induced-fit route, whereas ligands that bind primarily by short-range interactions will tend to bind by a conformational selection mechanism. Because the long-range interactions tend to be stronger and more typical of macromolecular binding rather than small-molecule binding, this in turn implies that proteins binding to small molecules tend to follow conformational selection, whereas proteins binding to macromolecules tend to follow induced fit. In the calculations discussed here, binding was always a mixture of the two: some molecules bound by one route whereas others bound by the other. Thus, again, it is likely that most binding events will not be pure induced fit or pure conformational selection, but some combination of the two. It is worth noting that the binding of ubiquitin to a range of ligands, discussed in Section 6.1.4, has indeed been shown to be a roughly equal mix of conformational selection and induced fit [54].

A further interesting implication of these calculations is that the energy surface is very likely to be different for an enzyme–substrate complex and an enzyme–product complex, because the long-range and short-range interactions formed by product and substrate are usually different [53]. It is thus perfectly possible that substrate binding and product release follow different routes.

6.2.4 Enzymes have a small population in an "activated conformation"

Enzymes need to have conformational mobility if they are to function. What is much less clear is whether they all need the same *degree* of mobility, and exactly what the mobility does. It is noteworthy that enzymes from thermophilic organisms, which normally function at elevated temperature, have low mobility at ambient temperature but "normal" mobility at high temperature, whereas enzymes from psychrophilic organisms have high mobility at ambient temperature but normal mobility at low temperature. There is thus apparently an optimum mobility that can be selected by evolution. Presumably, too low a mobility results in a slow enzyme (if the conformational change in the enzyme limits the catalytic rate, as discussed above), whereas too great a mobility could result in an unstable enzyme.

There is a limited amount of information on conformational dynamics in enzymes. However, it is a remarkable feature that for *all* the catalytic intermediates in DHFR (see Figure 6.16), the alternative state (that is, the next state in the cycle) has been shown to be populated to between 1.5 and 7% [42]. That is to say that in the DHFR.THF complex, for example, the conformation resembling the DHFR.NADPH.THF complex (and thus the state that is conformationally selected when NADPH binds) was populated to roughly 7% and is therefore about 6.5 kJ mol^{-1} higher in energy than the DHFR.THF complex. However, once NADPH has bound, the alternative state for this DHFR.NADPH.THF complex (which resembles the DHFR.NADPH complex, the next one in the cycle) is populated to 2.4%, or 9 kJ mol^{-1} higher in energy. Thus, to a first approximation, DHFR seems to have evolved so that at every stage in the catalytic cycle its local energy landscape looks similar (**Figure 6.20**). This is surely

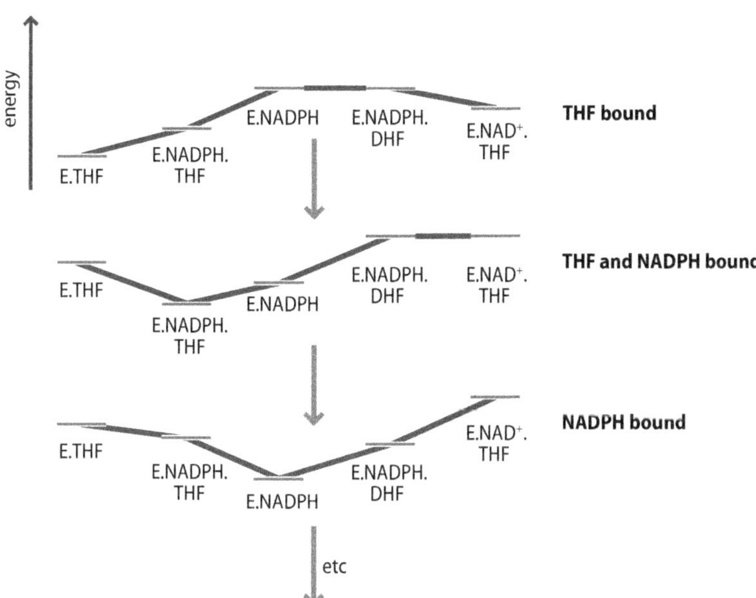

FIGURE 6.20
Diagrammatic energy landscape for dihydrofolate reductase. The energy of each conformation depends on what is bound to the enzyme. When only THF is bound, the energy required to interconvert to the E.NADPH.THF conformation is small; that is, the enzyme seems to have evolved so that the change in conformation to the next conformation in the catalytic cycle is energetically feasible. Similarly, once NADPH has bound and the enzyme has changed to the E.NADPH.THF conformation, the energy required for a change in conformation to the E.NADPH complex (the next in the cycle) is again small, and so on. Thus, each change in ligand binding sets up the enzyme for the next step.

a remarkable evolutionary achievement, at least as remarkable as the leveling of the hills and valleys in the free-energy surface of triosephosphate isomerase discussed in Chapter 5. Such a landscape has a clear biological advantage, because at each stage it allows a rapid transition to the next stage.

This behavior of DHFR does not seem to be an isolated finding. Ribonuclease A seems to do the same thing [55], with rather similar energies. One can speculate that this energy landscape is optimal because it gives acceptably fast reaction rates without a loss of stability or control, but there is much yet to learn.

Let us return for a moment to Figure 6.18. In this figure, the two energy minima were drawn in the same positions along the x axis in both parts. However, because ligand binding changes the shape of the energy landscape, there is no *requirement* that the minimum in the free protein should be in the same place as that in the bound protein, or even that there should be a minimum at all: in other words, in theory one could draw a diagram as in **Figure 6.21a** or Figure 6.21b, in which the free protein contains no energy minimum corresponding to the bound state. There are, however, two good reasons why the diagram must look something similar to Figure 6.18. One is essentially a kinetic argument: if the exchange between conformations is relatively slow, a diagram such as Figure 6.21 implies that the population of protein molecules that are in an appropriate conformation to bind to ligand is small, because it does not correspond to a minimum in the energy surface. Thus, the conformational selection would then require a significant time to repopulate the appropriate conformational region before further selection can occur, thereby negating the benefits of the conformational selection. The other reason is based on experimental results such as those discussed above for DHFR. In these experiments, measurements on one catalytic intermediate reveal the presence of a small percentage in a conformation characteristic of the next intermediate in the catalytic cycle but while still bound to the ligands of the first intermediate; for example, measurement of the E–NADPH–DHF complex reveals a small percentage of a structure resembling the E–NADP$^+$–THF complex, even though the

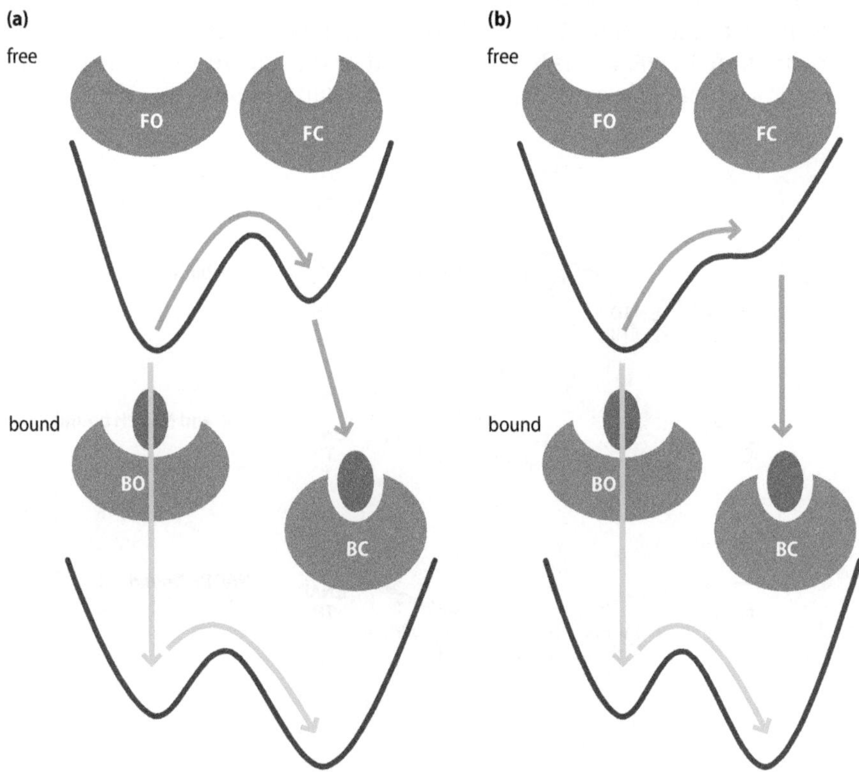

FIGURE 6.21
Two conceivable energy landscapes for the binding of a ligand to a protein. (a) The closed conformation in the absence of ligand is different from the closed conformation bound to ligand. (b) There is no minimum in the free state corresponding to the bound state; that is, the "selection" is from a continuum of sparsely populated states.

enzyme is in fact still bound to NADPH and DHF. The observation of such a relatively large population of this conformer must imply a conformational energy minimum very close to the actual conformation of the next intermediate. The clear implication is that for conformational selection to be a viable mechanism, there must be an energy minimum in the free protein that corresponds fairly closely to the bound conformation, as in Figure 6.18. This is an important conclusion, because it means that enzymes must have evolved not only a free conformation that can recognize substrate, but also a bound conformation that is in exchange with the free, and a dynamic process that ensures that these two conformations can exchange rapidly enough to be catalytically useful. Similar conclusions were obtained from a study on the bacterial two-component signaling protein NtrC [56]. In other words, conformational selection involves a choice between only two minima, rather than picking a conformation off a continuous energy landscape.

Little work has so far been done on the nature of these dynamic exchange processes, not surprisingly because they have populations too small to be observed directly. It has been suggested that the rate is speeded up by stabilization of the transition state by the formation of transient non-native hydrogen bonds [57]. This would be entirely consistent with what we have seen about the directionality and strength of hydrogen bonds in Chapter 1, and is similar to the way in which the potassium channel works, as described in Section 1.3.3. There, the selectivity filter for potassium is formed by a series of carbonyl groups; these form hydrogen bonds to the potassium ion, replacing hydrogen bonds from water and making the passage into the channel almost isoenergetic, and therefore very fast. A similar thing may happen here, in that transient hydrogen bonds from side chains may replace hydrogen bonds present in the ground state, to be replaced later by different hydrogen bonds in the activated conformation. It is therefore possible that some amino acids are selected by evolution not for their role in either the native state or the activated conformation, but for stabilization of the transition state between them.

6.3 FUNCTIONAL MOTION

6.3.1 Enzymes do not catalyze mobility along the reaction coordinate

In Chapter 5 we noted that when one draws an enzyme reaction using an energy diagram such as that in Figure 6.14, the x axis is generally called the "reaction coordinate." If the reaction under consideration is the breaking or making of a bond, the reaction coordinate is roughly equivalent to the length of the bond: it is a convenient single number to represent how far the reaction has proceeded. It does therefore represent a genuine distance, at least in some reactions.

One can then ask whether the motions occurring in enzymes are specifically directed along the reaction coordinate. In other words, in a reaction involving a bond formation between atoms A and B, does the enzyme move in such a way that it pushes A and B closer together? As the reaction approaches the transition state, does the enzyme move such as to push the reactants over the top of the transition state and thus increase the **transmission coefficient (*6.4)**? Does the motion of the enzyme thus directly push the substrate in the right direction to react? This is a very appealing idea, but the process would require some ingenuity on the part of the enzyme, since it would require a concerted rapid movement in the enzyme, ideally on the same timescale as the rate of crossing of the transition state, which is probably in the region of 10^{-11} s. Despite considerable effort, very little evidence has been obtained that enzymes actually do this, and it currently looks unlikely that this is a mechanism used by enzymes [58].

*6.4 Transmission coefficient

Classic transition-state theory says that the overall rate of a reaction is given by a modification of the Arrhenius equation:

$$\text{rate} = \kappa \frac{kT}{h} e^{-E_a/kT}$$

By comparison with the equation in Chapter 5, there is a minor change in that k is used instead of R (in other words, this is now measuring molecules not moles). This equation defines the pre-exponential factor A, and it introduces the symbol κ, which is called the transmission coefficient and normally has values between 0 and 1. It is thus a scaling factor that effectively says what the probability is of crossing the transition state once the energy is high enough. It was ignored in Chapter 5 because the value is predicted to vary over a rather small range (probably between 0.5 and 1 in enzymes) and therefore it has a much smaller effect on the overall rate than the activation energy. Furthermore, it is not directly measurable experimentally: it has to be calculated [96]. Effectively it measures how 'rough' the top of the energy curve is, and thus how strongly the substrate is hindered from going over the top. The value of κ is affected by dynamic fluctuations of the enzyme, which could act to hinder the crossing of the energy barrier, or (more interestingly) it is possible that enzymes have evolved such that their natural dynamics help to 'push' the substrate across the energy barrier; that is, that the natural thermal fluctuations of the enzyme cooperate in the direction of the reaction coordinate and assist the catalysis, for example by stretching a bond that gets broken during the reaction, or pushing together two molecules that need to react. This is an appealing idea, and one that has been much investigated, both theoretically and experimentally. However, so far it must be admitted that there is little evidence that such motions actually occur [58] (Section 6.3.1).

A possible exception to this is in enzymes that perform hydride transfer. In such enzymes the transfer of the hydrogen from cofactor to substrate could in principle occur not just by a simple movement of the hydride ion but at least partly by quantum tunneling, in which the quantum wave representing the hydride tunnels through the energy barrier rather than having to pass over it [59]. This tunneling serves to increase the transmission coefficient to a value greater than 1, thereby increasing the reaction rate reasonably significantly. The tunneling rate is very strongly dependent on the distance to be covered, and it has been suggested that enzyme dynamics may help to shorten the distance sufficiently to make tunneling a viable route. There is still considerable debate on this question, but it is reasonable to suggest that evolutionary pressure has been sufficient to drive enzymes in this direction.

In contrast, there is good evidence that enzymes use the slower hinge-bending or breathing motions to position active-site residues in the appropriate place, to block off the active site from water, and to organize the substrate for reaction. Thus, although the enzyme may not directly push the substrate over the top of the energy curve, it very probably brings it some of the way there, making it easier for rapid thermal motion to do the rest. Such motions have been characterized in lactate dehydrogenase [60] and cyclophilin A [61], for example, which also offers possible evidence for more rapid reaction coordinate motions [62]. Indeed, this is one of the main reasons why dynamic movement in enzymes is important, the other being the movement of flaps discussed next.

In a series of studies on the enzyme purine nucleoside phosphorylase, Schramm and coworkers have suggested that correct geometry of the active site, in a conformation that optimally stabilizes the transition state, is formed not just by the ground state of the enzyme but also by dynamic changes involving several parts of the enzyme structure, some of them very distant from the active site [63, 64]. Mutations of residues 25 Å from the active site produce no change in the geometry of the active site but do affect the structure of the transition state, and it is suggested that they do this by altering relatively slow breathing modes of the enzyme. They have no effect on the very rapid (about 10 fs) bond vibrations that actually push the substrate over the transition state. Thus, it seems that motions on several very different timescales are required for full enzyme function, including at least a slow segmental motion and rapid local vibrations. One suspects that similar motions are very common in enzymes but are just hard to detect experimentally.

6.3.2 Segmental motion is essential for binding and catalysis

In Chapter 4 we discussed the fact that binding interactions typically involve encounter complexes. In an encounter complex, protein and ligand first meet in an unproductive geometry. Because of the random nature of Brownian motion, they are in contact for some time, during which time some reorientation is possible. This time is also long enough for some conformational change to occur. Thus, many binding events are likely to occur in a pathway similar to that shown in Figure 6.19b: an initial encounter complex is formed, in which some binding occurs but with little or no conformational change (point 1), followed by some rearrangement of protein and ligand resulting simultaneously in some conformational change and some binding (point 2), followed by further binding. Such a mechanism is particularly likely for the binding of peptides and natively unstructured proteins [65], but could occur to some extent for any ligand.

On comparing a protein–protein or protein–ligand complex with the structures of the free components, it is very common to find that the protein conformation has changed on binding. Thus, at some point during the binding interaction the protein changes conformation. This has been studied using molecular dynamics, in which multiple docking trajectories were calculated [66]. Interestingly, the authors found that there were a large number of complementary protein:ligand interactions, many of which resembled neither the free nor the bound protein. In other words, it is suggested that there are multiple binding pathways, which cover a range between the conformational selection pathway of Figure 6.19a and the induced-fit type of pathway of Figure 6.19b. The authors therefore proposed a docking free-energy diagram similar to that in **Figure 6.22**, in which there is an initial rapid diffusional step to form an encounter complex, a second rearrangement step in which mutually

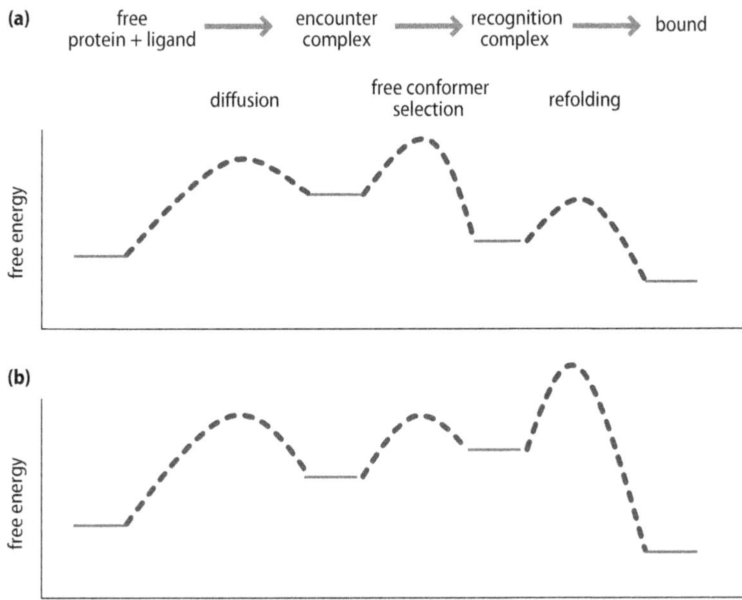

FIGURE 6.22
General scheme for the binding of a ligand to a protein. (a) If the highest barrier is the second one, this represents a conformational selection. (b) However, if the highest barrier is the final one, this represents an induced fit. Hence these two descriptions are two extremes of a continuum, depending on the relative heights of the barriers. Because the barrier heights may vary with ligand and solution conditions, the distinction between induced fit and conformational selection is very fluid.

complementary conformations of protein and ligand dock together, and a final step in which these "bound" complexes relax to form the final bound complex.

This route is consistent with almost all existing data, and in particular it allows for a continuum between induced fit and conformational selection. If the energy barrier to formation of the recognition complex is high, as in Figure 6.22a, then the slow step is the selection of appropriate conformers and we have a conformational selection process. But if the high energy barrier is the final refolding step, as in Figure 6.22b, then we have an induced-fit process. And if all the energy barriers are of similar height (as we have seen for triose-phosphate isomerase in a different context, for which it was argued that this is a natural result of evolutionary pressure), the route is a mixture of the two. Furthermore, the authors suggest that if the binding interaction is dominated by long-range electrostatic interactions, these will be important mainly in the initial binding step and will therefore lead to a mainly induced-fit binding. In contrast, if short-range interactions are dominant, these will come into play mainly in the final step, and will therefore produce a conformational selection process. This agrees with the conclusions reached above [53].

Finally, Grünberg et al. [66] point out that the conformational selection route (see Figure 6.22a) requires that there should be some minimum concentration of "bound" conformers in equilibrium with the free, for it to be kinetically viable. For a 100% conformational selection route, the rate is determined by the lifetime of the encounter complex and the rate of conformational equilibration, and the authors estimated that it would require 2–4% of "bound" conformer in equilibrium with free. This proportion matches well with the proportion of activated complex present in several dynamic studies, such as that of DHFR discussed above. It therefore seems likely that this represents a rather general situation.

It is obvious that for proteins in which the active site is completely buried, a 100% conformational selection route is not possible, because this would not allow access of the ligand into the active site [67]. However, a mixed route remains possible. In addition, the slowness of translational diffusion in comparison with conformational change implies that transient opening of flaps in what is essentially a bound conformation is probably possible. In an interesting computer simulation, it was shown that the lid enclosing the active site of triosephosphate isomerase opens and closes 10–20 times while the ligand is diffusing around close to the entrance: the flap therefore seems to have evolved dynamic properties that allow it to open and close often enough that it does not interfere with the rate at which the substrate can diffuse into the active site [68]. In the simulation, the flap was open about 50% of the time. Because the opening and closing is fast in comparison with diffusion, the presence of the flap has a surprisingly small effect on the overall on-rate [69]. In contrast, experimental results suggest that flap opening does limit the *turnover* rate, probably by limiting the (rate-limiting) rate at which the product is released [70]. A very similar conclusion was obtained for adenylate kinase [71]. This is probably another example of the effect described above: it is not the chemical reaction itself that limits the turnover but the rate at which the enzyme can rearrange itself in response to the reaction.

6.3.3 Buried waters are important for internal mobility

At the start of this chapter, we noted that water molecules are vital because they provide the thermal energy to allow proteins to move. Hans Frauenfelder and others have described the protein motions as being "slaved" to the motions of the water molecules, in that the direction and rate of protein motions are closely tied to the motions of the surrounding water, and that motions in dehydrated proteins are much restricted [72]. However, in this section we focus on a different group of water molecules, namely those water molecules that are buried inside the protein, with no direct contact with bulk water.

All proteins have small cavities inside them, which are a natural consequence of the difficulty of packing amino acid side chains together to fill all the available space completely. Some of these cavities are large enough to contain water molecules, and more than half of these larger cavities do have water molecules that can be seen by X-ray crystallography. The number of water molecules buried inside proteins is in general simply proportional to the size of the protein [73]. The locations of buried waters are well conserved in different crystal structures of the same protein, and even in homologous proteins, suggesting that they are structurally and/or functionally important (**Figure 6.23**) [74]. It is somewhat surprising that buried water molecules should be so common. When a water molecule is removed from bulk water and put into a cavity inside a protein, it should lose a considerable amount of its translational entropy. It forms hydrogen bonds in water, and hydrogen bonds inside the protein, which might be expected to be roughly equal in energy. Why, then, is it not highly energetically unfavorable to have a water molecule located in a cavity inside a protein?

The answer is a combination of enthalpy and entropy. Enthalpically, the answer is that water molecules very probably form more hydrogen bonds in a protein cavity than they do in water. A water molecule can form a maximum of four hydrogen bonds, as it does in ice. In liquid water, the average number of hydrogen bonds per water molecule is about 2.5, whereas in buried waters it is closer to 3 [75], thereby providing an enthalpic advantage for buried waters. One could also argue that the enthalpy is likely to be greater per hydrogen bond because of the lower dielectric constant (*4.2) inside a protein. Entropically, water molecules in protein cavities retain considerable mobility, although less than in bulk solvent. However, several studies, both computational [76–78] and experimental [18, 22, 23], have suggested the important result that the presence of a buried water molecule inside a protein provides the *protein* with significantly increased mobility. As the water molecule rotates within its cavity, it allows the protein to reorganize some of its hydrogen bonding and therefore allows it to move in ways that would be energetically unfavorable in the absence of the water. Thus, although the entropy of the water is decreased, the entropy of the protein is increased. It is possibly significant that

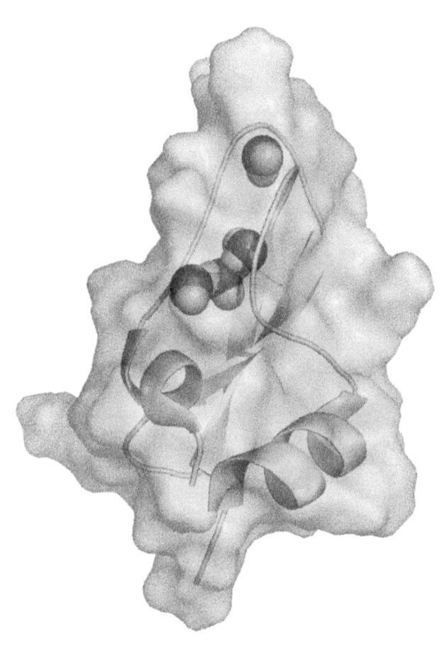

FIGURE 6.23
Buried water molecules inside proteins. This example shows the four water molecules buried inside basic pancreatic trypsin inhibitor (BPTI); that is, waters that have no accessible surface area. This structure (PDB code 5pti) is unusual because it is a joint refinement of X-ray and neutron diffraction (Section 11.4.10). This allows hydrogen atoms to be located accurately, as indicated by the protons on the water molecules, which shows the chain of three waters to be hydrogen bonding in the expected way. These water molecules have been particularly well characterized. The single water near the top of the structure (which is also close to the active site, namely the region of the protein that is inserted into the active site of trypsin) is unusually slow to exchange, with a residence time of about 170 μs at 27 °C. The other three exchange more rapidly (but still fairly slowly for buried waters) with residence times of between 10 and 1000 ns [98]. All four waters have low *B* factors (Section 11.4.5), comparable to the surrounding protein atoms, and therefore are no more mobile than the protein around them. The single water molecule rotates within its pocket, dragging the protein with it, and thus gives the protein significantly greater internal mobility [99].

many proteins have buried waters close to the active site, which may allow them extra mobility in this region. Interdomain interfaces also have a higher incidence of cavities, which again have been suggested to facilitate interdomain movement [79].

6.3.4 Internal dynamics can produce allostery

As we have seen in Chapter 3, allostery in proteins is normally readily explained by relative movements of domains, particularly in oligomeric proteins. However, there are some proteins for which such an explanation is not obvious.

Many years ago, it was suggested that allostery could occur via changes in protein dynamics rather than a change in structure. That is, if binding of an effector at one site changes the dynamic properties of an enzyme, this could affect its catalytic rate without any change in the average structure [80]. This idea has recently received both theoretical [81] and experimental [82, 83] support, and it may provide a possible explanation for long-range allosteric effects [46, 84, 85].

A bioinformatics approach has also provided support for this idea. On comparing a large set of homologous proteins, it was observed that some locations were "coupled," in that mutations to one residue tended to result in statistically significant correlated mutations in other residues. In many cases, these coupled pairs were adjacent in the three-dimensional structure. However, in some cases they were not, and further analysis suggested that they formed a dynamically linked pathway across the protein [86].

A related idea is that ligand binding may affect the degree of structural ordering, for example to increase the stability of a locally unfolded region; this increased stability could then propagate through the protein and affect more distant sites [85, 87, 88]. For example, there is convincing evidence that the signaling resulting from phosphorylation of NtrC (Chapter 8) acts more through dynamic changes than through structural changes [56].

It is clear that ligand binding to proteins does affect their mobility. The example of barnase discussed above is one such case, in which ligand binding leads to a decrease in mobility in the protein. This is the most common outcome of ligand binding, as one might expect, and affects both backbone and side chain, although not in any uniform or predictable manner [89]. However, there are several examples where ligand binding leads to an *increase* in mobility, particularly at sites distant from the ligand-binding site, and it has been argued that this increased mobility leads to an increase in the entropy of the protein and therefore to more favorable binding [11, 90]. It is still far from clear how large a role entropy has in the energetics of ligand binding [91], and the current consensus is that such entropic effects on stability are probably rather small [92].

All of these ideas are interesting and attractive. This does not necessarily make them true, and it remains to be established whether this is indeed a widespread feature of proteins.

6.4 SUMMARY

Local rapid motions in proteins are induced by the thermal motion of solvent water. They are uncorrelated and of small scale, and do not have direct functional relevance. As the motions become more correlated, they are also less frequent. These are the motions that lead to induced-fit movement, and they tend to occur on a timescale of about $10^4 s^{-1}$, which is much less frequent than the uncorrelated thermal motion. They often seem to be relayed through hydrogen bonds.

Proteins populate an energy landscape, and they can span a wide range of conformations even at ambient temperature. We therefore need to modify the induced-fit model to one of conformational selection, in which ligand binding does not induce a new conformation but selects one from the range already populated. The induced-fit and conformational selection models represent the two extreme models, and the reality for most proteins is probably somewhere in between, in that most proteins bind and restructure simultaneously. The balance between the two models is probably determined by how long-range the intermolecular forces are.

Enzyme motion is thus essential for their function. However, it seems that they do not push reactants together or pull them apart as they cross the transition state, attractive though this idea is. They do nevertheless have flaps and lids that open and close on timescales that have evolved to interfere with the reaction rate as little as possible. Some of this large-scale mobility probably arises from buried waters, which endow proteins with additional mobility.

Allosteric communication does not need to rely on structural change: it can be based on dynamic change. It is, however, too early to say how important this effect is.

6.5 FURTHER READING

There are no books on this subject, because it is too new. A recent review by Boehr et al. [11] is a good place to start. For the role of dynamics in enzyme function, try Goodey and Benkovic [46], and also a major review by Hammes-Schiffer and Benkovic [93].

6.6 WEBSITES

http://lorentz.immstr.pasteur.fr/nma/ *Calculates normal modes.*

http://sbg.cib.csic.es/Software/DFprot/ *Analyzes deformability/flexibility.*

http://services.cbu.uib.no/tools/normalmodes *Also calculates normal modes.*

http://wishart.biology.ualberta.ca/rci/cgi-bin/rci_cgi_1_e.py *Predicts flexible regions from chemical shifts.*

http://www.igs.cnrs-mrs.fr/elnemo/ *Also calculates normal modes.*

http://www.molmovdb.org/ *A directory of known or predicted movements, including movies.*

6.7 PROBLEMS

Hints for some of these problems can be found on the book's website.

1. Proteins have motions on timescales ranging from picoseconds to kiloseconds (see, for example, Figure 6.1). Do all of these motions have functional significance? If so, does this imply that these motions have evolved by natural selection? What does this imply about why residues are conserved; and conversely, does residue conservation allow us to explore particular types of motion?

2. If you could watch an enzyme undergoing slow ($10^3\,s^{-1}$) induced fit/conformational selection motions, what would these slow motions look like?

3. Find a website showing normal-mode motions of a protein. What do these look like? Does a real protein move like this, and if not, what is the relationship between normal mode motions and "real" motions?

4. Describe precisely what is meant by a "correlated" motion. How would you attempt to reveal such a motion from a molecular dynamics trajectory?

5. What effect would temperature have on the motions seen in a protein? And pressure? What are the implications for organisms that live at extremes of temperature and pressure?

6. From the arguments used in Section 6.2.3, describe the likely course of binding and conformational change in protein:protein interactions.

7. Toward the end of Section 6.3.4, it is argued that the increased mobility sometimes caused by the binding of a ligand "leads to an increase in the entropy of the protein and therefore to more favorable binding." Explain this argument. How could you test such a statement?

6.8 NUMERICAL PROBLEMS

N1. Let us put some numbers to Levinthal's paradox (Section 6.2.1). Assume that each amino acid can adopt only three backbone conformations. For a 200-residue protein, how many conformations is this altogether? If it can move between one and the next on a timescale of bond rotation ($10^{12} s^{-1}$), how long would it take to search through all possible conformations? By contrast, how long does a typical 200-residue protein take to fold?

N2. Study reference [86]. Describe the principle of the method. Does this explain the long-range effects observed for DHFR in [84]?

6.9 REFERENCES

1. FRN Gurd & TM Rothgeb (1979) Motions in proteins. *Adv. Protein Chem.* 33:73–165.
2. R Lumry & H Eyring (1954) Conformational changes of proteins. *J. Phys. Chem.* 58:110–120.
3. W Ferdinand (1976) The Enzyme Molecule. New York: Wiley.
4. GR Welch, B Somogyi & S Damjanovich (1982) The role of protein fluctuations in enzyme action—a review. *Progr. Biophys. Mol. Biol.* 39:109–146.
5. K Henzler-Wildman & D Kern (2007) Dynamic personalities of proteins. *Nature* 450:964–972.
6. K Lindorff-Larsen, RB Best, MA DePristo, CM Dobson & M Vendruscolo (2005) Simultaneous determination of protein structure and dynamics. *Nature* 433:128–132.
7. A De Simone, B Richter, X Salvatella & M Vendruscolo (2009) Toward an accurate determination of free energy landscapes in solution states of proteins. *J. Am. Chem. Soc.* 131:3810–3811.
8. JJ Chou, DA Case & A Bax (2003) Insights into the mobility of methyl-bearing side chains in proteins from $^3J_{CC}$ and $^3J_{CN}$ couplings *J. Am. Chem. Soc.* 125:8959–8966.
9. JC Hoch, CM Dobson & M Karplus (1985) Vicinal coupling constants and protein dynamics. *Biochemistry* 24:3831–3841.
10. RB Best, J Clarke & M Karplus (2004) The origin of protein sidechain order parameter distributions *J. Am. Chem. Soc.* 126:7734–7735.
11. DD Boehr, HJ Dyson & PE Wright (2006) An NMR perspective on enzyme dynamics. *Chem. Rev.* 106:3055–3079.
12. A Ansari, J Berendzen, SF Bowne et al. (1985) Protein states and protein quakes. *Proc. Natl. Acad. Sci. USA* 82:5000–5004.
13. RH Austin, KW Beeson, L Eisenstein, H Frauenfelder & IC Gunsalus (1975) Dynamics of ligand binding to myoglobin. *Biochemistry* 14:5355–5373.
14. G Wagner, A DeMarco & K Wüthrich (1976) Dynamics of aromatic amino acid residues in globular conformation of basic pancreatic trypsin inhibitor (BPTI). 1. ^{1}H NMR studies. *Biophys. Struct. Mech.* 2:139–158.
15. G Wagner (1983) Characterization of the distribution of internal motions in the basic pancreatic trypsin inhibitor using a large number of internal NMR probes. *Q. Rev. Biophys.* 16:1–57.
16. A Zhuravleva, DM Korzhnev, SB Nolde et al. (2007) Propagation of dynamic changes in barnase upon binding of barstar: an NMR and computational study. *J. Mol. Biol.* 367:1079–1092.
17. J Giraldo, L De Maria & SJ Wodak (2004) Shift in nucleotide conformational equilibrium contributes to increased rate of catalysis of GpAp versus GpA in barnase. *Proteins Struct. Funct. Bioinf.* 56:261–276.
18. DJ Wilton, R Kitahara, K Akasaka, MJ Pandya & MP Williamson (2009) Pressure-dependent structure changes in barnase on ligand binding reveal intermediate rate fluctuations. *Biophys. J.* 97:1482–1490.
19. KA Henzler-Wildman, M Lei, V Thai et al. (2007) A hierarchy of timescales in protein dynamics is linked to enzyme catalysis. *Nature* 450:913–916.
20. G Bouvignies, P Bernádo, S Meier et al. (2005) Identification of slow correlated motions in proteins using residual dipolar and hydrogen-bond scalar couplings. *Proc. Natl. Acad. Sci. USA* 102:13885–13890.

21. PRL Markwick, G Bouvignies & M Blackledge (2007) Exploring multiple timescale motions in Protein GB3 using accelerated molecular dynamics and NMR spectroscopy. *J. Am. Chem. Soc.* 129:4724–4730.

22. M Refaee, T Tezuka, K Akasaka & MP Williamson (2003) Pressure-dependent changes in the solution structure of hen egg-white lysozyme. *J. Mol. Biol.* 327:857–865.

23. MP Williamson, K Akasaka & M Refaee (2003) The solution structure of bovine pancreatic trypsin inhibitor at high pressure. *Protein Sci.* 12:1971–1979.

24. MP Williamson (2003) Many residues in cytochrome *c* populate alternative states under equilibrium conditions. *Proteins* 53:731–739.

25. DJ Wilton, RB Tunnicliffe, YO Kamatari, K Akasaka & MP Williamson (2008) Pressure-induced changes in the solution structure of the GB1 domain of protein G. *Proteins Struct. Funct. Bioinf.* 71:1432–1440.

26. OF Lange, NA Lakomek, C Farès et al. (2008) Recognition dynamics up to microseconds revealed from an RDC-derived ubiquitin ensemble in solution. *Science* 320:1471–1475.

27. RG Smock & LM Gierasch (2009) Sending signals dynamically. *Science* 324:198–203.

28. RM Daniel, RV Dunn, JL Finney & JC Smith (2003) The role of dynamics in enzyme activity. *Annu. Rev. Biophys. Biomol. Struct.* 32:69–72.

29. JH Tomlinson, VL Green, PJ Baker & MP Williamson (2010) Structural origins of pH-dependent chemical shifts in protein G. *Proteins Struct. Funct. Bioinf.* 78:3000–3016.

30. VJ Hilser, D Dowdy, TG Oas & E Freire (1998) The structural distribution of cooperative interactions in proteins: analysis of the native state ensemble. *Proc. Natl. Acad. Sci. USA* 95:9903–9908.

31. A Tousignant & JN Pelletier (2004) Protein motions promote catalysis. *Chem. Biol.* 11:1037–1042.

32. C Levinthal (1968) Are there pathways for protein folding? *J. Chim. Phys. Phys.-Chem. Biol.* 65:44–45.

33. S Meier, S Grzesiek & M Blackledge (2007) Mapping the conformational landscape of urea-denatured ubiquitin using residual dipolar couplings. *J. Am. Chem. Soc.* 129:9799–9807.

34. KA Dill & HS Chan (1997) From Levinthal to pathways to funnels. *Nature Struct. Biol.* 4:10–19.

35. CJ Tsai, BY Ma & R Nussinov (1999) Folding and binding cascades: shifts in energy landscapes. *Proc. Natl. Acad. Sci. USA* 96:9970–9972.

36. MI Page (1984) The energetics and specificity of enzyme-substrate interactions. In MI Page (ed.) The Chemistry of Enzyme Action. Amsterdam: Elsevier, 1–54.

37. CS Goh, D Milburn & M Gerstein (2004) Conformational changes associated with protein–protein interactions. *Curr. Opin. Struct. Biol.* 14:104–109.

38. BY Ma, S Kumar, CJ Tsai & R Nussinov (1999) Folding funnels and binding mechanisms. *Protein Eng.* 12:713–720.

39. J Gsponer, J Christodoulou, A Cavalli et al. (2008) A coupled equilibrium shift mechanism in calmodulin-mediated signal transduction. *Structure* 16:736–746.

40. D Tobi & I Bahar (2005) Structural changes involved in protein binding correlate with intrinsic motions of proteins in the unbound state. *Proc. Natl. Acad. Sci. USA* 102:18908–18913.

41. HR Bosshard (2001) Molecular recognition by induced fit: how fit is the concept? *News Physiol. Sci.* 16:171–173.

42. DD Boehr, D McElheny, HJ Dyson & PE Wright (2006) The dynamic energy landscape of dihydrofolate reductase catalysis. *Science* 313:1638–1642.

43. DD Boehr, HJ Dyson & PE Wright (2008) Conformational relaxation following hydride transfer plays a limiting role in dihydrofolate reductase catalysis. *Biochemistry* 47:9227–9233.

44. M Kurzyński (1998) A synthetic picture of intramolecular dynamics of proteins. Towards a contemporary statistical theory of biochemical processes. *Prog. Biophys. Mol. Biol.* 69:23–82.

45. M Kurzyński (1993) Enzymatic catalysis as a process controlled by protein conformational relaxation. *FEBS Lett.* 328:221–224.

46. NM Goodey & SJ Benkovic (2008) Allosteric regulation and catalysis emerge via a common route. *Nature Chem. Biol.* 4:474–482.

47. EZ Eisenmesser, DA Bosco, M Akke & D Kern (2002) Enzyme dynamics during catalysis. *Science* 295:1520–1523.

48. ED Watt, H Shimada, EL Kovrigin & JP Loria (2007) The mechanism of rate-limiting motions in enzyme function. *Proc. Natl. Acad. Sci. USA* 104:11981–11986.

49. DJ Heyes, P Heathcote, SEJ Rigby et al. (2006) The first catalytic step of the light-driven enzyme protochlorophyllide oxidoreductase proceeds via a charge transfer complex. *J. Biol. Chem.* 281:26847–26853.

50. DJ Heyes & CN Hunter (2005) Making light work of enzyme catalysis: protochlorophyllide oxidoreductase. *Trends Biochem. Sci.* 30:642–649.

51. R Wolfenden & MJ Snider (2001) The depth of chemical time and the power of enzymes as catalysts. *Acc. Chem. Res.* 34:938–945.

52. S Kumar, BY Ma, CJ Tsai, N Sinha & R Nussinov (2000) Folding and binding cascades: dynamic landscapes and population shifts. *Protein Sci.* 9:10–19.

53. K Okazaki & S Takada (2008) Dynamic energy landscape view of coupled binding and protein conformational change: induced-fit versus population-shift mechanisms. *Proc. Natl. Acad. Sci. USA* 105:11182–11187.

54. TW lodarski & B Zagrovic (2009) Conformational selection and induced fit mechanism underlie specificity in noncovalent interactions with ubiquitin. *Proc. Natl. Acad. Sci. USA* 106:19346–19351.

55. H Beach, R Cole, ML Gill & JP Loria (2005) Conservation of μs–ms enzyme motions in the apo- and substrate-mimicked state. *J. Am. Chem. Soc.* 127:9167–9176.

56. BF Volkman, D Lipson, DE Wemmer & D Kern (2001) Two-state allosteric behavior in a single-domain signaling protein. *Science* 291:2429–2433.

57. AK Gardino, J Villali, A Kivenson et al. (2009) Transient non-native hydrogen bonds promote activation of a signaling protein. *Cell* 139:1109–1118.

58. WR Cannon, SF Singleton & SJ Benkovic (1996) A perspective on biological catalysis. *Nature Struct. Biol.* 3:821–833.

59. MJ Sutcliffe & NS Scrutton (2000) Enzyme catalysis: over-the-barrier or through-the-barrier? *Trends Biochem. Sci.* 25:405–408.

60. L Young & CB Post (1996) Catalysis by entropic guidance from enzymes. *Biochemistry* 35:15129–15133.

61. PK Agarwal (2006) Enzymes: an integrated view of structure, dynamics and function. *Microbial Cell Factories* 5:2 (doi:10.1186/1475-2859-5-2).

62. PK Agarwal (2005) Role of protein dynamics in reaction rate enhancement by enzymes. *J. Am. Chem. Soc.* 127:15248–15256.

63. M Luo, L Li & VL Schramm (2008) Remote mutations alter transition-state structure of human purine nucleoside phosphorylase. *Biochemistry* 47:2565–2576.

64. S Saen-oon, S Quaytman-Machleder, VL Schramm & SD Schwartz (2008) Atomic detail of chemical transformation at the transition state of an enzymatic reaction. *Proc. Natl. Acad. Sci. USA* 105:16543–16548.

65. K Sugase, HJ Dyson & PE Wright (2007) Mechanism of coupled folding and binding of an intrinsically disordered protein. *Nature* 447:1021-U11.

66. R Grünberg, J Leckner & M Nilges (2004) Complementarity of structure ensembles in protein–protein binding. *Structure* 12:2125–2136.

67. SM Sullivan & T Holyoak (2008) Enzymes with lid-gated active sites must operate by an induced fit mechanism instead of conformational selection. *Proc. Natl. Acad. Sci. USA* 105:13829–13834.

68. RC Wade, BA Luty, E Demchuk et al. (1994) Simulation of enzyme–substrate encounter with gated active sites. *Nature Struct. Biol.* 1:65–69.

69. HX Zhou, ST Wlodek & JA McCammon (1998) Conformation gating as a mechanism for enzyme specificity. *Proc. Natl. Acad. Sci. USA* 95:9280–9283.

70. JP Loria, RB Berlow & ED Watt (2008) Characterization of enzyme motions by solution NMR relaxation dispersion. *Acc. Chem. Res.* 41:214–221.

71. M Wolf-Watz, V Thai, K Henzler-Wildman et al. (2004) Linkage between dynamics and catalysis in a thermophilic–mesophilic enzyme pair. *Nature Struct. Mol. Biol.* 11:945–949.

72. JA Rupley & G Careri (1991) Protein hydration and function. *Adv. Protein Chem.* 41:37–172.

73. MA Williams, JM Goodfellow & JM Thornton (1994) Buried waters and internal cavities in monomeric proteins. *Protein Sci.* 3:1224–1235.

74. U Sreenivasan & PH Axelsen (1992) Buried water in homologous serine proteases. *Biochemistry* 31:12785–12791.

75. S Park & JG Saven (2005) Statistical and molecular dynamics studies of buried waters in globular proteins. *Proteins Struct. Funct. Bioinf.* 60:450–463.

76. S Fischer & CS Verma (1999) Binding of buried structural water increases the flexibility of proteins. *Proc. Natl. Acad. Sci. USA* 96:9613–9615.

77. S Fischer, JC Smith & CS Verma (2001) Dissecting the vibrational entropy change on protein/ligand binding: burial of a water molecule in bovine pancreatic trypsin inhibitor. *J. Phys. Chem. B* 105:8050–8055.

78. LR Olano & SW Rick (2004) Hydration free energies and entropies for water in protein interiors. *J. Am. Chem. Soc.* 126:7991–8000.

79. SJ Hubbard & P Argos (1996) A functional role for protein cavities in domain:domain motions. *J. Mol. Biol.* 261:289–300.

80. A Cooper & DTF Dryden (1984) Allostery without conformational change. *Eur. Biophys. J.* 11:103–109.

81. RJ Hawkins & TCB McLeish (2006) Coupling of global and local vibrational modes in dynamic allostery of proteins. *Biophys. J.* 91:2055–206282.

82. RA Laskowski, F Gerick & JM Thornton (2009) The structural basis of allosteric regulation in proteins. *FEBS Lett.* 583:1692–1698.

83. K Gunasekaran, BY Ma & R Nussinov (2004) Is allostery an intrinsic property of all dynamic proteins? *Proteins Struct. Funct. Bioinf.* 57:433–443.

84. T Liu, ST Whitten & VJ Hilser (2006) Ensemble-based signatures of energy propagation in proteins: a new view of an old phenomonon. *Proteins Struct. Funct. Bioinf.* 62:728–738.

85. VJ Hilser & EB Thompson (2007) Intrinsic disorder as a mechanism to optimize allosteric coupling in proteins. *Proc. Natl. Acad. Sci. USA* 104:8311–8315.

86. SW Lockless & R Ranganathan (1999) Evolutionarily conserved pathways of energetic connectivity in protein families. *Science* 286:295–299.

87. E Freire (1999) The propagation of binding interactions to remote sites in proteins: analysis of the binding of the monoclonal antibody D1.3 to lysozyme. *Proc. Natl. Acad. Sci. USA* 96:10118–10122.

88. E Freire (2000) Can allosteric regulation be predicted from structure? *Proc. Natl. Acad. Sci. USA* 97:11680–11682.

89. AL Lee & AJ Wand (2001) Microscopic origins of entropy, heat capacity and the glass transition in proteins. *Nature* 411:501–504.

90. MJ Stone (2001) NMR relaxation studies of the role of conformational entropy in protein stability and ligand binding. *Acc. Chem. Res.* 34:379–388.

91. KK Frederick, MS Marlow, KG Valentine & AJ Wand (2007) Conformational entropy in molecular recognition by proteins. *Nature* 448:325–329.

92. R Grünberg, M Nilges & J Leckner (2006) Flexibility and conformational entropy in protein-protein binding. *Structure* 14:683–693.

93. S Hammes-Schiffer & SJ Benkovic (2006) Relating protein motion to catalysis. *Annu. Rev. Biochem.* 75:519–541.

94. H Remaut & G Waksman (2006) Protein–protein interaction through β-strand addition. *Trends Biochem. Sci.* 31:436–444.

95. H Wako, M Kato & S Endo (2004) ProMode: a database of normal mode analyses on protein molecules with a full-atom model. *Bioinformatics* 20:2035–2043.

96. M Garcia-Viloca, J Gao, M Karplus & DG Truhlar (2004) How enzymes work: analysis by modern reaction rate theory and computer simulations. *Science* 303:186–195.

97. RB Tunnicliffe, JL Waby, RJ Williams & MP Williamson (2005) An experimental investigation of conformational fluctuations in proteins G and L. *Structure* 13:1677–1684.

98. K Modig, E Liepinsh, G Otting & B Halle (2004) Dynamics of protein and peptide hydration. *J. Am. Chem. Soc.* 126:102–114.

99. S Fischer, JC Smith & CS Verma (2001) Dissecting the vibrational entropy change on protein/ligand binding: burial of a water molecule in bovine pancreatic trypsin inhibitor. *J. Phys. Chem. B* 105:8050–8055.

100. RJ Bingham, E Rudiño-Piñera, NAG Meenan et al. (2008) Crystal structures of fibronectin-binding sites from *Staphylococcus aureus* FnBPA in complex with fibronectin domains. *Proc. Natl. Acad. Sci. USA* 105:12254–12258.

CHAPTER 7
How Proteins Make Things Move

Everything is simpler than you think and at the same time more complex than you imagine.

Johann Wolfgang von Goethe (1749–1832), attrib.

We have already seen that much of the catalytic enhancement generated by enzymes derives simply from their binding substrates in close proximity, and having acidic and basic side chains placed at appropriate places. That is, enzymes do not in general need to do anything particularly 'clever,' and their remarkable rate enhancements are due not so much to subtle chemical reactivity or internal rearrangements as to rather straightforward physical principles, albeit arranged very precisely. This is an example of a major theme of this book, that most of what proteins do can be understood at a very simple 'macroscopic' level without needing to go into structural detail. This theme is pursued further in Chapter 8, where we will see that signaling depends mainly on an interplay of intramolecular and intermolecular protein interactions, and requires the same sort of physical principles. In other words, in these two aspects of protein function, the functionality derives mainly from the rates at which molecules bind and detach, and the rates at which they move randomly around the cell; the detailed inner workings of the proteins are interesting but do not need to be understood to get an idea of how the system works.

In this chapter we consider a protein function that should be a prime candidate for 'cleverness' requiring intramolecular dynamics of a very precise kind, namely directional movement: How do proteins make things move to where they need to go? We shall see that the key is ATP or GTP hydrolysis, which acts as a ratchet to force movement to go in one direction only. Many of the proteins that use this principle are homologous, implying that one can trace their history back to a single primitive GTPase that acted as a conformational switch, which is a good candidate for the most 'clever' protein in this book.

7.1 HOW PROTEIN MOTORS WORK

7.1.1 Most intracellular motion occurs by random diffusion

Cellular function frequently requires molecules to go from one part of the cell to another. To take an example from Chapter 8: signaling involves molecules going from the cytoplasm to the cell membrane, and others going from the cell membrane to the nucleus. Does this happen by random diffusion? In Chapter 4 we saw that a random walk is an inefficient way to get from one place to another, and that the cytoplasm is so crowded that free diffusion is significantly hindered. Does this mean that many proteins use some 'directed' process to get where they are going?

In general, the answer is no—almost all processes *do* occur by random diffusion. This may be inefficient, but most cells are small enough that this still happens rapidly.

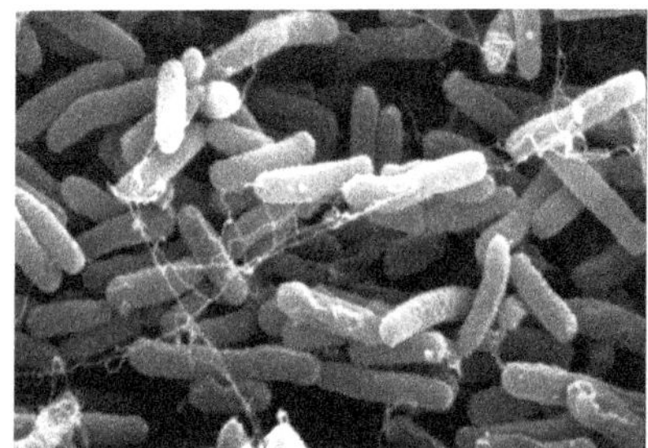

FIGURE 7.1
A group of *E. coli* cells, imaged by scanning electron microscopy. The image is 9.5 μm across, implying that an *E. coli* cell is approximately 1-2 μm long and 0.25 μm in diameter. (Courtesy of the Microbe Zoo and Michigan State University. http://commtechlab.msu.edu/sites/dlc-me/zoo/zah0700.html)

A good example is furnished by the system for controlling cell septum formation during cell division in *Escherichia coli*. *E. coli* is cylindrical (**Figure 7.1**) with a length of about 2 μm. When it divides, it forms a septum in the middle, whose location is determined by the assembly of a polymeric ring of the tubulin-like GTPase FtsZ. In turn, the location of the FtsZ ring is determined by a group of three proteins called MinC, D, and E. MinC inhibits FtsZ polymerization. These three proteins form a remarkable oscillating system in *E. coli* cells. MinC was

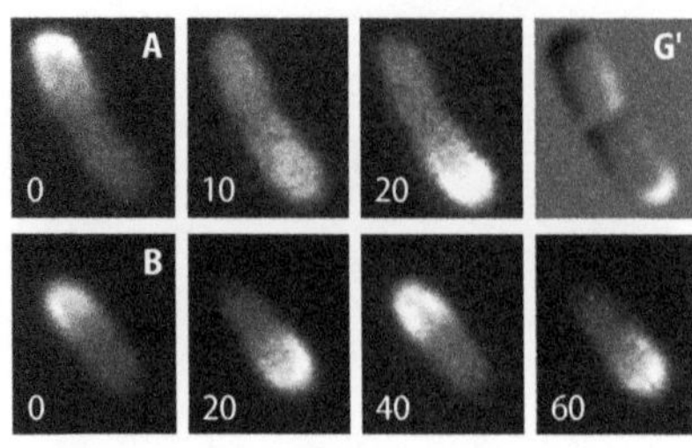

FIGURE 7.2
Dynamic properties of the *E. coli* cell division inhibitor MinD. These are fluorescence images of a fusion of green fluorescent protein (GFP) with MinD. The numbers are the times between frames, in seconds, for two different bacterial cells A and B. Frame G′ is a differential interference contrast image showing formation of the central septum. The width of the image is about 2 μm. (From D. M. Raskin and P. A. J. de Boer, *Proc. Natl. Acad. Sci. USA* 96: 4971–4976, 1999. With permission from the National Academy of Sciences.)

found to oscillate from one pole of *E. coli* to the other with a period of about 50 s [1, 2] (**Figure 7.2**). In fact it spends about 20 s at one end, and then 5 s later is observed entirely at the other end. Because on average MinC is found everywhere in the cell except in a ring around the middle, this is where FtsZ polymerization occurs.

The time taken for random diffusion from one end of *E. coli* to the other can be calculated from the standard equation for diffusion, which is known as Fick's law:

$$J = -D\,\partial\phi/\partial x$$

where J is the flux of molecules (that is, the amount of substance flowing across a given area) and has units of concentration.length^{-2}.time^{-1}, for example M μm^{-2} s^{-1}; D is the **diffusion coefficient (*7.1)**, in μm^2 s^{-1} (as discussed in Chapter 4); and ϕ is the concentration difference across a distance x. The symbol ∂ is used rather than the more normal d because this is a partial derivative.

In the cytoplasm of *E. coli*, D has been measured as roughly 8 μm^2 s^{-1} [3]. Thus we can immediately calculate that the flux of molecules from one end to the other has a maximum rate of 8/3 times the concentration (the 3 coming from the length of a cell): in other words, if we start off with all the protein at one end, it can all diffuse to the other end within about half a second (3/8 s). This is consistent with observations suggesting that it takes about 100 ms for proteins to diffuse 1 μm in *E. coli* [4]. Thus, the random diffusion rate of MinC is easily enough to explain this remarkable oscillatory behavior. In an elegant study, it was shown that some rather simple assumptions about interactions between MinC, D, and E can explain all of the oscillatory behavior without the need to invoke any other mechanism, and lead to a computer model that matches the observed timecourse [5].

If proteins can diffuse from one end of *E. coli* to the other within about 1 s, then diffusion rates over shorter distances, for example from the cell surface to DNA, will on average be faster. One second is a long time on a molecular scale: the mean velocity of a water molecule at room temperature is about 500 m s^{-1} or 1800 km h^{-1} (1100 miles per hour), so that unimpeded, a water molecule could get from one end of *E. coli* to the other in 6 ns. This observation reconciles the inefficiency of a random walk with the rapid diffusion rate: although a random walk is indeed very inefficient, each step in the walk is extremely fast, so overall the random diffusion still takes place at a reasonable rate.

The diffusion of small molecules is faster than that of proteins, and small molecules can move from one end of even large cells to the other in a time of roughly 0.1 s [6], or 1 μm in less than 1 ms [7]. This analysis also implies (as is indeed the case) that most movement within the cell, for example substrate movement toward enzymes, signaling, and recruitment of proteins for cellular processes, also happens by random diffusion and does not require any 'driving force' to make it happen [8].

***7.1 Diffusion coefficient**
The diffusion coefficient D is a number that describes how rapidly a molecule diffuses through a solution, and depends on temperature T, solution viscosity η, and the radius of the diffusing particle r according to the Stokes–Einstein equation, $D = kT/6\pi\eta r$, where k is the Boltzmann constant (Chapter 4). The units of D are area per unit time, which is somewhat confusing, because one might expect it to be either a distance per unit time (that is, how far it moves) or possibly a volume per unit time (that is, the mean volume occupied). And indeed diffusion does lead to the position of the molecule spreading out into a spherical volume that expands with time. The physics shows that the distribution is Gaussian, and that the half-width increases with a radius roughly equal to $\sqrt{(4Dt)}$ after time t [46], or equivalently as a sphere with an area $4\pi r^2$ or $16\pi Dt$; in other words, it is the area of the sphere that increases in a manner proportional to time. Hence the units of D are area per unit time.

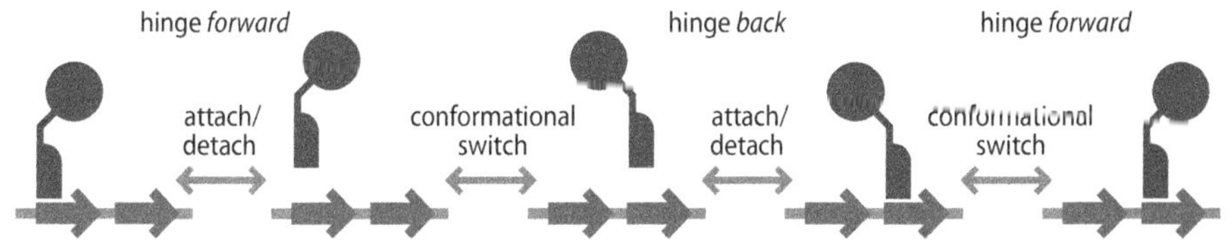

FIGURE 7.3
A simple stepping motion, requiring coordination of conformational change and binding or detachment. In this form, there is no energetic driving force to go in one direction or the other, and thus even though the fibrous substrate is directional, this represents a random movement.

7.1.2 Unidirectional movement requires a ratchet

Many of the directed movements that occur in the cell are motions of a motor protein along a one-dimensional fiber. This applies to the movements we consider below, such as movements of kinesin and dynein along microtubules, or actin along myosin. It applies also to the movement of a helicase along DNA, a ribosome along mRNA, or RNA polymerase along RNA (although in both these latter cases in fact the motion is more typically movement of the RNA along the enzyme).

In all these examples, the movement itself can be generalized as shown in **Figure 7.3**. The fiber contains multiple binding sites at evenly spaced intervals, and is directional, in that the motor can only bind one way round. The motor consists of two parts: a head that binds to the fiber, and a body. There is a hinge between the head and body, which can adopt one of two orientations, forward-facing and back-facing. A cycle of detachment, conformational switch of the hinge, attachment, and a second conformational switch serves to move the motor one step along the fiber. The movement drawn in Figure 7.3 is completely reversible, so that movement from left to right or from right to left is equally likely. Some movements in the cell, such as the searching of some transcription factors along DNA, do happen in exactly this way, and therefore search randomly forward or backward. Real systems also require some way of preventing the motor from completely detaching from the fiber at every step; different motors use different strategies, as we shall see shortly.

For movement in one direction, we require some kind of energy input. This is most clearly seen as a ratchet mechanism (**Figure 7.4**). In this example, the energy required is used to lift up the ratchet against gravity as the teeth move to the left. Once the ratchet falls, the energy required to go backward is too great, and the ratchet enforces unidirectional movement. Biological motor proteins work in the same way, with the energy input coming in the form of ATP hydrolysis. ATP has two big advantages as a gatekeeper. First, the energy obtained from ATP hydrolysis is large enough for hydrolysis to be essentially irreversible, and for the reaction to be linked to useful work; and secondly, the spontaneous rate of hydrolysis of phosphate monoesters is extremely slow, meaning that there is no risk of unwanted uncatalyzed hydrolysis. It has been estimated that phosphatases achieve the largest enzymatic rate acceleration of any enzyme identified so far, of up to a factor of 10^{21} [9].

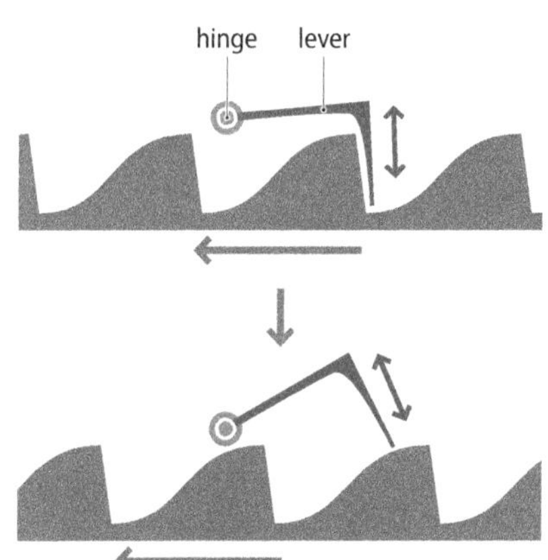

FIGURE 7.4
A ratchet mechanism. The lever can rise and fall, subject to gravity, but because of the shape of the teeth, the energy requirement for movement to the left (as indicated by the brown arrow) is much less than for movement to the right. The energy for movement of the teeth corresponds minimally to the energy needed to lift the lever against gravity.

For this mechanism to operate, we require the hinge conformation to depend on the nature of the bound ligand, such that (for example) the ATP-bound form has one conformation and the ADP-bound form has the other. Because ATP hydrolysis is essentially irreversible, this makes one of the four steps of Figure 7.3 irreversible, thereby forcing the motion to go in only one direction. There are two conformational changes in Figure 7.3: the first is from *forward* to *back*, and occurs while the head is detached; the second is from *back* to *forward*, and occurs while the head is attached. Either of these can be the step that is made irreversible by ATP hydrolysis. As we shall see, of the three main motor proteins (myosin, dynein, and kinesin), for myosin and dynein it is the first conformational change that is irreversible, while for kinesin it is less clear but may also be the first, but in a rather less direct way.

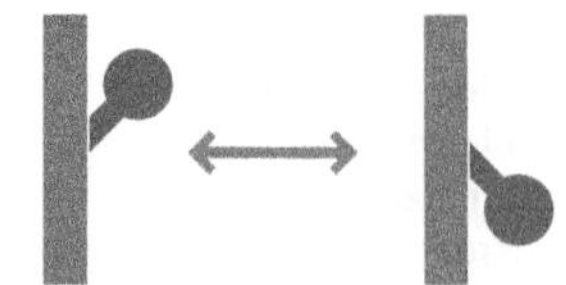

FIGURE 7.5
A switch should have only two stable positions.

The energy available from ATP hydrolysis is much greater than that required simply for a ratchet. It can therefore also be used to provide a force to drive motion. In the next sections, we see how this works.

7.1.3 Ras GTPase is the archetypal switch

As we have just seen, motor proteins work by having different conformations when bound to ATP and when bound to ADP. This is the key to their function. It is, however, easier to describe how this mechanism works by considering instead the conformational change in GTPases in response to the hydrolysis of GTP to GDP. The two are closely related and follow essentially an identical pattern. As we shall see, a GTPase switch must have evolved early, because it has been widely adopted.

The essence of a switch is that it must have two states that are both stable, but intermediate states are unstable (**Figure 7.5**). In Ras GTPase this is achieved by having different hydrogen bonding arrangements in the two states. In the GTP-bound state there is a set of hydrogen bonds that form a cooperative network; they strengthen each other, so that the formation of one stabilizes the next. The GDP-bound state is relatively disordered and 'relaxed.' In this way, any intermediate position is destabilized, because it would contain only a partial network and the amount of hydrogen bonding would not compensate for the loss of entropy. Switches are most easily constructed using hydrogen bonding, because hydrogen bonds are much more highly directional than other bonding interactions, as discussed in Chapter 1. Hydrogen bond networks form a crucial part of other switching systems, including hemoglobin (Chapter 3) and other allosteric proteins.

The basic function of GTPases is demonstrated by the signaling protein Ras (whose cellular function is considered in more detail in Chapter 8). Essentially all other GTPases and ATPases, such as the motor proteins discussed below, are similar but can have extra embellishments added on to this basic structure. The protein has a central β-sheet with helices on both sides (**Figure 7.6**) [10]. The nucleotide-binding site is highly conserved, with a highly conserved *P loop* (also called a Walker A motif) which binds to the α- and β-phosphates, and locks the GTP/GDP into place. When GTP is bound (which is the *on* state of the switch), the terminal γ-phosphate forms hydrogen bonds to two main-chain NH groups, Thr 35 and Gly 60, which are respectively in two loops known as *switch I* and *switch II*. These hydrogen bonds serve to pull these loops in toward the terminal phosphate in a 'spring-loaded' conformation (**Figure 7.7**). When the phosphate is hydrolyzed to GDP (creating the *off* state), these hydrogen bonds are lost, and the two switch loops spring back

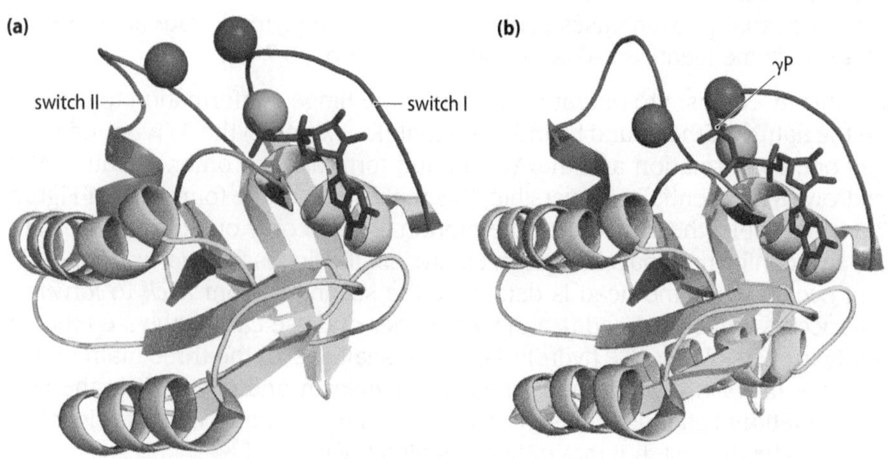

FIGURE 7.6
The basic switch mechanism of small GTPases, illustrated by the structures of (a) Ras bound to GDP (PDB file 4q21) and (b) Ras bound to a non-hydrolyzable GTP analog phosphomethylphosphonic acid guanylate ester (GCP) (PDB file 6q21). The two proteins are in roughly the same orientation, and have almost identical structures except for two loops known as switch I (magenta) and switch II (red). Also shown are GDP/GCP (blue), the magnesium ion (cyan), and the P loop (orange), which is the GAGGVG motif that binds the α- and β-phosphates. In the GDP-bound structure the two switch loops are relaxed and folded out, whereas in the GCP-bound structure they are folded in, with Thr 35 in switch I (magenta sphere, indicating the position of Cα) and Gly 60 in switch II (red sphere) making hydrogen bonds to the γ-phosphate (γP).

FIGURE 7.7
A diagrammatic version of Figure 7.6, emphasizing that hydrogen bonds from the γ-phosphate to residues in the two switch loops hold them 'spring-loaded,' ready to relax back again once the γ-phosphate has been removed. (Adapted from I. R. Vetter and A. Wittinghofer, *Science* 294: 1299–1304, 2001. With permission from AAAS.)

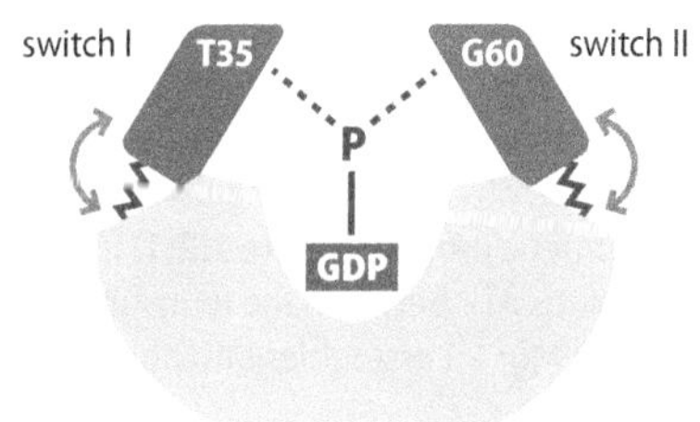

to a relaxed position. These switch loops are essentially the only structural changes between the GDP-bound and GTP-bound forms, supporting the argument made in Chapter 2 that it is much simpler if domains remain as rigid units, with the only conformational changes being to loops or linkers. The spring loading of switch I seems to be mainly provided by the Thr 35 hydrogen bonding, which also makes a hydrogen bond to the bound Mg^{2+} (see Figure 7.6b), where it replaces a water molecule that is found in the GDP-bound form.

It is of interest that in the many homologous proteins studied, the *on* state is basically the same in all of them, whereas the *off* state adopts a wide range of structures. In other words, we may picture the switch not so much as being between distinct *off* and *on* states, as being between a disordered GDP-bound state and a well-conserved cooperatively ordered GTP-bound state. Such a change is often referred to as a *disorder–order transition* and turns out to be a very common phenomenon. In evolutionary terms it is easier to evolve, because the structural constraints are onerous only for one of the two states.

The change in position of the two switch regions is relatively small. However, the structural change can be amplified, for example by having additional domains positioned next to switch I or switch II and hinged close to the switches (**Figure 7.8**). Further domains can be added to the structure so that they can sense and respond to these amplified changes.

In detail, the mechanism is of course rather more complicated than this. The spring loading requires changes in more than a single hydrogen bond. GTP hydrolysis is normally very slow, and these GTPases often require additional GTPase-activating proteins that catalyze the hydrolysis; they also require guanine exchange factors that catalyze the release of GDP so that GTP can re-bind.

It is interesting that the bases used in biology are limited almost entirely to guanine and adenine. These are purines, which are energetically more expensive to make than pyrimidines. They do, however, have the big advantage that they are larger and therefore presumably easier to recognize and bind to. Thus, evolution may tinker with whatever molecules are available, but it ends up using the ones that work best.

Having looked at a generic switch in this section, we are now ready to look at three real motors.

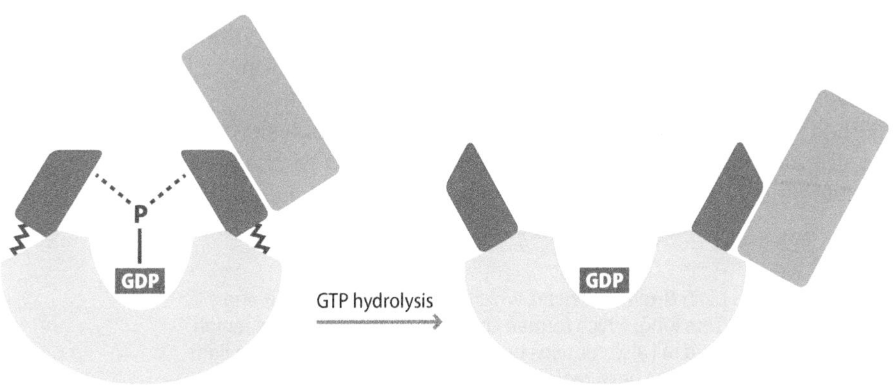

FIGURE 7.8
Amplification of the change in conformation of a switch region (here switch II) by addition of an extra domain that is bound to the switch loop.

*7.2 Actin

Actin is a very abundant protein in eukaryotic cells, particularly in muscle, often constituting 5% or more of total cell protein. The monomeric protein (**Figure 7.2.1**) is often called G (globular) actin, and binds ATP between the four subdomains. In the presence of ATP, K^+, and Mg^{2+} it assembles into polymeric actin filaments or fibers, where it is called F (filamentous) actin. The fibers have a fast-growing plus or barbed end, and a slow-growing minus or pointed end. After assembly, ATP hydrolysis to ADP is rapid. Fiber assembly is regulated by a large number of accessory proteins, as discussed in Section 7.3.2.

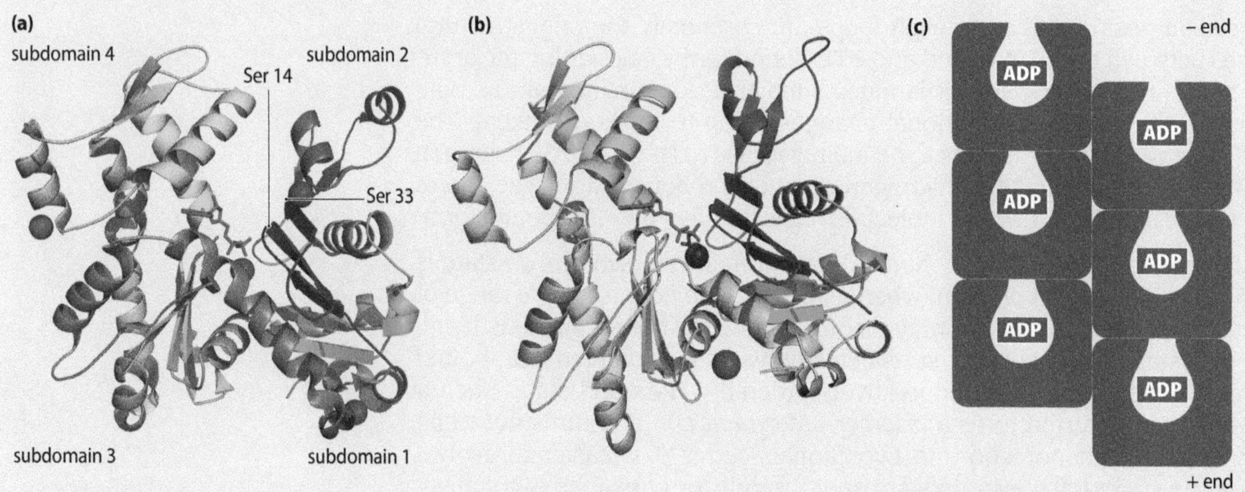

FIGURE 7.2.1
ATP-dependent assembly of actin fibers. (a) The structure of the actin.ADP complex (PDB file 1j6z). The protein is colored from blue at the N terminus to orange at the C terminus. Actin consists of four subdomains. (In Pfam actin is listed as having a single domain, in SCOP two, and in CATH 4.) Subdomains 1 and 3 have pseudo-two-fold symmetry, and the catalytic site for ATP hydrolysis is thought to be in the interface between these two subdomains, roughly where the Ca^{2+} ion (red, center) is modeled below the ADP (magenta). ADP is bound in a deep cleft in the interface between all four subdomains. (b) The structure of the actin.ATP complex (PDB file 1yag; actually a complex with gelsolin segment 1). The brown sphere is here modeled as a Mg^{2+} ion. Most of the structure is very similar to the ADP complex, but there has been an anticlockwise rotation of subdomain 2, leading to an alteration in the shape of the top surface. This is caused by a small change in the conformation of Ser 14, which hydrogen bonds to the γ-phosphate in the ATP complex but the β-phosphate in the ADP complex. This in turn leads to a roughly 1 Å movement in the main chain, causing rearrangement of Ser 33 (which forms a hydrogen bonding network to Ser 14), and thus to the orientation of subdomain 2 [49]. (c) Actin polymerizes to form filaments, in which the plus end is at the bottom of the figure, and the minus end is at the top.

7.2 MOTORS, PUMPS, AND TRANSPORTERS

7.2.1 Myosin is the linear motor of muscle

Myosin is the protein involved in muscle movement, and it tracks along a fiber made of **actin (*7.2)**. This form of myosin is called myosin II, and contains two head/hinges connected to a long coiled-coil body (**Figure 7.9**), of which only one contacts the actin at any one time. Myosin II is also involved in a range of other cellular activities, including vesicle transport and cytokinesis. The actin fiber (which in this context is usually called the actin filament) is directional, with two ends known as plus and minus (**Figure 7.10**); myosin II moves toward the plus (barbed) end. Muscle contraction occurs when myosin moves along the actin filament. Binding of ATP to myosin causes the head to detach from the actin filament. Subsequent hydrolysis of ATP produces the conformational switch of the hinge from *forward* to *back*, and constitutes the irreversible ratchet that prevents movement in a backward direction. After

actin-binding sites
ATP-binding sites
hinge

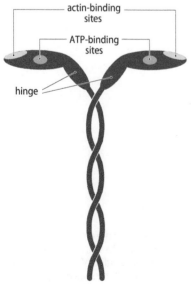

FIGURE 7.9
Myosin II consists of an N-terminal head, which binds to actin and ATP, and a C-terminal tail, 150 nm long, which forms a coiled coil dimer. The hinge region between the head and tail also contains two more molecules, the myosin light chains. The structure of the head region is shown in more detail in Figure 7.12.

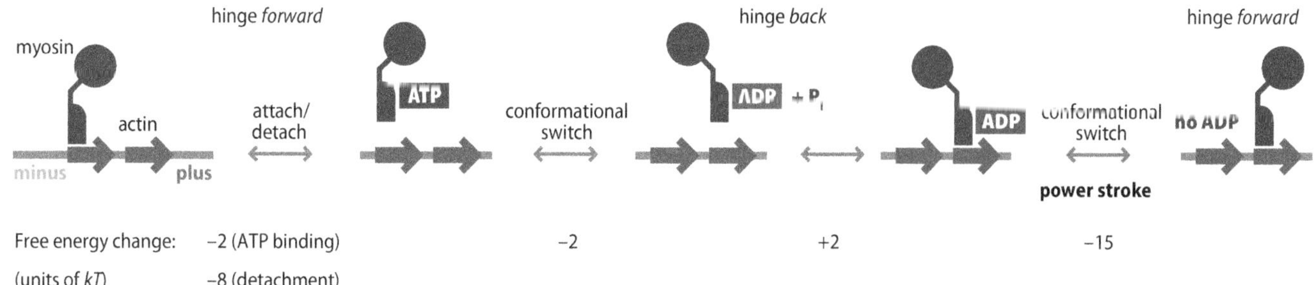

FIGURE 7.10
The mechanism of muscle movement, showing the conformational changes of myosin II, which are linked to ATP binding, hydrolysis, and release, as well as binding and detachment from actin filaments. The distance between *hinge forward* and *hinge back* corresponds to about 5 nm, which is the 'step size' of myosin II. The figure also shows the changes in free energy for each step. These are given in units of *kT*, to emphasize that most of the steps are reversible (a few *kT*), but one (the power stroke) has so large a change in free energy that it is essentially irreversible, and constitutes the ratchet that ensures unidirectional movement. The exact values for the free energy changes depend (among other factors) on the ratio of ADP to ATP around the muscle fiber and therefore can vary significantly in different situations.

this, phosphate dissociates, which enables the myosin head to bind to actin. Dissociation of ADP then leads to the second conformational switch. The conformation of the hinge therefore depends not so much on bound ATP as on bound ADP: when ADP is bound, the hinge is *back*, whereas when ADP is not bound (or when ATP is bound), the hinge is *forward*. Because this conformational change occurs with myosin bound to actin, this is the one that leads to a relative movement of actin and myosin and thus ultimately to muscle contraction; this step is therefore known as the *power stroke*. It is noteworthy that (despite being the 'power stroke') this step does not involve the hydrolysis of ATP, but merely the dissociation of ADP from myosin. How can it be that it is this step that generates the power for muscle movement?

To understand this question, we need to look more carefully at the energetics of the different steps. The exact energetics depend on the system studied, but approximate figures are given in Figure 7.10 [[11], cited in [12]]. These energies are given in units of *kT*, where the free energy for the hydrolysis of ATP under physiological conditions is roughly 25*kT*. Of this, about 15*kT* is actually used to cause muscle contraction, the remaining 10*kT* being lost as heat. (In other words, muscle has an efficiency of 60% as a motor—not a bad figure, compared for example with 30% efficiency for a gasoline/petrol car engine, or 45% for a diesel engine.)

We can see from these data that the ATP hydrolysis step itself is only favorable by 2*kT*; in other words, as part of the muscle mechanism it is almost at equilibrium, despite the fact that in free solution it is much more favorable. We can conclude that the free energy is being used to do something else: clearly, to drive the conformational switch. Thus, the ADP-bound *back* conformation is energetically unfavorable or 'strained.' This is clear from the fact that the dissociation of ADP during the power stroke liberates 15*kT*. So it is indeed true that the energy for the power stroke comes from ATP hydrolysis, although this energy is stored as a conformational strain in the ADP-bound state. It is no coincidence that the power stroke and loss of ADP is also the slowest step, taking about 80 ms under optimal conditions, thereby implying a maximum step rate per myosin head of $12\,s^{-1}$ [13].

As noted above, there needs to be some mechanism for preventing myosin from dissociating completely from actin at every step. In muscle, there are multiple myosin–actin interactions (**Figure 7.11**), thereby keeping the myosin head close to actin. Individual myosin heads dissociate frequently. In fact, this

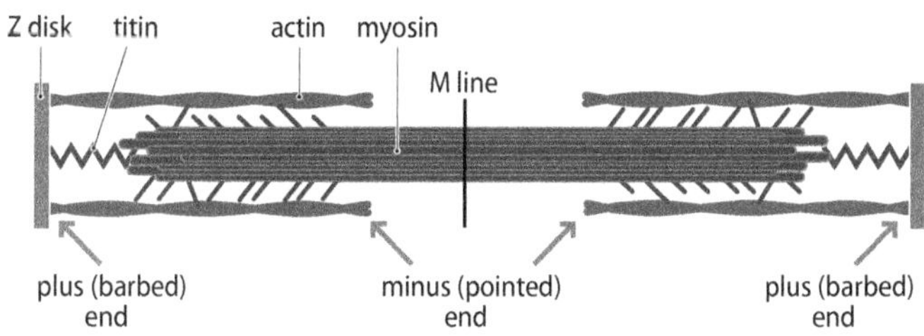

FIGURE 7.11
Actin filaments (see Figure 7.26) are attached to the Z disk via the plus or pointed end. Midway between two Z disks is the M line, which is the anchoring point for myosin. Myosin walks along actin filaments toward the plus end, which therefore leads to muscle contraction. Most of the myosin heads are not in contact with the actin at any one time; the heads not in contact therefore have time to get back into the *hinge forward* position ready for the next stroke. There is a giant protein called titin that attaches the myosin bundle to the Z disk and acts as an elastic connection to hold it correctly in place.

is an essential part of muscle function. The rapid movement of actin relative to the myosin heads occurs by multiple power strokes happening in quick succession. A single myosin head cannot shuttle backward and forward fast enough, and the rapid movement therefore requires that during the time that one head is detached, many other heads execute power strokes, each individual head contributing only a small fraction of the overall motion. It is therefore vital that the head is detached most of the time, and only attached actually during a power stroke.

Consideration of Figure 7.10 shows that ATP binding and hydrolysis has two effects: it causes a conformational change of myosin, and it also alters the affinity of myosin for actin. In particular, it ensures that during the power stroke myosin is bound, and during the other conformational change it is not, thereby maximizing the efficiency of the motor. This linking of binding affinity to conformational change is not absolutely essential, but it is such an obvious and easy route for improving efficiency that it is used (in different ways, as we shall see) by all motor proteins.

7.2.2 Myosin works by linking actin binding to head rotation

We have seen that there are two 'clever' parts of the myosin–actin system: the mechanism for making a large conformational change when ADP is bound, and the mechanism for weakening the affinity of myosin for actin when ATP is bound. These are made even more 'clever' when one considers that the binding site for ATP is some distance from the actin binding site and from the hinge. How does it occur? To understand this we need to look at the detailed structure of myosin.

The head of myosin contains four subdomains in the heavy chain. Following [14], these are called upper 50kDa, lower 50kDa, N-terminal, and converter (**Figure 7.12**). These subdomains are linked by small, flexible, highly conserved sections of linker peptide that are called switch I, switch II, strut, SH1 helix, and relay. Switch I and switch II have the same role as (and are homologous to) the regions with the same names in Ras GTPase [15]: switch I is part of the upper 50kDa domain, and switch II forms the connection between the upper 50kDa and lower 50kDa domains. The nucleotide binds in a site between upper 50kDa and N-terminal, and adjacent to a β sheet and to the P loop (again, as it does in Ras). Hydrolysis of ATP to ADP, or loss of ADP or P_i, results in changes to the conformations of switch I and switch II, as described in Section 7.1.3. By a series of linked changes in hydrogen bonding and side-chain conformation, these changes alter the relative positions of the four subdomains and of the linkers. By a mechanism closely related to that of GTPase discussed earlier, small changes close to the nucleotide-binding site can be

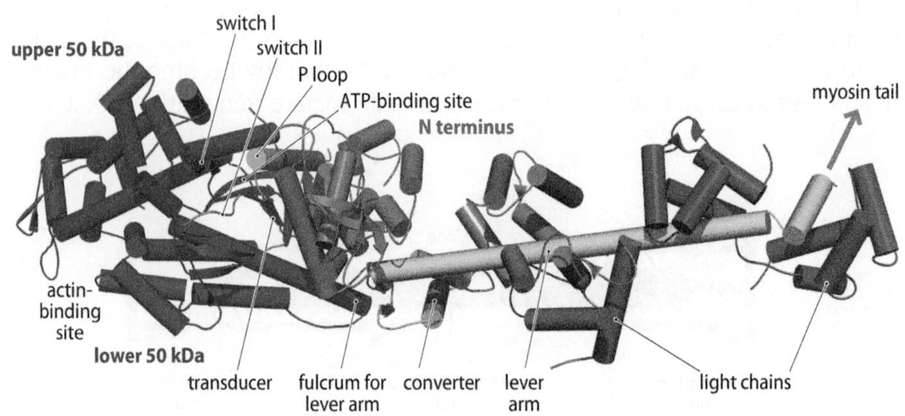

FIGURE 7.12
The myosin head in the rigor-like state. This is the state in which the myosin head is bound to ADP and in which it binds to actin (Figure 7.10), explaining why muscles become rigid (rigor mortis) after death. Coordinates are for squid myosin S1 (PDB file 3i5g). Regions discussed in the text are indicated. The lever arm is the first part of the long coiled-coil domain, which extends off to the right. ATP binds close to the P loop; as described for Ras, the γ-phosphate of ATP binds to residues in switch I and switch II. The conformational switch therefore consists of a change in switch I and switch II, leading to a change in the orientation of the brown upper 50 kDa domain relative to the blue lower 50 kDa domain. The domain boundaries used in this figure are: N-terminal domain, amino acid residues 1–203 and 665–707; upper 50 kDa, 217–465 and 603–625; lower 50 kDa, 466–602 and 643–664; converter, 708–778; P loop 175–183; switch I, 236–248; switch II, 464–473.

amplified into larger changes further away. In particular, the upper 50 kDa and lower 50 kDa domains form long extensions to the two switch loops, having the same rôle as the extra cyan colored domain in Figure 7.8. This means that loss of ADP leads to rotations in the positions of the upper 50 kDa and lower 50 kDa domains, which translates to a large change in the structure of the external junction of these two domains (top left of Figure 7.12). This is where actin binds; ATP hydrolysis and ADP dissociation therefore strongly affect actin binding.

Loss of phosphate from the ADP.P$_i$ complex leads to a change in the position of converter. The conformational change is mediated by a twisting in the β sheet in the transducer region linking the N-terminal and converter domains to the ATP-binding site, which is also thought to be where the energy liberated by the hydrolysis of ATP is stored before the power stroke. Converter is bound to a long helix called the lever arm, which is hinged from a fulcrum just above converter, between lower 50 kDa and N-terminal. A small change in the position of converter causes much larger movements in the lever arm, and therefore in the orientation of the myosin head, which is controlled by the lever arm. In myosin (as in dynein and kinesin), the distance moved by the filament-binding head exactly matches the spacing between binding sites. It is worth noting the general principle, described in Chapter 1, that rigid structural elements are carried by helices (such as the lever arm here), whereas flexible surfaces are carried by sheets.

In detail, the interactions between subdomains, and changes in the conformations and interactions of the linkers, are rather complicated. This is therefore one of the rather few examples in this book of functions that really do depend quite closely on detailed interactions. As noted above, it is also one of the few truly 'clever' functions of proteins. It is significant that the myosin head is homologous to the nucleotide-binding site of the small GTPases of the Ras superfamily (discussed above), another example of important functional consequences caused by replacement of a nucleotide triphosphate by a diphosphate. In other words, such 'clever' proteins do not evolve easily, and this relatively complicated mechanism has almost certainly been adapted and reused many times.

This description of muscle is very simplified. Muscle contains many more proteins than this, most of which have a regulatory function, for example to ensure that muscle contraction happens only in the presence of calcium ions (see Problem 5).

7.2.3 Dynein moves toward the minus end of microtubules

Dynein is a motor protein with two main functions. In the eukaryotic cytoplasm, it is required for mitosis and for the transport of vesicles along fibers made of **microtubules (*7.3)**. It is also the motor that generates the sliding motion of microtubules that leads to the beating of cilia and flagella. The microtubule, like the actin filament, is directional, with the two ends again being called plus and minus. Dynein moves toward the minus end, and seems to be evolutionarily unrelated to myosin. It is a relatively fast motor, being able to move along microtubules *in vitro* at a rate of 14 μm s^{-1}, or one 8 nm step (the spacing between tubulin binding sites) every 0.6 ms (pretty good, compared for example with the maximum *in vitro* ribosomal translation rate of one every 10–20 ms [16]). It is, however, also capable of larger steps, possibly up to 32 nm.

Dynein works in essentially exactly the same way as myosin (see Figure 7.10), with the power stroke linked to the dissociation of ADP.P$_i$, and the other conformational change occurring in an unbound state, stimulated by ATP binding. Its overall structure is, however, quite different from that of myosin. It

***7.3 Microtubule**

Microtubules are long stiff polymeric rods that extend through the cytoplasm of eukaryotic cells. They are made from the self-assembly of αβ dimers of the protein tubulin, which form parallel rods that pack together with 13 rods per microtubule (**Figure 7.3.1**). The two ends of the microtubule are different: the end that assembles and disassembles fastest is called the plus end, and the other one is the minus end. In general they grow out from the centrosome (a protein body close to the nucleus) to the cell membrane and form a dynamic transport network in the cell.

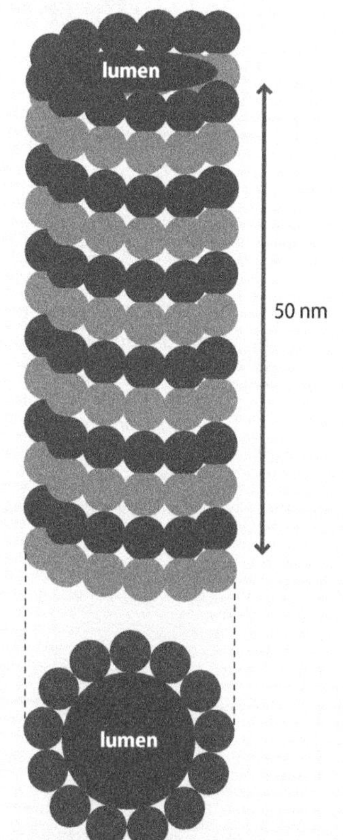

FIGURE 7.3.1
A microtubule is a stiff hollow tube made from 13 protofilaments running in parallel, each consisting of tubulin αβ heterodimers, aligned so that the β monomer (darker color) is always at the top or plus end. (Adapted from B. Alberts et al., Molecular Biology of the Cell, 5th ed. New York: Garland Science, 2008.)

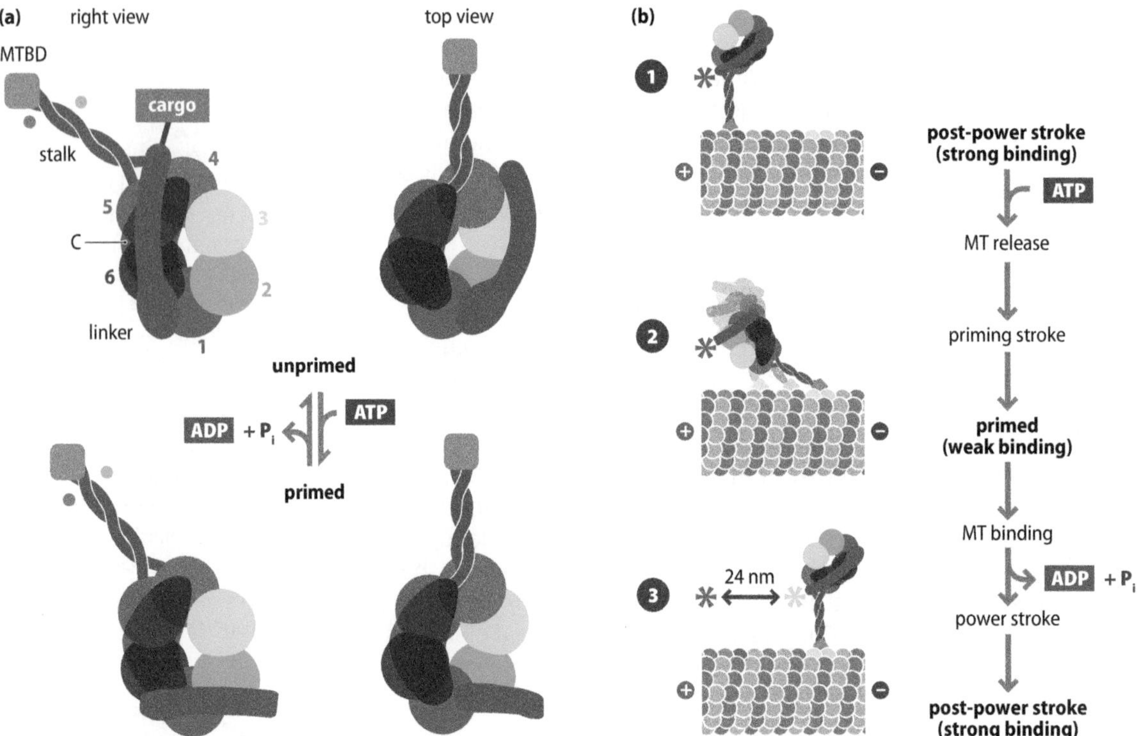

FIGURE 7.13
The mechanism of dynein, as suggested by Roberts et al. [17]. Dynein consists of a ring of six AAA+ domains, numbered 1–6, of which only AAA1 (magenta) is an active ATPase. A stalk extends away from the ring between domains 4 and 5, and consists of a coiled coil that leads to a microtubule-binding domain (MTBD). At the N terminus is a linker that connects to a tail and carries the cargo. In the unprimed state it runs past domain AAA4. At the C terminus is a rather poorly defined C-sequence, which is shown as a crescent-shaped shadow that connects AAA1 to AAA4 and AAA5, and therefore allows conformational changes in AAA1 arising from ATP hydrolysis or binding to be relayed to the stalk. (a) The main conformational change is a rotation of the linker, roughly in the plane of the paper in the top view shown on the right. There is a second conformational change, possibly a result of readjustments of the interfaces between AAA+ domains supported by the C-sequence, which causes the coiled-coil stalk to alter. A possible change in register of the two helices (brown and cyan dots) alters the linkage to the MTBD and therefore affects microtubule binding. (b) Suggested sequence of events. The asterisks denote the positions of the rest of the complex, which comprises not only the cargo (panel (a)) but also the other half of the dimeric complex (Figure 7.14). This anchors the hexameric ring to tubulin and keeps it in roughly the same place while reorientation of the linker is happening. Subsequent reattachment to microtubules followed by the power stroke causes a 24 nm movement along the microtubule (MT), which is shown as a cylinder of $\alpha\beta$ dimers (dark and light gray), with red and green dimers showing the initial and final binding sites, respectively. (Adapted from A. J. Roberts et al., *Cell* 136: 485–495, 2009. With permission from Elsevier.)

consists of a hexameric ring of proteins, to which are attached a long coiled-coil stalk with a microtubule-binding domain at the end, and an N-terminal linker and tail, which attaches to its cargo, for example vesicles (**Figure 7.13**). The hexameric ring is made up of six homologous AAA+ domains, a module that binds and hydrolyzes ATP [17]. However, only one of these (domain AAA1) is actually functional as the ATPase; three others bind nucleotide but are regulatory (and may possibly also be involved in regulating the function of dynein in response to external load), and the other two have lost the ability to bind nucleotide at all. It is thus one of many examples in this book of a system created by duplicated modules, most of which have lost their original function but may have gained new ones. There is also a C-terminal or C-sequence. The stalk is positioned between two of the AAA+ domains.

The suggested mechanism [17] is shown in Figure 7.13. The sequence starts off with the post-power-stroke (unprimed) conformation (as in Figure 7.10). Domain AAA1 is not bound to ATP or ADP, and the complex is bound tightly to microtubules. Binding of ATP causes detachment of the complex from microtubules. In a subsequent step, possibly requiring the hydrolysis of ATP, the conformation of the linker changes and the complex is now primed. It then binds to microtubules (possibly as a result of detachment of phosphate), after which a loss of ADP leads to the power stroke, in which the bound complex returns to its original conformation, thereby moving the complex about 24 nm along the microtubule.

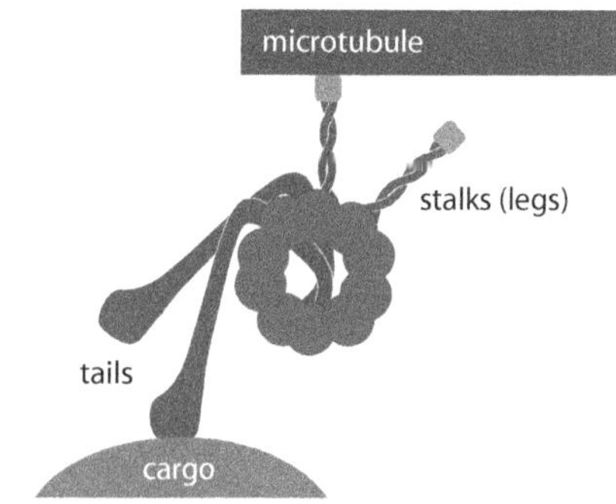

FIGURE 7.14
Model of intact dynein dimer. The dimerization region is near the point of attachment of the tail to the hexameric ring.

It is striking that ATP hydrolysis in domain AAA1 leads to changes to the angle of the stalk, which is between AAA4 and AAA5, and thus on the opposite side of the hexameric ring. It is not clear how the information occurs at the atomic level, but it is probably relayed by the C-sequence (dark region in Figure 7.13), which runs from AAA1 to the stalk, and quite possibly from changes in the relative angles of the AAA+ domains, although there is currently no information on this.

It seems that the ATP-bound form binds weakly to microtubules, whereas the ADP-bound form binds more strongly, as in myosin. Because the nucleotide-binding site resides in the hexameric ring, some 25 nm away from the microtubule-binding domain, this requires considerable long-range action. It is still unclear exactly how this occurs. However, the stalk consists of a **coiled coil**, and it is suggested that a change in the register of the two helices in the coiled coil, as indicated by the red and blue dots in Figure 7.13, could tilt the head and lead to altered affinity.

The structure described so far is only half of functional dynein, which is a dimer, linked by a dimerization domain in the tail. This means that dynein has two legs, which presumably enable it to walk along microtubules (**Figure 7.14**). There must be sophisticated regulatory mechanisms to ensure that one of the legs remains ADP-bound and firmly bound to the microtubule at all times, and therefore that one leg does not let go until the other has attached, but these details remain to be elucidated. In myosin V, which also has two heads that execute a walking movement, this process is probably controlled by the leading (front) arm changing the conformation of the dimerization region, which in turn causes a small rotation of the back arm that promotes ADP release. In this way dynein differs from myosin II, which frequently detaches from the actin filament; in contrast, dynein very rarely dissociates. This is functionally necessary, because single molecules of dynein are used to transport vesicles. If it were to detach frequently, the vesicles would fall off the microtubules and take much longer to reach their destination. The degree of binding is clearly under evolutionary control, because for example myosin V, which also transports vesicles along actin filaments, is also highly processive [18].

7.2.4 Kinesin moves toward the plus end of microtubules

Kinesin is the third motor protein to be discussed here. It looks similar to myosin II (**Figure 7.15**), and the key nucleotide-binding domain is distantly homologous to that of myosin, implying that at least some of the 'clever' details of generation of the bend and of binding to the fiber are similar to those discussed above for myosin. Like dynein, it binds to microtubules, but it moves toward the plus end, in the opposite direction to dynein. There are however exceptions to this generalization; a few myosins, kinesins, and dyneins move in the opposite direction to the majority. Details of how this is possible are given for example in [15]. There are clear structural reasons that explain the different directional motion: plus-end-directed kinesins have their motor domains at the N terminus, whereas minus-end-directed kinesins have their motor domains at the C terminus; and kinesins with motor domains in the middle do not move at all, but function to anchor microtubules to other structures. It is worth noting however that although the normal plus-end-directed kinesins are highly processive and very rarely detach from microtubules,

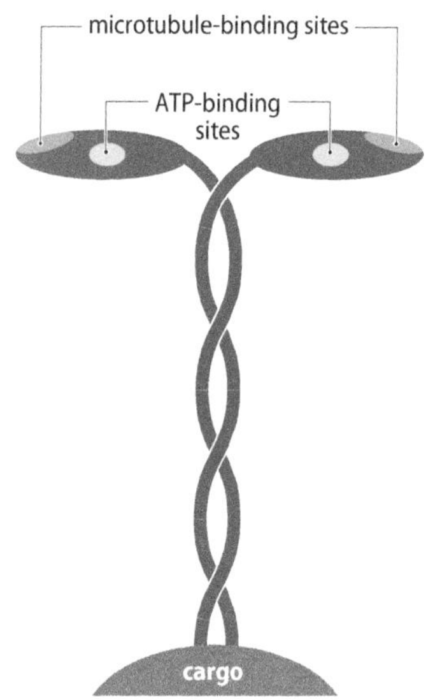

FIGURE 7.15
The structure of kinesin. Like myosin II it consists of a head that binds to a fiber and to ATP, and a coiled-coil tail. The tail carries a cargo, and the two heads are highly processive; that is, the rate of complete detachment of kinesin from a microtubule is very low.

minus-end-directed kinesins detach at every step and 'hop' along microtubules. This implies that several kinesins are required to ensure processive minus-end-directed movement.

Functionally, the kinesins are a more specialized motor, being mainly involved in **mitosis** and **meiosis**. There are many different kinesins, differing both in the cargo they carry and in the direction in which they move. In addition, some have two heads, whereas others have four and can therefore walk along two fibers at the same time, and lead to relative motion of the two fibers: either pulling together or pushing apart, depending on the type of kinesin.

How does kinesin work? It is agreed that, like myosin and dynein, conformational change and microtubule binding are related, and that the attachment and movement of the two heads are coordinated in some way by the linkage between them. This allows kinesin to walk along microtubules without detaching [18]. Other details of how kinesin moves along microtubules are still unclear. One possible model [19] suggests that kinesin works in a way roughly similar to myosin, in that the binding of ATP serves to detach a head, and that the irreversible ratchet is the loss of phosphate, with the hydrolysis of ATP again not directly providing energy but 'storing strain'. The novel feature of this model is that it sees the two heads as closely connected, in that in the absence of ATP the front head is not bound to microtubules, but stuck to the back head; binding of ATP releases the front head and allows it to scan forwards to find a binding site. If this model proves to be correct, it would imply that kinesin has more similarities than differences with myosin and dynein.

Although actin and tubulin homologs have been found in bacteria, so far no homologs of motor proteins have been identified. Possibly bacteria have never needed them because of their smaller size and reduced complexity.

7.2.5 ATP synthase is a circular motor

ATP synthase, also called F_OF_1 ATPase, is the justly famous protein complex that converts the proton gradient across the inner mitochondrial membrane into ATP. The elucidation of its mechanism by Paul D. Boyer and John E. Walker led to their being awarded the Nobel Prize in Chemistry in 1997. It is an abundant protein in all organisms, and is responsible for all ATP synthesis in eukaryotes. The basic mechanism, described in more detail below, is that flow of protons through a transmembrane channel leads to a rotation of a transmembrane assembly, which is attached to a rotor that passes through the center of a fixed head; this motion directly leads to ATP synthesis. The mitochondrial complex rotates at a maximum speed of about 130 Hz, generating 400 ATP molecules per second. Since for example the human body synthesizes more than half its weight in ATP every day (and much more in response to physical activity) it is a vital enzyme. (To put it another way, each molecule of ADP is rephosphorylated about 1000 times per day.)

The title of this section calls ATP synthase a motor (that is, a machine in which a chemical reaction leads to movement). However, its normal function in eukaryotic mitochondria is not as a motor at all but a dynamo, because it works in the opposite direction: rather than an energy source producing movement, it is movement of protons that causes energy to be stored. But it can work in both directions, and in some bacteria it does indeed work as a motor. The proteins that form the ATP-linked motor are very highly conserved in all forms of life, indicating that something similar was present in the last common ancestor, and it has been argued that the original function of the protein was as an ATP-driven motor rather than a proton gradient-driven dynamo [20]. We consider it here in the less common motor direction, in which the hydrolysis of ATP drives rotation of the motor, which pumps protons across the membrane, since it is simpler to describe in this direction. It is one of the

***7.4 Stoichiometry**

We may wonder why the number of components in the F_0 c ring is 10, 11, or 14, and how the organism knows how many components to use. A similar problem occurs for example in the photosynthetic light-harvesting 2 complexes (see Chapter 10), which has between 8 and 12 subunits in the ring. The answer seems to be a very simple geometrical one: subunits are added until a complete ring is made. The number of subunits in the ring is therefore determined by the shape of the subunit. For several multimeric proteins, site-directed mutagenesis has altered the interface slightly and has therefore altered the number of proteins in the assembly, with little or no effect on function.

very few occurrences in biology of truly rotary motion, and has been stated to have an efficiency of more than 90%, a truly remarkable figure compared for example with the already very respectable 60% efficiency of myosin [21].

ATP synthase consists of a membrane-embedded part called F_O (factor oligomycin) and a solvent-exposed part called F_1. Structural details for F_1 are fairly clear, whereas the details of F_O are still somewhat vague; we therefore know much more about how F_1 works than F_O, although the exact mechanism of both parts is still under discussion. Structural details are known for the eukaryotic mitochondrial complex and also for the bacterial complex. The two are very similar, and we will look at the mitochondrial complex.

F_1 consists of five subunits with **stoichiometry (*7.4)** $\alpha_3\beta_3\gamma\delta\epsilon$, and F_O consists of ab_2c_n, where n depends on the organism and has values between 10 and 15, typically being 10, 11, or 14 (**Figure 7.16**). In the description that follows, we shall assume that $n = 10$ (but see Problem N3). The subunits d, F_6, and OSCP (for oligomycin sensitivity-conferring protein) are also associated with F_1. In prokaryotes, δ, d and F_6 are absent, and the equivalent of OSCP is (confusingly) called δ. In eukaryotic mitochondria, n is 10. The $\alpha_3\beta_3$ proteins are arranged alternately in a hexamer, like segments of an orange. The rotor is composed of γ, which forms a rather bent α-helical coiled coil that runs through the center of $\alpha_3\beta_3$, and is attached to δε and the c_{10} ring. The membrane-bound c_{10} ring rotates past a fixed proton channel a, while the remaining parts form a fixed stator that holds the complex together and prevents the $\alpha_3\beta_3$ ring from moving round with the rotor.

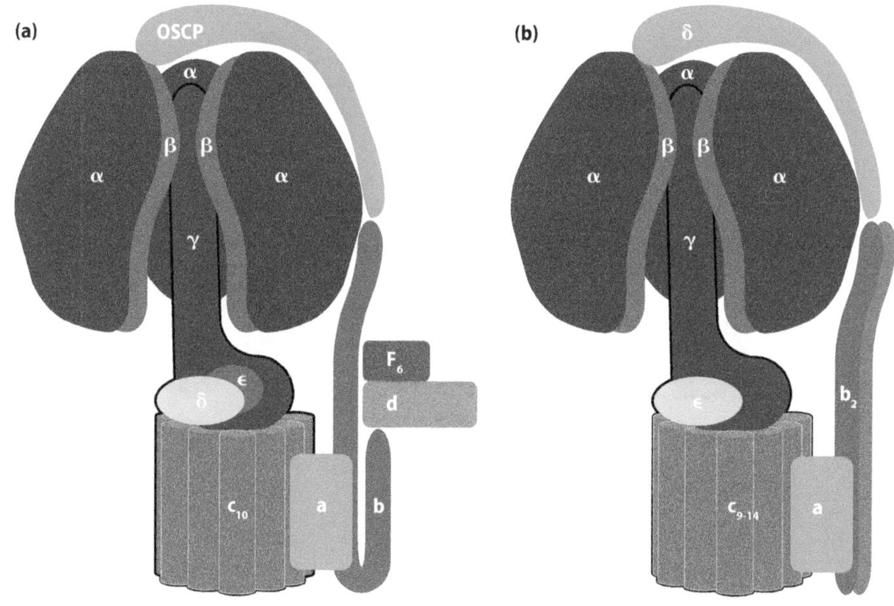

FIGURE 7.16
The structure of the F_oF_1 ATPase complex. (a) The mitochondrial complex. (b) The chloroplast and bacterial complex. Note that there is a δ subunit in both, but they are completely unrelated. The subunits enclosed by black lines (namely, the c ring, δ, ε, and γ in the mitochondrial enzyme) rotate together. There is a third β subunit that sits across the front of the $\alpha_3\beta_3$ hexamer, but this has been removed for clarity. (Adapted from D. Stock et al., *Curr. Opin. Struct. Biol.* 10: 672–679, 2000. With permission from Elsevier.)

The essential function is clear. Binding of ATP to the $\alpha_3\beta_3$ ring, followed by ATP hydrolysis and subsequent loss of ADP and P_i, cause structural changes to the ring that produce a rotation of the eccentrically oriented rotor γ. In turn, this leads to a rotation of the c_{10} ring past the proton channel a. The arrangement of charged side chains in c and a in the center of the membrane leads to a pumping of protons through the membrane, linked to the rotation of the ring. Each ATP-binding/hydrolysis cycle leads to a 120° rotation of the $\alpha_3\beta_3$ ring and thus of the rotor, while each c protein carries one proton. Every time a c protein passes the channel, a, it deposits its proton, thus pumping one proton across the membrane. Thus the hydrolysis of three ATP molecules leads to the pumping of 10 protons, an H^+–ATP ratio of 3.33. The symmetry mismatch between the hexameric $\alpha_3\beta_3$ ring and the decameric c_{10} ring is suggested to decrease any energy minima during the rotation and thus facilitate rotation [22].

There are models for how these steps work in detail. All models for the ATP motor are based on the 'binding change' principle described by Boyer [23], which was originally described in terms of three conformations for the ATP-binding sites in $\alpha_3\beta_3$ ring called tight, loose, and open. Following Weber & Senior [24], we use a different nomenclature of H (high affinity), M (medium affinity), and L (low affinity) plus O (open; that is, unoccupied). These affinities are very different: in *E. coli* the K_d values are 1 nM, 1 μM, and 30 μM, respectively.

The α and β proteins are homologous, and both contain the P loop and β sheet described for Ras and for myosin; they are therefore thought to be distantly related to these proteins. They do not, however, contain the switch regions. They each contain three domains. The binding site for nucleotide lies in the interface between α and β, though mainly on one of these. The α protein binds ATP but does not hydrolyze it: the active sites are on the three β proteins.

When ATP binds to the O form (**Figure 7.17**), the bottom half of the nucleotide-binding domain plus the C-terminal domain undergo a rotation of about 30° inward, toward the central axis [25]. This permits the formation of 15–20 hydrogen bonds to the ATP, plus it extends the central β sheet by another three strands, which are separated in the O form. The hydrogen bonding network forms two connected 'catches' that serve to lock the β conformation into

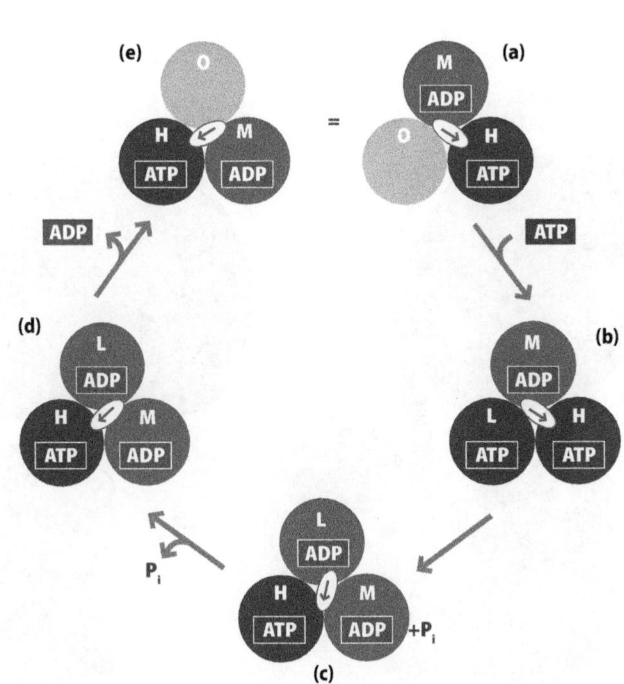

FIGURE 7.17
The rotary mechanism for ATP hydrolysis driving rotation of the central γ spindle. O, open (unoccupied); H, high affinity; M, medium affinity; L, low affinity. The central arrow represents the position of the γ spindle. Starting from top right, binding of ATP to the open site (panel (a) to panel (b)) causes hydrolysis of ATP in the adjacent anticlockwise H site by catalytic site cooperativity (panel (b) to panel (c)). The combined binding and hydrolysis of ATP produces an 80° rotation of γ, which also causes a switch in the conformation of the binding sites, and thus a change in affinity. This leads to dissociation of phosphate (panel (c) to panel (d)) associated with a further 40° rotation of the spindle, followed by ADP release (panel (d) to panel (e)), to regenerate the starting state, but rotated by 120°. In reverse, this is the mechanism for rotation of the γ spindle driving ATP synthesis. (Adapted from J. Weber and A. E. Senior, *FEBS Lett.* 545: 61–70, 2003. With permission from Elsevier.)

either a free or a bound state [26]. This rotation forces the γ stalk to rotate by about 80°, in what is essentially a mechanical push [24]. Both the γ stalk and the inner surface of the $\alpha_3\beta_3$ ring are hydrophobic (the latter being composed of proline-rich loops), presumably lubricating the rotor by removing any hydrogen-bonding interactions. Because the structural change to O alters the surface of the β domain, it is also communicated to the adjacent α domain and thence to the next β domain; that is, there are allosteric changes throughout the $\alpha_3\beta_3$ ring. In particular, it slightly alters the positions of E188 and R373 in the next β domain (anticlockwise, looking from the top of the complex down toward the membrane), which act respectively as general base and general acid to catalyze the hydrolysis of ATP, to produce ADP and P_i in the next β domain [26]. This hydrolysis, plus the rotation of the stalk, lead to a change in the relative affinities of the different sites (see Figure 7.17). Release of phosphate leads to a further 40° rotation of the γ stalk, producing a combined 120° rotation. Finally, loss of ADP from the now low-affinity third β subunit gives a structure E that is 120° different from the starting conformation A.

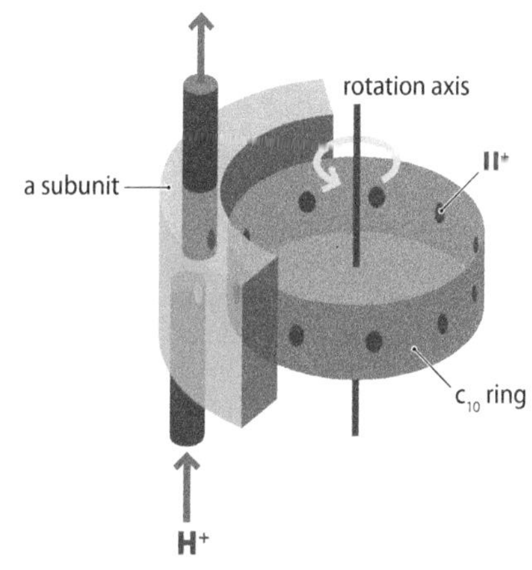

FIGURE 7.18
A model for the passage of protons through the a subunit and round the c_{10} ring as it rotates. A complete rotation of the ring translocates 10 protons. (Adapted from D. Stock et al., *Curr. Opin. Struct. Biol.* 10: 672–679, 2000. With permission from Elsevier.)

Because the rotation of the stalk is linked to proton transport in the F_O part, it is important that the rotation not be too jerky. It is also important that the force produced by the rotation be relatively uniform during the rotation. It is thought that this is achieved by the initial ATP binding occurring via a zippering up of the hydrogen bond network, which leads to a gradual change in the conformation of the β subunit, and thus a gradual rotation of the stalk [21]. (This is reminiscent of the zippering process that accompanies the dehydration of enzyme-binding sites discussed in Chapter 5, and is likely to be a very general feature of proteins.) It is also possible that the stalk is sufficiently flexible that it can 'store' some of the rotational energy as it rotates, and thus even out any energy barriers to rotation.

7.2.6 ATP synthase links the circular motor to a proton pump

We now turn to look at the proton pumping. The entire stalk composed of γ, δ, ϵ, and c_{10} rotates as one single unit. Each c subunit consists of two α helices in a hairpin arrangement. The internal helix is fixed, while the outer one is able to rotate. In the center of the membrane there is an acidic side chain, which in *E. coli* is D61. This residue has an unusually high pK_a of about 8, meaning that it prefers to be protonated while in the membrane. The c_{10} ring rotates past subunit a, which contains two half proton channels, asymmetrically placed (**Figure 7.18**), together with an arginine in the center of the membrane (R210 in *E. coli*, located on helix a_4). A plausible model [27] suggests that in what is drawn here as the starting conformation, two c subunits are symmetrically placed either side of subunit a, and are deprotonated, forming a bidentate salt bridge with R210. As the c_{10} ring rotates, the leading subunit (the right-hand helix dimer in **Figure 7.19**) moves away from subunit a and takes a proton from the adjoining periplasmic proton channel. It then carries this proton with it as it goes round the ring. Meanwhile, as the next c_2 passes sub-unit a, it rotates clockwise and the a_4 helix rocks clockwise, thus keeping the ion pair in good geometry. As the next c_2 subunit ($c2_L'$ in Figure 7.19c) approaches, it swivels round by about 180°, so that the protonated *c*D61 side chain points toward R210 in subunit a. The more polar environment caused by the R210–D61 salt bridge lowers the pK_a of c_3 D61 and therefore makes it energetically preferable for it to lose the proton. The proton can then exit through the cytoplasmic proton channel, and further rotation of the ring, and of a_4, restores the original position, with the c ring having moved on by one tenth of a complete revolution.

In the direction described here, ATP hydrolysis drives proton pumping; however, the standard direction in the mitochondrial enzyme is that proton pumping drives ATP synthesis, in which naturally everything happens backward

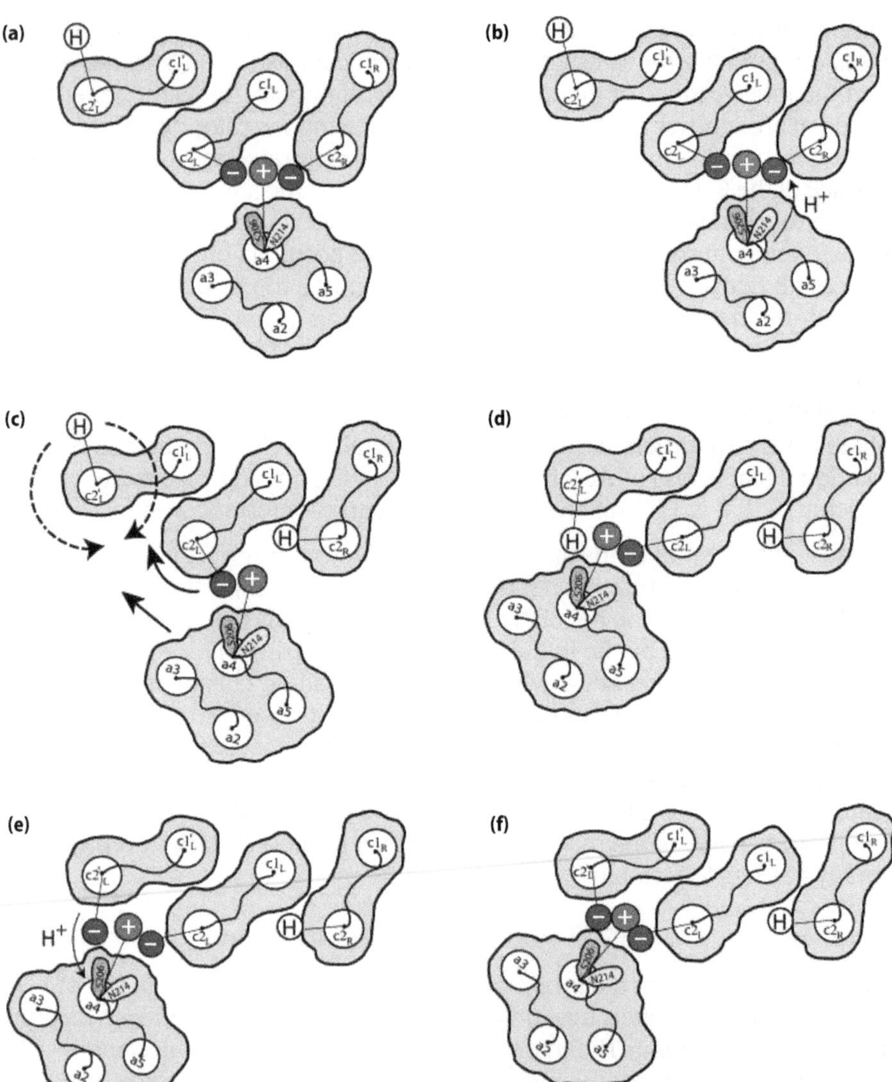

FIGURE 7.19
A detailed model of the helix and proton movements in subunits a and c as the c ring rotates past a. However, in this figure, for convenience, the c ring is drawn fixed while subunit a moves past it. Subunit a is at the bottom: the aR210 sidechain is drawn projecting out from a4 as the red positive charge. Three of the c helix pairs are shown. At the start (a), aR210 forms two salt bridges with cD61 on two adjacent c2 helices. (b) A proton moves from the 'bottom' proton channel onto the right-hand cD61. (c) The a subunit starts to move round to the left. As it goes, it rocks clockwise, and the central c2 helix also rotates clockwise to keep a good salt bridge geometry. As a approaches c2′, that helix rotates by roughly 180° so as to bring its proton round to the other side. (d) Subunit a has now moved round by 1/10 of a revolution. (e) The proton from c2′ is transferred to the 'top' proton channel and exits from the membrane, as shown in (f), which is the same as (a) but rotated by 1/10 of a revolution. (Reprinted from A. Aksimentiev et al., *Biophys J.* 86: 1332–1344, 2004. With permission from Elsevier.)

compared with what is described here. To ensure that the reaction goes to completion, there should be an excess of free energy in the forward direction. In an interesting hypothesis, it is suggested that the original enzyme was an ATP-driven motor, and that to ensure movement in the direction of ATP hydrolysis, the α subunits may have been catalytically active, thereby having an H^+–ATP ratio of 1.67 rather than the 3.33 described above which ensures that ATP hydrolysis pumps protons [20]. It is also noted that vacuolar ATPases, which are thought to be recently evolved pumps that use ATP to pump protons, evolved from the standard proton-driven dynamo described here by making every other c subunit inactive, thereby returning the ratio to 1.67. Thus, the driving force can be altered to suit the biological need by changing the ratio of H^+ to rotation. Note that the evolutionary direction is consistently toward *loss* of the original activity: see *3.5.

7.2.7 Bacterial flagella are related to ATP synthase

Bacterial flagella are attached to the cell surface via a circular ring, and their rotation is driven by proton flow. The mechanism for this is not well understood, but some of the proteins involved are homologous to those in F_O, and they are assumed to have in outline the same mechanism, namely proton flow through two half channels and a rotating unit.

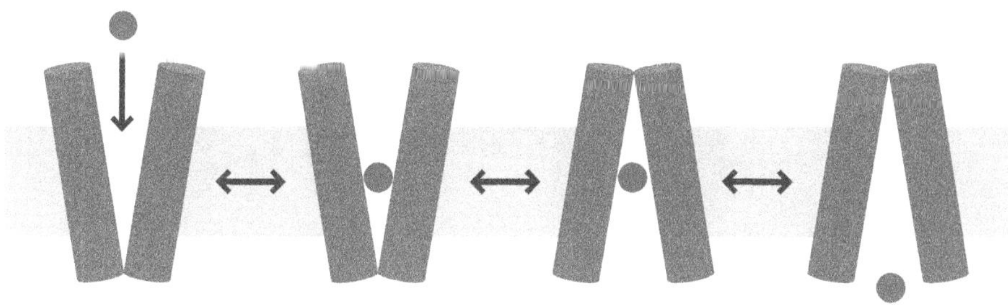

FIGURE 7.20
A simple model for how a membrane transporter works. The transporter consists of essentially two transmembrane sections, which can rotate within the membrane so that they point either in or out. In the simplest case, each step is an equilibrium, so that the direction of transport is determined simply by the concentration gradient across the membrane: in other words, this shows how passive transport occurs.

7.2.8 Many membrane pumps and transporters are based on a symmetric switch

Membranes are impermeable to water, ions, and charged and highly polar molecules including proteins, implying that any movement of such molecules across a membrane requires transport. The cell has many such transporters, which use different mechanisms. We have just seen one of these: the F_OF_1 ATPase is a proton pump driven by ATP, and several related systems exist for pumping either protons or ions such as sodium. In this section and the next we look at two other major classes, exemplifying different options available to proteins: the ATP-driven ABC transporter, and the light-driven proton pump rhodopsin.

Transport across the membrane can be classified as either active or passive. In *active* transport, the cell uses energy to drive transport against a concentration gradient. In *passive* transport, the cell makes no input of energy but uses an existing concentration gradient, and transports molecules down the concentration gradient. However, by linking this transport to transport of a second molecule, it can use a concentration gradient of the first to drive transport of the second. As we shall see, the basic mechanism for getting molecules across the cell is often similar: active transport requires linkage of the passive transporter to an ATP-driven energy source.

An outline mechanism for how a transporter might work was suggested more than 40 years ago [28], and it turns out that the suggestion was very largely correct. It was suggested that the transporter can switch between two conformations, in each of which it is open to one or other face of the membrane (**Figure 7.20**). The transporter has to have a central cavity with a binding site for the transported species. If the binding affinity in the two orientations is equal, then random opening and closing in the two orientations will transport the solute down an energy (concentration) gradient.

Many transporters do indeed seem to work in this way. For example, the *E. coli* LacY transporter, which is a **symporter (*7.5)**, transports lactose and a proton (or sodium ion) together in the same direction. Because there is an electrochemical proton gradient across the cell membrane, proton transport into the cell is energetically favorable. By contrast, the cell has more lactose inside than out, so transport of lactose into the cell is unfavorable, and in an uncoupled system, passive transport would transport lactose out. However, by transporting a proton and lactose at the same time, the cell can use the larger electrochemical gradient to transport lactose against its concentration gradient. The protein is α-helical and does indeed seem to adopt two conformations, facing in two different directions (**Figure 7.21**). Changes in free energy arising from a conformational change, coupled with the higher pH inside the cell, release a proton from the inward-facing conformation, providing the energy to drive the change [29]. It should, however, be added that despite a great deal of research on this transporter, it remains unclear exactly how the proton and lactose transport are coupled (although there are plausible models); this is an illustration of the difficulty of studying membrane proteins, discussed further in Section 11.4.8.

***7.5 Symporter**
A transporter that transports two solutes at the same time in the same direction (**Figure 7.5.1**). It generally uses a favorable concentration gradient for one of these solutes (proton, for example) to drive transport of the second solute against a (smaller) concentration gradient. Its opposite is antiporter.

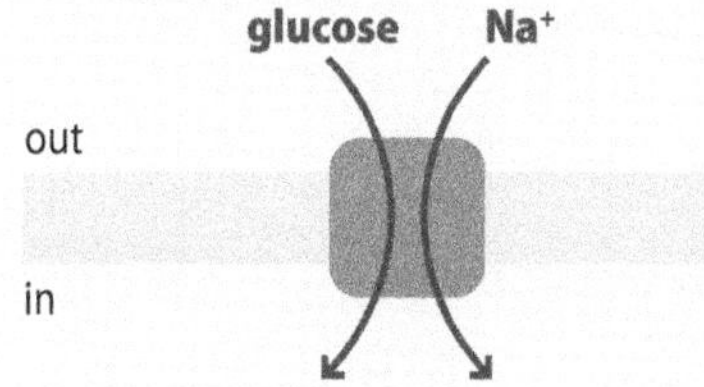

FIGURE 7.5.1
A symporter, here one that transports glucose against a concentration gradient, driven by the favorable concentration gradient for sodium. The SGLT1 transporter works in this way and therefore permits absorption of glucose from the intestine even against a concentration gradient.

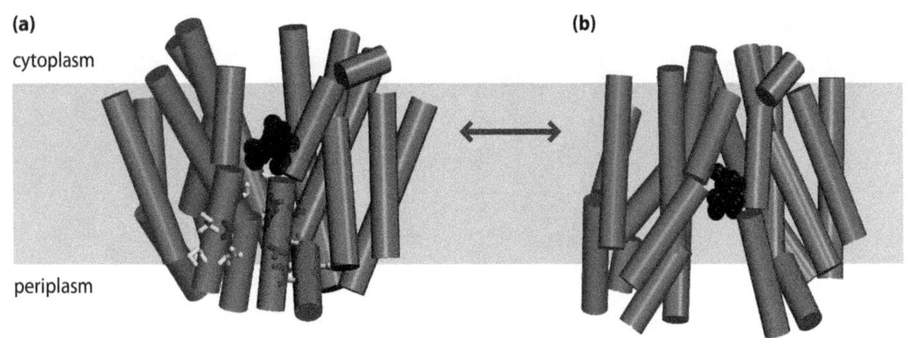

FIGURE 7.21
A model for the *E. coli* transporter LacY, which is a symporter that transports lactose and H^+. The two domains are shown in different colors. The protein has pseudo-two-fold symmetry, indicating an ancient duplication event. (a) The crystal structure (PDB file 1pv7). This is the protonated inward-facing conformation. A lactose analog, β-D-galactopyranosyl-1-thio-β-D-galactopyranoside (TDG) is shown in black, positioned centrally within the membrane, roughly centrally placed between the two domains. The periplasmic side (bottom) is tightly closed. The residues shown as sticks are implicated as being accessible to the periplasm in this conformation because when mutated to cysteine they are more reactive when substrate is bound. (b) A model for the outward-facing conformation, based on chemical modification and cross-linking experiments. Each domain was rotated by about 60° relative to its position in (a). (From J. Abramson et al., *Science* 301: 610–615, 2003. With permission from AAAS.)

Transporters can also operate as **antiporters (*7.6)**, in which they carry molecule A in one direction, down a concentration gradient, but then require the binding of a second molecule B to return to its starting position, pumping this against a concentration gradient.

Many ATP-driven transporters use a similar mechanism, coupling it to ATP hydrolysis. An important such class of transporters is the *ABC* or ATP-binding cassette transporters. This is the largest and most diverse group of ATP-driven pumps, used by bacteria for importing a range of substrates, and by bacteria and eukaryotes for export, including the export of drugs. ABC transporters are therefore largely responsible for drug resistance by pumping out the drugs as soon as they enter cells.

ABC transporters consist of the roughly two-fold symmetric structure shown opposite, which acts as the transmembrane transporter, linked to a dimeric ATPase domain via a 'coupling helix' that sits at the bottom of the transporter inside the cell (**Figure 7.22**) [30] and acts as a ball-and-socket joint to enable the transmembrane domain to swivel with respect to the ATPase domain. The mechanism is reminiscent of the GTPase mechanism described earlier in this chapter. The nucleotide binds via a P loop. The ATPase domain contains several other sequences with analogous functions to the switch domains in GTPases, which alter their conformation when the ATP is hydrolyzed and therefore lead to a domain reorientation. The effect of this is that in the ATP-bound conformation, the two coupling helices are brought together, being roughly 28 Å apart, leading to an outward-facing channel (see Figure 7.22), whereas in the free conformation, the coupling helices have moved apart by more than 10 Å and the channel is now inward facing [30]. For an *importer*, binding of ATP therefore flips the transporter into an outward-facing conformation ready to accept a ligand, and for an *exporter* it permits release of the substrate. Hydrolysis of ATP and release of ADP/P_i then flips the transporter in the opposite direction. It is likely that the conformational change also affects the affinity of the protein for its substrates.

***7.6 Antiporter**

A transporter that transports two solutes at the same time but in opposite directions (**Figure 7.6.1**). It generally uses a favorable concentration gradient for one of these solutes (sodium, for example) to drive transport of the second solute against a (smaller) concentration gradient. Its opposite is symporter.

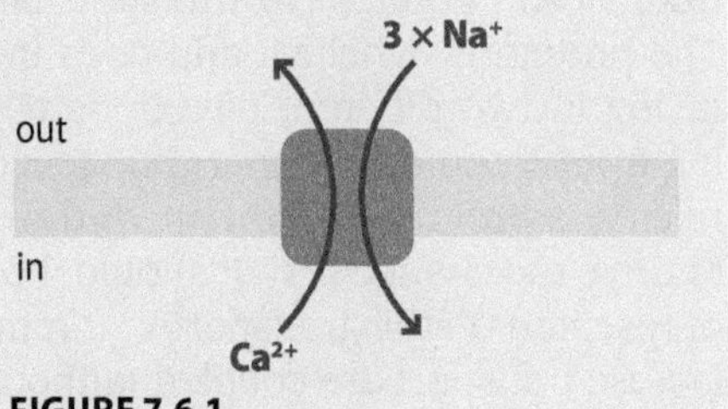

FIGURE 7.6.1
An antiporter, here the Ca/Na antiporter.

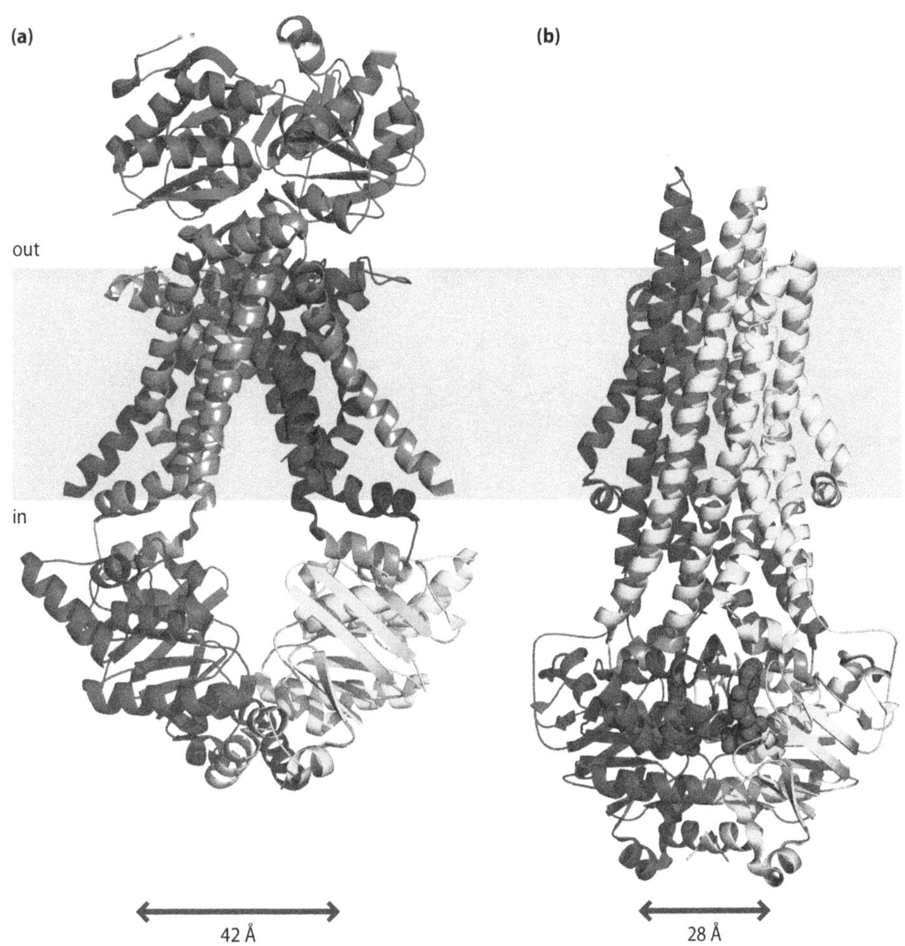

FIGURE 7.22
Conformational changes in ABC transporters. The approximate position of the membrane is indicated by the beige region. (a) The unbound molybdenum/tungsten transporter ModBC (PDB file 2onk). This protein consists of two pairs of polypeptides, constituting the ATP-binding domain (bottom; green and red) and the transmembrane domain (center; blue). In common with many such proteins, there is also a substrate-binding domain at the top, whose function is to bind the substrate and deliver it to the transporter. This is an inward-facing conformation, with an open ATP-binding site. In consequence, the coupling helices (orange) are rotated out, roughly 42 Å apart. (b) The ATP-bound state, shown by the *Staphylococcus aureus* transporter Sav1866, a homolog of human multidrug resistance transporter Mdr1, bound to the ATP analog AMP-PNP (magenta spheres) (PDB file 2onj). It is a dimer of identical chains. This is an outward-facing conformation, in which the coupling helices are now only approximately 28 Å apart. (Adapted from K. Hollenstein, R. J. P. Dawson and K. P. Locher, *Curr. Opin. Struct. Biol.* 17: 412–418, 2007. With permission from Elsevier.)

7.2.9 The light-driven proton pump rhodopsin is a seven-transmembrane-helix G-protein-coupled receptor

A different mechanism is exemplified by the light-driven proton pump bacteriorhodopsin (bR), which is related to the protein rhodopsin, the protein in the rod cells of the eye that contains retinal and traps light. In the halophilic bacterium *Halobacterium salinarum* it converts sunlight into energy by using light to make a proton gradient across the cell membrane. The protein is at such high concentration that it forms purple crystalline patches on the surface of the bacterium. This structure was the first to be determined to atomic resolution by electron microscopy (Chapter 11). It is similar in structure and function to rhodopsin, for which there are now several crystal structures. Rhodopsin consists of seven transmembrane helices, in the center of which is a retinal molecule, held by a Schiff base (see Figure 5.22) to lysine 296. The retinal sits in the middle of a proton channel and blocks it. Light causes a change in the structure of retinal from *cis* to *trans* (**Figure 7.23**), accompanied by changes in helix positions, which serve to block the channel in one direction and pump a proton from one side of the membrane to the other. In rhodopsin, the conformational change is transmitted to a heterotrimeric G-protein complex and triggers a signal: rhodopsin is a G-protein coupled receptor (Chapter 8), although a rather atypical one.

Rhodopsin has been investigated in enormous detail for many years. Many of the intermediates in the reaction have different spectroscopic signatures, enabling a characterization of many of them. However, the lack of a structure for some of the key intermediates has so far prevented the establishment of a complete picture of how it works. The current status is as follows [31, 32].

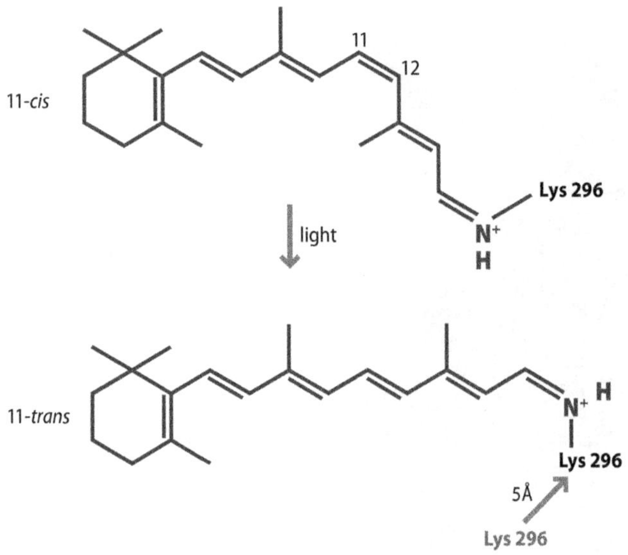

FIGURE 7.23
The pigment retinal undergoes isomerization between 11-*cis* and 11-*trans* forms on irradiation by light. Retinal is covalently attached to Lys 296, and in this diagram the isomerization moves the attachment point by 5 Å: in the protein Lys 296 stays in the same position but the chromophore moves by 5 Å. This is the basis for the detection of light in the eye.

Irradiation by light creates *trans*-retinal within less than 1 ps (**Figure 7.24**). This is followed by a series of much slower protein rearrangement steps, resulting in the key intermediate M-I. The Schiff base proton is lost from this to Glu 113 (blue in **Figure 7.25**) and can diffuse out of the bottom of the protein, resulting in the intermediate M-II. Steps after this are still unclear. A likely scenario is that the deprotonation of the Schiff base leads to a freeing up of transmembrane helix VI (cyan in Figure 7.25), which is able to move up and to the left. Because Gα is bound at the top end of helix VI, the freeing up of the helix allows it to exchange GTP for GDP and therefore initiate signaling within the cell. This further allows E134/R135 (red) to pick up a proton and transfer it back to the retinal. By far the slowest part of the photocycle (an enormous factor of 10^{11} slower than the isomerization!) is dissociation of the *trans*-retinal and rebinding of *cis*-retinal.

The mechanism has some interesting similarities to the mechanism of ABC transporters described previously, in that energy-dependent changes in helix orientations provide access of a central binding site to one side or the other: presumably, in the starting conformation, helix VI blocks access to the channel in the cytoplasmic direction. This is an example of convergent evolution: such an elegant mechanism is such a good solution that it has been created independently in these two systems.

It is noticeable that the proteins described in Section 7.2 are almost entirely helical. One good reason for this is that the proteins are required to be rather stiff, with two 'switch' positions. Such a requirement is satisfied by helical proteins, as discussed in Chapter 1.

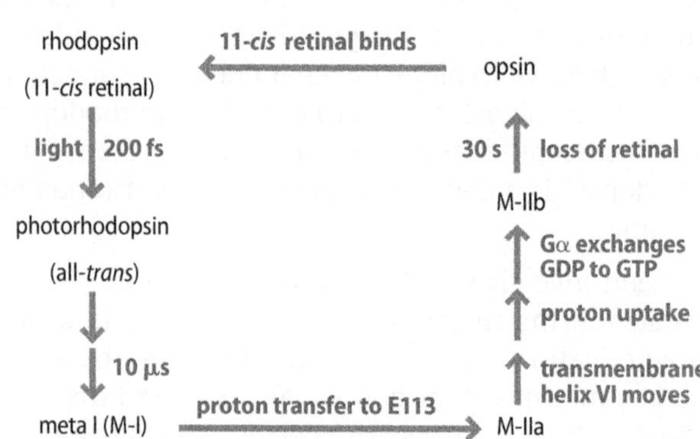

FIGURE 7.24
The rhodopsin photocycle. There are several intermediates between photorhodopsin and M-I. M-II shows only a single spectroscopic signal, but probably involves several forms in equilibrium.

FIGURE 7.25
The crystal structure of bovine rhodopsin (PDB file 1gzm), shown with the cytoplasmic face at the top. Retinal is shown in magenta, with the ring on the left. It is attached to K296, in pink, with the proton-bearing nitrogen shown as a dark pink sphere. The proton acceptor E113 is in blue. Proton uptake occurs via E134, R135, and Y136 (red). The G protein binds to a surface close to these residues, involving helices V (on the left) and VI (cyan). Helix VI moves in the direction of the arrow during the photocycle, as a consequence of structural rearrangements around the retinal, which passes close to it at a kink in the helix caused by P267 (indigo).

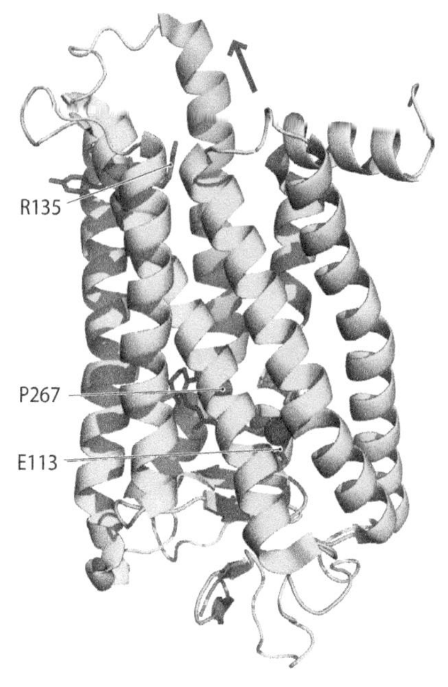

7.3 MOVEMENT ALONG ACTIN AND TUBULIN FIBERS

7.3.1 Actin and tubulin fibers continually assemble and disassemble

The shape and organization of eukaryotic cells are controlled by the positions of the fibers within them. These fibers come in three forms: microtubules, actin filaments, and intermediate filaments. The first two of these we have seen already; the intermediate filaments are less universal, and are different from the other two in that they are fibrous as monomers and form semi-permanent strengthening rather than fibers for movement. They are therefore more typical of the conventional permanent fibrous proteins (*1.14) and are less interesting than actin and tubulin. We shall not consider them further.

Actin filaments and microtubules form the highways on which all non-random intracellular transport occurs. As we shall see shortly, because the molecular motors move in a defined direction along their fibers, the function of each motor is more or less defined once the fibers are in place. Therefore the key to intracellular transport is the correct assembly of the fibers, and their growth and disassembly. A vital feature of both fibers is that they are not permanent: in fact they often only last a few minutes or even less before they are disassembled and rebuilt elsewhere. Alberts et al. [33] use a very apt analogy that the fibers are more like ant trails than highways.

Actin filaments are made of actin, whereas microtubules are made of tubulin. In both cases, the filaments are made of strings of globular proteins, rather than long fibrous proteins, which explains how they are able to assemble and disassemble so rapidly: the protein constituents are small and can diffuse rapidly wherever they are needed. Actin filaments are two-stranded twisted polymers (**Figure 7.26**), which makes them flexible, whereas microtubules are hollow cylinders made from 13 tubulin αβ dimers arranged in parallel, and are straight and stiff (**Figure 7.27**). However, in many other ways the assembly of the two fibers is very similar: I shall therefore use tubulin as an example.

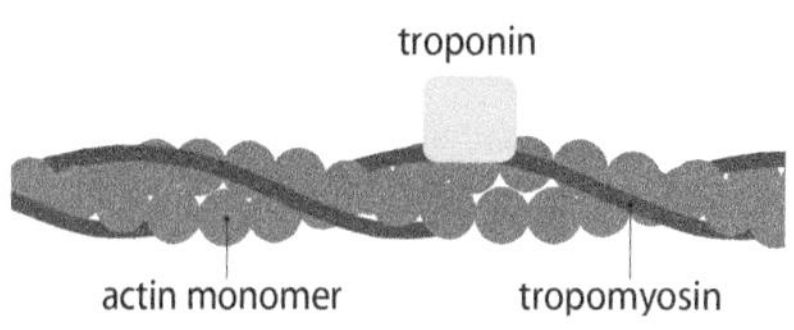

FIGURE 7.26
The structure of an actin filament. The basic actin filament consists of actin monomers (blue), which form twisted filaments with a period of about 37 nm. The figure shows the structure of actin filaments within muscle, which also have tropomyosin wound round them to stabilize them (Table 7.1) and are attached to complexes of troponin T, I, and C. The functions of these three proteins are respectively to bind near to ends of tropomyosin filaments, to bind actin, and to bind calcium, as well as troponins T and I.

The strong connections between the monomers in the fibers are longitudinal (along the fiber), but the transverse interactions also help to strengthen the fiber. This means that fibers are generally only made and destroyed at the ends. Assembly and disassembly are governed by rather simple principles, which can be summarized as follows.

The addition of a monomer to the end of a fiber can be described by a standard equilibrium:

$$\text{fiber}_n + \text{monomer} \underset{k_{\text{off}}}{\overset{k_{\text{on}}}{\rightleftharpoons}} \text{fiber}_{n+1}$$

The overall rate of formation per fiber is therefore k_{on}[monomer], whereas the rate of breakdown is k_{off}. At equilibrium these must be equal, and therefore the dissociation constant (*5.5) K is given by

$$K = k_{\text{off}}/k_{\text{on}}$$

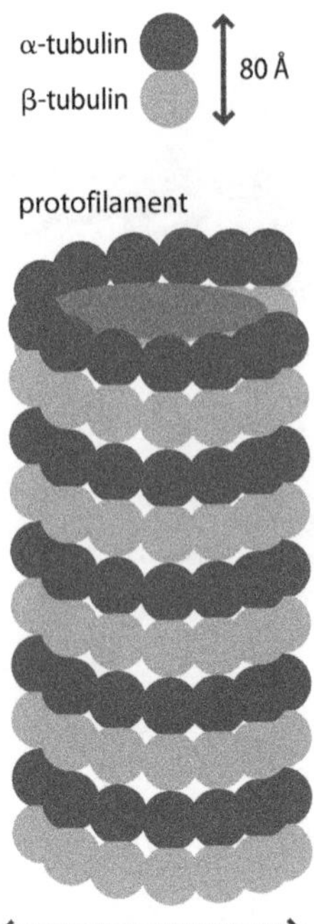

FIGURE 7.27
The structure of a microtubule, and of the constituent tubulin αβ dimer.

K has units of concentration. When the concentration of monomer is less than K, then k_{on}[monomer] is less than k_{off}, and the fiber dissociates. When the concentration is greater than K, the reverse is true and the fiber grows (**Figure 7.28a**). K is thus called the *critical concentration* of monomer.

Thus, the cell is able to change the rate of growth or shrinkage of the fiber by varying the monomer concentration. Usually this is done not by changing the total concentration of monomer (which is often very large) but by altering the activity of proteins that bind to monomer, such as thymosin, which binds monomeric actin and prevents it from assembling; or profilin, which binds monomers and presents them to the actin filament thus increasing the growth rate (see Tables 7.1 and 7.2 below).

The fiber has two different ends, and in a free fiber monomer can add on to either end. In many cases the fiber is attached to a nucleation center at one end, as discussed below, and thus the growth rate depends only on what happens at the free end. In a free fiber, the on- and off-rates are different for the two ends. However, because the equilibrium above looks the same whichever end the monomer attaches at, the equilibrium constant must be the same for both ends: in other words, the ratio of on- and off-rates must be the same at both ends. Thus, for any given monomer concentration, the fiber either grows or shrinks at both ends, but at different rates; and at the critical concentration there is no change at either end (Figure 7.28b). The end at which faster change occurs is called the plus end, and the other is the minus end; we have already seen that both actin filaments and microtubules have such ends, and that the motors move specifically in one direction.

7.3.2 Cells tightly regulate fiber growth

The cell needs to have more control over fiber growth than is provided by this simple mechanism. For actin, this is provided by ATP binding; for tubulin, it is provided by GTP binding. Tubulin is in fact distantly homologous to the GTPases discussed above such as Ras. A GTP-bound tubulin (t^T) binds much more strongly than a GDP-bound form (t^D), implying that K^T (the dissociation constant for the GTP form) is smaller (tighter binding) than K^D. Because K is the critical concentration, this also means that the critical concentration of monomer is lower when GTP is bound. The concentration of GTP or ATP in the

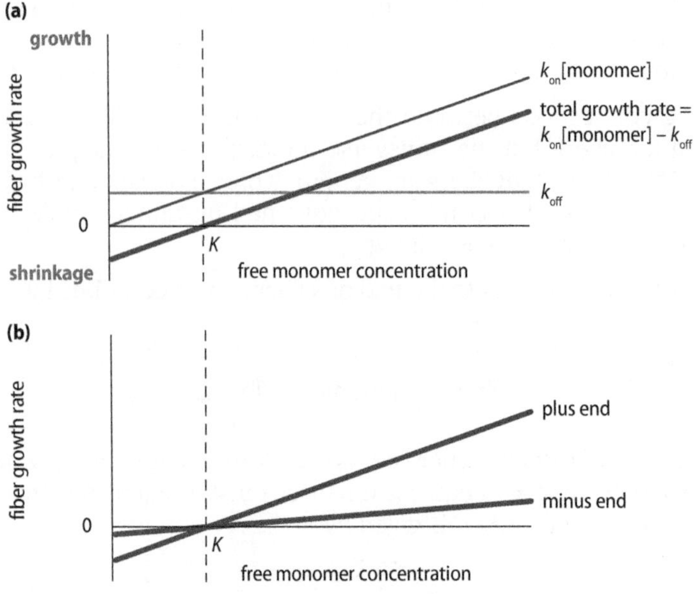

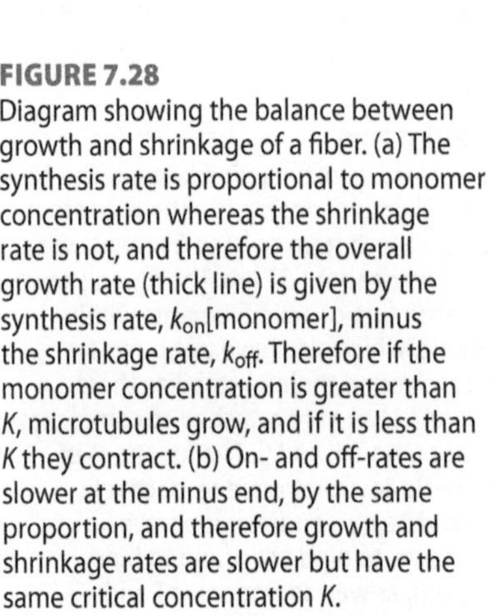

FIGURE 7.28
Diagram showing the balance between growth and shrinkage of a fiber. (a) The synthesis rate is proportional to monomer concentration whereas the shrinkage rate is not, and therefore the overall growth rate (thick line) is given by the synthesis rate, k_{on}[monomer], minus the shrinkage rate, k_{off}. Therefore if the monomer concentration is greater than K, microtubules grow, and if it is less than K they contract. (b) On- and off-rates are slower at the minus end, by the same proportion, and therefore growth and shrinkage rates are slower but have the same critical concentration K.

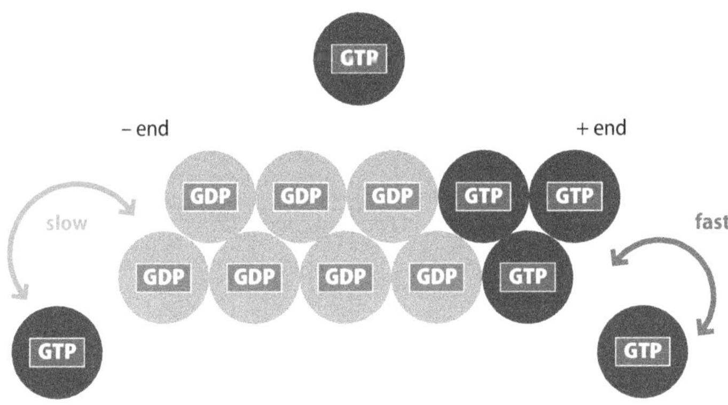

FIGURE 7.29
Each free tubulin monomer is bound to GTP. This is hydrolyzed slowly in free tubulin, but much more rapidly when the tubulin is part of a microtubule. This means that the plus end has one or two GTP-bound monomers (which get hydrolyzed to GDP at a rate dependent on the overall rate of growth) but the minus end usually only has GDP-bound monomer. Each circle is intended to represent an $\alpha\beta$ dimer.

cell is normally roughly 10 times the concentration of GDP or ADP. Therefore most of the free subunits are GTP-bound, and essentially all the subunits that add to the fiber are GTP-bound.

So far, this is not a mechanism for regulating fiber growth, because the ratio of free GTP to GDP in the cell is held almost constant. The regulation arises because in free monomer the hydrolysis rate of GTP to GDP is slow, but once the monomer has attached to the fiber the GTP hydrolysis rate is usually very fast: in fact, often faster than the rate of addition to the minus end, but not as fast as the rate of addition to the plus end. This means that the plus end usually has GTP bound, but the minus end has GDP bound (**Figure 7.29**). The growth rate at each end can therefore be differentially regulated by controlling the rate of hydrolysis of GTP within the fiber, which is done by the binding of regulatory proteins to the fiber.

The hydrolysis of GTP also makes possible the phenomenon known as *treadmilling* (**Figure 7.30**). We saw above that the critical concentration is lower when GTP is bound (K^T) than when GDP is bound (K^D) (**Figure 7.31**). This implies that at a monomer concentration greater than K^T but less than K^D, the plus end is growing, while the minus end is shrinking. There is some intermediate concentration, marked by the double-headed arrow in Figure 7.31, at which the two rates are equal. At this point, the fiber stays a constant length, but apparently moves in the direction of the plus end.

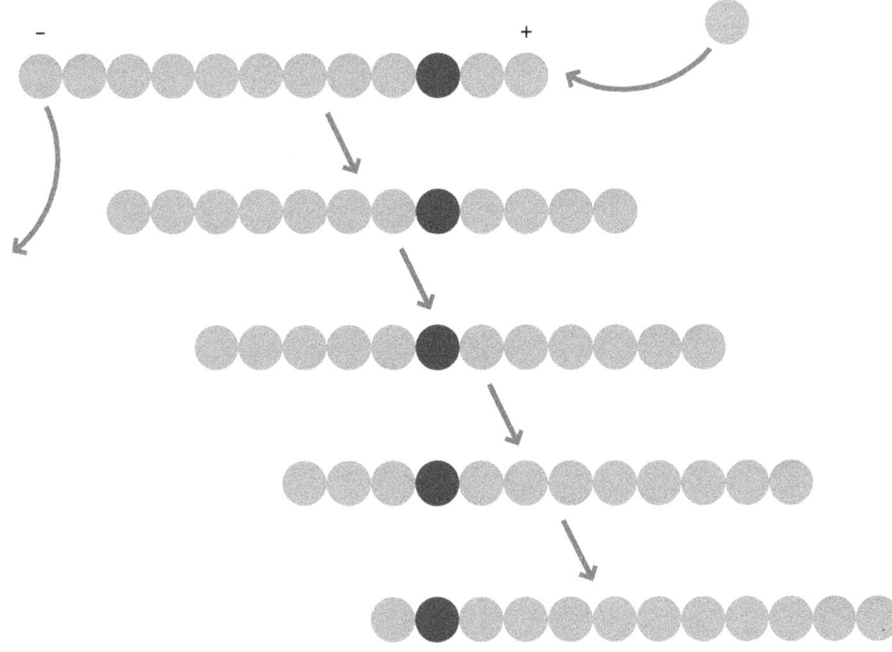

FIGURE 7.30
Treadmilling. The fiber seems to be moving to the right, but in fact the component monomers are staying in the same place (as indicated by the darker blue one), and new monomers are being added to the plus end at approximately the same rate as old ones are being removed from the minus end.

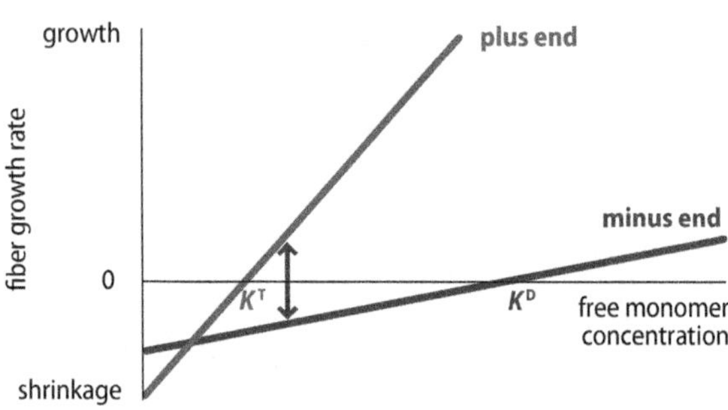

FIGURE 7.31
The mechanism of treadmilling. When the monomer concentration is in the range between K^T and K^D, there is growth at the plus end (where GTP is bound) but shrinkage at the minus end (where GDP is bound). There will be a concentration, marked by the double-headed arrow, at which the rate of growth at the plus end exactly matches the rate of shrinkage at the minus end: at this point the fiber will treadmill perfectly.

There is a further interesting consequence (**Figure 7.32**). As noted above, normally the plus end has GTP bound, because it is made from the addition of a GTP-tubulin unit. Once added to the fiber, the GTP gets hydrolyzed to GDP. Depolymerization of tubulin from the plus end is about 100 times faster from GDP-tubulin than it is from GTP-tubulin. Therefore, if the hydrolysis rate is similar to the rate of addition of new tubulin units, it can happen that the newly added but GDP-bound tubulin dissociates before a new GTP-tubulin can bind. Once this has happened, then the plus end will be terminated with a GDP-tubulin and yet more tubulin can dissociate. The consequence is that the growth rate can go very rapidly from a steady growth rate to a dramatic shrinkage (black double-headed arrow in Figure 7.32). Recall that hydrolysis and polymerization/depolymerization rates are averages of essentially random events: on an individual molecular scale they are unpredictable and subject to statistical fluctuations (Section 4.1.1). Thus, once the hydrolysis rate becomes similar to the rate of addition of new units, this reversal of growth can happen randomly, in either direction. It is called *dynamic instability*, and is most useful when the cell wants to remodel its fibers. The change between growth and shrinkage seems to occur often *in vivo* as part of the normal lifespan of fibers, every few minutes for microtubules and at even shorter intervals for actin. It is thought that dynamic instability predominates in microtubules, and treadmilling predominates in actin.

Thus, we have seen that the cell can regulate fiber growth at either end by controlling the concentration of free monomer. Both of these can be altered by a wide range of accessory proteins. However, there is another major control, which originates from nucleation of the fibers. Actin and tubulin fibers both require some kind of nucleation event to start them off. Actin uses two different proteins to do this: formin, which produces parallel bundles of actin, and the Arp complex, which produces a mesh. (See **Table 7.1**, which lists several proteins that interact with actin and thereby regulate its function, and **Table 7.2**, which does the same for tubulin.) These proteins are usually found close to the cell membrane, implying that actin produces bundles and meshes close

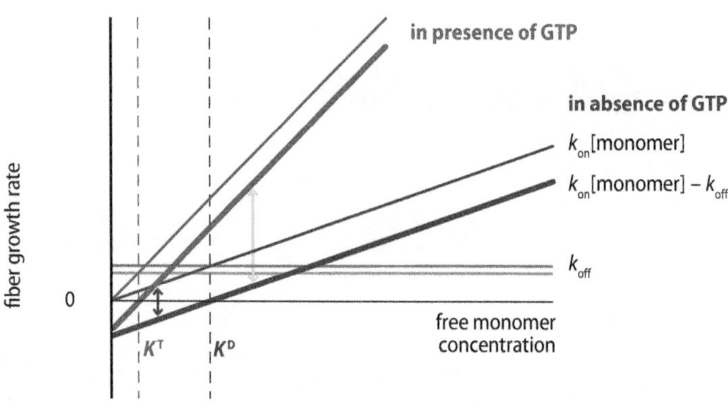

FIGURE 7.32
Dynamic instability. Growth at the plus end requires the addition of GTP-tubulin. Once bound to the fiber, the GTP is hydrolyzed to GDP. If the hydrolysis rate is similar to the rate of addition of new monomer, then sometimes the newly added monomer will become GDP-tubulin before the next monomer is added. Depolymerization of tubulin from the plus end is about 100 times faster from GDP-tubulin than from GTP-tubulin. Therefore, hydrolysis of GTP to GDP can result in depolymerization at the plus end. The result is either a marked slowing down of fiber growth (green double-headed arrow) or an even more marked sudden depolymerization (black double-headed arrow).

TABLE 7.1 Proteins that interact with actin

Protein	Function
Arp2/3, formin	Initiates filament formation
Nebulin, tropomycin	Stabilizes filaments
α-Actinin, fimbrin, villin	Cross-links filaments in parallel rows
Filamin	Cross-links filaments in meshes
Tropomodulin, CapZ	Caps ends
ADF/cofilin, gelsolin, thymosin	Cuts or depolymerizes filaments
Profilin, twinfilin	Binds monomers
Dystrophin, spectrin, talin, vinculin	Links filament to other proteins

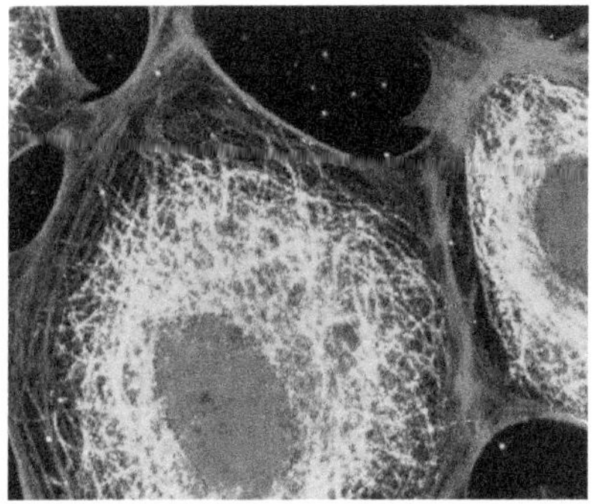

FIGURE 7.33
Fluorescence microscopy image of cells stained with a blue dye that binds to DNA and locates the nucleus; a green fluorescent antibody that binds to microtubules; and a red fluorescent antibody that binds to actin. The actin is located in bundles and meshes around the cell periphery, while the microtubules run in many directions but tend to radiate out from the nucleus. (Image from *Traffic, International Journal of Intracellular Transport*, virtual issue on cytoskeleton, 2010. With permission from Wiley-Blackwell.)

to the cell membrane (**Figure 7.33**). Microtubules are quite different: they are nucleated in most eukaryotes by a single *microtubule-organizing center*, called the centrosome in animals, which is usually located somewhere close to the nucleus. Microtubules radiate out from the centrosome in all directions, and it seems that there is a mechanism for evening out the lengths of microtubules, which has the effect of positioning the centrosome close to the center of the cell (**Figure 7.34**). Possibly this arises from microtubules pulling the centrosome toward the cell wall, which given enough microtubules pulling evenly will tend to pull the centrosome into the middle. There are also proteins that link actin filaments and microtubules to the cell membrane. This is important for both actin and tubulin, for different reasons, as we shall see shortly.

7.3.3 How cells move

Many cells move at some point in their life. Many bacteria move constantly; many cells in eukaryotes move during development; and circulating cells such as white blood cells are carried by blood flow but in addition need to stick, roll, or spread as required. Fibroblasts move around through tissue, remodeling and repairing it. They all do this by using tubulin and actin fibers.

Many cells have small hair-like structures on the outside, which they use either to propel themselves (**flagella (*7.7)** in sperm and some protozoa, or cilia in others) or to push fluid and nutrients past the cell. (The flagella are completely different from the bacterial flagella discussed above.) These structures consist

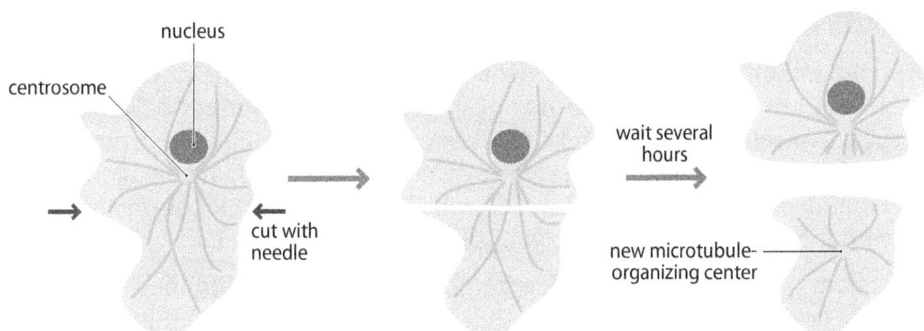

FIGURE 7.34
Microtubules can localize the centrosome in the center of the cell. After one end of a fish melanophore (pigment) cell is cut off with a needle, a new microtubule-organizing center is created in the detached fragment, and microtubules reorganize around it and position it approximately centrally; the other part also rounds up and relocates the centrosome. (Adapted from B. Alberts et al., Molecular Biology of the Cell, 5th ed. New York: Garland Science, 2008.)

TABLE 7.2 Proteins that interact with tubulin/microtubules

Protein	Function
γ-TuRC	Initiates filament formation
MAP, XMAP215	Stabilizes filaments
Tau, MAP-2	Cross-links filaments in parallel rows
Stathmin, kinesin 13, katanin	Cuts or depolymerizes filaments
+TIP, plectin	Links filament to other proteins

*7.7 Flagella

A flagellum (plural: flagella) is a hairlike protein assembly on the outside of a cell that is used to move it along. In eukaryotes, the flagella are fixed to the cell surface, being protrusions of the cell membrane supported by microtubules; eukaryotic flagella are long cilia, which move in a whip-like motion to push cells along (**Figure 7.7.1**). By contrast, the bacterial flagellum is a quite different structure made only of protein, which rotates within a circular assembly positioned in the cell membrane, using a mechanism probably related to the F_oF_1 ATPase.

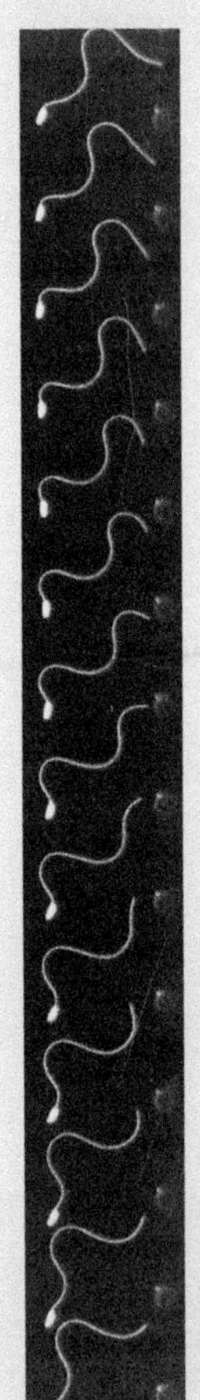

FIGURE 7.7.1
Spermatozoa use a single flagellum to move. Each image is 2.5 ms apart, showing a wavelike movement of constant amplitude moving from base to tip. (Image courtesy of C. J. Brokaw, CalTech.)

of microtubules, arranged in parallel in a remarkably complex organization that has an outer ring comprising nine pairs that are cross-linked together (**Figure 7.35**). The microtubules are covered with dynein at regular intervals, which link one microtubule pair to the next. On activation, dynein tries to walk along a microtubule, but the cross-linking prevents the sliding of one microtubule past another and leads instead to bending. This generates a beating motion of the flagellum, leading to movement.

Sliding or crawling motion of cells on a surface is achieved by using actin. Cells polymerize actin, and the growing plus end pushes the cell forward. Sometimes the actin is in parallel bundles and sometimes it is in sheets, depending on the cell type. At the same time, actin at the back of the cell is depolymerized, which serves to pull in the back of the cell. Some cells also use myosin to pull actin filaments. This is important for cells that also stick to the surface they crawl over, where the myosin II is mainly at the rear of the cell and involved in pushing the cell forward.

Microtubules do not generally 'push' the cell to the same extent. Microtubules are, however, involved in reorganizing the contents of moving cells. Because they emanate from a microtubule-organizing center close to the nucleus, they serve to keep the nucleus close to the center; as we shall see in the next section, they also keep other membrane-enclosed vesicles in the correct locations.

7.3.4 Vesicles are transported along microtubules

Microtubules run from the center of the cell (minus end) to the cell membrane (plus end), whereas actin fibers usually run along the cell membrane. Because most directed movement in the cell is either toward or away from the center, most movement happens along the microtubules. For movement away from the center, kinesin is required because it generally moves toward the plus end. A different kinesin is used for each type of cargo, differing both in the tail (to recognize the cargo) and in the legs (so that it goes to the correct place and stops there). Kinesins move membrane-enclosed vesicles from the **endoplasmic reticulum** (ER) out to the **Golgi apparatus** and from there into secretory pathways to the outside of the cell, and they disperse mitochondria around the cell. Kinesins were first observed in **axons**, of which the longest (from the bottom of the spinal cord to the foot) can be up to one meter long. A series of microtubules runs along axons, and kinesins are used to carry material to the tip of the axon, taking 2–3 days to travel a meter. (Contrast this with the time taken for diffusion over this distance, estimated at 3 years! [33]) The endoplasmic reticulum is continuous with the **nuclear envelope**, but is widely dispersed through the cell forming more than half the membrane content in the cell and containing about 10% of its volume. This dispersal is again carried out with kinesins.

By contrast, movement in to the center is performed by dynein, which moves vesicles from the cell surface (for example as a result of endocytosis), and recycles secretory vesicles and Golgi vesicles with their signal receptors back toward the center. This is most clearly demonstrated by adding an agent to the cell that depolymerizes microtubules, which leads to the ER contracting back toward the nucleus, and the Golgi dispersing around the cell.

For many organelles and vesicles, their position is determined by a balance between the actions of kinesin and dynein, which pull them in opposite directions. This is partly because there is apparently no direct mechanism to position a vesicle for example halfway between the centrosome and the cell membrane, other than by adjusting the rates of movement of kinesin and dynein; and partly because the continual balancing act means that when the cell needs to ensure rapid movement of a vesicle in one direction, it can easily do this by switching off the relevant motor.

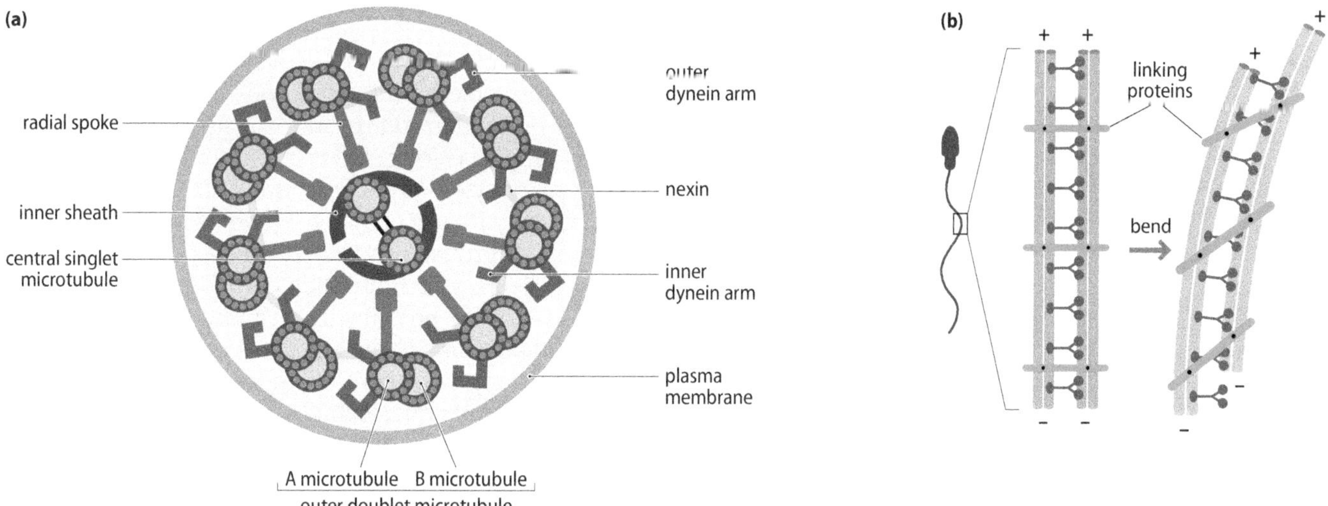

FIGURE 7.35
Bending of a flagellum is caused by dynein movement. (a) Cross-section through a eukaryotic flagellum or cilium. Microtubules are organized in nine pairs (blue rings) around a central pair. They are held together by occasional flexible cross-links made of nexin (light green). Dynein arms (blue/green) reach out from each pair to its neighbor. (b) On stimulation, the dynein arms on one face attempt to walk along their neighboring microtubule. Because of the cross-links, this creates a bend in the flagellum. (Adapted from B. Alberts et al., Molecular Biology of the Cell, 5th ed. New York: Garland Science, 2008.)

In some special cases, movement along the cell membrane is required, and here the cell uses actin and myosin. Pigment is contained in vesicles called melanosomes or melanophores, and skin color is altered (for example during sun tanning or in the color change in chameleons) by moving melanosomes out from the cell body into arms that extend outward from the cell; these are carried by myosin V along actin fibers.

All these movements of membrane-enclosed vesicles require labels on the vesicles to mark where they should go, and to permit recycling of the 'packaging' molecules back to their origin. These labels are provided mostly by small GTPases, called Rab proteins. There are many different Rab proteins, and the nature of their interactions with targets is very variable, although their basic function is straightforward, being for example '*cis* Golgi labels' (Rab2) or 'early endosome labels' (Rab5C). It is the Rab proteins that also act to attach the vesicle to its relevant motor protein.

Throughout this book (and particularly in the discussion of signaling mechanisms in Chapter 8), I note that evolution does not produce highly specific and effective signals or switches. Rather than producing a single 'perfect' system, it produces something that works, albeit clumsily, and then proceeds to refine it by adding more and more refinements until it ends up with something that does work. This process is called **embellishment**, and is exemplified in the targeting of proteins, where correct targeting relies on a sequence of rather weakly selective mechanisms, backed up by recycling systems that bring vesicles back for another try. Many Golgi enzymes, for example, actually cycle back and forth between ER and Golgi but spend most of their time in the Golgi; this is presumably the most efficient way of keeping them in the Golgi. Similarly, proteins that function in the ER do sometimes get carried off to the Golgi but then get recycled back to the ER, usually by using the C-terminal sequence KDEL or sometimes by other sequences such as a C-terminal KKXX, which marks them for binding to recycling receptors and packaging into vesicles for recycling.

7.3.5 Large cells require more directional intracellular transport

We noted at the start of this chapter that most movement of proteins and metabolites (as compared with the membrane-enclosed vesicles discussed in Section 7.3.4) is fast enough by random diffusion for it not to require extra assistance. However, this is only true for cells of 'normal' size. Large cells, such as plant cells, axons, or oocytes (egg cells), are orders of magnitude larger and are too large for everything to happen by diffusion: they require directed movement. Transport of proteins to the ends of axons requires the proteins to be bundled into vesicles, which are then carried by kinesin along the axon. Large cells can also localize proteins by ensuring that the relevant mRNA is localized, again using kinesin and dynein. This occurs in axons, and also in oocytes. In the *Drosphila* oocyte, the centrosome is not in the center but close to the anterior end. Consequently microtubules mainly have their minus ends at the anterior end of the oocyte and their plus ends at the posterior end. mRNAs can be clustered at the anterior or posterior end by attaching them to dynein or kinesin, respectively, often via intermediary carrier proteins. There is evidence that the mRNA molecules are anchored to actin once they reach their destination, to ensure that they do not diffuse back. Indeed, there is a significant body of evidence to suggest that the localization of mRNA in the correct part of the cell (and possibly also the ribosomes needed to translate it) is important for the correct location of proteins even in bacterial cells [35].

Plant cells can be very large, and unusually require some 'stirring' of the cytoplasm to assist diffusion and make sure that everything can get to where it needs to be. They undergo a remarkable process known as **cytoplasmic streaming**, which can easily be seen under a microscope as a vigorous movement of organelles in a circular motion around the edge of the cell [36, 37]. This is achieved by the bundling of actin filaments (using the specialist bundling protein villin; see Table 7.1) into parallel bundles along the cell wall. Myosin XI is attached to cargo vesicles and moves rapidly along the bundles, at speeds of several $\mu m\,s^{-1}$, with a maximum of about $100\,\mu m\,s^{-1}$ in the green alga *Chara*, generating very vigorous mixing of cell contents. Streaming also occurs in other large cells, such as oocytes, which in this case use microtubules to anchor the motor proteins [38].

7.3.6 Mitosis requires major intracellular movement

Very marked changes in the locations of the eukaryotic cell contents happen during **mitosis**, when the cell splits into two, pulling the chromosomes apart into two equal halves. This process is again facilitated by tubulin and actin motors. It is controlled by the **mitotic spindle**, a structure constructed from microtubules. First, the centrosome splits into two, and existing microtubules disassemble. Each half grows new microtubules, with the minus ends at the centrosome and the plus ends growing outward. This generates three classes of microtubule: *astral* microtubules, which project out toward the cell membrane and locate the spindle in the center of the cell; *interpolar* microtubules, which form antiparallel pairs with interpolar microtubules from the other centrosome; and *kinetochore* microtubules, which attach to chromosomes across the center of the spindle (**Figure 7.36**). Five motor proteins are involved in the correct formation, placement, and function of the spindle (**Figure 7.37**). Dynein attaches to the cell membrane; its minus-end-directed movement thereby pulls astral microtubules to the cell membrane, and therefore pulls the spindle itself toward the membrane. Kinesin-5 is located in the interface of the interpolar microtubules and is plus-end directed; it therefore pushes the two poles of the spindle apart, and works in the same general direction as dynein. By contrast, kinesin-14, which is also in the interface of the interpolar microtubules, is minus-end directed, and therefore pulls the poles together. A balance

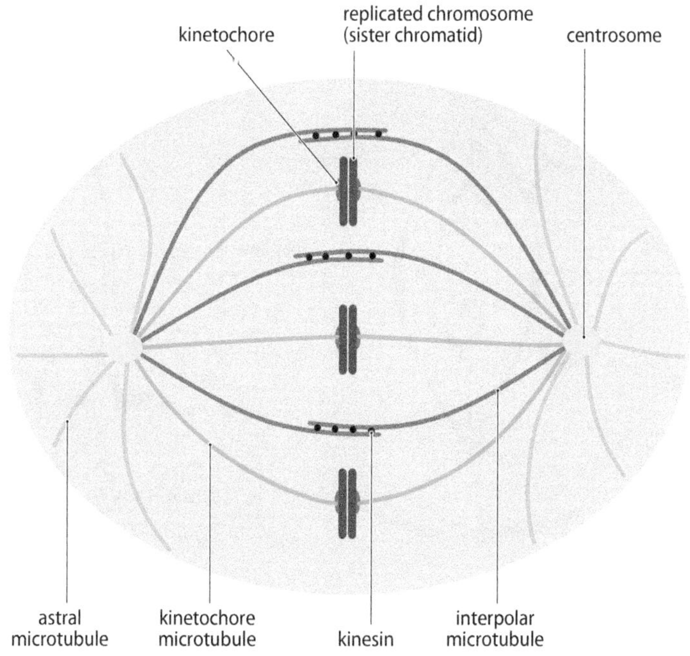

FIGURE 7.36
The arrangement of microtubules during mitosis. There are two centrosomes, one in each half of the dividing cell. Along the center of the cell are the chromosomes, aligned as pairs of sister chromatids. Each of these is attached to a kinetochore microtubule (blue) via a kinetochore (a protein assembly located at the chromosome centromere as the attachment point for microtubules). The astral microtubules (green) have a similar role to that of microtubules in normal cells, and locate the centrosome. The interpolar microtubules (red) reach out across the center of the cell and form pairs with interpolar microtubules extending from the other centrosome. The pairs are linked by kinesins (black dots), which act either to pull them together or to push them apart.

between these two kinesins apparently keeps the poles at the correct distance and centered on the midplane: it is not well understood how this balance is achieved. Finally, kinesin-4 and kinesin-10 attach to the chromosomes and are plus-end directed, therefore pushing the chromosomes away from the poles. There are several other mechanisms that operate on the chromosomes and lead to their correct placement in the center of the spindle.

During chromosome separation (anaphase), the protein cohesin, which holds the sister chromatids together, is degraded by a specific protease; the chromosomes are pulled toward the poles, and very shortly afterward the two poles are pulled apart. The pulling apart of the chromosomes occurs by depolymerization of the kinetochore microtubules at both ends (**Figure 7.38**). Depolymerization at the minus end is achieved while keeping the microtubule still attached to the centrosome, and therefore shortens the microtubule. Because the kinetochore is attached to the plus end of intact microtubules, depolymerization at the plus end keeps the kinetochore attached to the shrinking end, and it is pulled back along the microtubule. The pulling apart of the two poles is achieved with the dynein/kinesin-5 combination described above (see Figure 7.38).

At the end of mitosis, the cell splits into two, a process called **cytokinesis**. As we might expect from the discussion above, this is done by actin fibers, which make a ring around the center of the cell and are then pulled together by myosin II.

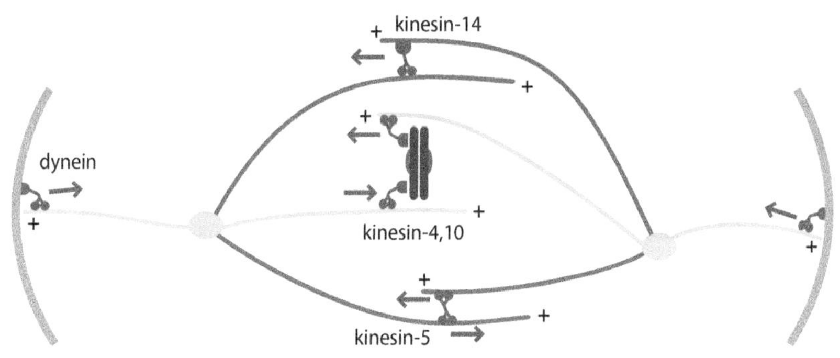

FIGURE 7.37
The five major motor proteins operating during mitosis. Dynein has its cargo-carrying tail (see Figure 7.14) attached to the cell membrane, and moves toward the minus end and the centrosome. It therefore pulls the centrosome toward the cell membrane. Thus, in presumably a similar way to many microtubules, the action of many such motors pulls the centrosome evenly in all directions and therefore locates it centrally, as well as pulling apart the two centrosomes. Kinesin-14 is an unusual minus-end-directed kinesin, which pulls two interpolar microtubules together. By contrast, kinesin-5 is a more conventional plus-end-directed kinesin but with two pairs of motor domains. It therefore works to push interpolar microtubules apart. These two kinesins therefore have opposite effects: an appropriate balance between the two presumably holds the centrosomes in the right place. Finally, kinesin-4 and kinesin-10 attach to chromosomes and push them away from centrosomes, thereby keeping them in the center of the cell. Several of these motors are also used in the final phases of mitosis to help separate the two halves (see, for example, Figure 7.38).

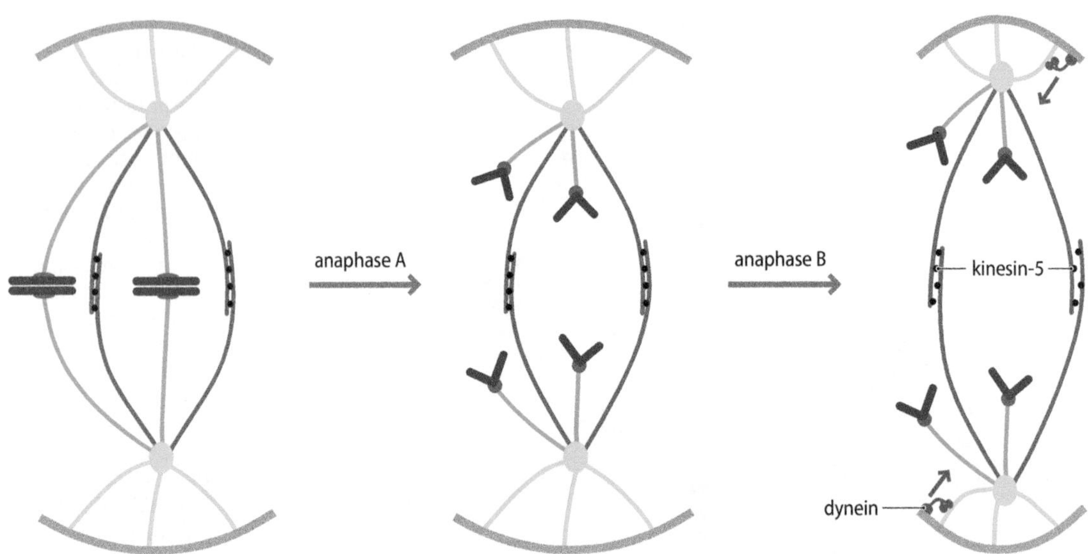

FIGURE 7.38
Anaphase, when the sister chromatids separate. In anaphase A, the kinetochore microtubules (blue) depolymerize at both ends, while still attached to the kinetochore and centrosome, thereby pulling the chromatids toward the centrosome. It is currently unclear exactly how this is achieved. In anaphase B, kinesin-5, now unfettered by kinesin-14, pushes the interpolar microtubules apart, and therefore helps to push the two poles apart. In addition, dynein, now with less resistance, is able to pull the astral microtubules (green) toward the cell membrane and help further to move the poles apart.

7.4 NUCLEAR TRANSPORT

The nucleus is surrounded by a double-membrane system called the nuclear envelope, which is covered by large nuclear pores (**Figure 7.39**). Nuclear pores are large enough for small proteins to be able to diffuse freely in and out. They are, however, lined by numerous *nucleoporins* or pore proteins, which are thought to be **natively unstructured** and contain a large number of repeats of the sequence Phe-Gly (FG repeats), which act as weak binding sites for the nuclear transporters. In the graphic analogy used by Alberts et al. [33], the pore proteins act rather like a kelp bed in the ocean, forming a curtain-like barrier to free diffusion. It seems that there are several weak interactions between nucleoporins that help to keep the curtains closed and prevent free diffusion of proteins across the nuclear pore. In particular, interactions between Asn residues (reminiscent of the interactions between glutamines in amyloid Gln-rich regions), as well as Phe–Phe and Phe–methyl interactions, help to keep the curtains closed. Interactions between Phe and nuclear transporters may function

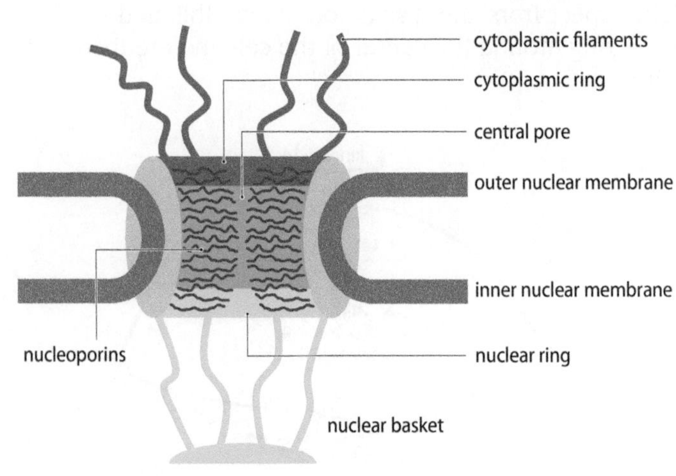

FIGURE 7.39
The structure of the nuclear pore complex. The nuclear envelope (brown) forms a double membrane studded with large holes, into which fit the nuclear pore complexes. These have eight-fold symmetry and have fibrils guarding the entrance on both sides; on the nuclear side they are gathered together into a nuclear basket. Nucleoporins form large natively unstructured curtains across the entrance. The structure is based mainly on electron micrographs.

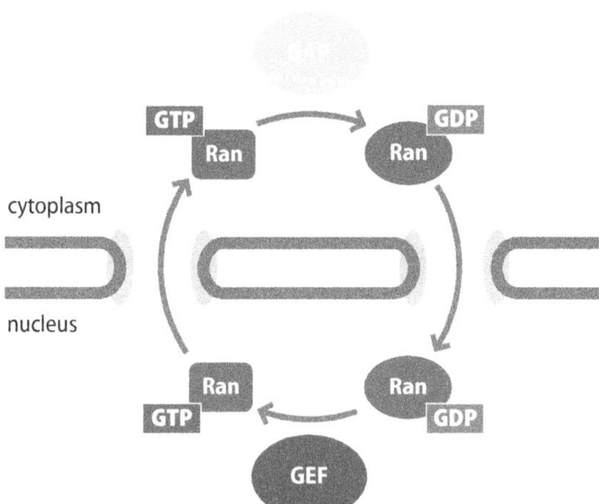

FIGURE 7.40
GTP-dependent nuclear transport. Ran can diffuse across the nuclear pore. In the nucleus, a GEF bound to chromatin ensures that all Ran is bound to GTP, while a GAP in the cytoplasm ensures that the GTP is hydrolyzed to GDP. This provides the directionality in nuclear transport, because nuclear export receptors will only bind to cargo in the presence of Ran-GTP, whereas nuclear import receptors will only bind to cargo in the absence of Ran-GTP (see Figure 7.41).

to unzip the curtains and allow transporters through [39]. Small proteins do not *require* any system to get them in and out of the nucleus, although larger proteins and complexes do. Nevertheless, most proteins do use facilitated transport, by containing a **nuclear localization signal** and/or its counterpart, a **nuclear export signal**, which can be exposed or hidden by regulatory events to control the location of the protein. Thus, toward the downstream end of signaling pathways in eukaryotes (see Chapter 8), there is usually some process such as dimerization or phosphorylation that exposes a nuclear localization signal (NLS) on a transcription factor or its transporter and leads to its localization into the nucleus. The NLS is typically a short sequence rich in positively charged lysines and arginines, and it can be at any exposed part of the protein, the archetypal sequence being that from the SV40 virus T-antigen, which is just KKKRK.

As we have seen many times already in this chapter, directed transport requires an input of energy. For nuclear transport, this takes the form of GTP, which binds to a small GTPase called Ran. Like other small GTPases, Ran forms a switch, in which Ran-GTP and Ran-GDP are recognized differently. There is a *GTPase-activating protein* or GAP in the cytoplasm, which stimulates the hydrolysis of Ran-GTP to Ran-GDP, and a *guanine exchange factor* or GEF in the nucleus (retained there by binding to chromatin), which stimulates the exchange of bound GDP back to the more abundant GTP (**Figure 7.40**). This means that most of the cytoplasmic Ran is bound to GDP, whereas most of the nuclear Ran is bound to GTP. This simple and elegant device permits directional transport, in that the import and export receptors bind only to Ran-GTP, not to Ran-GDP; in addition, export can only proceed in the presence of Ran-GTP, whereas import can only proceed in the absence of Ran-GTP.

This is explained in more detail in **Figure 7.41**. Transport requires the presence of nuclear export and import receptors, which are soluble proteins that bind reversibly to the nuclear pore and can push their way in either direction through the pore. We consider export first. In the nucleus, Ran-GTP binds to an export receptor. This complex can then bind to a protein, provided that it contains a nuclear export signal. The ternary complex is then free to diffuse randomly in any direction, including through the nuclear pore. However, once through the nuclear pore, GAP causes the bound GTP to be hydrolyzed to GDP, as a result of which Ran-GDP dissociates from the receptor, causing the exported protein to dissociate as well. The export receptor does not bind Ran-GDP, so Ran-GDP and the receptor both diffuse away. The receptor is kept close to the nuclear pore because of its binding to the pore and can therefore rapidly cycle back across the pore into the nucleus; Ran-GDP has an import

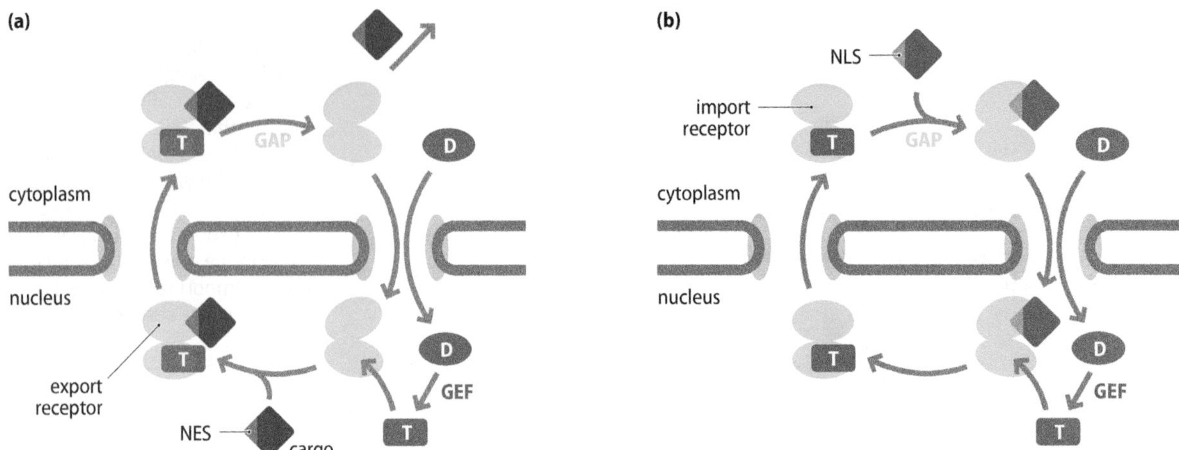

FIGURE 7.41
Nuclear transport. (a) Nuclear export. The export receptor (green) binds to Ran-GTP (red symbol labeled T) but not to Ran-GDP. It can only recognize the nuclear export signal (NES) on cargo to be exported from the nucleus if it is bound to Ran-GTP. Hence, it will only bind to cargo in the nucleus. Once it has diffused out through the nuclear pore, a GAP hydrolyzes Ran-GTP to Ran-GDP, as a result of which the Ran-GDP dissociates, shortly followed by the cargo. The cargo can then go off into the cytoplasm, while the receptor and Ran-GDP can both separately make their way back to the nucleus. (b) Nuclear import. The key difference from the export process is that the nuclear import receptor can bind to Ran-GTP or to cargo (containing a nuclear localization sequence, NLS) but not to both at the same time. In the cytoplasm, Ran-GTP is hydrolyzed to Ran-GDP, and hence it dissociates, leaving the receptor free to bind cargo. Once the cargo-laden receptor has arrived in the nucleus, the Ran-GTP there displaces the cargo and stays bound to the receptor until the receptor returns to the cytoplasm, where the GTP is again hydrolyzed, freeing the receptor for another cycle.

receptor that is specific for the GDP-bound form and therefore carries Ran-GDP back through the pore, where GEF exchanges it back to Ran-GTP ready to start again.

Import is similar (see Figure 7.41b). Import receptors in the cytoplasm will bind to any proteins that carry **nuclear localization signals**, and the bound receptors diffuse in the vicinity of the pore, including across it. However, once across it, the import receptor binds to Ran-GTP, which displaces the cargo protein. The receptor/Ran-GTP complex then diffuses back across the pore; once in the cytoplasm, Ran-GTP is hydrolyzed to Ran-GDP and dissociates. Because the import receptor can bind to either Ran-GTP or cargo but not both, it transports cargo in one direction only.

The receptors can cross the pore whether they are carrying cargo or not. This system is thus to some extent wasteful, in that non-productive movement occurs, and GTP is 'wasted' in the process. It is, however, still a very efficient mechanism. Depending on the ratio of useful to wasteful transport, it requires only a few high-energy phosphates to transport a protein across the nuclear pore. Compare this with protein synthesis for example, which requires at least four high-energy phosphates per amino acid added (ATP to AMP to link an amino acid to tRNA, plus two GTP for each elongation: one for EF-Tu and one for EF-G). The transport of a protein across a nuclear pore therefore takes about the same amount of energy as the addition of one amino acid to make a new protein!

7.5 TRANSPORT ACROSS AND INTO MEMBRANES

7.5.1 Transport into membranes requires a signal sequence

With the exception of a small number of proteins that are synthesized in the **mitochondrion (*7.8)** and the **chloroplast (*7.9)**, all eukaryotic proteins are synthesized on ribosomes in the cytoplasm. If their final location is within a membrane-enclosed organelle, in a membrane, or outside the cell, they need to cross a membrane barrier to get there. This process requires a tag, to label the protein with its target location, plus energy to make sure that it gets there. The label or **signal sequence** is made of a short peptide sequence, often at the N-terminal end of the protein, while the energy input can be a variety of sources: evolution has seized whatever materials were readily available and has made use of them. The N terminus is an obvious place for a label, because it is the first part of a protein to be synthesized and can therefore be recognized as soon as the protein starts to emerge from the ribosome. Once a protein has reached its location, it typically stays there. It is therefore common for the label to be removed during the transport process, because it is no longer needed.

*7.8 Mitochondrion

The mitochondrion (plural: mitochondria) is a membrane-enclosed organelle found in eukaryotes in which oxidative phosphorylation occurs. It is thought to have originated from a bacterial cell that was incorporated into an early eukaryotic precursor; it has its own DNA, and even a slightly different genetic code from the nucleus. However, most mitochondrial proteins originate from nuclear DNA and require transportation into the mitochondrion. It has two membranes, the inner membrane being highly folded into cristae, and enclosing the matrix (**Figure 7.8.1**). The matrix is very densely packed with protein, and is where the tricarboxylic acid cycle occurs.

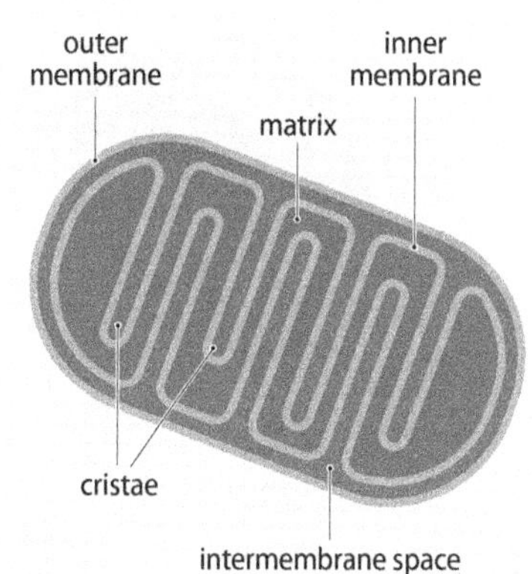

FIGURE 7.8.1
The structure of a mitochondrion.

Some typical signal sequences are given in **Table 7.3**. The cell export sequence, for example, typically comprises a short basic sequence, a hydrophobic stretch, and a short stretch containing short-chain amino acids. This last sequence is the recognition site for a signal peptidase, which removes the signal during the membrane translocation process. Note that the signal for recycling a protein back to the ER is C-terminal rather than N-terminal, because this signal may be required many times and should not be removed in the same way that the N-terminal sequences are.

Membranes form a major barrier to protein transport. Transport across membranes seems to have evolved rather early, because eukaryotic and prokaryotic codes are similar. Because the mitochondrion and chloroplast originated as distinct prokaryotic organisms that were incorporated into the young eukaryotic cell, they have double membranes surrounding them and are distinctly different from other membrane-enclosed organelles in the cell (although the mechanism for proteins crossing the membrane is rather similar, because it evolved before the eukaryote/prokaryote split). Most of the other organelles, such as the ER, the Golgi, secretory vesicles, lysosomes, and endosomes, as

TABLE 7.3 Typical signal sequences

Function	Example
Nuclear import	PP**KKKRK**V
Nuclear export	**L**A**L**K**L**AG**L**D**I**
Import into mitochondria	$^{+}$H$_3$N-MLS**LR**QS**IR**PP**K**PAT**R**TLCSS**R**Y**LL**
Import into plastid	$^{+}$H$_3$N-MVAMAMA**S**LQ**SS**M**SS**L**S**L**SS**N**S**FLGQPL**S**PI**T**L**S**PFLQG
Import into peroxisomes	**SKL**-COO$^-$ or $^{+}$H$_3$N-X$_n$-**RL**X$_5$**HL**
Import into ER	$^{+}$H$_3$N-MMSFVS**LLLVGILFWA**T**E**A**E**QLT**K**C**E**VFQ
Return to ER	**KDEL**-COO$^-$
Export from cell	$^{+}$H$_3$N-MATGS**R**TS**LLLAFGLLCLPWL**QEGSA
Insert into cell membrane	$^{+}$H$_3$N-**MLL**Q**AFLFLLA**G**FAAKI**S**A**

Important residues are highlighted in bold.

*7.9 Chloroplast

The chloroplast is the organelle in plants in which photosynthesis occurs. In many ways it is similar to the mitochondrion: it originated from a bacterium that was engulfed by a primitive prokaryote, it has its own DNA, it has an inner and outer membrane, and it creates a membrane proton gradient. The inner membrane, however, is not folded like the mitochondrial one; inside it there is a third membrane-enclosed system, the **thylakoid membrane** (**Figure 7.9.1**), where photosynthesis occurs.

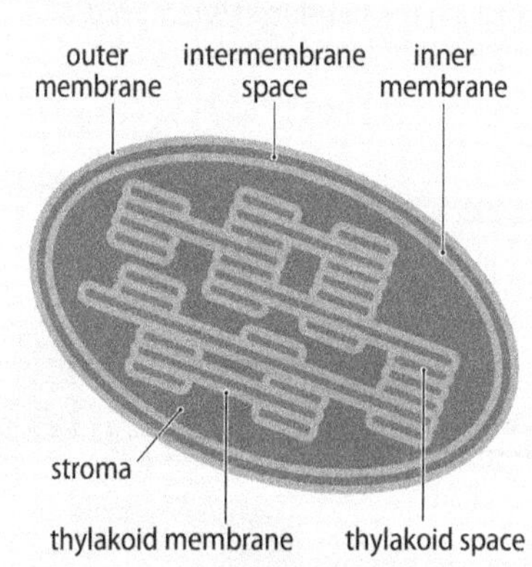

FIGURE 7.9.1
The structure of a chloroplast.

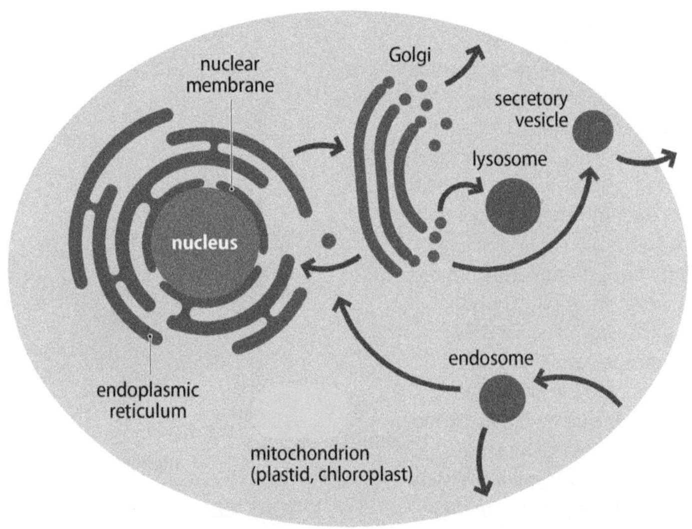

FIGURE 7.42
Traffic between eukaryotic cellular compartments. Most compartments form part of a single transport system, so that once a protein is (for example) inside the ER lumen, it can be transported from there to the Golgi, and from there it can be secreted or inserted into the cell membrane. The nuclear membrane is continuous with the ER. This also implies that once a protein is inserted into the ER membrane, its orientation and transmembrane pattern are determined for any future destination. The only major system that is different, and therefore requires an independent transport system, is mitochondria (and plastids such as chloroplasts in plants).

well as the cell membrane, form essentially one membrane system, within which proteins can be transported by budding off vesicles and fusion with the membrane of another organelle, or with the cell membrane (**Figure 7.42**). Therefore once a protein is inside a membrane or membrane-enclosed organelle, normally first the ER, its progress to its final location is determined mainly by vesicle transport and membrane fusion events.

7.5.2 The channel in the ER membrane is Sec61

Proteins cannot cross membranes in a folded state. They must be threaded through the membrane in an unfolded state and refolded on the other side. Thus, the majority of proteins that have to be either inserted into a membrane or passed through a membrane do so at the same time as their synthesis, to avoid having to fold, unfold, and then refold. Proteins destined for membrane location, or within the secretory system (such as the ER or the Golgi)—in fact any target except within mitochondria and chloroplasts—are therefore recognized by a signal recognition particle (*1.20) (SRP) as soon as they start to emerge from the ribosome. The SRP is a very early evolutionary system, as mentioned in Chapter 1. The SRP halts protein synthesis until it has attached the ribosome to a receptor in the ER membrane (**Figure 7.43**). This enables the end of the protein to be inserted into a translocator called Sec61 [40], after which the SRP and receptor detach, regulated by the hydrolysis of GTP that is bound to the receptor. Ribosomal synthesis then resumes, pushing the protein into the ER.

Sec61 has a hydrophilic channel. In the resting state there is a plug in the channel, to prevent leakage (**Figure 7.44**). This plug is displaced on binding of

FIGURE 7.43
Insertion of the nascent peptide chain into the Sec61 transporter. (1) As the nascent peptide chain (red) emerges from the ribosome, if it contains an appropriate signal sequence it is recognized by the signal recognition particle, SRP (see Figure 1.63), which also binds to the ribosome. This halts translation. (2) The ribosome–SRP complex then binds to a receptor in the ER membrane, and positions the nascent chain close to the Sec61 entrance. (3) Binding to the receptor causes hydrolysis of GTP bound both to SRP and the receptor. (4) This allows the SRP to detach from the receptor. The plug blocking the translocon is removed, and protein translation restarts.

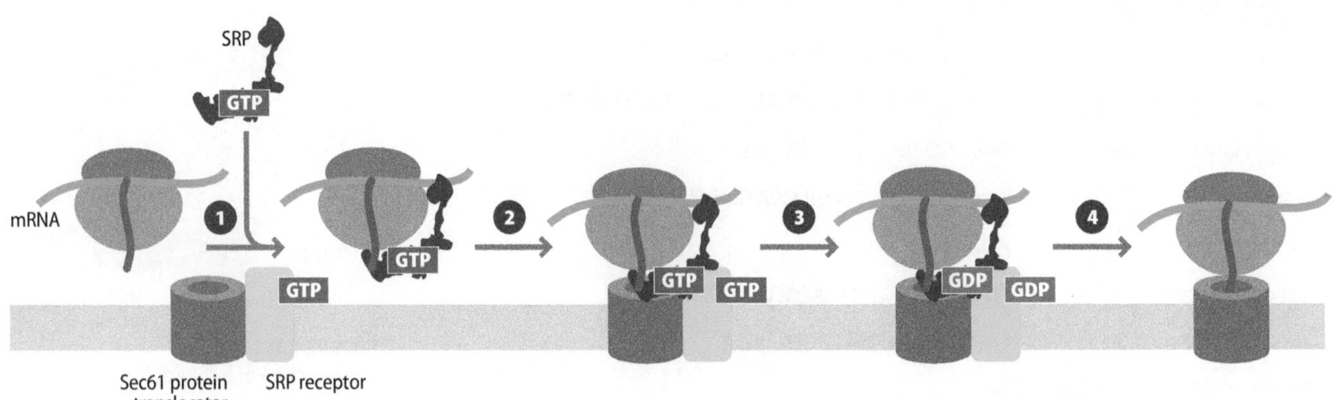

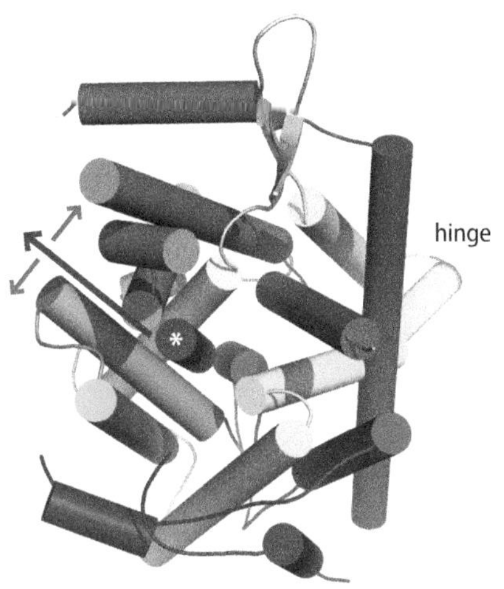

FIGURE 7.44
The structure of Sec61/SecY from *Methanococcus jannaschii* (PDB file 1rh5), looking from the top (a ribosomal view). The plug in the channel is shown in magenta. When proteins are being secreted through the translocon, it is suggested that the plug is removed (that is, the purple helix detaches from the protein and moves away from the core) and the growing chain moves roughly through where the plug is here, while the signal or stop/start-transfer sequence is held in jaws and is located roughly where the red asterisked helix is positioned (not part of Sec61). The protein consists of two semicircular rings of helices, with a hinge on the right; the region between the brown and blue helices can hinge open as shown by the thin red arrows, allowing the transfer sequence to squeeze out into the membrane (thick arrow). (After B. van den Berg, W. M. Clemons, I. Collinson et al., *Nature* 427: 36–44, 2004. With permission from Macmillan Publishers Ltd.)

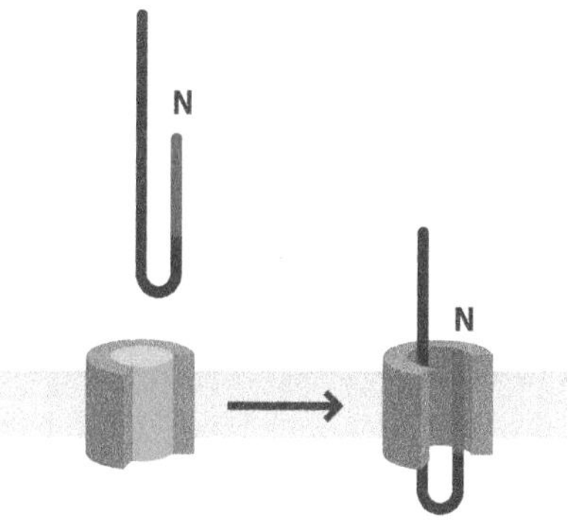

FIGURE 7.45
Location of nascent peptide within the Sec61 translocon. The nascent chain has a signal peptide (red). Once the plug (blue) has been removed by the SRP, the signal peptide is folded back to form a hairpin and inserted into Sec61 so that the hydrophobic signal peptide is held within the jaws of Sec61.

ribosome plus signal peptide. The growing peptide chain inserts into Sec61 as a loop (**Figure 7.45**). Hydrophobic regions of sufficient length (roughly 20 amino acids) are recognized as transmembrane spans by the translocator, in a similar way that **hydropathy plots (*7.10)** are used for the same purpose,

*7.10 Hydropathy plot

A hydropathy plot is a graph of amino acid hydrophobicity against sequence (**Figure 7.10.1**). Hydrophobic regions have a positive hydropathy index, and sufficently long continuous hydrophobic regions have a high probability of forming transmembrane helices.

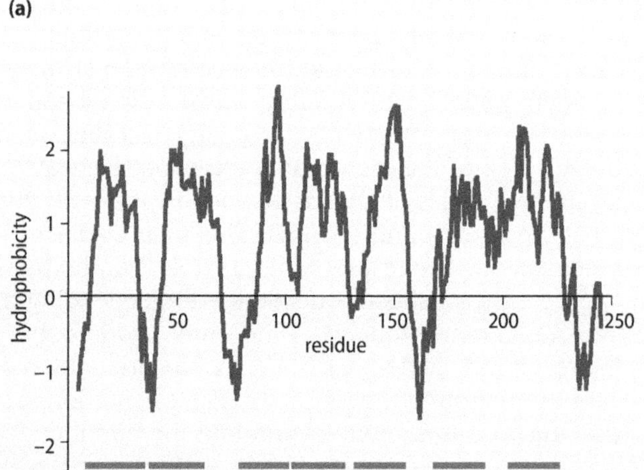

(b)

FIGURE 7.10.1
A hydropathy plot. (a) A hydropathy plot of bacteriorhodopsin from *Halobacterium salinarum*. The data were calculated using the Expasy server (http://www.expasy.ch/cgi-bin/protscale.pl) using the Kyte & Doolittle scale with a window size of nine residues. Regions at least 20 residues long with a hydropathy score of greater than 1 have a high probability of being transmembrane helices. (b) The crystal structure of the same protein (PDB file 2ntu), showing the approximate position of the membrane. This structure was used to determine the actual positions of the transmembrane regions, shown in magenta in (a). There is a good, although by no means perfect, correlation.

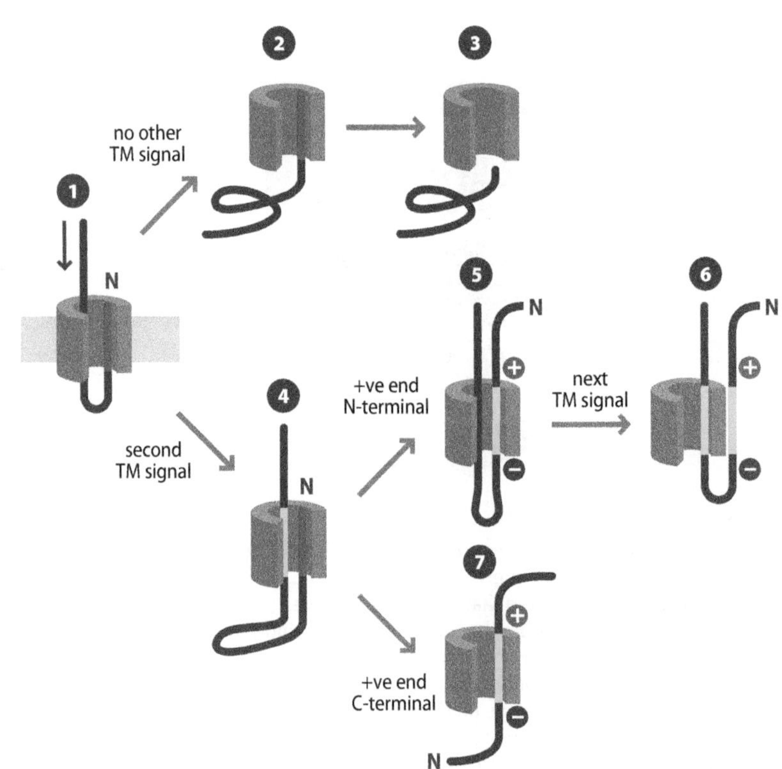

FIGURE 7.46
The placement of proteins as they translocate through Sec61. (1) The signal peptide (red) is held between the jaws of Sec61, and translation pushes the growing peptide chain through the Sec61 channel. (2) If no further transmembrane (TM) signal is detected, the entire protein passes through the channel until the C-terminal end. (3) At this point, the signal peptide is cleaved off, leaving the protein within the ER. (4) However, if a second transmembrane sequence is detected during translation, this acts as a preferred sequence for the jaws of the channel, because of the short length of the N-terminal signal sequence (see the text). This second (stop-transfer) sequence is held so that the positively charged end is on the outside of the membrane (5, 7). If the positive end is at the N-terminal side of the stop-transfer sequence, then the N terminus of the protein is on the outside (5). Translation then continues. If there is no subsequent transmembrane sequence, the protein ends up as a single-span transmembrane protein with the N terminus on the outside. However, if a second TM sequence is encountered (6), this then replaces the previous one in the jaws, and the first sequence is squeezed out sideways into the membrane. Translation continues, and any further TM sequences behave in the same way, squeezing the previously held one out into the membrane, with the intervening loops being alternately inside and outside. (7) However, if the positively charged end of the stop-transfer sequence is at the C-terminal side, the protein orients so that the N terminus is on the inside. If no further TM sequence is encountered, the result will be a single-span TM protein with the N terminus on the inside (7); if additional TM sequences are detected, they will behave in the same way as described above, by displacing the previous sequence, and pushing the old one out into the membrane.

and retained within the membrane. They act alternately as *start-transfer* and *stop-transfer* signals (that is, signals to start and stop translocating a peptide through the membrane, respectively) for the transfer of peptide through the Sec61 transmembrane channel. The details are still not entirely clear, but the following is a plausible model.

The first of these hydrophobic sequences to be recognized is the signal peptide itself, which acts as a start-transfer signal and is held within Sec61 as the protein is translated by the ribosome (**Figure 7.46(1)**). Actually the hydrophobic region of the signal peptide is shorter than the standard transmembrane helix length, which means that it is held less tightly than a normal transfer signal. If no other hydrophobic sequence is encountered, the entire protein is translated through into the ER, the signal peptide is cut off, and the protein folds within the ER (Figure 7.46(2) and (3)). If a second transmembrane (TM) sequence is identified (Figure 7.46(4)), then it acts as a stop-transfer signal and its orientation within the membrane is determined by the charge distribution of the peptide sequence on each side of the membrane: the peptide always orients so that the negatively charged end is on the inside and the positively charged end is on the outside (Figure 7.46(5) and (7)) [41]. This then displaces the signal peptide, which is cleaved off. The result of this stage is that the N terminus of the protein can end up either inside or outside the ER, depending on the orientation of the transmembrane sequence. If no other hydrophobic sequence is encountered, the result is a single-pass TM helix, the orientation depending on the charge distribution around the TM helix. The orientation produced by this process is vital, because in a multi-TM protein the orientation of subsequent TM stretches is determined entirely by this first insertion, which serves to guide all subsequent insertions.

If a third hydrophobic sequence is then encountered, this acts as another start-transfer signal, and leads to insertion of the sequence plus subsequent protein through the membrane. Each new hydrophobic sequence acts alternately as a start-transfer and stop-transfer sequence, implying that the orientation of protein sequences either side of the membrane is determined entirely by the order of start-transfer and stop-transfer sequences within the protein. Because the same orientation is then maintained throughout subsequent transfers through the secretory pathway, the final orientation of all membrane proteins is determined in this remarkably simple manner, although subsequent glycosylations may well serve to lock the positions.

Proteins that contain multiple transmembrane helices need to accumulate the helices within the membrane as they pass through Sec61. This is probably achieved by Sec61 opening at the side and allowing transmembrane helices to slide sideways into the membrane as they arrive in the translocator (arrows in Figure 7.44). Such a model is also consistent with the little that is known about the folding of helical transmembrane proteins, in which it is thought that the transmembrane helices form first, with organization of the helices coming later [42] [see **folding of membrane proteins, *7.11**].

7.5.3 Transport into mitochondria and chloroplasts is similar

By contrast, proteins destined for mitochondria or chloroplasts are synthesized in the cytoplasm, where they are kept unfolded by being coated with other proteins. Some of these are the classic **chaperone** proteins (*4.12); others are proteins that are specialized for binding to mitochondrially targeted proteins. They are then recognized (within minutes at most) by their signal sequences (which can be either N-terminal or internal), and directed to their target. At the target organelle, the N terminus inserts into a translocator complex, which in mitochondria is called TOM, the energy being supplied by the hydrolysis of ATP bound to the chaperone.

What happens next depends on the eventual location of the protein. For proteins targeted to the mitochondrion, the N terminus always contains positively charged residues, and in most cases the membrane potential across the inner membrane is sufficient to pull the N terminus through a channel in the inner membrane. It is thought that the protein then binds to ATP-bound Hsp70 chaperones in the mitochondrial matrix, which serve to pull the protein through the inner membrane. Once there, hydrolysis of bound ATP again works to remove the chaperone and eventually allow the protein to fold; this type of ATP-driven directional movement has an early evolutionary origin and is very common. Proteins destined for the outer mitochondrial membrane (or the bacterial cell membrane) are pulled through the outer membrane only, by specialized chaperones, which then attach them to the membrane surface, and by ATP

***7.11 Folding of membrane proteins**

Remarkably little is understood about the way in which membrane proteins fold. As discussed in the text above, the orientation of transmembrane helices is determined largely by the distribution of positive charges, whereas their positions are determined by their hydropathy. It seems that the helices are formed first, either before insertion into the membrane or during it, and that association of the helices into a folded state occurs subsequently. NMR studies in micelles of membrane proteins containing multiple transmembrane helices usually show very unstable or fluctuating interactions between the helices, again suggesting that interactions between helices in the membrane are rather easily disrupted and therefore presumably occur after the helices are positioned in the membrane. Calculations suggest that the intermolecular association between transmembrane sequences is greatly strengthened by their location in a membrane [47], particularly if the membrane already contains a high concentration of protein. Thus, the overall picture is that once the helices are in place within the membrane, their association is rapid and relatively weak, and requires no special help. This is, however, not true for the assembly of membrane protein complexes, which seem to assemble in a particular sequence, and often seem to require the assistance of extra 'chaperone' proteins. Indeed, several transmembrane complexes contain 'extra' proteins whose function seems to be solely to help them assemble properly [48] (see, for example, Chapter 10).

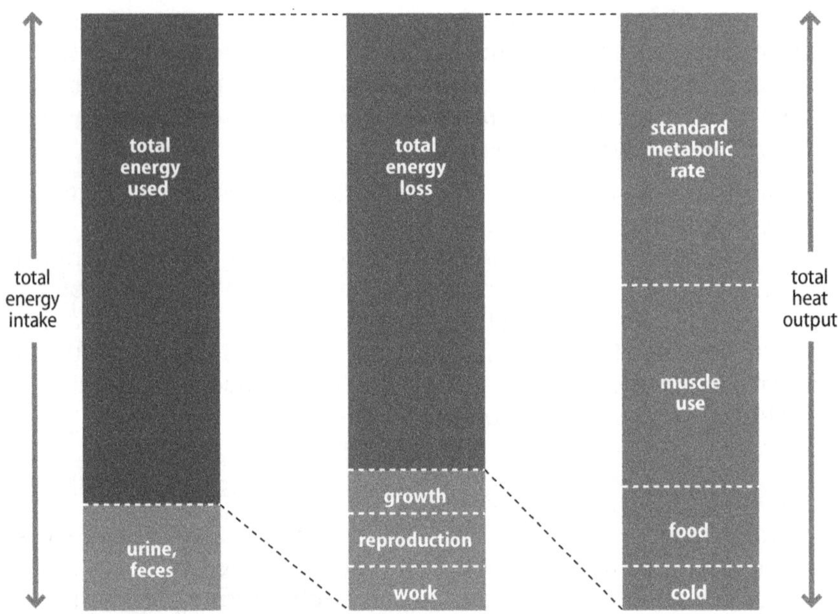

FIGURE 7.47
Approximate usage of energy in an adult human. The first column shows the total energy intake; the second is a breakdown of the total energy used; and the third is a further breakdown of the total energy lost (that is, not used for anything 'useful' and thus eventually converted into heat). (Redrawn from D. F. S. Rolfe and G. C. Brown, *Physiol. Rev.* 77: 731, 1997. With permission from the American Physiological Society.)

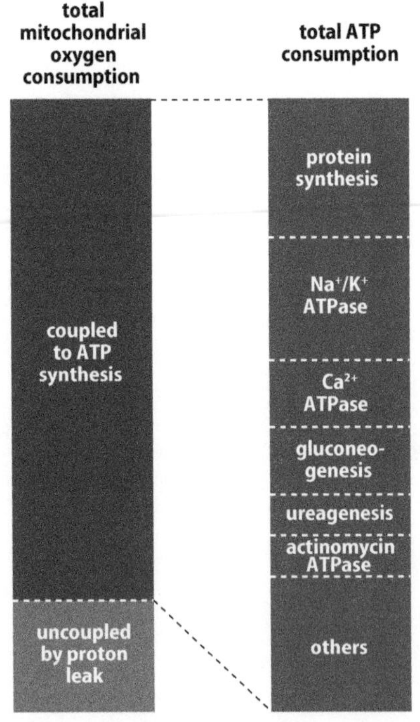

FIGURE 7.48
Estimated contribution of different processes to energy consumption in humans in the standard state. The first column shows that 20% of mitochondrial oxygen consumption is lost through proton leakage. The second column shows contributions of ATP-consuming processes to total ATP consumption, of which the largest is the Na^+/K^+ pump. (Redrawn from D. F. S. Rolfe and G. C. Brown, *Physiol. Rev.* 77: 731, 1997. With permission from the American Physiological Society.)

hydrolysis 'push' them to insert into the membrane. Chloroplasts do not have a membrane potential across their inner membrane, and so they have to use ATP-linked or GTP-linked hydrolysis for this purpose.

7.5.4 Transport requires energy

Almost all the processes described in this chapter require energy. Moreover, we have seen that many transport processes occur as a balance between two opposing movements, and are therefore continually consuming energy just to stand still. It is of interest to ask what proportion of the cell's energy output is consumed in such processes. The answer is not entirely clear; however, it has been investigated in detail in mammals, and the proportion is probably rather small [43]. In an animal at rest, most of the energy taken up as food is used, with about 15% being wasted in feces and urine (**Figure 7.47**). Of the energy input, only about 20% is used for 'work' (that is, growth, reproduction, and doing work), the rest being eventually converted into heat. And from the 80% 'heat' output, about 40%, or roughly one-quarter of the food ingested, goes into standard metabolic turnover, including the transport processes discussed here. Mammals need to maintain a constant body temperature, of course, so some of this turnover is required to achieve this.

A more detailed look at energy consumption shows that remarkably little is used for transport, most of this being used in ion pumps, with in particular roughly 25% of cellular ATP consumption (65% in neurons) being used for the sodium/potassium pump (**Figure 7.48**). Of course, much of the work done by Na^+, K^+, H^+, and Ca^{2+} pumps is to replenish losses incurred during function, but the authors estimate that about 20% of cellular energy is 'wasted' by leakage of protons across the mitochondrial membrane. However, transport processes themselves contribute an insignificant amount to total cellular energy requirements, except of course for muscle movement. And even muscle movement is a relatively low-energy process, as witnessed by the small quantity of calories consumed even during prolonged exercise—unfortunately for those of us whose calorie intake has resulted in the deposition of unwanted adipose tissue!

7.6 SUMMARY

Most motor proteins are in some way based on the small GTPases, in which the binding of GTP leads to the formation of a spring-loaded conformation. The energy from this can be released as a ratchet to prevent backward motion and to drive forward motion. In muscle, the ATPase myosin II in its ADP-bound form binds to actin filaments. Dissociation of ADP produces the power stroke that rotates the myosin head; subsequent binding of ATP releases the head from actin, and hydrolysis to ADP rotates it back to the original position, setting it up for the next power stroke. Dynein and kinesin work in roughly similar ways, except that they move along microtubules, in opposite directions.

Mitochondrial ATPase is a circular motor with a long central asymmetric spindle that turns a circular membrane-enclosed proton pump. Membrane pumps and transporters such as the ABC transporters are based on a rather simple in/out pivoting system that can be linked to an ATPase to provide the driving force for transport against a gradient.

Actin and tubulin fibers are constantly being formed and dissolved, allowing the cell to respond rapidly to environmental changes. Microtubules run outward from the cell center, whereas actin coats the cell membrane; their functions are therefore different. Microtubules are used to transport vesicles in and out from the cell center, whereas actin is used to attach and move molecules around the cell membrane, and to move cells across surfaces.

Nuclear transport again relies on GTPases to provide the energy for the directionality of movement. Transport into and across membranes uses a range of energy sources, including the membrane potential and ATP-driven chaperone binding. Insertion of proteins into membranes follows a rather simple ordered mechanism, the first insertion depending on the distribution of basic residues on each end of the transmembrane helix.

7.7 FURTHER READING

For all the cell biology details, I recommend Molecular Biology of the Cell, by Alberts et al. [33]: this is a stunning book.

I also recommend three reviews: two reviews of motor proteins [15, 18], and a review of transmembrane transport [40].

7.8 WEBSITES

http://valelab.ucsf.edu/ *A movie of kinesin in action (nicely shows how 'random walk' it is!).*

http://wolfpsort.org/ *A subcellular location predictor.*

http://www.cbs.dtu.dk/services/SignalP/ *Identification of signal sequences (see [44] for details of how to use these sites).*

http://www.cbs.dtu.dk/services/TargetP/ *Prediction of subcellular location of eukaryotic proteins on the basis of signal sequence.*

http://www.cco.caltech.edu/~brokawc/Demo1/BeadExpt.html *A movie of the beating of a flagellum from sea urchin.*

http://www.res.titech.ac.jp/~seibutu/ *A movie of the celebrated experiment demonstrating the circular movement of F_oF_1 ATPase by attaching fluorescently labeled actin to an immobilized ATPase.* (Go to "ATP synthase – the rotary engine in the cell", which is the third bullet in the ATP synthase section.)

7.9 PROBLEMS

Hints for some of these problems can be found on the book's website.

1. What difference does length scale (the distance to be moved) make to motor-driven movement? In particular, how does it affect what drives the movement?

2. Related to Question 1: if you have movement without a ratchet, will anything move anywhere at all? Is this a useful form of movement, either for a molecule or for a macroscopic object—and if there is a difference between these two, why?

3. In this chapter, it is argued that the small GTPase was a very early mechanism for a recognizable conformational switch. Suggest reasons why the hydrolysis of GTP to GDP might have been selected for this function. In particular, why a nucleotide? Why a phosphate hydrolysis? Why guanosine? And why GTP to GDP and not for example GTP to GMP or GDP to GMP?

4. This chapter (Section 7.2.1, etc.) discusses only one member of the myosin family, namely myosin II. Describe the structures and functions of the other members of the myosin family.

5. The model of muscle function described here is very simplified. What are the roles of tropomyosin, troponins T, I, and C, and titin?

6. Until fairly recently, it was thought that bacteria have no homolog of actin—which is fairly reasonable, because they contain no muscle. However, it is now realized that they do. What is/are these homologs, and what is their function? What does this tell us about the evolution of actin?

7. Research and describe the role of intermediate filaments. What proteins make up intermediate filaments, and how do their structures fit them for this role?

8. If a residue is on the outside of the ER in an ER-embedded membrane protein, and that protein goes through the Golgi to the cell membrane, does the residue end up on the inside or the outside of the cell?

9. The system described in Section 7.5.2 for getting proteins across the eukaryotic ER membrane is often called co-translational import (or translocation), because the protein translocates at the same time as it is being translated. Several proteins are expressed in the cytoplasm but are then subsequently transported into the ER by a process called post-translational import. Describe how this works, and in particular where the energy comes from. How much less efficient is this as a transport system than co-translational import, for example in ATP phosphates per amino acid?

7.10 NUMERICAL PROBLEMS

N1. The unhindered diffusion of a smallish molecule takes about 100 μs to diffuse 1 μm. The solution of Fick's law in one dimension shows that the time taken to travel a distance is proportional to the square of the distance. How long would it take to diffuse across an oocyte of diameter 100 μm? And along an axon of length 1 m?

N2. Dynein is described here as being able to move along a microtubule at up to 14 μm s^{-1}. How long would it take to get from one end of a microtubule to the other; that is, from the cell membrane to the centrosome in a typical cell?

N3. ATP synthase has a membrane-embedded F_o part containing a ring of n c subunits. What is the relationship between n and the number of ATPs produced per proton? Specifically, if n changes from 10 to 14, what is the effect on the ratio of ATPs per proton? Is it 'better' to have $n = 10$ or 14? In which case, why are both observed?

N4. I show above that actin/myosin is a remarkably fuel-efficient motor. How about power-to-weight ratio? The myosin hexamer has a molecular weight of about 5×10^5 Da, whereas actin is about 43 kDa. The force of interaction between one actin/myosin pair is roughly 4 pN, moving the myosin head relative to actin by about 11 nm in about 5 ms [45]. By comparison, a typical turbocharged V8 diesel engine weighs 380 kg and has an engine power of 250 kW (330 horsepower), and thus has a power-to-weight ratio of 0.65 kW kg^{-1}, and the Space Shuttle's main engines have a power-to-weight ratio of 153 kW kg^{-1}. Is this a fair comparison?

7.11 REFERENCES

1. DM Raskin & PAJ De Boer (1999) MinDE-dependent pole-to-pole oscillation of division inhibitor MinC in *Escherichia coli*. *J. Bacteriol.* 181:6419–6424.
2. ZL Hu & J Lutkenhaus (1999) Topological regulation of cell division in *Escherichia coli* involves rapid pole to pole oscillation of the division inhibitor MinC under the control of MinD and MinE. *Mol. Microbiol.* 34:82–90.
3. CW Mullineaux, A Nenninger, N Ray & C Robinson (2006) Diffusion of green fluorescent protein in three cell environments in *Escherichia coli*. *J. Bacteriol.* 188:3442–3448.
4. MB Elowitz, MG Surette, PE Wolf et al. (1999) Protein mobility in the cytoplasm of *Escherichia coli*. *J. Bacteriol.* 181:197–203.
5. H Meinhardt & PAJ de Boer (2001) Pattern formation in *Escherichia coli*: a model for the pole-to-pole oscillations of Min proteins and the localization of the division site. *Proc. Natl. Acad. Sci. USA* 98:14202–14207.
6. G Albrecht-Buehler (1990) In defense of 'nonmolecular' cell biology. *Int. Rev. Cytol.* 120:191–241.
7. HP Kao, JR Abney & AS Verkman (1993) Determinants of the translational mobility of a small solute in cell cytoplasm. *J. Cell Biol.* 120:175–184.
8. HV Westerhoff & GR Welch (1992) Enzyme organization and the direction of metabolic flow: physicochemical considerations. *Curr. Top. Cell Regul.* 33:361–390.
9. C Lad, NH Williams & R Wolfenden (2003) The rate of hydrolysis of phosphomonoester dianions and the exceptional catalytic proficiencies of protein and inositol phosphatases. *Proc. Natl. Acad. Sci. USA* 100:5607–5610.
10. IR Vetter & A Wittinghofer (2001) Signal transduction: the guanine nucleotide-binding switch in three dimensions. *Science* 294:1299–1304.
11. J Howard (2001) Mechanics of Motor Proteins and the Cytoskeleton. Sunderland, MA: Sinauer.
12. RS MacKay & DJC MacKay (2006) Ergodic pumping: a mechanism to drive biomolecular conformation changes. *Physica D* 216:220–234.
13. EM de la Cruz, AL Wells, SS Rosenfeld et al. (1999) The kinetic mechanism of myosin V. *Proc. Natl. Acad. Sci. USA* 96:13726–13731.
14. YT Yang, S Gourinath, M Kovács et al. (2007) Rigor-like structures from muscle myosins reveal key mechanical elements in the transduction pathways of this allosteric motor. *Structure* 15:553–564.
15. EP Sablin & RJ Fletterick (2001) Nucleotide switches in molecular motors: structural analysis of kinesins and myosins. *Curr. Opin. Struct. Biol.* 11:716–724.
16. M Lovmar & M Ehrenberg (2006) Rate, accuracy and cost of ribosomes in bacterial cells. *Biochimie* 88:951–961.
17. AJ Roberts, N Numata, ML Walker et al. (2009) AAA+ ring and linker swing mechanism in the dynein motor. *Cell* 136:485–495.
18. RD Vale & RA Milligan (2000) The way things move: looking under the hood of molecular motor proteins. *Science* 288:88–95.
19. MC Alonso, DR Drummond, S Kain et al. (2007) An ATP gate controls tubulin binding by the tethered head of kinesin-1. *Science* 316:120–123.
20. RL Cross & V Müller (2004) The evolution of A-, F-, and V-type ATP synthases and ATPases: reversals in function and changes in the H^+/ATP coupling ratio. *FEBS Lett.* 576:1–4.
21. I Antes, D Chandler, HY Wang & G Oster (2003) The unbinding of ATP from F_1-ATPase. *Biophys. J* 85:695–706.
22. D Stock, C Gibbons, I Arechaga et al. (2000) The rotary mechanism of ATP synthase. *Curr. Opin. Struct. Biol.* 10:672–679.
23. PD Boyer (2002) Catalytic site occupancy during ATP synthase catalysis. *FEBS Lett.* 512:29–32.
24. J Weber & AE Senior (2003) ATP synthesis driven by proton transport in F_1F_0-ATP synthase. *FEBS Lett.* 545:61–70.
25. AGW Leslie & JE Walker (2000) Structural model of F_1-ATPase and the implications for rotary catalysis. *Phil. Trans. R. Soc. Lond. B* 355:465–471.
26. JP Abrahams, AGW Leslie, R Lutter & JE Walker (1994) Structure at 2.8 Å resolution of F_1-ATPase from bovine heart mitochondria. *Nature* 370:621–628.
27. A Aksimentiev, IA Balabin, RH Fillingame & K Schulten (2004) Insights into the molecular mechanism of rotation in the F_o sector of ATP synthase. *Biophys. J.* 86:1332–1344.
28. O Jardetzky (1966) Simple allosteric model for membrane pumps. *Nature* 211:969–970.

29. L Guan & HR Kaback (2006) Lessons from lactose permease. *Annu. Rev. Biophys. Biomol. Struct.* 35:67–91.

30. K Hollenstein, RJP Dawson & KP Locher (2007) Structure and mechanism of ABC transporter proteins. *Curr. Opin. Struct. Biol.* 17:412–418.

31. KD Ridge & K Palczewski (2007) Visual rhodopsin sees the light: structure and mechanism of G protein signaling. *J. Biol. Chem.* 282:9297–9301.

32. B Knierim, KP Hofmann, OP Ernst & WL Hubbell (2007) Sequence of late molecular events in the activation of rhodopsin. *Proc. Natl. Acad. Sci. USA* 104:20290–20295.

33. B Alberts, A Johnson, J Lewis et al. (2008) Molecular Biology of the Cell, 5th ed. New York: Garland Science.

34. R Phillips, J Kondev & J Theriot (2009) Physical Biology of the Cell. New York: Garland Science.

35. A Danchin (1996) By way of introduction: some constraints of the cell physics that are usually forgotten, but should be taken into account for *in silico* genome analysis. *Biochimie* 78:299–301.

36. T Shimmen (2007) The sliding theory of cytoplasmic streaming: fifty years of progress. *Curr. Top. Plant Res.* 12:31–43.

37. T Shimmen & E Yokota (2004) Cytoplasmic streaming in plants. *Curr. Opin. Cell Biol.* 16:68–72.

38. WE Theurkauf (1994) Premature microtubule-dependent cytoplasmic streaming in cappuccino and spire mutant oocytes. *Science* 265:2093–2096.

39. C Ader, S Frey, W Maas et al. (2010) Amyloid-like interactions within nucleoporin FG hydrogels. *Proc. Natl. Acad. Sci. USA* 107:6281–6285.

40. TA Rapoport (2008) Protein transport across the endoplasmic reticulum membrane. *FEBS J.* 275:4471–4478.

41. B van den Berg, WM Clemons, I Collinson et al. (2004) X-ray structure of a protein-conducting channel. *Nature* 427:36–44.

42. DM Engelman, Y Chen, C-N Chin et al. (2003) Membrane protein folding: beyond the two stage model. *FEBS Lett.* 555:122–125.

43. DFS Rolfe & GC Brown (1997) Cellular energy utilization and molecular origin of standard metabolic rate in mammals. *Physiol. Rev.* 77:731–758.

44. O Emanuelsson, S Brunak, G von Heijne & H Nielsen (2007) Locating proteins in the cell using TargetP, SignalP and related tools. *Nature Protocols* 2:953–971.

45. JT Finer, AD Mehta & JA Spudich (1995) Characterization of single actin-myosin interactions. *Biophys. J.* 68 (4 Suppl.):291s–297s.

46. OG Berg & PH Von Hippel (1985) Diffusion-controlled macromolecular interactions. *Annu Rev. Biophys. Biophys. Chem.* 14:131–160.

47. B Grasberger, AP Minton, C Delisi & H Metzger (1986) Interaction between proteins localized in membranes. *Proc. Natl. Acad. Sci. USA* 83:6258–6262.

48 DO Daley (2008) The assembly of membrane proteins into complexes. *Curr. Opin. Struct. Biol.* 18:420–424.

49. LR Otterbein, P Graceffa & R Dominguez (2001) The crystal structure of uncomplexed actin in the ADP state. *Science* 293:708–711.

CHAPTER 8
How Proteins Transmit Signals

I am never content until I have constructed a mechanical model of the subject I am studying. If I succeed in making one, I understand; otherwise I do not.

Lord Kelvin (1904), Baltimore Lectures on Molecular Dynamics and The Wave Theory of Light

Everything that can be, is. [Tout ce qui peut être, est.]

Georges-Louis Leclerc (1707–1788), Comte de Buffon

This book is concerned with the *principles* of how proteins work: their structure, evolution, oligomerization, binding, dynamics, and so on. It is not helpful to try to describe *all* the functions that proteins have, or to present a comprehensive list of all the types of structure that there are, not least because a wide range of information is available in books and on the web. However, in Chapters 7 and 8 we are looking at two key functional areas, because these illustrate the ways that proteins behave, and how their structures allow them to achieve their functions. In Chapter 7 we looked at movement and showed that a single protein, the small GTPase, has spawned a large range of systems that make use of that switch to generate motors: complexity and function coming from a fairly simple origin, but branching out in many directions. In this chapter we look at another major functional area, that of signaling, and examine whether the same is true here. That is, are there simple principles that we can identify that make sense of the bewildering complexity that is signaling? We shall see that there are indeed. Signaling is a relatively recent evolutionary need, which means that the embellishments of evolution are strikingly apparent. However, there are rather few basic mechanisms, which develop naturally from the principles that we have seen in earlier chapters. In particular, we shall see that specificity, a key requirement for signaling, arises not from a single strong interaction but from a collection of weak interactions. Evolution has added extra domains and sequences to achieve the specificity needed.

8.1 AN OUTLINE OF THE PROBLEMS AND SOLUTIONS

8.1.1 Signaling pathways have to overcome several problems

A **multicellular (*8.1)** organism has regulatory problems to solve that a single-celled organism does not. It needs to coordinate the activities of its different cells and organs such that they work together. This coordination must span a range of timescales (from immediate to very long, the latter being for developmental regulation) and a range of distances between cells. It does this mainly by signaling pathways, in which molecules are produced by one cell, secreted, and transported to other cells, and then evoke a response from the target cell. This is typically a change in gene transcription, which often affects several genes simultaneously, although not necessarily all to the same extent. In most cases the signaling molecule (typically a hormone or cytokine) cannot cross the cell membrane, and the signaling pathway therefore has to overcome a major problem: how can a molecule binding to a receptor on one side of a membrane generate a signal on the other side? We need to bear in mind that, on a molecular scale, a membrane is a wide barrier, and it is no easy matter to transmit a physical signal across such a long distance. On the basis of what we have seen so far (Section 1.3.2), it should be no surprise that all transmembrane signaling proteins are helical, because helices are much better than sheets at transmitting mechanical signals. A signaling pathway needs the minimum elements shown in **Figure 8.1**, namely a receptor that can transmit a signal across the membrane, and a second protein or *messenger* inside the cell that can recognize the intracellular signal and lead to further changes within the cell, such as activation of an enzyme. Alternatively, the messenger may move to the nucleus and alter transcription.

***8.1 Multicellular**
Single-celled eukaryotic organisms such as yeast have far fewer signaling mechanisms than multicellular organisms. The human genome codes for roughly 87 proteins that contain 95 SH2 domains, whereas there is only one protein with an SH2 domain in *Saccharomyces cerevisiae*, which essentially has no tyrosine kinase signaling system [61]. Similarly, humans have 27 PTB domains, whereas *S. cerevisiae* has none.

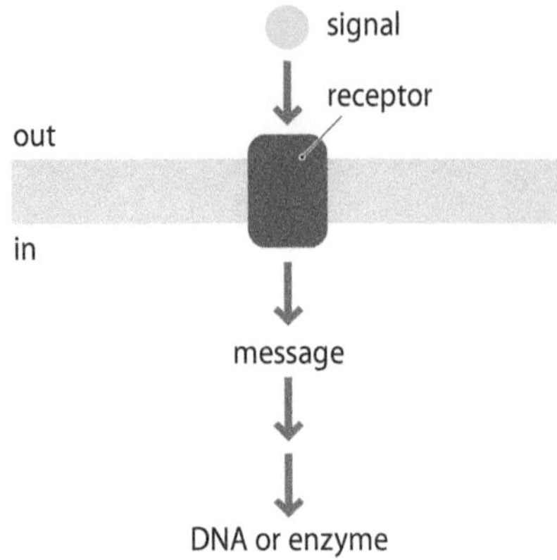

FIGURE 8.1
The key need for a signaling pathway is to transmit a signal across the cell membrane. The initial signal is passed on either to alter an enzyme activity or to alter DNA transcription.

This process has to have evolved from whatever lay to hand in the prokaryote-like ancestor of multicellular organisms. It needs a high degree of specificity, to avoid excessive cross-talk between one signal and another. However, it also requires some flexibility, to allow one pathway to amplify or inhibit another. And it needs to be readily adaptable and expandable, so that if the need for a new response arises, it is possible to evolve one without having to start from scratch, and without messing up existing pathways.

The other problem to be addressed in a signaling system is how ideal to make the switch. For most applications, the ideal switch is a perfect one, which is 100% off in the absence of signal and 100% on in the presence of signal. As discussed in Section 3.3.1, the 'ideality' of a switch depends on the binding energy between proteins: tighter binding allows more perfect discrimination between on and off. This implies that for an ideal signaling system, there should be tight binding for all the protein:protein interactions throughout the pathway. However, this makes the signaling pathway slow to turn on and very slow to turn off (because tight binding implies a slow off-rate); and also makes it hard to evolve a new system because it requires major remodeling of the interacting proteins. Thus, a 'perfect' switch is not compatible with useful on/off rates of switching. The standard evolutionary solution of adding frills to a basic and nonideal system serves very well here, essentially because adding extra proteins as scaffolds or modulators adds to the cooperativity of the system, and thus increases the specificity without slowing down the pathway unduly.

8.1.2 The membrane barrier can be crossed by lipophilic signals

One way of transmitting a signal across a large barrier such as a cell membrane is to avoid the problem by allowing the signaling molecules to cross the membrane themselves. This is the solution adopted by the gas nitric oxide (NO), which dissolves readily in membranes and so is able to travel directly from the site of its synthesis, for example in endothelial cells, to its target cells, for example in smooth muscle cells (**Figure 8.2**). It is also used by lipophilic hormones, as discussed in Chapter 3. This solution, however, creates new problems of its own, because the lipophilic hormones then do not readily dissolve in water, and they therefore need a transport system and possibly a receptor system to dock hormone carriers onto target cells. I do not discuss these systems further (but see Problem 2).

8.1.3 The membrane barrier can be overcome by receptor dimerization

Eukaryotic cells have evolved three main nonlipophilic systems, representing alternatives to the problem of transmitting a signal across a membrane. These are: (1) kinase-linked receptor dimerization, (2) G-protein-coupled receptors, and (3) ion channels. Prokaryotic cells generally use a system similar to the kinase-linked receptor dimerization, although in detail it works completely differently and is probably unrelated through evolution. We will consider each of these in detail, and start by summarizing these systems in the next three sections. There is a fourth set of mechanisms related to proteolysis, discussed at the end of the chapter.

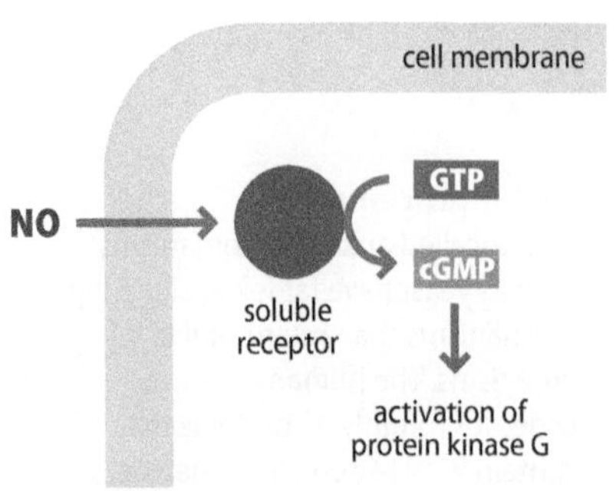

FIGURE 8.2
Nitric oxide does not need a cell surface receptor because it can cross the cell membrane. Steroids also cross the cell membrane without needing a membrane receptor.

In the receptor dimerization system (**Figure 8.3**), the receptor has an extracellular domain, which recognizes the hormone, and an intracellular kinase domain. These two domains can be part of a single protein or two proteins associated together. The kinase is (weakly) constitutively active: that is, it is permanently turned on, and when presented with a suitable substrate it will phosphorylate it, albeit slowly. The crucial aspect of the signaling system is that its substrate is a second receptor. In the absence of a hormone, two receptor molecules do not bind together, and so there is no phosphorylation of one

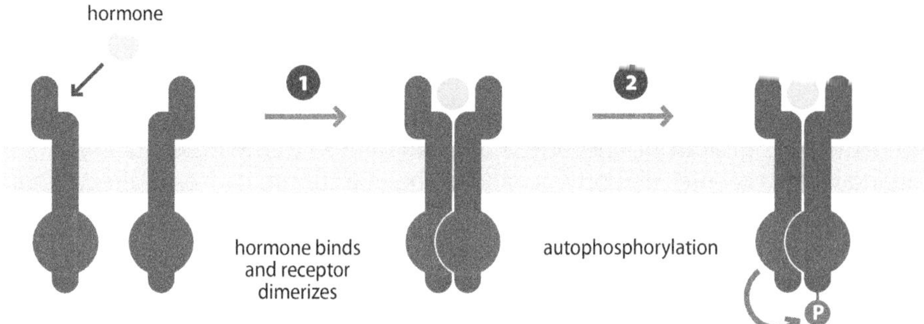

FIGURE 8.3
The heart of the receptor kinase mechanism. (1) Binding of an extracellular signal to the receptor leads to dimerization of the receptor. (2) Because the receptor carries a constitutively active kinase, this leads to transphosphorylation of the receptor. There is thus an intracellular change (phosphorylation of the receptor), which can be recognized intracellularly and passed on.

receptor by another. However, the hormone has two receptor-binding sites, so that when the hormone binds to a receptor, it leads to receptor dimerization. This brings the intracellular kinase domains close enough together for long enough that one of them can phosphorylate the other. This constitutes the signal: hormone binding on the outside leads to phosphorylation on the inside. Of course, the signal needs transmission to downstream components, but this phosphorylation is the all-important initial signal, after which the rest of the pathway is 'merely' recognizing the signal and passing it on.

This system has an inherent problem that the cell needs to solve: if the kinase is constitutively active, what is to stop it working in the absence of signal? In the previous paragraph, I wrote that "in the absence of a hormone, two receptor molecules do not bind together": why not? Or, to put it in more scientific terms, how does binding of a hormone to a receptor alter the change in free energy for dimerization sufficiently for there to be 0% dimerization in the absence of ligand and 100% in its presence (that is, a perfect switch)? As we should have come to expect, the solution adopted as a result of evolution is not to make an ideal monomer/dimer switch (which is too costly in terms both of binding energy and evolution), but to add extra tweaks so that the switch works as well as needed. But a key element of the solution is that the signal should not be activated until there is a genuine hormone binding event: in other words, it is better if the ligand-bound state is not 100% on, if this can ensure that the ligand-free state really is 100% off (see Problem N1).

A key part of the receptor dimerization mechanism is that the phosphorylated kinase is much more active than the unphosphorylated form. The mechanism for this was presented in Chapter 4, but is restated briefly here. Phosphorylation of the kinase occurs at a tyrosine that is part of the 'activation loop,' sometimes called an activation lip (**Figure 8.4**). In the unphosphorylated state, this loop is disordered and incapable of binding substrate; however, in the phosphorylated state it becomes fixed in a conformation that is able to bind substrate in the correct orientation to react.

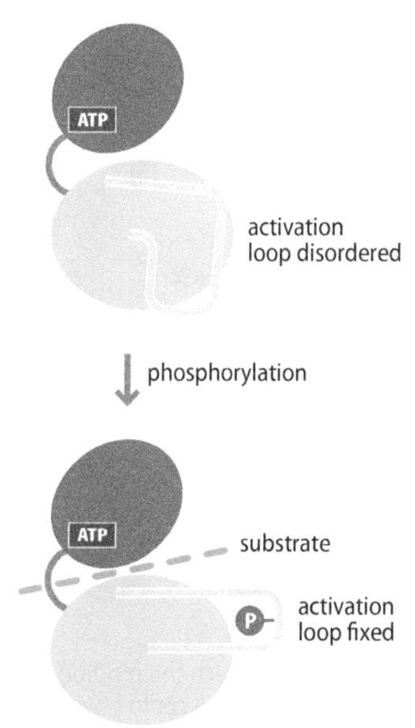

FIGURE 8.4
Activation of kinases by phosphorylation. The phosphorylation takes place on a residue within the activation loop, and functions to fix the loop into a rigid and active conformation, capable of binding the substrate in the correct orientation.

In the following text, I often draw the inactive and active kinases with frowning and smiley faces (**Figure 8.5**), as shorthand for this change in the structure of the activation loop.

8.1.4 The membrane barrier can be overcome by helix rotation

The G-protein-coupled receptor system is a completely different solution to the same problem. Here, there really is a mechanical transmission of signal from one side of the membrane to the other, which is transmitted via a helix. Because a helix is relatively rigid, one can imagine several possible ways in which the mechanical transmission could work (**Figure 8.6**). There could be a piston-like movement along the length of the helix. This would require a rather precise placement of side chains and is rare, although minor movements like this certainly occur in globular proteins. A prominent example of probable

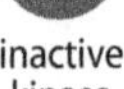

FIGURE 8.5
Inactive and active kinases: a shorthand for the kinase structures having disordered and ordered activation loops, respectively.

piston-like movement is found in rhodopsin (Chapter 7). There could be a leverlike movement, as occurs in hemoglobin (Chapter 3), or a rotation. What actually happens is a combination of both of these: rotation of a bent helix leads to a large displacement at the end of the helix, which is quite adequate to convey a signal from one side of the membrane to the other. Binding of a ligand to the receptor site on one side of the membrane causes a direct structural change on the other side.

8.1.5 The membrane barrier can be crossed by opening a channel

In a ligand-gated ion channel, there is a channel that is normally closed. Binding of a ligand causes the channel to open (**Figure 8.7**), and there is then a rapid diffusion of ions through the channel down a concentration gradient, leading to a change in their intracellular concentration, which constitutes a signal because it can be picked up by conformational changes in ion-selective proteins (for example calmodulin, which specifically binds and changes its conformation in response to calcium ions). The physical mechanism for the channel opening is actually rather similar to that found in G-protein-coupled receptors, namely rotation of a helix.

Every signal needs to be turned off again at some point. For the receptor dimerization and G-protein-coupled receptors, this is achieved rather simply: once the ligand dissociates, the receptor returns to its original conformation. Phosphorylated receptors also need to be dephosphorylated, which is typically done by the receptor itself through its autophosphorylase activity. However, an opened ion channel takes rather more work to rectify, in that the ions have to be pumped back across the membrane. This of course takes energy, as discussed in Chapter 7.

FIGURE 8.6
Possible ways in which the binding of an extracellular ligand could physically transmit a signal through the membrane. (a) A piston-like movement of one helix into the membrane. (b) A lever-like movement of one helix. (c) A rotation of one or more helices about their long axes, leading to a change in the relative positions of amino acids at the other ends of the helices.

8.1.6 Signaling pathways make use of specialized protein modules

In Section 8.1.1, we noted that specificity in signaling is enhanced by a major embellishment, namely the addition of pairs of interacting modules [1, 2]. There are many such modules, some of which are listed in **Table 8.1**. However, most signaling is performed by only a few of these, summarized in **Figure 8.8**. In typical manner, these modules tend to have their N and C termini close together and on the opposite face to their binding sites, which is the ideal structure for modules that can be plugged into host proteins as required.

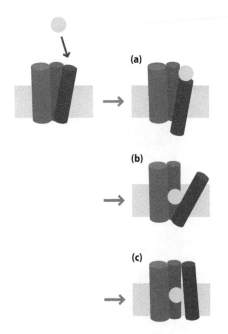

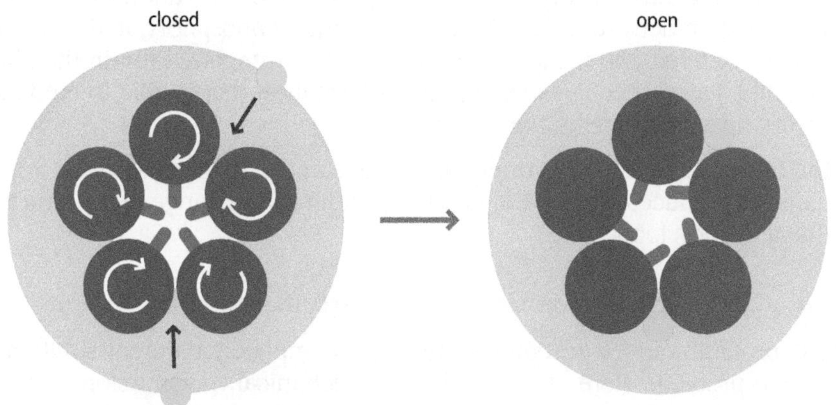

FIGURE 8.7
An ion channel (such as the acetylcholine receptor) is closed because side chains project into the channel and block it. When a ligand binds, it leads to a clockwise rotation of the helices, which rotates the side chains and opens the channel. This is a view down onto the membrane. In the acetylcholine receptor, only two of the five subunits can bind acetylcholine.

TABLE 8.1 Ligand recognition modules

Protein name	Specificity	Function
SH2	Phosphotyrosine	Signaling
SH3	Proline-rich region	Signaling
WW	Proline-rich region	Signaling
PH	Membranes, pT, pS, etc.	Signaling, cytoskeleton
PTB	Phosphotyrosine	Signaling, focal adhesion
FHA	Phosphoserine/phosphothreonine	Signaling
14-3-3	Phosphoserine/phosphothreonine	Signaling
ARM	Extended acidic peptides	Signaling
WH1 (EVH1)	Proline-rich region	Signaling, nuclear transport
bZIP	DNA	Transcriptional regulation
PDZ	G(S/T)XVI at C terminus; phospholipids	Ion channels
PX	Phosphoinositides	Signaling/membrane binding
kringle	Lysine, phospholipids, etc.	Regulation of blood clotting
Bromo	Acetylated lysine	Transcriptional control
Tudor	Methylated lysine	Transcriptional control
PHD	Trimethylated lysine	Transcriptional control
Zinc finger	DNA	Transcription factor
RING finger	Various	Ubiquitin-dependent degradation
CARD	Other CARD domains	Apoptosis
DED	Other DED domains	Apoptosis
EF-hand	Calcium	Signaling
FYVE	Mainly phosphatidylinositol 3-phosphate	Membrane traffic
IQ	Calmodulin (Ca-independent)	Signaling
SAM	Other SAM	Development/signaling
UBA	Ubiquitin	Degradation, signaling

Not all of the domains have (so far at least) been found in signaling pathways. They are, however, grouped here for convenient reference.

SH2 (Src homology 2) domains recognize phosphotyrosines, and are the most common domains for this purpose. They typically recognize phosphotyrosine plus the three succeeding residues, as described in Table 2.2. They are an extremely useful domain in signaling pathways, because phosphorylation and dephosphorylation are so common and easy to achieve. The SH2 domain binds roughly 1000 times more strongly to a phosphorylated tyrosine than to the unphosphorylated version, making this an effective switch. In the receptor dimerization mechanism, the first intracellular signal is often the phosphorylation of a tyrosine residue, either on the receptor or on an associated protein. This is usually recognized by an SH2 domain. There are many such domains,

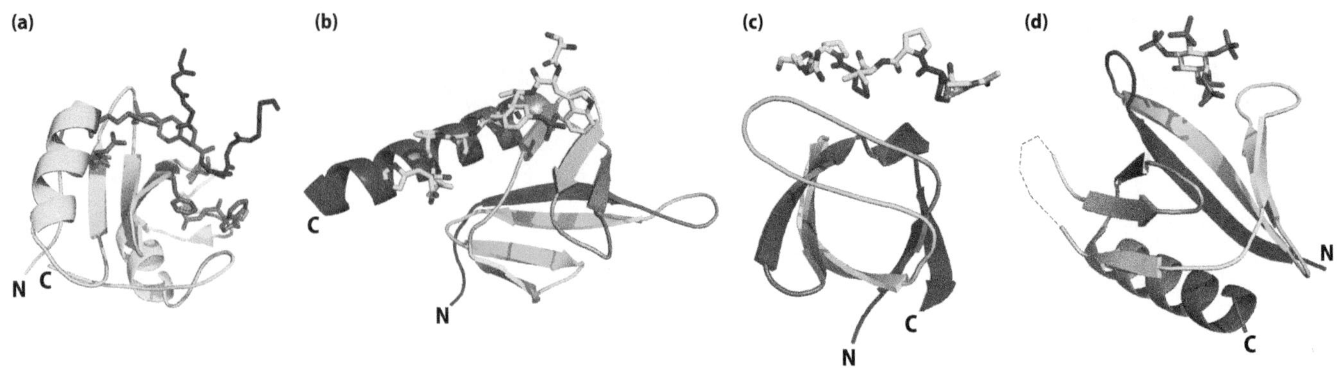

FIGURE 8.8
The four most common domains involved in signaling pathways, particularly those involving receptor tyrosine kinases. (a) The SH2 domain, which binds to phosphotyrosines. The figure shows the SH2 domain from Grb2 (PDB file 1jyr) complexed to a phosphotyrosine peptide (red). The phosphotyrosine (magenta) points into a pocket containing two arginines (orange, left) which is therefore positively charged and good at discriminating between tyrosine and phosphotyrosine, while the residues C-terminal to the phosphotyrosine sit in a hydrophobic pocket (orange, right) containing hydrophobic and aromatic residues. (b) The PTB domain, which also recognizes phosphotyrosine. The figure shows a PTB domain from talin (integrin-binding protein) bound to a phosphotyrosine peptide (PDB file 2g35). The phosphotyrosine is at top center: this protein has a significant binding interface to both the N and C terminus of the phosphotyrosine. The major binding interface is, however, more commonly on the N-terminal side. (c) The SH3 domain, which recognizes proline-rich sequences. Notice how the peptide has a polyproline II conformation, with the two PXXP motif recognition prolines (red) on the same face of the polyproline helix. The figure shows the Sem-5 protein from *C. elegans* (PDB file 1sem). (d) The PH domain, which recognizes phosphoinositides, lipids, and other ligands such as proteins. The figure shows the PH domain from pleckstrin in complex with D-*myo*-inositol (1,2,3,4,5)-pentaphosphate. Notice that part of the protein chain is missing (dashed line): this is because the electron density was absent because of disorder in the crystal in this region (see Section 11.4.5).

with their own (but overlapping) specificities. Remarkably, there are almost no SH2 domains in lower eukaryotes such as yeast [3]. SH2 domains have two binding pockets: one for the phosphotyrosine and one mainly for the residue that comes three after the tyrosine, which is usually a hydrophobic residue. Binding to SH2 domains resembles the binding of a two-pronged plug into a socket, in which the major affinity comes from the phosphotyrosine binding, but the specificity comes from the hydrophobic interaction.

PTB (phosphotyrosine-binding) domains also recognize phosphotyrosines, but they recognize in addition the three residues on the N-terminal side of the phosphotyrosine, most commonly NPXpY. They are much less common than SH2 domains but complementary because of their different sequence recognition.

SH3 (Src homology 3) domains recognize proline-rich sequences, typically PXXP with two or three additional residues at one end, including a basic residue. This function is discussed further in Chapter 4. SH3 domains and proline-rich sequences are also found in a range of proteins associated with the actin cytoskeleton and myosins. It therefore seems that SH3:proline-rich protein interactions are important for locating signaling systems close to the cytoskeleton, and for linking signaling to cell remodeling and dynamics. The binding of SH3 to proline-rich ligands can be modified by phosphorylation of adjacent serines or threonines, although the effect is less marked than for phosphorylation of tyrosines on SH2 binding. The binding of SH3 domains is therefore most often regulated by intramolecular autoinhibition, as we shall see later.

PH (pleckstrin homology) domains are also found frequently in signaling pathways. They bind to phosphoinositides (of which there are many different types), including to phosphoinositides in cell membranes. In signaling pathways, PH domains are often used to attach a protein to the membrane, or to orient a protein correctly with respect to the membrane [4]. However, PH domains are used in a more versatile way than the other domains, since they can also bind

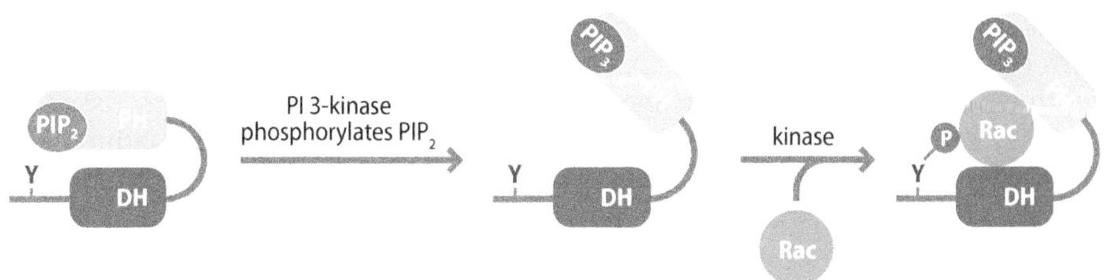

FIGURE 8.9
PH domains can bind both to phospholipids and to other proteins, particularly the Dbl homology (DH) domain. Binding to DH often acts as a switch, giving increased flexibility to the inputs to a pathway. In this example, binding of the Vav PH domain to phosphatidylinositol 3,4,5-triphosphate (produced by phosphorylation of the diphosphate) leads to its dissociation from DH, followed by phosphorylation of a tyrosine by a signaling kinase and binding of Rac. PI 3-kinase, phosphoinositide 3-kinase. (Adapted from B. Das, X. D. Shu, G. J. Day et al., *J. Biol. Chem.* 275: 15074–15081, 2000. With permission from the American Society for Biochemistry and Molecular Biology.)

other proteins. One of these is the Dbl homology (DH) domain, a domain that occurs only in combination with PH domains and acts as a guanine nucleotide exchange factor [5]. For example, in Tiam1 (a member of the Dbl family of guanine nucleotide exchange factors that activate Rho-family GTPases), one of the PH domains is suggested to affect the activity of the adjacent Dbl homology domain by binding to membrane and thereby altering the binding interface of the PH domain with the Dbl homology domain [6], whereas in the related Vav guanine nucleotide exchange factor, a direct interaction between PH and Dbl homology domains is weakened by the binding of PH to phosphatidylinositol 3,4,5-triphosphate, thereby activating the protein (**Figure 8.9**) [7], and also creating a link between this signaling pathway and the activity of phosphoinositide 3-kinase. PH domains are therefore useful for linking one signaling pathway to another.

8.1.7 Signaling pathways use these modules to achieve specificity

We have already seen that the genome of a multicellular eukaryote is not vastly larger than a typical prokaryotic genome, nor are there significant numbers of protein folds or functions found in eukaryotes that are not present in prokaryotes. Rather, the small number of module types outlined in the previous section are connected together in different ways to enhance the specificity of binding [8]. The small number of modules used does of course potentially make the problem of cross-talk worse, and means that the evolution of *specificity* is crucial for an effective signaling system. As we shall see, this is an excellent example of embellishment, in which elaborate frills are added to what is basically a simple system so as to produce specificity.

Two types of addition are common (**Figure 8.10**). The first is to pick a pair of interacting proteins or peptides, such as an SH3 domain and a proline-rich peptide, or an SH2 domain and a phosphotyrosine (see Table 8.1), and attach one to the upstream protein and the other to the downstream protein. In this way, the receptor:messenger interaction is made more specific. This concept is explored in Chapter 2, which forms a useful background to this chapter. Effectively the additional modules convert an intermolecular recognition into an intramolecular recognition, which makes it both faster and stronger.

The second type of addition is similar and is also discussed in Chapter 2: to use a **scaffold** protein, which achieves the same effect but requires an additional protein. Again, it effectively converts an intermolecular interaction into an intramolecular one.

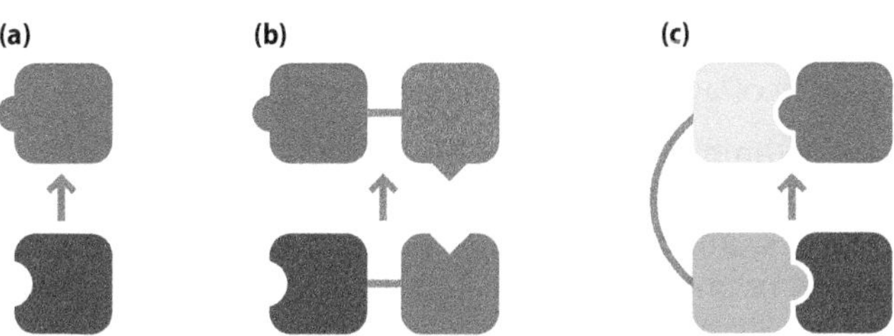

FIGURE 8.10
Two ways to increase the specificity of protein:protein interactions. (a) A simple and relatively nonspecific recognition. In (b), extra domains are added onto each partner, for example an SH2 domain on one and its complementary phosphotyrosine sequence on the other. In (c), a third scaffold protein is used that can bind to both partners and bring them close together. The extra interactions can improve the affinity also, but their main purpose is to increase the specificity.

We shall see that these additions then create an additional problem: how to turn off the interactions in the absence of signal. The solution to this requires yet further embellishments in the form of **autoinhibition**, also discussed in Chapter 2. In most cases, there is some intramolecular interaction within each signaling molecule that inactivates it and stops it from inadvertently transmitting a signal. This interaction is weak, and can be readily released by the incoming signal. The autoinhibition often requires the addition of yet more modules or peptides.

8.1.8 Signaling pathways make use of colocation to achieve specificity

Evolution adapts and tinkers: it also prefers to use multiple weak controls rather than one single strong one. Therefore, in addition to the embellishments described above, it makes extensive use of one more feature: it increases the specificity of the signal transmission by locating all the components in the same place. This makes the chance of passing on the correct signal much greater, and correspondingly decreases the chance of an incorrect signal. For example, kinases are generally rather unspecific in their substrate selectivity. However, if the only substrate present is the one required in the signaling pathway, then the problem diminishes markedly. (There is still a problem that even a single substrate protein has multiple potential phosphorylation sites; in general, cells seem to have very little selectivity over phosphorylation, because eventually all of the sites get phosphorylated, although not all at the same rate. Many of the phosphorylation events probably have little biological effect, but this is still far from clear.)

This effect of colocation (or *context*) is important, because it means that a simple analysis of protein sequence, to identify a phosphorylation or recognition site, often yields a very inaccurate picture. It is necessary to consider what other modules are present in order to work out which sites are likely to be important. To use a computing analogy, nature uses *fuzzy logic*, and integrates several effectors to result in the desired outcome. Colocation also implies that overexpression of a signaling protein may well have effects that are irrelevant to its function *in vivo*, because effectively the protein is being forced to act in places where it does not normally function. Overexpression of scaffold proteins can have very contradictory effects, because in the correct stoichiometry they help assemble the proteins within a pathway and enhance signaling, whereas in the wrong stoichiometry (for example, too much) they competitively disassemble proteins and thus inhibit the signaling [9].

In an important paper, Kuriyan and Eisenberg [10] convincingly argue that colocation is an important first step in the evolution of a range of signaling systems. The idea is that once two proteins are brought together, either by being on the same peptide chain or by binding to a scaffold, the evolution of regulatory mechanisms is greatly speeded, because their high mutual effective concentration amplifies the effect of mutations. Thus, effects such as allostery and autoinhibition are readily evolved once the proteins are already tethered together. A very interesting corollary of this is that the evolution of an allosteric or autoinhibitory mechanism is to a large extent dependent on the 'accident' of the prior tethering event. This has the important implication that allostery and autoinhibition will not be conserved across related systems or organisms, as indeed they are not. As we shall see, evolution has developed a bewildering array of autoinhibitory mechanisms.

An interesting example is provided by Pawson and Kofler [11]. The Abl and Src tyrosine kinases both have the domain structure SH3-SH2-kinase, and in both cases the SH3 and SH2 domains produce an autoinhibition of the kinase domain which is relieved on activation. However, the mechanism is different. In Src, regulation is achieved largely by phosphorylation of a tyrosine close

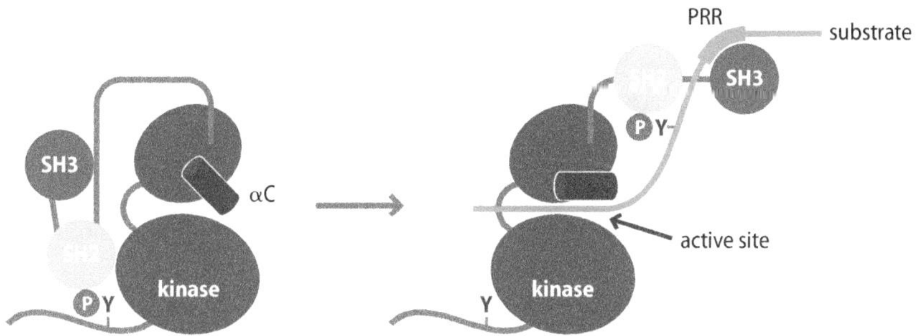

FIGURE 8.11
Regulation of the activity of Src kinase. In the autoinhibited state, a phosphotyrosine close to the C terminus binds to the SH2 domain, which fixes the linker that attaches the SH2 domain to the kinase domain. In turn, this allows the SH3 domain to bind to a proline-rich region in the linker. The orientation of the SH2-kinase linker also perturbs the αC helix, which is part of the active site, and deactivates the kinase. Dephosphorylation of the phosphotyrosine allows unwinding of these interactions and opens potential ligand-binding sites on the SH2 and SH3 domains to bring substrate close to the active site. As an alternative to a phosphotyrosine on the substrate, the substrate could present a proline-rich site for binding to the SH3. PRR, proline-rich region.

to the C terminus, which binds intramolecularly to the SH2 domain (**Figure 8.11**) [12]. In turn, this rigidifies the linker between SH2 and kinase, which contains a proline-rich sequence, and therefore leads to a second intramolecular binding event between the SH3 domain and proline-rich sequence. The binding is crucially dependent on the linker between the SH2 and SH3 domains, which has been described as an inducible 'snap lock' that clamps the two domains together in the presence of phosphorylation of the C-terminal tyrosine but allows them flexibility when it is dephosphorylated [13, 14]. This structure inhibits the kinase by distorting the active site, in a similar way to that described for cyclin-dependent kinases in Chapter 4. Activation of Src occurs by a stronger intermolecular binding to SH3 or SH2 domains, and/or by dephosphorylation of the C-terminal tyrosine. By contrast, in Abl, the SH2 binds directly to the back of the kinase domain, in an interaction that does not involve phosphotyrosine; however, this binding also distorts the kinase to prevent activity, but in a different way. The SH3 also binds to the restructured SH2-kinase linker. Intermolecular binding of SH2 to a phosphotyrosine ligand activates the kinase (**Figure 8.12**). The anticancer drug Gleevec (imatinib) works by specifically inhibiting Abl kinase but not other kinases. Such specificity is a challenge, because Gleevec, like almost all kinase inhibitors, is a competitive inhibitor at the ATP-binding site. How, then, can it specifically recognize Abl and not other kinases? Pawson and Kofler point out that it binds to the distorted (autoinhibited) Abl kinase but not to the distorted Src kinase, and hence achieves specificity in its action, not by recognizing the active kinase but by stabilizing the inactive kinase. Because of the system-specific nature of autoinhibition, this provides a good potential target for drug specificity [11].

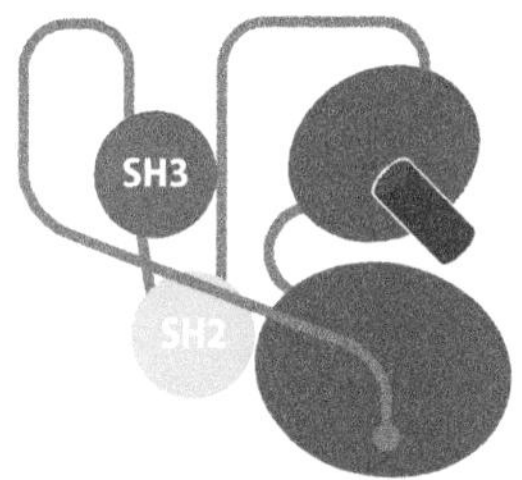

FIGURE 8.12
Regulation of the activity of Abl kinase. The active state is very similar to that of Src kinase (Figure 8.11). The inactive state is also similar to the inactive state of Src, except that the SH2 domain now binds to the kinase C-terminal domain rather than to a phosphotyrosine. This pulls the C-terminal domain (bottom half) of the kinase back toward the SH2 domain, keeping the active site open and inactive. This structure is probably strengthened by an interaction between a myristoyl group attached to the N terminus (gray circle) and the C-terminal kinase domain. Gleevec binds into the opened active site, under the αC helix.

8.2 DIMERIZING RECEPTOR KINASE SYSTEMS

8.2.1 The Jak/Stat system is a simple pathway

Most receptor dimerization systems lead ultimately to phosphorylation of DNA-binding proteins, and thus to alteration of transcription. Such systems therefore tend to be relatively long-term in their activity, and they are the main systems responsible for regulating cell development and differentiation. This of course makes them medically important. We begin with the Jak/Stat class of receptor dimerization mechanism because of its simplicity (and its likely early evolutionary origins, as discussed later).

***8.2 Cytokine**

Cytokines are small globular proteins or peptides (usually less than 10 kDa) that act as messengers between cells (**Figure 8.2.1**). The word derives from the Greek for 'cell movement.' They are effectively local hormones, and are found in blood and in the immune system. For example, interleukins and interferons are generally secreted by helper T cells and target other immune cells such as B or T cells, stimulating proliferation or maturation. They also act as chemoattractants (hence the name), leading to movement (chemotaxis) of the target cell toward the source of the cytokine. Such proteins are often called chemokines. Cytokines also include transforming growth factors TGF-β and factors leading to cell differentiation such as erythropoietin and thrombopoietin.

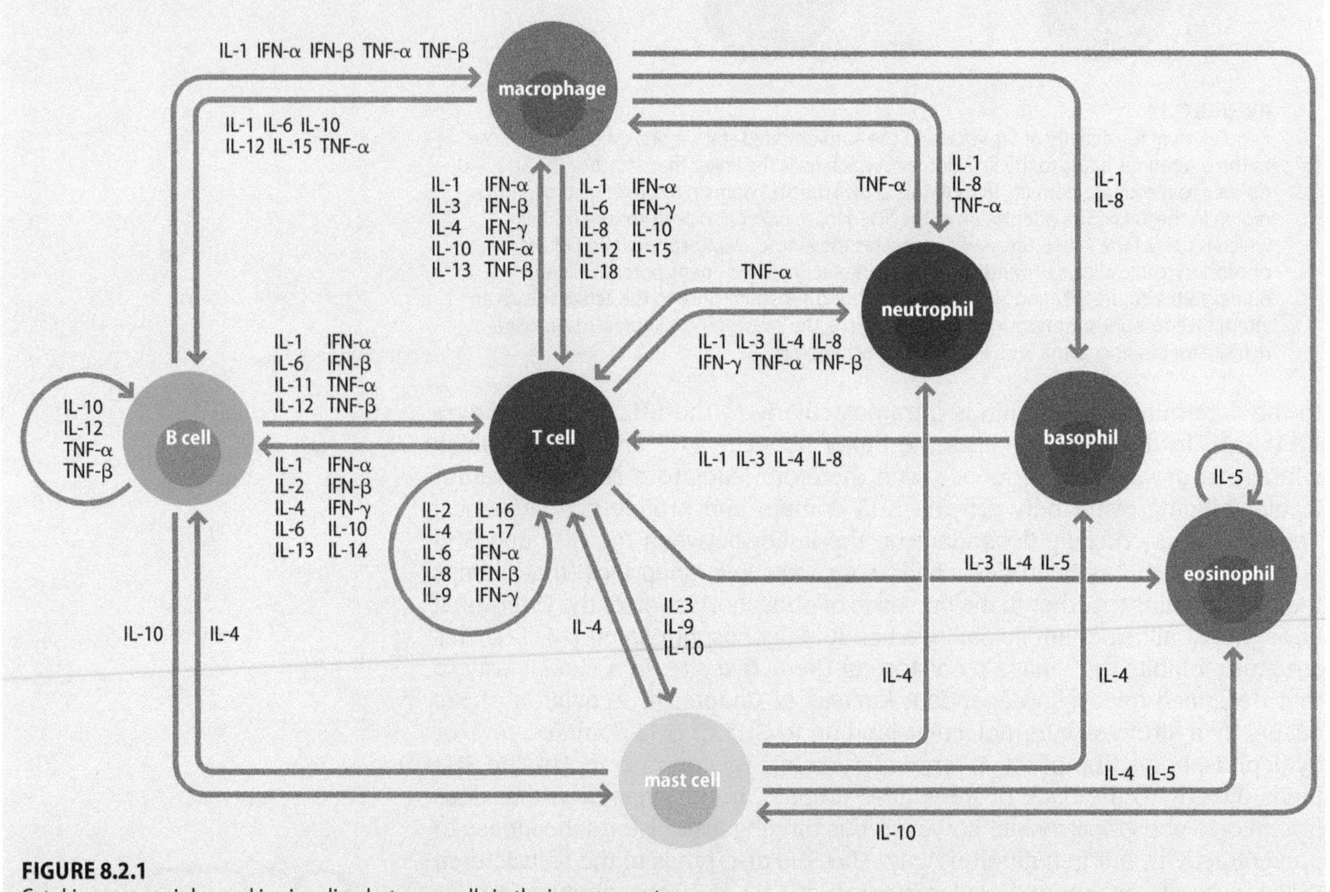

FIGURE 8.2.1
Cytokines are mainly used in signaling between cells in the immune system.

The Jak/Stat pathway is typically turned on by a cytokine. A **cytokine (*8.2)** is a small extracellular protein produced by one group of cells and acting on another group. Because cytokines cannot cross the membrane, the signal carried by the cytokine has to be mediated by a receptor. The receptor for many cytokines (including α- and γ-interferon and growth hormone) is a simple monomeric molecule with a single transmembrane helix, although normally it self-associates to form a dimer or trimer, even in the resting state (which makes it different from most of the other dimerizing systems discussed later). Intracellularly, each receptor is bound to a tyrosine kinase, which is called a Janus kinase (Jak), after Janus, the Roman god of doorways, who has two faces to look in both directions at the same time (hence January, the first month of the year): Jaks have two catalytic domains. The receptor–Jak complex is a permanent and stable complex. Binding of a cytokine to the receptor causes conformational changes that lead to a rearrangement of the structure (**Figure 8.13**). This rearrangement brings the Jak domains together and allows them to phosphorylate each other; that is, the substrate for the kinase activity of Jak is Jak itself.

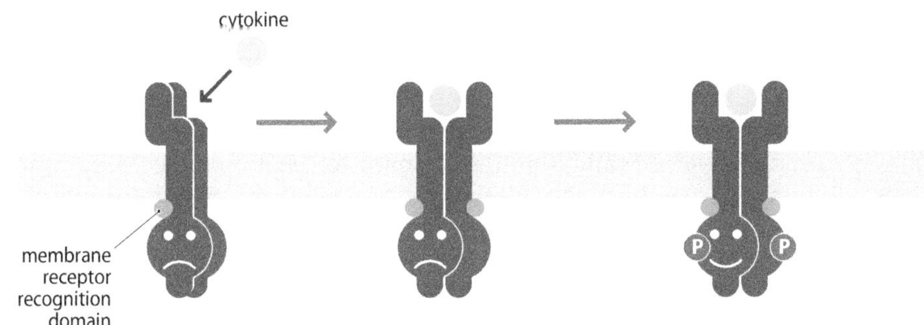

FIGURE 8.13
Activation of the Jak/Stat signal. Binding of a cytokine to the receptor leads to a rearrangement of the receptor structure, which brings the active sites of the two Jak kinases close to their partners. This allows them to phosphorylate, and thus activate, each other.

Essentially, this completes the difficult part of the signaling pathway: how to get a signal from one side of the membrane to the other. The extracellular binding of a ligand has produced an intracellular change.

This very simple mechanism has thus found one solution to the problem described above: how to prevent the constitutive activity of the kinase from leading to unwanted turning on of the signal. By pre-assembling the receptor–Jak complex so as to keep different Jaks apart until the receptor is bound to ligand, unwanted phosphorylation is greatly decreased, although not eliminated.

All that remains is to complete the pathway: how can the phosphorylation event be recognized and used to alter DNA transcription? The solution in this system relies on a third protein, Stat (signal transducer and activator of transcription). Stats bind to phosphorylated tyrosines by using an SH2 domain (Section 8.1.6). They also have a second domain, which binds to DNA. Completion of the signaling pathway then requires that binding of Stat to the phosphorylated tyrosine leads to some change in Stat that alters the state of the DNA-binding domain to allow it to bind to DNA. As we have seen already in Chapter 3, a very common way of doing this is to dimerize the DNA-binding domain; that is, the 'change in Stat' that is needed is a dimerization. And this is done in a very simple way: by phosphorylation of Stat, using the existing kinase activity of Jak.

In fact, Stats do not recognize the phosphorylated tyrosine on Jak directly. What happens is that the initial phosphorylation of Jak activates it so that it then phosphorylates a tyrosine residue on the *receptor* (**Figure 8.14**), to which it is of course close, because it is permanently bound to one receptor and held close to the other by the dimerization of the receptor. This trick is used repeatedly in signaling pathways, as we shall see. Stat then binds to this phosphorylated tyrosine. This binding of course positions Stat close to Jak, and means that Stat also gets phosphorylated by Jak, on another tyrosine (see Figure 8.14). In other words, Stat gets phosphorylated not because it is an inherently good substrate for Jak, or because Jak is an inherently good or specific kinase, but mainly because Stat is held close to the kinase. This is such a key element of so many signaling mechanisms that it is worth repeating: the specificity in the signaling is obtained by bringing the sequential partners in the signal pathway

FIGURE 8.14
Transphosphorylation of the Jak kinases activates them to phosphorylate the receptor. The phosphorylated receptor is then recognized by the SH2 domain of a Stat. In turn, this leads to phosphorylation of Stat. Finally, the phosphorylated Stats bind together. This dimeric Stat is then able to go into the nucleus and activate DNA transcription.

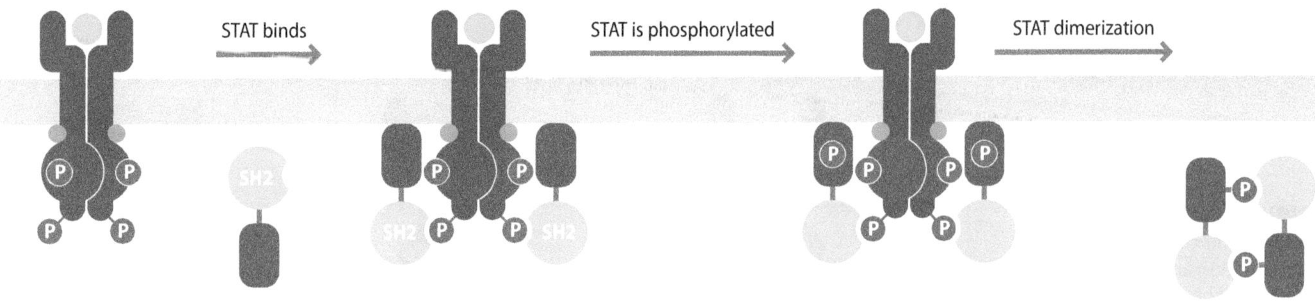

together, and allowing a relatively nonspecific signaling mechanism to transmit the signal between the colocalized partners.

Thus, Jak has to perform three different phosphorylation events. The third phosphorylation achieves the final goal of the signaling system (namely to pass on the signal specifically from the receptor to the messenger, Stat), and does two things in one. First, the phosphorylation weakens the binding of Stat to the receptor so that it dissociates. And second, it provides a phosphorylated tyrosine on Stat, which can then be recognized by the second Stat protein that is conveniently close, and produce the desired Stat dimer. Formation of the dimer uncovers a nuclear localization sequence (Section 7.4), which relocates the protein into the nucleus, after which the dimer is capable of binding cooperatively to DNA.

As we shall see repeatedly in this chapter, the evolution of such a system does not require any fundamentally new domains or functions: it simply requires the putting together of existing domains in a new way, followed of course by tinkering to make the system work properly.

8.2.2 Receptor dimerization takes a variety of forms

The basic mechanism for receptor dimerization is that a single monomeric receptor has an extracellular ligand recognition domain and an intracellular receptor tyrosine kinase (RTK) domain. In general, the RTK domain is part of the same polypeptide chain rather than being a different protein as in Jak/Stat. Binding of a ligand to the recognition domain leads to dimerization of the receptor, which brings two RTKs together intracellularly, so that they can phosphorylate each other. This generates the intracellular signal, which is then transmitted further: by the binding of Stats, as discussed above; or by the binding of enzymes such as phospholipase C or phosphoinositide 3-kinase, which generate small-molecule **second messengers**; or by passing the signal through to the activation of a small GTPase such as Ras, and from there to a kinase cascade that eventually phosphorylates DNA-binding transcription factors. This third system is the most common form of RTK signaling and is discussed in detail in the following sections.

The system described in Sections 8.2.2 to 8.2.5 is commonly known as the RTK (receptor tyrosine kinase) system. This is a somewhat confusing name because the Jak/Stat system is also an RTK; however, Jak/Stat is conventionally considered a different type of system and is not usually called an RTK because the kinase is on a separate polypeptide chain. It is a widely used signaling system, and is the pathway used by epidermal growth factor (EGF), platelet-derived growth factor (PDGF), interleukins, insulin, colony-stimulating factors, and many growth hormones. We can view it as a more sophisticated version of Jak/Stat.

The simplest mechanism for receptor dimerization is exemplified by several receptors, including probably those for vascular endothelial growth factor (VEGF), stem cell factor (SCF), PDGF, and neurotrophins (the Trk receptor). For all of these, the ligand is a dimer, which interacts identically with two receptors and therefore generates dimerization of the receptor (**Figure 8.15**) [15].

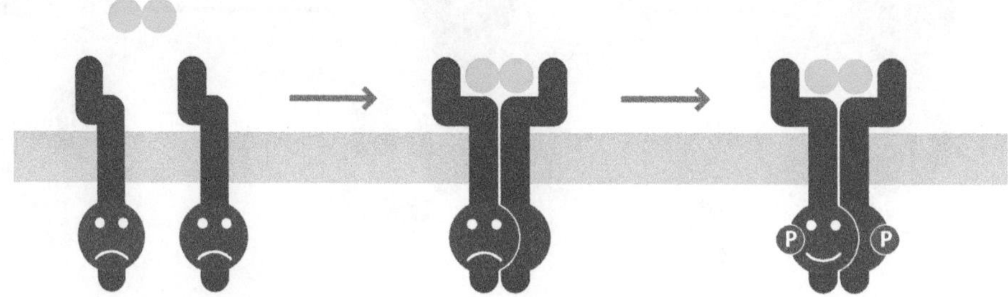

FIGURE 8.15
A simple mechanism for receptor dimerization and activation: the ligand (here VEGF) is also a dimer.

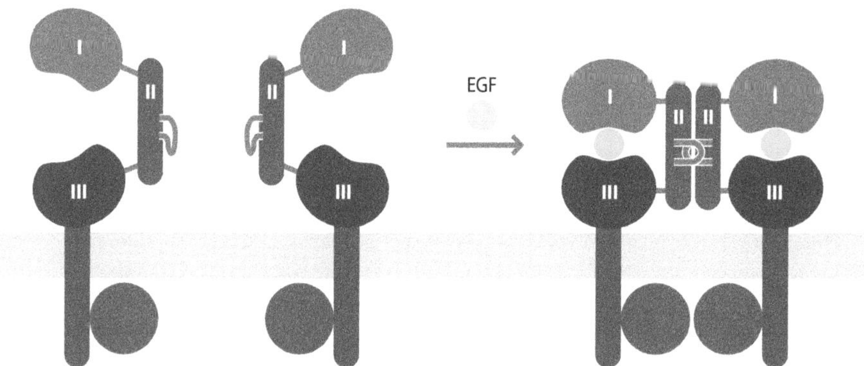

FIGURE 8.16
The EGF receptor binds monomeric EGF, which leads to a domain reorganization that pushes a loop out from domain II (also called the cysteine-rich domain). Two loops on adjacent receptors interact and produce a dimeric receptor.

It is also possible to have a monomeric ligand, which binds to two identical receptors but has different interactions with each. This mechanism creates some interesting possibilities, because the affinities and on-rates for the two sites can be quite different. Such a situation occurs for the binding of growth hormone to its (Jak-associated) receptor.

A somewhat more complicated mechanism is used by EGF and transforming growth factor-α (TGF-α) [16]. EGF binds as a monomer to a monomeric receptor. The binding causes a conformational change in the receptor, which means that it binds strongly to a second EGF-bound receptor molecule (**Figure 8.16**). This brings together the two intracellular kinase domains which can go on to autophosphorylate.

The bound monomeric EGF receptor can also bind to other receptors to form heterodimers. As we have seen in Chapter 3, this provides the potential for a much greater variation in function. In particular, it can dimerize with the transmembrane proteins ErbB2, ErbB3, and ErbB4. ErbB3 and ErbB4 are receptors for the protein heregulin, implying that this pathway only gets switched on when *both* EGF and heregulin are present. On the other hand, ErbB2 may not need a ligand at all, in which case signal transmission depends strongly on the concentration of ErbB2.

Another possible mechanism for receptor dimerization is illustrated by interleukin-4 (IL-4), which initiates a Jak/Stat pathway. Here, the receptor is a heterodimer (**Figure 8.17**) [17]. IL-4 binds to the α subunit of the receptor, and this complex then binds the γ subunit, a much shorter molecule that probably has a similar role in the signaling of a large number of interleukins (an interesting example of the 'common parts' tool discussed in Chapter 2). The **insulin receptor (*8.3)** is already dimeric in the absence of ligand. As one might expect, evolution has experimented with other multimeric structures. For example, the tumor necrosis factor (TNF) receptor is a trimer, matching the trimeric structure of TNF.

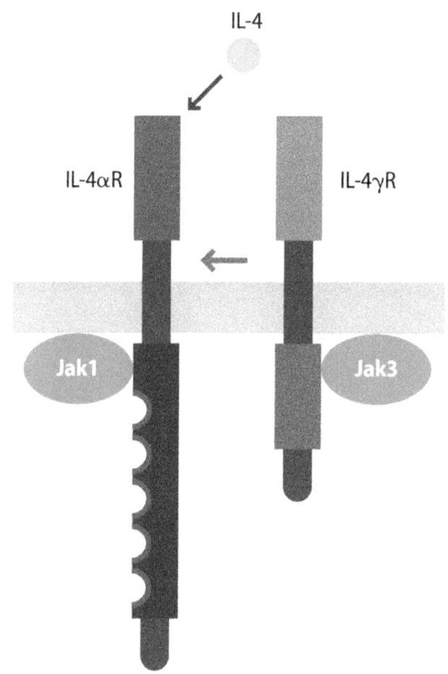

FIGURE 8.17
The interleukin-4 receptor. The red semicircles on IL-4αR represent potential phosphorylation sites.

There are two general points worth making about receptor dimerization. First, the transfer of the signal across the membrane requires the association of two transmembrane receptors. The association of two receptors is of course a search on a two-dimensional membrane surface. As discussed in Chapter 4, this is a rather rapid process.

Second, the fact that signal transmission requires a dimeric receptor means that anything that prevents dimerization will inhibit the signal. Therefore, for example, ligands with only a single binding site can often prevent signal transduction, as can monomeric antibodies that bind only to a single ligand. This presents a good opportunity for drugs to inhibit the activation of specific pathways. Conversely, anything that leads to dimerization will produce a signal. In

*8.3 The insulin receptor

The insulin receptor is unusual in several ways. First, the dimeric structure is already preformed in the absence of insulin, although it is of course inactive, as are dimeric Jak kinase receptors. The insulin receptor (IR) actually consists of four peptide chains held together by extracellular disulfide bridges (**Figure 8.3.1**; compare with a selection of receptor structures in Figure 2.11). Structural studies of the receptor have led to a model for IR function (**Figure 8.3.2**), in which insulin binds asymmetrically to the two symmetry-related halves of the receptor [62]. The two halves of the receptor are positioned by the disulfide bridges, which effectively provide a hinge. When insulin binds, the hinge closes, allowing the kinase domains to approach close enough to autophosphorylate.

Second, a large number of the tyrosine autophosphorylation sites are not on the receptor at all, but on a scaffold protein (also referred to as a multidocking protein) associated with the receptor, called IRS1 (insulin receptor substrate 1). This provides a significant increase in the range of possible interaction partners, and thus facilitates the integration of signals from other pathways (**Figure 8.3.3**).

Third, there is another scaffold called IRS2, which further modulates the activity of the receptor. IRS2 is anchored to the activated receptor because the activation involves phosphorylation at four sites on the receptor. Three of these lead to the removal of an inhibitory chain from the active kinase site, and the fourth (Y972) forms a binding site for IRS2. Like IRS1, IRS2 contains a PH domain, to position it close to the membrane, together with a PTB domain, which binds to pY972 on the activated receptor (**Figure 8.3.4**) [63]. IRS2 contains a short peptide sequence known as KRLB (kinase regulatory-loop binding), which binds to the active site of the receptor kinase in a similar manner to the receptor's own inhibitory chain [64]. This blocks the active site and therefore inhibits further kinase activity. Whether this is inhibitory or stimulating to the overall function of the receptor is as yet unclear, but it illustrates the complexity that can arise from the binding of additional scaffolds, not least because all of these interactions are regulated by phosphorylation and dephosphorylation.

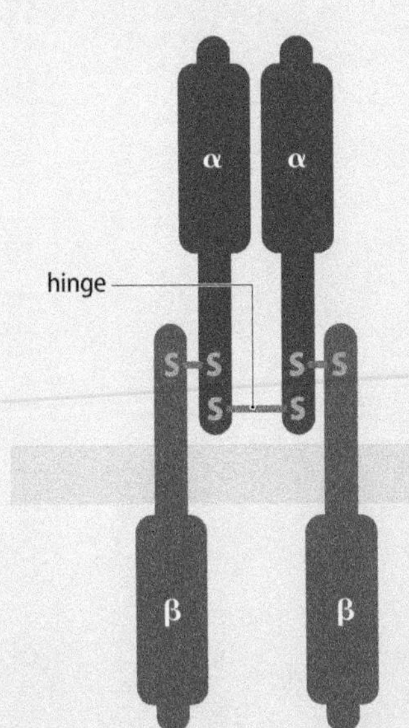

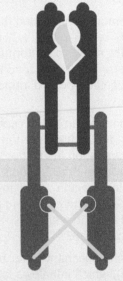

FIGURE 8.3.1
The insulin receptor consists of four polypeptide chains (two α and two β), held together by disulfide bridges.

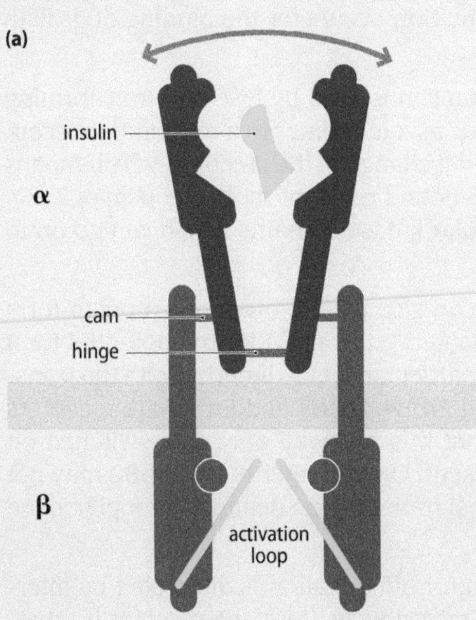

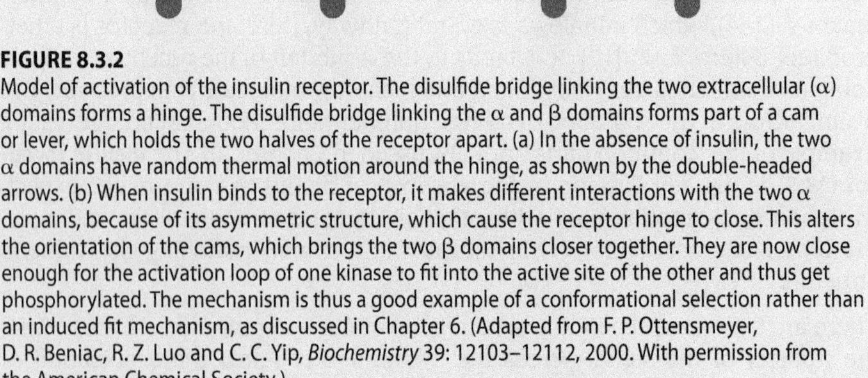

FIGURE 8.3.2
Model of activation of the insulin receptor. The disulfide bridge linking the two extracellular (α) domains forms a hinge. The disulfide bridge linking the α and β domains forms part of a cam or lever, which holds the two halves of the receptor apart. (a) In the absence of insulin, the two α domains have random thermal motion around the hinge, as shown by the double-headed arrows. (b) When insulin binds to the receptor, it makes different interactions with the two α domains, because of its asymmetric structure, which cause the receptor hinge to close. This alters the orientation of the cams, which brings the two β domains closer together. They are now close enough for the activation loop of one kinase to fit into the active site of the other and thus get phosphorylated. The mechanism is thus a good example of a conformational selection rather than an induced fit mechanism, as discussed in Chapter 6. (Adapted from F. P. Ottensmeyer, D. R. Beniac, R. Z. Luo and C. C. Yip, *Biochemistry* 39: 12103–12112, 2000. With permission from the American Chemical Society.)

particular, normal dimeric antibodies with two identical Fab arms that recognize and bind to two receptors simultaneously can often lead to signal activation. Indeed, this is the mechanism for signaling mast cell degranulation in the inflammatory response caused by IgE in response to allergens.

***8.3 The insulin receptor (cont.)**

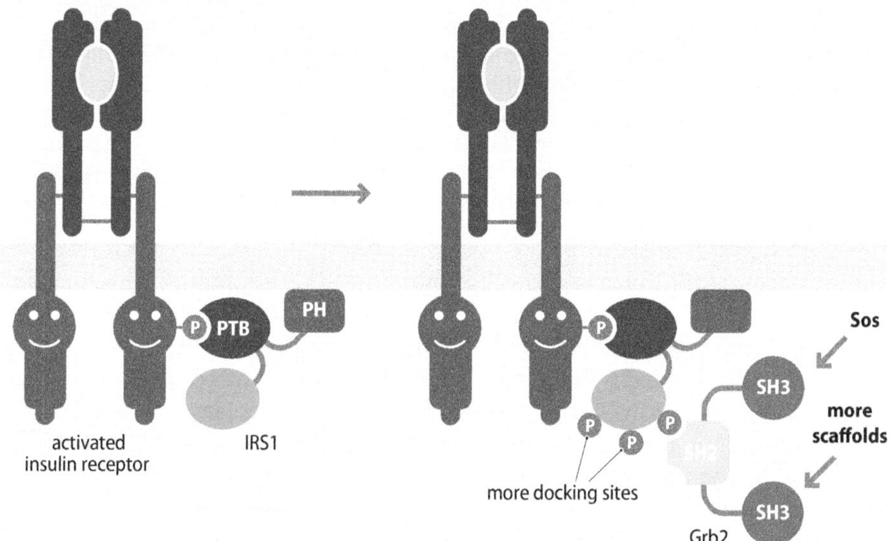

FIGURE 8.3.3
The activated insulin receptor binds to a scaffold or multidocking protein called IRS1 (insulin receptor substrate 1), via a PTB domain that recognizes the phosphorylated receptor. IRS1 also has a PH domain that attaches it to the cell membrane, and a third domain that gets phosphorylated by the receptor kinase and acts as a docking site for a range of other proteins, including Grb2 (discussed below).

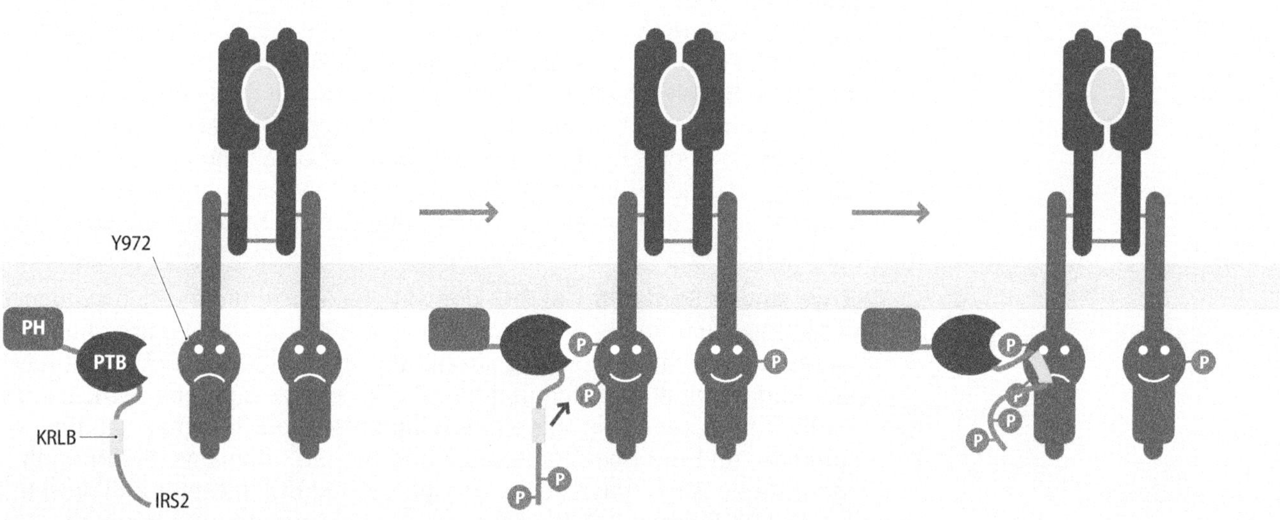

FIGURE 8.3.4
IRS2 contains a PH domain that recognizes inositol phosphates in the cell membrane, and a PTB domain, which recognizes Tyr 972 in the insulin receptor once it has been autophosphorylated. This binding leads to the phosphorylation of several tyrosines in the C-terminal end of IRS2, which probably act as binding sites for further proteins. A short sequence within IRS2 called KRLB binds to the active site of the receptor kinase and inhibits it.

8.2.3 Ras is the immediate target of the receptor tyrosine kinase (RTK) system

Once the initial intracellular phosphorylation has occurred, how is the signal passed to the nucleus? Essentially, there are two parts. In the first part, the specific signal is linked to a generic intracellular activation system, by leading to the replacement of GDP by GTP bound to the small GTPase protein Ras, which activates Ras. And in the second part, the activated Ras initiates a **cascade (*8.4)** of phosphorylation reactions leading ultimately to the phosphorylation of transcription factors, which consequently have altered DNA binding affinity and so lead to changes in gene expression. One can see that this two-part pathway could have major benefits over the more simple Jak/Stat system. For

FIGURE 8.18
Electrical adaptors. These two are different: they have one end that recognizes a British 3-pin plug, and another end that plugs into a 2-pin socket, and has either flat pins (left) or round ones (right). They thus have one 'common' end, and differ at the other end. Similar arrangements are very common in signaling adaptor proteins.

*8.4 Kinase cascade

A kinase cascade is a series of kinases, such as that in the MAPKKK/MAPKK/MAPK group (Figure 8.26), in which the first kinase acts on the second to phosphorylate it and thereby activate it; the second phosphorylates and activates the third, and so on. In theory, this can lead to a very large enhancement of the initial signal, because one kinase can phosphorylate a large number of substrates. In practice, it is not clear how much of an amplification actually occurs, because for the correct specificity, the kinases need to be held together in a complex by a scaffold protein. In such cases, amplification can only occur if the downstream kinases exchange on and off the scaffold rapidly compared with loss of the upstream kinase. There are other cascades such as a proteolytic cascade in blood clotting factors.

example, the more sophisticated generic signaling system leads to a larger signal (because of the cascade), and one that can alter expression of many genes simultaneously. However, it creates some major problems: if essentially the same system is used by a large number of ligands, how can it be made specific? Here the answer again is to add extra bits until it *is* sufficiently specific, as we shall see.

The first step is very similar to that used by Jak/Stat: the phosphorylated tyrosine is recognized by an SH2 domain (see Table 2.2). However, now the SH2 domain has to feed in toward the generic Ras pathway. The SH2 domain forms half of an **adaptor** protein. This works much like a universal electrical adaptor, used for example to plug a range of electrical devices from different countries into a British socket, or vice versa (**Figure 8.18**): there is one 'socket' end (the SH2 domain) that is specialized for each signaling system and recognizes only a single phosphorylated receptor, and the other end is a general 'plug' that connects the receptor into a more generic system. Typically, the adaptor contains an SH2 domain at one end and one or more SH3 domains at the other end (**Figure 8.19**).

As we saw in Section 8.1.6, SH3 domains recognize proline-rich sequences. As discussed in Chapter 4, proline-rich sequences are ideal parts of a signaling pathway, because they bind specifically and reasonably tightly but have a small surface area, meaning that they can recognize and release their targets rapidly. The protein Grb2 acts as the adaptor for the EGF signaling pathway in mammals, and has one SH2 domain and two SH3 domains (**Figure 8.20**). Its essential role is to recognize the phosphorylated EGF receptor and present its SH3 domain for recruitment of the next element downstream, Sos.

Sos (**Figure 8.21**), like Grb2, has two ends: a proline-rich sequence at one end, and an active site at the other, which functions as a **guanine nucleotide exchange factor** (GEF, also known as guanine nucleotide releasing protein,

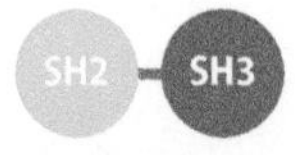

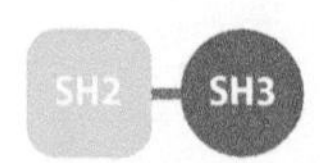

FIGURE 8.19
Signaling adaptors. Like the electrical adaptors in Figure 8.18, these have one common end, in this case an SH3 domain, and two different ends, in this case SH2 domains that recognize different phosphotyrosine sites.

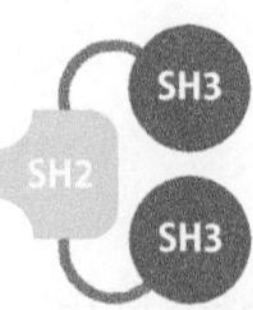

FIGURE 8.20
The adaptor protein Grb2 (growth factor receptor-bound protein 2), which has one SH2 and two SH3 domains. In *Drosophila* the equivalent protein is known as Drk (which boringly stands for downstream of receptor kinase), while in *C. elegans* it is called Sem-5.

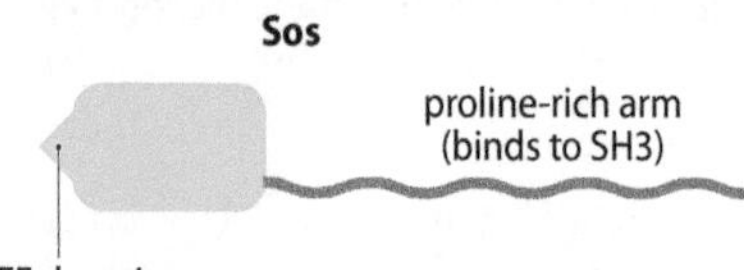

FIGURE 8.21
Sos, which has a guanine exchange factor (GEF) domain at one end and a proline-rich tail at the other. The protein is in fact considerably more complex than this, as discussed shortly. Sos stands for son of sevenless. It is a *Drosophila* protein that occurs downstream of a gene named sevenless, ablation of which leads to loss of the seventh photoreceptor in the *Drosophila* eye and consequent inability to see ultraviolet light. There are close homologs in many eukaryotes, including humans.

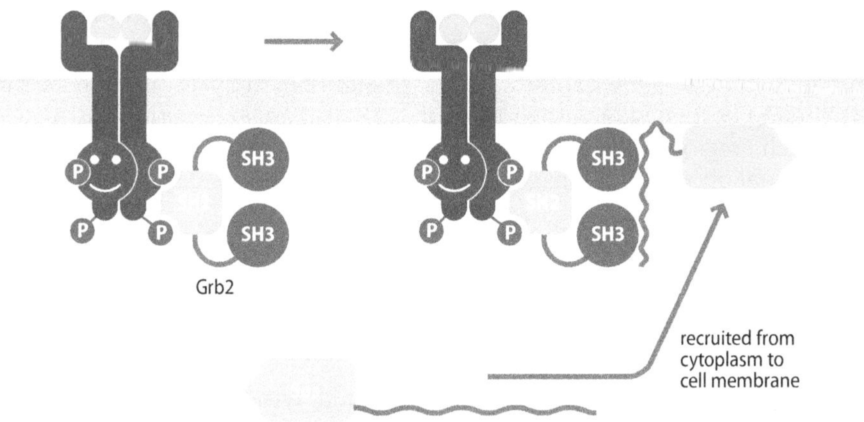

FIGURE 8.22
In a typical RTK pathway, the activated receptor binds to the SH2 domain of the adaptor Grb2, which recognizes a phosphotyrosine that has been phosphorylated by the action of the receptor kinase. The SH3 end of Grb2 then binds to the polyproline tail of Sos, and thereby brings it from the cytoplasm next to the membrane.

GNRP) and catalyzes the exchange of GTP for GDP on Ras. It is another constitutively active protein, and the main reason why it does not act as a GEF all the time is simply due to its location: normally Sos is present in the cytoplasm, whereas Ras has a membrane anchor and is thus firmly attached to the membrane. Therefore effectively the Grb2 adaptor merely brings Sos up to the membrane surface so that it can interact with Ras (**Figure 8.22**). Once Sos is present at the membrane surface, another two-dimensional search to bind Ras to Sos is then required.

Ras is a classic small GTPase (Section 7.1.3). A large number of variants on Ras are found in different pathways, as illustrated in **Table 8.2**. It contains the switch I and switch II regions characteristic of small GTPases, and adopts different conformations when bound to GDP or GTP. It is active as a signaling molecule only when bound to GTP. However, it has an intrinsic GTPase activity, which means that the signal gets switched off at a rate determined by the GTPase activity, and it gets switched on by detachment of GDP and binding of GTP. In the cell, the concentration of GTP is about 10 times higher than that of GDP. Ras can therefore be activated simply by a **guanine nucleotide exchange factor (*8.5 GEF)**, which opens up the binding site to allow the equilibration of GDP and GTP; conversely, it can be deactivated by **GTPase-activating proteins** or **GAPs (*8.6)**, which accelerate the rate of GTP hydrolysis, by a factor of about 100 relative to the intrinsic hydrolysis rate (or more, depending on the inherent activity of the GTPase). Ras works not unlike the classic spring-loaded mouse trap (except in reverse): it remains in the GDP-bound state until prodded in the right place by the GEF Sos, which displaces part of the structure and allows it to spring open, thereby allowing GDP to be released and GTP to bind, after which it re-closes into its 'tense' active state (**Figure 8.23**) [18]. The conformational change in Ras is the typical change between the spring-loaded GTP-bound

TABLE 8.2 The Ras superfamily of small monomeric GTPases

Family	Examples	Function
Ras	H-Ras, K-Ras, N-Ras	RTK signaling
	Rheb	Neural plasticity
Rho	Rho, Rac, Cdc42	Signaling to cytoskeleton
ARF	Arf1-6	Assembly of vesicle coats
Rab	Rab1-60	Vesicle traffic
Ran	Ran	Nuclear transport and mitotic spindle assembly

***8.5 Guanine nucleotide exchange factor**
A GEF is a protein that stimulates the loss of GDP from a Ras-like small GTPase protein, and therefore allows activation by binding to the more abundant GTP (**Figure 8.5.1**). Such proteins are very common in signaling systems, and act to turn on the GTPase switch. Many GEFs contain a Dbl domain followed by a PH domain. Examples of GEFs discussed in this book are Sos and Vav, and G-protein-coupled receptors. The complementary off switch is provided by a GTPase-activating protein (GAP).

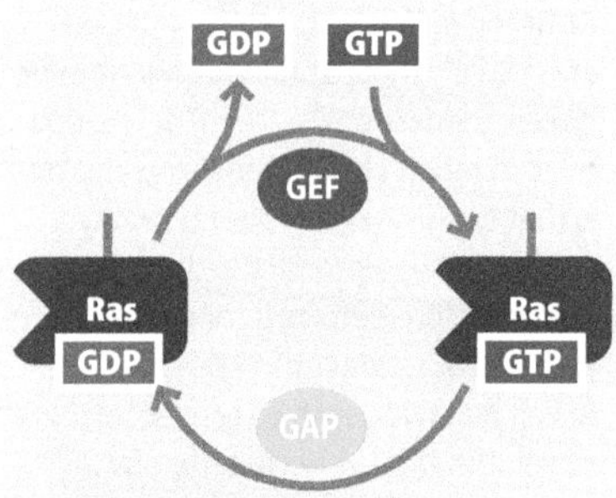

FIGURE 8.5.1
The function of GAPs and GEFs. GEFs generally act by binding to the 'spring-loaded catch' in the G protein (for example Ras), indicated by the indentation on the left, and releasing the spring to allow GDP to be released.

***8.6 GTPase-activating protein**
A GAP is a protein that stimulates the hydrolysis of GTP to GDP when bound to a Ras-like small GTPase. This turns off the signal. GAPs are therefore opposite in their action to guanine nucleotide exchange factors (GEFs).

state and the relaxed GDP-bound state discussed in Chapter 7. In Chapter 7 the change was described as a conformational change driven by GTP/GDP exchange: here it is the conformational change that happens first (pushed by the binding of Ras), which then leads to GTP exchange. Sos in fact displaces the switch I region, which sits at one end of the nucleotide-binding site, where the terminal phosphate binds: the incoming GTP therefore displaces Sos [18].

Once Sos has done its job and activated Ras, it is no longer needed and can be switched off and removed (essentially a reverse of the activation process described in Section 8.2.7). Indeed, it *has* to do so, because modeling studies imply that the upstream and downstream partners, Sos and Raf, bind to the same part of Ras, and thus Sos has to leave before Raf can bind [19].

It is striking that with the exception of the initial tyrosine kinase (which was also part of the Jak/Stat pathway), none of the proteins involved in this pathway so far is an enzyme. The evolution of a new enzyme is harder than the evolution of a binder (basically because an enzyme has to bind *and* catalyze a reaction), which means that variations on the pathway are somewhat easier to produce when there are no new enzymes needed.

8.2.4 Ras activates the kinase Raf

Ras forms a key intersection for many signaling pathways. Many signals converge on Ras, and after Ras there is a classic kinase cascade (*8.4), in which the phosphorylation of a protein activates it as a kinase, which then phosphorylates the next protein. Effectively, activation of Ras lets loose an almost unstoppable process. It is therefore critical that Ras be activated correctly.

The activated Ras-GTP transmits the signal by binding to Raf. Just as activated Grb2 binds to Sos and relocates Sos to the membrane, so Ras-GTP binds Raf and relocates it to the membrane. This act of relocation is the main function of Ras-GTP. The binding of Ras-GTP to Raf probably also causes a conformational change in Raf that relieves an autoinhibition. This aspect has proved very difficult to characterize, because of the involvement of a number of scaffold proteins, kinases, and phosphatases, which add considerably to the complication. As a result, the details of Raf activation are still unclear. It is, however, evident that Raf activation requires the involvement of a protein called 14-3-3.

There are several isoforms of 14-3-3, and it is one of the most abundant proteins in the cell. Its function is to bind to phosphorylated serines and threonines (pS/T), and it recognizes and binds to a very wide range of proteins [20]. Remarkably, all the discussion so far in this chapter has concentrated on phosphotyrosine, despite the fact that **Ser/Thr kinases (*8.7)** constitute roughly 92% of the protein kinases in humans [21]. 14-3-3 is a dimer with a rather rigid structure (**Figure 8.24**), in which the pS/T-binding site lies within a long extended channel. The most plausible model for its function suggests that the protein functions by forcing pS/T-containing ligands (generally as dimers) to

***8.7 Ser/Thr kinases**
As noted in the text, Ser/Thr kinases are actually much more abundant than tyrosine kinases. Why, then, do we not hear so much about them? The answer is that tyrosine kinases are typically involved in receptor activation, and therefore defects in tyrosine kinases cause drastic alterations in cell behavior, such as cancers or developmental abnormalities. By contrast, Ser/Thr kinases regulate almost everything else: enzyme activity, transport, degradative processes, cell cycle, and so on. They are therefore also extremely important but individually tend to have less marked effects.

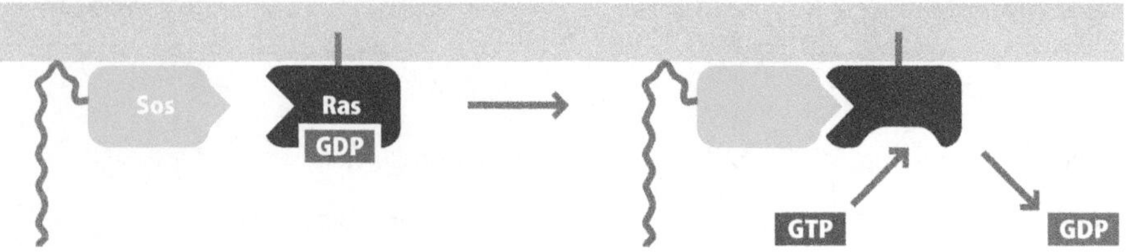

FIGURE 8.23
Sos acts as a GEF on Ras, and releases the bound GDP, which is replaced by GTP. The GTP-bound Ras is now capable of activating the kinase Raf.

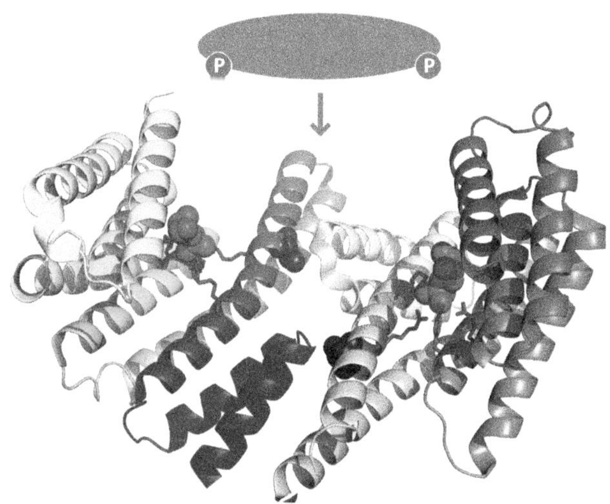

FIGURE 8.24
The structure of 14-3-3. The protein is a dimer, the dimer interface being in the center of the figure. Each monomer has one binding site for phospho-Ser/Thr, indicated by the red residues, binding to phosphoserines (magenta spheres). The protein is rigid, and therefore requires the two phospho-Ser/Thr residues to be a fixed separation apart, as indicated on the gray protein. 14-3-3 itself can also be phosphorylated; for example, phosphorylation of Ser 58 (gray spheres) disrupts the dimer interface and therefore prevents 14-3-3 binding correctly, whereas phosphorylation at Ser 185 disrupts the binding site. 14-3-3 is so called because of its migration on gels. Drawn using PDB file 2br9.

undergo a conformational change so as to fit within the 'molecular anvil' that is 14-3-3. Because of the rigid dimeric structure of 14-3-3, its binding to a correctly oriented dimeric ligand is highly cooperative (Chapter 2). It is therefore suggested [20] that 14-3-3 works by binding to a pS/T site on Raf; cooperative binding to a second pS/T site then acts either to clamp Raf shut in an inactive state or to clamp it open in an active state, depending on which S/T sites are phosphorylated. Thus, 14-3-3 works both as an inhibitor and as a scaffold and thus activator for Raf, depending on the location and phosphorylation state of Raf. This dual function has of course not helped in understanding how Raf activation occurs.

Raf has at least four suitable binding sites for 14-3-3 [21], of which the two most important ones seem to be S621 and S259. Beyond this, the details are somewhat unclear. A plausible hypothesis is drawn in **Figure 8.25**. In the inactive state (bottom left), Raf is cytosolic and is clamped shut by the binding of 14-3-3 to both pS259 and pS621. Activation of Ras by binding to GTP increases its affinity for Raf, which binds via its Ras-binding domain (RBD) and possibly also via its cysteine-rich domain (CRD). This binding brings Raf to the cell membrane, which is the key to all subsequent steps. In some way and in some order, this displaces 14-3-3 from pS259, and allows protein phosphatase 2

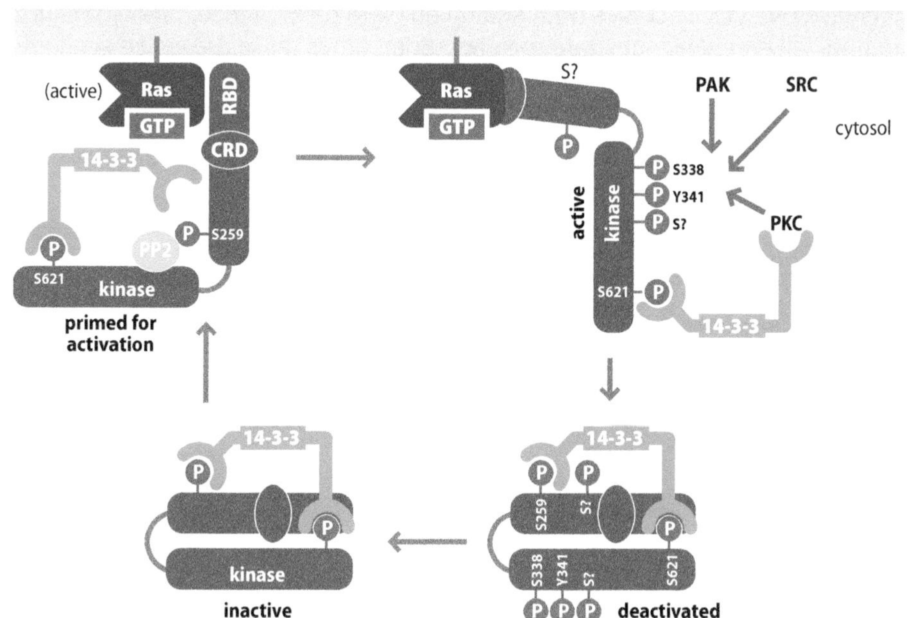

FIGURE 8.25
A plausible model for the activation of Ras, for which the key step is phosphorylation of Tyr341. See the text for details. PP2 is a phosphatase; CRD is the cysteine-rich domain, also called the zinc finger; RBD is the Ras-binding domain; PAK, SRC, and PKC are the kinases p21cdc42/rac1-activated serine/threonine kinase, Src kinase, and protein kinase C. (Adapted from W. Kolch, *Biochem. J.* 351: 289–305, 2000. With permission from the Biochemical Society.)

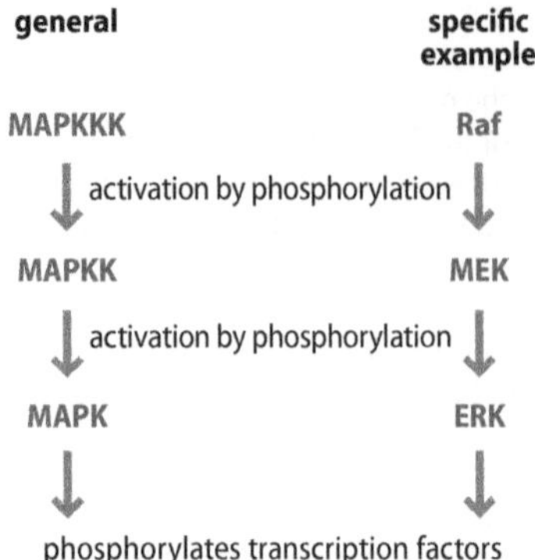

FIGURE 8.26
The downstream kinase cascade. Binding of GTP to Ras leads to activation of the kinase Raf. Raf then phosphorylates MEK and activates it, which in turn phosphorylates ERK and activates it. Activated ERK then goes on to phosphorylate several other proteins, many of which are transcription factors. This is a specific example of the more general MAPKKK/MAPKK/MAPK cascade.

(PP2, also called PP2A) to dephosphorylate S259. 14-3-3 is thus freed at one end, uncovering the kinase site on Raf and enabling it also to act as a scaffold and bind to other proteins such as a range of other kinases. In part this probably involves activation of the Rho family of GTPases, a large family related to Ras but involved in cytoskeletal signaling (see Table 8.2) [22]. Subsequent phosphorylations of Raf serve mainly to activate it further.

This is clearly not the complete picture: there are many more proteins in the Ras–Raf complex; dimerization of both Ras and Raf has been suggested; and there are several other important phosphorylation sites on Raf. In addition, full activity of Raf requires the presence of chaperone (*4.12) proteins such as Hsp90 and Hsp50, whose roles are as yet unclear. There is a very complicated series of interactions, presumably necessary for the regulation of Raf and the integration of signals from different pathways.

8.2.5 The downstream pathway from Raf is a kinase cascade

Activated Raf is now a functional kinase, and initiates a kinase cascade (**Figure 8.26**). In the standard cascade there are three levels, in which the first is a MAP (mitogen-activated protein) kinase kinase kinase or MAPKKK, the next is a MAPKK, and the final stage is a MAP kinase. It is called mitogen-activated because some of the extracellular ligands that initiate the signaling pathway are mitogens. There are at least six versions of this cascade, but in the one that concerns us here, the ERK (extracellular-signal-regulated kinase) pathway, the unique initial kinase [23] MAPKKK (here also known as MEK kinase) is Raf, the MAPKK is called MEK (MAPK/ERK kinase), and the MAPK is called ERK. ERK phosphorylates a range of proteins, notably transcription factors such as Fos, Jun, and Myc (Chapter 3), as well as a range of cytoplasmic enzymes (for example kinases) and other proteins such as the inflammatory mediator phospholipase A_2 (**Figure 8.27**).

We saw in Chapter 4 that kinases do not recognize their substrates very specifically. Thus, one kinase can often phosphorylate several substrate proteins; and many proteins can get phosphorylated by several kinases, often at different sites. Evolution has of course tinkered with this, by taking what is basically a defect in the system and turning it to an advantage. There is a bewildering number of proteins that recognize different stages of phosphorylation and either activate or inhibit the signaling pathway, or else link it to other pathways. The key to understanding what is going on is the recognition that increased specificity comes from adding on extra pairs of recognition domains to bring enzyme and substrate together. Sometimes these are called scaffolds or adaptors, but their function is simply to bring the active site of the kinase (for example) close to a potential substrate so that it produces phosphorylation. This is equally true for phosphatases.

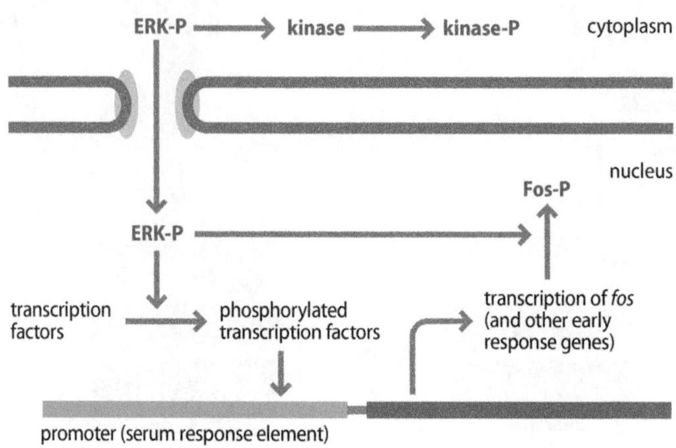

FIGURE 8.27
The activated ERK can remain in the cytoplasm and phosphorylate other proteins, including other kinases, or it can translocate to the nucleus and phosphorylate several transcription factors. In particular, it can phosphorylate transcription factors that bind to a DNA sequence called the serum response element and activate the transcription of several 'early response genes,' including the *fos* gene encoding the Fos protein (Section 3.3.6), which together with Jun comprises the AP1 transcription factor. Full activation of Fos requires a further phosphorylation by ERK, presumably as a mechanism for ensuring that activation of Fos does not occur as a result of phosphorylation by a nonspecific kinase.

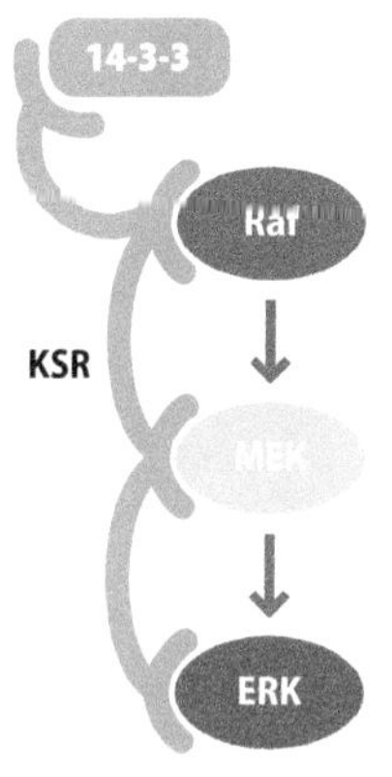

FIGURE 8.28
KSR is a scaffold protein that holds the ERK kinase cascade components close together, thus ensuring that they phosphorylate each other and not other potential substrates. Note that KSR is probably not a rigid scaffolding but a series of binding domains with flexible linkers. Scaffolds are discussed in more detail in Section 2.2.10.

In particular, specificity in this kinase cascade is increased by several scaffold proteins. For example, the scaffold KSR (kinase suppressor of Ras) has been suggested to bind activated Raf, 14-3-3, MEK, and ERK (**Figure 8.28**). There is another scaffold called CNK (connecter-enhancer of KSR), whose role is not completely clear, plus a range of other mechanisms; for example in mammals but not in yeast, some MEK isoforms contain proline-rich sequences that are regulated by phosphorylation and are used to attach them to Raf [24].

The core substrate recognition site for ERK is the sequence (S/T)P surrounded by other amino acids; however, this alone is not enough to provide complete specificity, and as we might expect by now, specificity is provided by an additional recognition between the MAP kinase and another domain on the substrate (**Figure 8.29**) [25].

The final protein in the cascade, ERK, must translocate to the nucleus if it is to phosphorylate transcription factors. As described in Chapter 7, transport into and out of the nucleus is regulated by nuclear location sequences (NLS) and nuclear export sequences (NES), respectively. Typically these sequences are short continuous sequences exposed on the surface of the protein that are recognized by the nuclear import and export machinery. ERK5 has an NLS that is exposed and permanently active. However, in the unphosphorylated state, it also has an NES that is more active than the NLS and keeps ERK5 in the cytoplasm most of the time. Phosphorylation of ERK5 disrupts the NES and therefore leads to translocation of ERK5 into the nucleus (**Figure 8.30**).

8.2.6 Colocation provides extra control

The basic RTK and Jak/Stat systems require further evolutionary tinkering to allow them to function *in vivo*. As suggested at the start of this chapter, they need the addition of **embellishments** to give them specificity, to reduce (or allow) cross-talk between related systems, and to prevent signals from being transmitted in the absence of ligand. Evolution has produced a proliferation of different frills, which we are still discovering. It has tried almost every variation one can imagine, and many more that one might not have imagined. It would be fair to say that the enormous number of variations on a theme that

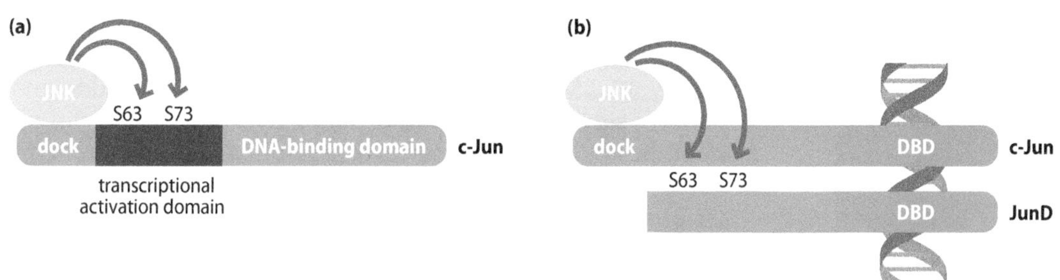

FIGURE 8.29
The MAPK c-Jun N-terminal kinase (JNK) phosphorylates Jun at Ser 63 and Ser 73, within the transcriptional activation domain. (a) Its recognition of Jun is made more specific by binding of JNK at the JNK docking region (dock). (b) Remarkably, an isoform of Jun lacking the docking region, such as JunD, can still be phosphorylated by JNK as long as it can be recruited by a full-length c-Jun and both Jun proteins are bound adjacently on DNA. DBD, DNA-binding domain.

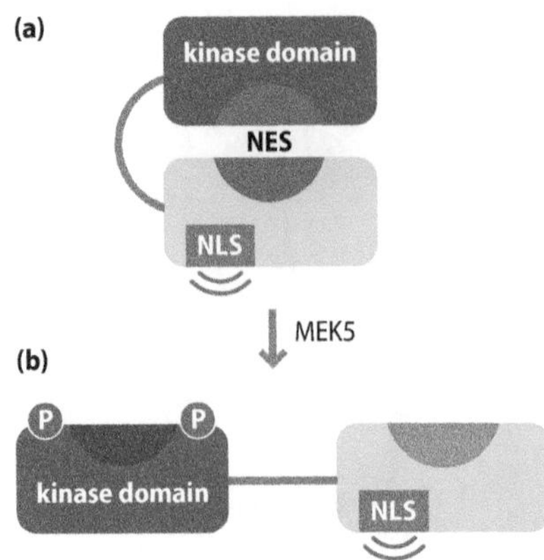

FIGURE 8.30
The location of ERK5 is determined by a balance of the activities of its nuclear localization sequence (NLS) and its nuclear export sequence (NES). (a) The NES is only formed when the two domains of ERK5 can bind together: in this state the NES outweighs the NLS and the protein mainly stays outside the nucleus. (b) However, on phosphorylation the NES is disrupted and the NLS acts to transport the protein into the nucleus. (Adapted from K. Kondoh, K. Terasawa, H. Morimoto and E. Nishida, *Mol. Cell Biol.* 26: 1679–1690, 2000. With permission from the American Society for Microbiology.)

are observed in signaling systems is not so much a reflection on their inherent cleverness, but more a result of their inherent faults, which need patching up by whatever means that evolution can find. In this section we discuss a few of these.

All of the action described here has taken place at the membrane surface. The membrane has been essential to the efficiency of the signaling, in that intermolecular recognition both of receptor and of Ras is a two-dimensional search and is therefore rapid. However, in other ways the membrane has been a somewhat passive part of the system. Naturally, evolution has tinkered with this part of the system too.

Membranes are composed largely of phospholipids, and different membranes contain different phospholipids. Therefore a more targeted system can be constructed by adding domains such as PH domains that recognize specific lipids. PH domains both attach proteins to the membrane and orient them correctly. They can also bind to specific lipids, and link RTK signaling into other pathways. Another example is the different isoforms of Ras. Ras has three main isoforms, H-Ras, K-Ras, and N-Ras (see Table 8.2), which have very similar structures and regulation. However, K-Ras is targeted to the membrane by farnesylation, whereas in addition N-Ras is palmitoylated and H-Ras is doubly palmitoylated. This means that K-Ras is found only at the plasma membrane, while the other two are also found on the Golgi membrane, in different ratios. Repeated cycles of cleavage and reattachment of the palmitoyl groups serve to redirect their location continually [26]. This gives each of the isoforms a distinct cellular function.

Of course, if the signaling system itself creates a modified lipid, then this will help in the specificity of the signal. A common mechanism is to attach membrane-modifying enzymes to the activated receptor. This is simple to do: evolution merely needs to take the relevant enzyme and join it to an SH2 domain, so that phosphorylation of the receptor leads to binding of the enzyme. Because the RTKs normally have rather low substrate specificity, they generally phosphorylate at several sites on the receptor, so that there is no shortage of phosphotyrosines to bind to. The common enzymes are phosphoinositide 3-kinase, which as its name suggests phosphorylates phosphoinositides, and phospholipase C-γ (PLC-γ). The modified lipids can then act as additional binding and regulatory sites. PLC-γ generates diacylglycerol and inositol phosphate from membrane lipids, which is part of a different signaling system that for example releases calcium and so links RTK signaling to calcium signaling. It also activates protein kinase C, another major signaling component.

There is increasing evidence (although not uncontested) that membranes are far from uniform. In particular, membranes probably contain regions known as **lipid rafts**, which are enriched in cholesterol and sphingolipids, making them more rigid and ordered than the rest of the membrane. It seems that lipid rafts contain different lipids and different proteins from the rest of the membrane; in particular, they seem to be organizing centers for cell signaling, in that they contain large numbers of receptors, and intracellularly form anchoring points for cytoskeletal elements such as actin and microtubules. Not surprisingly, it is likely that there are a range of mechanisms for targeting signaling molecules to such regions. It is currently rather too early to go into much detail, as many of the details of lipid rafts are still unclear.

We have seen that specificity in signaling is obtained not by having exquisitely specific binding sites but by tacking on additional scaffolds and binding sites until the required specificity is obtained. This creates something of a problem for experimental probing of signaling mechanisms, because the system that emerges has a good deal of built-in redundancy and cross-talk. Hence, overexpression or even knockout of a protein is unlikely to produce a clean single effect [27]; similarly, removal of a single phosphorylation site is very unlikely to have an all-or-nothing effect. The analysis presented here suggests that the removal of scaffolds and 'extra' binding domains will make the picture clearer and could be a useful experimental strategy.

8.2.7 Autoinhibition provides extra control

It seems that essentially every element in the RTK pathway is autoinhibited; that is, it is turned off in the absence of signal. A very wide range of mechanisms has been identified to accomplish this. These regulatory mechanisms have very probably evolved as additional embellishments once the main pathway had evolved. They therefore show little common pattern [10], and are often unique or specific to particular systems. Some examples are presented in this section, showing that all the components from receptor through to Ras can be autoinhibited.

The EGF receptor was described above as a typical example of a dimeric receptor. However, on the intracellular face it is far from typical: it forms an asymmetric dimer (**Figure 8.31**) [28]. The kinase in the monomeric state is inactive, because a glutamate residue in helix αC that is required for activity is blocked from forming the correct interactions. A similar effect was described in Chapter 4, in which the equivalent crucial glutamate in the PSTAIRE helix of the cyclin-dependent kinase is incorrectly positioned and thus inactivates the kinase. In the cyclin-dependent kinase, activity is regained by the binding of a cyclin; in the EGF receptor, activity is regained in an exactly equivalent way by the binding of the C-terminal domain of the kinase. This of course implies that only one of the two kinase domains in the dimeric receptor is active.

Receptor kinases are normally activated by the phosphorylation of residues in an activation loop. This basic system has been subject to a wide range of embellishments, as described in Chapter 4. Signaling systems also contain other kinases, as part of adaptors and scaffolds. A particularly well-studied scaffold system is the Src kinase. *Src* (pronounced 'sarc') was originally

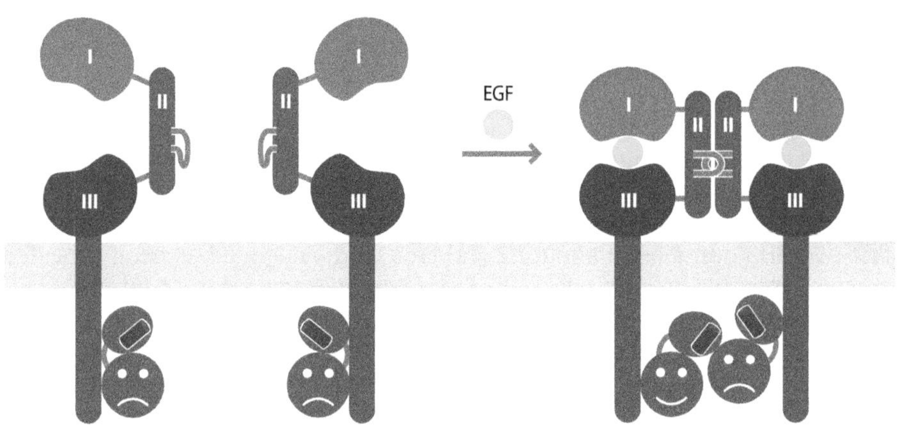

FIGURE 8.31
Activation of the EGF receptor. The kinase is inhibited because helix αC (brown) is incorrectly positioned, and hence the ATP cannot be bound in the correct orientation. On dimerization of the receptor, an asymmetric dimer is formed, in which the C-terminal domain of one kinase (in this diagram, the one on the right) interacts with the αC helix of the other and reorients it, in a similar manner to the activation of cyclin-dependent kinases by cyclins. Only the left-hand kinase is therefore active. It is, however, possible that the two kinases can subsequently swap places to activate the other kinase.

*8.8 Oncogene

Oncogenes are genes whose expression leads to cancer. They come in four classes. Many are viral; infection by the virus therefore increases the risk of cancer. Because cancer is a disease of unregulated cell division, most oncogenes code for proteins involved in signaling systems. Class 1 are genes that code for growth factors, such as the *sis* gene that codes for platelet-derived growth factor, PDGF. Class 2 are genes that code for receptors, such as the epidermal growth factor receptor, EGFR. Class 3 are genes that code for cytosolic proteins downstream of the receptor, such as *ras*, *src*, and *raf*, described elsewhere in Chapter 8. Finally, class 4 are genes that code for nuclear proteins, mainly transcription factors such as *jun* and *fos*.

An interesting group of oncogenes code for signaling pathway components that differ from the host proteins in that they cannot be turned off. An important example is the viral *ras* oncogene, whose protein differs from the human protein in that Gly 12 is replaced by an alanine residue. This makes the protein much slower at hydrolyzing bound GTP to GDP, and consequently the signal stays on for much longer than it should. A similar defect is found in about one-fifth of human cancers. Because this mutation makes the protein hyperactive, it is a dominant effect. The human *ras* gene is called a proto-oncogene, in that some alterations to the gene can cause cancers. Viral oncogenes are thought to have been derived from host proto-oncogenes.

described as a viral gene responsible for a mesodermal cancer called sarcoma (of which leukemia and lymphoma are examples), thus giving the protein its name. *Src* is therefore an **oncogene (*8.8)**. Its discovery led to the award of the Nobel Prize in Physiology or Medicine in 1989 to J. Michael Bishop and Harold E. Varmus. The protein Src contains three domains that were recognized by sequence comparisons, and later named Src homology 1 (now always referred to as the tyrosine kinase), Src homology 2 (SH2) and Src homology 3 (SH3) (Section 8.1.6). These domains are in the order SH3-SH2-kinase, and in the inactive protein they form a set of mutually inhibitory interactions, which are described in Section 8.1.8. The length of the linkers between domains is important to the degree of cooperative autoinhibition in this and other systems [29] (Section 2.2.8).

Adaptor proteins are also regulated by autoinhibition, as exemplified by Grb2. It would clearly be of benefit to the system if Grb2 were only able to bind to its downstream ligand Sos once Grb2 was bound to the receptor: there needs to be a safety switch on the adaptor so that it functions only when ready. This is achieved with a neat example of intramolecular regulation (Chapter 2) (**Figure 8.32**). Grb2 has an SH2 domain with an SH3 domain on each side. In the crystal, the two SH3 domains are folded back and contact each other, to partly block the SH3 binding sites [30]. This is, however, a very weak interaction, because for example in solution, in the presence of short peptide ligands for the SH3 domains, the protein is flexible [31]. When the SH2 domain binds to its cognate phosphorylated receptor, the SH3:SH3 interactions are weakened, thereby shifting the equilibrium so that Grb2 is now turned on and able to bind to Sos. Grb2 has a second, quite different, autoinhibitory system involving a protein called Cbl, which has a proline-rich sequence that binds to the SH3 domain of Grb2 and thereby turns off Grb2. The inhibition is removed by phosphorylation of Cbl by the RTK. In a further twist, the phosphorylated Cbl then binds to the SH2 domain of the adaptor Crk, to initiate further signaling events [32].

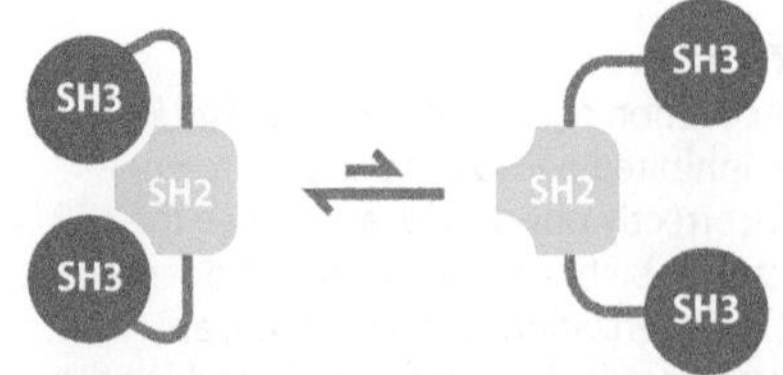

FIGURE 8.32
The Grb2 adaptor is in equilibrium between a closed form and an open form. In the unactivated protein, the closed form is the predominant form, thereby autoinhibiting it. On binding of the SH2 domain, for example to a phosphorylated receptor, the SH3 domains are then freed to bind elsewhere, for example to Sos.

In a similar manner, the adaptor protein Crk is autoinhibited. However, the mechanism is different from that found in Grb: Crk has a short proline-rich insert in the SH2 domain that is capable of binding SH3 domains [33]. Another interesting example is provided by the adaptor protein Nck, which contains three SH3 domains followed by an SH2 domain. There is a linker between the first two SH3 domains of about 52 residues, and evidence has been provided showing that this linker binds to the following SH3 domain in an intramolecular interaction, masking the region that binds to proline-rich sequences and thus autoinhibiting it [34]. The sequence within the linker that binds to the SH3 is -(K/R)x(K/R)RxxS-, and thus contains no prolines at all. As an intramolecular interaction (27 residues away from the SH3 domain) it binds essentially

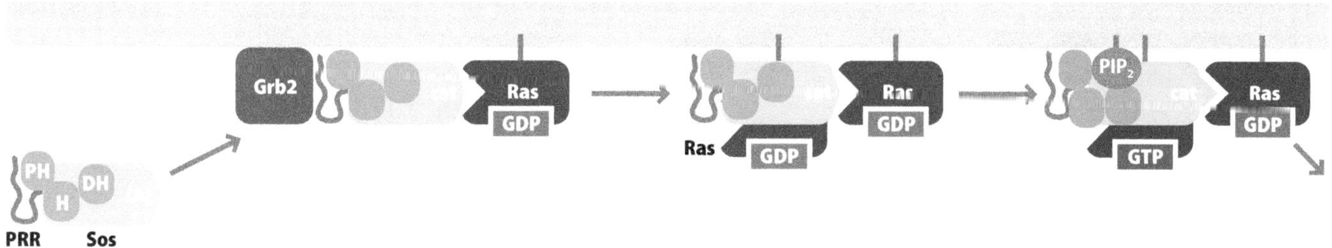

FIGURE 8.33
Activation of Sos. Sos consists of five domains: a PH domain, a histone fold domain (H), a Dbl homology domain (DH), the catalytic domain (cat), and the C-terminal proline-rich region (PRR). Although the key signaling connection is the bringing of Sos up to the membrane by Grb2, the autoinhibited Sos also requires activation in several other ways before it can become fully active as a GEF. Binding to Ras-GDP serves to activate it, whereas binding to Ras-GTP is even more effective, implying that activation of Sos is autocatalytic, because once the first Ras has been activated, the newly created Ras-GTP can replace the activating Ras-GDP, further enhancing the activity of Sos, so that activation of subsequent Ras molecules is faster. Sos is also activated by the binding of PH to PIP_2, and by phosphorylation.

100% to the SH3; in contrast, as an intermolecular interaction it has an affinity of only about 2 mM. Thus, as discussed in Chapter 2, the intramolecular nature of the interaction contributes greatly to the strength of the interaction.

Sos is also autoinhibited, preventing the activation of Ras except when both Sos and Ras are properly located at the membrane [35]. Sos contains several domains (**Figure 8.33**): an N-terminal regulatory section with two histone folds, a central regulatory section containing a Dbl homology (DH) domain followed by a PH domain, the guanine nucleotide exchange catalytic section, and the C-terminal proline-rich region that interacts with Grb2. In the inactive state, the histone domain binds to the linker between PH and catalytic domains, while the DH domain binds to the catalytic domain so as to allosterically inhibit it. When both Sos and Ras are at the membrane surface, Ras displaces the DH domain and binds to DH's allosteric site on the catalytic domain. This molecule of Ras is in addition to the one that binds to the active site. The allosteric effect produces a roughly 500-fold increase in the catalytic activity of Sos. Once Sos has catalyzed nucleotide exchange of Ras, Ras-GTP can replace Ras-GDP in binding to the allosteric site, further increasing the activity of Sos by about another factor of 10. In this way, a positive feedback loop is created and the activity of Sos is autocatalytic. A further enhancement of activity arises from the binding of phosphatidylinositol 4,5-bisphosphate (PIP_2) to the PH domain, which alters the relative position of the DH, PH, and catalytic domains and releases the histone domain, thereby linking Ras activation to other pathways that generate PIP_2, such as EGF signaling. Almost inevitably, Sos is also further activated by phosphorylation.

Raf is autoinhibited in the absence of Ras, which binds to a zinc finger on Raf [23]. This interaction is strengthened by prenylation of Ras, which presumably works to locate Ras more exactly next to the membrane.

8.2.8 The bacterial two-component signaling system has a histidine kinase

The tyrosine receptor kinase system described above is found only in eukaryotes. Prokaryotes mainly use a related but simpler system known as the two-component system, which is also found in eukaryotes but only to a limited extent, and not at all in animals. It is different in that the kinase phosphorylates a histidine residue, and the subsequent step is transfer of this phosphate from the histidine kinase (HK) to an aspartate residue on the receiver domain of a response regulator (RR) (**Figure 8.34**). Often the RR is attached to an effector domain, which becomes activated once the RR is phosphorylated. Both the HK and the RR have phosphatase activities which turn off the signal. In different systems this can occur at rates ranging from seconds to hours. Bacteria typically have a separate sensor, HK, and RR for each type of signal. The HK and RR domains are well conserved across different systems, but the sensor and effector domains are often entirely different in structure and mechanism. *Escherichia coli* has more than 60 such systems, representing roughly 1% of the genome, which regulate a wide range of processes such as chemotaxis,

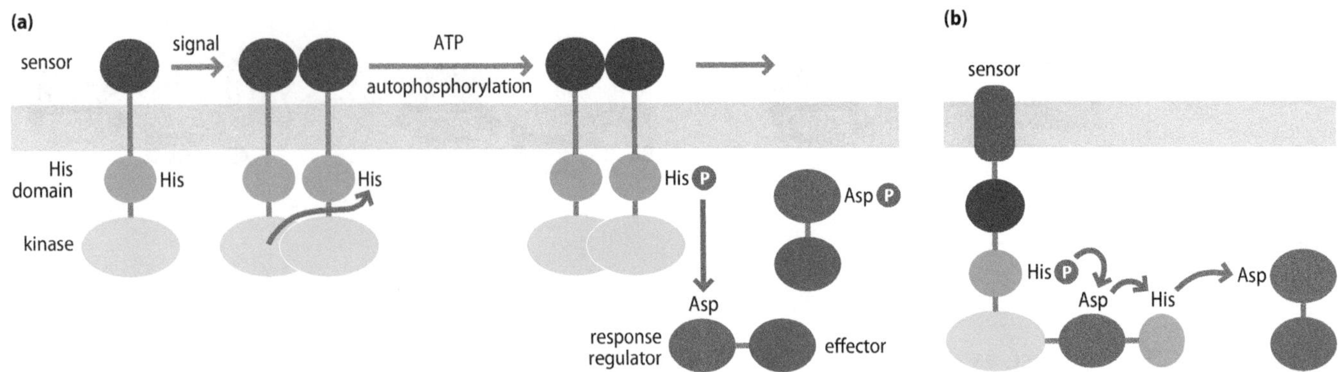

FIGURE 8.34
The two-component system. (a) A simple two-component system as found in the osmoregulator OmpR. The sensor is in the periplasm; the rest of the protein, and the response regulator, are in the cytoplasm. (b) A more complex system as found in the redox regulator ArcA.

osmoregulation, redox response, metabolism, and transport [36, 37]. Because such systems are unique to bacteria, they represent interesting targets for antibacterial drugs.

The basic system is very simple, although embellishments are common, particularly in eukaryotes, such as the addition of extra domains (see Figure 8.34b) [36]. Often the effect of the external signal is to initiate signaling by dimerization or tetramerization of the sensor; as with the RTKs, the histidine kinase is typically constitutively active. The downstream effector domain typically functions as a transcription factor and is able to bind to DNA only when the RR is phosphorylated. However, the way in which this occurs differs. In NarL, the RR physically blocks the binding of the effector to DNA; phosphorylation leads to release of the effector domain, which binds to DNA as a symmetric dimer. In contrast, the phosphate starvation response protein PhoB and the sporulation protein Spo0A bind as two tandem domains. The redox sensor PrrA is able to bind DNA in its inactive state, but only very weakly; phosphorylation of the RR leads to release of the effector from the RR, which is then able to dimerize and bind much more strongly [38]. Interestingly, PrrA binds to a wide range of DNA regulatory sites, which leads to the activation of genes involved in photosynthesis, electron transport, nitrogen fixation, and carbon fixation, and to the deactivation of genes involved in aerobic respiration. This range of effects seems to be mediated by the DNA sequences of the symmetric binding sites, which differ in the spacing between the two halves and so presumably have differing binding affinities [39]. The unphosphorylated nitrogen fixation regulator NtrC is dimeric, and it forms oligomers on phosphorylation, leading to ATP hydrolysis by an additional central domain, which in turn leads to a conformational change and hence to DNA binding. The chemotaxis RR CheB has a methylesterase as its effector; unphosphorylated RR blocks access to the active site.

The key switch in the two-component system is the RR, which changes its conformation when phosphorylated. RRs have a structural relationship to the small GTPases discussed in Chapter 7, and they can be presumed to have a common evolutionary origin [40]. However, the relationship is rather distant. A somewhat oversimplified version of the mechanism [41] is as follows. When the RR is phosphorylated, a highly conserved serine or threonine moves to bind to the phosphate group. This creates a hydrophobic hole in the protein, which is filled by a conserved phenylalanine or tyrosine residue, which moves from a surface-exposed location into the newly created hole (**Figure 8.35**) [42]. In this way, phosphorylation of the protein leads to the loss of a large hydrophobic residue from the surface of the protein. In turn, this produces rearrangements of local secondary structure elements which constitute the signal, for example altering the binding interface to the effector domain.

The difference in species distribution between two-component and RTK signaling is striking. The reason is unlikely to be the simplicity of two-component

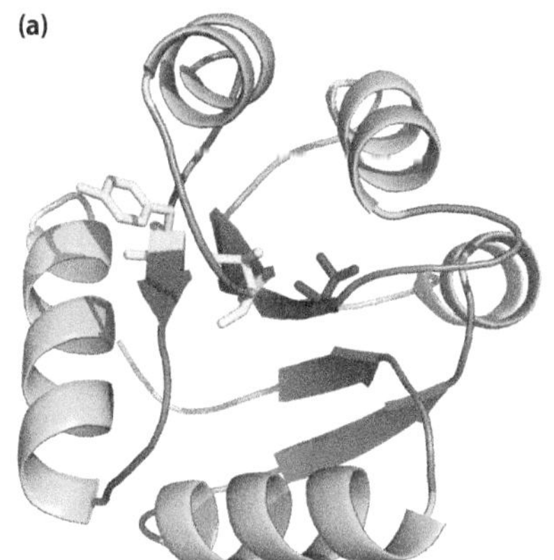

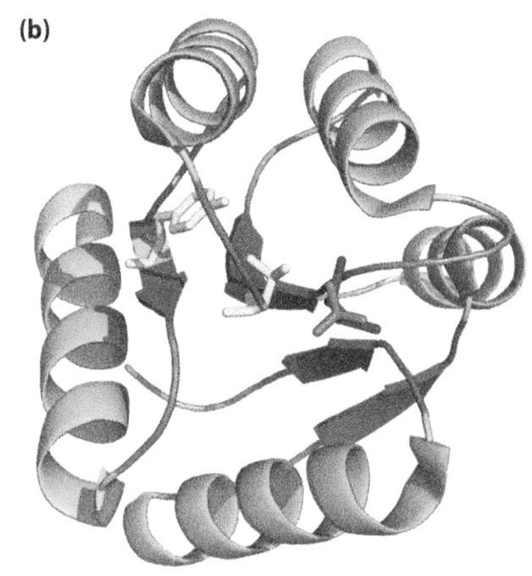

FIGURE 8.35
The mechanism of activation of the response regulator CheY, responsible for bacterial chemotaxis. CheY is phosphorylated at Asp 57 (red), changing the protein from its inactive state (a) to its active state (b). The phosphate makes a hydrogen bond with Thr 87 (green), causing the Thr 87 side chain to rotate and the methyl group to move, leaving a hydrophobic hole behind. This hole is filled by Tyr 106 (green), which rotates inward to do so. In turn, this creates quite a large change to the protein surface, which can be recognized by proteins that bind to CheY, in particular FliM (a regulator of the bacterial flagellum). Drawn using PDB files 3chy (free CheY) and 1fqw (in a complex with BeF_3^-). BeF_3^- (brown) is used experimentally because it forms a good mimic of the phosphate group, and is more stable.

signaling, because eukaryotes are certainly capable of introducing embellishments if required. It may be that one reason is the efficiency of the on/off switch. The small GTPase switch represented by Ras has well over a 1000-fold difference in the ratio of inactive to active conformation. However the RR switch is less good, with estimates of between 200-fold [38] and as low as 20-fold [43]. A number of reports have identified fluctuations between 'on' and 'off' states of RRs in solution, in either the phosphorylated or unphosphorylated forms [44], these being particularly clear examples of conformational selection (Chapter 6). Although embellishments can improve imperfect switches, it helps to have a good system to start with, which is presumably why the small GTPase switch is so widespread throughout biology. It is therefore possible that Ras-type small GTPase switches supplanted two-component switches in early eukaryotes because their switches were less leaky.

8.2.9 An evolutionary perspective provides a unifying explanation

Throughout this chapter, I have tried to show that the evolution of signaling pathways is, like any other evolutionary process, opportunistic. An initial attempt will be developed and added to if it works well enough, or it may be replaced if a sufficiently good alternative can be found. This leads to a large and complicated overall system, with considerable overlap of function, redundancy, cross-talk, and a large number of mechanisms that are unique to individual signaling systems [10].

In other words, evolution works by taking something simple, copying and transplanting it, and then tinkering with it to add embellishments until it does what is required. In this section we strip out all the embellishments to appreciate the underlying structure.

For a receptor kinase pathway involving signaling across the membrane, more or less the simplest pathway is as shown in **Figure 8.36**. The key element is that an extracellular signal should be converted into an intracellular event, in this case a phosphorylation. Some kind of recognition of this phosphorylation is then needed, leading to a change in an intracellular protein. I have suggested that the Jak/Stat pathway is almost as simple as this. Interestingly, *Caenorhabditis elegans* has Stat homologs but no Jak homologs, implying that

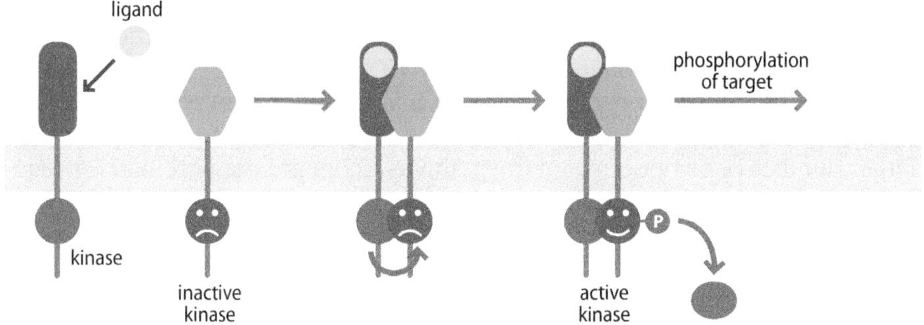

FIGURE 8.36
A simple dimerizing receptor kinase mechanism. The two kinases can be the same or different, and the two extracellular receptors can also be the same or different. The key requirement is that binding of the ligand leads to dimerization, and hence brings one kinase close to the other.

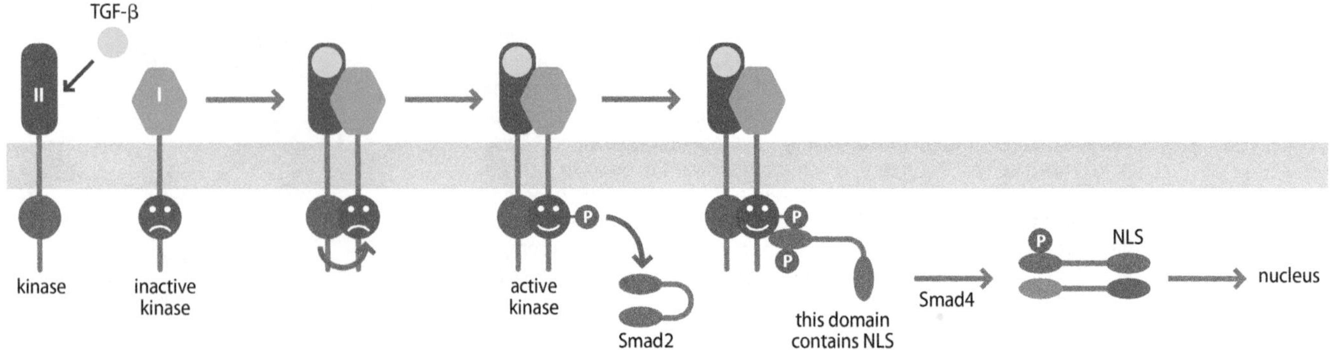

FIGURE 8.37
The Smad signaling system. This important family controls a variety of functions including cell death, tissue repair, proliferation, and differentiation.

the Stat system is more widespread and therefore presumably older than Jak. In other words, signaling via phosphorylation of Stat-like proteins is an ancient mechanism, but it is not necessary to phosphorylate Stat using a receptor-associated tyrosine kinase such as Jak. Again, it is interesting that prokaryotes have very few tyrosine kinases, but many serine/threonine kinases. Similarly, plants have very few receptor tyrosine kinases, but many serine/threonine kinases. It is thus very likely that the first receptor kinases were serine/threonine kinases. The tyrosine kinases are structurally related to serine/threonine kinases and are presumably more recent adaptations that arose by changes to serine/threonine kinases.

It is therefore relevant that there is a eukaryotic signaling system that is even simpler than Jak/Stat, namely the Smad system, which looks like a good candidate for an 'early' signaling system [45]. The receptor is in fact tetrameric, but its core is a heterodimer (**Figure 8.37**), consisting of a type II receptor, which has an intracellular serine/threonine kinase domain, and a type I receptor, which has a different kinase domain. The ligand (for example transforming growth factor-β, TGF-β) binds to the type II receptor and leads to its heterodimerization with the type I receptor. This permits the type II kinase to phosphorylate the type I kinase and activate it. The type I kinase then phosphorylates a ligand called Smad. The phosphorylated Smad then dissociates and translocates to the nucleus, where it affects gene expression. Usually this involves the association of the phosphorylated Smad with another different Smad. This system has of course evolved further embellishments, including inhibitory Smads and other regulatory proteins such as the attractively named Smurfs, discussed further below.

One can thus imagine an evolutionary route in which an initial simple serine/threonine kinase system like that shown in Figure 8.36 diverged into either Smads, by rather minor modifications to the kinase substrate, or to Jak/Stat, by replacing the kinase substrate by a pre-existing Stat, followed by some tinkering with the substrate for the receptor kinase.

Why, then, should the more complicated RTK system emerge? First, because it could: it is likely that the small GTPase is a rather early evolutionary invention, and the adaptors connecting to it would not be that difficult to pick up and re-use. Again, one can imagine a pre-existing pathway from Ras to transcription factors that could be picked up and re-used with minimal modification. It has been suggested that the main benefit of the RTK system is not signal amplification, not least because once scaffolds get involved, the possibilities for amplification are severely reduced [46]. The reason is most likely to be that having a larger number of components in the pathway offers greater potential for regulation, control, cross-regulation, and branching.

Signaling within the apoptotic pathway has both striking similarities to, and differences from, the mechanisms described above, and possibly is an indication of the way in which the evolution of signaling pathways could have gone.

Binding of the signal leads to trimerization of the receptor rather than dimerization (a solution also adopted by NFκB signaling), and many of the interactions made are homo- rather than hetero-associations, in that they occur between structurally related domains, in particular death domains and death effector domains (DD and DED domains, respectively). However, in other respects the essential mechanisms are very similar, with a variety of adaptor and scaffold proteins, few enzyme functions, and a kinase cascade, although the major end product is a proteolytic cascade rather than a kinase cascade [47].

8.2.10 Switching off the signal

Equally as important as the creation and transmission of the signal is how the signal is switched off. More medical implications arise from signals that are not turned off properly rather than signals that do not start properly, emphasizing the importance of turning off the signal correctly.

Because the intracellular signal is transmitted via protein phosphorylation, it is in general turned off by a corresponding dephosphorylation. As discussed in Chapter 2, and also in Sections 4.4.2 and 8.2.5, there are only about one-third as many phosphatases as there are kinases. This must imply that phosphatases are even less specific in their activity than kinases. They achieve specificity in the same way as kinases, by being located close to their substrate. Thus, a common mechanism for turning off the signal is the presence of a phosphatase, which is joined to a domain (such as SH2) that recognizes the phosphorylated residue that initiated the pathway. It can then bind and remove the phosphate, shutting down the pathway.

As an example, the Jak/Stat pathway can be turned off by a phosphatase that binds to an activated receptor via an SH2 domain and dephosphorylates Jak. This phosphatase is present in all cells but is inactive; it is turned on by binding to the receptor, which uncovers the phosphatase's active site. The Smad pathway can be turned off by a phosphatase directed to the receptor, which is one of the gene products produced by activation of the pathway.

There are of course multiple routes for turning off the signal. The signal is often turned off in a cruder but effective manner, by degradation of the signaling proteins. In Jak/Stat, one of the genes activated by the pathway is called SOCS, which has an SH2 domain that binds to the phosphorylated receptor, and a second domain (a 'SOCS box') that recruits an E3 ubiquitin ligase, leading to proteasomal degradation of the receptor. Similarly, Smad signaling can also be turned off by the production of *Smad ubiquitylation regulatory factors* or Smurfs, which bind to several Smads and other Smad-associated proteins and recruit E3 ligases, leading to their proteasomal degradation [48]. Depending on the target proteins, this can lead to enhancement or repression of signal; it can also down-regulate signal intensity in resting cells.

8.3 G-PROTEIN-COUPLED RECEPTOR SIGNALING

G-protein-coupled receptors (GPCRs) also constitute a major signaling pathway. They form the largest family of cell-surface receptors in humans (there are more than 700 in the human genome), and it is estimated that 50% of drugs in use today work by interfering with GPCRs. By contrast with receptor dimerizing systems, GPCRs tend to be used where rapid responses are required, leading to direct regulation of intracellular processes rather than transcription.

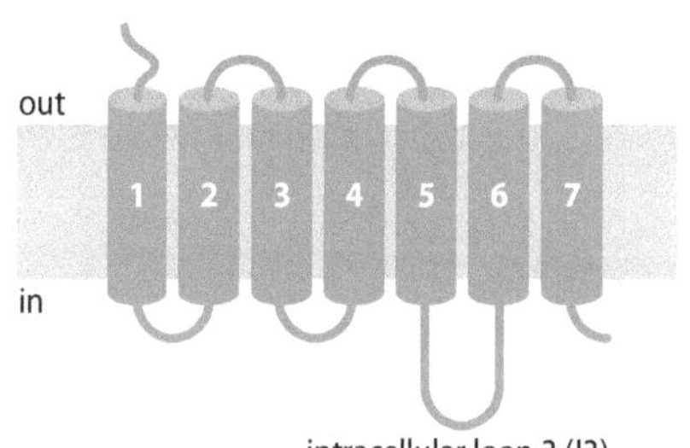

FIGURE 8.38
A G-protein-coupled receptor. This figure shows the typical diagrammatic representation, with the seven transmembrane helices and the loops connecting them, particularly the long loop I3 connecting helices 5 and 6 on the intracellular side.

The signaling mechanism is shorter and simpler than that of receptor kinases. The receptor consists of seven transmembrane helices, with an extracellular domain at the N terminus and an intracellular part composed of loops that link together the different helices (**Figure 8.38**). A detailed mechanism has been lacking until recently, because of the lack of a suitable high-resolution

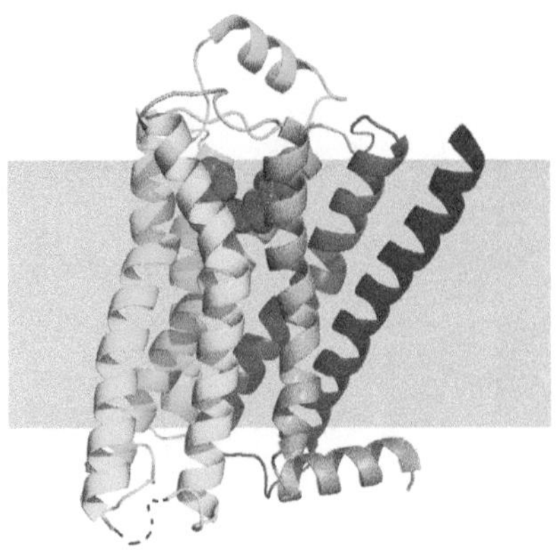

FIGURE 8.39
A G-protein-coupled receptor in higher resolution, here the β_2-adrenergic GPCR (PDB file 2rh1). This structure includes a ligand, the inverse agonist carazolol (magenta). Distinct bends are clearly visible in the three helices at the front (green/yellow). Loop I3 (dashed line) is not visible in the electron density because it is disordered (that is, mobile) but would be in front at the bottom, between the two green helices.

structure. The first structure at atomic detail was that of rhodopsin [49], which is a rather specialized member of the GPCR family and not really representative. However, more recent crystal structures of the human β_2-adrenergic receptor (**Figure 8.39**), the β_1-adrenergic receptor, opsin, and the adenosine A_2 receptor have opened the way to a greater understanding (**Figure 8.40**) [50–52]. These are all reasonably close homologs and do not cover the spread of all GPCRs, so it is as yet unclear how general will be the conclusions obtained from these receptors.

The ligand-binding site is buried well within the transmembrane region, and is in a similar place in all the receptors. In β_2-adrenergic receptor, binding of the ligand leads to a rearrangement of some amino acid side chains, in particular a tryptophan described as a toggle switch [52], which in turn leads to the rotation of a helix that is kinked owing to the presence of a proline. The repacking of helices that ensues is facilitated by several water molecules that lubricate the interfaces. This mechanically transmits a signal to the inside of the cell, by altering the shape of the receptor. It is likely that something similar happens in opsin, leading to the tilting of one of the transmembrane helices and the rotation of another, thereby presenting a different binding surface for the Gα. In detail, ligand specificity is not well understood and seems to depend on the details of helix:helix interactions within the membrane.

Inside the cell, there is a complex of three proteins that make up the G protein: Gα, Gβ, and Gγ. Gβ and Gγ are permanently associated, and essentially form one heterodimeric protein, which is attached to the membrane by a lipid anchor. Gα also has a lipid anchor and is a GTPase, which works in the same way as the small GTPases discussed in Chapter 7 and earlier in this chapter; it also has structural similarities, suggesting a common evolutionary origin [53]. It is inactive when bound to GDP and is activated by a nucleotide exchange factor, which permits the release of GDP and the binding of GTP. In the inactive state, Gα is bound to the Gβγ complex (**Figure 8.41**). The nucleotide exchange factor (*8.5) is the activated GPCR: sometimes the heterotrimeric G-protein complex is already attached to the GPCR and sometimes it is freely diffusible within the membrane. Note again that this requires only a two-dimensional search because Gα is attached to the membrane. The binding stimulates the release of GDP and binding of GTP to Gα (by now we should be seeing some similarities with the RTK pathway!), which then usually causes the dissociation of Gα from Gβγ and from the receptor.

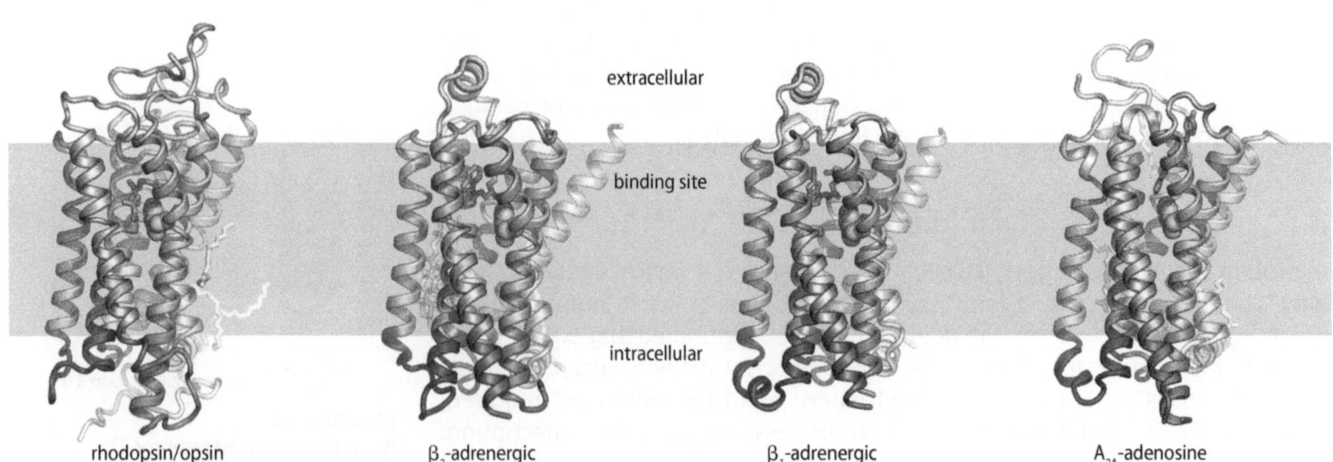

FIGURE 8.40
Comparison of the structures of several GPCRs. Transmembrane regions are in light blue, intracellular regions darker blue, and extracellular regions brown. The ligands are shown as orange sticks, bound lipids are yellow, and the conserved toggle switch tryptophan is green spheres. These GPCRs are reasonably closely similar in sequence, and have very similar structures and ligand-binding sites. (From M. A. Hanson and R. C. Stevens, *Structure* 17: 8–14, 2009. With permission from Elsevier.)

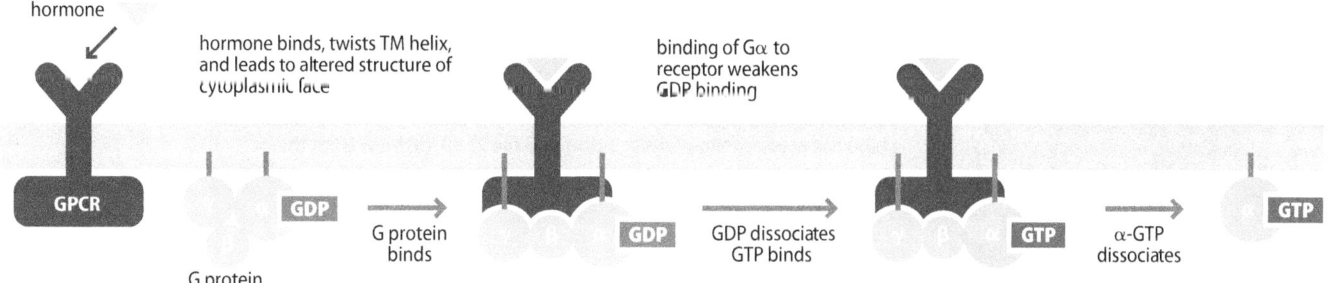

FIGURE 8.41
The mechanism of activation of the heterotrimeric G protein by a G-protein-coupled receptor. Note that both Gα and Gγ are held at the membrane surface by lipophilic anchors.

The Gα.GTP complex is then free to search around on the membrane surface, and interacts with its target. In many cases this is the enzyme adenylate cyclase (also attached to the membrane), which is activated by binding to Gα and starts converting ATP to cyclic AMP (cAMP). The cAMP then acts as a second messenger within the cell and has further effects. Gα also has some GTPase activity (note again the parallels with receptor kinase signaling), which can be modulated by GAPs (*8.6) and is enhanced by binding to adenylate cyclase, meaning that the signal is turned off relatively quickly once it has fed through to its target. (Gα in fact contains a GAP domain and is therefore inherently rapidly self-deactivating, by contrast with Ras for example, which requires external GAPs to deactivate it at any appreciable rate.) Once the GTP has been hydrolysed to GDP, the Gα.GDP is inactive and reassociates with Gβγ, to return to the starting state.

Cholera is caused by a toxin produced by the bacterium *Vibrio cholerae*, which binds to a specific Gα and prevents GTP hydrolysis, thereby leaving a prolonged signal that in this case causes the excretion of excessive sodium and water into the gut. There are of course a range of other embellishments of the basic G-protein system that have been tacked on during the course of evolution, such as proteins that either stimulate or inhibit adenylate cyclase activity. One of the enzymes activated by Gα is phospholipase C-β, which has similar activity to PLC-γ and so links the GPCR and RTK pathways.

8.4 ION CHANNELS

The third class of transmembrane signaling system is ion channels. The channel opens in response to a change in the voltage across the membrane (a voltage-gated channel), to a ligand binding either intracellularly or extracellularly (a ligand-gated channel), or to mechanical force (a mechano-gated channel). The transmission of a nerve impulse is achieved by the sequential opening and closing of voltage-gated channels along the nerve axon, creating an electrical **action potential** that travels rapidly along the axon. When a nerve impulse reaches the end of an axon, it stimulates the exocytosis of acetylcholine contained in vesicles at the end of the axon, which diffuses the short distance across the synapse and binds to receptors in muscle cells. These receptors are called **nicotinic acetylcholine receptors (*8.9)**. Binding of acetylcholine opens the channel, allowing the flow of more than 10^7 potassium and sodium ions per second, which stimulates contraction of the muscle. Acetylcholine in the synapse is rapidly removed by acetylcholinesterase, thereby switching off the channel.

Structural details of the acetylcholine receptor have been obtained largely through electron microscopy (Chapter 11). It is a pentamer (**Figure 8.42**),

***8.9 Nicotinic acetylcholine receptor**

There are two completely different kinds of acetylcholine receptor. The *nicotinic* receptor is an ion channel that is normally found in muscle and nerve cells, particularly in synapses, and is used to transmit signals via the binding of acetylcholine, which opens the channel. This channel can also be opened by the binding of nicotine, which is responsible for the physiological effect and addictive properties of nicotine. By contrast, the *muscarinic* receptor binds to the alkaloid muscarine and is involved in the parasympathetic nervous system used in smooth muscle control and glandular secretions. This is a G-protein-coupled receptor.

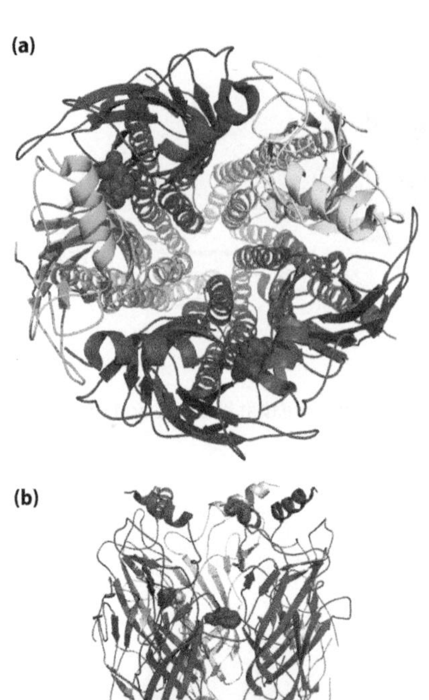

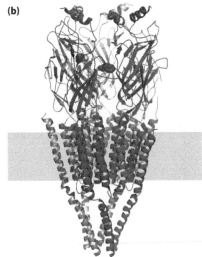

FIGURE 8.42
The structure of the acetylcholine receptor. (a) View from outside the cell. The protein is a pentamer; clockwise from the top, the subunits are α (red), β (green), δ (magenta), α, and γ (cyan). Only the α subunits bind acetylcholine, indicated by the highlighted residues αTrp 149 (blue), which bind the acetylcholine. Drawn using PDB file 2bg9. (b) View from the membrane. Only the α and δ subunits are colored.

consisting of chains (clockwise from above) $\alpha\beta\delta\alpha\gamma$. The four subunits are homologous, but only the α subunit binds to acetylcholine: presumably having only two subunits directly binding to the ligand provides the desired response, in terms of the concentration of acetylcholine needed to trigger opening. The channel protrudes a long way from the membrane surface on both faces, and the binding site for acetylcholine is about 40 Å above the membrane [54]. In the closed state, the two α chains are distorted or (to use the terminology commonly used for allosteric change discussed in Chapter 3) 'tense' [55]. Binding of acetylcholine leads to changes in the positions of loops around it, which in turn leads to a distortion or rotation of a small β sheet. This sheet is attached to a helix, which therefore also rotates. The helix is the gate of the ion channel, forming a hydrophobic channel that is too small for hydrated ions (**Figure 8.43**). Binding of acetylcholine to the α subunits therefore leads to rotations of their gate helices. These two rotations destabilize the interactions between the α subunits and their neighbors, which permits a concerted rotation of the helices in all five subunits, all in the same direction by about 15°. This moves the large hydrophobic leucine side chains that were blocking the pore out of the way and replaces them by smaller hydrophilic side chains, thereby allowing ions to pass though. The overall change is not unlike the change in aperture of a camera lens.

The switch represented by this channel is not 100% on/off. In both the ligand-bound and ligand-free states it oscillates between open and closed states. However, in the ligand-free state it is almost 100% closed, whereas in the ligand-bound state it is about 10% closed. This is a 'safer' position than having a partly open ligand-free state.

The acetylcholine receptor is homologous to other ligand-gated ion channels such as those for serotonin, γ-aminobutyric acid, and glycine, and rather more distantly to glutamate-gated and ATP-gated channels, which all probably work in the same way.

Once the ion has entered the cell, there need to be mechanisms for sensing the ion and also for sequestering it once it has done its job. A common mechanism for sensing ions is calmodulin, which senses calcium ions; this is not discussed further here but is the subject of Problem 3.

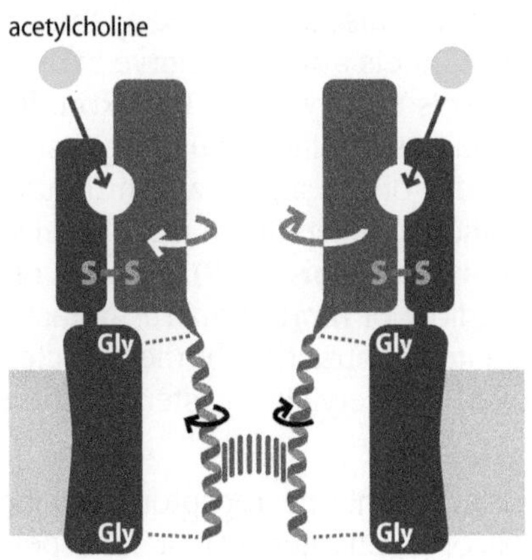

FIGURE 8.43
A likely mechanism for opening of the ion channel in the acetylcholine receptor. See the text. (Adapted from A. Miyazawa, Y. Fujiyoshi and N. Unwin, *Nature* 423: 949–955, 2003. With permission from Macmillan Publishers Ltd.)

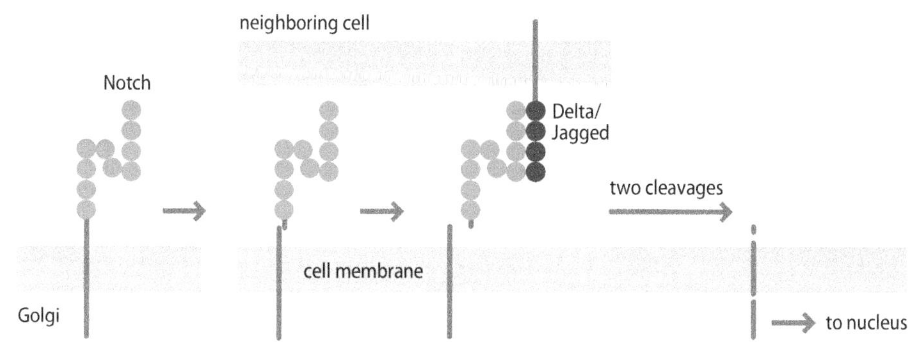

FIGURE 8.44
The mechanism of Notch function. Notch receptor is processed in the Golgi by proteolytic cleavage to give two polypeptide chains. On binding to a ligand, the extracellular peptide is released. This results in two further proteolytic cleavages, thereby releasing the intracellular part of the receptor, which relocates to the nucleus and leads to gene activation.

8.5 SIGNALING VIA PROTEOLYSIS OF A LATENT GENE REGULATORY PROTEIN

8.5.1 Notch receptor directly activates gene transcription

There is yet another common signaling mechanism, which is completely different from anything discussed so far. It is irreversible, which limits its use to cell patterning during the development of cells and embryos, because it uses proteolytic cleavage to release an active protein into the cell. It is the simplest mechanism yet found for passing a signal from the cell surface down to the nucleus, and may therefore represent a very primitive mechanism. The Notch system is a good example of this mechanism, so called because it was first identified by a mutant producing a notch in the wings of *Drosophila*.

The Notch receptor is a single-pass transmembrane protein. During maturation of the receptor in the Golgi, there is a proteolytic cleavage, so that the mature receptor actually consists of two separate chains (**Figure 8.44**): a large extracellular part and a shorter part consisting of a small extracellular sequence, a transmembrane region, and a rather longer intracellular region. The extracellular part of the Notch1 receptor consists of 36 repeats of the EGF module (Chapter 2). This receptor binds to a ligand, which is another cell-surface protein and therefore has to be on a closely neighboring cell; the ligand also consists of multiple repeats of the EGF module. (There are many examples in this book of this type of behavior, where the two parts of a protein:protein interaction system are clearly related. Presumably this arises because evolution picked up the easiest thing that came to hand, which was a self-recognition.) Binding of the ligand to the receptor leads, in some poorly understood way, to detachment of the extracellular part of the receptor and thus to exposure of the small extracellular sequence of the transmembrane part. The exposed sequence gets cut by two successive proteases, which reside on the cell membrane and presumably are only prevented from acting in the resting cell because they cannot access the cleavage site. The fragment released can then enter the cell and move to the nucleus, where it binds to a transcriptional repressor and converts it into an activator (see Figure 8.44).

Because the Notch ligand and receptor have to be on the surfaces of neighboring cells, Notch signaling is typically used for one cell to signal to all its neighbors, very often to make them different from the cell carrying the Notch ligand (**Figure 8.45**). Thus, for example, it is used in the development of nerve cells. Nerve cells originate as single cells within a sheet of epithelial precursor cells. Once one cell develops into a neuron, it expresses Notch ligand and therefore signals to all its neighbors *not* to become neurons, a process known as lateral inhibition. Notch is involved in many cell patterning processes.

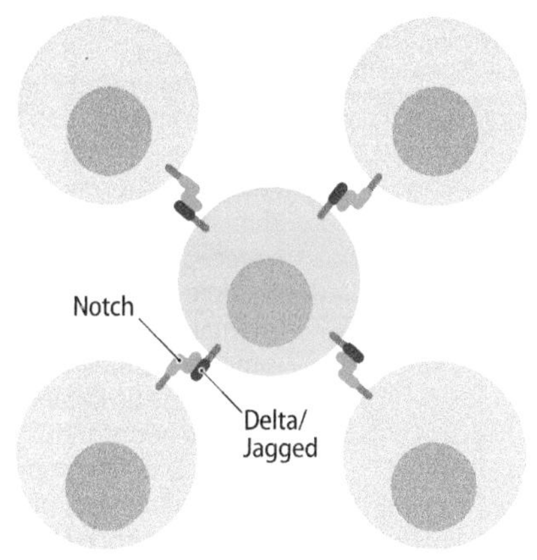

FIGURE 8.45
Notch signaling is often involved in patterning. A differentiated cell (green) expresses Notch ligand on its surface. Adjoining cells carrying the Notch receptor are activated, often to prevent them from differentiating in the same way, and thus to establish alternating patterns of cells.

***8.10 Hedgehog**

It will be apparent from Section 8.5 that developmental biologists tend to give their proteins somewhat whimsical names. A further example is the ligands for the Notch receptor, three of which are called Delta, Serrate, and Jagged. The system described in Section 8.5.2 is that in *Drosophila*, which is moderately well understood. In humans, it is far less well understood. There are three Hedgehog gene products in humans, known as Sonic, Desert, and Indian. Unusually, these proteins are covalently attached to cholesterol. Hedgehog gets its name from the spiky appearance of *Drosophila* larvae in which the gene is mutated.

8.5.2 Hedgehog prevents proteolysis of an intracellular signal

Another important type of developmental signal is provided by **Hedgehog (*8.10)**. Whereas Notch ligand is attached to the cell membrane, and therefore leads to patterning of cells that are in contact, Hedgehog is a protein ligand that is only able to diffuse a small distance and leads to patterning of cells in zones. It is encoded by a key gene involved in the development of the segmental body plan in flies and worms, and also in the development of many tissues in mammals. Defects in Hedgehog signaling give rise to many diseases, including cancers and developmental abnormalities. A particularly striking such abnormality is the presence of a single central eye (a condition called *cyclopia*) in mammals in which Hedgehog is inhibited.

Hedgehog binds to a receptor called Patched (**Figure 8.46**). The normal function of Patched is to inactivate a second transmembrane protein called Smoothened, by an unknown mechanism. When Hedgehog binds, it leads to internalization of Patched and degradation of Patched by lysosomes. This then leaves Smoothened free to function: it gets phosphorylated, which allows it to bind two further proteins, in particular a scaffold protein called Costal2. Costal2 then releases a protein called Ci, which is free to translocate to the nucleus and activate transcription.

When Hedgehog is not present, Costal2 binds to Ci and to microtubules and recruits several other proteins, including the kinases mentioned above that phosphorylate Smoothened. This complex leads to the phosphorylation of Ci, which subsequently leads to its tagging by ubiquitin (*4.9) and its eventual degradation. One of the Ci fragments so produced acts as a transcriptional inhibitor of many of the genes that are activated by full-length Ci.

There are thus several similarities between Notch and Hedgehog. There is proteolysis involved, and the signaling leads to a switch from gene repression to gene activation. Hedgehog signaling has some similarities with yet another signaling pathway, that by the ligand family Wnt (which includes Wingless), which binds a receptor called Frizzled and again prevents the proteolysis of an intracellular protein, leading to a switch from repression to activation.

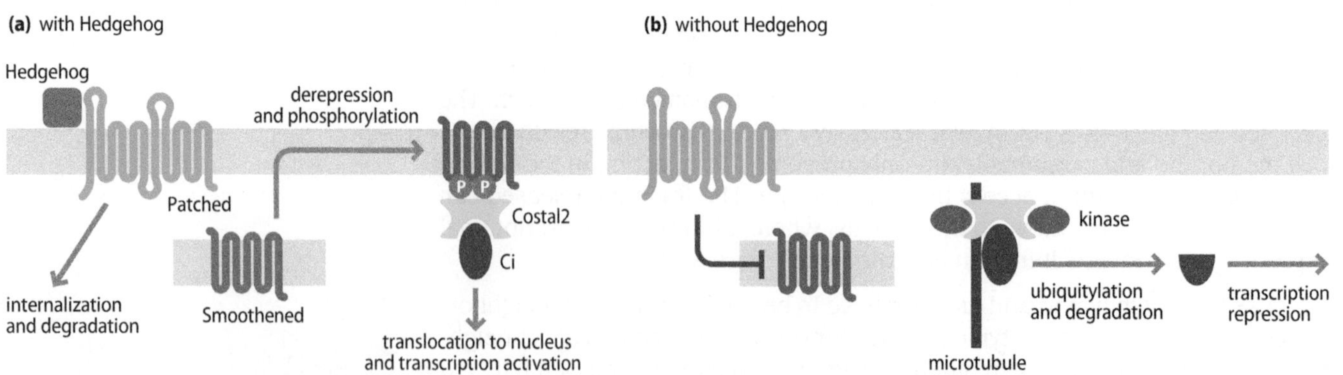

FIGURE 8.46

The mechanism of signaling by Hedgehog. (a) Hedgehog binds to the receptor Patched. Binding normally involves a co-receptor, iHog (not shown), which is a typical receptor in that it is a single-pass transmembrane receptor containing multiple immunoglobulin and fibronectin type III modules (compare Figure 2.11). Binding of Hedgehog to Patched leads to internalization and degradation of Patched, which in turn removes the repression of Patched on Smoothened. Smoothened is then phosphorylated and moves from an internal vesicle to the cell membrane, where it interacts with Costal2 and leads it to release Ci. Ci can then move to the nucleus and activate transcription. (b) In the absence of Hedgehog, Patched inhibits Smoothened. Costal2 is therefore able to bind to Ci and recruit kinases, which eventually leads to the ubiquitylation of Ci and its degradation. One of the fragments of Ci moves to the nucleus and acts as a transcriptional repressor. Note that this mechanism means that in the (more normal) absence of Hedgehog, Ci is continuously degraded. This is not an uncommon mechanism, despite its rather wasteful appearance.

8.6 SUMMARY

The basic challenge faced by signaling is how to transmit a signal across the cell membrane. The simplest method is to use a hydrophobic signal which can pass directly through the membrane.

Many receptors are linked to an intracellular kinase; ligand binding leads to dimerization of the receptor and therefore autophosphorylation. A simple such system is Smad, where an autophosphorylated heterodimeric receptor is activated as a kinase and phosphorylates its ligand Smad, which then translocates to the nucleus and alters DNA transcription. A slightly more complicated system is Jak/Stat, in which the autophosphorylated receptor recruits Stat and phosphorylates it, leading to dimerization and nuclear translocation. However, most systems are receptor tyrosine kinases, in which the phosphorylated receptor is recognized by an adaptor protein such as Grb2, which has an SH2 domain to recognize phosphotyrosine, and an SH3 domain to bind a proline-rich region on Sos. Sos is a constitutive guanine nucleotide exchange factor, and catalyzes the exchange of GDP to GTP on Ras. GTP-bound Ras recruits Raf and activates it, aided by a range of additional proteins. Activated Raf is a kinase that initiates a kinase cascade, ending in the phosphorylation of transcription factors.

At each stage, the signaling intermediate is autoinhibited when unwanted. Activation requires the assistance of additional proteins such as scaffolds and additional interaction domains to increase the specificity of signaling. There is a wide range of embellishments in different systems.

Prokaryotes have a related mechanism called two-component signaling, in which a histidine residue is phosphorylated and passes the phosphate on to an aspartate residue on a receiver domain. This then activates a covalently attached effector domain, again in a wide range of mechanisms.

Three other widely used signaling methods are G-protein-coupled receptors, ion channels, and the release of proteolytic fragments. In the first, the binding of a ligand leads to helix rotations that alter the intracellular surface, which in turn activates a Gα protein to stimulate the production of second messengers. In the second, ligand binding also leads to helix rotation, this time to open a channel and allow ions to pass. And in the third, ligand binding leads to alterations in proteolytic degradation of intracellular proteins that directly affect gene transcription.

8.7 FURTHER READING

Signaling pathways are discussed in several excellent textbooks. My particular favorites are Molecular Biology of the Cell, by Alberts et al. [56], and Molecular Cell Biology, by Lodish et al. [57]: there is also good coverage in Cell Signalling, by Hancock [58].

There are excellent descriptions of many of the systems discussed here in Protein Structure and Function, by Petsko and Ringe [59], which comes from a structural angle rather than cell biology.

All the biological context discussed here is described elegantly in Molecular Biology of the Cell, by Alberts et al. [56].

A thought-provoking paper by Kuriyan and Eisenberg [10] sets out the argument that proximity is the key first requirement for the evolution of further embellishments.

8.8 WEBSITES

http://smart.embl-heidelberg.de/ *Protein interaction motifs (and much more).*

http://www.cellsignal.com/reference *Intracellular protein interaction domains.*

8.9 PROBLEMS

Hints for some of these problems can be found on the book's website.

1. This chapter starts by noting that signaling across a membrane barrier presents problems that need special solutions to overcome them. A eukaryotic cell has numerous membrane-enclosed internal organelles, which would imply the need to develop signaling mechanisms for each organelle. Is this true? For example, are there signaling systems for the cytoplasm to signal to the mitochondrion?

2. A signaling system that avoids the need for transmembrane receptors altogether, such as the NO system (Section 8.1.2), seems a much simpler solution than most discussed in this chapter. Describe the benefits and limitations of signaling by NO. In what situations is it used, and why is it a good solution in these cases? Your answer should include a short account of the drugs developed to interfere with NO signaling.

3. This chapter almost completely ignores calcium signaling, including the role of calmodulin. (a) Describe the structures of calmodulin in the free and calcium-bound states. (b) Why is the calcium-bound form able to bind ligands, whereas the unloaded form is not? (c) What is the major amino acid in the calmodulin-binding site? Suggest why this might have been chosen. (d) Calcium has been shown to bind to several different classes of ligand. Describe the main mode of interaction with ligands. In particular, what is the difference in binding mode between skeletal muscle myosin light-chain kinase and melittin?

4. Describe how the anticancer drug Gleevec (imatinib) binds to Abl kinase. Why does it not bind equally well to other kinases? What does this tell us about the design of drugs that act against kinases (currently the most popular drug target for the pharmaceutical industry)?

5. What would you expect would be the biological effect of an antibody directed against RTK signal receptors? Does it make a difference if the antibody has one or two antigen recognition sites?

6. Discuss what you would expect to observe if you performed the following experiments: (a) knock out a receptor, (b) knock out the kinase domain of a receptor, (c) alter an RTK so that it dimerizes constitutively, (d) alter an RTK so that it cannot dimerize, (e) knock out a scaffold, (f) overexpress a receptor, (g) overexpress a scaffold, (h) knock out a phosphatase that removes the phosphate from an activated receptor, (i) overexpress such a phosphatase, and (j) alter a dimeric ligand to prevent it dimerizing.

7. What is the evidence that lipid rafts are involved in signaling pathways?

8. A significant fraction of cancers are linked to mutations in Ras that prevent it from binding to a GAP. Explain why this might lead to cancer.

9. A rare mutation in the gene encoding the receptor for erythropoietin (TGF-β family) leads to truncation of the transcript and deletion of the C-terminal domain, which is an inhibitor of the receptor. This leads to over-activation of the receptor. What would be the consequences of such a mutation?

10. The nicotinic acetylcholine receptor has five subunits, but only two of them bind to acetylcholine. How might you expect the channel to behave differently if all five subunits were to bind to acetylcholine?

8.10 NUMERICAL PROBLEMS

N1. A good biological switch (such as Ras or SH2) has about 99.9% efficacy; that is, the inactive state is about 1000 times worse at transmitting a signal than the active state. If the activated switch is in fact 'on' only 10% of the time, for what fraction of the time is the inactivate switch 'on'? Standard risk assessment for a nuclear power plant says that the risk of a severe core damage accident should be less than 10^{-4} per plant per year (http://canteach.candu.org/library/19990102.pdf), and that the overall health risk is the risk of simultaneous severe core damage *and* inadequate containment. Furthermore, the overall risk is the frequency of an event multiplied by the consequences of the event. Compare this with the risk of transmission of an unwanted signal. Assuming that evolution has arrived at the best solution, what does this tell us about the risk to the host organism of an incorrect signal?

N2. Within the human genome there are about 530 kinases and 180 phosphatases. Of the kinases, 67% are serine/threonine kinases and 17% are tyrosine kinases; of the phosphatases, 54% are tyrosine phosphatases, 19% are serine/threonine phosphatases, and 27% have dual specificity [60]. Calculate the number of serine/threonine and tyrosine kinases and phosphatases. It is thought that roughly one-half of the 25,000 genes in the human genome are phosphorylated (although the exact number is still unclear; in Chapter 4 I quote a figure of one-third), and there may be an average of up to 40 sites of phosphorylation per protein. Of these, roughly 90% are on serine, 10% on threonine, and only 0.05% on tyrosine. How many potential phosphorylation sites is this? Calculate the number of sites phosphorylated and dephosphorylated (on average) on serine/threonine and tyrosine by each kinase or phosphatase. What do you conclude?

8.11 REFERENCES

1. J Schultz, F Milpetz, P Bork & CP Ponting (1998) SMART, a simple modular architecture research tool: identification of signaling domains. *Proc. Natl Acad. Sci. USA* 95:5857–5864.
2. I Letunic, T Doerks & P Bork (2008) SMART 6: recent updates and new developments. *Nucleic Acids Res.* 37:D229–D32.
3. GB Cohen, RB Ren & D Baltimore (1995) Modular binding domains in signal-transduction proteins. *Cell* 80:237–248.
4. N Blomberg, E Baraldi, M Nilges & M Saraste (1999) The PH superfold: a structural scaffold for multiple functions. *Trends Biochem Sci* 24:441–445.
5. RA Cerione & Y Zheng (1996) The Dbl family of oncogenes. *Curr. Opin. Cell Biol.* 8:216–222.
6. AM Crompton, LH Foley, A Wood et al. (2000) Regulation of Tiam1 nucleotide exchange activity by pleckstrin domain binding ligands. *J. Biol. Chem.* 275:25751–25759.
7. B Das, XD Shu, GJ Day et al. (2000) Control of intramolecular interactions between the pleckstrin homology and Db1 homology domains of Vav and Sos1 regulates Rac binding. *J. Biol. Chem.* 275:15074–15081.
8. RR Copley, J Schultz, CP Ponting & P Bork (1999) Protein families in multicellular organisms. *Curr. Opin. Struct. Biol.* 9:408–415.

9. W Kolch (2000) Meaningful relationships: the regulation of the Ras/Raf/MEK/ERK pathway by protein interactions. *Biochem. J.* 351:289–305.

10. J Kuriyan & D Eisenberg (2007) The origin of protein interactions and allostery in colocalization. *Nature* 450:983–990.

11. T Pawson & M Kofler (2009) Kinome signaling through regulated protein–protein interactions in normal and cancer cells. *Curr. Opin. Cell Biol.* 21:147–153.

12. M Huse & J Kuriyan (2002) The conformational plasticity of protein kinases. *Cell* 109:275–282.

13. MA Young, S Gonfloni, G Superti-Furga et al. (2001) Dynamic coupling between the SH2 and SH3 domains of c-Src and Hck underlies their inactivation by C-terminal tyrosine phosphorylation. *Cell* 105:115–126.

14. JD Faraldo-Gómez & B Roux (2007) On the importance of a funneled energy landscape for the assembly and regulation of multidomain Src tyrosine kinases. *Proc. Natl. Acad. Sci. USA* 104:13643–13648.

15. SW Cowan-Jacob (2006) Structural biology of protein tyrosine kinases. *Cell. Mol. Life Sci.* 63:2608–2625.

16. J Schlessinger (2002) Ligand-induced, receptor-mediated dimerization and activation of EGF receptor. *Cell* 110:669–672.

17. K Nelms, AD Keegan, J Zamorano et al. (1999) The IL-4 receptor: signaling mechanisms and biologic functions. *Annu. Rev. Immunol.* 17:701–738.

18. PA Boriack-Sjodin, SM Margarit, D Bar-Sagi & J Kuriyan (1998) The structural basis of the activation of Ras by Sos. *Nature* 394:337–343.

19. P Aloy & RB Russell (2006) Structural systems biology: modelling protein interactions. *Nat. Rev. Mol. Cell Biol.* 7:188–197.

20. MB Yaffe (2002) How do 14-3-3 proteins work? Gatekeeper phosphorylation and the molecular anvil hypothesis. *FEBS Lett.* 513:53–57.

21. E Wilker & MB Yaffe (2004) 14-3-3 proteins: a focus on cancer and human disease. *J. Mol. Cell. Cardiol.* 37:633–642.

22. WQ Li, HR Chong & KL Guan (2001) Function of the Rho family GTPases in Ras-stimulated Raf activation. *J. Biol. Chem.* 276:34728–34737.

23. J Avruch, A Khokhlatchev, JM Kyriakis et al. (2001) Ras activation of the Raf kinase: tyrosine kinase recruitment of the MAP kinase cascade. *Recent Prog. Hormone Res.* 56:127–155.

24. AD Catling, HJ Schaeffer, CWM Reuter et al. (1995) A proline-rich sequence unique to Mek1 and Mek2 is required for Raf binding and regulates Mek function. *Mol. Cell. Biol.* 15:5214–5225.

25. L Chang & M Karin (2001) Mammalian MAP kinase signalling cascades. *Nature* 410:37–40.

26. O Rocks, A Pekyer, M Kahms et al. (2005) An acylation cycle regulates localization and activity of palmitoylated Ras isoforms. *Science* 307:1746–1752.

27. J Downward (2001) The ins and outs of signaling. *Nature* 411:759–762.

28. XW Zhang, J Gureasko, K Shen et al. (2006) An allosteric mechanism for activation of the kinase domain of epidermal growth factor receptor. *Cell* 125:1137–1149.

29. WA Lim (2002) The modular logic of signaling proteins: building allosteric switches from simple binding domains. *Curr. Opin. Struct. Biol.* 12:61–68.

30. S Maignan, JP Guilloteau, N Fromage et al. (1995) Crystal structure of the mammalian Grb2 adapter. *Science* 268:291–293.

31. S Yuzawa, M Yokochi, H Hatanaka et al. (2001) Solution structure of Grb2 reveals extensive flexibility necessary for target recognition. *J. Mol. Biol.* 306:527–537.

32. L Buday (1999) Membrane-targeting of signalling molecules by SH2/SH3 domain-containing adaptor proteins. *Biochim. Biophys. Acta* 1422:187–204.

33. M Anafi, MK Rosen, GD Gish et al. (1996) A potential SH3 domain-binding site in the Crk SH2 domain. *J. Biol. Chem.* 271:21365–21374.

34. K Takeuchi, Z-YJ Sun, S Park & G Wagner (2010) Autoinhibitory interaction in the multidomain adaptor protein Nck: possible roles in improving specificity and functional diversity. *Biochemistry* 49:5634–5641.

35. J Gureasko, WJ Galush, S Boykevisch et al. (2008) Membrane-dependent signal integration by the Ras activator Son of sevenless. *Nat. Struct. Mol. Biol.* 15:452–461.

36. AM Stock, VL Robinson & PN Goudreau (2000) Two-component signal transduction. *Annu. Rev. Biochem.* 69:183–215.

37. AH West & AM Stock (2001) Histidine kinases and response regulator proteins in two-component signaling systems. *Trends Biochem. Sci.* 26:369–376.

38. C Laguri, RA Stenzel, TJ Donohue et al. (2006) Activation of the global gene regulator PrrA (RegA) from *Rhodobacter sphaeroides*. *Biochemistry* 45:7872–7881.

39. C Laguri, MK Phillips-Jones & MP Williamson (2003) Solution structure and DNA binding of the effector domain from the global regulator PrrA (RegA) from *Rhodobacter sphaeroides*: insights into DNA binding specificity. *Nucleic Acids Res.* 31:6778–6787.

40. PJ Artymiuk, DW Rice, EM Mitchell & P Willett (1990) Structural resemblance between the families of bacterial signal-transduction proteins and of G-proteins revealed by graph theoretical techniques. *Protein Eng.* 4:39–43.

41. M Simonovic & K Volz (2001) A distinct meta-active conformation in the 1.1-Å resolution structure of wild-type ApoCheY. *J. Biol. Chem.* 276:28637–28640.

42. HS Cho, SY Lee, DL Yan et al. (2000) NMR structure of activated CheY. *J. Mol. Biol.* 297:543–551.

43. A Bren & M Eisenbach (1998) The N terminus of the flagellar switch protein, FliM, is the binding domain for the chemotactic response regulator, CheY. *J. Mol. Biol.* 278:507–514.

44. BF Volkman, D Lipson, DE Wemmer & D Kern (2001) Two-state allosteric behavior in a single-domain signaling protein. *Science* 291:2429–2433.

45. CH Heldin, K Miyazono & P ten Dijke (1997) TGF-β signalling from cell membrane to nucleus through SMAD proteins. *Nature* 390:465–471.

46. J Avruch, XF Zhang & JM Kyriakis (1994) Raf meets Ras: completing the framework of a signal transduction pathway. *Trends Biochem. Sci.* 19:279–283.

47. A Ashkenazi & VM Dixit (1998) Death receptors: signaling and modulation. *Science* 281:1305–1308.

48. L Izzi & L Attisano (2004) Regulation of the TGFβ signalling pathway by ubiquitin-mediated degradation. *Oncogene* 23:2071–2078.

49. K Palczewski, T Kumasaka, T Hori et al. (2000) Crystal structure of rhodopsin: a G protein-coupled receptor. *Science* 289:739–745.

50. DM Rosenbaum, V Cherezov, MA Hanson et al. (2007) GPCR engineering yields high-resolution structural insights into β_2-adrenergic receptor function. *Science* 318:1266–1273.

51. P Scheerer, JH Park, PW Hildebrand et al. (2008) Crystal structure of opsin in its G-protein-interacting conformation. *Nature* 455:497–502.

52. MA Hanson & RC Stevens (2009) Discovery of new GPCR biology: one receptor structure at a time. *Structure* 17:8–14.

53. SR Sprang (1997) G protein mechanisms: insights from structural analysis. *Annu. Rev. Biochem.* 66:639–678.

54. A Miyazawa, Y Fujiyoshi & N Unwin (2003) Structure and gating mechanism of the acetylcholine receptor pore. *Nature* 423:949–955.

55. N Unwin (2005) Refined structure of the nicotinic acetylcholine receptor at 4 Å resolution. *J. Mol. Biol.* 346:967–989.

56. B Alberts, A Johnson, J Lewis et al. (2008) Molecular Biology of the Cell, 5th ed. New York: Garland Science.

57. H Lodish, A Berk, CA Kaiser et al. (2008) Molecular Cell Biology, 6th ed. New York, W.H. Freeman.

58. JT Hancock (2010) Cell Signalling, 3rd ed. Oxford: Oxford University Press.

59. GA Petsko & D Ringe (2004) Protein Structure and Function. New York. Sinauer Associates.

60. S Arena, S Benvenuti & A Bardelli (2005) Genetic analysis of the *kinome* and *phosphatome* in cancer. *Cell. Mol. Life Sci.* 62:2092–2099.

61. MB Yaffe (2002) Phosphotyrosine-binding domains in signal transduction. *Nat. Rev. Mol. Cell Biol.* 3:177–186.

62. FP Ottensmeyer, DR Beniac, RZ-T Luo & CC Yip (2000) Mechanism of transmembrane signaling: insulin binding and the insulin receptor. *Biochemistry* 39:12103–12112.

63. SY Park & SE Shoelson (2008) When a domain is not a domain. *Nat. Struct. Mol. Biol.* 15:224–226.

64. JH Wu, YD Tseng, CF Xu et al. (2008) Structural and biochemical characterization of the KRLB region in insulin receptor substrate-2. *Nat. Struct. Mol. Biol.* 15:251–258.

CHAPTER 9

Protein Complexes: Molecular Machines

If I have seen further it is only by standing on the shoulders of giants.

Isaac Newton (15 February 1676), letter to Robert Hooke

In the rush to be first, it is important to avoid the tendency to do so on the backs of our predecessors, rather than, as used to be the case, on their shoulders.

Paul Srere (2000), [1]

Thou shalt not believe something just because you can explain it.

Arthur Kornberg (2003), Commandment III [2]

The nature of science is such that experimentalists push their experimental data to the limits of reliability, and sometimes beyond this point.

Alan Fersht (1985), [3]

In the first few chapters we covered more or less all of the general principles that guide how proteins work. In Chapters 7 and 8 we started to investigate the kind of roles that proteins have in cells, and we have seen for example that many complicated systems, such as motors, have a common and fairly simple core, but that the workings of evolution tend to generate a **patchwork** or **mosaic** system, with components brought in from elsewhere wherever needed. This can generate systems of considerable complexity as well as redundancy. In Chapter 8 we saw how interactions between one domain and another build up specificity but can also facilitate cross-talk between one signaling system and another.

In this chapter and Chapter 10 we continue this theme by seeing how proteins combine together to generate molecular machines of considerable sophistication. In fact, *most proteins in vivo associate with other partners*. In this context, it is relevant to quote another passage from Kornberg [2], which essentially forms the subject matter of this chapter:

Beyond the catalytic face, enzymes have two additional faces: regulatory and social. The regulatory site binds a ligand that modifies the rate and specificity of the enzymes. The social face associates the enzyme with other components, such as a membrane or a scaffold, or complexes with other enzymes.

We have already seen (mainly in Chapter 2) that, to a remarkably large extent, different functions of a protein can be identified with different domains. In many proteins, one can readily identify for example a catalytic domain, a ligand-binding domain, or a membrane attachment domain. In this chapter we see how the different functions of a molecular machine can similarly be identified with its component proteins. Thus, we shall see how the multiprotein exosome complex (which digests RNA in a tightly regulated manner) has a catalytic core, a series of substrate-recognition proteins, a group of regulatory proteins, and sets of transient partner proteins that are plugged onto the complex when particular substrates are bound.

The word 'transient' in the previous sentence is vital. The exosome digests a range of different RNAs, in different ways. It does this by having a core or hub comprising the essential digestive machinery, and then interacting with different protein subcomplexes depending on the specific RNA to be digested. Some interactions are strong and long-lasting; others are weak and highly transient. This chapter discusses the structure and formation of such interactions, a system often described as the *interactome* of the cell.

Chapter 10 also discusses protein complexes. The difference between the two chapters is that the complexes described in this chapter are transient, whereas those in Chapter 10 typically have a fixed structure and stoichiometry. Most of the complexes in this chapter have a single function, such as digesting or polymerizing RNA. This may be a complex function that requires regulation, but there is basically just one 'enzyme.' By contrast, the complexes in Chapter 10 have more than one enzymatic function, typically two or more sequential reactions in a pathway.

Falling in between these two are the metabolons that are discussed at the end of this chapter, which carry out a series of enzyme reactions but do it in the form of transient and regulated complexes. It seems more appropriate to

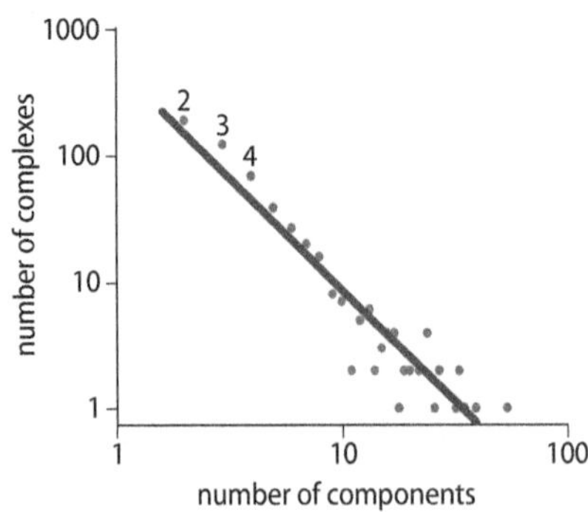

FIGURE 9.1
Distribution of the number of components per complex in the yeast interactome. On a double logarithmic scale, the correlation is close to being linear, indicating a scale-free network. There are about 190 complexes with only two members, 120 with three, and so on. (Adapted from N. J. Krogan, G. Cagney, H. Yu et al., *Nature* 440: 637, 2006. With permission from Macmillan Publishers Ltd.)

discuss them here, because the experimental problems and techniques are similar to those of 'true' molecular machines.

9.1 THE CELLULAR INTERACTOME

9.1.1 Interactomes have similar structures

The material covered in this chapter is in many ways the least well understood of anything in this book. There have been many publications recently that describe the interactome—the networks of interactions that proteins form within the cell. These studies use a variety of techniques. Mainly they have used either the two-hybrid screen or pull-down techniques such as tandem affinity purification (TAP)-tag, in which one component of a complex is tagged with an additional domain or sequence that allows it to be purified, bringing with it any other proteins with which it may be associated [4]. These techniques are discussed further in Section 11.8. In the next section we consider how reliable the information is: here we survey what these screens have found.

Probably the most detailed description of the interactome comes from studies on the baker's yeast *Saccharomyces cerevisiae*. Several independent TAP-tag screens have been conducted [5–8]: all potential proteins encoded by the genome were tagged individually and used to identify complexes. Of course, not all proteins worked. Strikingly, around two-thirds of proteins had at least one partner: that is, a large majority of proteins associate with at least one other in the cell, or, viewed alternatively, very few proteins go it alone. Similarly, many proteins take part in more than one complex. From the 6500 ORFs of *S. cerevisiae*, around 500 complexes were identified in each screen. However, the authors predicted that the total number of complexes in yeast should be closer to 800, because of the incompleteness of the protein sampling. The number of protein components involved in each complex ranged from 2 to about 100, with an average of about 5, although most proteins were in complexes of 2–4 members (**Figure 9.1**). The largest complexes were for proteasomes, the ribosome, and RNA transcription (RNA polymerase II and Mediator), presumably reflecting the fact that these are more complex activities that need careful regulation. Complexes represented all areas of cellular activity. It therefore seems that essentially all cellular activities benefit from collaboration between proteins, although many require only two or three. This observation is probably a natural extension of what we have seen in earlier chapters: specificity is increased by the addition of extra domains or proteins, and by colocation of related functions.

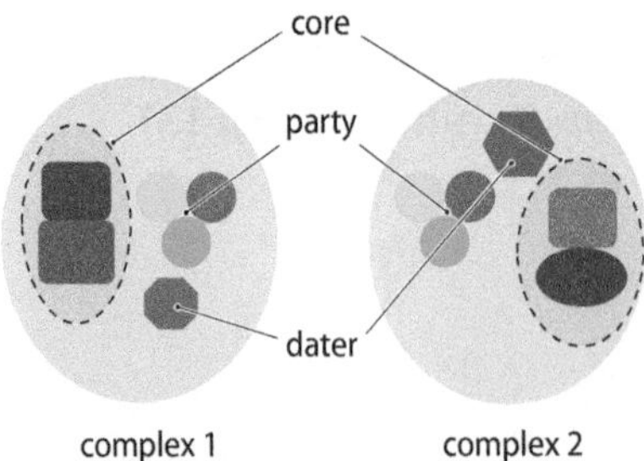

FIGURE 9.2
The sociology of complexes. The cores of the complexes constitute their essential catalytic activity and are always present. These two complexes contain the same group of three 'party' proteins, which typically perform a defined function, for example binding to single-stranded RNA; these proteins tend to be found as the same group in a range of complexes, and may sometimes not be present when the complex is isolated. They also contain 'dater' proteins, which are often found transiently attached to a range of complexes and often have a more generic function such as protein interactions.

Closer analysis of the complexes in *S. cerevisiae* shows that complexes tend to consist of 'core' or 'hub' proteins, which are always present whenever the complex is isolated, and a range of other proteins that probably bind transiently (**Figure 9.2**). Many of the proteins that bind do so in groups. We can envisage the hub complexes being visited by little clusters of other proteins. Many of these clusters can visit a number of core complexes. These two types of protein have been nicely described as 'party' hubs [9, 10], meaning that they always go around with the same group of partners, and 'daters,' who frequently change their partners, although as noted above many seem to go around essentially permanently with the same one or two other proteins. The structures and dynamics of complexes has been aptly called the *molecular sociology* of the cell [11].

Each complex seems to have a defined function, which (where we know the details) is made up in a fairly simple way by the constituent proteins: the whole is (more or less) just the sum of its parts. It is thus a straightforward modular assembly, in which the necessary functions are contributed by individual component proteins. The modularity of the complexes mirrors the modularity of multidomain proteins themselves. Thus, complexes can be labeled as being responsible for *RNA degradation* or *membrane fusion*, within which the component proteins can be generally seen to have independent roles. The authors comment, "Modularity might very well represent a general attribute of living matter, with *de novo* invention being rare and reuse the norm," which certainly fits within the general theme of this book [6]. Thus there is no evidence for subtle interplay between proteins and complexes (except for a few special cases that are considered in the next chapter): they have clearly defined single functions, although in many cases there is redundancy, with several proteins having apparently the same role. This is what we might expect from evolutionary arguments: a system involving cross-talk between components is much harder to modify, as any computer programmer can tell you! However, complexes that one might expect to be involved in two related functions, such as cell cycle control and transcription, do indeed have proteins in common. This is presumably the rationale behind the division into 'party' and 'dater' components: the 'party' hub is the specialist function, and the 'daters' are the ancillary and more general-purpose components.

This is what we might expect from everything we have seen so far: evolution tends to work by the accretion of proteins and domains, tinkering by adding extra little functions, rather than by the evolution of a new single functional unity that does everything. Indeed, as we have seen, evolution is probably incapable of re-engineering on such a large scale, because there is no evolutionary mechanism to start off such a major reorganization. Evolution is, however, good at adapting, and there is strong evidence that different components in complexes coevolve [7].

All interactome studies find a power-law or scale-free network, in which a small number of proteins make many connections, but most proteins make very few (see Figure 9.1). (The same is incidentally true of domain:domain interactions within multidomain proteins, implying that proteins and their domains are selected for combination because of their function and not because of any specific structural features [12].) There are a small number of vital 'hubs' that are highly conserved and tend to be essential genes (that is, genes whose knockout prevents growth) [13], especially where those hubs interact directly with other hubs [14]. A yeast two-hybrid study found that these essential genes are more likely to interact within large networks than are non-essential genes [15].

So far we have looked only at the yeast interactome. However, yeast seems to be fairly typical. A study of the *Escherichia coli* interactome found that only 18% of proteins were not in some kind of complex; again, a scale-free network was found [16], in which there are a small number of 'party' hubs with a large number of connections, and many proteins with only one or two. A two-hybrid screen of *Drosophila* had similar overall features [17].

9.1.2 The picture is still far from clear

It is still very early days, and it is not yet clear how reliable these global interactomes are. There is a hand-curated database of protein interactions assembled by the Munich Information Center for Protein Sequences (MIPS; http://mips.helmholtz-muenchen.de/genre/proj/corum), which is often cited as the 'gold standard' of interacting proteins [18, 19]. In 2006, it listed 217 known complexes, of which between one-half and one-third were not identified in the TAP-tag screens. There could be many reasons for this. It is certainly true that many complexes are formed transiently, depending on the stage of the cell cycle or growth conditions; for example, the TAP-tag studies may have not used the conditions appropriate for some complexes. Addition of the tag may have disrupted the complex, although this is unlikely because each protein in the complex is tagged in turn, and proteins are often tagged twice in two separate screens, once at the N terminus and once at the C terminus, precisely to pick up this kind of problem.

The two TAP-tag studies had little overlap in detail, even though they were studying the same organism by the same technique [9, 20, 21]. This can be viewed either as complementary coverage [9] or as problems that still need ironing out. However, comparisons suggest that they have a reliability as good as non-high-throughput methods.

The results from the two TAP-tag screens have been combined [22] to produce a set that is suggested to be as accurate as traditional low-throughput methods. The conclusion is that complexes probably exist in a continuum of components: some components are always associated, whereas others come and go at different rates and with different affinities. This combined analysis concludes that 1622 proteins (out of a genome total of around 6500) were involved in 'high-confidence interactions' (probably equating to strong or stable interactions). Similar conclusions were reached by a study that attempted to include structural information [23]. This study noted that some proteins apparently had hundreds of possible partners. This is structurally impossible: one globular protein has a maximum of about 14 partners with which it can interact directly at the same time. The study concluded that of 1269 high-quality interactions, 438 were mutually exclusive. This means that these interactions cannot occur at the same time and that they must therefore reflect alternative partners within a complex, confirming the idea that the 'party' hubs can combine with a range of other complexes depending on the needs of the cell.

Analysis of TAP-tag screens is developing. It is clear that they produce a significant proportion of false positives (that is, proteins that copurify but are not in fact in a complex), and we are starting to learn how to identify these [24]. Abundant proteins, particularly ribosomal proteins, tend to be picked up more often than they should, and membrane proteins are under-represented.

The yeast interactome has also been probed with two-hybrid technology, which picks out binary protein–protein pairs only [25]. There have been a number of criticisms of two-hybrid screens, mainly on the grounds of the large number of false positives that they generate (that is, pairs of proteins that are suggested to interact but in fact probably do not). The set of complexes resulting from the two-hybrid screen is very different from that identified by the TAP-tagging. This may reflect real differences in the types and durations of interactions [26], and it has been suggested that two-hybrid screens are more suited to identifying transient and intercomplex interactions [15], implying again that many proteins interact weakly and transiently, sometimes with more than one complex.

Interactome technology is still new. In any new technology, the protagonists tend to exaggerate, and it takes a while until it is clear what the technology can really do [27]. We are now entering the second stage, where methodical hard work and careful analysis will eventually reveal what is really going on. It is nevertheless clear that the broad outlines revealed so far are reliable, even if the details are not.

9.1.3 Interacting complexes have defined but transient structures

Do the complexes described above have a defined three-dimensional structure? It certainly looks as though the 'core' elements, and probably also a considerable proportion of the accessory elements, do have a well-defined structure. We shall look at two of these shortly. It may well be that some of the individual components are natively unstructured, and only become structured when bound within the complex. In Chapter 4 we discussed the concept of

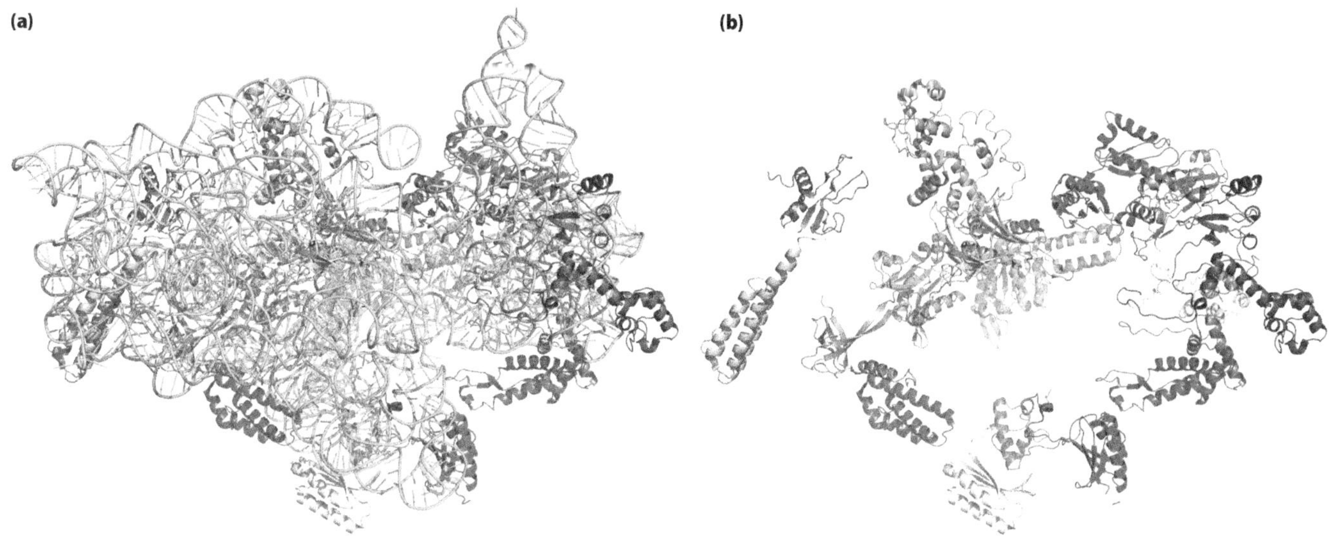

FIGURE 9.3
The structure of the 30S ribosomal subunit of *Thermus thermophilus* (PDB file 1j5e) (compare Figure 1.62). (a) The entire 30S subunit. RNA is in green. (b) The same view, but without the RNA. The proteins form a protective coat around the RNA. Some are more or less normal globular proteins (such as the blue protein at 7 o'clock, ribosomal protein S15), while others have very non-globular regions that are only structured because they form part of the interface with RNA (such as the magenta protein at 5 o'clock, ribosomal protein S11).

fly-casting, in which disordered regions of proteins are used as ways of attracting and binding to a partner. Many ongoing studies of complexes identify such features as important for the formation of complexes. It is probably relevant to compare this picture to the structure of the ribosome, in which many of the component proteins are structured, but not globular: it is only their interactions with RNA or other proteins that provides them with structure (**Figure 9.3**). Thus, although many of the complexes are dynamic, in that components come and go as needed, they are in general probably of a defined structure, with defined interactions between the different components. In general, proteins undergo conformational changes on binding to their partners, ranging from minor rearrangements of side chains to disorder/order transitions. This makes the task of computational docking very difficult, and means that for example predicting binding energies is as yet virtually impossible [28]. We therefore still rely almost entirely on experimental observations.

9.1.4 Interactomes constitute molecular machines

The complexes that perform the major functions in the cell (DNA, RNA, and protein polymerization and degradation; DNA repair; RNA processing; movement within the cell; membrane fusion; regulation of the cell cycle; mitosis; and many others) have been described as 'molecular machines' [29] in that they consist of multiple coordinated components. They retain the features identified above—each component has a clearly identifiable function, and the functions are largely sequential and independent—but because they are in one complex, their activity can be more coordinated. A considerable effort is under way to identify and understand these machines.

In his 1998 review, Alberts noted that the assembly of such machines seems often to be driven by energy-requiring catalytic factors (**Figure 9.4**) [29]. These factors are needed for the assembly of the machines, but after doing this they are released to allow them to initiate another cycle. Interestingly, in several membrane protein complexes, the assembly factor seems to be still present (that is, the assembly factor here is not catalytic, and is not recyclable); either the assembly of membrane protein complexes is more difficult, or it never evolved to recycle the assembly factors.

In Sections 9.2 and 9.3 we discuss two reasonably well studied and well understood complexes, as exemplars of the way such complexes may work.

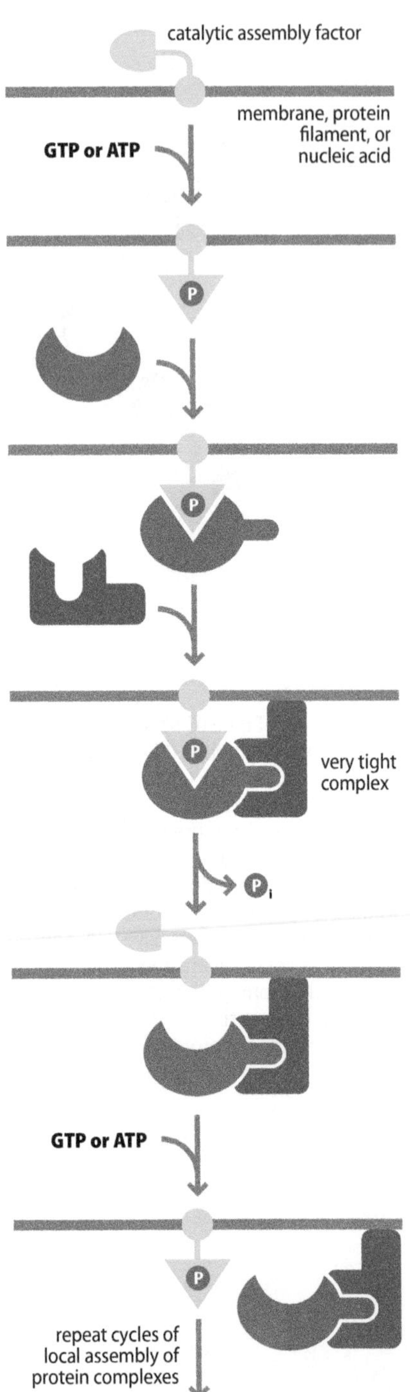

FIGURE 9.4
A general scheme showing how the hydrolysis of nucleotide triphosphates provides the energy for assembly of protein complexes. The catalytic assembly factor (green) is activated either by binding to GTP or by phosphorylation. The extra phosphate enables the assembly factor to bind to the red protein; in turn this induces a conformational change in the red protein that enables it to bind to the blue protein. This produces a very tight complex, from which dissociation of any of the partners is very slow. Loss of the phosphate by hydrolysis provides the energy needed to liberate the green assembly protein and allow it to take part in another catalytic assembly cycle. (From B. Alberts, *Cell* 92: 291–294, 1998. With permission from Elsevier.)

9.2 THE EXOSOME

The exosome is a relatively small (600 kDa) eukaryotic complex composed of 10 core proteins (**Table 9.1**) whose function is to degrade RNA in the 3′ → 5′ direction from one end (in other words, it is an exoribonuclease not an endoribonuclease), a function often termed RNA decay [30, 31]. It is also involved in the processing of other RNAs, including ribosomal RNA and small nucleolar and nuclear RNAs (snoRNAs and snRNAs). This is a quite different function from RNA decay, because here it is required only to trim the RNA, not to destroy it. It is therefore in this context an RNA hydrolase. However, to describe its function as simply a hydrolase is a large understatement, because it needs to be able to recognize which RNA to hydrolyze, and to stop as appropriate. To a large extent this recognition is done not by the exosome itself but by associated complexes such as cytoplasmic Ski (an abbreviation of 'superkiller', so named because it is involved in destroying viral RNA) complexes and nuclear TRAMP,

TABLE 9.1 Proteins in the yeast exosome

Yeast	Human	Archaeal related	Bacterial related	Function
Rrp44	Dis3		RNAse R, RNAse II	Processive hydrolytic exoribonuclease
Rrp41		Yes	RNAse PH, PNPase PH	Processive phosphorylytic exoribonuclease
Rrp42		Yes	RNAse PH, PNPase PH	
Rrp43	OIP2		RNAse PH, PNPase PH	
Rrp45	PM/Scl-75		RNAse PH, PNPase PH	
Rrp46			RNAse PH, PNPase PH	
Mtr3			RNAse PH, PNPase PH	
Csl4		Yes		S1/KH domains bind RNA
Rrp4		Yes		S1/KH domains bind RNA
Rrp40				S1/KH domains bind RNA
Rrp6	PM/Scl-100		RNAse D	Distributive hydrolytic exoribonuclease

Polynucleotide phosphorylase (**PNPase**) is a bifunctional enzyme with a phosphorolytic 3′ → 5′ exoribonuclease activity and a 3′-terminal oligonucleotide polymerase activity. **RNase PH** is a 3′ → 5′ exoribonuclease and nucleotidyltransferase present in archaea and bacteria that is involved in tRNA processing. In contrast with hydrolytic enzymes, it is a phosphorolytic enzyme, meaning that it uses inorganic phosphate as a cofactor to cleave nucleotide–nucleotide bonds, releasing diphosphate nucleotides.

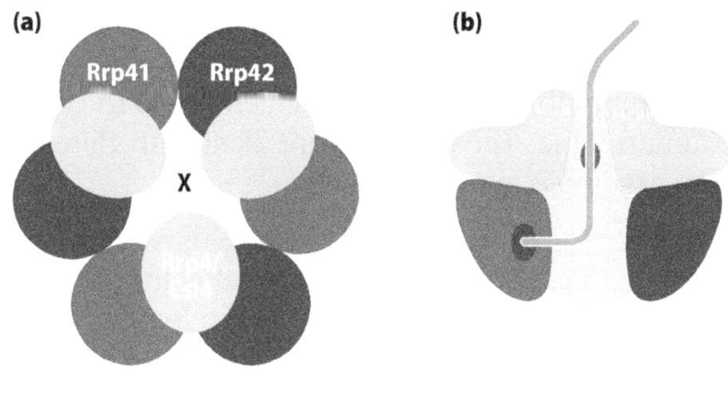

FIGURE 9.5
The archaeal exosome consists of a hexameric ring composed of alternating Rrp41 (red) and Rrp42 (blue), capped by a trimer in which the monomers are either Rrp4 or Csl4 (green). (a) Top view. X marks the entry point for RNA into the complex. (b) Side view, showing the route of RNA in through the central cavity to the active site on Rrp41. In a crystal structure of the complex (PDB file 2jea), two RNA fragments can be identified bound to the protein, indicated by ovals: one at the blue/green position and one at the active site (brown).

the first of which probably bind to non-standard RNA ends and unwind them, and the second tags RNA that is to be degraded by adding poly(A). Hence the exosome is a good example of a molecular machine because its primary function is RNA degradation, but this is regulated by the transient complexes within which the degradative function sits, turning it into either an indiscriminate nuclease or a precise trimmer.

Exosomes are found only in eukaryotes and archaea, not prokaryotes. However, in a common pattern, the proteins within the exosome have similarities to bacterial proteins, but there are more of them. They are more closely related to archaeal proteins, and indeed the archaeal exosome (**Figure 9.5**) seems to be a simpler version of the eukaryotic complex (**Figure 9.6**), in which there are fewer different proteins but the same overall arrangement. Thus, the basic structure consists of heterodimers (Rrp41 and Rrp42 in archaea; Rrp41/Rrp42, Rrp46/Rrp45, and Mtr3/Rrp43 in yeast), with clear resemblance to the bacterial PNPase heterodimer, in which the single active site is between the two components of the dimer (and there is therefore presumably an evolutionary relationship). These are assembled into a hexameric ring, and capped by Csl4 or Rrp4 in archaea, and Rrp4, Rrp40, and Csl4, respectively, in yeast. Mass spectrometry (Chapter 11) was able to demonstrate that this architecture is mirrored by the stability of the interactions, with the heterodimers as the strongest interaction and thus the basic building blocks. The trimeric cap has a small hole through which only single-stranded RNA can enter. It is relevant to note that the intact archaeal complex has weak exonucleolytic activity (although the yeast complex probably does not), whereas a complex lacking the cap has distributive activity (meaning that it cuts the RNA randomly along its length, as opposed to cutting in a processive manner from one end). This suggests that the cap acts to grip RNA and ensure processivity (Chapter 4). It also has a function in specifically recognizing mRNA, which probably resides in Csl4, the cap protein that interacts least with the hexameric ring [32]. A structure of the archaeal complex was obtained with RNA bound in the channel [33], which shows that the RNA is bound by the cap but then is unstructured until it reaches one of the active sites within the hexameric ring (see Figure 9.5). Csl4 in the cap also seems to be the binding site for the Ski7 complex.

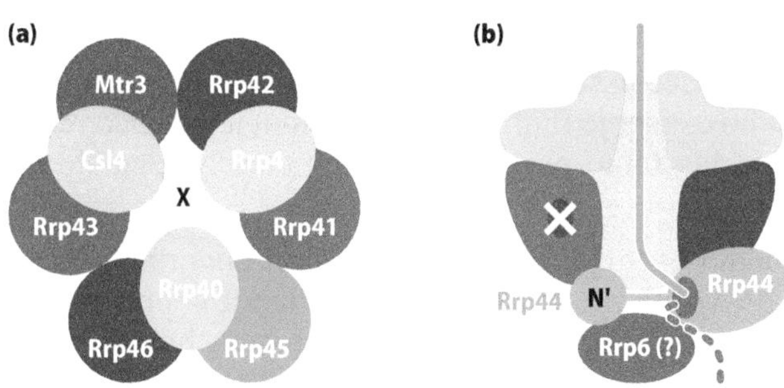

FIGURE 9.6
The eukaryotic (yeast) exosome is similar to the archaeal one, except that each subunit is now different. The three Rrp41 homologs (Rrp41, Rrp46, and Mtr3) are in shades of red, and the three Rrp42 homologs (Rrp42, Rrp45, and Rrp43) are in shades of blue. The three Rrp4 homologs (Rrp4, Rrp40, and Csl4) are in shades of green. (a) Top view. (b) Side view. The hexameric ring no longer contains an active exonuclease: the active site of the exonuclease is located in Rrp44, which can potentially be accessed without going through the hexameric ring (brown dashed route). There is a second nuclease activity present in some complexes, provided by the Rrp6 domain. The location of this domain is not yet clear; one possible location is shown here.

As we might expect, the cap proteins contain domains whose function is to bind to the proteins in the hexameric ring. In fact, at least two of the cap proteins have two binding domains, each of which binds to one of the two proteins in the hexameric ring. The domains are all slightly different, allowing the yeast complex to assemble in only one way.

The yeast hexameric ring is composed of proteins with homology to bacterial phosphorolytic exonucleases (that is, they use inorganic phosphate as a cofactor to cleave nucleotide–nucleotide bonds, releasing diphosphate nucleotides) and yet the exosome has hydrolytic exonuclease activity (that is, it uses water, not inorganic phosphate, to break the phosphate bond). The proteins in the hexameric ring have, however, no phosphorolytic activity, and the key hydrolytic activity resides in Rrp44, which binds at the 'bottom' face (opposite the cap), along with the final component, Rrp6 (see Figure 9.6b). This binding causes very little perturbation to the ring [34]. The function of the proteins in the hexameric ring is still not exactly clear. They act as sites of recognition and binding for further proteins, for example to improve selectivity in substrate binding, and they also both restrict the access of the degraded RNA to the active site of Rrp44 and act as a second binding site for RNA, thereby improving the processivity of the enzyme. It is also possible that they act as a spacer between the entrance hole and the degradative enzyme, allowing control over the length of the RNA that is trimmed. It seems that the original bacterial degradative hexameric complex has evolved in eukaryotes by losing almost all its original enzymic activity, converting the original enzymes into regulators, and regaining activity by adding another protein.

Rrp44 contains the processive hydrolytic activity. Thus, in a sense it *is* the essential exosome. It is active in the absence of the rest of exosome, and probably functions in this form. It can be dissociated from the core exosome with 0.5 M $MgCl_2$ and has not been found associated with the exosome in human or *Trypanosoma brucei* exosomes. It therefore probably has other activities that are quite separate from those that it has in the exosome. Within the exosome, its activity is essentially the same as the free enzyme against single-stranded RNA, but is much decreased against stem-loop-containing RNA. Electron microscopy (Chapter 11) showed that Rrp44 sits across the bottom of the hexameric ring. Rrp44 contains a catalytically active 'body' plus a 'head' (the N' domain in Figure 9.6), which in free Rrp44 is unconstrained but in the intact exosome has a fixed position and limits access to the active site [34]. Rrp44 is homologous to *E. coli* RNAse II, which contains a catalytic (RNB) domain, in which the active site is shielded by three other domains: two cold-shock domains and an S1 domain, all of which belong to the common OB (oligonucleotide/oligosaccharide-binding) fold. However, in yeast Rrp44, these domains are oriented differently, and restrict the RNA entering the active site to being single-stranded [35]. They may also be responsible for the fact that Rrp44 is able to hydrolyze regions of structured RNA given a long enough overhang, whereas RNAse II stalls on encountering secondary structure [35]. This discussion makes it clear that the exosome machine is essentially a complex 'front end' that regulates the input and action of the key catalytic function, similarly to the way in which most of a DVD player is a 'front end' that controls what the actual reading device does.

The discussion above assumes that Rrp44 receives RNA via the channel that runs through the exosome ring. It is possible that it can also receive RNA from the other side, this route presumably being closely regulated by additional proteins.

The head domain of Rp44 is also active as an endoribonuclease [32], preferentially to RNA with a 5′ phosphate, although it is unclear how this fits in with other activities of the exosome; it may possibly help to coordinate 3′ and 5′ cleavage. This activity further supports the idea that RNA may be able to enter the active site of Rp44 from the bottom as well as the top. Unusually, this domain seems to have two functions: the endonuclease activity plus one in attaching the Rp44 head to the hexameric ring [36].

The final component, Rrp6, has distributive ribonuclease activity (it cuts the RNA randomly along its length). It binds to the exosome via a C-terminal domain, but also functions independently of the exosome: it digests different substrates in the presence of the exosome [37]. It can interact with the oligomeric yeast protein Rrp47 via an N-terminal binding domain, which may function to allow Rrp6 to degrade structured RNAs [38]. However, only about one-sixth of Rrp6 complexes are bound to Rrp47, so this is not its only activity. It is also suggested that a function of Rrp6 within the intact exosome is to remove the poly(A) tail added by the TRAMP complex, after which the RNA is passed on to Rrp44 (presumably by other proteins that form part of the TRAMP complex) for further digestion [39]. Rrp6 has been located at the top of the exosome in a low-resolution electron microscopy study, but biochemical results suggesting that it binds at the bottom next to Rrp44 are possibly more trustworthy, so its exact location is not yet clear: it is shown at the bottom in Figure 9.6 [31].

We have seen that the archaeal and yeast exosomes look rather similar (a hexameric ring capped by three proteins), but functionally they are fundamentally different in that the nucleolytic activity in the yeast complex is not contained within this assembly at all; instead it is in a separate protein, Rrp44. In the language introduced in Section 9.1, we can describe the yeast exosome as having a core consisting of Rrp44 plus the nine associated proteins, which feed in the substrate and regulate the activity of Rrp44. Rrp6 is a 'dater', because it can also be found in a range of other complexes. In addition, as mentioned briefly above, there are a large number of other 'party' components that come and go as needed, including the Ski complex (which recognizes non-standard RNAs and presents them) and the TRAMP complex, which is restricted to the nucleus and adds poly(A) tails, but also contains a helicase activity. Other components also include various RNA-binding proteins, which recognize specific types of RNA and present them appropriately; helicases; and the Nrd1–Nab3 complex, which recruits the exosome to snRNA/snoRNA precursors and can also preprocess them. The yeast interactome study also identified several members of the small-subunit processome that is involved in the processing of pre-18S rRNA as another set of 'party' proteins [6].

This account is a short summary of what is known about the exosome; further details, particularly of interacting partners in both nucleus and cytoplasm, which include helicases, polyadenylases, and a range of other RNA- and protein-binding factors, can be found elsewhere [30].

In summary, the exosome is an interesting product of evolution. Its core structure is similar to a bacterial RNA-degrading complex, yet this hexameric core has lost all the original RNA-degrading activity; it is now used to thread single-stranded RNA into the active enzymes, and to bind extra regulatory proteins. The two enzymic activities can both dissociate from the exosome, probably having other degradative roles within the cell independent of the exosome; when attached to the exosome they have restricted and much more specific activity, controlled not only by binding to the exosome but by additional proteins that bind transiently. Indeed, in many ways the whole of the rest of the exosome is an accretion that has collected around the active enzyme to moderate and regulate it. The two enzymes can probably also act in concert, substrates being passed from one to the other, again by the help of additional transient components. A plausible scenario is that at some point in evolution, binding of both degradative enzymes simultaneously to the exosome turned out to be synergistic and thus worth keeping. Furthermore, there is some redundancy within the exosome, with for example some endonuclease activity. It is not yet clear whether this is a feature that is important for the overall function of the exosome or whether it is some evolutionary relic (not unlike the appendix) that has no particular value to the cell: it certainly seems that the exosome is something of a tinker's product—it works, but it is not an elegant solution. Its similarity to prokaryotic enzymes (and the fact that it works on RNA, the 'original' biopolymer) has suggested that the exosome may represent a very early complex [30]. Perhaps this early origin (and essential role) contributes to its rather 'messy' construction, in that the pressure on it to work properly has restricted it from evolving into a more streamlined machine.

9.3 THE RNA POLYMERASE II COMPLEX

9.3.1 Pol II assembles sequentially

RNA polymerases (Pol) generate RNA based on a double-stranded DNA template. Bacteria have only one RNA polymerase, whereas eukaryotes have three: Pol I makes 18S, 5.8S, and 28S rRNA, Pol III makes tRNA and 5S rRNA, while Pol II makes mRNA and also heterogeneous nuclear RNA. Pol II is the most complicated of the three enzymes, containing a large number of subunits, listed in **Table 9.2**.

RNA polymerization has three stages. *Initiation* is the search for a suitable promoter site, together with a helicase action to unwind DNA and uncover the single-stranded template. *Elongation* is the polymerization itself, in a completely processive manner (Chapter 4), and *termination* is the recognition of a termination sequence and subsequent detachment.

TABLE 9.2 The constituents of yeast RNA polymerase II

Protein	Number of components	Role
Pol II	12	Polymerase
TFIIA	2	Stabilizes TBP and TFIID binding. Blocks transcription inhibitors. Positive and negative gene regulation
TFIIB	1	Binds TBP, Pol II, and DNA. Helps determine start site
TFIID TBP	1	Binds TATA element and bends DNA. Platform for assembly of TFIIB, TFIIA, and TAFs
TFIID TAFs	14	Binds INR and DPE promoters. Target of regulatory factors
Mediator	24	Binds cooperatively with Pol II. Kinase and acetyltransferase activity. Stimulate basal and activated transcription
TFIIF	3	Binds Pol II and is involved in Pol II recruitment to PIC and in open complex formation
TFIIE	2	Binds promoter near transcription start. May help open or stabilize the transcription bubble in the open complex
TFIIH	10	Transcription and DNA repair. Kinase and two helicase activities. Essential for open complex formation
SAGA TAFs	5	Unknown
SAGA Spts, Adas, Sgfs	9	Structural. Interact with TBP, TFIIA, and Gcn5
SAGA Gcn5	1	Histone acetyltransferase
SAGA Tra1	1	Large activator protein. Part of the NuA4 HAT complex
SAGA Ubp8	1	Ubiquitin protease

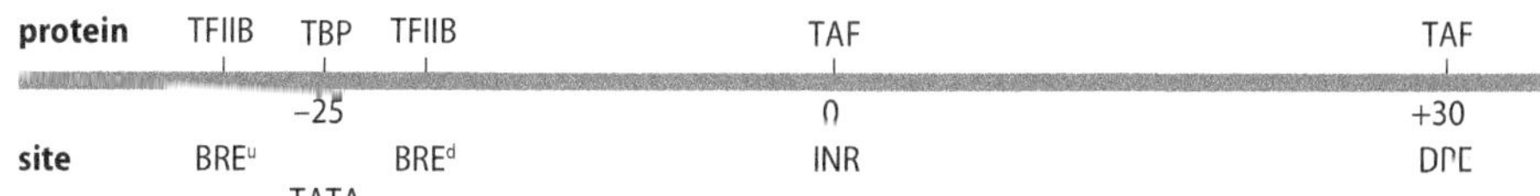

FIGURE 9.7
Binding sites on DNA for components of the Pol II complex. Numbering is relative to the transcription start site. Above the DNA are indicated the proteins that bind to DNA: TBP (TATA-binding protein) and the 14 proteins of the TAF complex together comprise the TFIID complex (see Table 9.2). There are two possible BRE sequences, either upstream or downstream (u and d, respectively) of the TATA sequence. The TATA sequence is symmetrical: correct orientation onto the DNA therefore comes (partly) from binding to the other elements. Several other recognition sites have also been described, most notably the DCE and MTE sites just downstream of the transcription start site.

As we should expect, the eukaryotic proteins are homologous to the prokaryotic ones, and in addition they have a large number of extra proteins, whereas archaea have an intermediate complexity, having only two general transcription factors, TBP (TATA-binding protein) and TFB (similar to Pol II TFIIB), whose function is to attach the polymerase to DNA. In all cases, initiation requires the binding of transcription factors and Pol II to the correct DNA site in the correct orientation. All the specific recognition is done by transcription factors, which bind first and prepare the way for Pol II to bind later. There are generally four DNA sequences that are recognized, these being the TATA element (recognized by TBP and containing the sequence TATA, about 25 nucleotides upstream of the transcription start), BRE (recognized by TFIIB, about 5 nucleotides away from the TATA element), Inr (initiator, at the start site and thus 25 nucleotides downstream from TATA), and DPE (downstream promoter element, found about 30 nucleotides into the transcription sequence) (**Figure 9.7**). It is not necessary for all four to be present.

Normally the first protein to bind to DNA is TBP (part of the TFIID complex; see Table 9.2), which is roughly saddle-shaped and binds to the TATA element, bending it by about 80° (**Figure 9.8**). This interaction requires the presence of the TATA element and does not occur in its absence (**Figure 9.9**) [40]. TBP is usually associated with the TAF (TBP-associated factor) complex, which also interacts with the Inr and DPE DNA elements and helps to orient it correctly: the TBP–TAF complex is called TFIID (see Figure 9.9). The TAFs form an example of the 'dater' components described earlier, in that they can also be found associated with other complexes, for example those involved in chromatin modification (the SAGA complex; see Table 9.2). Many of these form protein heterodimers.

TBP then interacts with TFIIB on one side and TFIIA on the other, which also help to make sure that it is bound in the correct orientation. TFIIA and TFIIB recognize both TBP and the bent DNA.

Interestingly, the assembled TFIID complex has an autoinhibitory mechanism similar to those discussed in Chapter 8. A TAF domain binds at the DNA recognition surface of TBP and inhibits DNA binding. Binding of TFIIA to this domain of TBP therefore removes the inhibition and enhances the binding to DNA [41].

Thus, binding at the correct location is achieved not by a single highly specific interaction but by several weaker interactions that work cooperatively. As we have seen numerous times already, this achieves equally strong overall binding, but does it with a faster on- and off-rate.

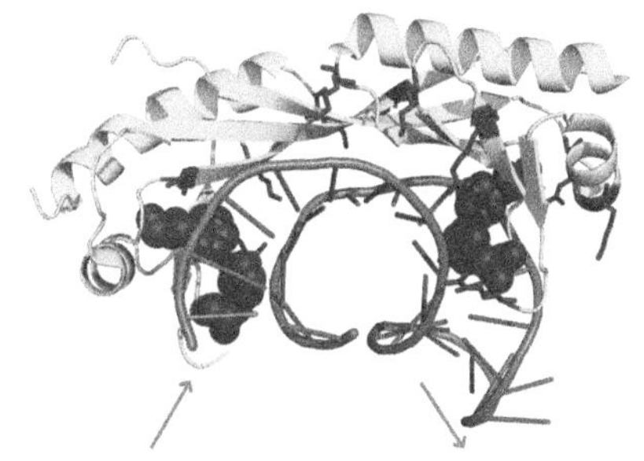

FIGURE 9.8
The structure of the complex of TBP with the TATA site (PDB file 1ytb). The DNA double helix bends by roughly 80°, as indicated by the gray arrows. TBP interacts nonspecifically with the DNA using a number of arginines and lysines (red). The bending in the DNA is induced by the insertion of four phenylalanine rings (blue) in between DNA bases. The sequence-specific recognition is due very largely to the unusually high deformability of the 5′-TATA sequence. The TBP has pseudo-twofold symmetry, presumably as a result of ancestral gene duplication.

At this point the polymerase itself attaches, as a complex with TFIIF, itself a heterodimer. Most of the binding interactions involve TFIIF rather than Pol II itself. We can thus think of Pol II as being dragged into the complex, and onto the DNA, by the transcription factors. Pol II has 12 subunits: 5 of these are homologous to subunits of the bacterial polymerase as well as Pol I and Pol III, 4 more have close homologs in Pol I and Pol III, and the remaining 3 are present only in Pol II, and are the least tightly bound. Two of them are required only for formation of the initiation complex, and dissociate during elongation. TFIIF is homologous to the bacterial σ factor.

This is followed by TFIIE and TFIIH, the latter of which is a large protein complex that contains both a helicase (to unwind DNA) and the CDK7–cyclin H kinase complex. This important complex helps to attach the polymerase to

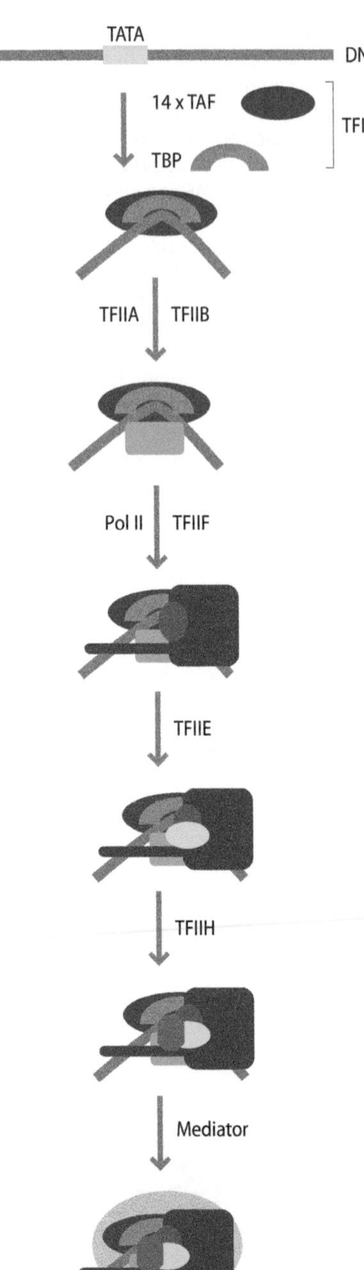

FIGURE 9.9
Assembly of the pre-initiation complex. When a TATA box is present (only in about 10–20% of genes), the TBP binds first. If it is not present, TAFs bind other recognition elements (Figure 9.7), and recruit the TBP. TBP binding is essential because it bends the DNA (Figure 9.8). This complex then binds to TFIIA, which recognizes TBP; and to TFIIB, which recognizes bent DNA, particularly the BRE elements. This TFIID–TFIIA–TFIIB complex then binds to a complex containing Pol II (brown; a 12-protein complex) plus TFIIF (blue/green; a two-protein complex related to bacterial σ factors). Pol II has a long C-terminal domain (CTD), which is extended. Being proline-rich it interacts with several other transcription factors. TFIIE then binds and may help to unwind the DNA at the promoter site. Next, TFIIH binds, interacting with TFIIE. This is a large complex containing helicase activity (to unwind the DNA) and kinase activity (to phosphorylate the CTD). Finally, the Mediator complex binds, wrapping itself around most of the rest of the Pol II complex. Mediator also interacts with other transcription factors, some of which can be many kilobases away. This completes the pre-initiation complex; priming occurs by the binding and hydrolysis of ATP, which opens up the transcription bubble, and by phosphorylation of the CTD, which releases it from binding to DNA–transcription factor complexes.

the template DNA strand, unwinds the DNA, phosphorylates serine residues in the C-terminal domain (CTD) of Pol II, and performs nucleotide excision repair on damaged DNA. Finally the protein complex Mediator binds, wrapping itself around the entire assembly and forming the site of interaction with enhancers, which can be far upstream or downstream from the binding site. Mediator is composed of a large number of different proteins, probably with different compositions depending on growth conditions. This entire assembly is generally known as the pre-initiation complex (PIC) (see Figure 9.9).

Thus, the assembly of the PIC is an ordered process, in which different components come together in roughly a fixed order. However, it is not completely ordered. The existence of non-canonical promoters (which for example do not contain the TATA element) implies that there must be several alternative routes of assembly, and probably a flexible interface between some of the components to allow this variety in assembly.

9.3.2 There is an electron microscopy structure of the pre-initiation complex

There is no high-resolution structure of the PIC. It is probably inherently rather flexible. Partly this is because the fact that Pol II can attach at sites lacking the TATA element means that there must be some flexibility in the way the complex interacts with DNA. It is also clear that some of the TAFs bind initially to the promoter but then move to allow other proteins to interact during formation of the PIC. There are, however, crystal structures of Pol II (free, and bound to TFIIB and DNA/RNA hybrids) [42] and of several component complexes, and electron microscopy structures of several complexes including TFIID and the Pol II–Mediator complex: see **Figure 9.10**, for example [43]. Because of the way in which the electron microscopy images are generated (as described in Chapter 11), this must mean that the complex is a reasonably defined structure. Remarkably, all of the 'preparation' of the DNA ready for strand separation is done by the general transcription factors; there are no direct interactions of promoter DNA with Pol II [44].

Pol II itself has a large cleft in the center into which double-stranded DNA enters, where it is bent sharply (presumably helping the separation and pairing with RNA), separated, and paired with the growing RNA strand (**Figure 9.11**) [45]. The cleft thus houses the entire transcription bubble. This cleft is covered by a 'clamp' structure formed from several subunits, which is able to

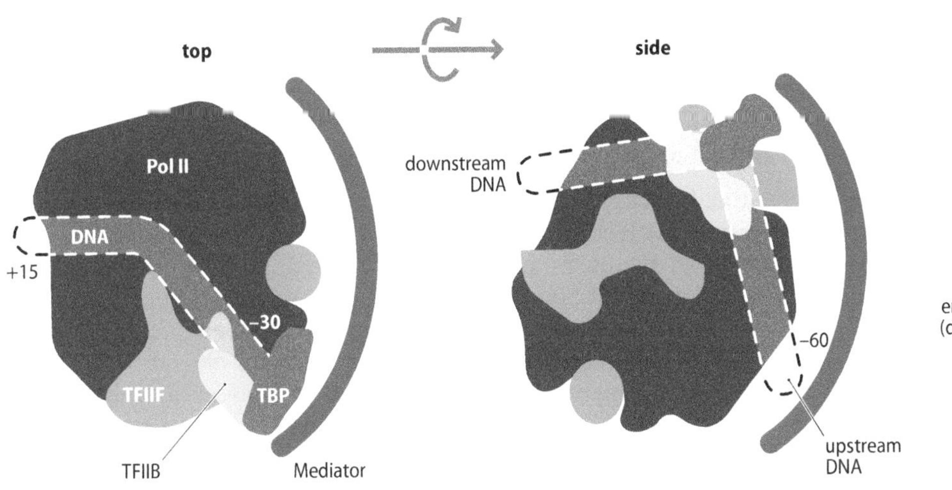

FIGURE 9.10
A model of the complex of Pol II–TFIIF with TFIIB and TBP. Pol II is shown in brown and TFIIF in cyan: the structure of these two comes from electron microscopy. The identification of TFIIF comes partly from structures of σ factors in the bacterial holoenzyme. The structures of TFIIB, TBP, and DNA come mainly from the crystal structure, while the rest of the DNA (dashed lines) is modeled in. The position of Mediator is based on electron microscopy.

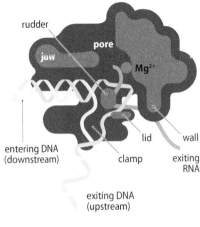

FIGURE 9.11
Top view of Pol II. DNA is in dark green (template strand) and light green (nontemplate strand) and RNA is blue. (From S. Hahn, *Nat. Struct. Mol. Biol.* 11: 394–403, 2004. With permission from Macmillan Publishers Ltd.)

move up to 30 Å, not least so as to bind onto DNA at initiation and release the DNA strand at termination. The remaining proteins within the PIC seem to be grouped around Pol II in a fairly tight bundle.

9.3.3 The C-terminal domain is a key component during elongation

Pol II contains an unusual section known as the C-terminal domain (CTD). The CTD contains a large number of repeats of the seven-residue sequence YSPTSPS (52 repeats in mammals and 26 in yeast; in general, more complex organisms have a longer CTD). The prolines occur every three or four residues and help to form it into what is presumed to be a **polyproline II helix**, which would therefore have the prolines roughly all on one face, with Ser 2 and Ser 5 adjacent on the same face of the helix. This is its conformation when bound to the peptidyl-proline isomerase Pin1, for example [46]. As an extended sequence, this makes the CTD roughly 1200 Å long, or eight times the length of the rest of Pol II [47]. It is thought to form a proline-rich platform or scaffold (Chapter 4) onto which a variety of transcription factors, initiation, elongation, termination, and mRNA-processing proteins (capping, 3′ end processing, and splicing) can bind, which has been called a 'transcription factory.' It is also very likely to bind to proteins that interact with DNA, for example to disrupt chromatin structure so as to allow access for the polymerase complex.

TFIIH contains two kinases, which can phosphorylate Ser 5 all along the CTD. This destabilizes interactions between the CTD and several of the transcription factors (including Mediator, which not only wraps around the PIC but also binds several of its components), thereby allowing Pol II to detach from the PIC and start the process of elongation (**Figure 9.12**). It is thought that many of these factors are left behind at the promoter site, and may remain there in a scaffold complex ready to start another round of assembly.

Once elongation is in progress, the active complex contains Pol II, TFIIB, and TFIIF, and seems to have a well-defined structure. It is able to bind to a range of new proteins, many via the Ser 5-phosphorylated CTD. Interestingly, in the complex with the mRNA capping enzyme Cgt1 the CTD does not have a

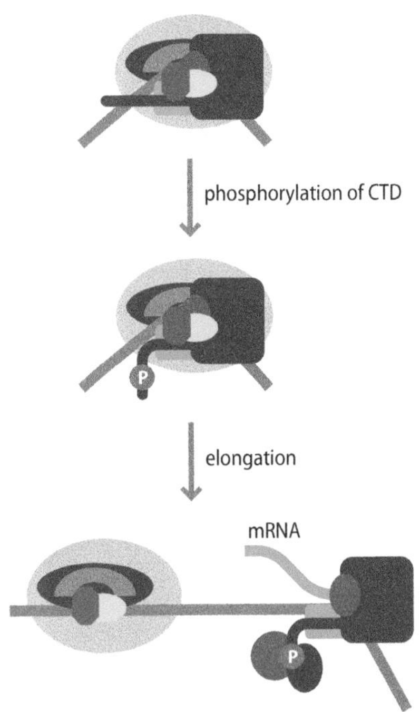

FIGURE 9.12
Elongation of Pol II. The pre-initiation complex is liberated by formation of the transcription bubble and phosphorylation of the CTD. The Pol II–TFIIF complex, together with TFIIB and TFIIF, then detach from the rest of the complex and move along DNA, transcribing mRNA. In this diagram the CTD is shown interacting with additional proteins.

polyproline II helix, but an extended β structure. This has a roughly 180° alternating structure per residue, putting the side chains in a quite different orientation from that in the Pin1 complex [48]. Thus, the CTD is able to adapt its structure to bind to different targets (or, more exactly, the structure is flexible enough to allow it to recognize a number of targets: see Chapter 6).

The proteins that bind to the Ser 5-phosphorylated CTD after release from the PIC include those involved in mRNA 5′ capping. It seems likely that there is an ordered series of events, in metazoans although possibly not in yeast [49]. Elongation proceeds about 30 nucleotides, long enough to generate a free 5′ end, and then pauses until the capping proteins are recruited and have added the 5′ cap. The 5′ cap then recruits a kinase, which further phosphorylates the CTD at Ser 2 once capping is complete. It also phosphorylates another protein that holds the CTD onto the promoter, which weakens the interaction, thus facilitating the release of Pol II from the promoter region and allowing the binding of new proteins, including those involved in RNA splicing and polyadenylation and in termination. There is also a phosphatase that removes Ser 5 phosphates and helps to free the CTD for further elongation. There must also be at least one other phosphatase linked to termination that removes all the phosphates ready for another round.

The CTD thus acts as a temporary holding bay where useful proteins can be kept until required, regulated by phosphorylation [50]. It has been suggested to be the platform that holds the proteins that allow a coupling between transcription and RNA processing. In common with many of the interactions mediated by proline-rich sequences, the binding is mediated by multiple weak interactions and is therefore not strongly dependent on the exact sequence and details of phosphorylation. There is no single region of the CTD that is crucial for its function: instead, it is the overall length and number of repeats that matter.

In summary, the Pol II complex is not unlike the exosome: a core complex associates first and then binds to the enzyme, and acts to direct and regulate the action of the enzyme. The complex assembles a large group of other enzymes, which come and go regulated by phosphorylation of the CTD. The 'core' is relatively small, consisting of only Pol II, TFIIB, and TFIIF, but the number of other proteins that can be attached is very large. Like the exosome, it has essentially just a single enzyme function, but a large assortment of extra activities. This complex truly is a molecular machine.

9.4 THE METABOLON CONCEPT

9.4.1 Metabolons are controversial

The complexes discussed so far are justly called molecular machines, because they perform a single function that involves considerable coordination. In this final section, I discuss a rather different type of complex, namely one in which a group of related enzymes function together to conduct a series of reactions. This has been given the name of *metabolon*.

In a very cogent and compelling review in 1987, Paul Srere argued that many groups of enzymes within common metabolic pathways are physically associated together into metabolons [51] (as did Welch in a review 10 years earlier [52]). This argument makes a great deal of sense. First, there is a 'theoretical' reason, which is based on metabolic pathway charts such as that shown in **Figure 9.13**. The point is that many pathways have several sequential steps with no side branches: the pathway exists only to make the end product. There is therefore no value to the cell in making the intermediates and allowing them to diffuse around the cell: it would be far more economical to have some

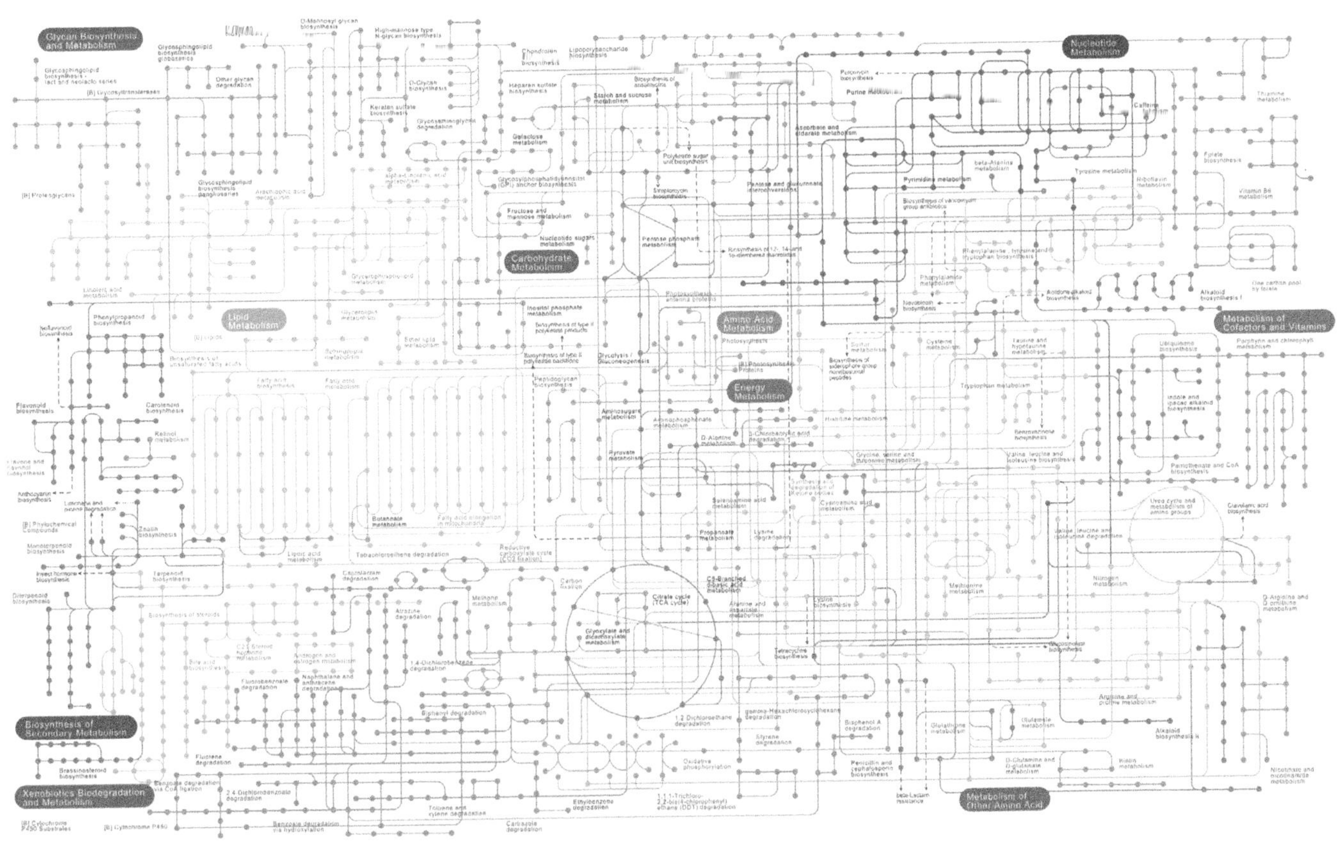

FIGURE 9.13
A representation of the main human metabolic pathways. Carbohydrate metabolism is in blue; the circle in the lower center represents the TCA cycle. (Taken from KEGG, the Kyoto Encyclopedia of Genes and Genomes (http://www.genome.jp/kegg/) [81].)

mechanism whereby the intermediates were channeled directly through the pathway without being lost. About 80% of the metabolites in Figure 9.13 fall into this category. Such a metabolon would also facilitate tighter regulation of the biosynthetic pathway, and presumably make it go faster. Analyses using artificial gene fusions have indeed shown rate enhancements [53]. The same also applies to degradative (catabolic) pathways, which are mainly not shown in Figure 9.13. However, an important counter-argument has been made [54], namely that *"we must be careful not to presume that effects we find attractive must have been adopted by nature"*: merely because it would be attractive for cells to do this does not mean it actually happens (see also the quotes at the head of this chapter, particularly the third one).

Second, there are a number of experimental observations. At the time of the Srere review in 1987, there were relatively few relevant results to cite, but there were a wide range of different types of result. There are many more now. The most compelling relate to colocalization or copurification of related enzymes, and to evidence for channeling. Colocalization of enzymes is necessary but not sufficient evidence for a metabolon. For example, there is clear evidence that the two enzymes responsible for the final steps in cysteine biosynthesis associate in the cell; however, this now seems to be a regulatory mechanism, because in the complex one of the enzymes is almost inactive [55]. I discuss these two arguments in turn.

The metabolon concept has generated considerable controversy. As noted above, it is such an attractive idea that it 'has' to be true. On the other hand, experimental evidence has been hard to produce. For observation both of colocalization and of channeling, the fragility of metabolon complexes makes it necessary to perform the experiments in whole cells *in vivo*, which is usually what makes the experiments difficult. In addition, metabolons are very likely to be metabolically regulated assemblies, and thus transient and easily dissociated, which may explain why they are difficult to see.

9.4.2 Colocalization provides evidence for metabolons

There are six enzymes (three of which are multifunctional) that catalyze purine biosynthesis in eukaryotes. *In vitro*, no colocalization is observed. However, fluorescent labeling demonstrated that these enzymes colocalize *in vivo* and that this localization can be disrupted in response to purine levels [56]. This provides strong evidence not only that some association exists, but also that the general lack of evidence for association in other systems need not be seen as compelling negative evidence. Similar colocalization has been seen for a range of other metabolic systems after nutrient starvation, although the phenomenon was attributed to 'depots' rather than metabolons [57]. The suggestion of some kind of association is supported by some kinetic evidence of channeling. Photobleaching of mitochondria containing TCA enzymes tagged with green fluorescent protein suggested that the TCA enzymes existed as a high-molecular-mass complex [58], which was immobilized in about half of the cases. Many other papers have reported complexes of TCA enzymes, for example that by Sumegi et al. [59].

There have been numerous unrelated suggestions that complexes of metabolic enzymes in eukaryotic cells may be localized to membranes, in some cases at **lipid rafts** [60], and in others associated with membrane proteins. As in many areas of this topic, the results are controversial [55, 61]. Three of these are described below.

1. Srere has shown that a model can be constructed of three enzymes from the TCA cycle that has an electrostatic channel connecting the active sites, which explains the kinetics seen in a fusion protein [62, 63]. Significantly, these enzymes have been shown to bind to the inner surface of the mitochondrial inner membrane and have been copurified (together with two more TCA enzymes) as a catalytically active complex. (See also [64], which provides a much earlier example of purification of a complex of five TCA enzymes.)
2. The enzymes of the Calvin cycle, the enzymes involved in the biosynthesis of cyanogenic glucosides (which have toxic intermediates), and enzymes involved in the biosynthesis of phenylpropanoids have also been suggested to associate (separately) at a membrane surface *in vivo* [55].
3. Remarkably, centrifugation of intact *Neurospora* or *Euglena* cells resulted in the stratification of cell contents, in which *all* the enzymes of central metabolic pathways were found not in the cytoplasmic layer but in the membrane layer (discussed in [1]). The most obvious explanation of this result is that all the central metabolic enzymes are not in fact floating freely in the cytoplasm but are in some way attached to membrane surfaces.

Are these disparate observations that have alternative explanations, or are they part of a coherent picture showing regulated membrane colocalization of metabolons? It is not yet clear.

9.4.3 Channeling provides evidence for metabolons

Channeling is clearly observed in some multienzyme complexes, as discussed in Chapter 10. In such complexes, the substrate is passed directly from one enzyme to the next and is never free to diffuse within the solution. This is the essential feature that constitutes channeling, and provides prima-facie

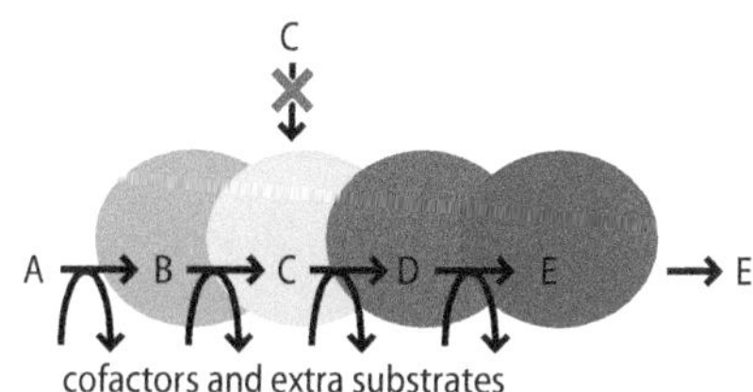

FIGURE 9.14
Channeling as evidence for a metabolic complex or metabolon. Substrate (compound A) is fed in at one end of the complex, and product (compound E) exits from the other. A highly channeled complex has no other entry or exit routes (except, of course, for other substrates and products), and therefore feeding intermediate C exogenously has no effect, either on the rate or, for example, on the isotopic labeling pattern of the product.

evidence of a complex. Channeling can be observed in several ways, most obviously using isotope labels. For example, labeled substrate (for the first enzyme) is added, together with a high concentration of unlabeled intermediate. If the intermediates are completely channeled, there will be no dilution of the isotopic label in the final product; however, if it diffuses freely, the label will be diluted (**Figure 9.14**). Evidence for channeling has been reviewed [55, 61, 65, 66]. As one would expect, the best evidence for channeling is found in multienzyme complexes, where the structures included tunnels that facilitate such channeling. In contrast, for metabolons the evidence is more patchy. Significantly, the best evidence for channeling comes from membrane-bound complexes. This applies not only to TCA cycle enzymes, but also to the urea cycle [67, 68].

There are many other lines of evidence that suggest metabolic complexes. Two are presented here, not because they are especially convincing but because I like them.

1. One way to detect interactions between proteins is to add a protease, and see whether the rate of proteolytic digestion is affected by addition of the partner. Such changes are indeed observed *in vivo*: for example, the digestion of cytochrome *c* is slowed in the presence of cytochrome $a.a_3$, implying an interaction [69].
2. The brine shrimp lives in tidal waters, and at low tide it can get dehydrated. It can be dehydrated to only 35% of normal water content, at which point there is effectively no free water at all inside the cell, only water of hydration coating the molecules inside. Yet the cells still function [70, 71]. If there is no free water to permit the diffusion of metabolites, how is this possible, unless in fact diffusion on a large scale is not required because everything is organized into metabolons?

9.4.4 High-throughput methods provide no evidence for metabolons

The advent of genome-wide tools for measuring protein interactions allows us to look for metabolons in a much more methodical way. Now that we have a much larger array of experimental tools, do these arguments still stand up? An argument made forcefully by Srere, and by many others before and since, is that it is difficult to experimentally detect complexes that are present inside cells. This is because the environment inside the cell is very crowded, which promotes the association of proteins (Chapter 4). As soon as you break open the cell to examine complexes, the complexes dissociate. Thus, even if such complexes are there *in vivo*, they will not be observed experimentally.

This is a good argument (although difficult to argue against!), but we now know that intracellular protein complexes *can* be observed in this way, using for example the TAP-tag method, which involves the formation of complexes *in vivo* but still requires cells to be broken open for the purification and detection of the complexes. So the obvious question is: do we observe metabolons with TAP-tag technology?

Basically, the answer is no. In the yeast whole-genome experiments described at the start of this chapter, there were certainly some metabolic enzyme complexes observed, but these were mostly specialized (and known) multienzyme complexes such as tryptophan synthase, carbamoyl phosphate synthase, pyruvate dehydrogenase, and 2-oxoglutarate dehydrogenase (discussed in the next chapter). In [6], out of the 491 complexes identified with confidence, the only complexes that could make any claim to be metabolons were trehalose-6-phosphate synthase/phosphatase, anthranilate synthase, phosphoribosyl diphosphate synthase, α-aminoadipate-semialdehyde dehydrogenase, and ribonucleoside diphosphate reductase, which are all single-function enzymes. There were no biosynthetic or catabolic complexes, with the exception of a single complex that included the glycolytic (and related) enzymes TIM, enolase, fructose-1,6-bisphosphate aldolase, phosphoglycerate mutase, phosphoglycerate kinase, and glycerol-3-phosphatase, together with several unrelated proteins. The TAP-tag method has an identified bias in that it over-represents complexes that contain high-abundance proteins, and the glycolytic enzymes are among the most abundant proteins in the cell. Therefore the observation of the glycolytic complex could either be an artifact due to the high abundance of these proteins, or it could imply that there are many other such metabolons present, which were not detected because they fell apart before they could be identified. Of course, TAP-tagging does identify many other complexes, so if the latter argument is true, it must be the case that metabolons are held together much more weakly than the molecular machines discussed above. This is very likely to be so, because their formation is quite likely to be reversibly stimulated by a range of metabolic and cellular factors. Weak complexes are certainly physiologically relevant: it has been estimated that physiologically important interactions can have affinities as weak as 100 μM *in vivo*, and potentially even weaker once the cell has been broken open [72]. Problem N3 shows that such complexes will not be observed by TAG-tag experiments.

It is possibly significant that in TAP-tag experiments probing the exosome, no association has been found between the exosome and the catalytic nuclease Rrp44 in human or *T. brucei* exosomes. For reasons given above, it seems hard to believe that these exosomes do not associate with Rrp44 (or some other nuclease) *in vivo*, because otherwise they would have no catalytic function and therefore no reason to exist. So it is certainly true that TAP-tagging does not observe all the complexes that there are. As discussed above, two-hybrid screens are inherently better at finding weaker interactions, and have indeed found many relevant examples; however, they also have a tendency to find false positives, so it is currently hard to tell whether or not these are real.

Thus, the tentative conclusion to be drawn from TAP-tagging is that metabolons are not widespread, if they exist at all; however, this may be an erroneous conclusion.

9.4.5 There is reasonably good evidence for a glycolytic metabolon

In Section 9.4.4 we noted that the only 'real' metabolon identified from TAP-tagging in yeast contained glycolytic enzymes. Theoretically, glycolysis would be a good system to integrate into a metabolon. It is such a major pathway that one might imagine that the evolutionary pressure to optimize it would be strong enough to have created a metabolon. However, unlike some of the other pathways discussed above, many of the intermediates *are* used for other purposes, implying that a completely channeled metabolon would not be biologically sensible. A strong theoretical argument does not mean that such a metabolon *must* exist. But there is in fact considerable evidence that glycolytic enzymes in a range of organisms may operate in a metabolon.

Most experiments to detect the association of glycolytic enzymes have been negative. However, some have found association. In erythrocytes, antibody staining implied that various glycolytic enzymes associate to the cell membrane, attached to the transmembrane protein band 3, this assembly being regulated by oxygenation and phosphorylation [73]. Notably, such interactions are not measurable *in vitro* at physiological ionic strength. The authors rationalize this difference as being due to macromolecular crowding. In this context it is worth noting that crowding was also invoked as being necessary for producing the rapid transfer of metabolites through enzymes in the *E. coli* phosphotransferase system [74].

In some organisms, the gene for triose phosphate isomerase (TIM) is covalently attached to the gene for phosphoglycerate kinase (PGK) as a result of a frameshift mutation, in which the stop codon of PGK is lost, being replaced by the first codon of a slightly overlapping TIM gene [75]. PGK is in fact not the next enzyme in the glycolytic pathway, it is the one after that, which implies that for this fusion to improve rates, one would need a complex of at least three enzymes within this metabolon. This is an interesting result because, as shown in Chapter 5, TIM is an evolutionarily perfect enzyme, whose rate is limited by diffusion; in other words, the only way to make it faster would be to increase the diffusion rate on or off. Attaching it in a complex with other enzymes in the pathway would do exactly this. Interestingly, however, the PGK activity of the fusion protein was measured to be about three times slower than in isolated PGK, possibly as a consequence of reduced mobility or accessibility in the fusion. This makes the useful point that the assembly of multiple enzymes into a complex does not have unmixed benefits. Nevertheless, it has been observed that artificially created gene fusions do indeed show the expected enhancements of rate and reduced transient time [65], although again this is not proof that they do associate *in vivo*. Of possible relevance is the observation that cases of gene fusion are overwhelmingly found in metabolic enzymes [76]. One presumes that these cases have been at least partly selected by natural selection and therefore have at least a small benefit to the cell; the most obvious benefit is to assist in some type of channeling.

More convincingly, there have also been observations of channeling in glycolytic pathways. In *E. coli*, feeding of cells with fructose 1,6-bisphosphate labeled with ^{14}C gave rise to almost 100% channeling of the carbons through to $^{14}CO_2$, even in the presence of ^{12}C-labeled intermediates, suggesting a complex that includes all the glycolytic enzymes [77]. In a particularly interesting set of experiments, Graham et al. showed using ^{13}C labeling that, in *Arabidopsis* and potato, channeling is close to 100% for three enzyme pairs, and about 50% for two more, and that the ^{13}C label is carried through to TCA intermediates; this requires channeling all the way through to pyruvate and then coordinated transport across the mitochondrial outer membrane [78]. In support of this conclusion, they also showed colocalization of the glycolytic enzymes, which bound to the outer mitochondrial membrane protein VDAC (voltage-dependent anion channel). Most strikingly, they showed that the glycolytic enzymes existed in two pools: one that was apparently free in the cytosol, and one that was attached to the mitochondrial membrane. Enzymes moved from the cytosolic pool to the membrane-bound complexes according to respiratory demand, so that increased respiration rate attracted more enzyme to the membrane, the response occurring within minutes. They therefore suggest that the complexes bound to the mitochondrial membrane could be metabolically regulated to act as channeled sources of pyruvate to feed across the membrane into the subsequent TCA cycle.

Glycolysis presents the best evidence so far that metabolons do exist, and agrees with a growing body of evidence that metabolons are weak and metabolically regulated complexes, most probably forming on membrane surfaces. (Compare Chapter 4, which shows that membrane binding is a good way to assemble complexes. Srere's 1987 review also concluded that many metabolons are likely to be attached to membranes.) The success of cell-free systems for protein expression, for example, suggests that even systems known to be

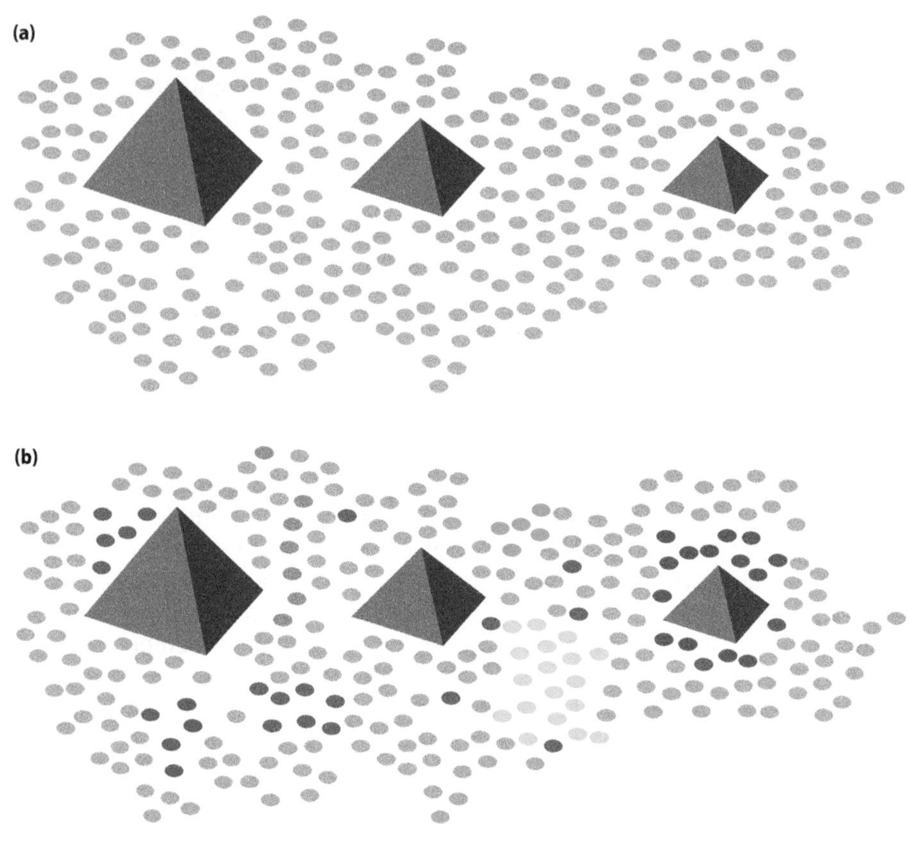

FIGURE 9.15
Do metabolons exist? (a) Tourists grouped around the pyramids may appear to be completely uncorrelated, but (b) after being labeled with a suitable label (language, country of origin, next destination, hotel, and so on) they can be seen to be grouped together in various ways; not all of these groupings may be relevant or reflect genuine commonality.

physically associated *in vivo* function quite happily as a 'bag of enzymes' (to quote a term used derogatively by Srere): metabolons are probably not vital for cell function. Nevertheless, the extensive recent work on molecular machines certainly provides convincing evidence that metabolic processes are spatially coordinated, and it would be surprising if metabolons do not turn out to be more common than suspected so far; in other words, that the TAP-tag results probably do not detect metabolons because their association is indeed weak and metabolically regulated, and possibly because many of the complexes are bound to membrane proteins and therefore are of low abundance in soluble fractions.

I conclude with an analogy borrowed and adapted from [67] (**Figure 9.15**). If we look at proteins within the cell they seem to resemble tourists at a busy tourist site, moving round apparently at random. However, if we label them with an appropriate probe, we can see that in fact most of the tourists are moving in groups, with the groups constantly forming, dissolving and re-forming, using a variety of attractive forces to hold them together. It remains to be seen how valid this analogy is. One need is to develop appropriate labels that can follow the proteins *in vivo*.

9.5 SUMMARY

Genome-wide studies suggest that at least 65% of proteins in the cell work as part of a complex. The complexes typically contain from two to four members, although some are much larger. Larger complexes make up molecular machines, which tend to contain 'hubs' surrounded by 'daters' who go round in groups. These correspond to the core catalytic function and associated regulatory functions, respectively.

In this chapter, we have looked at two molecular machines in detail. The exosome contains a single main catalytic protein, and several other proteins whose role is to limit its proteolytic activity to particular substrates and chain lengths. Part of the complex consists of proteins that have homology with bacterial exonucleases but have now lost this function. The complex has therefore evolved in a far from linear manner. RNA polymerase II contains a large number of components that are assembled in a consistent, although not entirely fixed, order; many of these components later dissociate once elongation begins, to be replaced by other components. Many of these are held on the C-terminal domain of Pol II, to be called on as needed, with their binding regulated by phosphorylation of the CTD.

Finally, we consider the concept of the metabolon, a weakly associated and metabolically regulated group of metabolic enzymes. This has been a controversial area, and I conclude that the evidence for some metabolons (for example in glycolysis) is convincing, particularly from channeling. Metabolons probably associate with membrane proteins. Further experimental work is needed.

9.6 FURTHER READING

For a good introduction to the interactome, see the paper by Devos and Russell [9].

The review by Srere [51] presents a fascinating (if slightly one-sided) argument for the existence of metabolomes.

9.7 WEBSITES

http://3dcomplex.org/ *Three-dimensional complex.*

http://3did.irbbarcelona.org/ *3DID (interacting domains).*

http://cluspro.bu.edu/login.php *ClusPro (docking).*

http://consurftest.tau.ac.il/ *ConSurf (functional regions in proteins).*

http://dunbrack.fccc.edu/ProtBuD/ *ProtBuD (identifies asymmetric units).*

http://nic.ucsf.edu/asedb/ *ASEdb (alanine scanning energetics: protein hotspots).*

http://ppidb.cs.iastate.edu/ppidb/ *Protein:protein interactions database.*

http://prism.ccbb.ku.edu.tr/prism/ *PRISM (protein interactions).*

http://www.bioinformatics.sussex.ac.uk/protorp/ *PROTORP (protein interfaces).*

http://www.boseinst.ernet.in/resources/bioinfo/stag.html *Proface (several programs).*

http://www.ebi.ac.uk/thornton-srv/databases/cgi-bin/valdar/scorecons_server.pl *Scorecons (scores residue conservation).*

http://www.piqsi.org/ *PiQSi (quaternary structures database).*

9.8 PROBLEMS

Hints for some of these problems can be found on the book's website.

1. **When performing TAP-tag screening, it is common to attach the tag at the N terminus and again independently at the C terminus. Why?**

2. **Look at the list of complexes identified in yeast in Supplementary Table 2 of [6]. You will need to go to the journal's website and find**

the link to Supplementary Information. For the complexes with names (that is, an identified and/or known function, not just for example 'Complex 5'), identify their function, and classify the functions into suitable groups. What are the main functional groups you have identified? Find a list of the main cellular functions: one source of this information is [79]. Do these groups represent the main cellular functions roughly in proportion? If not, why not?

3. In the two papers referenced here on the yeast interactome as studied by TAP-tag [6, 7], what measures were used to check accuracy and completeness of coverage?

4. The global yeast two-hybrid study [15] found that 'daters' were the main type of interacting protein identified and were important for maintaining the integrity of cellular metabolism, whereas the TAP-tag studies found that the dominant types were 'party' hubs. Is there a conflict between these two results?

5. Define false positive and false negative interactions. Scientists are usually more concerned about false positives. Why?

6. What are the two main functions of the exosome? How does its structure facilitate these?

7. Is RNA polymerase II homologous to polymerases I and III? Which components?

8. Almost all of the metabolons identified so far are anabolic not catabolic. Why should this be so, and is this likely to be 'real' or an experimental artifact?

9. Assuming it is true that metabolons assemble on membranes only in response to cellular need, suggest experiments that might be used to identify them.

9.9 NUMERICAL PROBLEMS

N1. The graph of (number of complexes) versus (size of complex) described in [7] looks similar to that drawn in Figure 9.1 and has the equation $y = 2556.9x^{-1.8992}$. How many complexes would you expect of size 4 (that is, composed of four proteins)? How many proteins in a complex does it take to make it unlikely to find one this large (that is, with $y < 1$)?

N2. The equation given in question N1 denotes a scale-free network with a rather small exponent (namely, less than 2). What is the significance of a low value for the exponent? (You cannot readily answer this without searching the literature!)

N3. It is argued that TAP-tagging does not pick up weak complexes because the washing steps wash away weakly bound components; in essence, if a component dissociates during the time taken to complete a wash, it will be lost. In Section 9.4.4 it is stated that physiologically important complexes can have dissociation constants as weak as $100\,\mu M$. To be on the cautious side, if a complex has a dissociation constant of $1\,\mu M$, and I can identify 10% of a complex remaining after washing, roughly how short does my washing step need to be? Is the washing argument valid? Following the arguments of Chapter 4, how could I increase the affinity of my complex components during the wash?

9.10 REFERENCES

1. PA Srere (2000) Macromolecular interactions: tracing the roots. *Trends Biochem. Sci.* 25:150–153.
2. A Kornberg (2003) Ten commandments of enzymology, amended. *Trends Biochem. Sci.* 28:515–517.
3. AR Fersht (1985) Enzyme Structure and Mechanism, 2nd ed. New York: WH Freeman.
4. A Dziembowski & B Séraphin (2004) Recent developments in the analysis of protein complexes. *FEBS Lett.* 556:1–6.
5. AC Gavin, M Bösche, R Krause et al. (2002) Functional organization of the yeast proteome by systematic analysis of protein complexes. *Nature* 415:141–147.
6. AC Gavin, P Aloy, P Grandi et al. (2006) Proteome survey reveals modularity of the yeast cell machinery. *Nature* 440:631–636.
7. NJ Krogan, G Cagney, HY Yu et al. (2006) Global landscape of protein complexes in the yeast *Saccharomyces cerevisiae*. *Nature* 440:637–643.
8. Y Ho, A Gruhler, A Heilbut et al. (2002) Systematic identification of protein complexes in *Saccharomyces cerevisiae* by mass spectrometry. *Nature* 415:180–183.
9. D Devos & RB Russell (2007) A more complete, complexed and structured interactome. *Curr. Opin. Struct. Biol.* 17:370–377.
10. J-DJ Han, N Bertin, T Hao et al. (2004) Evidence for dynamically organized modularity in the yeast protein–protein interaction network. *Nature* 430:88–92.
11. CV Robinson, A Sali & W Baumeister (2007) The molecular sociology of the cell. *Nature* 450:973–982.
12. G Apic, J Gough & SA Teichmann (2001) Domain combinations in archaeal, eubacterial and eukaryotic proteomes. *J. Mol. Biol.* 310:311–325.
13. NN Batada, LD Hurst & M Tyers (2006) Evolutionary and physiological importance of hub proteins. *PLOS Comp. Biol.* 2:748–756.
14. NN Batada, T Reguly, A Breitkreutz et al. (2006) Stratus not altocumulus: a new view of the yeast protein interaction network. *PLOS Biol.* 4:1720–1731.
15. HY Yu, P Braun, MA Yildirim et al. (2008) High-quality binary protein interaction map of the yeast interactome network. *Science* 322:104–110.
16. G Butland, JM Peregrín-Alvarez, J Li et al. (2005) Interaction network containing conserved and essential protein complexes in *Escherichia coli*. *Nature* 433:531–537.
17. L Giot, JS Bader, C Brouwer et al. (2003) A protein interaction map of *Drosophila melanogaster*. *Science* 302:1727–1736.
18. HW Mewes, D Frishman, U Guldener et al. (2002) MIPS: a database for genomes and protein sequences. *Nucleic Acids Res.* 30:31–34.
19. HW Mewes, C Amid, R Arnold et al. (2004) MIPS: analysis and annotation of proteins from whole genomes. *Nucleic Acids Res.* 32:D41–D4.
20. J Gagneur, L David & LM Steinmetz (2006) Capturing cellular machines by systematic screens of protein complexes. *Trends Microbiol.* 14:336–339.
21. J Goll & P Uetz (2006) The elusive yeast interactome. *Genome Biol.* 7:223.
22. SR Collins, P Kemmeren, XC Zhao et al. (2007) Toward a comprehensive atlas of the physical interactome of *Saccharomyces cerevisiae*. *Mol. Cell. Proteomics* 6:439–450.
23. PM Kim, LJ Lu, Y Xia & MB Gerstein (2006) Relating three-dimensional structures to protein networks provides evolutionary insights. *Science* 314:1938–1941.
24. R Jansen, HY Yu, D Greenbaum et al. (2003) A Bayesian networks approach for predicting protein–protein interactions from genomic data. *Science* 302:449–453.
25. B Suter, S Kittanakom & I Stagljar (2008) Two-hybrid technologies in proteomics research. *Curr. Opin. Biotechnol.* 19:316–323.
26. LJ Jensen & P Bork (2008) Not comparable, but complementary. *Science* 322:56–57.
27. RB Russell & P Aloy (2008) Targeting and tinkering with interaction networks. *Nat. Chem. Biol.* 4:666–673.
28. D Reichmann, O Rahat, M Cohen et al. (2007) The molecular architecture of protein–protein binding sites. *Curr. Opin. Struct. Biol.* 17:67–76.
29. B Alberts (1998) The cell as a collection of protein machines: preparing the next generation of molecular biologists. *Cell* 92:291–294.
30. M Schmid & TH Jensen (2008) The exosome: a multipurpose RNA-decay machine. *Trends Biochem. Sci.* 33:501–510.
31. E Lorentzen, J Basquin & E Conti (2008) Structural organization of the RNA-degrading exosome. *Curr. Opin. Struct. Biol.* 18:709–713.
32. D Schaeffer, B Tsanova, A Barbas et al. (2009) The exosome contains domains with specific endoribonuclease, exoribonuclease and cytoplasmic mRNA decay activities. *Nat. Struct. Mol. Biol.* 16:56–62.
33. E Lorentzen, A Dziembowski, D Lindner et al. (2007) RNA channelling by the archaeal exosome. *EMBO Rep.* 8:470–476.
34. HW Wang, J Wang, F Ding et al. (2007) Architecture of the yeast Rrp44-exosome complex suggests routes of RNA recruitment for 3′ end processing. *Proc. Natl. Acad. Sci. USA* 104:16844–16849.
35. E Lorentzen, J Basquin, R Tomecki et al. (2008) Structure of the active subunit of the yeast exosome core, Rrp44: diverse modes of substrate recruitment in the RNase II nuclease family. *Mol. Cell* 29:717–728.
36. C Schneider, E Leung, J Brown & D Tollervey (2009) The N-terminal PIN domain of the exosome subunit Rrp44 harbors endonuclease activity and tethers Rrp44 to the yeast core exosome. *Nucleic Acids Res.* 37:1127–1140.
37. KP Callahan & JS Butler (2008) Evidence for core exosome independent function of the nuclear exoribonuclease Rrp6p. *Nucleic Acids Res.* 36:6645–6655.
38. JA Stead, JL Costello, MJ Livingstone & P Mitchell (2007) The PMC2NT domain of the catalytic exosome subunit Rrp6p provides the interface for binding with its cofactor Rrp47p, a nucleic acid-binding protein. *Nucleic Acids Res.* 35:5556–5567.
39. QS Liu, JC Greimann & CD Lima (2006) Reconstitution, activities, and structure of the eukaryotic RNA exosome. *Cell* 127:1223–1237.
40. NA Woychik & M Hampsey (2002) The RNA polymerase II machinery: structure illuminates function. *Cell* 108:453–463.
41. T Kokubo, MJ Swanson, JI Nishikawa et al. (1998) The yeast TAF145 inhibitory domain and TFIIA competitively bind to TATA-binding protein. *Mol. Cell. Biol.* 18:1003–1012.
42. AL Gnatt, P Cramer, JH Fu, DA Bushnell & RD Kornberg (2001) Structural basis of transcription: an RNA polymerase II elongation complex at 3.3 Å resolution. *Science* 292:1876–1882.
43. WH Chung, JL Craighead, WH Chang et al. (2003) RNA polymerase II/TFIIF structure and conserved organization of the initiation complex. *Mol. Cell* 12:1003–1013.
44. DA Bushnell, KD Westover, RE Davis & RD Kornberg (2004) Structural basis of transcription: an RNA polymerase II-TFIIB cocrystal at 4.5 angstroms. *Science* 303:983–988.
45. S Hahn (2004) Structure and mechanism of the RNA polymerase II transcription machinery. *Nat. Struct. Mol. Biol.* 11:394–403.
46. MA Verdecia, ME Bowman, KP Lu et al. (2000) Structural basis for phosphoserine-proline recognition by group IV WW domains. *Nat. Struct. Biol.* 7:639–643.

47. SM Carty & AL Greenleaf (2002) Hyperphosphorylated C-terminal repeat domain-associating proteins in the nuclear proteome link transcription to DNA/chromatin modification and RNA processing. *Mol. Cell. Proteomics* 1:598–610.

48. C Fabrega, V Shen, S Shuman & CD Lima (2003) Structure of an mRNA capping enzyme bound to the phosphorylated carboxy-terminal domain of RNA polymerase II. *Mol. Cell* 11:1549–1561.

49. G Orphanides & D Reinberg (2002) A unified theory of gene expression. *Cell* 108:439–451.

50. CD Lima (2005) Inducing interactions with the CTD. *Nat. Struct. Mol. Biol.* 12:102–103.

51. PA Srere (1987) Complexes of sequential metabolic enzymes. *Annu. Rev. Biochem.* 56:89–124.

52. GR Welch (1977) On the role of organized multienzyme systems in cellular metabolism: a general synthesis. *Prog. Biophys. Mol. Biol.* 32:103–191.

53. J Ovádi & PA Srere (1992) Channel your energies. *Trends Biochem. Sci.* 17:445–447.

54. JR Knowles (1991) Calmer waters in the channel. *J. Theor. Biol.* 152:53–55.

55. BSJ Winkel (2004) Metabolic channeling in plants. *Annu. Rev. Plant Biol.* 55:85–107.

56. SG An, R Kumar, ED Sheets & SJ Benkovic (2008) Reversible compartmentalization of de novo purine biosynthetic complexes in living cells. *Science* 320:103–106.

57. R Narayanaswamy, M Levy, M Tsechansky et al. (2009) Widespread reorganization of metabolic enzymes into reversible assemblies upon nutrient starvation. *Proc. Natl. Acad. Sci. USA* 106:10147–10152.

58. PM Haggie & AS Verkman (2002) Diffusion of tricarboxylic acid cycle enzymes in the mitochondrial matrix *in vivo*: evidence for restricted mobility of a multienzyme complex. *J. Biol. Chem.* 277:40782–40788.

59. B Sumegi, AD Sherry & CR Malloy (1990) Channeling of TCA cycle intermediates in cultured *Saccharomyces cerevisiae*. *Biochemistry* 29:9106–9110.

60. SW Martin, BJ Glover & JM Davies (2005) Lipid microdomains—plant membranes get organized. *Trends Plant Sci.* 10:263–265.

61. K Jørgensen, AV Rasmussen, M Morant et al. (2005) Metabolon formation and metabolic channeling in the biosynthesis of plant natural products. *Curr. Opin. Plant Biol.* 8:280–291.

62. C Velot, MB Mixon, M Teige & PA Srere (1997) Model of a quinary structure between Krebs TCA cycle enzymes: a model for the metabolon. *Biochemistry* 36:14271–14276.

63. AH Elcock & JA MaCammon (1996) Evidence for electrostatic channeling in a fusion protein of malate dehydrogenase and citrate synthase. *Biochemistry* 35:12652–12658.

64. SJ Barnes & PDJ Weitzman (1986) Organization of citric acid cycle enzymes into a multienzyme cluster. *FEBS Lett.* 201:267–270.

65. RJ Conrado, JD Varner & MP DeLisa (2008) Engineering the spatial organization of metabolic enzymes: mimicking nature's synergy. *Curr. Opin. Biotechnol.* 19:492–499.

66. M Milani, A Pesce, M Bolognesi et al. (2003) Substrate channeling—molecular bases. *Biochem. Mol. Biol. Educ.* 31:228–233.

67. P Mentré (1995) L'eau dans la Cellule [Water in the Cell]. Paris: Masson.

68. C-W Cheung, NS Cohen & L Raijman (1989) Channeling of urea cycle intermediates *in situ* in permeabilized hepatocytes. *J. Biol. Chem.* 264:4038–4044.

69. DA Pearce & F Sherman (1995) Diminished degradation of yeast cytochrome *c* by interactions with its physiological partners. *Proc. Natl. Acad. Sci. USA* 92:3735–3739.

70. AB Fulton (1982) How crowded is the cytoplasm? *Cell* 30:345–347.

71. JS Clegg (1981) Metabolic consequences of the extent and disposition of the aqueous intracellular environment. *J. Exp. Zool.* 215:303–313.

72. ML Dustin, DE Golan, DM Zhu et al. (1997) Low affinity interaction of human or rat T cell adhesion molecule CD2 with its ligand aligns adhering membranes to achieve high physiological affinity. *J. Biol. Chem.* 272:30889–30898.

73. ME Campanella, HY Chu & PS Low (2005) Assembly and regulation of a glycolytic enzyme complex on the human erythrocyte membrane. *Proc. Natl. Acad. Sci. USA* 102:2402–2407.

74. JM Rohwer, PW Postma, BN Kholodenko & HV Westerhoff (1998) Implications of macromolecular crowding for signal transduction and metabolite channeling. *Proc. Natl. Acad. Sci. USA* 95:10547–10552.

75. H Schurig, N Beaucamp, R Ostendorp et al. (1995) Phosphoglycerate kinase and triosephosphate isomerase from the hyperthermophilic bacterium *Thermotoga maritima* form a covalent bifunctional enzyme complex. *EMBO J.* 14:442–451.

76. S Tsoka & CA Ouzounis (2000) Prediction of protein interactions: metabolic enzymes are frequently involved in gene fusion. *Nat. Genet.* 26:141–142.

77. G Shearer, JC Lee, J Koo & DH Kohl (2005) Quantitative estimation of channeling from early glycolytic intermediates to CO_2 in intact *Escherichia coli*. *FEBS J.* 272:3260–3269.

78. JWA Graham, TCR Williams, M Morgan et al. (2007) Glycolytic enzymes associate dynamically with mitochondria in response to respiratory demand and support substrate channeling. *Plant Cell* 19:3723–3738.

79. EA Winzeler, DD Shoemaker, A Astromoff et al. (1999) Functional characterization of the *S. cerevisiae* genome by gene deletion and parallel analysis. *Science* 285:901–906.

80. A Barabási & ZN Oltvai (2004) Network biology: understanding the cell's functional organization. *Nat. Rev. Genet.* 5:101–113.

81. M Kanehisa, S Goto, M Furumichi et al. (2010) KEGG for representation and analysis of molecular networks involving diseases and drugs. *Nucleic Acids Res.* 38:D355–D360.

CHAPTER 10
Multienzyme Complexes

It ain't what you do, it's the way that you do it.

Jazz song, first recorded in 1939 under the title *T'ain't what you do (It's the way that you do it)*

In earlier chapters, we have seen the increased regulation and complexity that is possible as a result of protein oligomerization. We saw that many enzymes have evolved from domains that bind the individual substrates, which are then put together and tinkered with. In Chapter 8 we saw how combining domains in different ways can produce signaling of remarkable complexity and control. And in Chapter 9 we saw that actually most proteins function as part of complexes rather than as individuals. However, sometimes this is not enough, and in a rather small number of cases evolution has had to come up with something special. These cases are discussed in this chapter on the grounds that by comparing 'special' enzymes with normal ones, it gives a better idea of what normal ones can and cannot do. They are special because of the high degree of communication between subunits, and the reactions catalyzed fall into two main categories. There are some reactions in which the product of a reaction is so reactive and/or expensive that it cannot be allowed to be released into the cell, and it has to be held on to and converted in a second reaction to something less dangerous or reactive. And then there is a larger group of reactions in which the product of one reaction forms the substrate for a second reaction, *and* for which the gain in linking together the separate reactions outweighs the evolutionary cost. These are typically biosynthetic pathways, in particular cyclic ones.

The proteins discussed in this chapter are multienzyme complexes (MECs). An MEC is defined here as an assembly of enzymes, of reasonably well defined stoichiometry and structure (for example, well enough that it should be possible to crystallize it), that catalyzes a series of related reactions. Some exist as a set of separate polypeptides, while others are composed of one or two multiprotein chains. However, they all exist as stable and essentially permanent complexes. This forms the major distinction between the MECs discussed in this chapter and the more transient complexes discussed in the previous one.

The other distinction, and what makes an MEC something much more interesting than merely a set of enzymes bound together, is that the MEC provides an additional benefit beside bringing related proteins together. The whole is more than just the sum of its parts. This is because (as defined here) the component proteins within the MEC affect each other. There may for example be allosteric communication between proteins, as in tryptophan synthase, or functions that require the proteins to be assembled in a particular way, such as channeling, active-site coupling in pyruvate dehydrogenase or termination in fatty acid synthase.

This chapter also discusses several systems that have the first property, namely that of a set of enzymes with defined stoichiometry (although possibly not a defined structure), but not the second (cooperativity). These are described here as *almost* MECs. As discussed at the end of the chapter, the explanation for the remarkable properties of MECs is evolutionary pressure, which has put a premium on cooperative behavior. For the 'almost' MECs, the evolutionary pressure has probably not been quite so great or extensive, and consequently the complexes exist but do not display any obvious cooperativity. A prime example of this is polyketide synthases, which are nevertheless included in this chapter because of their similarities to fatty acid synthase. Nonribosomal peptide synthases are much the same. Similarly, the *arom* complex is an 'almost'

MEC because it does not seem to have a defined structure or (probably) any cooperativity between the component enzymes, although this is still unclear. Finally, we consider some integral membrane protein complexes, and conclude that they certainly have fascinating mechanical properties but do not have the organization required for an MEC, possibly because two-dimensional searching within the membrane is so fast that further organization is unnecessary. This allows us to place limits on the situations in which MECs are found.

We begin this chapter with the small number of MECs that need to be special because the product produced by the first reaction is too dangerous or reactive to allow it to diffuse freely within the cell.

10.1 SUBSTRATE CHANNELING

10.1.1 Tryptophan synthase is the best example of substrate channeling

For genetics students, it may come as something of a surprise to find that the genes in the *trp* operon actually have a function. The *trp* operon is one of the classic examples of prokaryotic operons, and in particular of the phenomenon known as **attenuation (*10.1)**. It codes for five enzymes, which perform the last five sequential steps in the biosynthesis of tryptophan (**Figure 10.1**). (In passing, note that chorismate mutase at the bottom of the figure is almost the only known example of an enzyme that performs a common and very useful organic chemistry transformation: a six-electron pericyclic reaction typified by the Diels–Alder reaction. The other examples occur in the formation of unusual secondary metabolites, although many of these may have stepwise rather than concerted mechanisms [1, 2].) However, there is not simply one gene per enzyme: one of these genes codes for two enzymes, and the last two code for a single multienzyme complex. It is these latter that are discussed here. *trpB* codes for the β subunit of tryptophan synthase, while *trpA* codes for the α subunit. These two subunits form a single $\alpha_2\beta_2$ tetramer and conduct sequential reactions:

$$\alpha\text{: indole-glycerol phosphate} \rightarrow \text{indole} + \text{glyceraldehyde 3-phosphate}$$

$$\beta\text{: indole} + \text{serine} \rightarrow \text{tryptophan}$$

***10.1 Attenuation**

Most regulation of prokaryotic gene translation is done by repressors that block translation, forming an on/off switch. However, regulation can also occur at the transcriptional level by attenuation, which is more like a volume control. The *trp* operon forms a classic example (**Figure 10.1.1**). The RNA sequence upstream of the *trp* gene contains four regions: 1, 2, 3, and 4. Region 1 codes for a 14-residue peptide containing two adjacent tryptophans, a rare amino acid. Region 3 can form secondary structure with either region 2 or region 4. If tryptophan is abundant, the ribosome moves rapidly through regions 1 and 2, pausing before region 3 long enough to allow region 3 to base-pair with region 4 to form a hairpin that blocks further transcription of the downstream gene. However, if tryptophan is scarce, the ribosome can only move slowly through region 1, allowing region 2 to base-pair with region 3. This is a weaker binding, and allows the ribosome to move through it, therefore allowing transcription of the downstream *trp* gene. Thus, the control derives from the rate of ribosome movement compared with the rate of formation of secondary structure in the RNA. Similar mechanisms have been found in several other amino acid biosynthetic genes. (Note again that these sophisticated systems are found in biosynthetic pathways, generally of the more expensive products!)

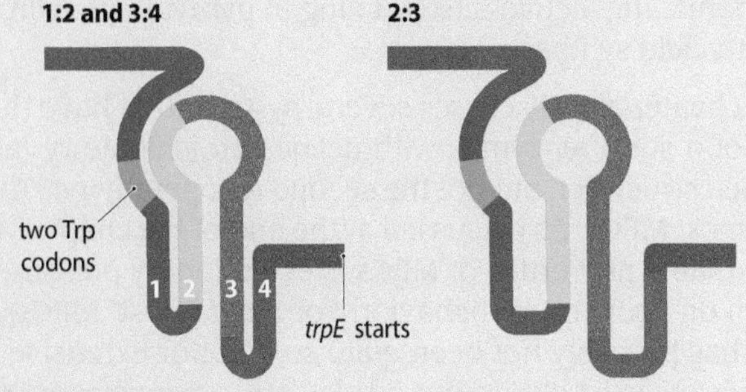

FIGURE 10.1.1
The amount of TrpE produced varies according to how much tryptophan is present.

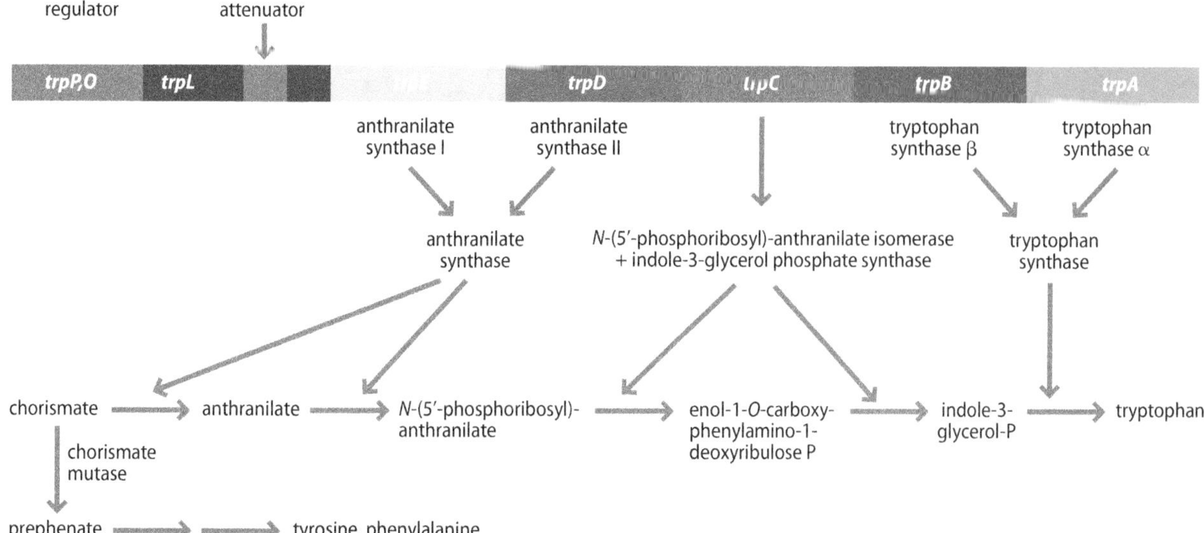

FIGURE 10.1
The genes encoded by the *trp* operon. Five genes code for enzymes that catalyze the last five steps in the synthesis of tryptophan. In particular, *trpA* and *trpB* code for the α and β subunits of tryptophan synthase, respectively.

Indole (**Figure 10.2**) is a reactive compound, which can undergo both oxidations and electrophilic substitutions. However, by itself this hardly seems a strong enough reason for nature to have evolved the complex synthetic system described below. It is fairly hydrophobic (and interestingly one of the few uncharged metabolites in the cell [3]), and therefore can diffuse out of cells: being planar and aromatic it can also bind to many enzyme active sites and intercalate into DNA. So not only is it reactive, it is also toxic. It is interesting to note its chemical similarity with the plant hormone indoleacetic acid, the most common auxin (**Figure 10.3**), which diffuses freely across the plant cell wall and has a wide range of biological effects. Richard Perham has described it as a 'hot potato' by analogy with baking potatoes in a fire: the sort of thing that you would not want to hold on to for long, and would want to pass on as quickly as possible [4]. It is possibly also relevant that it is an expensive compound for the cell to make, and so is not one that you would wish to allow to be lost. The biosynthetic cost of tryptophan is the greatest of any amino acid: depending on how it is measured, it is 25% more expensive than the next amino acid, phenylalanine, and 8–10 times as expensive as the cheapest. It is probably relevant also to note that animals do not attempt to make tryptophan or the other aromatic amino acids, preferring to rely on plants to make them for them. Finally, the β subunit on its own in the absence of indole also conducts the related reaction

$$\text{serine} \rightarrow \text{pyruvate}$$

which in this context would be an unwanted side reaction, possibly providing another reason why the cell has worked so hard to assemble both activities into a single MEC. The chemistry of the two individual reactions, together with some details of the conformational changes occurring during catalysis, have been reviewed recently [5].

Each of the two components alone does have enzyme activity, but it is much lower than the activity in the intact tetramer: the isolated α subunit is about 100 times less active, whereas the β subunit is 30 times less active [6]. It seems a reasonable speculation that the reason for this difference is that both separate reactions are wasteful (and, in the case of the α reaction, dangerous) on their own, so the evolution of a system in which the two separate components are almost inactive makes a lot of sense.

The striking feature about the enzyme structure (**Figure 10.4**) is that there is a hydrophobic 'tunnel' approximately 25 Å long leading from the α active site to

tryptophan
COOH
NH_2
N
H
indole

FIGURE 10.2
The structure of indole (brown) and tryptophan (entire structure).

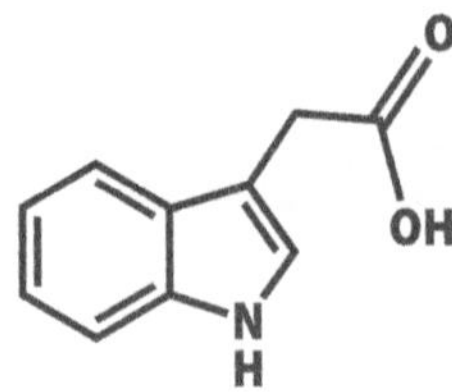

FIGURE 10.3
The structure of indoleacetic acid (a plant auxin; that is, a hormone for plant growth).

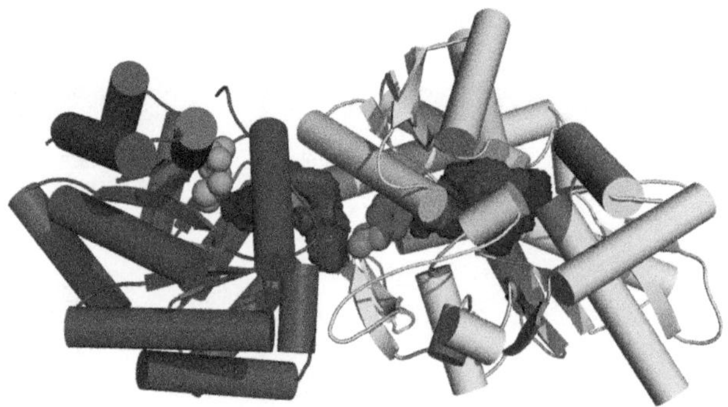

FIGURE 10.4
The structure of tryptophan synthase (PDB file 2j9x). The α subunit is in red and the β subunit in green. The blue regions show the internal channel, which runs from the active site of the α subunit (cyan balls) to the active site of the β subunit at the right-hand end, and allows the indole intermediate to move from one active site to another without going into bulk solution. In this structure the channel is blocked by the sidechain of βPhe280 (magenta spheres) which needs to fold out of the way to allow the intermediate to pass. The channel was drawn using the program Voidoo [63].

the β active site, which is just wide enough to allow indole to pass through. This means that the indole product of the α reaction is able to diffuse directly to the β active site without leaving the enzyme. This feature is known as *substrate channeling*. Clearly the major benefit is that the substrate is never free to act as a hot potato elsewhere.

A secondary benefit is that the rate of delivery of substrate to the second enzyme is much faster. Consider two sequential enzymes that are free in the cytoplasm (**Figure 10.5**). On leaving the first active site, the product (which is also the substrate for the second enzyme) then has to diffuse through the cytoplasm in a random walk until it finds the second enzyme. As discussed in Chapter 4, this can be a very slow process, particularly when the product has a tendency to bind to other molecules within the cell.

Is substrate channeling a useful mechanism for speeding up a reaction? Only if the inherent enzyme turnover is fast compared with the rate at which substrate diffuses into the active site. We can perform an approximate calculation to see if this is true in this case. The turnover of tryptophan synthase is actually very slow ($24\,s^{-1}$) [7]. The rate of diffusion of a substrate S into an enzyme's active site is given by k[S], where k is the pseudo-first-order rate constant for diffusion. k is proportional to the diffusion constant and is therefore slower for large, hydrophobic, or highly charged compounds that are likely to bind to other cell components. This is partly true here: indole is hydrophobic, as discussed above. Let us assume a diffusion rate that is 10 times slower than the standard figure for unhindered diffusion in water; that is, a diffusion rate of $10^8\,M^{-1}s^{-1}$ (see Chapter 4). In addition, if [S] is low, the diffusion rate is slow. A concentration of one molecule of indole in a bacterial cell corresponds to a concentration of roughly 1 nM. Therefore the diffusion rate of that one molecule to the enzyme is about $10^8 \times 10^{-8}$ or $1\,s^{-1}$. Thus, in this case, substrate channeling probably does speed up the reaction, although not by an enormous amount: the big benefit is likely to be the effective removal of indole from the cell cytoplasm.

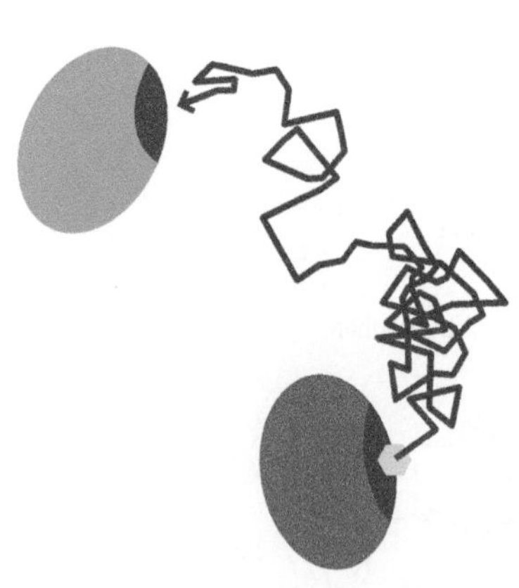

FIGURE 10.5
Random diffusion of the product of one reaction from one active site to another can be a slow process.

The other striking feature of tryptophan synthase is that the reaction catalyzed by the α subunit (that is, the reaction that creates the hot potato) does not

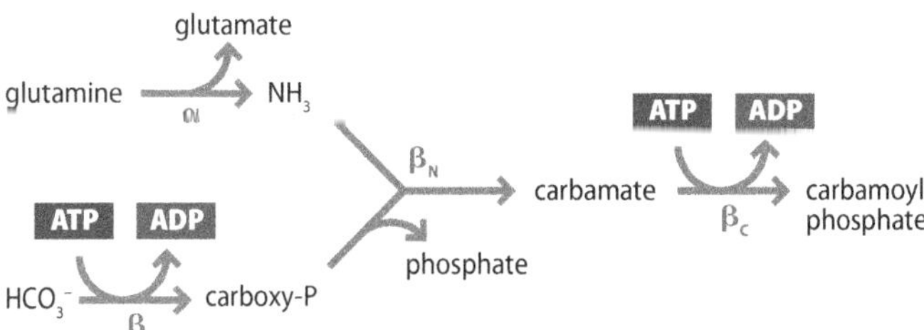

FIGURE 10.6
The reaction catalyzed by carbamoyl phosphate synthetase.

occur until the β subunit is bound to serine and ready to accept the hot potato [7]. This makes a lot of sense. It is, however, a remarkable allosteric interaction, because of the distance between the two active sites, and requires several concerted structural changes to achieve it [8].

This book comments frequently on the little tricks that evolution can use to improve its tinkering and come up with new enzymes fairly easily. On looking at tryptophan synthase, however, one has to pause for a while and marvel at its complexity, which must have required considerable evolutionary pressure to make something so complex.

10.1.2 Most other examples of substrate channeling involve toxic intermediates [9]

Substrate channeling has been suggested for a large number of enzymes but proven in rather few, and indeed has been disproved in several [10]. The rate analysis above suggests that channeling does not have a major effect on rate, at least for a typical slowish enzyme: it seems to have evolved mainly when the intermediate is particularly toxic and/or expensive.

The other clear example of substrate channeling is for several enzymes that generate ammonia as an intermediate. Ammonia is very soluble in water, basic, and apt to bind to proteins. Being small, it is also a difficult molecule to "grab hold of." Ammonia is produced by carbamoyl phosphate synthetase, an enzyme that catalyzes the synthesis of carbamoyl phosphate from bicarbonate and glutamine, using ATP in two separate phosphorylation steps (**Figure 10.6**). There are four reactions, catalyzed by three active sites. The product of the first reaction is ammonia, and the products of the second and third reactions are carboxy-phosphate and carbamate, also reactive intermediates. The enzyme has a remarkable structure and contains a long hydrophilic tunnel that links the three active sites (**Figure 10.7**) [11]. Again there is evidence for allosteric communication between the active sites. The product of the reactions, carbamoyl phosphate, is itself unstable and is used as an intermediate in two pathways: pyrimidine nucleotide biosynthesis, and synthesis of arginine and urea. There is evidence for further partial channeling of carbamoyl phosphate into the pyrimidine synthesis pathway, via the enzyme aspartate transcarbamylase [12].

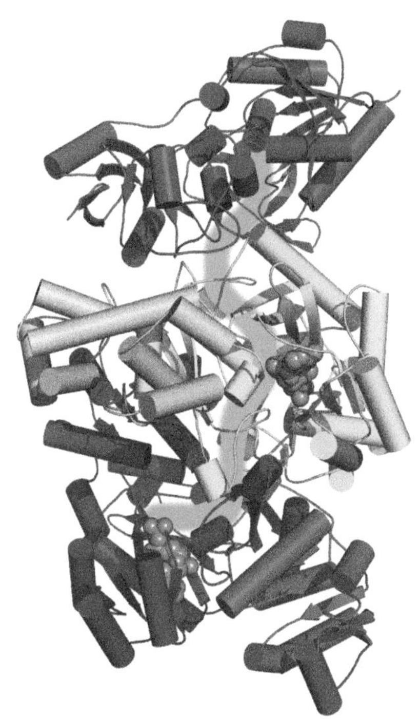

FIGURE 10.7
The structure of carbamoyl phosphate synthetase (PDB file 1jdb) [11]. The α or small subunit (magenta) catalyzes the release of ammonia from glutamine. The β or large subunit catalyzes the other three reactions: the *N*-terminal domain (green) catalyzes the phosphorylation of bicarbonate plus its condensation with ammonia, while the blue domain catalyzes the formation of carbamoyl phosphate. The other two domains in the β subunit are an oligomerization domain (yellow) and an allosteric domain (red). The tunnel for ammonia is shown in cyan and is 96 Å long, with an average radius of 3.3 Å. Two molecules of ADP (orange) are bound in the two active sites of the β domain. The β subunit has pseudo-twofold symmetry.

Ammonia is also produced by glutamine phosphoribosylphosphate amidotransferase, which has a 20 Å hydrophobic tunnel that is formed only once substrates are bound [13, 14]. The enzyme again has an allosteric regulatory mechanism that prevents the unwanted formation of ammonia until the phosphoribosylpyrophosphate is ready to accept it [15].

A small number of other enzymes also display substrate channeling, including the related enzymes asparagine synthase, glutamine amidotransferase, and imidazole glycerol phosphate synthase, all of which have a tunnel for channeling ammonia.

Another well-supported example of substrate channeling is less dramatic, because it involves not a tunnel through the enzyme but a 'highway' across the enzyme surface [16], which probably acts not so much to restrict the motion of the substrate as to guide it. The enzymes thymidylate synthase and dihydrofolate reductase catalyze two sequential reactions in the biosynthesis of thymidine:

$$\text{dUMP} + \text{5,10-methylene-THF} \rightarrow \text{dTMP} + \text{dihydrofolate}$$

$$\text{dihydrofolate} + \text{NADPH} \rightarrow \text{THF} + \text{NADP}^+$$

The tetrahydrofolate produced is not particularly toxic, but is certainly expensive: like tryptophan, it is an essential dietary ingredient for humans, in the form of vitamin B_9 (folic acid). It is also highly charged, because of the four glutamates that are added onto the folate unit; indeed, it seems highly likely that these glutamates are added purely to make it easier to handle and retain in the cell, in the same way that many metabolites (glucose for example) get phosphorylated as the first stage in their metabolism.

In most organisms these two enzymes are separate. However, in protozoa they are linked into a single polypeptide chain. This system provides kinetic evidence for channeling: with two separate enzymes there is a lag time of about 60 s before a steady state is reached, whereas in the linked enzyme there is no observable lag time [17]. In the crystal structure of the enzymes from *Leishmania major*, there is a 40 Å-long positively charged highway across the enzyme, and both calculations and experiments suggest that it acts to retain tetrahydrofolate on the enzyme surface. There is also evidence of allosteric linkage between the two active sites [18], making this a good candidate for genuine MEC-type channeling. However the linked protozoal enzymes from *Cryptosporidis hominis* do not seem to show channeling or allostery; and, as noted above, in most organisms the two enzymes are not linked at all, although it is of course quite possible that they are involved in a regulated metabolon, as discussed in Chapter 9. Thus in most cases the complex arrangement needed for true substrate channeling does not seem to be necessary.

Electrostatic channels have been suggested in several cases, but experimental evidence is thin and such channels are probably very rare. Calculations show that the escape of a charged molecule from an enzyme surface is usually much faster than the catalytic turnover time, and therefore that true channeling is likely to require some sort of barrier to prevent loss of the intermediate [19]. Thymidylate synthase/dihydrofolate reductase may represent an exception because of the very high charge density on the substrate.

10.2 CYCLIC REACTIONS

10.2.1 Cyclic reactions require coordination

The MECs discussed so far fulfill a role of ensuring that toxic and/or expensive metabolites do not leak away into the cell, and this is presumably why they evolved in this way. By contrast, the remaining examples in this chapter catalyze complicated sequential and usually cyclic reactions, where the rationale for their evolution is that linkage and coordination of the different steps is so important that it is worth evolving a system in which all the enzymes are held close together in fixed arrangements. In different ways, the substrate of one reaction is passed on directly to the next enzyme. For most of these systems, the obvious analogy is a production line, in which each enzyme in the complex does its specialist task, and then the substrate is moved on to the next one. The movement of substrate is reminiscent of substrate channeling, except that the substrate here is covalently bound to a domain whose function is to transport it between active sites.

The first and most elaborate of these is exemplified by pyruvate dehydrogenase (PDH), an enzyme aptly described by Richard Perham as being of 'almost gothic complexity' [20]. PDH catalyzes five separate reactions using three

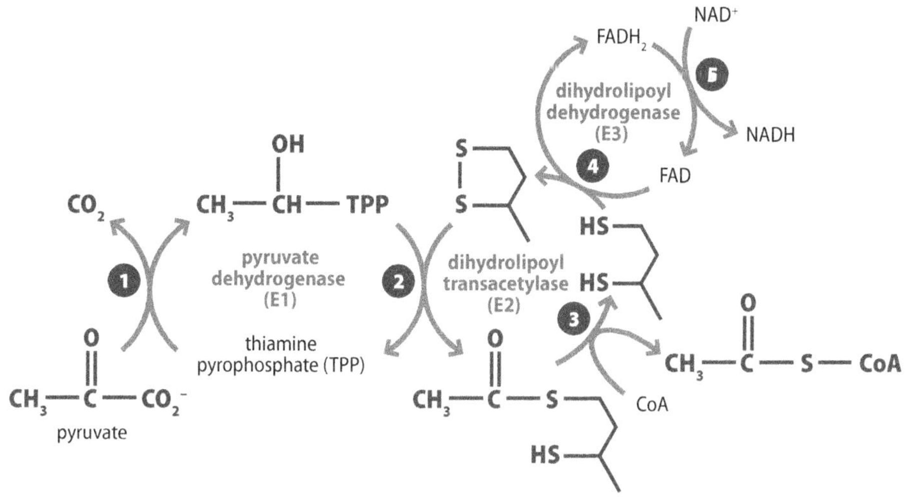

FIGURE 10.8
The five reactions catalyzed by PDH. The E1 component catalyzes reactions 1 and 2, using thiamine pyrophosphate as a cofactor (the active form of thiamine, vitamin B_1). The E2 component catalyzes reaction 3 using lipoamide. The E3 component catalyzes reactions 4 and 5 using FAD (the active form of riboflavin, vitamin B_2) and NAD^+ (the active form of niacin, vitamin B_3). Reaction 3 also involves coenzyme A (the active form of pantothene, vitamin B_5). The involvement of so many essential cofactors indicates the chemical challenge of the reaction for the cell.

enzyme components, usually simply called E1, E2, and E3. These are all somewhat complicated reactions, involving expensive cofactors, but probably the 'reason' for the existence of PDH is not the expense or complication, but the cyclic nature of the reactions and therefore the benefit from keeping the three reactions fixed in close proximity so that products are not lost from the system but recycled [20]. Indeed, this seems to be the rationale for most MECs.

10.2.2 PDH has a large and complicated structure

It is certainly true that PDH has an important metabolic role: it converts pyruvate to acetyl-CoA, and so acts as the exit from glycolysis into the TCA cycle. This effectively commits the carbon atoms to leaving glycolysis, because reentry is either energetically demanding or impossible (for humans at least), and the decarboxylation of pyruvate itself is irreversible. It is regulated both by product inhibition (by NADH and acetyl-CoA) and in eukaryotes by phosphorylation of the E1 subunit, also controlled by NADH and acetyl-CoA, which are products of fatty acid oxidation as well as of PDH, thereby inhibiting PDH if fatty acid oxidation has already provided enough acetyl-CoA.

The reactions conducted by PDH are shown in **Figure 10.8**. The overall reaction is

$$\text{pyruvate} + \text{CoA} + \text{NAD+} \rightarrow \text{acetyl-CoA} + CO_2 + \text{NADH}$$

The first enzyme component, E1, is **pyruvate dehydrogenase (*10.2)**, which catalyzes two reactions: the decarboxylation of pyruvate with the formation of hydroxyethyl-TPP (TPP being thiamine pyrophosphate), followed by the transfer of the hydroxyethyl group to a lipoyl group that is attached to E2 via a lipoyl arm. This second reaction regenerates TPP to be used again. The second component, E2 or dihydrolipoyl transacetylase, catalyzes a thioester exchange reaction, the transfer of the acetyl group to coenzyme A, to give acetyl-CoA and dihydrolipoamide. And finally, the third component, E3 or lipoamide dehydrogenase, reoxidizes dihydrolipoamide back to the lipoyl group, and then in turn is oxidized back to its original form by NAD^+.

Note that a crucial part of the whole process is the lipoyl arm (**Figure 10.9**), which is physically attached to the E2 subunit and interacts sequentially with the active sites of E1, E2, and E3. The arm is on the end of a lysine side chain, and is about 14 Å long altogether. It is this sequential movement that allows the enzyme to operate cyclically, without dissociation of any intermediates from the enzyme, and this is what provides the main benefit to be obtained by assembling everything together into one complex.

***10.2 Pyruvate dehydrogenase**
Pyruvate dehydrogenase enzyme E1 is also known as pyruvate decarboxylase, although confusingly so because there is another quite different enzyme with the same name. Because E1 and the whole complex are called pyruvate dehydrogenase, it is common to call the complex *pyruvate dehydrogenase complex* or PDHC.

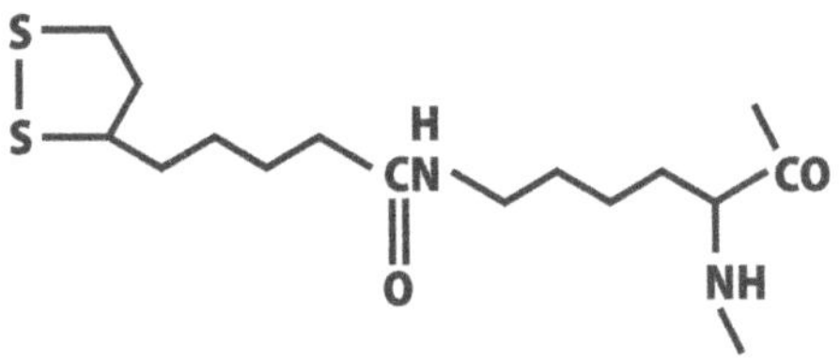

FIGURE 10.9
The structure of the lipoyl arm, which is covalently attached to the end of a lysine side chain.

The subunits are physically associated together in a remarkably complex manner. The construction is different in different species. In all species, the core of the enzyme is an oligomer of E2 subunits, which in *Escherichia coli* consists of 24 subunits. This core is surrounded by a cage comprising the E1 and E3 subunits; in *E. coli*, there are 24 E1 and 12 E3 subunits, arranged approximately in a cube as shown in **Figure 10.10**. This gives an MEC of 4600 kDa, a particle roughly the size of a ribosome. The arrangement is regular and stable enough to be amenable to electron microscopy (**Figure 10.11**) [21], but is also flexible enough to tolerate missing or damaged subunits, so that a 'complete' 60-protein assembly is not required for function, which is clearly an advantage. This structure creates a space inside the cube, within which the active sites of the three enzymes can face each other (**Figure 10.12**) [21]. Apart from this feature, there seems to be no particular rationale to the number and detailed arrangement of the subunits in different species. The mammalian enzyme has different numbers of subunits, for example typically 60 E2 subunits, and looks like a dodecahedron rather than a cube; but the internal arrangement of the components is probably similar [22, 23].

Each E2 subunit also carries the lipoyl arm, as explained above. The lipoyl arm is attached to a lysine residue on a specialized lipoyl domain, which is linked to the E2 core by a flexible linker. In *E. coli*, there are three tandem lipoyl domains (**Figure 10.13**), whereas other organisms have different numbers (two in mammals and one in yeast). There is also a very small peripheral subunit-binding domain (PSD), which binds to both E1 and E3. Presumably, the PSD helps to anchor the lipoyl arms next to the active site for long enough to allow the reaction to occur.

We saw in Chapter 1 that the TCA cycle contains another enzyme that is homologous with PDH, namely α-ketoglutarate dehydrogenase. This enzyme conducts a very similar reaction, the only difference being that the acetyl group

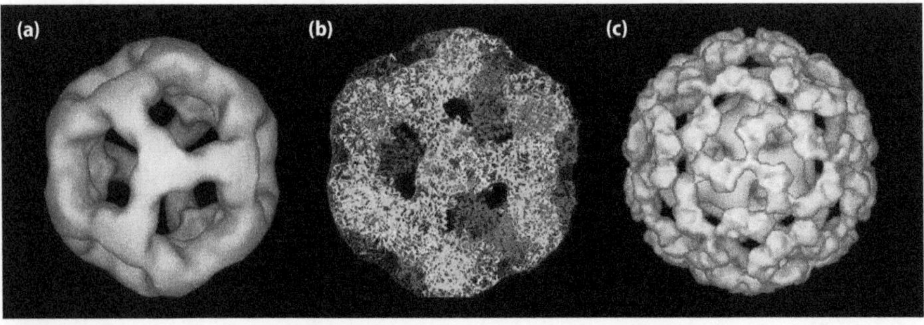

FIGURE 10.11
Electron microscope images of the PDH complex from *Bacillus stearothermophilus*, which has an icosahedral E2 core containing 60 copies. (a) The E2 core, which is roughly 225 Å in diameter. (b) The same structure, into which has been modeled crystal structures of E2. The crystal structures very nearly match the electron density, which does not by itself prove that the model is correct, but does show that it is at least consistent. (c) The electron microscopy structure of the E2 core (as in part (a)) surrounded by 60 E1 $\alpha_2\beta_2$ heterotetramers. (From J. L. S. Milne, D. Shi, P. B. Rosenthal et al. *EMBO J.* 21: 5587–5598, 2002. With permission from Macmillan Publishers Ltd.)

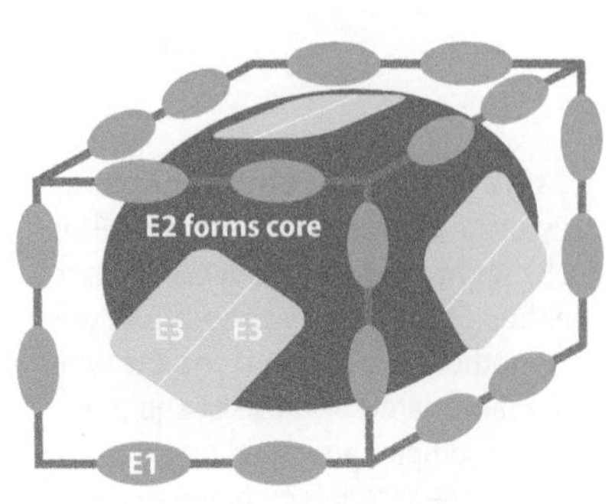

FIGURE 10.10
The arrangement of the E1, E2, and E3 subunits within the *E. coli* PDH complex.

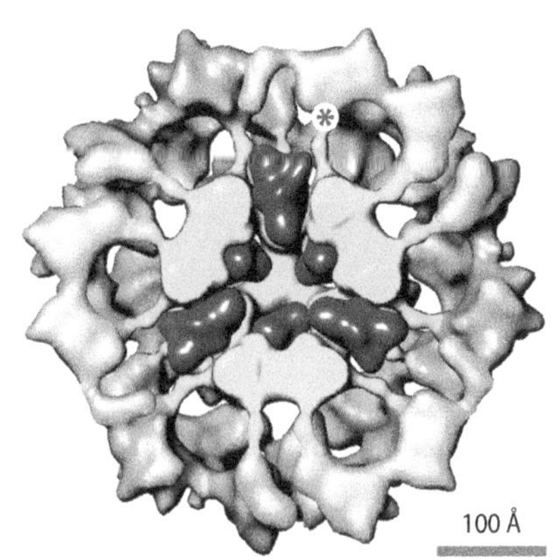

FIGURE 10.12
A cut-away model (front half removed) of the intact PDH complex, based on a combination of electron microscopy and X-ray crystallography. The green regions comprise the E2 core from *S. cerevisiae* determined by electron microscopy, and the yellow regions are E1. In blue are shown the linkers that connect E1 and E2. Into the core were docked homodimers of E3 (red), obtained by X-ray crystallography and filtered to produce structures of the same resolution as the electron microscopy (20 Å). The location marked by an asterisk (and its equivalent position on all the other linkers) is suggested to be the pivot around which the linker to the lipoyl domain rotates, which is roughly 50 Å from the three active sites on E1, E2, and E3. In many models of the PDH complex (such as Figure 10.10), E3 is docked onto the outer E1 ring rather than, as here, onto the inner E2 core. It may of course in reality contact both E1 and E2. (From Z. H. Zhou et al. *Proc. Natl. Acad. Sci. USA* 98: 14802–14807, 2001. With permission from the National Academy of Sciences.)

(CH_3CO) is replaced by $^-O_2C\text{-}CH_2\text{-}CH_2CO$: the chemistry that is performed is otherwise the same. We also saw there that there is good evidence that α-ketoglutarate dehydrogenase was recruited from PDH. Its E1 and E2 domains are homologous with those of PDH, and its E3 domain (which regenerates E2 and therefore does not interact directly with the ketoacid) is identical; thus is a nice example of the 'Common parts with variable ends' tool design described in Chapter 2. Furthermore, there is a third homologous enzyme, branched-chain α-ketoacid dehydrogenase, which is used in the degradation of isoleucine, leucine, and valine. Here the acetyl group is replaced by RCO, the R being the side chains of these amino acids; and again the E3 domain is identical.

Interestingly, in aerobic archaea, the conversion of pyruvate to acetyl-CoA is performed by a different enzyme, pyruvate ferredoxin oxidoreductase, which uses TPP to perform the decarboxylation but does not involve lipoyl groups. It is a much smaller and simpler enzyme. This raises the question of why evolution should have produced such a remarkably complicated system in all other branches of life. There is no really convincing answer to this question. A possible reason is that under most growth conditions the concentration of CoA is low, and so a system that 'concentrates up' the available CoA is preferred; by itself this does not seem a strong enough reason.

10.2.3 PDH shows active-site coupling

The lipoyl arm carries the lipoyl group between the three active sites (**Figure 10.14**), in the space between the E2 core and the E1/E3 outer surface (compare Figures 10.14 and 10.12). For optimum efficiency, it should be restricted in its movements, so that it is most likely to be in the places it is wanted (namely, at the active sites) and less likely to be anywhere else. In other words, the linker should be fairly stiff, but hinged. It is often described as a 'swinging arm' [24]. As we have already seen in Chapter 4, this role is nicely fulfilled by proline-rich sequences, and the linkers in PDH are indeed proline-rich.

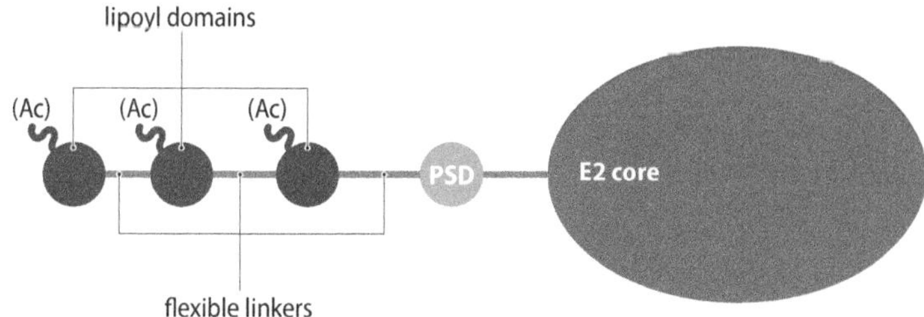

FIGURE 10.13
The structure of the E2 polypeptide chain in *E. coli* PDH. Each lipoyl domain carries a lipoyl arm, which may or may not be acetylated.

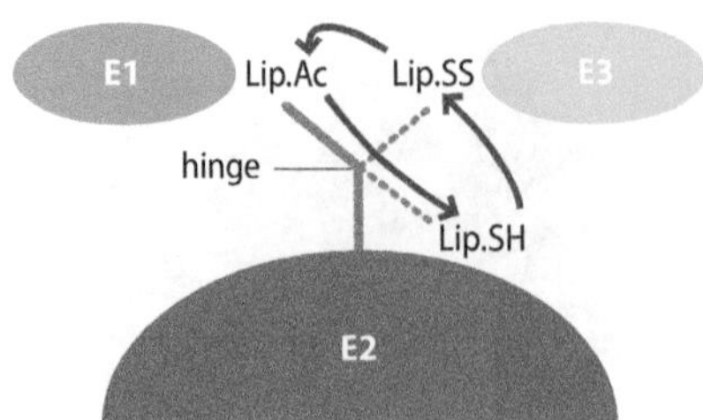

FIGURE 10.14
The linker with its attached lipoyl arm has to cycle round the three active sites. At E1 the pyruvate is decarboxylated and attached to the lipoyl arm (drawn here as Lip.Ac). The lipoyl arm then swings round to E2, where the acetyl group is transferred to coenzyme A. Finally it swings round to E3, where it is oxidized back to the disulfide form (compare Figure 10.8). To achieve this in an efficient manner, the linker should be relatively rigid and hinged (at the point marked with an asterisk in Figure 10.12).

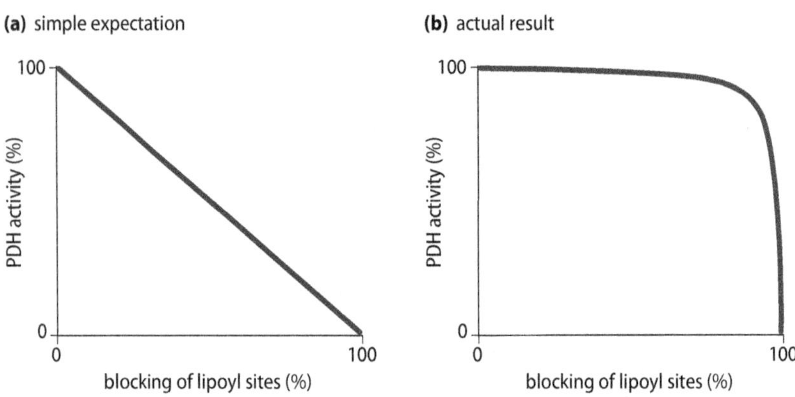

FIGURE 10.15
Active-site coupling. Removal or deactivation of lipoyl arms does not lead to a proportionate loss in enzyme activity, because substrate can be passed around from one arm to the next.

PDH has been studied by NMR. Normally a molecule of this size (4.6 MDa) would not be expected to have any sharp signals at all, because slowly tumbling molecules produce broad lines in NMR (Section 11.3.9). However, a few sharp signals can be observed, which can be assigned to the linkers between the lipoyl domains, showing that the linkers have considerable mobility [20, 25]. The linkers are composed of prolines and alanines. A linker composed only of prolines adopts a rigid **polyproline-II helix** structure, whereas a linker composed entirely of alanines forms an α helix, which again has a rigid structure. However, a linker formed from a mixture of alanines and prolines forms an extended structure that has significant internal flexibility, which is presumably why such a sequence was selected by evolution [25].

PDH has a fascinating and almost unique enzymatic property known as *active-site coupling*. We have seen that the lipoyl arm is the crucial part of the machinery and is involved in all three reactions. One might therefore expect that the removal of a lipoyl arm (either chemically or genetically) would lead to a proportional loss of activity (**Figure 10.15a**). However, this is not actually true: one can remove a significant fraction of the lipoyl arms before there is any apparent change in the reaction rate (Figure 10.15b) [26, 27]. The explanation for this lies in the complicated oligomeric structure of PDH. The flexibility of the linkers that connect the lipoyl domains to E2 allows the lipoyl arm from an *adjacent* E2 to take the place of the 'normal' one (**Figure 10.16**). The slowest reaction is the one catalyzed by E1. This gives adjacent lipoyl domains time to pick up the acetyl group from E1 and take it to different E2 active sites. It is even possible to pass the substrate 'hand over hand' from one lipoyl group to another, either within a polypeptide chain or from one chain to another. Presumably this happens in normal undamaged PDH: substrate and the reduced lipoyl

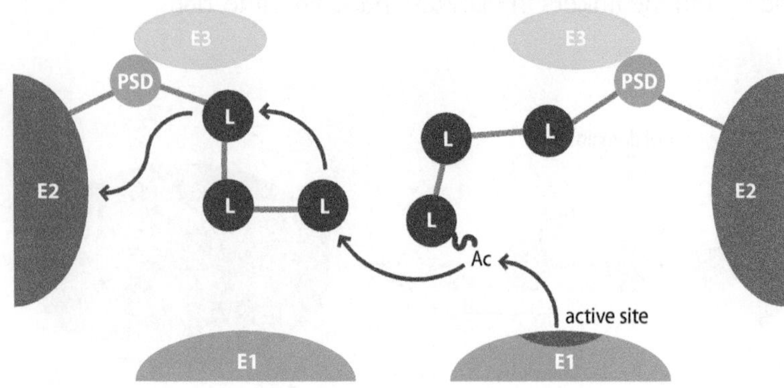

FIGURE 10.16
The explanation for active-site coupling. Substrate can be passed around from the active site on E1 to a lipoyl domain and then from one lipoyl domain to another until it reaches an E2 active site, not necessarily on the same polypeptide chain.

groups are shuttled around different E2 and E3 active sites until they find one that is free. This is reminiscent of parallel computing, where exactly the same thing occurs: calculations are shuttled around to whichever processor is free, thereby avoiding bottlenecks and very significantly speeding up calculations. Thus, the oligomeric structure can lead to very significant rate enhancements, at least in some circumstances. Active-site coupling effectively provides PDH with a 'backup' set of E2 and E3 enzymes that can be used if needed [28].

We saw above that *E. coli* has three lipoyl domains, whereas in other species the number is one or two. Why is this? To examine this question, mutants of *E. coli* PDH were created that contained from one to nine lipoyl domains, and it was found that the mobility of the lipoyl domain for transferring acyl groups was optimal at three [29]. In other words, when there were *more* than three lipoyl domains, the different lipoyl arms from different E2 domains started to get tangled up with each other; but when there were *fewer* than three, the arms had difficulty in reaching from one E2 to its neighbor. Thus, in *E. coli* at least, three lipoyl domains provides the best active-site coupling.

10.2.4 Fatty acid synthase involves multiple rounds of a cyclic reaction

Long-chain fatty acids are synthesized from acetyl-CoA in a series of reactions, in which two-carbon units are added sequentially (**Figure 10.17**). At the

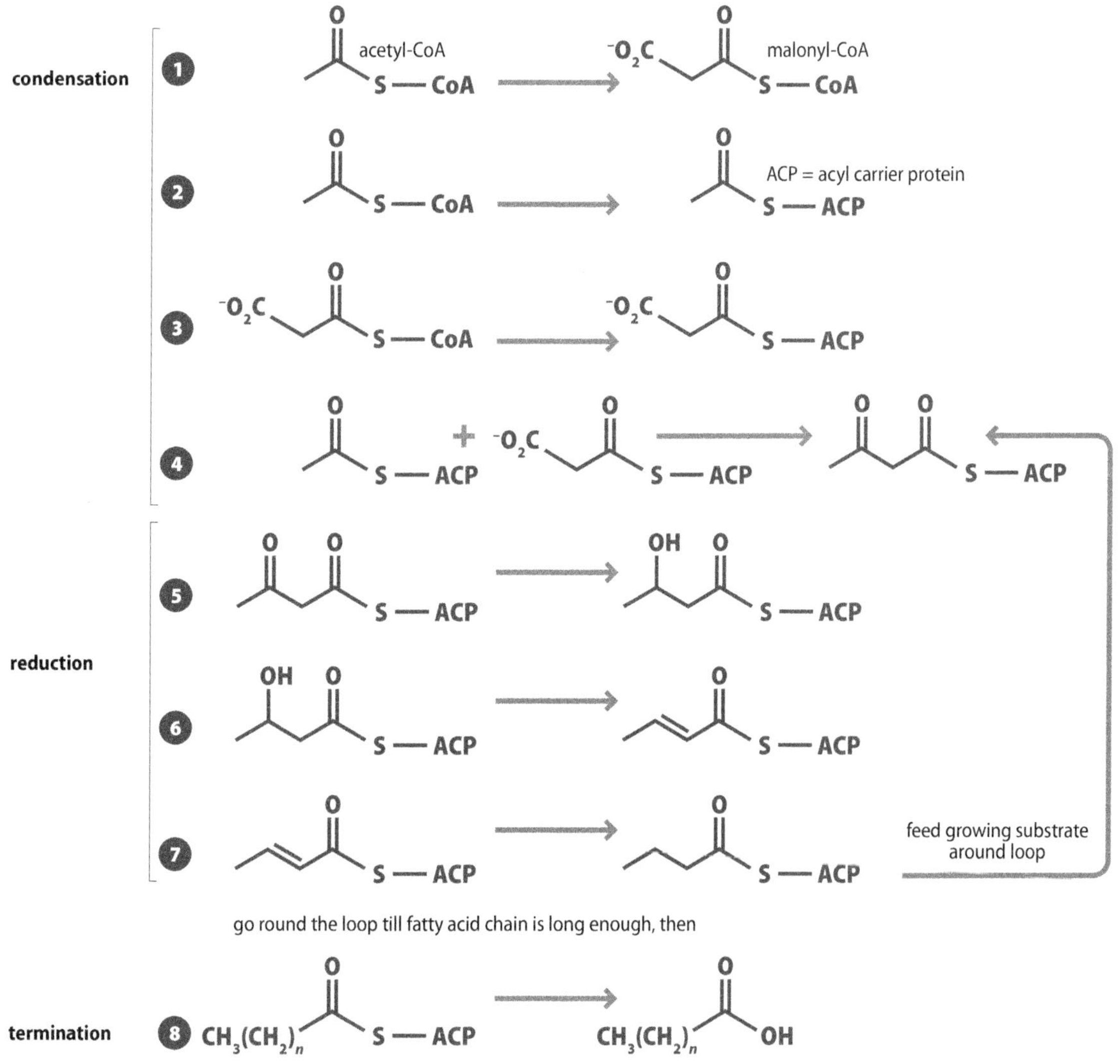

FIGURE 10.17
The reactions catalyzed by fatty acid synthase. The first three reactions load two-carbon units onto acyl carrier protein (ACP), after which comes the key *condensation* reaction that joins a two-carbon unit onto the growing fatty acid chain. The next three reactions reduce the carbonyl group to a hydrocarbon, and the reaction then cycles round reactions 3–7 multiple times until the hydrocarbon chain is of the required length, at which point the termination reaction occurs and the chain is transferred off the ACP.

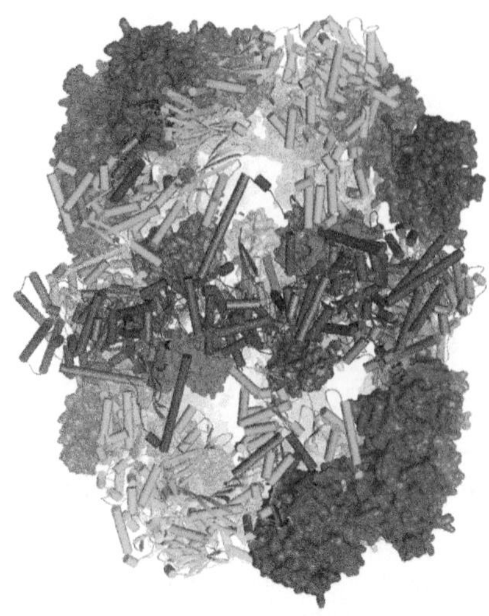

FIGURE 10.18
The structure of FAS from *Saccharomyces cerevisiae* (PDB file 2uv8) [64]. The complex has composition $\alpha_6\beta_6$, where the α chain contains (in order from N terminus to C terminus) part of malonyl/palmitoyl transferase (MPT; reactions 3 and 8; reaction numbering is as in Figure 10.17), the acyl carrier protein ACP, ketoacyl reductase (KR; reaction 5), ketoacyl synthase (KS; reaction 4), and phosphopantetheine transferase (required for activation of apo-ACP); the β chain contains acetyltransferase (AT; reaction 2), enoyl reductase (ER; reaction 7), dehydratase (DH; reaction 6), and the majority of MPT; reactions 3 and 8). The order of enzymes within the protein sequence therefore bears no obvious relation to the order of reactions. There are two chambers, each formed by an α_3 trimer, which forms the base, and a β_3 trimer, which forms a dome. Two α_3 trimers associate together to form the complete complex. Each chamber therefore contains three complete sets of all enzyme domains. The α domains are shown in brown, red, and pink; the β domains are shown in shades of blue. Two of the three β domains are shown as surfaces, while the third is shown as a cartoon to permit a view into the interior. The structure has openings in the side walls and in the central disk to allow substrates to diffuse into the reaction chambers. The ACP is shown as brown, red, and pink surfaces sitting on the base of the chamber. It is anchored to the rest of the structure by two flexible arms, shown for the brown chain as dashed purple lines, attached at the sites shown by purple spheres. This allows the active site of the ACP (green spheres) to move around between different active sites inside the dome, as shown in more detail in Figure 10.19.

start of the process, the first acetyl group is attached to fatty acid synthase (FAS) via a specialized binding domain known as acyl carrier protein (ACP), a situation analogous to the attachment of the acyl group to the lipoyl arm in PDH. As in PDH, the carrier protein has a long flexible arm attached to it, in this case a phosphopantetheine arm which is about 20 Å long [24]. A second acetyl group is then activated to make it more reactive by adding a carboxylate group, to make malonyl-CoA, which is then attached to a second ACP. These three reactions complete the preliminary stages, after which the *condensation* reaction can occur, which joins the two units together, keeping the growing (now four-carbon) chain attached to ACP.

There then follow three *reduction* reactions, to reduce the carbonyl group to a hydrocarbon. After this, the chain is ready to accept another two-carbon unit in the form of another malonyl-ACP; the growing fatty acid goes round this loop until the chain is long enough. Most fatty acids are an even number of carbon atoms long, because they are made from multiple two-carbon additions, and are typically 16, 18, or 20 carbons long, requiring respectively 8, 9, or 10 loops around the circuit described above. The elegance of the FAS MEC is that it (1) keeps the growing chain attached to the enzyme at all times, and therefore can perform the whole process much more quickly; (2) performs each circuit with the same set of enzymes, and so requires far fewer enzymes; and (3) does all this within an enclosed space, which helps to control the length of the final product. Once the chain has reached the required length, a *termination* reaction occurs, in which the fatty acid is liberated from ACP. In mammals, this is achieved with an additional enzyme; in yeast, the termination enzyme (which transfers the mature fatty acyl chain from ACP to CoA) is the same as the malonyltransferase that performed the initial transfer of the malonyl group from CoA to ACP, and is known as malonyl/palmitoyl transferase.

In *E. coli*, these enzymes are all present as separate peptide chains within a complex. In yeast, there are two polypeptides, α and β, which are associated together in a hexamer of dimers: $\alpha_6\beta_6$. The different enzymes are scattered rather randomly around the two chains: the malonyl/palmitoyl transferase enzyme is even split between both chains. In mammals, all the activities are on the same polypeptide, which is arranged as a dimer [30]. This tendency for multiple reactions to be on a single chain in eukaryotes and on multiple chains in prokaryotes has been noted several times in this book, and is presumably related to the greater difficulty of assembling multiple polypeptide chains in the larger eukaryote, and possibly to the fact that in prokaryotes the separate chains tend to be within the same operon and therefore can be expressed together, making them easier to assemble.

10.2.5 Structures of FAS contain a large cavity where the cyclic reaction occurs

There are crystal structures of both the yeast and mammalian complexes, which show how these reactions are organized (Figures 10.18 and 10.20, respectively). In both there is a large cavity lined with the different enzymes, which also contains the ACP. In the yeast enzyme structure (**Figure 10.18**), the ACP could be observed stalled at the active site of the ketoacyl synthase enzyme, whereas in the mammalian enzyme it was not visible at all as electron density because it was too mobile to see. However, it must be in a position that allows it access to the other domains, which are arranged roughly in a ring around it, as indicated for the yeast enzyme in **Figure 10.19**. Surprisingly, the domains do not occur within the protein sequence in the order that the reactions happen, and the three-dimensional arrangement of the domains is also not so as to put sequential active sites adjacent [31]. Because the trajectory from one domain to another is largely random, possibly this is immaterial.

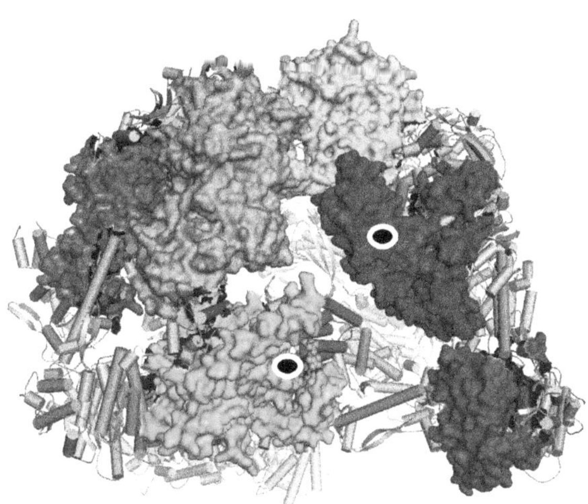

FIGURE 10.19
Detailed three-dimensional arrangement of catalytic domains in yeast FAS (PDB file 2uv8). The view is from the outside of one $\alpha_3\beta_3$ dome (that is, the top half of the structure shown in Figure 10.18), with the α_3 base at the bottom. The ACP, which has to carry the growing fatty acid chain from one domain to another, is shown in magenta, with the location of the fatty acid chain in green. The ACP is attached to the inside of the dome by two attachment points, located roughly where the black-and-white ovals are positioned. These presumably allow the ACP to swing around to visit the different sites. The order of sites (for reactions 2, 3, 4, 5, 6, 7, and 8, with 8 being the same as 3 because MPT performs both) is shown respectively by the green, purple, cyan, blue, brown, olive, and purple domains. The route of the ACP is therefore not particularly regular. The different domains come from a total of four different chains (ACP and KR from chain α_2; KS from α_1; AT, DH, and ER from β_1; and MPT from β_2), implying that the trimerization is essential for enzyme function.

It is striking that the enzyme again has a modular construction: each domain has its own function, and the whole enzyme is constructed very simply just by joining the different functions together [32].

The mammalian enzyme (**Figure 10.20**) is X-shaped and has two equivalent halves, each of which contains a full set of enzyme activities, with the condensing enzymes on one arm and the reducing and termination enzymes on the other (**Figure 10.21**). The complex seems to have a good deal of flexibility, possibly even allowing a rotation of the top half relative to the bottom half. Indeed, such flexibility is probably necessary to allow the fatty acid-ACP

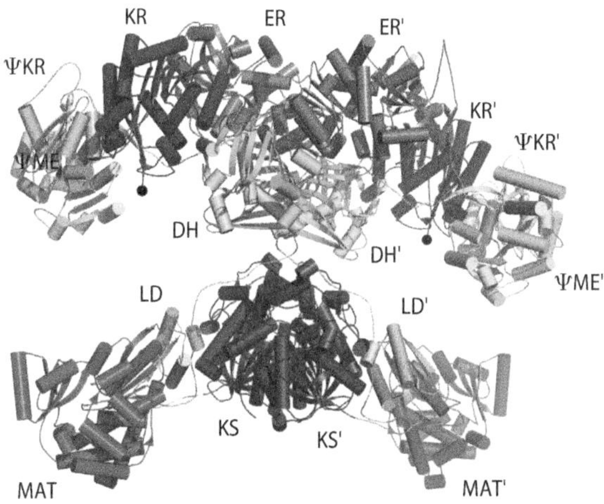

FIGURE 10.20
The crystal structure of the pig FAS (PDB file 2vz9) [65], which is closely similar to the human enzyme. The different functional domains are shown in different colors, in a rainbow order from violet to red. The mammalian enzyme is a single polypeptide chain containing all the enzyme activities, arranged as an X-shaped dimer. The left and right halves of the X each contain a complete set of enzymes. In the mammalian enzyme, the two activities of reactions 2 and 3 (Figure 10.17) are conducted by a single enzyme, malonyl-CoA/acetyl-CoA ACP transacetylase (MAT). The order of reactions around the cycle is therefore as indicated by the numbers on Figure 10.21: MAT, KS, KR, DH, ER. The path is therefore again not particularly linear. The C-terminal end of the polypeptide contains ACP and the thioesterase termination reaction (reaction 8). These two domains could not be located in the crystal structure, presumably because of their mobility, but they are attached at the points marked by black balls. The ACP is therefore presumably able to swing around within the cavity to visit the different active sites. The mammalian enzyme contains three non-enzymatic domains: a linker domain (LD), a domain with homology to methyl esterase (ΨME), and a domain with homology to KR (ΨKR). The second chain in the dimer is indicated with superscript primes. The hinge below the DH domains is thought to be very flexible: flexible enough to allow rocking of the top part with respect to the bottom part and thus both alter the size of the cavity and move ACP around it, and possibly flexible enough to allow the top part to swing round through 180° and thus permit the ACP of one half to interact with condensation active sites on the other.

(a)

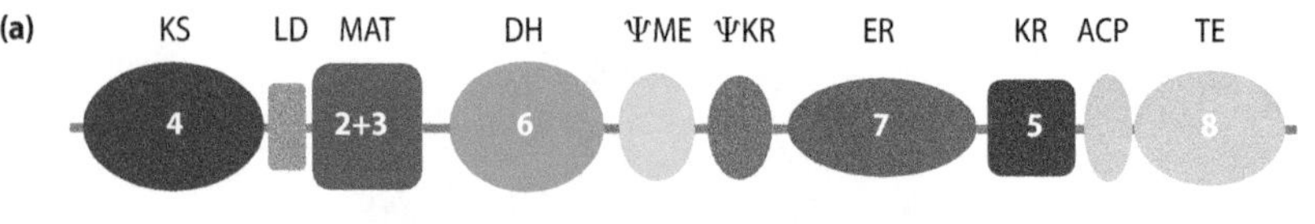

(b)

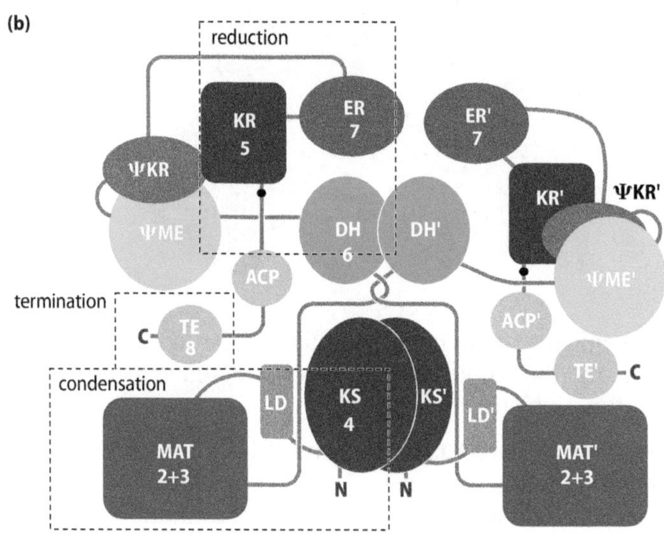

FIGURE 10.21
Diagrammatic structure of pig FAS. (a) Linear arrangement along the sequence, roughly to scale. The color scheme is the same as in Figure 10.20, and the order of reactions is indicated by the numbering, which matches that in Figure 10.17. (b) Three-dimensional arrangement, laid out in the same orientation and with the same colors as in Figure 10.20. The second domain (marked 2 + 3) is a combined malonyl-CoA/acetyl-CoA ACP transacetylase (MAT). The ACP and TE domains are not visible in the crystal structure. The order of domains within the chain is therefore very far from the order of reactions, although there is a clear grouping of condensation and reduction reactions.

to reach all the active sites. There is some evidence to suggest that the two halves communicate, so that the reaction occurs only on one side of the complex at any time, with some kind of alternating mechanism; the details are, however, far from clear as yet.

By contrast, the fungal enzyme is more rigid. The six α chains form a central flat 'wheel' that is capped on both sides by a dome of β chains. There are therefore three complete sets of enzymes in each half, as well as three ACPs which are held at both ends by flexible linkers. In a manner highly reminiscent of PDH, the N-terminal ACP linker is Ala/Pro-rich, which is presumed to limit its movement and stop the three ACPs in each half from getting tangled up [31]. Examination of the structure shows that each ACP probably circulates round its own 'local' set of enzymes, with the three sets acting independently (see Figure 10.19). The 'local' group, however, derives from more than one α and β chain.

Recognition of the appropriate chain length for termination works somewhat differently for mammalian and yeast enzymes [33]. In mammals, the condensation reaction becomes inefficient for long chains, possibly by simple steric hindrance. In addition, the termination domain contains a 'ruler' for fatty acid length. It has three fatty acid binding pockets of different lengths. The longest one matches the length of the desired fatty acid, and when the fatty acid binds, it adopts a conformation that is favorable for acyl group transfer to the termination domain (**Figure 10.22**). By contrast, in yeast the malonyl/palmitoyl transferase enzyme has two binding sites: one for the malonyl group, and one that can accommodate any length up to C_{18}. Normally the malonyl site is occupied by malonyl, preventing the binding of longer fatty acid-ACP chains, so that the enzyme functions exclusively to transfer malonyl groups to ACP. However, a C_{16}- or C_{18}-ACP binds strongly enough to the enzyme that it can displace malonyl, and the enzyme can then work in the 'reverse' direction, transferring the fatty acid back to CoA. Thus, the chain length for termination is more easily varied in yeast than in mammals; in addition, longer chains than malonyl can act as initiators. The resultant chain length is thus regulated by both substrate and product concentrations. This metabolic flexibility may be important for yeast, which is more versatile in the fatty acids it can make.

10.2.6 β Oxidation is approximately the reverse of fatty acid synthesis

The previous section discussed the elegant enzyme that synthesizes long-chain fatty acids, passing the growing chain around the same cycle several times until it is long enough to be released. It makes good sense to assemble such a system into an MEC. What about the oxidative breakdown of fatty acids, which is essentially the reverse process?

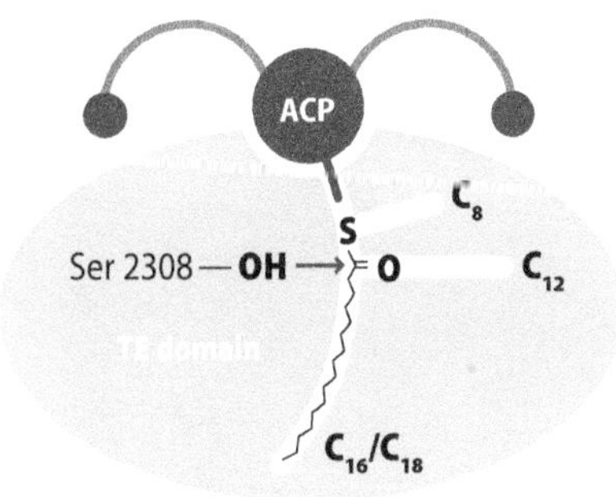

FIGURE 10.22
The termination reaction catalyzed by the thioesterase (TE) domain in mammals. ACP visits the TE domain as it cycles round the different active sites. The growing fatty acid chain can bind in one of three sites, which bind preferentially to C_8, C_{12}, and C_{16}/C_{18} chain lengths. If an acyl chain binds within the C_{16}/C_{18} site, it is positioned such that the thioester bond can be attacked by Ser 2308, and the ester is subsequently hydrolyzed by water to produce the free fatty acid. (Adapted from M. Leibundgut, T. Maier, S. Jenni, and N. Ban, *Curr. Opin. Struct. Biol.* 18: 714–725, 2008. With permission from Elsevier.)

The breakdown of fatty acids is known as β-oxidation and occurs in the mitochondrion in animals [34]. It is indeed approximately the reverse of fatty acid synthesis: in particular, the three reductive steps of FAS are mirrored in three oxidative steps of β-oxidation. In other respects it is different. There is no single enzyme complex in the same sense, and there is no ACP; cleavage of each two-carbon unit is performed directly by coenzyme A. In part, this contrast with the cyclic nature of fatty acid synthesis is because fatty acids can have *cis* or *trans* double bonds in different places, and can also be odd numbers of carbons long. β-Oxidation has to be able to handle these molecules. This means that it has to be able to allow branching in the oxidation process: if there is a double bond, a different route is followed. Odd-numbered chains are dealt with in the final stage, where the three-carbon propionyl-CoA is handled differently. Degradation has to be able to handle more variability in its substrates. Nevertheless, the elegance of fatty acid synthesis contrasts strangely with the messiness of β-oxidation. There is one set of enzymes for handling the oxidation of long-chain fatty acids (C_{12} to C_{18}) (which do indeed form an MEC, as discussed below), and two additional sets of enzymes for shorter chains down to C_4.

The long-chain fatty acid oxidase complex in bacteria and humans is organized as an MEC, known as the mitochondrial trifunctional protein. It is an $\alpha_4\beta_4$ hetero-octamer in humans, in which the α chain contains two of the oxidative activities while the β chain carries out the cleavage of the two-carbon unit (**Figure 10.23**) [35]. In bacteria, it is an $\alpha_2\beta_2$ heterotetramer, essentially half of the mammalian version. The complex does display channeling, but in a

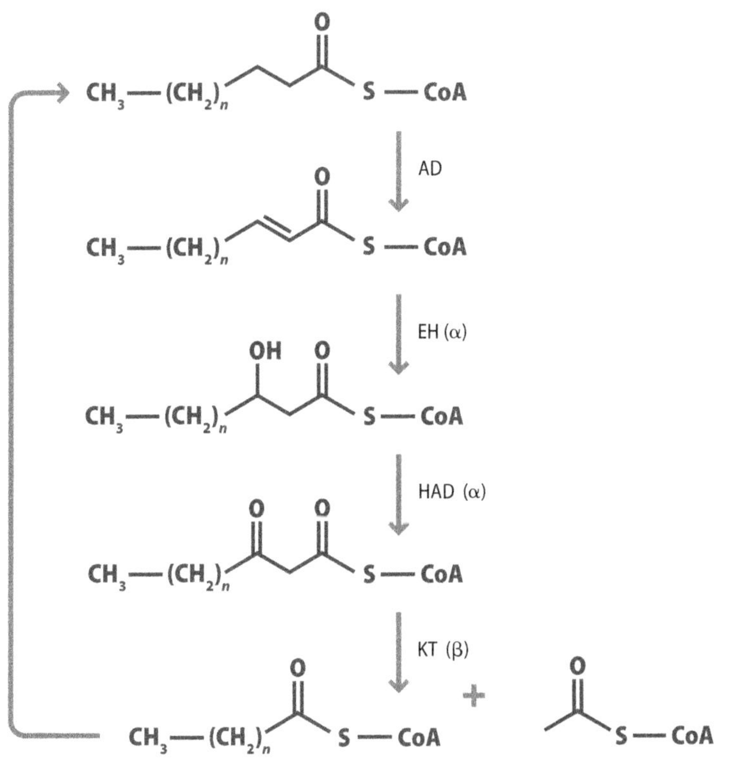

FIGURE 10.23
The reactions catalyzed by the mitochondrial long-chain fatty acid oxidase complex. The reactions are more or less the reverse of the FAS reactions, except that the intermediates are attached to coenzyme A rather than being covalently attached to the enzyme. The individual enzyme components are: (1) acyl-CoA dehydrogenase (AD) (there are four different forms of this enzyme, specializing in different chain lengths, and It is not part of the MEC); (2) enoyl-CoA hydratase (EH), which is part of the α subunit; (3) 3-L-hydroxyacyl-CoA dehydrogenase (HAD), which is also part of the α subunit; and (4) β-ketoacyl-CoA thiolase (KT), which is part of the β subunit.

rather leaky form. The first enzyme activity of the cycle, acyl-CoA dehydrogenase, is not part of either the mammalian or the bacterial complex, and is carried on a separate enzyme. This is possibly because four different versions of fatty acid oxidase are required for substrates of different lengths, and different ones again for branched substrates; it may therefore be more efficient to have detachable versions of acyl-CoA dehydrogenase that can be used as required [34]. (Compare the analogy of enzymes with tools in Chapter 2: this is another example of the 'common parts with variable ends' theme.)

Although β-oxidation does have some problems with having to handle unsaturated fatty acids, this does not seem to be a strong reason why it is so much less elegant as an MEC. One possible reason is to do with the relative energetic cost of synthesis and degradation, and thus with evolutionary pressure. Biosynthesis is an energetically demanding process, and there is considerable evolutionary pressure to streamline it. We reasoned earlier in this chapter that a major factor in the evolution of the tryptophan synthase MEC was the energetic cost of making tryptophan. By contrast, β-oxidation is an energy-producing process. Naturally an organism will want to extract the maximum energy possible from its foodstuffs, but it does not matter quite so much whether the process is efficient or not, and there is therefore less evolutionary pressure for an elegant and coordinated system. It is no coincidence that many of the true MEC systems discussed in this chapter are biosynthetic.

10.3 ENZYME COMPLEXES THAT ARE ALMOST MECS

10.3.1 Type I polyketide synthase is chemically similar to FAS but is not an MEC

Many bacteria synthesize **polyketides** (**Figure 10.24**) in a series of reactions related to fatty acid synthesis [36]. However, in this case some of the reductive steps are omitted. The result is a polymer with varying degrees of oxidation along the chain. In addition, bacteria can add three- and four-carbon units as well as two-carbon units (propionyl-CoA and butyryl-CoA as well as acetyl-CoA), so that the polymers can also have methyl and ethyl groups attached. These polymers can then fold up and cyclize in a variety of ways, to form polyketides. Many of these have useful properties as antibiotics, such as

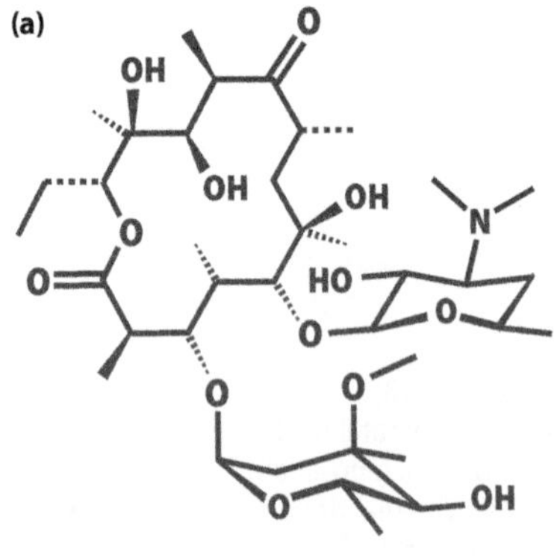

FIGURE 10.24
Some polyketides: (a) erythromycin; (b) rifamycin B; (c) amphotericin B. All three are biosynthesized by linking together a series of two- or three-carbon units (acetyl-CoA or propionyl-CoA) with a variable number of reduction steps, leading to ketone, double bond, hydroxyl, or hydrocarbon groups on alternate carbons along the backbone.

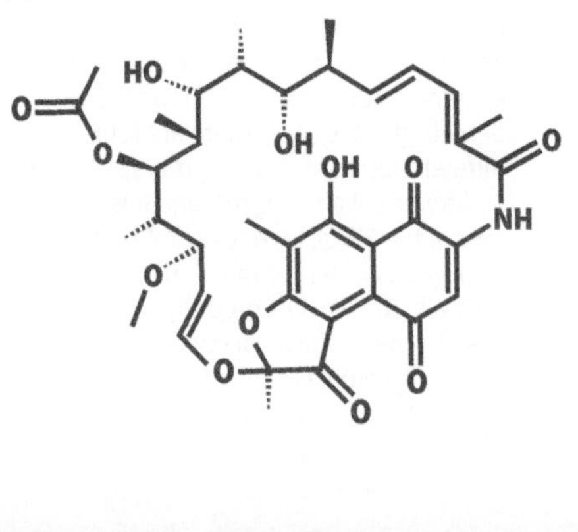

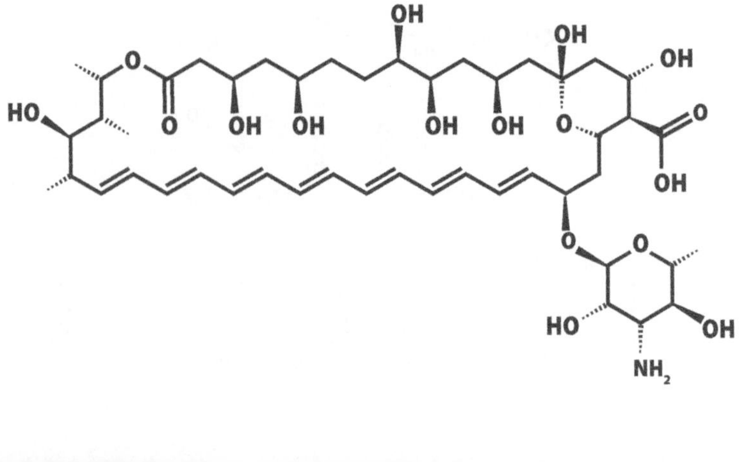

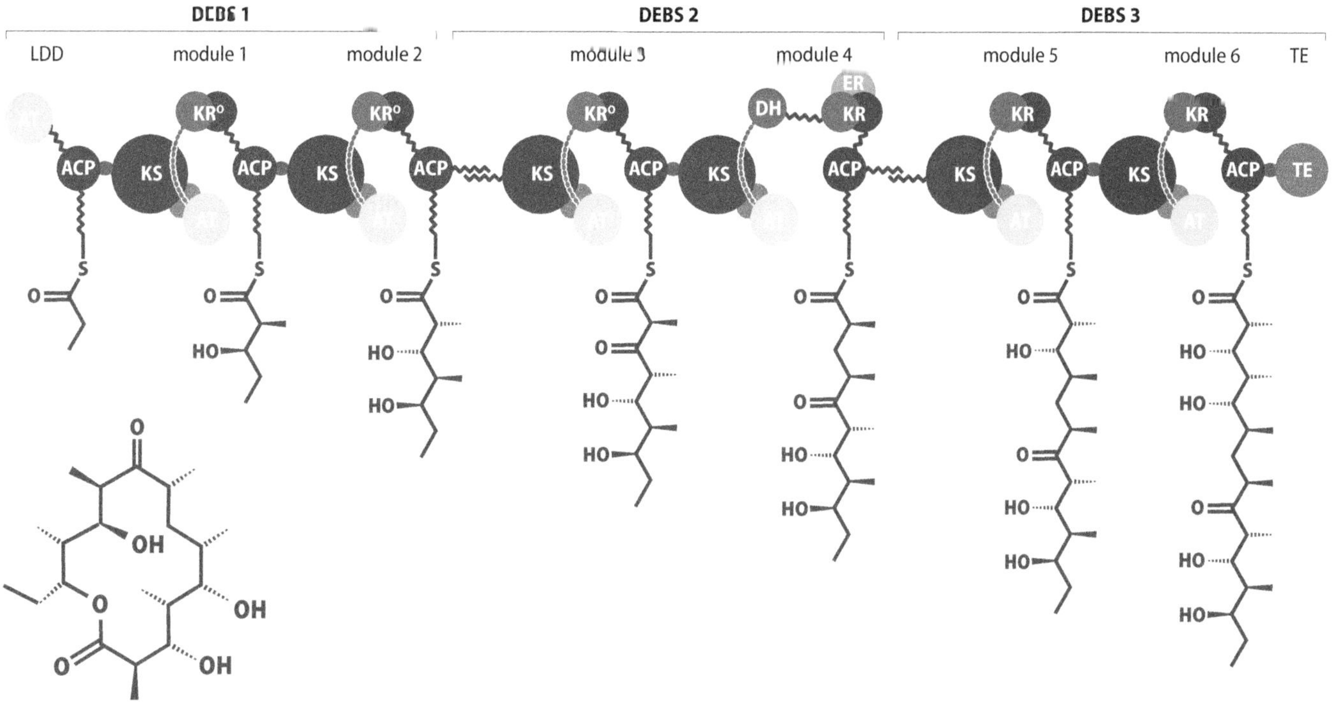

FIGURE 10.25
Modular organization of the type I polyketide synthase 6-deoxyerythronolide B synthase (DEBS). The enzyme contains three polypeptide chains (DEBS 1, 2, and 3), each of which is a dimer in the intact complex. Each of these chains consists of two modules, and each module is responsible for the attachment of one appropriately modified two-carbon unit. A 'complete' module would contain the functions acetyltransferase (AT), ketosynthase (KS), ketoreductase (KR), dehydratase (DH), and enoyl reductase (ER), plus the acyl carrier protein ACP, which is the same set of reactions as conducted by FAS. This complete set would produce a fully saturated two-carbon unit as in fatty acids. The omission of one or more functions results in a partly functionalized unit. The minimal set (AT, KS, and ACP) results in a ketone, as in module 3, in which the KR domain is present but non-functional. Addition of KR produces an alcohol, as in modules 1, 2, 5 and 6. The final domain is the thioesterase termination, which also cyclizes the product. Note that the order of domains within the chain is not identical to the order of reactions. The structure is drawn in what may be the functional organization. In particular there is a cleft between the KS and AT domains, which is likely to be the docking site for the ACP; from here the ACP can deliver the growing polyketide chain to the KR, DH, and ER domains before passing on the chain to the next module. It is interesting to note a close structural similarity between DEBS KR module 1 and the KR/ΨKR pair in pig fatty acid synthase [65], indicating a clear evolutionary relationship between polyketide synthase and fatty acid synthase. LDD, loading domain. The ketoreductase domain in module 2 is nonfunctional. (Adapted from Y. Y. Tang et al., *Chem. Biol.* 14: 931–943, 2007. With permission from Elsevier.)

the tetracyclines, erythromycin, azithromycin, rifamycin, and clarithromycin. Together with the immunosuppressant rapamycin and the antifungal amphotericin B and many others, annual sales of polyketides exceed US$10 billion.

The best-characterized synthases are the type I polyketide synthases. These use the same sequence of enzymes as in fatty acid synthesis, and also have an ACP, but by contrast with the elegant cyclic system used for fatty acid synthesis, *all* the enzyme activities (including the ACP) are used only once. Thus, the biosynthesis of a large polyketide requires a very large number of enzymes (**Figure 10.25**). These are often assembled into one polypeptide, or a small number of polypeptides, of immense length, often as long as 2 MDa; a particularly striking enzyme that produces the polyketide ECO-02301 has 122 domains contained on nine proteins, with a total size of 4.7 MDa [36]. The different subunits are linked strictly in the order in which the reactions occur, and are grouped into blocks (confusingly, these are often called modules), each of which has its own ACP, which adds on and modifies a two-carbon unit; sometimes, however, additional enzymes such as acyltransferase are required that are not part of the main polypeptide [37]. The linkers between enzymes are generally composed mainly of alanines and prolines, like those in PDH, and presumably work in the same way, having limited flexibility to optimize transfer of the ACP from one active site to the next. There is probably some organization within each block, imposed by the nature of the linkers between enzymes, which in turn leads to interactions between domains [38, 39]. Evidence is also emerging for 'docking domains,' possibly based on coiled-coil interactions, between one block and the next [37]. A crystal structure of a fragment from 6-deoxyerythronolide B synthase provides suggestions of the interactions between an ACP and its associated ketosynthase and acyltransferase domains, implying a matched set of interactions within each block [40]; it is, however, clear that within each block there must be substantial mobility [41]. We may view these enzymes as a more evolutionarily primitive version of FAS, with which they share many structural and functional similarities [37].

10.3.2 Some polyketide synthases are proper MECs

There are, however, iterative type I polyketide synthases, found in eukaryotes, that do use the same set of enzymes for more than one cycle. 6-Methylsalicylic acid synthase, lovastatin synthase, and aflatoxin synthase have this property. For these systems, the key question is how the enzyme produces the correct set of modifications and cyclizations, which in the more common bacterial systems (described above) are conducted by separate enzymes. The answer seems to reside in an extra module found only in these iterative synthases, which has been called the 'product template' domain [42]. It is as yet unclear how this domain works.

Type II polyketide synthases tend to be bacterial and have the same relationship to type I synthases as bacterial fatty acid synthases do to eukaryotic fatty acid synthases (and presumably for similar reasons): they consist of multiple monofunctional proteins assembled into a complex. They tend to be limited to the synthesis of rather simple polyketides such as actinorhodin, tetracenomycin, or doxorubicin.

There is also a group of type III polyketide synthases. These do not use a carrier protein, and just use coenzyme A derivatives directly. Possibly in compensation, however, they do have a defined structure, which is a homodimer, each half of which can in some synthases catalyze multiple cycles.

Polyketide synthases are thus closely related to FAS, and probably have a common evolutionary origin [43]. They are, however, not as well organized; they are a sort of primitive MEC. Why are they not as organized? Again, presumably, the answer lies in evolutionary pressure: the pressure to evolve something really efficient has not been sufficient. It is clear that fatty acid synthesis is a vital part of primary metabolism, and it therefore really needs to be efficient. However, polyketides are *secondary metabolites*: compounds that are not essential parts of an organism's function. There is a fascinating literature on secondary metabolites, concerning their role and importance. Most of our antibiotics are derived from secondary metabolites, so they are certainly important for us. But they are arguably less important for the organism that produces them, at least for normal purposes. The consensus view is probably that secondary metabolites began as a way of disposing of things that the organism did not need, or were produced by 'unwanted' **moonlighting** reactions, but because the product turned out to have some value, the synthetic routes were refined and extended. Secondary metabolites have enabled symbiotic relationships to coevolve, and probably help to define niches in which different species can live side by side without undue competition. There has certainly been less evolutionary pressure on secondary metabolism than on primary metabolism. One piece of evidence is the observation that there has apparently been comparatively little horizontal transfer of genes involved in secondary metabolism, something that is certainly not true for more essential genes (Chapter 1) [44].

10.3.3 Nonribosomal peptide biosynthesis is similar to polyketide synthase

Nonribosomal peptide synthesis (NRPS) is, as the name implies, the synthesis of peptides but without using a template RNA and without using ribosomes. It has a lot in common with type I polyketide biosynthesis. The products are again secondary metabolites. It is a pathway for the synthesis of polymeric compounds (in this case peptides) that uses a series of enzymes whose sequence determines the structure that is produced. Each amino acid is added on and modified if necessary by a group of enzymes that have a highly modular nature; the structure of the intact peptide can often be predicted from the sequence of the synthase. As in PKS, there is a starter module, a series of elongation modules, and a termination module. There is a carrier protein with a phosphopantetheine arm that carries the growing chain, and it seems that the linkers between the modules affect the nature of the interactions between them [36, 45]. There may be 'tailoring' enzymes docked onto particular modules to carry out specific reactions.

Some NRPS enzymes contain polyketide modules to add on such components when required, sometimes even on the same polypeptide chain. There is a relatively recently characterized group of *trans*-acyltransferase systems that contain both polyketide and NRPS modules and seem to have arisen from extensive horizontal gene transfers, with highly mosaic structures and unusual domain orders. They differ from the other polyketide and NRPSs discussed so far because there is no acyltransferase domain within each extension module; instead, there are one or more *trans*-acting acyltransferases that must associate with the rest of the complex to perform their function.

It seems that there is little structural organization to NRPS, and rather less than in PKS: they are very much a fledgling MEC (or, alternatively, possibly an extremely primitive system that never had the pressure or opportunity to evolve further).

10.3.4 Aromatic amino acid synthesis is a poor MEC

We have already discussed tryptophan synthase, which consists of the final two enzymes in the pathway for the synthesis of tryptophan from chorismate. The likely reason for it to have evolved to such an intricate system is the high metabolic cost of tryptophan. In fact all the aromatic amino acids are expensive: after tryptophan, the next most expensive amino acids are phenylalanine and tyrosine, followed by histidine or isoleucine, depending on exactly how it is calculated (**Table 10.1**), which are only half the cost of tryptophan. Tryptophan, phenylalanine, and tyrosine are synthesized from chorismate, which in turn is synthesized from shikimate, which is synthesized from sugars. Chorismate is also the starting point for folate (vitamin B_9) and for vitamin K.

The biosynthetic pathway for chorismate is shown below in Figure 10.26 [46]. This pathway is present in bacteria, microbial eukaryotes, and plants, but not in animals, hence the occurrence of the vitamins and essential amino acids in the list above.

In bacteria, the enzymes in this pathway are all present as separate polypeptides; in eukaryotes they exist as a single polypeptide, which has been given the name AROM (corresponding to the *arom A* gene, *arom B* being chorismic acid synthase), which forms a dimer. The order of genes within the AROM polypeptide does not follow the sequence of the reactions within the biosynthetic pathway: the reaction sequence is B, A, L, D, E, but the actual sequence runs (using the *E. coli* gene names) *aroB*, *aroD*, *aroE*, *aroL*, *aroA*. This should not be a surprise, because the genes in the much more organized FAS complex do not follow the order of the reaction either. Presumably the fungal AROM complex arose by gene fusion from individual bacterial-like genes.

TABLE 10.1 Metabolic cost of amino acids

Amino acid	Cost (number of ATP molecules)
Ala	20
Arg	44
Asp	21
Asn	22
Cys	19
Glu	30
Gln	31
Gly	12
His	42
Ile	55
Leu	47
Lys	51
Met	44
Phe	65
Pro	39
Ser	18
Thr	31
Trp	78
Tyr	62
Val	39

The figures are based on the number of ATP molecules required to make the amino acid from metabolic intermediates.

Is this merely a multifunctional polypeptide, or could it qualify as a genuine MEC? Or, to put it another way, how much does a polypeptide need to do to qualify it as an MEC? First, it needs to have some three-dimensional organization; that is, a contact between the domains. There is little evidence of this, although cross-linking studies have indicated some globular structure to the AROM complex [46]. Second, there must be communication between enzymes. Expression of single components or of pairs of components in most cases produces monofunctional or difunctional enzymes with properties essentially identical to full-length AROM. This is not completely true, however, because there is evidence for altered activity of some components in the intact complex [46–48].

An MEC would be expected to show some kind of property that is different from what one would get just by joining the enzymes together, for example allostery or substrate channeling. There is no evidence of allostery. There is, however, some evidence for channeling, for which we need to consider a second pathway, the quinate pathway.

In the fungus *Aspergillus nidulans* there is a pathway that allows it to grow on leaf litter, called the quinate utilization (qut) pathway (**Figure 10.26**) [46]. Quinate comprises about 10% of leaf litter by weight, and is a major foodstuff for many fungi. A comparison of the genes involved in quinate utilization and the AROM genes (**Figure 10.27**) suggests that all the genes used in quinate utilization were adapted from pre-existing genes [49]. The *qutD* and *qutG* genes, encoding respectively a permease and a presumed phosphatase, have similarities to bacterial sugar transporters and to inositol monophosphatase. The *qutE* gene encoding the catabolic dehydroquinase has no similarity to the anabolic dehydroquinase in AROM, but fungi have two completely different non-orthologous dehydroquinases, and *qutE* is related to the other (type II) enzyme (of which more later). It has been suggested that the type II enzyme was recruited as a catabolic enzyme by horizontal transfer, once the quinate substrate became abundant [50]. In several fungal species the (presumed) original type I enzyme has been lost, meaning that the type II enzyme now fulfills both anabolic and catabolic functions in different species. The *qutB* gene encoding quinate dehydrogenase is related to the bacterial *aroE* shikimate dehydrogenase gene.

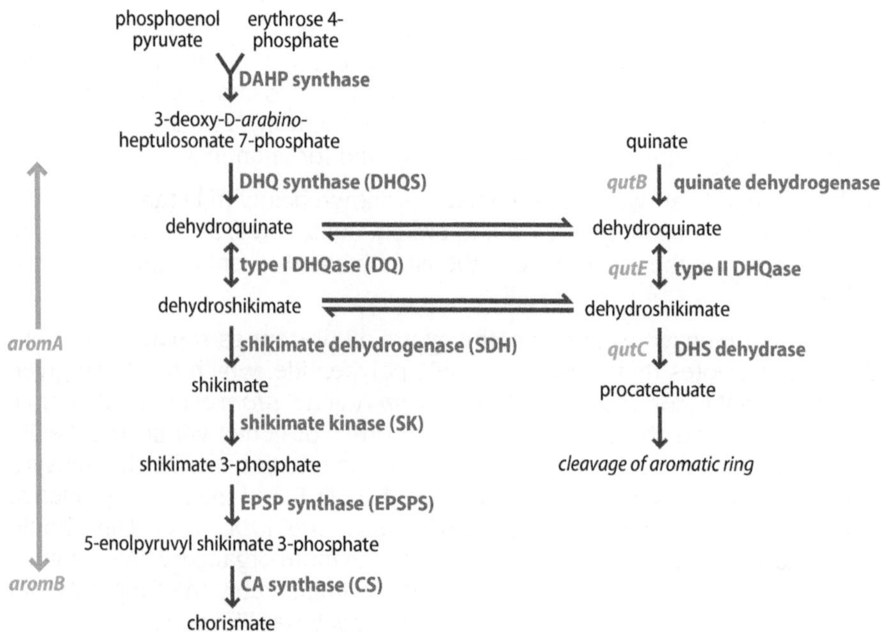

FIGURE 10.26
The enzymes, genes, and intermediates in the shikimate biosynthetic pathway (left) and quinate degradative pathway (right) in the fungus *Aspergillus nidulans*. Dehydroquinate and dehydroshikimate are in both pathways: the horizontal arrows denote possible different locations (channeling) for the two pathways.

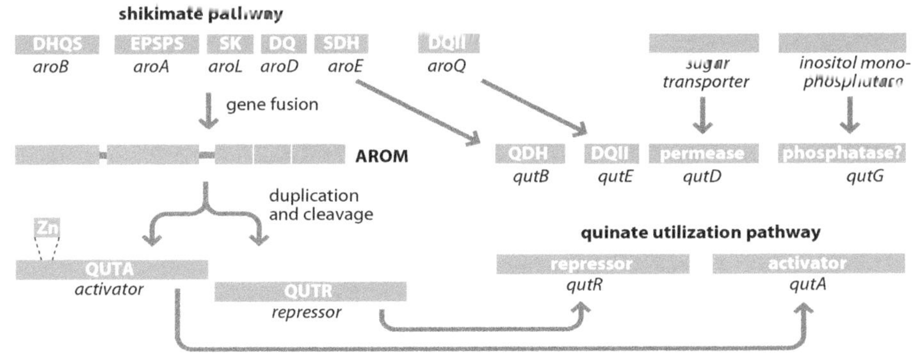

FIGURE 10.27
The modular structure of the enzymes and regulatory proteins comprising the shikimate and quinate pathways in *A. nidulans*. The boxes at top left that are named *aro* describe genes from bacteria encoding monofunctional shikimate pathway enzymes (see Figure 10.26). These genes probably fused to produce the *aromA* cluster encoding AROM. The genes of the quinate utilization (qut) pathway are *qutB* and *qutE* (Figure 10.26) together with a permease and a possible phosphatase, plus the activator *qutA* and the repressor *qutR*. Note that although there is a dehydroquinase gene in both pathways, they are structurally unrelated: some bacteria have a type II gene with a similar sequence to the quinate utilization pathway gene. *qutA* and *qutR* are likely to have arisen by gene duplication of *aromA*, splitting into two parts (within the EPSPS gene), and insertion of a DNA-binding zinc finger sequence. (Adapted from A. R. Hawkins et al., *J. Gen. Microbiol.* 139: 2891–2899, 1993. With permission from the Society for General Microbiology.)

However, the most interesting genes are two regulatory genes: a repressor *qutR* and an activator *qutA*. Both of these are likely to have evolved from a duplication and subsequent splitting of AROM, with at some point a loss of the original catalytic function. *qutA* is similar to the final three proteins in AROM, whereas *qutR* is similar to the first two, with the incorporation of a zinc binuclear cluster near the N terminus, which helps to provide the regulatory function. This provides a nice example of the way in which evolution borrows whatever it needs from elsewhere: obviously it is easier if the source is close at hand! And it seems very reasonable that quinate utilization should have taken genes from AROM, because there is no point in having a *qut* gene cluster unless there is quinate in leaf litter to digest, and quinate is a major product of the shikimate pathway in plants. (About 20% of the carbon fixed by plants goes through the shikimate pathway, mainly for the synthesis of lignin-related compounds, which are degraded to quinate.) There are many other examples of regulatory domains that have most probably started off as duplicates of catalytic domains and subsequently lost the catalytic function but retained binding ability. We have already seen in Chapter 1 that binding ability is much easier to modify than catalytic activity.

The quinate utilization pathway (see Figure 10.26, right) involves the conversion of quinate to dehydroquinate, and then the conversion of dehydroquinate to dehydroshikimate by the type II dehydroquinase mentioned above [46]. Thus, dehydroquinases are involved in both a biosynthetic pathway and a degradative pathway in the same cell. One could therefore expect some kind of channeling to separate these two pathways, which otherwise would tend to go round in a 'futile' (energetically wasteful) cycle. When the *qutE* type II dehydroquinase gene was deleted in *A. nidulans*, the fungus lost its ability to grow on quinate. However, a fivefold overexpression of AROM (containing the type I dehydroquinase) restored growth [51]. This suggests that AROM does channel its metabolites, but not very well: it is a leaky channel. The channeling may be most effective when flux through the pathway is low; that is, when the organism is growing in its natural environment rather than in a laboratory.

We conclude that AROM is a poor MEC, although it has started along that path. It is always dangerous to suggest reasons for evolution to have taken the path it has, but one can speculate that there is some pressure to evolve an MEC, because of the expense of the pathway. However, it is not as expensive as tryptophan biosynthesis (a later stage in the same pathway), and it needs to be leaky because shikimate feeds into a range of biosynthetic pathways that do not involve chorismate.

10.3.5 Integral membrane complexes are not MEC-like

In this section, we consider a quite different type of complexes: those formed by membrane proteins. This also gives us the opportunity to examine whether complexes formed from membrane proteins are significantly different from

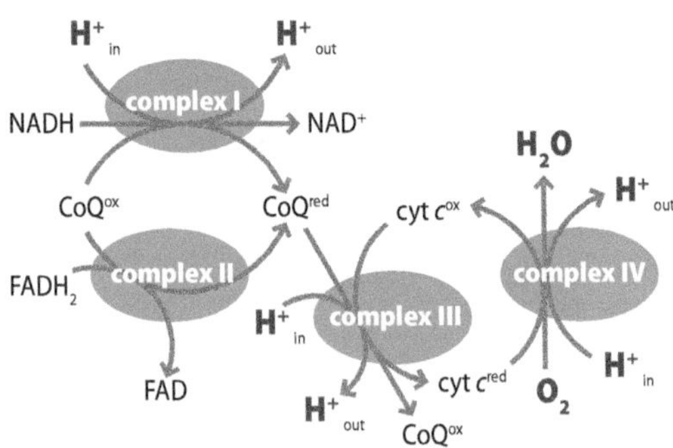

FIGURE 10.28
The mitochondrial electron transport chain. Complexes I, III, and IV pump protons out of the mitochondrion and thereby create a proton gradient used to make ATP. Coenzyme Q (CoQ) is also known as ubiquinone. Cyt *c* is cytochrome *c*. Complex II does not pump protons; its main function is to inject the electrons from $FADH_2$ into the electron transport chain.

MECs formed by globular proteins. We consider three main complexes: the electron transport chain, F_OF_1-ATPase, and the photosynthetic reaction center.

The electron transport chain is the end of the respiratory chain in eukaryotes. NADH and $FADH_2$ are oxidized by O_2 to create a proton gradient across the mitochondrial membrane, which is used to make ATP. The electrons generated by the oxidation pass through four successive protein complexes, known as complexes I, II, III, and IV. Essentially each of these complexes is a single enzyme, which catalyzes a two-substrate redox reaction in which one substrate reduces the other. In the process, they also move protons from one side of the membrane to the other. This is a linear system, in which the product of one reaction is passed directly to the next enzyme (**Figure 10.28**). Complex II is, however, a branch, feeding in electrons from a different source. There would therefore be considerable advantage to be gained from having all the components assembled into a single MEC.

However, as far as we know, they are not: each of the complexes is thought to exist as a separate independent entity. Each individual component (each of the four complexes) has a defined structure: they need to, because the rate of electron transport is heavily dependent on the distance between redox centers. Therefore it is important that the redox centers should all be close enough to pass on the electrons rapidly. The structure of complex IV (better known as cytochrome *c* oxidase) is shown in **Figure 10.29**. It is remarkably complicated, considering that it performs a relatively straightforward oxidation reaction. It consists of between 8 and 13 subunits with four redox centers (two heme

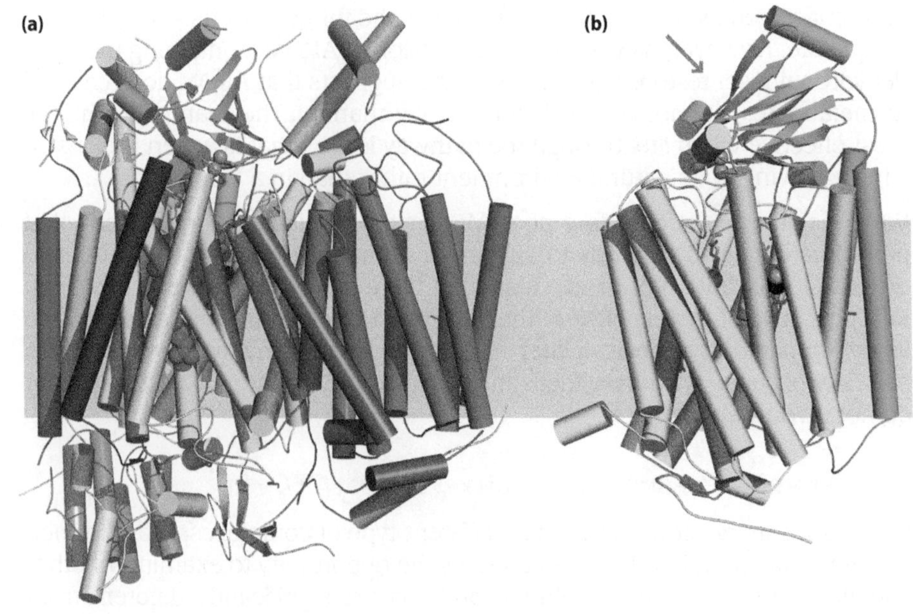

FIGURE 10.29
Cytochrome *c* oxidase from bovine heart (PDB file 3ag1). Different peptide chains are colored differently. There are 13 chains, 10 encoded by nuclear DNA and 3 by mitochondrial DNA. The approximate location of the membrane is shown. The key parts of the structure are two heme groups, which are called cytochrome *a* and cytochrome a_3, and two copper centers called Cu_A and Cu_B. Oxygen binds to the cytochrome a_3/Cu_B binuclear center; the structure depicted here contains bound carbon monoxide, shown in gray, to the right of the red spheres of cytochrome a_3. The oxygen is reduced by an electron originating from reduced cytochrome *c*, which docks close to the Cu_A center (orange spheres at the top); the electron gets to cytochrome a_3 via cytochrome *a*. (a) View of the entire complex. The intact complex is a homodimer, but only one monomer is shown here. (b) View from a different angle of subunits 1 and 2 only, using the same color scheme as in (a) except that cytochrome *a* is shown in salmon pink. The two cytochromes are shown as sticks rather than spheres, except for the Fe atom in the center. The docking site for cytochrome *c* is indicated by an arrow.

cytochromes and two coppers). Some subunits are encoded by mitochondrial DNA and some by nuclear DNA; some are integral and some peripheral. Many have no known function.

The least well understood of the four complexes is complex I. This again catalyzes a fairly simple reaction (though again associated with proton movement). The complex from bovine heart has 46 different subunits, of which 14 are homologous to bacterial subunits and seem to be 'core,' and probably 9 cofactors. Again, some are nuclear and some mitochondrial, and many have unknown function.

The electrons are carried from complexes I and II to complex III by ubiquinone (coenzyme Q) (see Figure 10.28). We might expect this to be attached to a swinging arm or to diffuse through a channel; however, as far as is known it merely diffuses within the membrane, and there is no fixed spatial relationship between the complexes. Similarly, electrons are carried from complex II to complex IV by cytochrome *c*. Cytochromes are sometimes attached to the membrane by a hydrophobic tail, but otherwise there is again no spatial linkage. It is worth pointing out that the electron transport chain in photosynthetic plants (photosystems I and II, together with cytochrome b_6f) similarly has no obvious organization within the membrane, and no covalent channeling of carriers between one complex and the next: it merely relies on diffusion through the membrane.

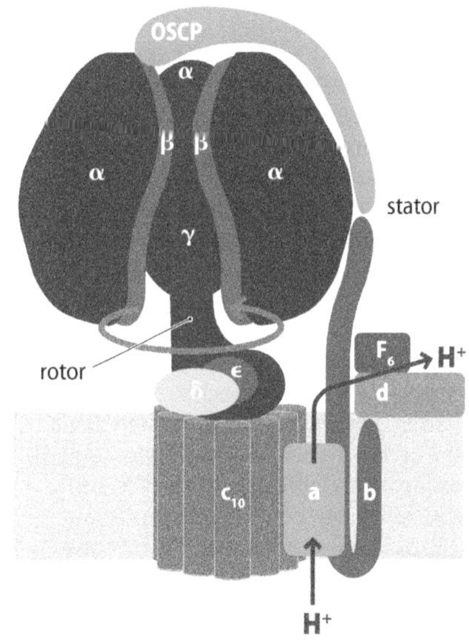

FIGURE 10.30
The structure of mitochondrial F_OF_1-ATPase. Compare with Figure 7.16, where the structure and function of this molecular machine are discussed in more detail.

The electron transport chain is thus rather a disappointment as regards MECs. It is instructive to compare eukaryotes with prokaryotes, where we find similar complexes used for energy generation. However, prokaryotes have a wide range of alternative electron donors and acceptors, which are usually inducible. They are also modular, in that different starts and ends can be used depending on what is available. Therefore it is fair to say that within prokaryotes there has probably been little opportunity for evolution to create super-efficient production-line energy generators. The situation can be compared with that seen above for fatty acid synthesis and β-oxidation: there was more pressure for an efficient synthesis (energy-consuming) system than for an energy generation system.

By contrast, F_OF_1-ATPase (Chapter 7) has the complexity and structure we would expect for a genuine MEC. It converts a proton gradient to mechanical movement (of an α-helical protein, as we might expect), and thence to the synthesis of ATP. Its remarkable structure (**Figure 10.30**) is essentially a motor, where the flow of a current of H^+ rotates a spindle and in this case generates ATP. In comparison with the genuine MECs we have looked at in this chapter, it has little regulatory sophistication, or feedback from one part to another (so far as we know). It cannot qualify as an MEC despite its high degree of structural organization; it does, however, contain a quite new macroscopically mechanical aspect. We do not discuss it further here because it was discussed in detail in Chapter 7.

Finally, we look at the photosynthetic reaction center. We shall consider the bacterial reaction center, which is rather better understood. This converts light into chemical energy in the form of a proton gradient. The reaction center itself is, like the mitochondrial electron transport complexes, a complicated structure with three protein chains and about 10 redox cofactors. The arrangement of the cofactors has almost complete two-fold symmetry (**Figure 10.31**), yet electrons travel down only one arm; the other arm does not seem to be involved. The final destination of the electrons produced is a quinone, namely ubiquinone or Q_B, which is shown at the bottom left of Figure 10.31b.

The reaction center itself does not have a large enough surface area to trap a very high proportion of the light falling on the bacterium. Light is therefore trapped by an antenna complex consisting of bacteriochlorophyll molecules

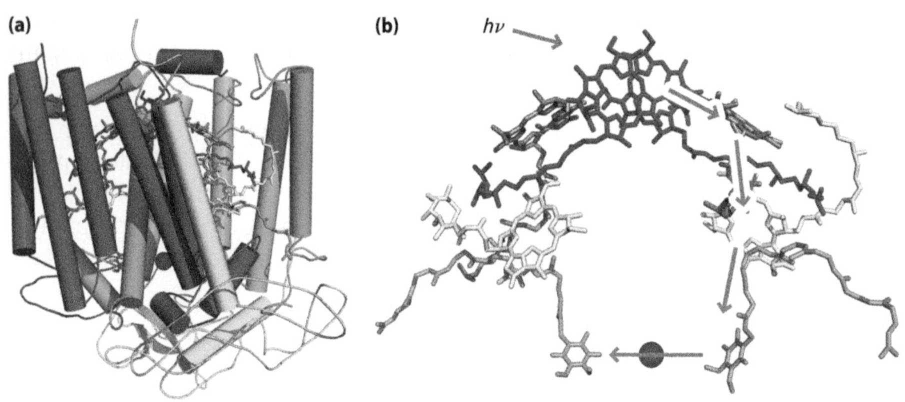

FIGURE 10.31
The photosynthetic reaction center from the bacterium *Rhodobacter sphaeroides* (PDB file 2rcr). There are three polypeptide chains, called L (cyan), M (magenta), and H (green). The L and M subunits show pseudo-two-fold symmetry. (a) Intact complex. (b) The same view, but cofactors only. The cofactors are arranged with almost exact two-fold symmetry, but electrons are transferred down only one arm. Photons ($h\nu$) are trapped by the 'special pair' of bacteriochlorophylls (blue). Within 3 ps, the energy is transferred via the accessory bacteriochlorophyll (green) to the bacteriopheophytin (yellow) on the right-hand or L subunit arm. About 200 ps later it has reached the quinone Q_A (orange). It then takes a time approaching 100 μs to move to the other quinone, Q_B, on the left, which is also called ubiquinone. The intervening iron atom (red) seems to take little part in this electron transfer. Because of the large distances between some of the chromophores, it is assumed that aromatic rings within the protein help to carry electrons. Once Q_B has received a second electron (and is therefore a relatively unreactive hydroquinone rather than a very reactive radical) it is free to diffuse away from the center and transfer the electrons to a cytochrome bc_1.

held by small 'light harvesting' (LH) proteins, which surround the reaction center (RC). In the photosynthetic bacterium *Rhodobacter sphaeroides*, there are two types of LH protein, each of which is ligated to a bacteriochlorophyll, which is what actually traps the light. LH1 αβ dimers surround the reaction center and pass energy directly to it, while LH2 proteins form separate rings and pass energy to LH1. The LH1 rings are always larger than the LH2 rings, partly because they have to be large enough to surround the RC. The number of αβ dimers in each ring depends on the species and does not seem to be important: it is presumably fixed by the shape of the dimer, and thus by the number that pack neatly into a complete circle (see *7.4). The LH rings and RC can be seen by atomic force microscopy (AFM; Section 11.5.2) in native membranes (**Figure 10.32**).

There are crystal structures of LH2 rings and of the RC (see Figure 10.31), but so far the native LH1–RC complex has not been crystallized. The structure of the LH1–RC complex has been pieced together by using a combination of X-ray crystallography, electron microscopy, and NMR. The overall structure of the complex was determined by cryo-electron diffraction (Section 11.4.11), and is shown in **Figure 10.33**. Rather than being a simple ring surrounding the RC, it is actually an S-shaped dimer containing two RCs. One reason for this structure is that the Q_B that becomes reduced by light energy (see Figure 10.31b) has to leave the RC and diffuse to cytochrome bc_1. If the LH ring were complete, there would be no gap through which Q_B could exit. Formation of the dimeric structure requires the presence of a small single-transmembrane protein called PufX, which is located at the green circle in Figure 10.33 but which has an N-terminal arm that reaches out and pairs with a second PufX at the dimer interface.

An atomic-level model of the structure has been generated by using the following components: the crystal structure of the RC; a crystal structure of LH2, suitably modified by expanding the ring size and replacing the LH2β peptide by the NMR structure of LH1β [52]; and the NMR structure of PufX [53]. Both the NMR structures show a transmembrane helix that has a significant bend

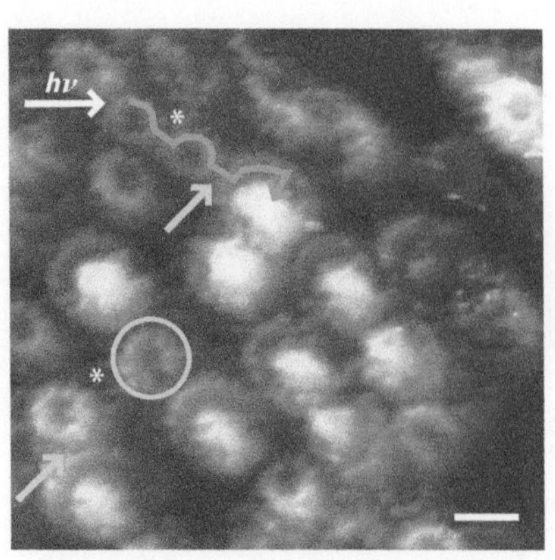

FIGURE 10.32
Atomic force microscopy image of native membranes from *R. sphaeroides*. LH2 rings are small circles, two of which are marked by asterisks. They can be clearly seen to be composed of nine αβ subunits. The LH1 rings are larger and contain a reaction center (RC), which is the brighter (that is, taller) region, particularly the RC H subunit (compare Figure 10.31a, in which the RC H subunit is at the bottom). The LH2 ring within the green ring has an LH1 either side. Light energy passes from LH2 to LH2 to LH1 to RC (blue line). The scale bar is 10 nm. (Adapted from S. Bahatyrova, R. N. Frese and C. A. Siebert, *Nature* 430: 1058–1062, 2004. With permission from Macmillan Publishers Ltd.)

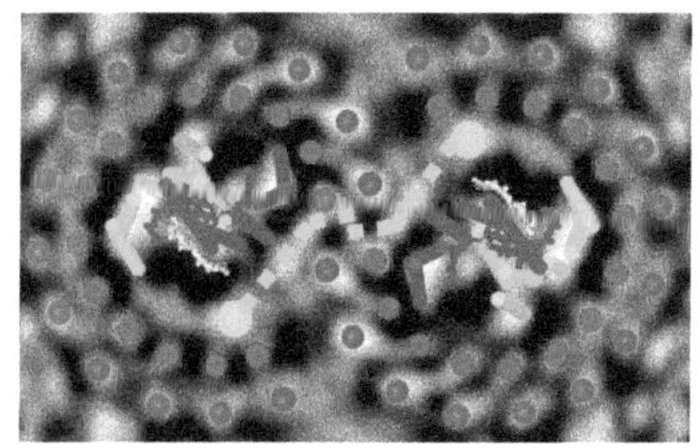

FIGURE 10.33
The projection structure of the LH1/RC/PufX dimer, determined by cryo-electron diffraction. The white regions indicate high electron density. Superimposed on this density is the crystal structure of the RC, in the center of each ring, with Q_B indicated in yellow and the special pair of bacteriochlorophylls in blue; the L, M, and H subunits of the RC are in magenta, green and cyan respectively. The figure also shows the LH1 α and β peptides, which are single-transmembrane peptides and are indicated by red and blue circles respectively. An additional intense region of electron density is shown in green and has been attributed to PufX. The N-terminal end of PufX extends along the surface of the membrane, roughly along the dashed line, and the two PufX N termini contact each other at the dimer interface. (From P. Qian, C. N. Hunter and P. A. Bullough, *J. Mol. Biol.* 349: 948–960, 2005. With permission from Elsevier.)

near the top of the membrane. In LH1β this leads to an assembled LH ring in which the LH1β ends form a domed arch encasing the RC; for PufX the N-terminal helix runs almost parallel to the membrane (see the dashed line in Figure 10.33). These structures were modeled into the electron density.

The final structural detail comes from single-particle cryo-electron microscopy (Section 11.5.1). A large number of LH1–RC–PufX dimers were studied, from which a three-dimensional structure could be calculated (**Figure 10.34**). The result shows a significantly bent structure, which fits the cryo-electron diffraction data well: for example, the RC H helix is almost perpendicular to the membrane plane in this model, matching well the very sharp intense electron density found in the electron microscopy structure.

In photosynthetic bacteria, the LH1–RC complexes tend to occur in spherical membrane invaginations called chromatophores, and evidence is emerging that the structure of the chromatophore is determined by the shape of the LH1–RC complex: because the complex is significantly bent, this makes the membrane curved (**Figure 10.35**) [54]. It seems likely that the cytochrome bc_1 complexes are not located in the chromatophore at all, but outside it. Therefore, the quinones have to shuttle back and forth a significant distance to offload their electrons. Despite this surprisingly inefficient structure, it manages to be remarkably rapid and energy efficient. No doubt this is partly because the two-dimensional diffusion required is still very rapid (Section 4.2.2).

So far little is known about the assembly of these structures. The proteins are expressed in a fixed time order: RC first, then PufX, and then the LH proteins. It is certainly possible that the complexes are entirely self-assembling—once PufX has located to the correct position on RC, it initiates formation of the LH ring. Isolated LH peptides can be completely dissociated and will then spontaneously reassociate into intact rings. It is therefore possible that the entire chromatophore, with a diameter of about 70 nm, can assemble entirely spontaneously. There is thus considerable organizational complexity encoded within the protein structures.

Our brief survey of transmembrane complexes suggests that they are indeed different from soluble proteins. Although in some cases they have a highly organized structure, there is little communication between subunits (disqualifying them as MECs). The initial assembly of the complex seems to be more difficult than for soluble proteins, despite the fact that two-dimensional diffusion is rapid and there is much less difficulty in achieving the correct orientation [55]; perhaps this is because lipids are larger and less mobile than water. However, once assembled, the two-dimensional surface makes transport to and from the complexes easier, and also permits the assembly of more 'mechanical' structures than we have seen so far for soluble proteins.

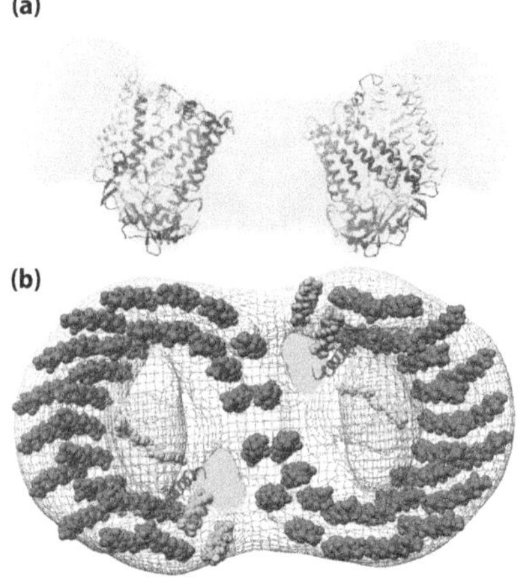

FIGURE 10.34
The three-dimensional structure of the LH1–RC–PufX dimer, determined by single-particle cryo-electron microscopy. (a) A side view of the dimer. The green density shows the structure, into which is placed the crystal structure of the RC. (b) Top view of the dimer. The green mesh shows the boundary of the cryo-EM density. LH1α and β are shown in red, except for the fourteenth pair, which is in blue. PufX is the magenta ribbon, and Q_B is the yellow space-filling representation. The flat yellow regions contain no protein density and are ideally placed to act as pools for replacement quinone molecules. (Courtesy of Neil Hunter, Sheffield.)

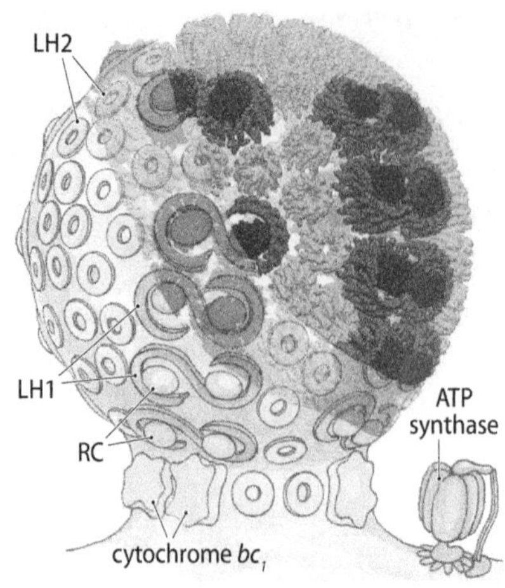

FIGURE 10.35
A model of the photosynthetic chromatophore from *R. sphaeroides*. LH2 is green, LH1 red, and RC blue. The locations of cytochrome bc_1 and F_oF_1-ATP synthase are speculative, but there is no evidence that they are located within the chromatophore itself. AFM (Figure 10.32) and electron microscopy suggest that the LH1/RC dimers are organized in rows as shown here, interspersed by LH2. Modeling suggests that this arrangement permits the rapid transfer of light energy from anywhere on the chromatophore to an RC. (From M. K. Şener et al., *Proc. Natl. Acad. Sci. USA* 104: 15723–15728, 2007. With permission from the National Academy of Sciences.)

10.4 POSSIBLE ADVANTAGES OF MULTIENZYME COMPLEXES

Multienzyme complexes, in the sense used in this chapter, are rare. The examples given here are almost all the well-attested cases that there are. It seems that the cell can get on quite happily without needing such complex organization most of the time; or, to put the same thing more scientifically, evolutionary pressure has not been sufficiently strong to lead to the evolution of many MECs. The examples that we have seen have mainly been in a small number of very intensively energy-requiring biosynthetic processes, particularly cyclic ones in which it is particularly advantageous to be able to hold on to the substrate. We should therefore be careful not to overemphasize their importance. In this section, we consider the advantages that have been suggested to arise from MECs.

10.4.1 Cycling of substrates

In most of the true MECs considered in this chapter, namely PDH, FAS, the long-chain fatty acid β-oxidation system, and some polyketide synthases, the clear rationale for their existence is the need to hold on to the substrate (or, in the case of PDH, the lipoyl group) throughout what is essentially a cyclic series of reactions.

10.4.2 Substrate channeling

This advantage is clear, although limited to a small number of reactions that generate particularly expensive or harmful intermediates. The channeling from one active site to another must be functionally important, because it is linked to allosteric effects that do not permit the first reaction until the other substrate has already bound in the second site; this is the 'hot potato' scenario. The complexes that benefit from substrate channeling are in fact those that do *not* benefit from substrate cycling (Section 10.4.1): together, these two advantages therefore seem to be the main benefits of MECs.

The AROM complex is different, because the intermediates are not harmful and are less expensive; if there is channeling, its justification is more likely to be that the intermediate(s) are involved in both biosynthetic and degradative pathways. Although substrate channeling does seem to occur in the AROM system, it is not very effective or apparently important. Indeed, in eukaryotes at least, the separation of one cellular component from another is generally done using membrane compartmentation rather than protein complexes. Obvious examples are lysosomes and peroxisomes, and one could also argue that mitochondrial membranes have essentially the same function. The nuclear membrane is very different because it has numerous large pores, but then the nucleus does not contain metabolites that need separating in the same sense, so channeling for this reason is apparently not necessary.

10.4.3 Speeding up reactions

It has often been suggested that substrate channeling helps to make reactions faster. In particular, when a second reaction uses a substrate that is the product of the first, their assembly into an MEC should make the overall reaction faster. As discussed above, this is only true to a limited extent, and except for very fast enzymes it probably has little effect. Certainly the relative paucity of MECs in multistep reactions does not suggest that rate enhancement is a major effect. (Chapter 9 suggests, however, that there may be many more complexes in cells than we are yet aware of, so this argument may become weaker with time.)

10.4.4 Faster response time

It has also been suggested that the assembly of a group of sequential enzymes into one spatially defined complex should help to make the response time of the system faster. If there is a sequence of enzymes, such as a biosynthetic pathway, it is desirable that the pathway can be stopped or started rapidly in response to an excess or lack of the end product. It has been estimated for example that a sequence of 10 reactions catalyzed by enzymes placed randomly within the cell would take 1 hour to switch off, because of the time taken for substrates to diffuse from one enzyme to the next, whereas the same 10 reactions could respond within 10 seconds if all the enzymes were spatially colocated [56]. This is a compelling argument, and several authors have presented similar theoretical arguments to suggest that channeling 'must' be important [57, 58]; however, there is very little evidence that it is in fact true, and several authors have presented evidence that this is at best not a strong argument [17].

In Chapter 9, we discussed arguments that suggest that enzymes within the same pathway tend to be spatially close within the cell. The balance of evidence so far suggests that this is 'good enough,' and that there is no need to hold sequential enzymes physically within the same complex in order to enhance the response time.

10.4.5 Active-site coupling

Active-site coupling undoubtedly occurs in PDH, and is a remarkable phenomenon, made possible by the oligomeric structure of the PDH MEC. PDH (and its close relatives) is, however, the only enzyme with this property, and it is not obvious that it is actually all that useful. In the study that looked at *E. coli* PDH with different numbers of lipoyl groups (Section 10.2.3), the overall rate was unaffected by the number of lipoyl groups, implying that, at least under laboratory conditions, active-site coupling has little overall effect on reaction rate. FAS is a dimer and uses a carrier protein, but it has no obvious active-site coupling; indeed, its structure suggests that such coupling is physically impossible. One can therefore speculate that active-site coupling is an adventitious phenomenon made possible by the structure of the PDH complex, but is not such an important feature that it is worth evolving for its own sake.

10.4.6 Increase in solvent capacity

It is argued that there is very little free water in the cell, and therefore that the cell needs to work hard to reduce the concentrations of both metabolites and enzymes so as to 'release' water to allow it to solvate all the solutes present. Thus, enzymes that contain their metabolites within channels require fewer solvent molecules and are presumably beneficial. It is certainly true that there is very little free water (see Chapter 4). However, considering how little free water there is, there is very little evidence (as discussed in Chapter 9) that this presents much of a problem to the cell. It has also been demonstrated that the concentration of a pathway intermediate is not greatly affected by being in a multienzyme complex [59, 60] (this conclusion has, however, been challenged [61]). And, as discussed above, there are much stronger arguments for why channeling has evolved. It therefore seems that, attractive as this idea is, it is not strong enough to produce any evolutionary pressure.

10.4.7 Conclusion

We can conclude that the only advantages of MECs that have actually operated as a force for natural selection are the first two in this list: the need to hold on to intermediates in cyclic reactions, and the need to prevent unusually toxic or expensive intermediates from escaping. All the other reasons are good, but they do not actually seem to be strong enough to produce MECs. In this sense they fall into the same category as Horowitz's retrograde evolution suggestion (Chapter 1): great idea, but not actually true.

10.5 SUMMARY

Multienzyme complexes are special in that there is communication between the components. The total is therefore more than the sum of its parts. MECs require significant evolutionary pressure to evolve.

A few MECs have evolved tunnels between sequential active sites, and allosteric mechanisms to ensure that the different reactions are coordinated, in a process known as substrate channeling. These all generate toxic or reactive intermediates. Other MECs conduct cyclic reactions, in which the coordination of the reactions, and the handing over of an intermediate from one enzyme to another, greatly increase the efficiency of the reaction. These are mostly biosynthetic processes such as fatty acid synthesis. Pyruvate dehydrogenase is also an MEC, both because of the need to recycle cofactors and (possibly) because of its key metabolic role. PDH also displays active-site coupling, in which the substrate is efficiently passed around within the complex. AROM is a poor MEC, in that it displays leaky channeling.

Several other enzyme systems are almost MECs but not quite. These include polyketide biosynthesis and nonribosomal peptide biosynthesis. Membrane complexes are often complicated multiprotein structures but show no evidence for communication between sites. It is possible that diffusion within the membrane is so fast that such communication is not necessary.

Possible reasons for the existence of MECs are discussed, and it is concluded that the only convincing ones are the cycling of substrates and substrate channeling of 'hot potatoes.' Evolutionary pressure sufficient to produce MECs is therefore rare.

10.6 FURTHER READING

Details of most of the systems discussed in this chapter can be found in any biochemistry textbook. I particularly recommend Biochemistry, by Voet and Voet [62], for this material.

The paper by Perham [4] is mainly about PDH, but is an entertaining read on MECs in general.

10.7 WEBSITES

http://blanco.biomol.uci.edu/Membrane_Proteins_xtal.html *Membrane proteins of known structure.*

http://www.mpdb.tcd.ie/ *Membrane Protein Data Bank.*

10.8 PROBLEMS

Hints for some of these problems can be found on the book's website.

1. **What is the experimental evidence for channeling in tryptophan synthase?**

2. **Chapter 10 argues that tryptophan synthase may have evolved into such a complex system because of the cost of tryptophan synthesis. The next most expensive amino acids are Phe, Tyr, and Lys or Ile (see Tables 1.7 and 10.1). Are their syntheses achieved in a similarly complex way?**

3. Is aspartate kinase–homoserine dehydrogenase a multienzyme complex?

4. Draw out the substrates of PDH and of related enzymes (α-ketoacid dehydrogenase, etc.). Discuss the evolution of this enzyme.

5. Describe the regulation of PDH by phosphorylation.

6. Describe the mechanisms of medium-chain and short-chain fatty acid oxidases. Could it be argued that they form multienzyme complexes?

7. Review the evidence that nonribosomal peptide synthases are structurally organized into complexes and therefore may merit a description as multienzyme complexes.

8. The only true integral membrane enzymes pump something across the membrane (although there are many enzymes anchored to membranes, but for which the enzyme part is effectively a soluble globular domain.) One should therefore not expect to find multienzyme complexes within the membrane. Is this fair?

10.9 NUMERICAL PROBLEMS

N1. In the absence of channeling, is the rate of *E. coli* carbamoyl phosphate synthetase likely to be limited by the concentration of ammonia in solution; in other words, is channeling likely to speed up the reaction significantly? The turnover rate of carbamoyl phosphate synthetase can be found in the BRENDA enzyme database (http://www.brenda-enzymes.org/). Note that we are interested in ammonia, not the ammonium ion. The intracellular concentration of ammonium is roughly 150 μM.

10.10 REFERENCES

1. G Pohnert (2001) Diels-Alderases. *ChemBioChem* 2:873–875.
2. H Oikawa & T Tokiwano (2004) Enzymatic catalysis of the Diels–Alder reaction in the biosynthesis of natural products. *Nat. Prod. Rep.* 21:321–352.
3. RM Stroud (1994) An electrostatic highway. *Nat. Struct. Biol.* 1:131–134.
4. RN Perham (1975) Self-assembly of biological macromolecules. *Phil. Trans. R. Soc. Lond. B* 272:123–136.
5. MF Dunn, D Niks, H Ngo et al. (2008) Tryptophan synthase: the workings of a channeling nanomachine. *Trends Biochem. Sci.* 33:254–264.
6. P Pan, E Woehl & MF Dunn (1997) Protein architecture, dynamics and allostery in tryptophan synthase channeling. *Trends Biochem. Sci.* 22:22–27.
7. KS Anderson, EW Miles & KA Johnson (1991) Serine modulates substrate channeling in tryptophan synthase: a novel intersubunit triggering mechanism. *J. Biol. Chem.* 266:8020–8033.
8. TR Schneider, E Gerhardt, M Lee et al. (1998) Loop closure and intersubunit communication in tryptophan synthase. *Biochemistry* 37:5394–5406.
9. EW Miles, S Rhee & DR Davies (1999) The molecular basis of substrate channeling. *J. Biol. Chem.* 274:12193–12196.
10. MK Geck & JF Kirsch (1999) A novel, definitive test for substrate channeling illustrated with the aspartate aminotransferase malate dehydrogenase system. *Biochemistry* 38:8032–8037.
11. JB Thoden, HM Holden, G Wesenberg et al. (1997) Structure of carbamoyl phosphate synthetase: a journey of 96 Å from substrate to product. *Biochemistry* 36:6305–6316.
12. V Serre, H Guy, X Liu et al. (1998) Allosteric regulation and substrate channeling in multifunctional pyrimidine biosynthetic complexes: analysis of isolated domains and yeast-mammalian chimeric proteins. *J. Mol. Biol.* 281:363–377.
13. SH Chen, JW Burgner, JM Krahn et al. (1999) Tryptophan fluorescence monitors multiple conformational changes required for glutamine phosphoribosylpyrophosphate amidotransferase interdomain signaling and catalysis. *Biochemistry* 38:11659–11669.
14. JM Krahn, JH Kim, MR Burns et al. (1997) Coupled formation of an amidotransferase interdomain ammonia channel and a phosphoribosyltransferase active site. *Biochemistry* 36:11061–11068.
15. JL Smith (1998) Glutamine PRPP amidotransferase: snapshots of an enzyme in action. *Curr. Opin. Struct. Biol.* 8:686–694.
16. AH Elcock, MJ Potter, DA Matthews et al. (1996) Electrostatic channeling in the bifunctional enzyme dihydrofolate reductase-thymidylate synthase. *J. Mol. Biol.* 262:370–374.
17. DR Knighton, CC Kan, E Howland et al. (1994) Structure of and kinetic channeling in bifunctional dihydrofolate reductase-thymidylate synthase. *Nat. Struct. Biol.* 1:186–194.
18. PH Liang & KS Anderson (1998) Substrate channeling and domain-domain interactions in bifunctional thymidylate synthase: dihydrofolate reductase. *Biochemistry* 37:12195–12205.

19. HV Westerhoff & GR Welch (1992) Enzyme organization and the direction of metabolic flow: physicochemical considerations. *Curr. Top. Cell. Regul.* 33:361–390.
20. RN Perham (1991) Domains, motifs, and linkers in 2-oxo acid dehydrogenase multienzyme complexes: a paradigm in the design of a multifunctional protein. *Biochemistry* 30:8501–8512.
21. JLS Milne, D Shi, PB Rosenthal et al. (2002) Molecular architecture and mechanism of an icosahedral pyruvate dehydrogenase complex: a multifunctional catalytic machine. *EMBO J.* 21:5587–5598.
22. ZH Zhou, DB McCarthy, CM O'Connor et al. (2001) The remarkable structural and functional organization of the eukaryotic pyruvate dehydrogenase complexes. *Proc. Natl. Acad. Sci. USA* 98:14802–14807.
23. XK Yu, Y Hiromasa, H Tsen, et al. (2008) Structures of the human pyruvate dehydrogenase complex cores: a highly conserved catalytic center with flexible N-terminal domains. *Structure* 16:104–114.
24. RN Perham (2000) Swinging arms and swinging domains in multifunctional enzymes: catalytic machines for multistep reactions. *Annu. Rev. Biochem.* 69:961–1004.
25. SL Turner, GC Russell, MP Williamson & JR Guest (1993) Restructuring an interdomain linker in the dihydrolipoamide acetyltransferase component of the pyruvate dehydrogenase complex of *Escherichia coli*. *Protein Eng.* 6:101–108.
26. ML Hackert, RM Oliver & LJ Reed (1983) A computer-model analysis of the active-site coupling mechanism in the pyruvate dehydrogenase multienzyme complex of *Escherichia coli*. *Proc. Natl. Acad. Sci. USA* 80:2907–2911.
27. LC Packman, CJ Stanley & RN Perham (1983) Temperature-dependence of intramolecular coupling of active sites in pyruvate dehydrogenase multienzyme complexes. *Biochem. J.* 213:331–338.
28. MJ Danson, AR Fersht & RN Perham (1978) Rapid intramolecular coupling of active sites in the pyruvate dehydrogenase complex of *Escherichia coli*: mechanism for rate enhancement in a multimeric structure *Proc. Natl. Acad. Sci. USA* 75:5386–5390.
29. RS Machado, JR Guest & MP Williamson (1993) Mobility in pyruvate dehydrogenase complexes with multiple lipoyl domains. *FEBS Lett.* 323:243–246.
30. T Maier, S Jenni & N Ban (2006) Architecture of mammalian fatty acid synthase at 4.5 Å resolution. *Science* 311:1258–1262.
31. S Jenni, M Leibundgut, D Boehringer et al. (2007) Structure of fungal fatty acid synthase and implications for iterative substrate shuttling. *Science* 316:254–261.
32. C Khosla & PB Harbury (2001) Modular enzymes. *Nature* 409:247–252.
33. M Leibundgut, T Maier, S Jenni & N Ban (2008) The multienzyme architecture of eukaryotic fatty acid synthases. *Curr. Opin. Struct. Biol.* 18:714–725.
34. WH Kunau, V Dommes & H Schulz (1995) β-Oxidation of fatty acids in mitochondria, peroxisomes, and bacteria: a century of continued progress. *Prog. Lipid Res.* 34:267–342.
35. M Ishikawa, D Tsuchiya, T Oyama et al. (2004) Structural basis for channelling mechanism of a fatty acid β-oxidation multienzyme complex. *EMBO J.* 23:2745–2754.
36. MA Fischbach & CT Walsh (2006) Assembly-line enzymology for polyketide and nonribosomal peptide antibiotics: logic, machinery, and mechanisms. *Chem. Rev.* 106:3468–3496.
37. KJ Weissman & R Müller (2008) Protein–protein interactions in multienzyme megasynthetases. *ChemBioChem* 9:826–848.
38. RS Gokhale, SY Tsuji, DE Cane & C Khosla (1999) Dissecting and exploiting intermodular communication in polyketide synthases. *Science* 284:482–485.
39. C Khosla, Y Tang, AY Chen et al. (2007) Structure and mechanism of the 6-deoxyerythronolide B synthase. *Annu. Rev. Biochem.* 76:195–221.
40. YY Tang, AY Chen, CY Kim et al. (2007) Structural and mechanistic analysis of protein interactions in module 3 of the 6-deoxyerythronolide B synthase. *Chem. Biol.* 14:931–943.
41. YY Tang, CY Kim, II Mathews et al. (2006) The 2.7-Å crystal structure of a 194-kDa homodimeric fragment of the 6-deoxyerythronolide B synthase. *Proc. Natl. Acad. Sci. USA* 103:11124–11129.
42. JM Crawford, PM Thomas, JR Scheerer et al. (2008) Deconstruction of iterative multidomain polyketide synthase function. *Science* 320:243–246.
43. H Jenke-Kodama, A Sandmann, R Muller & E Dittmann (2005) Evolutionary implications of bacterial polyketide synthases. *Mol. Biol. Evol.* 22:2027–2039.
44. CP Ridley, HY Lee & C Khosla (2008) Evolution of polyketide synthases in bacteria. *Proc. Natl. Acad. Sci. USA* 105:4595–4600.
45. DE Cane & CT Walsh (1999) The parallel and convergent universes of polyketide synthases and nonribosomal peptide synthetases. *Chem. Biol.* 6:R319-R25.
46. AR Hawkins, HK Lamb, JD Moore et al. (1993) The pre-chorismate (shikimate) and quinate pathways in filamentous fungi: theoretical and practical aspects. *J. Gen. Microbiol.* 139:2891–2899.
47. JD Moore & AR Hawkins (1993) Overproduction of, and interaction within, bifunctional domains from the amino-termini and carboxy-termini of the pentafunctional AROM protein of *Aspergillus nidulans*. *Mol. Gen. Genet.* 240:92–102.
48. AR Hawkins & M Smith (1991) Domain structure and interaction within the pentafunctional AROM polypeptide. *Eur. J. Biochem.* 196:717–724.
49. AR Hawkins & HK Lamb (1995) The molecular biology of multidomain proteins: selected examples. *Eur. J. Biochem.* 232:7–18.
50. DG Gourley, AK Shrive, I Polikarpov et al. (1999) The two types of 3-dehydroquinase have distinct structures but catalyze the same overall reaction. *Nat. Struct. Biol.* 6:521–525.
51. HK Lamb, JPTW van den Hombergh, GH Newton et al. (1992) Differential flux through the quinate and shikimate pathways: implications for the channeling hypothesis. *Biochem. J.* 284:181–187.
52. MJ Conroy, WHJ Westerhuis, PS Parkes-Loach et al. (2000) The solution structure of *Rhodobacter sphaeroides* LH1β reveals two helical domains separated by a more flexible region: structural consequences for the LH1 complex. *J. Mol. Biol.* 298:83–94.
53. RB Tunnicliffe, EC Ratcliffe, CN Hunter & MP Williamson (2006) The solution structure of the PufX polypeptide from *Rhodobacter sphaeroides*. *FEBS Lett.* 580:6967–6971.
54. P Qian, PA Bullough & CN Hunter (2008) Three-dimensional reconstruction of a membrane-bending complex: the RC-LH1-PufX core dimer of *Rhodobacter sphaeroides*. *J. Biol. Chem.* 283:14002–14011.
55. DO Daley (2008) The assembly of membrane proteins into complexes. *Curr. Opin. Struct. Biol.* 18:420–424.
56. FH Gaertner (1978) Unique catalytic properties of enzyme clusters. *Trends Biochem. Sci.* 3:63–65.
57. GR Welch & JS Easterby (1994) Metabolic channeling versus free diffusion: transition-time analysis. *Trends Biochem. Sci.* 19:193–197.
58. J Ovádi & PA Srere (1992) Channel your energies. *Trends Biochem. Sci.* 17:445–447.
59. A Cornish-Bowden (1991) Failure of channelling to maintain low concentrations of metabolic intermediates. *Eur. J. Biochem.* 195:103–108.
60. A Cornish-Bowden (2004) Fundamentals of Enzyme Kinetics, 3rd ed. London: Portland Press.
61. P Mendes, DB Kell & HV Westerhoff (1992) Channelling can decrease pool size. *Eur. J. Biochem.* 204:257–266.
62. DJ Voet & JG Voet (2004) Biochemistry, 3rd ed. New York: Wiley.
63. GJ Kleywegt & TA Jones (1994) Detection, delineation, measurement and display of cavities in macromolecular structures. *Acta Cryst.* D50:178–185.
64. M Liebundgut, S Jenni, C Frick & N Ban (2007) Structural basis for substrate delivery by acyl carrier protein in the yeast fatty acid synthase. *Science* 316:288–290.
65. T Maier, M Leibundgut & N Ban (2008) The crystal structure of a mammalian fatty acid synthase. *Science* 321:1315–1322.

CHAPTER 11

Techniques for Studying Proteins

The problem of biology is not to stand aghast at the complexity but to conquer it.

Sydney Brenner (2004), [1]

This book is mainly concerned with the principles of how proteins work, rather than the experiments that uncovered the principles. But clearly experiments are vital, and the results (and the papers that present them) can only be appreciated if one has some understanding of the experimental techniques underlying protein science. This topic could easily be a textbook on its own (and indeed there are many such books), so I have necessarily had to be somewhat selective in the material presented in this chapter. In particular, I have only described the outlines of how the techniques work, my guiding principle being whether the descriptions enable a reader to understand the current research literature at least roughly. I have therefore given more detail about some topics (for example, NMR) than others. I have also tried to evaluate the techniques, rather than merely describe them. My evaluations are obviously biased by personal experience, which I hope is not atypical. Inevitably, some readers will be very familiar with some of these techniques, and completely baffled by others. There is no substitute for going into a lab and discussing methods with the practitioners!

11.1 EXPRESSION AND PURIFICATION

In virtually all cases, any protein to be studied has to be **overexpressed** and purified before it can be studied. In this context it is worth repeating Kornberg's fourth commandment: Don't waste clean thinking on dirty enzymes [2]. In other words, it is always worth purifying your protein because you could get very misleading results otherwise. The degree of purity depends on the methods to be used, but as a working rule it needs to be clean enough to produce a single band on an **SDS-PAGE gel (*11.1)**.

*11.1 SDS-PAGE

A gel is made from polyacrylamide into which is incorporated sodium dodecyl sulfate (SDS). The SDS helps to denature proteins, coating them evenly with this negatively charged detergent, so that all proteins have a similar appearance, being extended and hydrophobic with a negative charge proportional to their length. A voltage is placed across the gel (**Figure 11.1.1**), with the positive voltage at the bottom, so that proteins run down the gel at a rate dependent on their size; the fastest (smallest) appear at the bottom of the gel. The method forms a reliable and quick way to separate proteins according to their size. Proteins are most commonly located by staining with Coomassie brilliant blue, which stains proteins blue. If greater sensitivity is needed (for example to detect proteins in two-dimensional gels), they can be stained with silver salts.

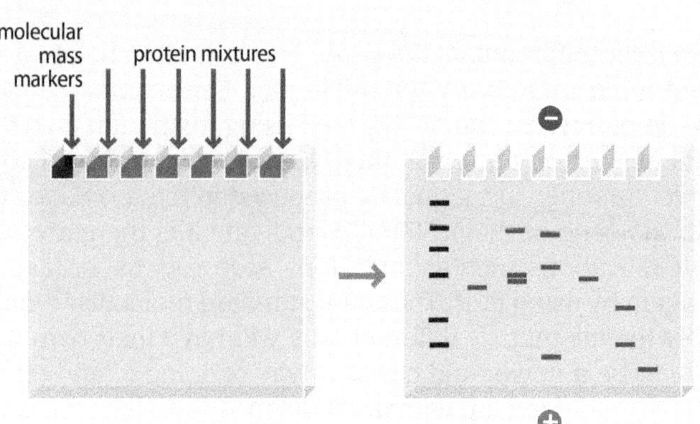

FIGURE 11.1.1
Typically, an SDS-PAGE gel will have several lanes, one of which contains a commercially available set of molecular mass marker proteins. Proteins run with the smallest proteins fastest (at the bottom of the gel). The relationship between size and mobility is logarithmic rather than linear: the larger proteins bunch up together at the top of the gel.

*11.2 PCR

Polymerase chain reaction (PCR) is a very common and useful tool in molecular biology, used for amplifying double-stranded DNA. It relies on a heat-stable DNA polymerase, usually the Taq polymerase from the thermophile *Thermus aquaticus*, which can withstand high temperature. In this technique (**Figure 11.2.1**), double-stranded DNA is heated to more than 90°C, which causes it to melt (become single stranded). This is done in the presence of Taq polymerase and deoxynucleotide triphosphates (dNTPs), as well as two primers, which are short sequences of DNA that are complementary to the two ends of the DNA to be amplified. The solution is then cooled to about 60°, at which temperature the primers bind (anneal) to the DNA template, and the polymerase adds dNTPs to the ends of the primers, making a second copy of each strand. This procedure is then repeated many (for example 20–30) times; in principle, each cycle doubles the amount of DNA present. The method is used widely to produce enough DNA for sequencing or cloning, and for example in forensic DNA fingerprinting. It resulted in a Nobel Prize for Kary Mullis in 1993.

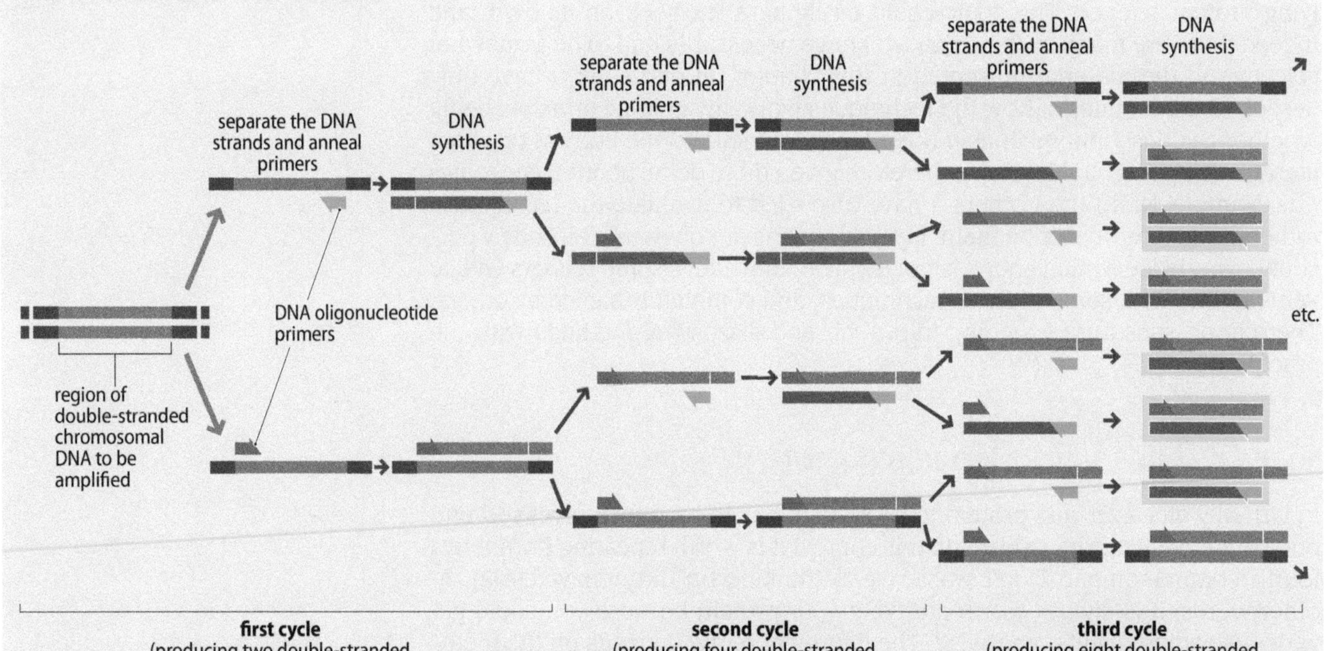

FIGURE 11.2.1
The polymerase chain reaction (PCR) uses 20–30 repeated cycles, each of which in principle doubles the number of DNA strands present. At the start is present the double-stranded DNA to be amplified, a heat-stable DNA polymerase, and an excess of dNTPs (dATP, dGTP, dCTP, and dTTP) plus two primers. These are short single-stranded DNA sequences: one is complementary to the 5′ end of the DNA sequence to be amplified, while the other is complementary to the 5′ end of the other strand, so that together they serve as primers for both DNA strands. Each cycle consists of a brief heating to separate the double-stranded DNA, followed by cooling in the presence of primers to allow annealing of the primers to newly revealed single-stranded DNA, and then an incubation to allow the polymerase to synthesize a new DNA strand starting from the primer. After a few cycles, nearly all of the double-stranded DNA present consists exactly of sequences running from one primer to the other (highlighted in green at the right). (From B. Alberts et al. Molecular Biology of the Cell, 5th ed. New York: Garland Science, 2008.)

The overexpression of a protein requires the gene. This can often be obtained using **PCR (*11.2)** from a plasmid library (or colleague). For example, a plasmid containing the gene of interest can be digested using restriction enzymes and cloned into a suitable vector, for which there are a large number of commercial sources. Very commonly, the sequence produced in this way does not correspond exactly to the gene or the protein desired, and it is therefore necessary to introduce modifications such as mutations, stop codons, or tags (of which more below), again by using PCR. These are standard molecular biology tools, which inevitably means that (1) different labs will have their own little tricks, (2) there is sure to be a newer and cleverer way of doing it, and (3) if it doesn't work, it is not easy to come up with good diagnostic tests to work out what went wrong. Graduate students in my lab spend at least half their time in trying to produce good-quality overexpressed protein. Nevertheless, in experienced hands, this part of the process is usually successful.

Before proceeding to 'real' experiments, it is advisable to check that the gene being used has the desired sequence. PCR introduces mutations at a significant frequency, and it is particularly annoying to discover that you are working with a mutated protein only after a year of unsuccessful experiments.

The molecular biology is almost always carried out in *Escherichia coli*, because it is easy to do so. There are a wide variety of host strains that can be used for protein expression. *E. coli* BL21(DE3) is usually the host of choice, because it grows quickly and is well understood: the BL21(DE3) strain is protease-deficient and is therefore less likely to digest the target protein. However, if the protein to be expressed comes from a eukaryotic organism, expression in *E. coli* may not be successful, because **codon usage** is different in eukaryotes and prokaryotes. The gene can be modified to remove rare codons, and there are strains of *E. coli* that are particularly designed for expressing eukaryotic genes. Expression of disulfide-rich proteins can also be difficult. There are many tricks available, such as the use of N-terminal signal sequences directing the expressed protein to the periplasm; this is also useful for proteins that are toxic to *E. coli*. Expression of membrane proteins can also be problematic. One can also use different hosts. Eukaryotic genes can be expressed in yeast, such as *Saccharomyces cerevisiae* or *Pichia pastoris*, both of which have good genetic systems for transformation and expression. These systems will glycosylate eukaryotic glycoproteins, although not usually in the same way as higher eukaryotes, by contrast with expression in *E. coli*, in which glycosylation usually does not occur. Eukaryotic genes can also be expressed in animal cells, or in insect cells transfected by genes carried by a baculovirus vector. These methods are more difficult, slower, and more expensive, and are only to be considered as a last resort.

Overexpression is most commonly initiated by putting the gene under the control of a *lac* promoter and inducing with isopropyl β-thiogalactopyranoside (IPTG). Excessively rapid expression may lead to the desired protein's being produced in **inclusion bodies**, in which the protein aggregates into insoluble clumps. These can be isolated by centrifugation and can sometimes be resolubilized with denaturants, followed by gradual removal of the denaturant by dialysis.

A protein to be studied is often expressed as a **fusion protein**, linked to a tag. Commonly this would be a **histidine tag**, consisting of six histidine residues in a row. This sequence binds tightly to some metal ions such as nickel, and it is usually straightforward to purify histidine-tagged proteins with a metal affinity column, a technique also known as immobilized metal affinity chromatography or IMAC (**Figure 11.1**). Alternatively, the protein could be fused to a protein affinity tag, such as maltose-binding protein (MBP) or glutathione S-transferase (GST), for which commercial affinity columns are available. Fusion to another protein can sometimes help with solubility and folding of the target protein. Heterologous expression can also benefit from an N-terminal fusion, because if the fusion protein comes from the host cell, expression levels are often increased.

When using a fusion protein, one always has to consider whether the tag could interfere with the subsequent experiments. Each case has to be considered separately. A histidine tag is often not a problem, except sometimes in crystallography, where it can prevent crystallization. GST tends to dimerize, implying that a protein fused to GST will often in fact be dimeric. Many commercial expression vectors contain not only the fusion tag but also a protease-specific cleavage site, to permit cleavage of the tag. Naturally, cleavage does not always work.

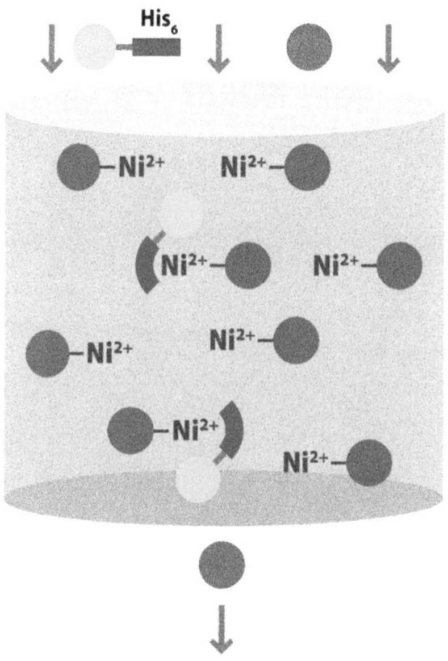

FIGURE 11.1
Purification of His-tagged proteins by immobilized metal affinity chromatography. The column is packed with beads (red) that have nickel ions (Ni^{2+}) bound to them. When a mixture of proteins (green) is passed down the column, the His-tagged proteins will stick to the nickel ions, whereas other proteins pass straight through. The column can then be washed, and His-tagged proteins can subsequently be eluted from the column by washing with a high concentration of imidazole, which has the same structure as the histidine ring.

*11.3 Ion-exchange chromatography

A form of chromatography based on charge:charge interactions. To purify a protein that has a negative charge at neutral pH, one would typically use anion-exchange chromatography, in which the resin making up the column bed is positively charged, for example a Q or quaternary resin, or DEAE (**Figure 11.3.1a**). When a mixture containing the protein is run down the column, the protein will bind (Figure 11.3.1b). It is typically eluted off the column by applying a gradient of salt: once the salt concentration is high enough, it screens the charge sufficiently for the protein to detach and wash off the column (see Electrostatic screening, *4.4). Conversely, a positively charged protein would be purified by cation-exchange chromatography, in which the resin is negatively charged, for example by having sulfate groups attached (for example an SP resin).

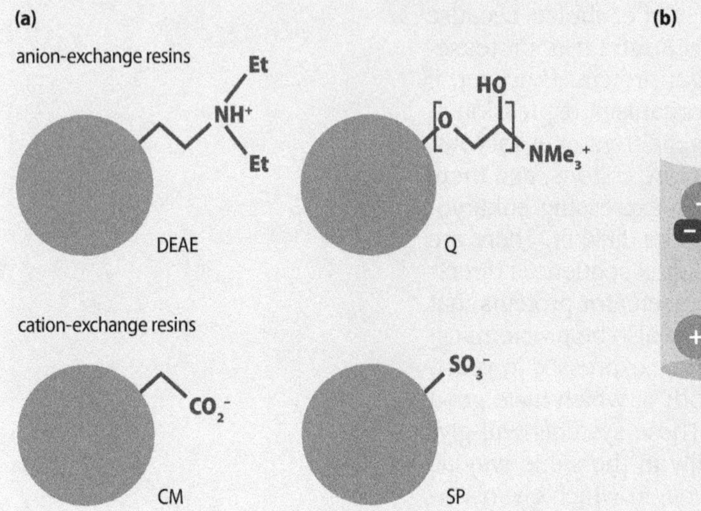

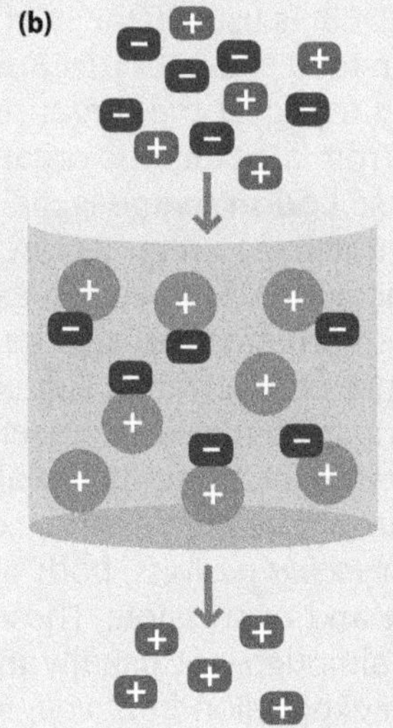

FIGURE 11.3.1
Ion exchange chromatography. (a) Ion-exchange resins. (b) An anion-exchange resin binds to negatively charged proteins, which can subsequently be washed off by increasing the salt concentration or lowering the pH.

A single IMAC step is not usually sufficient to generate a protein clean enough for further study. There are a range of further options, of which the most common is some form of **ion-exchange chromatography (*11.3)**, which is useful because not only can it be readily tailored to the protein under study, but it can also often be used as a way of concentrating the protein. **Gel filtration** is also common, which separates proteins according to their radius. It is less likely to lead to unexpected loss of protein but is somewhat less selective as a purification step; it is, however, also a good way to identify potential problems with aggregation at an early stage.

A survey of protein production methods for high-throughput structural genomics has been published; it provided a helpful description of problems and solutions, and found that only around 30% of bacterial proteins and 20% of eukaryotic proteins could be expressed and purified with standard high-throughput methods. This is despite the fact that the proteins were selected to be purifiable and therefore did not include membrane proteins, which are always difficult [3]. With more attention and a greater range of conditions, no doubt these percentages could be increased significantly, but these numbers show that it is far from straightforward to pick a gene at random and produce usable protein. Among the most useful conclusions are:

- It is often difficult to predict domain boundaries. Therefore it is advisable to produce a number of constructs in which the start and end sites are systematically altered.
- Use ligation-independent cloning rather than restriction enzymes: it is usually quicker to generate multiple clones.

Once the protein has been purified, it is important to check that it is pure and functional. Purity is most simply checked by SDS-PAGE (*11.1), which can also provide the molecular mass of the monomer, although for unusual proteins such as hydrophobic or proline-rich ones, the apparent molecular mass on

SDS-PAGE can often be significantly in error. It is often quick and simple to use mass spectroscopy to check the molecular mass. Many spectroscopic methods are also quick. We turn to these next.

Molecular biology provides a vast toolbox of methods for sequencing genes and proteins, performing mutagenesis and fusions, introducing proteins into non-natural places, and identifying genes and proteins. These steps are usually quick and easy in comparison with many of the other procedures described here, and are well described elsewhere.

11.2 SPECTROSCOPIC METHODS

11.2.1 An introduction to spectroscopic methods

Spectroscopy consists of the analysis of a sample by irradiating it with electromagnetic radiation, and then detecting some interaction of the radiation with the sample. This covers a very wide range of frequencies and wavelengths (**Figure 11.2**), which are related by the equation $f = c/\lambda$, where f is the frequency, λ the wavelength, and c the speed of light ($3 \times 10^8\,\mathrm{m\,s^{-1}}$). Although radio waves are used for NMR and X-rays for crystallography, they are not used in the same way as in classical spectroscopy and are discussed separately in this chapter. For protein studies, the only radiation types we need to consider are infrared, visible, and ultraviolet; visible and ultraviolet (UV) radiation are so close together that they are generally handled together under the heading UV/vis. For humans, there is of course a big difference between visible and ultraviolet radiation, but not for proteins!

Infrared (IR) is most useful in the range 30–120 THz. For historical reasons, IR frequencies are always measured in wavenumbers ($\mathrm{cm^{-1}}$), where the frequency in hertz is 3×10^{10} times the number in reciprocal centimeters ($\mathrm{cm^{-1}}$). These frequencies are characteristic of bond vibrations. This means that IR radiation is absorbed by molecules when its frequency matches that of a particular bond (**Figure 11.3**). A particularly prominent band in IR spectra of proteins is the carbonyl (C=O) stretch at about $1650\,\mathrm{cm^{-1}}$, whose position is subtly different for α-helical and β-sheet amides. This means that IR can be used to distinguish helix from sheet, but the interpretation is not easy, and it is much more common to use circular dichroism for this purpose, as discussed below.

By far the most useful radiation is UV/vis. This radiation corresponds to the energy needed to push an electron into an excited orbital. Irradiation of a protein with UV/vis promotes an electron to an excited orbital, after which the electron usually falls back again very rapidly. The spectrum therefore carries information about electrons and their energies. Because electrons are what chemical bonds are made of, the wavelength and intensity of UV/vis spectra are related to the bonding in proteins. In particular, they are absorbed by **aromatic (*11.4)** systems. Proteins do not contain many aromatic systems: they contain only tryptophan, phenylalanine, and tyrosine rings (and histidine, which is only weakly aromatic), plus any aromatic cofactors. DNA contains many aromatic systems, and one simple application of UV is to detect contamination of proteins by DNA, because protein and DNA absorb maximally at different wavelengths. The applications of UV/vis are considered next.

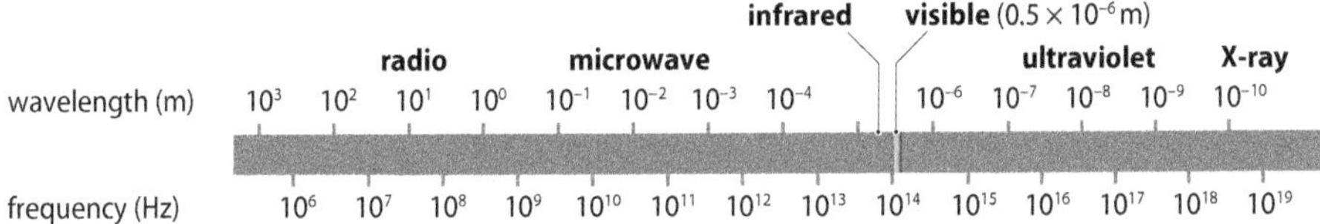

FIGURE 11.2
The electromagnetic spectrum.

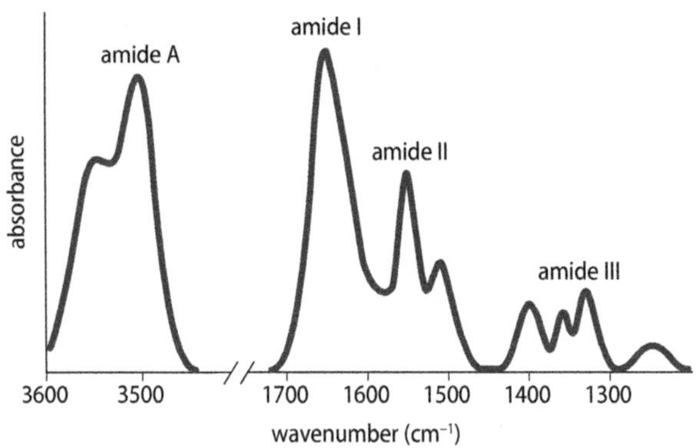

FIGURE 11.3
The infrared spectrum of a protein. The most commonly used regions of the IR spectrum are the amide I and amide II bands, due respectively to C=O bond stretching and N–H bond bending. The resolution of IR spectra is often enhanced by calculating second-derivative spectra ($\partial^2 A/\partial \nu^2$), which makes it possible to pick out different signals within the bands and therefore for example to identify different secondary structure elements, and with luck, perseverance, and the use of difference techniques (for example using isotopes to shift particular bands), specific interactions.

11.2.2 UV/vis absorbance

UV/vis is absorbed by **chromophores**, these being molecular groups, often aromatic. Each chromophore has a different ability to absorb UV/vis, this being expressed by the **molar extinction coefficient** and $\lambda_{\mathbf{max}}$, the wavelength at which the maximum radiation is absorbed. These are characteristic of the chromophore (**Table 11.1**). By far the most common application of UV/vis is as a simple method for quantitating the concentration of a given chromophore: UV radiation is passed through a measuring cell, usually with a 1 cm pathlength, and the amount of light absorbed is measured, usually as the difference from that of a reference beam that did not pass through the cell. The concentration of a protein can be worked out reasonably accurately (as long as there are no other chromophores present) by measuring the absorbance at 280 nm in a 1 cm pathlength cell, and using the **Beer–Lambert** law based on Table 11.1, $c = A_{280}/\epsilon$; ϵ is calculated as $5540n_{\mathrm{Trp}} + 1480n_{\mathrm{Tyr}} + 134n_{\mathrm{S\text{-}S}}$, where n is the number of these groups in the protein. Absorbances can be measured accurately over a range of about 0.05–1, meaning that protein concentrations can be measured down to the low micromolar range, depending on the composition of the protein.

Water absorbs UV at wavelengths below about 180 nm, placing a limit on the wavelengths available for studying proteins.

Because many cofactors are aromatic and absorb UV, they can also be readily detected. In particular, NADH absorbs strongly with a λ_{max} of 340 nm, well

***11.4 Aromatic**
In organic chemistry, this has nothing to do with smell: it denotes systems of conjugated electrons, in which there are $4n + 2$ π electrons (that is, $2n + 1$ double bonds) arranged in a closed loop. Therefore benzene and its derivatives are aromatic ($n = 1$), as are chlorophyll and heme ($n = 4$) (**Figure 11.4.1**). Such systems have strong UV absorbance and give characteristic NMR spectra.

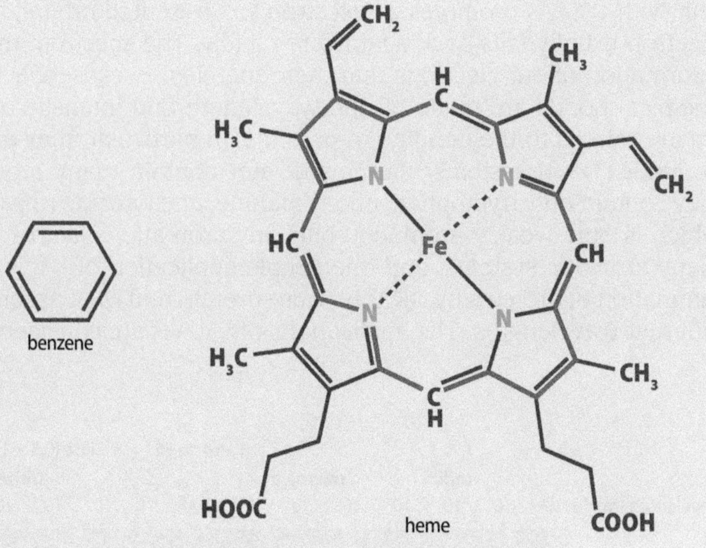

FIGURE 11.4.1
Benzene and heme are examples of aromatic ring systems, with $4n + 2$ π electrons.

***11.5 Enzyme linked assay**

The simplest way to measure the rate of an enzyme-catalyzed reaction is to follow it spectrophotometrically, by measuring a change in absorbance as a function of time. This of course requires that a substrate or product absorb light in some characteristic way. However, very often the reaction to be measured involves no suitable chromophore. In this case, it can prove very convenient to set up an assay in which the product of the reaction is the substrate for a second reaction, which *does* involve a suitable chromophore. The NADH/NAD^+ pair is very convenient, because NADH absorbs strongly at 340 nm but NAD^+ has zero absorbance. It is therefore convenient where possible to link a reaction to a reduction by NADH. For example, the reaction catalyzed by hexokinase is the attachment of a phosphate group at the 6 position of glucose:

$$\text{glucose} + \text{ATP} \longrightarrow \text{glucose 6-P} + \text{ADP}$$

None of these compounds is useful as a chromophore. However, the reaction can be linked to the NAD^+-catalyzed oxidation of glucose 6-phosphate to gluconate 6-phosphate by the enzyme glucose-6-phosphate dehydrogenase:

$$\text{glucose 6-P} + \text{NAD}^+ \longrightarrow \text{gluconate 6-P} + \text{NADH} + \text{H}^+$$

Therefore, to assay hexokinase, one adds an excess of NAD^+ and glucose-6-phosphate dehydrogenase, so that any glucose 6-phosphate produced is immediately used up in the second reaction; in this way, the rate of the second reaction is necessarily equal to the rate of the first.

TABLE 11.1 Average molar extinction coefficients at 280 nm for different groups

Group	Molar extinction coefficient, ϵ280
Trp	5540
Tyr	1480
Phe	2
S–S bond	134

In folded proteins, ϵ_{280} can vary over a reasonably wide range because of local interactions such as hydrophobic burial. Values are given in $M^{-1}\,cm^{-1}$, which would give the absorbance of a 1 M solution in a cell with a path length of 1 cm. The λ_{max} for phenylalanine is about 250 nm, where it has an ϵ_{250} of about 200 $M^{-1}\,cm^{-1}$.

away from any protein absorbance, whereas NAD^+ has zero absorbance at this wavelength. This makes it particularly easy to measure the rate of conversion of NADH to NAD^+ or vice versa, and therefore to measure the rate of an enzyme-catalyzed reaction that performs this conversion. By judicious choice of reaction, it is often possible to measure the rate of many enzyme-catalyzed reactions by conducting them as **enzyme-linked assays (*11.5)**, with the enzyme that performs the NADH conversion in excess (**Figure 11.4**).

UV/vis spectra are rather broad and featureless, and convey little other information. The values for λ_{max} and ϵ are dependent on the environment, meaning that they can be used to follow protein folding and unfolding. Circular dichroism and fluorescence are, however, often more discriminating techniques for following folding.

11.2.3 Circular dichroism (CD)

CD is a technique that uses UV radiation to detect **chiral** molecules or structures. The UV is passed through a linear (plane) polarizer. This plane-polarized radiation can be considered as the sum of two circularly polarized waves,

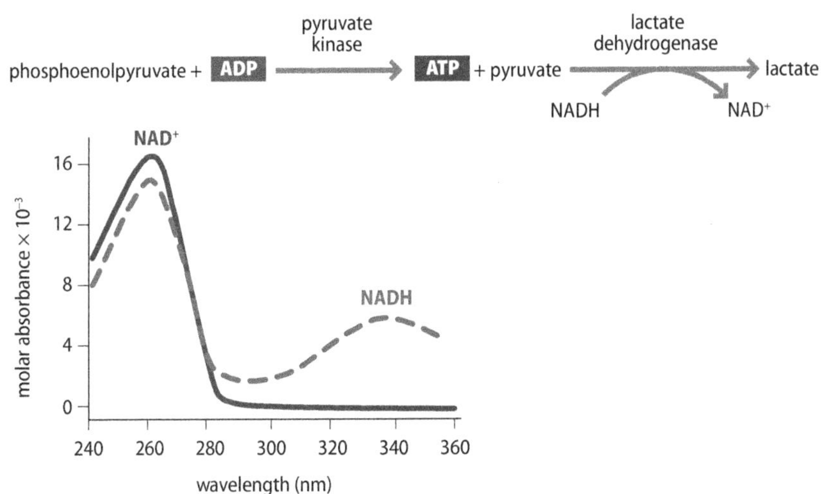

FIGURE 11.4
An enzyme-linked assay for pyruvate kinase. The substrates and products for this reaction are difficult to detect spectroscopically, so the reaction is linked to lactate dehydrogenase, whose activity can be easily detected because it oxidizes NADH to NAD^+. This leads to large changes in absorbance at 340 nm, which are directly proportional to the amount of NADH consumed, and thus to the amount of pyruvate. Provided that the LDH reaction is going rapidly compared with the pyruvate kinase reaction (that is, if LDH and NADH are in excess), this rate is also equal to the rate of the LDH reaction.

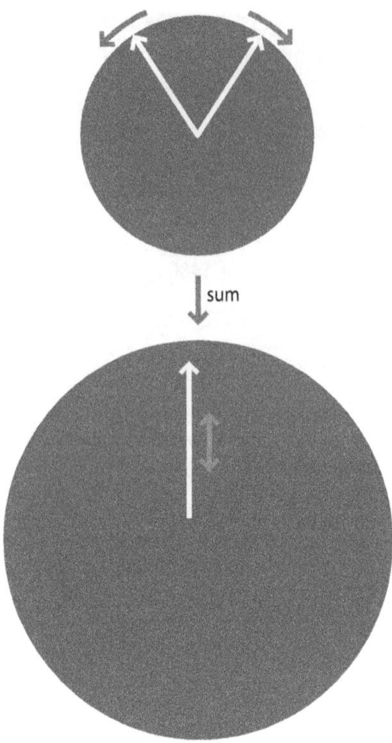

FIGURE 11.5
Plane polarized light is the sum of two circularly polarized waves rotating in opposite directions. Because the two rotating vectors (top) are going at the same speed but in opposite directions, their sum has zero horizontal component at all times and is thus an oscillating plane wave (bottom).

one going clockwise and the other going anticlockwise (**Figure 11.5**), and CD machines usually generate left- and right-circularly polarized light in rapid alternation using a piezoelectric oscillator. Chiral molecules will absorb the right- and left-circularly polarized light to slightly different extents, and this can be detected. The difference in absorbance of the two components is not great, and therefore CD is a less sensitive technique than UV.

UV is absorbed by proteins at wavelengths of about 200 nm (2000 Å). CD is therefore not sensitive to the chirality of individual amino acids (which are too small), but it does interact with protein secondary structure. The main application of CD is to characterize the secondary structure content of proteins (**Figure 11.6**). Typically, a CD spectrum can be deconvoluted to obtain the relative fractions of helix, sheet, and 'coil' (that is, everything else). The relative fractions depend heavily on the shape of the reference spectra for 100% helix, sheet, and coil, which are inherently variable. It is therefore difficult to get exact values from deconvolution of CD spectra, but approximate values are nevertheless extremely useful. CD is simple to do, and it is used widely to see whether proteins are folded (that is, whether they contain helix or sheet), to detect structural change, and to measure the kinetics of folding. Changes in CD at around 200 nm ('far-UV CD,' where the peptide backbone absorbs) reveal changes in secondary structure, whereas changes around 280 nm ('near-UV CD') reveal changes in the environment of Trp and Tyr, and thus tertiary structure.

11.2.4 Fluorescence

When a molecule absorbs UV radiation, an electron is promoted into a different orbital. Almost all molecules have all their bonding orbitals full and their antibonding orbitals empty. Therefore, excitation of an electron takes it out of a bonding orbital and puts it into an antibonding orbital. As a result, the excited state is less tightly bonded, and the bond length is longer. This gives rise to the situation shown in **Figure 11.7**, in which the structure of the excited state is different from that of the ground state. Movement of an electron is much faster than movement of nuclei, and so the molecule after absorption of UV is in an excited vibrational state. In most molecules, the excited electron can drop back down again by giving off the excess energy as heat. However, depending on the rate at which the molecule rearranges its structure and the rate of electron transition, relaxation of the electron can involve a second transition (*emission*), as shown in Figure 11.7. This transition is necessarily smaller in energy than the first one, and therefore the molecule emits light at a lower energy (longer wavelength) than the excitation wavelength. This is fluorescence. In most applications, the emitted light is in the visible range, and the excitation is often in the UV.

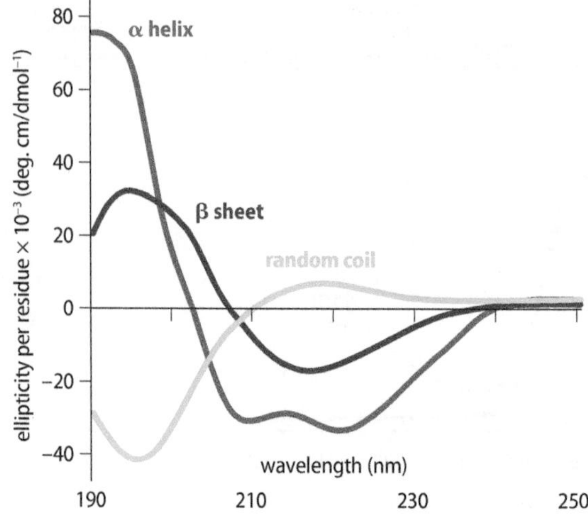

FIGURE 11.6
Typical CD spectra for the main secondary structures in proteins. Note that the spectrum of an α helix is easily recognizable in having a double minimum. This means that quantitation of helix content in proteins is reasonably reliable with CD, but quantitation of β-sheet content is much more prone to error.

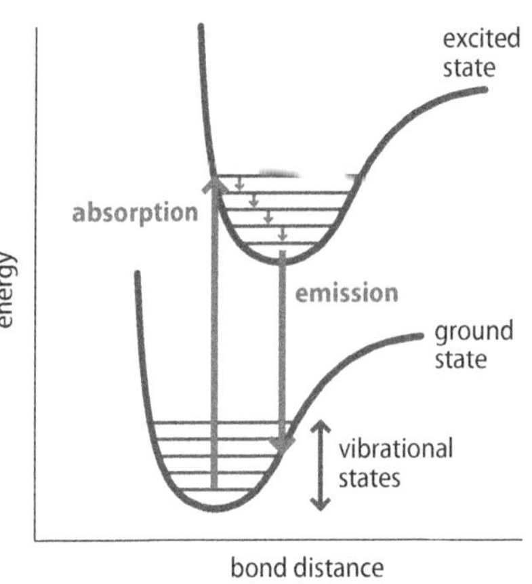

FIGURE 11.7
The principle of fluorescence. A molecule in the ground state absorbs UV radiation and an electron is promoted to a different orbital, producing an excited state. For fluorescence to occur, the re-emission must be slow compared with the timescale of vibration, allowing radiationless vibrational transitions to occur before the emission, which is therefore at a lower energy than the absorption.

The fraction of incident radiation that gets re-emitted is called the **quantum yield** and is usually much less than 1. Consequently, the signal detected in fluorescence is weaker than that detected in UV. However, UV measures a difference between the amount of incident and transmitted light, whereas fluorescence measures only the transmitted light. Thus, in UV the detector cannot be very sensitive because it has to be able to measure the incident beam as well as the transmitted beam, whereas in fluorescence it is designed to measure only the weak fluorescent signal. This in practice makes fluorescence much more sensitive than UV. In favorable cases, it is even possible to see a *single fluorescent molecule*, as discussed below.

Fluorescence has many applications. Its drawback is that rather few molecules fluoresce well, and most proteins fluoresce very weakly. However, this is also its big advantage, because it means that fluorescent tags can be seen clearly even when many untagged proteins are present. There are a very large number of fluorescent markers now available, which can be covalently attached to proteins. For example, one can buy antibodies with different fluorescent labels attached. It is perfectly possible to have three different antibodies with different fluorescent markers, fluorescing red, green, and blue, which can be detected selectively by using a microscope with filters at appropriate wavelengths. One can then use them to detect different cellular features. For example, **Figure 11.8** shows a cell stained for nucleus, microtubules, and actin. Fluorescence is thus the best way to visualize localization and colocalization of cellular components, time-resolved if necessary. One can use a **confocal microscope (*11.6)** to identify specific locations more precisely.

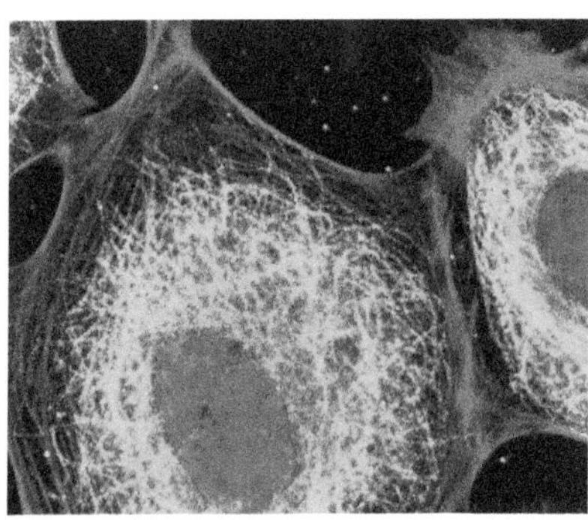

FIGURE 11.8
The use of fluorescence to detect cellular components. Fibroblasts were fixed and stained using antibodies, to which fluorescent probes emitting at different wavelengths were attached: a green probe for microtubules, and a red probe for actin. The nucleus was stained blue with a fluorescent blue dye that binds to DNA. It is clear that microtubules radiate out from the nucleus, and actin tends to lie parallel to the cell membrane, as discussed in more detail in Chapter 7. (Image from *Traffic, International Journal of Intracellular Transport*, virtual issue on cytoskeleton, 2010. With permission from Wiley-Blackwell.)

***11.6 Confocal microscope**

A special setup of the microscope involving a point illumination and a pinhole aperture in the detector which is in front of the specimen (**Figure 11.6.1**). This arrangement eliminates signals from out-of-focus objects and is therefore good at seeing fluorescence from a very narrow plane within the sample. By changing the focus, one can scan up and down through the sample. The disadvantage is that very little of the emitted light reaches the detector, and so the sensitivity is low.

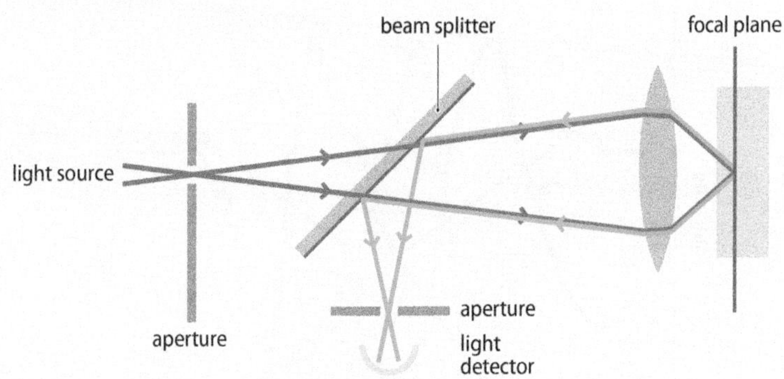

FIGURE 11.6.1
The arrangement for a confocal microscope.

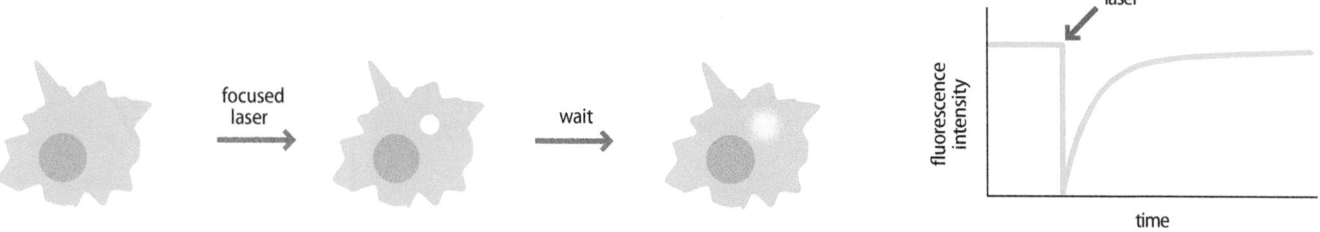

FIGURE 11.9
Fluorescence recovery after photobleaching (FRAP). A small region of the sample is photobleached using a laser, after which the rate of recovery of the fluorescence in this region can be measured, giving a measure of the mobility of the fluorescent probe.

After excitation of fluorescence by UV irradiation, the fluorescence intensity decays away. The rate of decay depends on many factors, and is affected by the environment of the protein. For example, ligand binding or protein folding and unfolding can affect the rate of decay. Measurement of fluorescent lifetimes therefore provides information on different environments or interactions. This measurement is called fluorescence lifetime imaging microscopy or **FLIM**.

Fluorescence can be bleached by intense UV radiation, for example from a laser. If this bleaching is focused on a small region of a cell, it produces a small region that does not fluoresce and appears black (**Figure 11.9**). If the molecules producing the fluorescence then diffuse away and are replaced by unbleached molecules, one can measure recovery of the fluorescence, in a technique called fluorescence recovery after photobleaching (**FRAP**). This is a powerful technique for measuring protein mobility.

The Nobel Prize in Chemistry 2008 was awarded to Osamu Shimomura, Martin Chalfie, and Roger Y. Tsien for the discovery and development of **green fluorescent protein** (**GFP**). This is a protein from jellyfish that is fluorescent. Remarkably, the **fluorophore** is formed spontaneously from naturally occurring amino acids without the need for external cofactors. This means that heterologous expression of GFP, in particular as a tag for a target protein, results in a fluorescent protein. This is a very powerful tool for locating and characterizing proteins, and for studying their movement, regulation, and interactions. Several variants of GFP are now available (such as cyan and red fluorescent protein; CFP and RFP), with different emission wavelengths, again allowing the simultaneous detection of two or three fluorescent proteins.

An important application of fluorescence is known as fluorescence resonance energy transfer or **FRET**. If two fluorophores are close together, and if the emission spectrum of one overlaps with the absorption spectrum of the other (**Figure 11.10**), the fluorescent energy can be transferred directly from one fluorophore to the other, resulting in a decrease in fluorescence of the donor

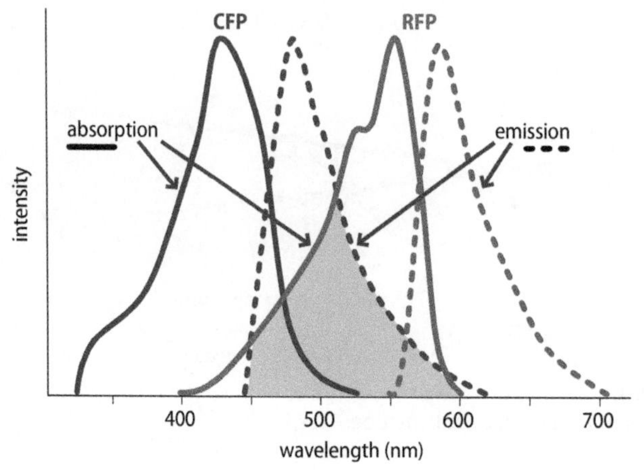

FIGURE 11.10
Fluorescence resonance energy transfer (FRET). This example shows the absorption (solid lines) and emission (dashed lines) spectra of cyan fluorescent protein (CFP) and red fluorescent protein (RFP). Where the emission spectrum of CFP overlaps the absorption spectrum of RFP, energy can be passed from CFP to RFP, provided that the two chromophores are close enough and in an appropriate orientation. This is detected as an increase in the fluorescence of RFP coincident with a decrease in the fluorescence of CFP. If the sample can be irradiated with light filtered so that it excites only CFP and not RFP, there is no red fluorescence unless FRET occurs. Normally filters are not so precise, and there is weak fluorescence from RFP even in the absence of FRET.

and an increase in fluorescence of the acceptor. The transfer depends on r^{-6}, where r is the distance between fluorophores, and also on their relative orientation, and operates over distances of up to about 100 Å. This method can therefore be used to detect the proximity of two fluorophores. For example, if two suitable fluorophores are attached to two different proteins, FRET can be used to detect their direct interaction; or if two fluorophores are attached to different parts of the same protein, it can detect a conformational change that alters the distance between them. The two fluorophores can be chemically attached or they can be attached genetically, for example as CFP and RFP. This is a very powerful technique, particularly because it can (albeit with considerable difficulty) be used at the single molecule level.

If the sample is irradiated at an angle steeper than the critical angle for total internal reflection, the light is reflected, but some light is able to penetrate the sample for a short distance: this is called an evanescent field and is the same idea that lies behind surface plasmon resonance (Section 11.6.1). This gives rise to fluorescence from fluorophores close to the surface (within about 100 nm), in a technique called total internal reflection fluorescence or **TIRF**. This can be a very powerful method for seeing proteins at the cell surface, for example actin.

Finally, fluorescence can be used to measure the rate of molecular reorientation, in a technique known as fluorescence depolarization. The idea is that if a fluorophore is irradiated with polarized light, it will also fluoresce polarized light. However, it will only remain polarized if it does not move between irradiation and emission. Loss of fluorescence polarization therefore gives information on the rotational correlation time: it is rapid for a small and rapidly tumbling molecule but slow for a large molecule. Hence the major use of this technique is in the detection of binding of small fluorophores to receptors, for example receptor–ligand or protein–DNA binding.

11.2.5 Single-molecule methods

A discussion of fluorescence leads on naturally to the exciting field of single-molecule methods, because almost the only technique with the sensitivity to see single molecules is fluorescence. Observations of FRAP and FRET can be made on single molecules. This is an important current area of research. Almost all other experimental methods measure averages over large numbers of probes. This means that if different molecules behave in very different ways, this will not be detected by most methods, which will only describe the average. A well-discussed example is protein folding (Chapter 4). It is now believed that proteins fold on an energy landscape (Section 6.2.1). That is, the pathway along which they fold is different for different molecules, depending on the unfolded structure that they started with. However, most measurements of protein folding detect only what looks like a single pathway, because they can only detect the average. By contrast, single-molecule methods show how each individual molecule folds, and therefore they provide details on the folding landscape that cannot be obtained by any other experimental technique.

Some other methods should be mentioned here, which also provide information at a single-molecule level. These include single-particle cryo-electron microscopy and atomic force microscopy, discussed further below. Another is **patch clamping**. This is a method for measuring the current flow through a transmembrane channel. A very small electrode is pushed onto a membrane, for example a single cell. Either by applying a small amount of suction or by pulling, the electrode can be sealed onto the cell in different geometries (**Figure 11.11**). One can then measure either the voltage or the current. The method is highly sensitive and can easily detect the opening and closing of single channels (**Figure 11.12**), permitting measurements of voltage gating, drug

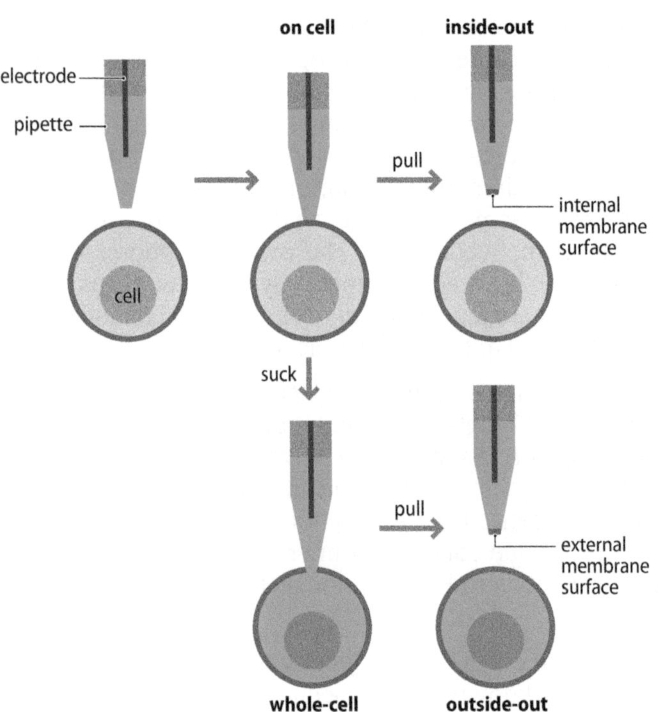

FIGURE 11.11
Patch clamping. The technique measures a voltage or current between an electrode contained inside a very narrow pipette tip and a second electrode on the other side of a cell membrane. Different geometries can be probed depending on how the membrane is attached to the pipette tip. The first stage is to push the tip onto the cell surface. Gentle suction makes a seal and attaches the membrane to the tip in the 'on cell' geometry. A rapid pull of the tip away from the cell breaks the membrane and makes an 'inside-out' patch in which the side of the membrane that was originally inside the cell is now on the outside. This allows, for example, different intracellular ligands to be tested. In contrast, harder suction applied to the 'on cell' geometry breaks the membrane, making the whole intracellular volume continuous with the pipette in the 'whole-cell' geometry. This allows the measurement of all channels on the cell surface simultaneously (rather than just the small patch sealed to the end of the pipette). Finally, gentle pulling from the 'whole-cell' geometry creates an 'outside-out' patch, which allows experiments with different extracellular ligands.

binding, and so on. One of the key findings is that channels open and close randomly, although with probabilities dependent on the overall opening probability. Erwin Neher and Bert Sakmann received the Nobel Prize in Physiology or Medicine 1991 for the development of patch clamping.

11.2.6 Hydrodynamic measurements

There is a group of methods that are used to measure the size and shape of proteins; these are important features, which in addition to shape give information on oligomerization and complexation. The simplest and one of the most informative is gel filtration, which is an excellent preliminary way of detecting and characterizing oligomerization. One can also use native PAGE (that is, in the absence of SDS), which is equally quick and requires very little sample, although it is rather more prone to artifacts. *Dynamic light scattering* is also straightforward. In this method, a laser is used to illuminate a protein solution, and a detector is placed to measure light scattered at a fixed angle. The light is scattered by reflecting off particles in the solution. The intensity of the light fluctuates as a consequence of tumbling of the particles: from an analysis of the fluctuations one can calculate the sizes of the particles. The method requires very clean solutions, and is less successful if there is a distribution of particle sizes present, but it is simple to do and fairly simple to analyze. Shape can also be determined by low-angle X-ray scattering in solution, which has been used very effectively to analyze conformations of multidomain proteins.

The other technique that can be very useful is *analytical ultracentrifugation* (AUC). The principle is very simple: a larger molecule will move faster down a centrifuge tube. Movement of the protein is normally detected by UV/vis absorbance, or alternatively by refractive index. The sedimentation rate

FIGURE 11.12
A typical patch clamp result. The pipette is in either the 'inside-out' or 'outside-out' geometry, so that the area of the membrane is small enough that only a small number of channels are present on the membrane. The measurement shows an alternation between three conductance states: closed (top level), and one or two channels open. Note that the transition between open and closed is very rapid. The measurement gives the current in the open state, as well as the fraction of time spent open, and thus the equilibrium constant for opening. Two channels open produces twice the current produced by one channel.

***11.7 Magnetization**

In NMR, each individual nucleus behaves like a tiny bar magnet, which rotates around the applied magnetic field. The signal that is observed from a sample inside an NMR spectrometer is the vector sum of all these individual magnetic fields, and is called the magnetization. For most applications it can be treated like a conventional magnet.

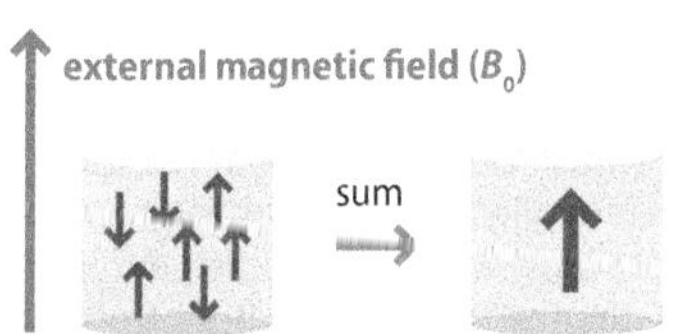

FIGURE 11.13
In the presence of a magnetic field (usually referred to as B_0), individual nuclear spins will have a slight preference for the α (*up*) state rather than the β (*down*). The sum of all the individual spins is the observable macroscopic magnetization, which is therefore pointing upward.

depends on both size and shape, although the shape effect is not strong. AUC can also be run as a sedimentation equilibrium measurement, so that the protein reaches an equilibrium in a density gradient. In this form, the position depends only on size and not on shape. This form is more accurate but takes longer to do (at least overnight to reach equilibrium). AUC requires protein concentrations of about 0.1 μM or greater, so it is not suitable for looking at very strong binding.

11.3 NMR

11.3.1 Nuclear spin and magnetization

In Sections 11.3 and 11.4, we look at the two techniques that provide the most information about protein structure and dynamics, NMR and X-ray crystallography. Because of their importance, these topics receive rather more detailed coverage.

NMR is a spectroscopic technique, and like the other spectroscopies described here it relies on the absorption of energy to produce a transition between two states. Here, the states are two different orientations of the nuclear spin, normally described as *up* (α) and *down* (β). These states only have a different energy if the nucleus is in a magnetic field, and hence NMR spectroscopy requires a magnet. The energy difference, and consequently the frequency, is proportional to field strength. Most NMR spectrometers use a strong superconducting magnet, in which the magnetic field runs vertically. Thus, at equilibrium the **magnetization (*11.7)** of the sample is also oriented vertically, with a net *up* orientation because the α state is of lower energy (**Figure 11.13**). The energy difference between these states is very small, which is equivalent to saying that the frequency measured is relatively low (see Figure 11.2): between 100 and 1000 MHz, depending on the strength of the magnet. From the Boltzmann distribution (*5.2) $N_1/N_2 = \exp(-\Delta G/kT)$, this means that at 500 MHz, for example, the excess population in the lower-energy state is only about 1 in 10^4. Hence, a major limitation of NMR is its low sensitivity. NMR typically requires sample concentrations of 0.5–1 mM, which is two or three orders of magnitude greater than for UV, and many orders of magnitude more than for fluorescence or mass spectroscopy. It is also typically a much higher concentration of protein than the physiological concentration, and NMR suffers more than most methods from protein aggregation and insolubility because of the high concentrations needed.

Because of the small population difference, NMR cannot be measured in the same way as UV or IR, by measuring the fraction of energy absorbed, because this fraction is too low. Instead, it is measured by a resonance technique. The sample is irradiated by a short pulse of radiation at a frequency that matches the energy difference. This frequency depends on the nucleus being observed and the field strength: in a 21-tesla (21 T) magnet, it is 900 MHz for ^{1}H. (A 21 T magnet is therefore normally described as a '900 MHz magnet' or just as 'a 900.') The radiofrequency (rf) pulse generates a magnetic field that acts to rotate the nuclear magnetization, and it is turned on for a time sufficient to rotate it by exactly 90°. It is therefore called a *90° pulse*. The nuclear magnetization vector

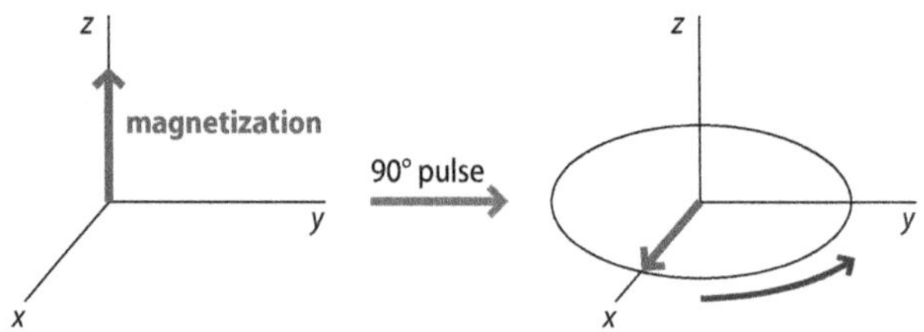

FIGURE 11.14
A radiofrequency field (here applied along the *y* axis) rotates the magnetization, and is applied long enough for it to rotate it by 90°, onto the *x* axis. After the radiofrequency pulse has been turned off, the magnetization then precesses in the *xy* plane at a frequency determined by its chemical shift.

is now in the *xy* or transverse plane (**Figure 11.14**), and under the influence of the applied magnetic field it rotates around the *z* axis. Technically, this rotation is usually described as a *precession*, which is motion like that of a gyroscope or child's top, which spins about its own axis as well as a vertical axis. This rotating magnetic field generates a voltage in a detector coil placed around the sample, which can be detected and amplified. The resultant signal is called a *free induction decay* or FID, and is a function of intensity against time. To produce the spectrum, which is a function of intensity against frequency, a **Fourier transform (*11.8)** is used (**Figure 11.15**).

There are two features of NMR that have made it a uniquely powerful spectroscopic method. One is its very high resolution. In UV, a typical signal is about 50 nm wide, in a total usable spectral width of about 150 nm. One can therefore resolve at best about three signals. By contrast, in ^{1}H NMR a resolution of 0.5 Hz can be obtained across a spectral width of about 5000 Hz: one can therefore in principle resolve about 10,000 signals. The second is that NMR theory, which is based on quantum mechanics, is very detailed and matches the experimental spectra with remarkable precision. One can interpret these high-resolution data using very sophisticated treatments, and relate the spectra to physical phenomena to a very high level.

*11.8 Fourier transform (FT)

A mathematical technique, used extensively in both NMR and X-ray crystallography, originally proposed by **Joseph Fourier (*11.9)**. It is used to express a function as a sum of cosine waves of different frequencies, and can therefore be used to convert a time-dependent function (such as the NMR signal) into a frequency-dependent signal (the NMR spectrum). Mathematically it is expressed as $f(\omega) = \int f(t)e^{-i\omega t}dt$. Essentially, it obtains the intensity at each frequency by multiplying the time-dependent function by a cosine wave at that frequency (**Figure 11.8.1**). In most cases the integral or sum of this multiplication is zero, because any positive value is always cancelled by a negative value somewhere else. However, over a small range of frequencies the multiplication consistently gives a positive value and therefore the sum is non-zero: the function returns a peak at this frequency.

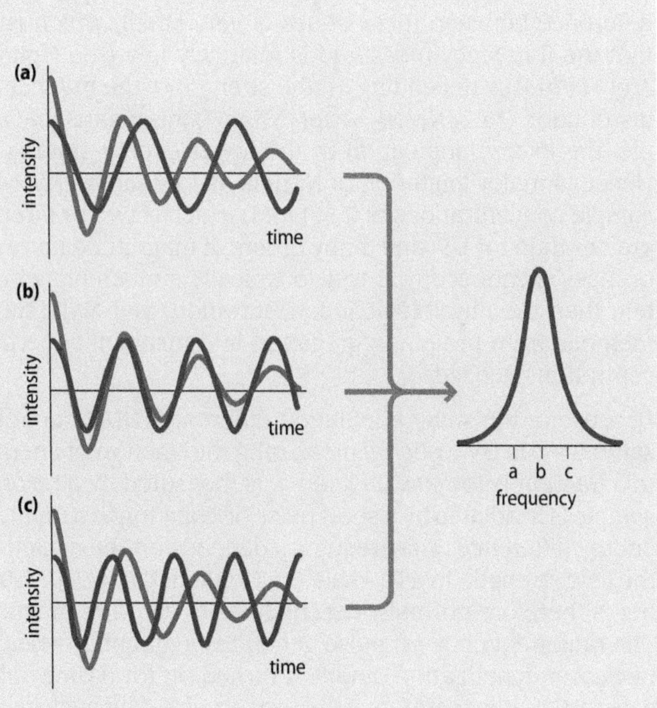

FIGURE 11.8.1
To carry out a Fourier transform of the red function (a typical NMR FID for example), it is multiplied by test cosine waves at different frequencies (blue). A test wave at the right frequency (b) gives a product that is always positive, and the summed intensity of the product is large, as indicated by the peak at point b on the right. Test frequencies that are too low (a) or too high (c) give products that tend to cancel out and thus have low intensity. The result is a peak centered at frequency b.

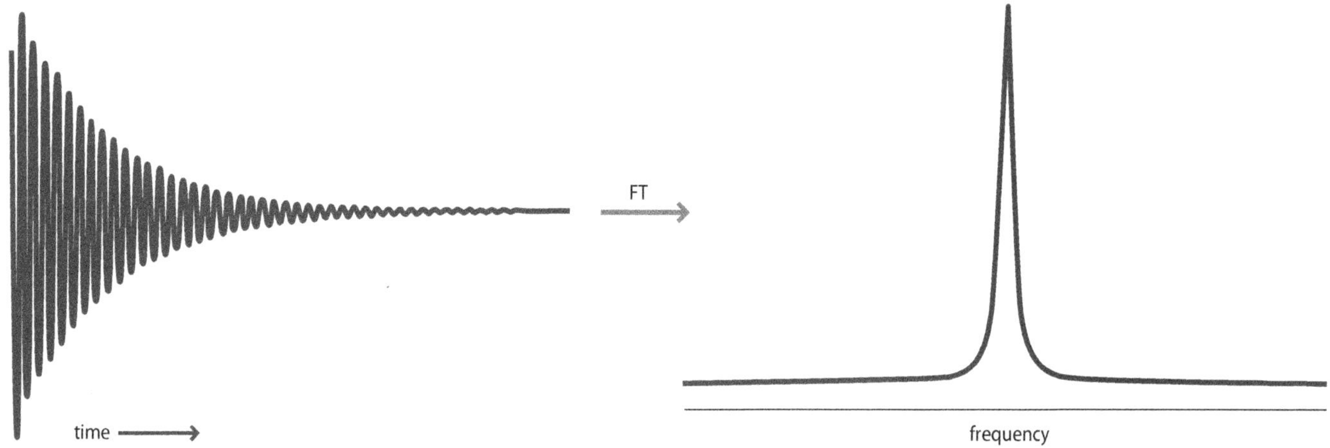

FIGURE 11.15
The rotating magnetization is detected as a damped oscillating voltage in the *x* and *y* directions, known as the FID, which is a function of time, typically lasting about 0.5 second. It is converted to a spectrum (a function of frequency) by a Fourier transform (FT).

11.3.2 Chemical shift

NMR measures the signal from nuclear spins. Not all nuclei have spins. NMR can observe ^{1}H, ^{2}H, ^{13}C, ^{15}N, ^{19}F, and ^{31}P but not ^{12}C, ^{14}N, or ^{16}O. The signals from different nuclei come at sufficiently different frequencies that one always selectively pulses and observes a single nuclear type. Normally this is ^{1}H (proton), because it is both abundant and sensitive (**Table 11.2**). Note that the normal carbon and nitrogen isotopes are not observable. However, it is fairly easy and cheap to grow *E. coli* in minimal medium made up using ^{15}N and/or ^{13}C for all nitrogen and carbon sources, implying that the production of labeled protein is very common.

The nucleus is partly shielded from the external magnetic field by the electrons around it. This means that each nucleus experiences a slightly different magnetic field and therefore resonates at a slightly different frequency. These frequencies are always called the **chemical shift**. Chemical shifts are always described as frequencies relative to a reference compound. The relative frequencies are very small, and hence chemical shifts are measured in the somewhat confusing unit of parts per million (ppm) (**Figure 11.16**). For ^{1}H and ^{13}C, the reference is the compound DSS (**Figure 11.17**), which is soluble and inert and has a chemical shift lower than almost all the others in proteins.

The chemical shift of both ^{1}H and ^{13}C varies in a fairly predictable way with structure. In recent years there have been many studies to try to improve the calculation of shifts from structures, and conversely to be able to calculate structures by using chemical shifts as restraints. These are looking very

TABLE 11.2 Observation frequencies of common nuclei in a 14.1 T magnet

Nucleus	Frequency (MHz)	Natural abundance (%)
^{1}H	600	99.99
^{2}H	92	0.015
^{13}C	151	1.1
^{15}N	61	0.37
^{19}F	565	100
^{31}P	243	100

A higher frequency corresponds to a higher sensitivity of detection.

*11.9 Joseph Fourier

Jean Joseph Fourier (1768–1830) (**Figure 11.9.1**) was a mathematician interested in the flow of heat. As part of this work he developed the Fourier series analysis, showing that a continuous function can be expressed as a sum of sine waves. He kept his rooms very hot and wore a heavy coat, even indoors, believing that this was good for health. This may have dated from the 3 years he spent in Egypt, employed by Napoleon. Fourier subsequently became an efficient bureaucrat, was sacked by Napoleon in 1815, and became a professor and Permanent Secretary of the French Academy of Sciences. In his lifetime Fourier analysis was treated with a good deal of scepticism. He made several other contributions to science. He developed dimensional analysis; that is, the use of the dimensions of an equation to verify and relate physical properties. He performed many investigations into heat transfer and is credited with the discovery that gases in the atmosphere may be responsible for warming the Earth: the greenhouse effect.

FIGURE 11.9.1
Joseph Fourier. (Wikimedia Commons.)

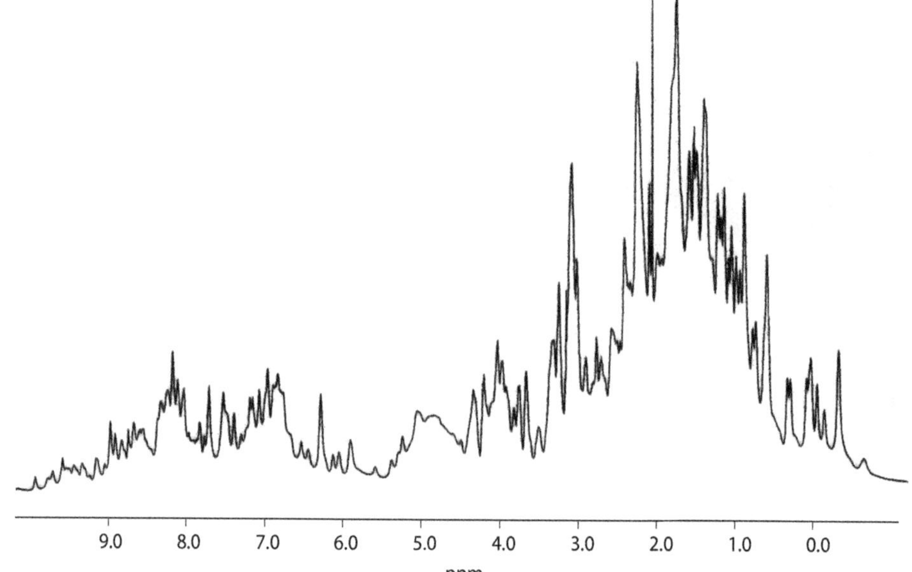

FIGURE 11.16
The ^1H NMR spectrum of a small protein in water. The chemical shifts are given in ppm values, calculated as $10^6 \times$ [(frequency of signal) – (frequency of reference)]/(frequency of reference). This makes the chemical shift independent of the strength of the NMR magnet. The strange shape around 5 ppm is an artifact arising from the way in which the very strong signal from the H_2O solvent was digitally removed.

promising, although so far for accurate structures they need to be supplemented by additional information. It is for example possible to use computational tools to predict protein structure, and then use chemical shifts as a filter to select the 'best' predictions. This method works well for small proteins and has considerable potential [4].

11.3.3 Dipolar coupling

Chemical shift, as we have seen, is governed by the magnetic field experienced at the nucleus. One source of magnetic fields is neighboring nuclei. Their magnetic fields are small in comparison to the applied field, but they are only a short distance away, and hence the effect can be very strong: up to 10 kHz, which makes it larger than any other effect on chemical shift. The effect is orientation dependent, because nuclear magnetic fields are orientation dependent (**Figure 11.18**), and hence the effect changes as molecules tumble in solution. In fact, in most cases it averages completely to zero as a result of molecular tumbling.

Recently, however, it has been realized that if a protein can be induced to have a slight preference to point in one particular direction, the dipolar coupling will *not* average to zero. The partial orientation required can be very slight, typically only 0.1%, and can be induced by placing the protein in an anisotropic solution

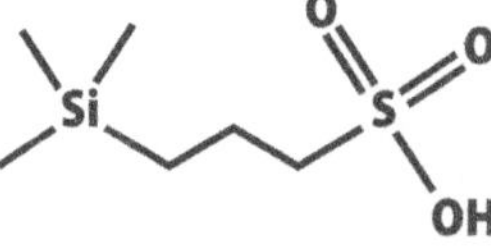

FIGURE 11.17
The structure of DSS (2,2-dimethyl-2-silapentane-5-sulfonic acid), which is the standard reference compound for both ^1H and ^{13}C NMR. The signal from the methyl groups is defined to have a frequency of 0 ppm. In organic chemistry, the reference is tetramethylsilane ($Si(CH_3)_4$ or TMS). This is not suitable for proteins because it does not dissolve in water. The methyl groups in DSS and TMS have almost identical chemical shifts.

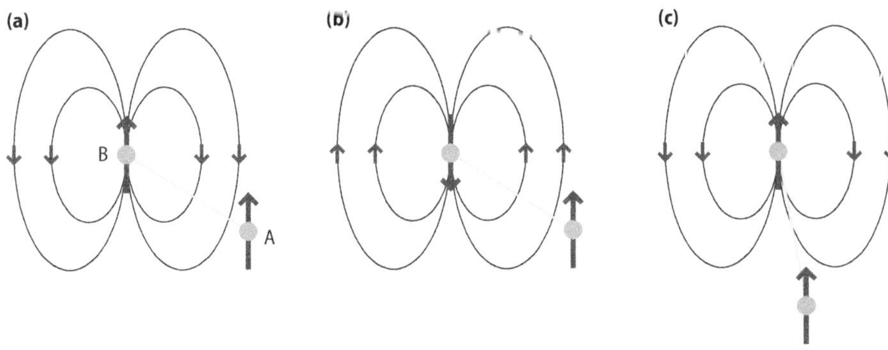

FIGURE 11.18
Dipolar coupling: the alteration in magnetic field at one nucleus caused by the relative position (and spin state) of another. For a molecule (containing nuclei A and B) tumbling randomly in solution, this effect averages to zero, but is the main source of relaxation for most nuclei. In solids, the effect causes a wide spread of resonance frequencies and is usually removed by magic angle spinning (that is, rapid rotation of the sample at the 'magic angle' of 54.7°; this angle corresponds to the diagonal of a cube, and rotation at this angle therefore averages all orientations equally, making the sample behave as if it were tumbling rapidly and oriented randomly). In (a), nucleus A experiences a total magnetic field smaller than B_0 as a result of the local magnetic field created by the spin (dipole) B. (Nuclear spins are often called dipoles because they have a north and south magnetic pole, like any other magnet.) In (b) and (c), the total magnetic field experienced by A is greater than B_0. For a randomly tumbling molecule, the average effect is zero.

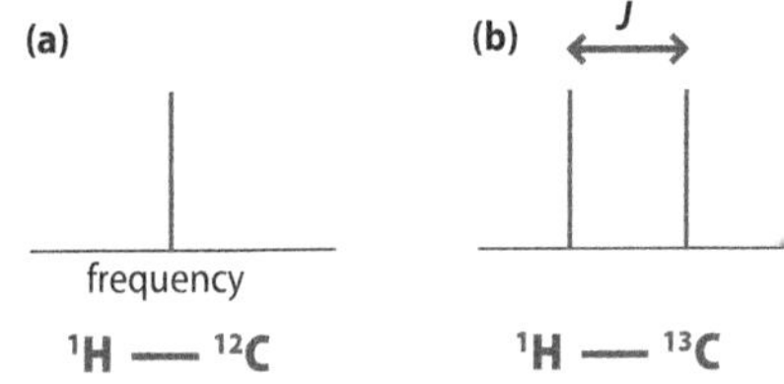

FIGURE 11.19
J coupling. If a ^1H nucleus is bonded to an NMR-active nucleus such as ^{15}N or ^{13}C, its signal will be split into two, corresponding to the α and β spin states of the attached nucleus. Thus, the ^1H NMR spectrum of $^1H–^{12}C$ consists of a single signal (a), whereas the $^1H–^{13}C$ spectrum (b) consists of two signals, in which the higher-frequency signal comes from molecules with the ^{13}C in the α state, and the lower-frequency signal from molecules with ^{13}C in the β state. The splitting of the signals is measured in Hz and is given the symbol *J*.

such as a liquid crystal or a polyacrylamide gel that has been squeezed in one direction. The *residual dipolar couplings* that remain provide information about the angles between the internuclear vector and the alignment axis. This information is a valuable source of structural information.

11.3.4 *J* coupling

In the previous section, we saw that a neighboring spin affects chemical shifts by dipolar coupling, which is a through-space interaction. It can also do it by *J* coupling, also known as scalar coupling, which is a through-bond interaction. This is a much weaker effect, but unlike dipolar coupling, *J* coupling is not averaged to zero by tumbling. The electron distribution around a nucleus affects the electron distribution in its chemical bonds, which means that nuclear transitions affect each other via the bonds that connect them. In other words, *J* coupling produces a splitting of the NMR signal into two (**Figure 11.19**). For a $^1H–^{15}N$ amide pair connected by a single bond the *J* coupling is about 94 Hz, in contrast with a ^1H chemical shift distribution of several thousand Hz. It is thus relatively small, but easily measurable. Couplings can be transmitted through several bonds: $^1H–^1H$ couplings through three bonds (for example H–C–C–H, written ^3J) are in the range 0–12 Hz, depending on the angle between the bonds (the **Karplus curve**: **Figure 11.20**).

J couplings are useful for two reasons. One is that the Karplus curve relates the magnitude of a coupling to angle; examples are backbone dihedral angles and side-chain orientation. However, by far the more important reason is

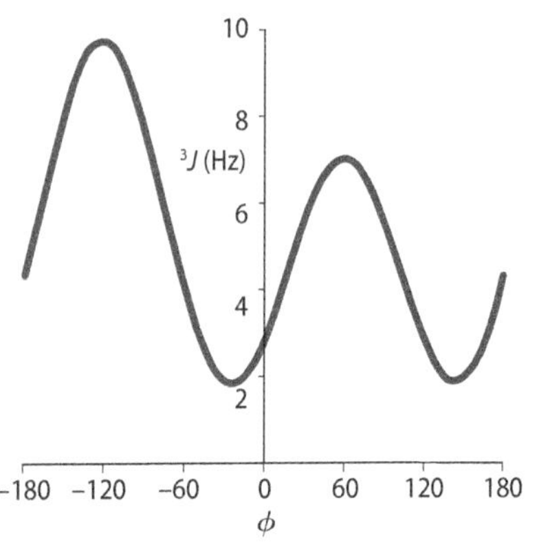

FIGURE 11.20
The dependence of the three-bond H–X–X–H coupling on the dihedral angle, usually called the Karplus curve. This curve shows the dependence of the $H_N–H_\alpha$ coupling constant on the backbone dihedral angle ϕ. Curves for other three-bond couplings are similar. The coupling is largest when the two protons are as far apart as possible, with an H–N–C–H dihedral angle of 180°: this corresponds to a ϕ angle of –120°.

that J coupling can be used to transfer magnetization from one nucleus to its J-coupled neighbor. The mechanism of this transfer will be described shortly, when we shall see how the connectivity between one nucleus and its neighbor can be used for the all-important step of *assigning* the NMR spectrum.

11.3.5 Two-dimensional, three-dimensional, and four-dimensional spectra

Despite the excellent resolution of NMR, it cannot resolve the several thousand signals from all the protons in a protein. To increase the resolution, NMR makes extensive use of higher-dimensional spectra.

As we have seen, NMR data are acquired by a simple **pulse sequence** involving a pulse followed by a period during which data are recorded during the time t, and constitute the FID. The FID is processed by a Fourier transform (FT) to yield a spectrum. A two-dimensional spectrum is a development of this method, and consists of the scheme preparation–t_1–mixing–t_2 (**Figure 11.21**). The preparation and mixing each consist of combinations of pulses and delays that affect spins in different ways, while the time t_2 is equivalent to t in the one-dimensional (1D) spectrum: Fourier transformation of the t_2 data yields spectra. A series of 100–500 experiments is performed, in which the delay t_1 is incremented regularly. A second Fourier transform in the t_1 direction produces a two-dimensional spectrum, in which both of the time axes have been converted to frequency. There is a difference between the two axes, in that the frequencies occurring during t_2 are directly detected, whereas those occurring during t_1 are detected only indirectly, by their effect on the signal measured during t_2. A two-dimensional (2D) spectrum is normally presented as a contour plot, similar to the representation of height in a map (**Figure 11.22**). A 2D spectrum typically takes longer to acquire than a 1D spectrum because of the need to acquire multiple 1D experiments. The development of 2D NMR won a Nobel Prize for Richard Ernst in 1991.

A three-dimensional (3D) spectrum is an obvious development of 2D NMR, consisting of the scheme preparation–t_1–mixing–t_2–mixing–t_3, followed by three Fourier transforms; a four-dimensional (4D) spectrum involves yet

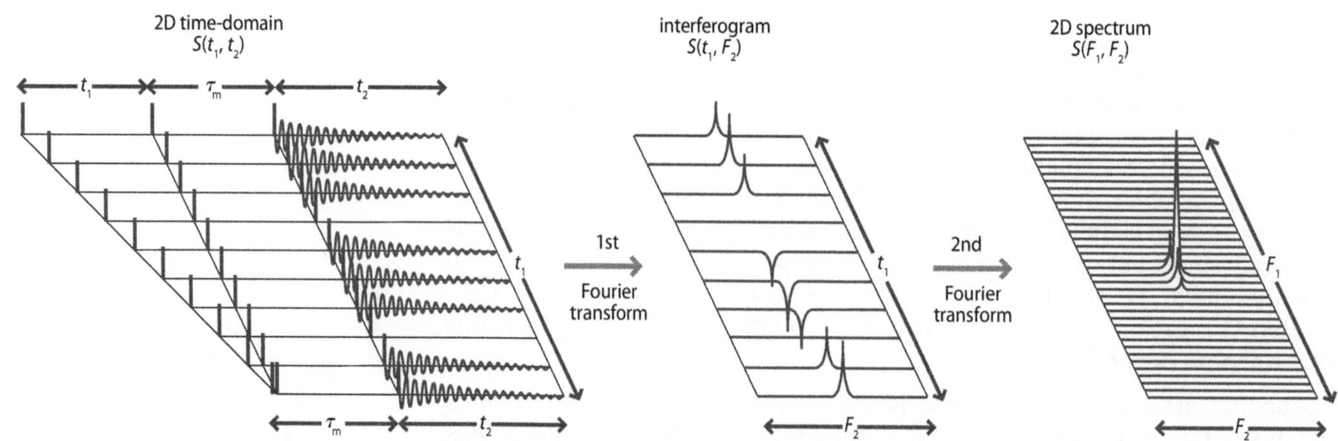

FIGURE 11.21
An example of a two-dimensional spectrum, here illustrated by the homonuclear NOESY spectrum, used to measure NOEs (see *11.10). The pulse sequence (left) consists of: relaxation delay, 90° pulse, t_1 delay, 90° pulse, τ_m delay, 90° pulse, acquire (t_2). By comparison with the standard 2D scheme in the text, it can be seen that the *preparation* in this experiment consists merely of the relaxation delay followed by a pulse; and the *mixing* consists of pulse, τ_m, pulse. The time τ_m in a NOESY experiment is normally referred to as the mixing time and is the time during which the NOE builds up. The pulse sequence is repeated many times using incrementally different values for t_1, and the resultant FIDs are stored in a 2D time-domain matrix. Fourier transformation of the directly acquired data (that is, the FIDs, which are a function of t_2) gives an 'interferogram' containing a series of 1D spectra in which each peak undergoes an intensity variation as a function of t_1. The corresponding frequencies are extracted by performing a second Fourier transformation of the data along t_1.

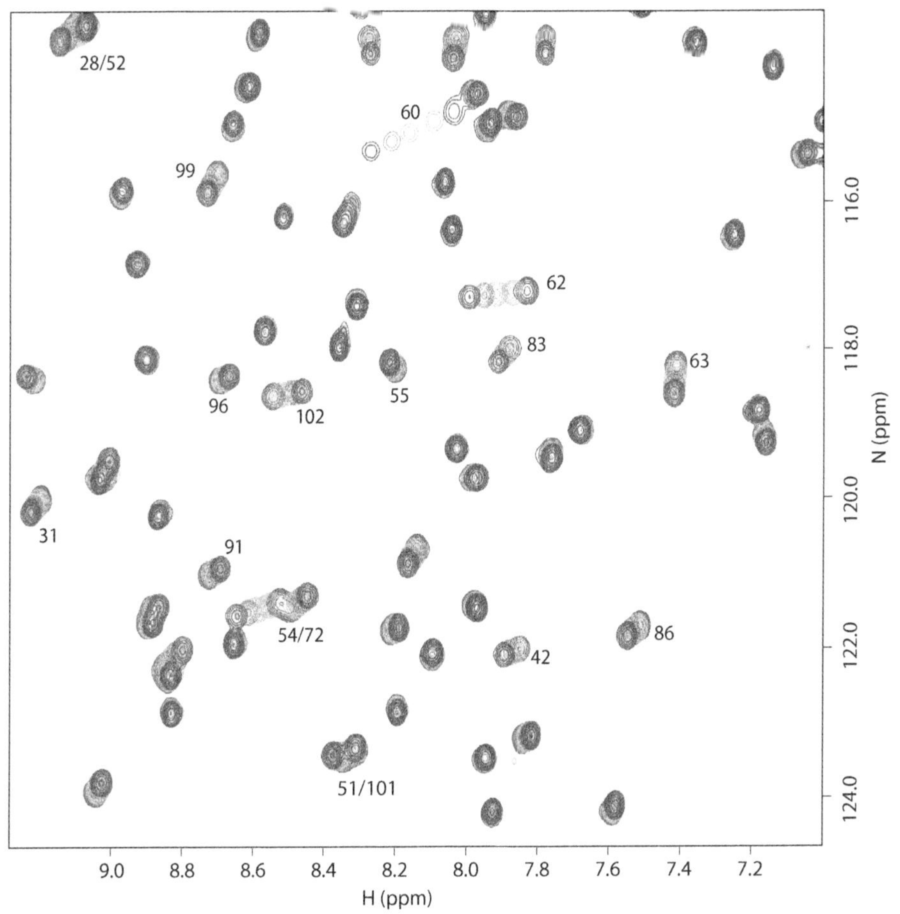

FIGURE 11.22
A two-dimensional spectrum. This spectrum is part of a ^{1}H–^{15}N HSQC spectrum (see Section 11.3.6, where the closely related ^{1}H–^{13}C HSQC spectrum is described) of the ribonuclease protein barnase. Each peak is shown as a series of contours, in the same way as a contour map. This spectrum contains one peak for the backbone H–N pair in each amino acid residue, and shows a titration of barnase with a ligand. Some peaks move during the titration (blue through red, indicated by numbers showing the residue number), indicating the parts of the protein close to the ligand-binding site.

another mixing and acquisition time. Each increase in dimensionality adds exponentially to the time required, and usually leads to a reduction in signal-to-noise ratio, meaning that 4D spectra are not often done. 3D and 4D spectra are normally presented as 2D planes from the higher-dimensional data. Planes in different directions can be selected depending on the application.

11.3.6 An example: the heteronuclear single-quantum coherence experiment

As an example of a 2D experiment, we discuss the heteronuclear single-quantum coherence (HSQC) experiment. Not only is this a relatively quick (about 1 hour) and simple heteronuclear experiment, it also forms the basis of almost all other protein NMR experiments. For technical reasons, we present the ^{13}C HSQC rather than the more common ^{15}N HSQC;[1] they are, however, almost identical.

[1]First, because nitrogen has a negative **gyromagnetic ratio**, which makes the α spin state confusingly of higher energy; and second, because the ratio of ^{1}H and ^{15}N gyromagnetic ratios is close to 10, whereas for carbon it is close to 4. Energy diagrams for ^{15}N are therefore less clear.

The pulse sequence is shown in **Figure 11.23**, and involves a series of pulses on both ^{1}H and ^{13}C. Some of these are 90° pulses, and some are 180° pulses:

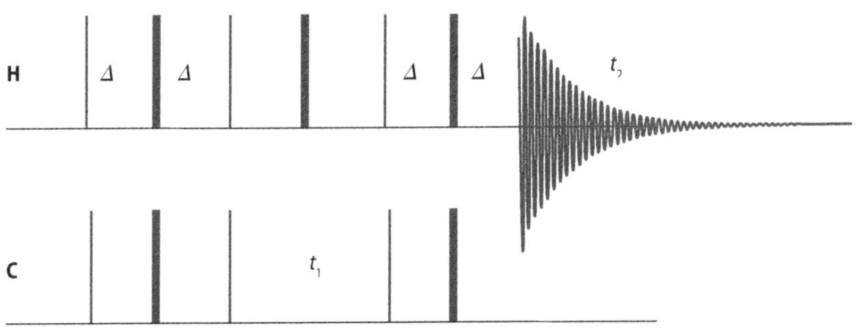

FIGURE 11.23
The HSQC pulse sequence. The horizontal axes are time; thin lines represent 90° pulses and thick lines 180° pulses. The experiment therefore uses pulses on both ^{1}H and ^{13}C, often simultaneously. As discussed for the NOESY pulse sequence, the experiment consists of the collection of several hundred FIDs, each using a different incremented value of t_1. The value of the delay Δ is $1/(4J)$, where J is the coupling constant between H and C, as discussed in the text.

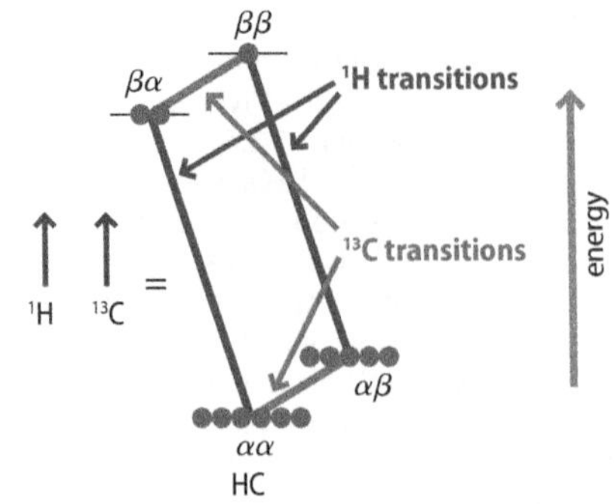

FIGURE 11.24
An isolated ^{1}H–^{13}C pair has four possible nuclear energy levels: $\alpha\alpha$, $\alpha\beta$, $\beta\alpha$, and $\beta\beta$, where the first letter indicates the spin state of the ^{1}H and the second that of the ^{13}C. The gyromagnetic ratio for ^{13}C is approximately one-quarter of that for ^{1}H (compare Table 11.2, in which observation frequency is proportional to gyromagnetic ratio), which means that the energy difference between α and β spin states for ^{13}C is about one-quarter of that for ^{1}H, and therefore at equilibrium (both spins along the $+z$ axis) the differences in population across the ^{13}C transitions are only one-quarter of those for the ^{1}H transitions, here shown as 1 unit compared to 4. The population is greater in the lower (α) energy level for each transition.

There is J coupling between H and C. This means that the two H transitions differ in energy by J Hz, as do the two C transitions (necessarily, because the total difference between $\alpha\alpha$ and $\beta\beta$ must be constant). The effect on the diagram, and on the populations, is tiny. For example, in an 11.7 T magnet, the energy of the ^{1}H transition is 500 MHz, whereas a typical J coupling is 130 Hz.

as discussed above, a 90° pulse rotates magnetization by 90°, and a 180° pulse rotates magnetization by 180°. Some of the pulses are applied along the x axis, and so rotate magnetization around the x axis (from z to $-y$, according to the right-hand rule), whereas y pulses rotate it around the y axis (from z to x). Pulses are selective, in that ^{1}H pulses affect only ^{1}H, and ^{13}C pulses affect only ^{13}C.

At the start of the pulse sequence, both ^{1}H and ^{13}C magnetizations are at equilibrium, along the $+z$ axis. This corresponds to equilibrium population differences across the ^{1}H and ^{13}C α/β transitions, as shown in **Figure 11.24**. The experiment starts with a ^1H_x 90° pulse, which rotates ^{1}H z magnetization to $-y$. During the subsequent time Δ, ^{1}H magnetization rotates under the influence of both its chemical shift and its J coupling to ^{13}C. Fortunately, it is possible to consider these two effects separately. Chemical shift corresponds to a rotation around the z axis, at a rate dependent on the chemical shift. **Figure 11.25** shows that a 180_y pulse on ^{1}H moves the ^{1}H magnetization such that, after a second period Δ, it is *refocused* back to the $-y$ axis, no matter what chemical shift it has. We can therefore ignore ^{1}H chemical shift evolution. J coupling corresponds to a splitting of the ^{1}H signal by an amount J Hz. This means that we need to split the ^{1}H magnetization vector into two components. One moves clockwise (as a result of J coupling to the ^{13}C β spin) at a rate $-J/2$ Hz, and the other moves anticlockwise (as a result of J coupling to the ^{13}C α spin), at a rate $+J/2$ Hz, corresponding to two components centered at zero chemical shift and separated by J Hz. The period Δ is chosen to be of duration $1/4J$ seconds, which means that it is long enough for each component to move through 45° (**Figure 11.26**). The ^1H_y 180° pulse interchanges the two components, which has the effect of making them change direction. However, the ^{13}C 180° pulse (which is simultaneous with the ^{1}H 180° pulse) acts on the ^{13}C z magnetization, and turns α into β, and β into α. This interchanges the spin state labels on the two ^{1}H components, which therefore change direction again: the net effect of the two 180° pulses is thus that the two components carry on going in the original direction (see Figure 11.26). The consequence is that after the second delay Δ, each component has moved through 90°, and thus one ^{1}H component is along

FIGURE 11.25
Refocusing of chemical shift by the 90-Δ-180-Δ pulse sequence. The initial H_{-y} magnetization created by the 90_x pulse precesses around the z axis, and after a time Δ it has moved an angle $\theta = \Delta\omega$, where ω is its chemical shift. The 180_y pulse rotates it through 180° around the y axis, which places it at an angle $-\theta$ to the $-y$ axis. Therefore after another time Δ it has come back to the $-y$ axis; in other words, chemical shifts are all refocused. The 90–Δ–180–Δ pulse sequence is often called a *spin echo* sequence because of this refocusing effect.

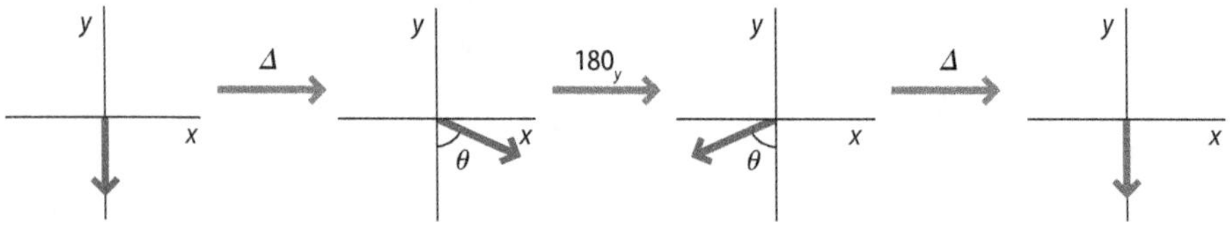

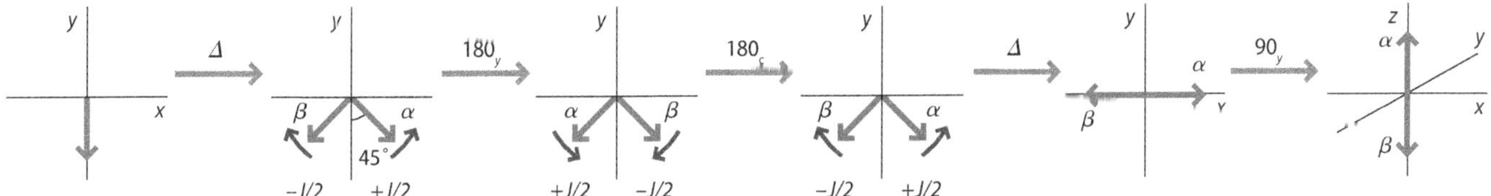

FIGURE 11.26
Creation of antiphase ^{1}H magnetization by the 90–Δ–180–Δ pulse sequence. The movement of magnetization is described in the text. Note that in the final panel, the axes are shown in three dimensions, whereas all the others are shown looking down onto the *xy* plane from the *z* direction.

+*x* while the other is along –*x*. At this point there is a 90°_y pulse, which rotates these two components onto the –*z* and +*z* axes, respectively (see Figure 11.26).

Now comes the really clever part. We have already seen that a vector along +*z* corresponds to a population distribution in which there is an excess of spins in the lower-energy α spin state. But now we have one of the *J*-coupled components along +*z* and the other along –*z*. This is shown in **Figure 11.27**, in which one component is up and the other down. We have therefore succeeded in inverting the population across one of the two ^{1}H transitions (the one coupled to the ^{13}C β spin state). However, because the ^{1}H and ^{13}C spins form one connected system, this means we have *also* affected the population difference across the ^{13}C transitions: the differences are larger than they were initially (by an amount equal to the ^{1}H spin population difference, implying that the population differences are greater by a factor of γ_H/γ_C, or very close to 4), and one is up while the other is down. We have thus coherently transferred magnetization from ^{1}H to ^{13}C, effectively with 100% efficiency. This elegant step is the core of the HSQC experiment, and of almost all heteronuclear NMR experiments.

The first 90° pulse on ^{13}C converts the *z* population difference into *y* magnetization. During the subsequent t_1 time, the ^{13}C magnetization rotates under the influence of its chemical shift. (The 180° ^{1}H pulse in the middle of t_1 interconverts the ^{1}H α and β spin states and therefore removes any effect from *J* coupling during t_1: it is said to *decouple* the ^{13}C–^{1}H coupling during this period.) This means that by the end of t_1, the ^{13}C magnetization has moved by an angle dependent both on its chemical shift and on the length of t_1. This affects the amount of signal that is reconverted to ^{1}H by the second half of the pulse sequence, and provides an indirectly detected measure of the ^{13}C chemical shift. The second half of the pulse sequence is the exact reverse of the first half, and reconverts the ^{13}C magnetization to ^{1}H via their *z* populations; the final t_2 period is used to measure the ^{1}H precession frequency.

Because the final signal that is measured during t_2 has an intensity that is modulated by the position that the ^{13}C magnetization had reached by the end of t_1, Fourier transformation with respect to t_1 generates ^{13}C frequencies, and Fourier transformation of the t_2 signals generates ^{1}H frequencies. The net result is a spectrum that correlates the ^{1}H and ^{13}C frequencies of H–C ^1J-coupled pairs (**Figure 11.28**).

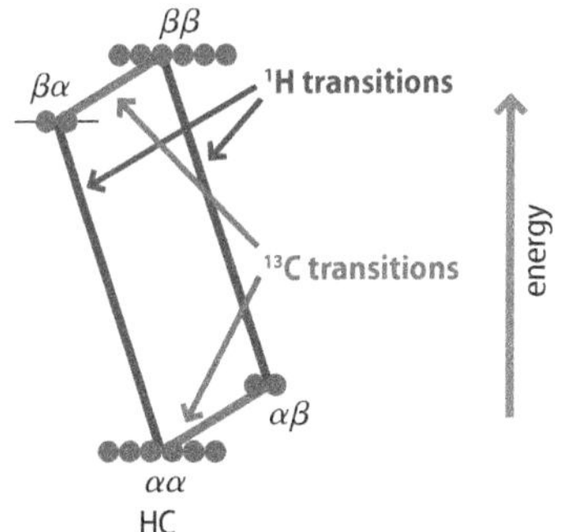

FIGURE 11.27
Spin populations at the end of the 90–Δ–180–Δ–90 pulse sequence. One ^{1}H transition is inverted, implying that the populations across this transition are also inverted. This necessarily also means that the population differences across the two ^{13}C transitions are of the same magnitude (that is, the ^{1}H population differences have now been transferred to ^{13}C) and that one is inverted.

11.3.7 Assignment of protein NMR spectra

Assignment is the essential step of working out the frequency of each ^{1}H, ^{13}C, and ^{15}N nucleus in the protein sequence. This cannot be predicted accurately enough to be useful, and must be worked out experimentally. It is done by using a set of 3D **triple resonance** experiments.

The 90°–Δ–180°–Δ block described above is found in virtually all heteronuclear experiments, and it can be used to transfer magnetization sequentially from one nucleus to its *J*-coupled neighbor. In one experiment, called the HNCO experiment, magnetization is transferred in this way from ^{1}HN to ^{15}N, then from ^{15}N to ^{13}CO, and then back from ^{13}CO to ^{15}N, and finally back to ^{1}H (**Figure 11.29**). On the way, there are incremented t_1 and t_2 periods, during

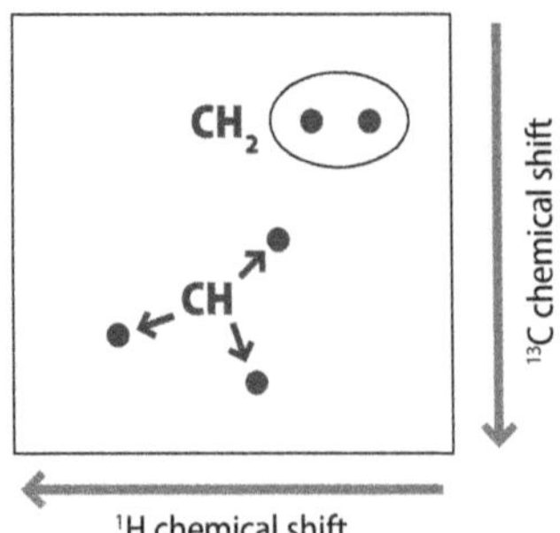

FIGURE 11.28
A $^{13}C/^{1}H$ HSQC spectrum. Each peak shows the correlation between a ^{1}H nucleus and its attached ^{13}C. Note that a CH_2 group therefore in general has two peaks, with different ^{1}H frequencies but the same ^{13}C frequency. In the conventional representation, the ^{1}H axis is horizontal, with increasing frequency from right to left, whereas the ^{13}C axis is vertical, increasing from top to bottom.

which the chemical shifts of ^{13}CO and ^{15}N are measured, in the same way as described for the HSQC experiment above. The result is a three-dimensional experiment in which each peak is characterized by three frequencies, these being the $^{1}H_{i+1}$, $^{15}N_{i+1}$, and $^{13}C'_{i}$, where for convenience we represent the carbonyl carbon by C′ (see Figure 11.29). The $i/i+1$ nomenclature indicates that the carbonyl carbon comes from the amino acid residue preceding that of the H and N.

One can also do a second three-dimensional experiment called HN(CA)CO. As one might expect from the name, in this experiment magnetization is passed along in the sequence $^{1}H_i$–$^{15}N_i$–$^{13}C\alpha_i$–$^{13}C'_i$, and then comes back along the same route (see Figure 11.29). The t_1 and t_2 periods are organized so that the three frequencies measured are $^{1}H_i$, $^{15}N_i$, and $^{13}C'_i$.

These two spectra form a pair, in which each C'_i nucleus along the backbone is connected either to its following HN_{i+1} (in the HNCO) or its intraresidue NH_i (in the NH(CA)CO), or conversely, each NH_i unit along the protein backbone is connected to either $^{13}C'_i$ or $^{13}C'_{i-1}$. By matching up carbonyl frequencies in the two spectra, it is therefore possible to walk along the protein sequence, and thus assign all backbone H, N, and C′ signals.

One can do a similar thing for Cα, using the pair of experiments called HNCA and HN(CO)CA; and also for Cβ. These experiments together provide assignments of all backbone H, N, Cα, and Cβ nuclei, as well as giving considerable redundancy in the inevitable case of signal overlap of some C′, Cα, or Cβ nuclei. Much of this procedure can be automated, making backbone assignment a relatively rapid procedure. Assignment of side chains, however, remains a much more difficult matter because the spectra are much more crowded, and usually requires significant human intervention.

11.3.8 Chemical shift mapping

Once assignments have been made, a particularly simple and powerful experiment is to measure $^{1}H,^{15}N$ HSQC spectra over the course of a titration with a ligand or protein interaction partner. Chemical shifts are extremely sensitive

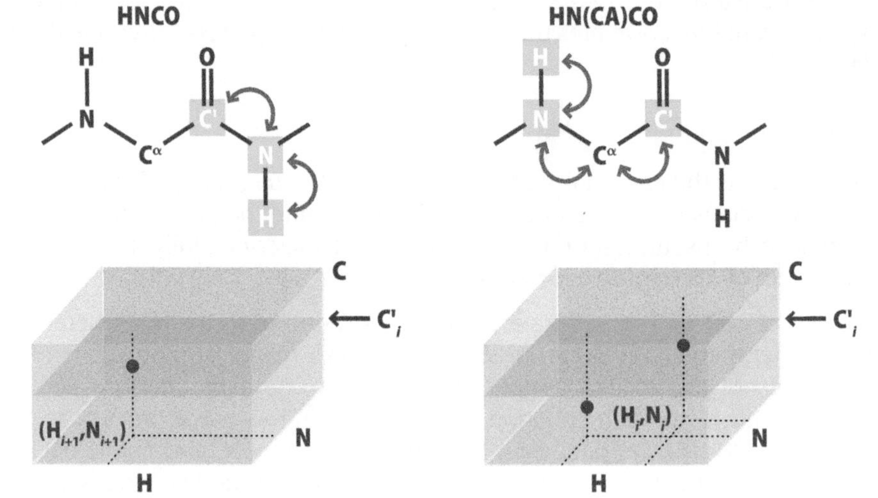

FIGURE 11.29
Sequential assignment using the HNCO and HN(CA)CO experiments. The HNCO experiment is a 3D experiment that correlates the frequencies of $^{1}H_{i+1}$, $^{15}N_{i+1}$, and $^{13}C'_i$, as shown by the green squares: the brown arrows show the route of the magnetization as it is passed from one nucleus to the next (H to N to C′ and back). In the 3D HNCO spectrum this correlation is shown as a peak at these three frequencies: the peak (black circle) is found on the $^{13}C'_i$ plane. By contrast, the HN(CA)CO experiment correlates the three intraresidue frequencies of $^{1}H_i$, $^{15}N_i$, and $^{13}C'_i$. This means that any particular C'_i frequency (as indicated by the shaded plane) is correlated to different (H,N) frequencies in the two spectra. Therefore, by matching up C'_i frequencies in the two spectra, one can 'walk' from one HN to the next along the sequence.

to structural effects, and ligand binding leads to changes in the HSQC spectrum that can be followed very easily (see Figure 11.22). This provides a good indication of the binding site of the ligand, and very often also permits measurement of the affinity. This experiment is called chemical shift mapping or complexation-induced shift (CIS).

11.3.9 Relaxation

The equilibrium position of magnetization is along the $+z$ axis. Pulses move magnetization away from equilibrium (for example, a 90° pulse moves it into the xy plane), after which nuclei undergo a process of **relaxation** to return to their equilibrium position. There are two main relaxation processes possible. One describes the recovery of z magnetization, and occurs at a rate $R_1 = 1/T_1$. The other describes the loss of xy magnetization, and occurs at a rate $R_2 = 1/T_2$ (**Figure 11.30**). In proteins, R_2 is very much faster than R_1. This is because R_1 relaxation requires transitions between α and β states; that is, flips in z magnetization. By contrast, R_2 relaxation can arise from flips in any magnetization (x, y, or z). There are therefore more processes that can cause R_2 relaxation, and hence R_2 relaxation is faster.

A single isolated spin relaxes extremely slowly: spins need something to stimulate their relaxation. The most common source of relaxation is another spin. Other spins produce local magnetic fields, and (as we have seen already in the context of pulses) magnetic fields can cause magnetization to rotate, but only if they are changing at a frequency corresponding to the nuclear transition that is to be altered. This means that the effectiveness of the local field to cause relaxation depends on the rate of relative motion of the two spins, in other words on the local mobility within the protein. It also depends on the distance between the nuclei.

If spin A relaxes spin B, then it is also possible (although not necessary) that spin B relaxes spin A at the same time. For example, the two spins can exchange magnetization. This process is also distance-dependent in the same way as the other relaxation effects described above. This effect is known as the **nuclear Overhauser effect (*11.10)** or NOE, and provides the single most powerful structural parameter for calculating protein structure. The effect is proportional to r^{-6}, where r is the distance between the two nuclei, and has a

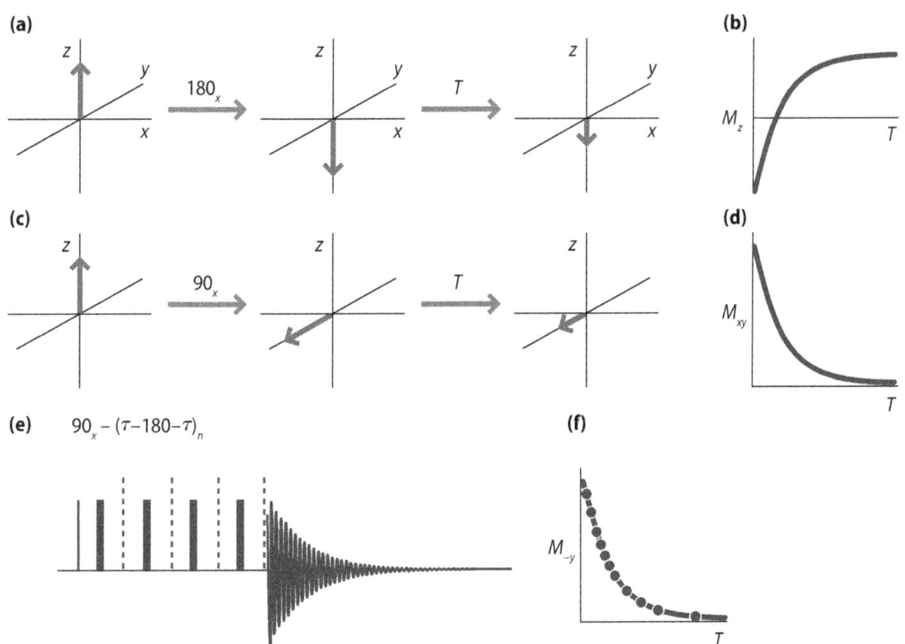

FIGURE 11.30
T_1 and T_2 relaxation. After a 180° pulse (a), magnetization relaxes back to equilibrium (b) (that is, along the $+z$ axis) following the equation $M_z = M_z(0)(1-2e^{-t/T_1})$; that is, an exponential return to equilibrium with a time constant of T_1. By contrast, after a 90° pulse (c), two relaxation processes must occur. Magnetization in the $+z$ direction recovers in a similar way to that in (a), but also magnetization in the xy plane is lost, in an exponential decay (d) with time constant T_2, $M_{xy} = M_{xy}(0)e^{-t/T_2}$. The measurement of T_2 is often performed with repeated *spin echoes* (see Figure 11.25), because this compensates for some instrumental imperfections and enables magnetization to be measured along the same axis in all measurements (e), in this example along the $-y$ axis. M_y can then be measured only at the peak of each echo (f), where $T = 2n\tau$.

*11.10 Nuclear Overhauser effect (NOE)

This is a very important NMR effect because it is used to measure the distance between two nuclei, usually two protons. If two protons I and S relax each other, then a change in the intensity of S (for example saturation) can also result in a change in the intensity of I, the rate of change being proportional to r^{-6}, where r is the distance between them. This can be understood fairly simply. The intensity of an NMR signal is proportional to the difference in population between the lower-energy α state and the higher-energy β state. For the two-spin IS system, there are four energy levels, as shown in **Figure 11.10.1a**, where for example the label $\beta\alpha$ means that I is in the β state and S is in the α state. At equilibrium, there is an excess of spins in the $\alpha\alpha$ state and correspondingly a deficit in the $\beta\beta$ state (see Figure 11.10.1a). This is represented in the figure by populations of +1/2 in $\alpha\alpha$ and −1/2 in $\beta\beta$, because the total I signal (summed over both I transitions) is the sum of the population differences $(\alpha\alpha - \beta\alpha) + (\alpha\beta - \beta\beta) = 1$. When S is continuously saturated, the populations of $\alpha\alpha$ and $\alpha\beta$ are equalized, as are the populations of $\beta\alpha$ and $\beta\beta$ (Figure 11.10.1b). As yet, this has no effect on the net population of I, which is still 1. However, in proteins an efficient relaxation route is an exchange of magnetization between I and S, often described as a 'flip-flop' transition from $\alpha\beta$ to $\beta\alpha$ and vice versa, or from (↑↓) to (↓↑). In Figure 11.10.1c this is indicated as the W_0 (zero quantum) transition. This leads to an exchange of spin population between these two energy levels; because the population of $\alpha\beta$ is greater than that of $\beta\alpha$, the net result is a decrease in the population of $\alpha\beta$ and an increase in $\beta\alpha$. These population differences get equalized over the S transition because of the continuing saturation of S. The effect on the net population of I is to decrease it by an amount δ. Thus, saturation of S results in a decrease in the intensity of I: in other words, an NOE. The relaxation route between $\beta\beta$ and $\alpha\alpha$ is also possible, but in proteins it is much less efficient; this leads to an increase in the intensity of I, and is observed in small molecules.

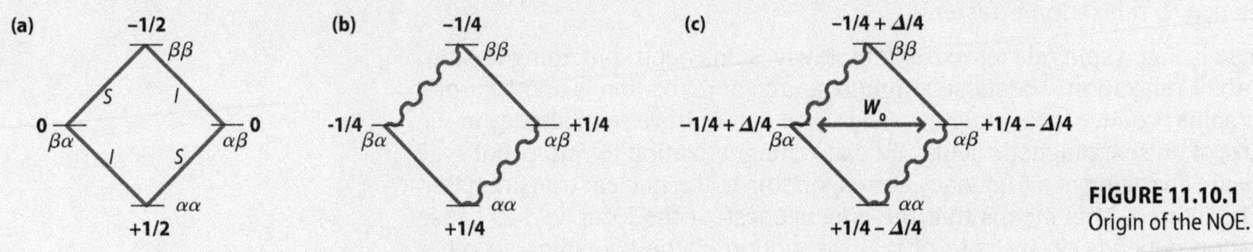

FIGURE 11.10.1
Origin of the NOE.

very strong similarity to FRET, because its physical origins are very similar. It is, however, a much more short-range effect, being limited to about 6 Å.

The rates of the NOE, R_1, and R_2 relaxations all depend both on internuclear distance and on the rate of local tumbling in the protein, implying that for a fixed distance, such as the relaxation of ^{15}N by its attached ^1H, the rates depend only on local tumbling. These rates can therefore provide a detailed measure of local mobility, and have been a powerful method for describing rapid local motions, as described at the start of Chapter 6. They can be analyzed to calculate an **order parameter**, given the name S^2, which describes what proportion of local motion is due to tumbling of the whole protein as a rigid body, and conversely what proportion is due to rapid local motions.

On a rather less quantitative level, R_2 relaxation rates increase fairly linearly with the **correlation time** of the molecule. This means that as a molecule gets larger, the R_2 relaxation rate increases and so does the linewidth; in other words, large and slowly tumbling molecules have broad NMR signals. In the NMR spectrum of a protein, small-molecule impurities can usually be spotted very simply by their sharper lines.

One of the origins of R_2 relaxation is the exchange of a nucleus between two different environments in which it has different chemical shifts. This could be due for example to conformational change or ligand binding. Such effects can be detected by using a modified R_2 measurement designed to measure *relaxation dispersion* (**Figure 11.31**). This provides information on dynamic processes happening on a timescale between 10^{-4} and 10^{-2} second, which is of great interest because it is the typical timescale of enzyme catalytic turn-over and induced-fit motions. This experiment provided the data for the slow dynamics of dihydrofolate reductase discussed in Chapter 6.

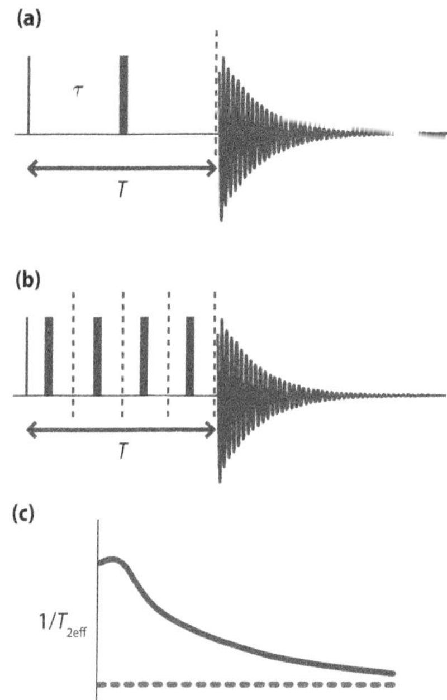

FIGURE 11.31
Relaxation dispersion. T_2 decay is measured with a spin echo pulse sequence over a fixed time T, but with different numbers of echoes (here 1 and 4 in (a) and (b), respectively). If there is no exchange occurring during time T, the measured T_2 will be the same in both experiments (c, dashed line). However, if exchange occurs, the spin echo will not refocus magnetization fully, and the result is an apparently faster relaxation; that is, a larger effective value of $1/T_2$ (c, solid line). The measurement is normally performed on ^{15}N relaxation, and actually uses a 2D experiment more like that shown in Figure 11.23.

Relaxation can also be induced by *unpaired electrons* (free radicals). Free radicals can be deliberately introduced into proteins, for example by the attachment of nitroxide radicals or paramagnetic metal ions, where they give rise to distance-dependent relaxation and hence provide very useful long-range distance information in the form of *paramagnetic relaxation enhancement* (PRE). This relaxation is due to the electron spin, which has an energy transition nearly 2000 times that of the ^1H nuclear spin, and is therefore much more sensitive. Electron spins are observed in *electron paramagnetic spectroscopy* (EPR, also known as ESR), which provides useful information about the immediate vicinity of the electron spin. This requires completely different apparatus from NMR, because the frequencies are so much higher, in the microwave range.

11.3.10 Protein structure calculation from NMR data

Protein assignment, described above, merely provides a list of the chemical shifts of all ^1H, ^{13}C, and ^{15}N nuclei in the protein. To calculate structure, one needs structural information. This comes from three main sources, which have been described above: the NOE, residual dipolar couplings, and chemical shifts. The NOE is the most powerful of these because it provides the approximate distance between two specific nuclei. Dipolar couplings provide angles, but because the dipolar coupling is proportional to $(1 - 3\cos^2\theta)$, which typically has four possible solutions for each angle, these are not unique. Chemical shifts also provide angles, but not very precisely: the relationship between chemical shift and structure remains somewhat imprecise.

For a structure calculation on a typical small protein of 150 residues, one might expect to have around 2000 NOEs, supported by around 300 chemical shifts and possibly another 120 residual dipolar couplings. Such measurements are still often collected rather laboriously by hand, although efforts to automate this process are likely to provide a substantial reduction in time and labor [5]. A protein of this size has roughly 900 independent heavy atoms (that is, heavy atoms whose position cannot be worked out from other known atoms), and therefore requires (to a first approximation) around 2700 ($3n$) pieces of information to characterize x, y, and z coordinates. This would suggest that the information available from NMR is very nearly enough. However, none of the data are precise. In addition, many of the NOEs are likely to be ambiguous (that is, the assignment is not certain because several protons share the same chemical shift value). This means that, in practice, the information from NMR is nothing like enough to define the structure properly. Therefore the NMR information is always supplemented by other information, such as bond lengths, bond angles, van der Waals radii, and planarity of aromatic rings and peptide bonds: all the things we know about the covalent structure of proteins.

Essentially, the structure calculation is a complicated optimization problem: given the known covalent structure, determine the three-dimensional structure that best satisfies the experimental restraints. There are several computational approaches that can be used to solve this problem, but by far the most widely used is known as *restrained molecular dynamics* or *simulated annealing*.

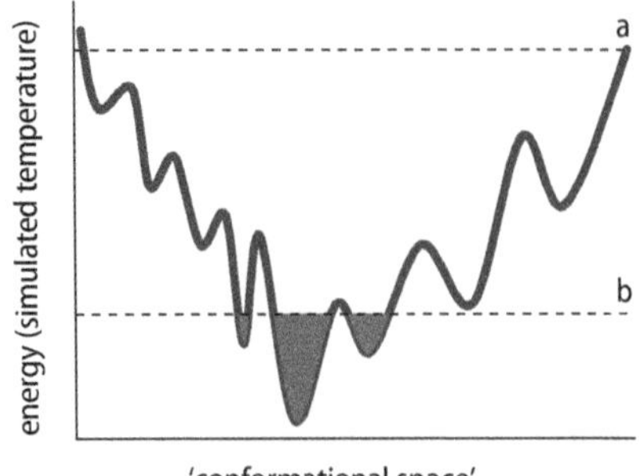

FIGURE 11.32
Simulated annealing. The graph represents the conformational space available to the protein, and is essentially the same as the energy landscape discussed in Chapter 6. A simulated annealing calculation starts as a simulation at high temperature, and therefore high kinetic energy, where the molecule has sufficient energy to search the conformational space widely (a). The temperature is then gradually lowered. By temperature (b), for example, only three regions of conformational space (shown in red) are populated significantly (although the molecules have a distribution of velocities given by the Boltzmann distribution (*5.2), and therefore some molecules will have higher energies and populate other regions). The aim is that by the time the temperature has been reduced to its final value, all molecules will have settled into the lowest energy well.

A standard molecular dynamics calculation is performed (as described in Section 11.9.2 below), to which the structural restraints are added as additional forces. For example, NOEs are added as a force between the two atoms concerned, to bring them to the correct distance. The calculation is performed initially at a high simulated temperature (often several thousand kelvins), to give the protein atoms enough kinetic energy to search conformational space rapidly. The system is then gradually cooled ('annealed'), to allow the conformation to settle down into the global energy minimum (**Figure 11.32**). Typically, the calculation uses a limited set of energy terms, to speed it up, and is thus in no way to be thought of as representative of real folding. The full set of energies can then be added back at the end to improve the local geometry.

Such a calculation represents an approximate solution to the correct structure. Typically, many such calculations (about 50) are performed, and the ensemble is used to describe how well the structure can be determined from the NMR data (**Figure 11.33**).

In practice, the experimental NOE list is not only ambiguous but is also likely to contain noise and errors. Such structure calculations are therefore performed iteratively: at each iteration the list is tidied, so that by the final iteration a stable solution has been reached.

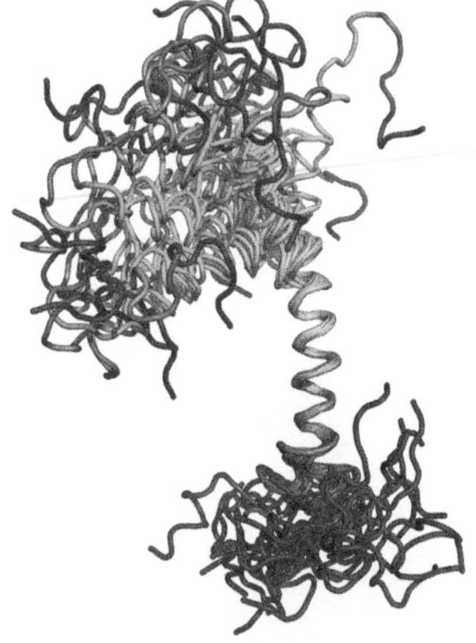

FIGURE 11.33
A typical NMR protein structure. The structure is represented by a bundle of 10–50 structures, each of which represents a good solution to the input restraints, composed of NOEs, chemical shifts, RDCs, etc. This example is a transmembrane protein PufX (Section 10.3.5), colored from blue at the N terminus to red at the C terminus. The central helical section is well defined, and there is also a well-defined helical section N-terminal from it, but the relative orientations of the two helical sections are not well defined. The N and C termini are completely undefined. In this case the well-defined green and yellow helical regions are within the membrane in the native protein, while the termini extend out beyond the membrane.

11.4 DIFFRACTION

11.4.1 Microscopy and the diffraction limit

In conventional light microscopy, light is used to illuminate a sample. Some of the light is absorbed by the sample, leading to a darker region in the image or, if the absorption is dependent on the wavelength of the light, a colored image. For many biological samples, the amount of light absorbed is very small, leading to poor contrast and thus a weak image. The contrast is improved greatly by using *phase-contrast microscopy*. This technique (which led to the awarding of the Nobel Prize in Physics to Frits Zernike in 1953) is based on the realization that samples not only absorb light: they also slow it down (because of their different refractive index) and therefore lead to a small change in the phase of the light. The phase-contrast microscope splits the incident light up into two parts, one of which acts as a reference beam and is made to undergo a 180° phase change, so that when recombined with the beam going through the sample it exactly cancels out if there has been no phase change in the latter. The only light that produces net intensity is light that has had its phase altered by going through a sample with different refractive index. Thus, the small phase variation is converted into an amplitude variation.

The ultimate limit to the resolution achievable by microscopes arises from the fact that light undergoes diffraction if the detail to be seen becomes smaller

*11.11 Numerical aperture

The numerical aperture of a lens is given by NA = $\eta\sin\alpha$, where η is the refractive index of the medium and α is the angle at which the light enters the lens (**Figure 11.11.1**). The maximum resolution attainable is 0.61λ/NA, implying that a larger numerical aperture gives a better resolution. The maximum value of α is 90°, at which $\sin\alpha = 1$, while the refractive index of air is 1. With an oil-immersion lens, η rises to 1.4, explaining why such an arrangement can give better resolution.

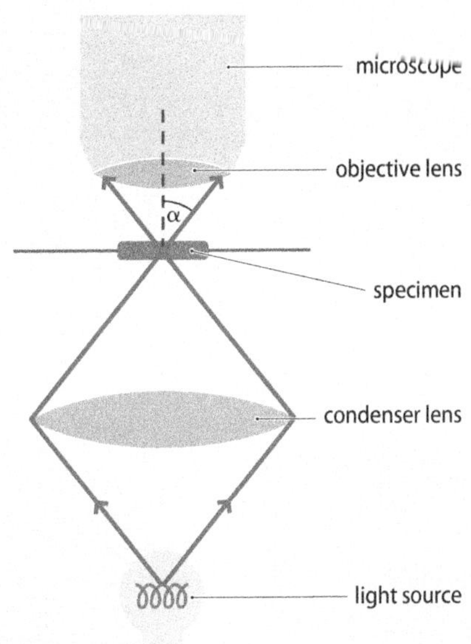

FIGURE 11.11.1
Definition of α.

than about half the wavelength of the radiation. The wavelength of blue light is roughly 400 nm, implying that objects smaller than about 200 nm cannot be seen with a light microscope. It is therefore necessary to use other techniques. The diffraction limit also depends on the size of the **numerical aperture (*11.11)** of the objective lens. There are some very interesting techniques emerging that get round this limitation.

11.4.2 X-ray diffraction

X-ray crystallography is by far the most productive method for the determination of protein structures. As of September 2010, the protein data bank (http://www.rcsb.org/) contains 68,000 structures, of which 59,000 come from X-ray diffraction, 8600 from NMR (13% of the total, a fairly consistent figure over the past few years), and only 306 from electron microscopy (almost all of which are at very low resolution), reflecting the greater experimental difficulties associated with that technique. Without structures we can understand relatively little of how a protein works, and it is true to say that crystallography has contributed more to our understanding of proteins than any other technique. There have been many Nobel prizes in this area, starting with the prizes in physics awarded to von Laue in 1914 and the Braggs (father and son) in 1915, and most recently the prize in chemistry awarded to Venkatraman Ramakrishnan, Thomas A. Steitz, and Ada E. Yonath in 2009 for their remarkable structural work on the ribosome.

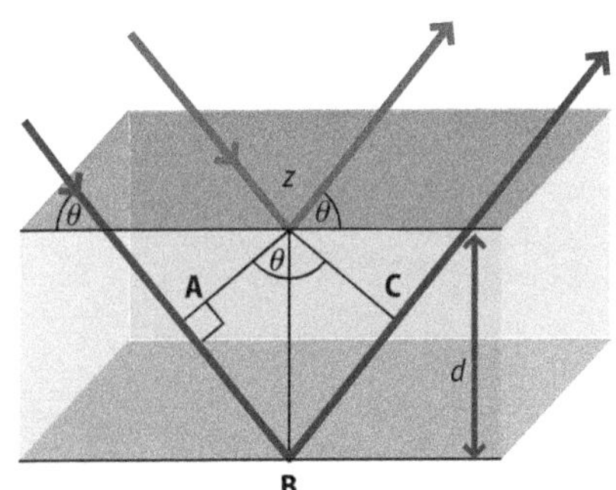

FIGURE 11.34
The principle of X-ray diffraction from crystals. A beam of light can be reflected from a crystal plane. The figure shows two light rays, of which one travels a distance AB + BC farther than the other. AB = BC = $d\sin\theta$. If this distance is equal to an exact multiple of the wavelength of the light (that is, if $2d\sin\theta = n\lambda$), the diffracted beams will be in phase and will add up constructively, creating a diffraction spot.

X-ray crystallography requires protein crystals. These are arrays of protein molecules, arranged regularly in three dimensions. The smallest repeating unit, which may contain one or more protein molecules, is called the *unit cell*. X-rays have wavelengths in the order of ångströms, which makes them ideal for studying molecular structure, because this is the same order as bond lengths. X-ray diffraction is based on the principle that when X-rays pass through matter, a small proportion will interact with the electrons in it and will be scattered by them. The scattering is in all directions, but in most directions the scattered X-rays from different atoms will interfere with each other and give no coherent signal. However, constructive interference will occur when the X-rays are scattered off a crystal lattice, such that the X-ray beams scattered from adjacent crystal planes are in phase (**Figure 11.34**). The Braggs

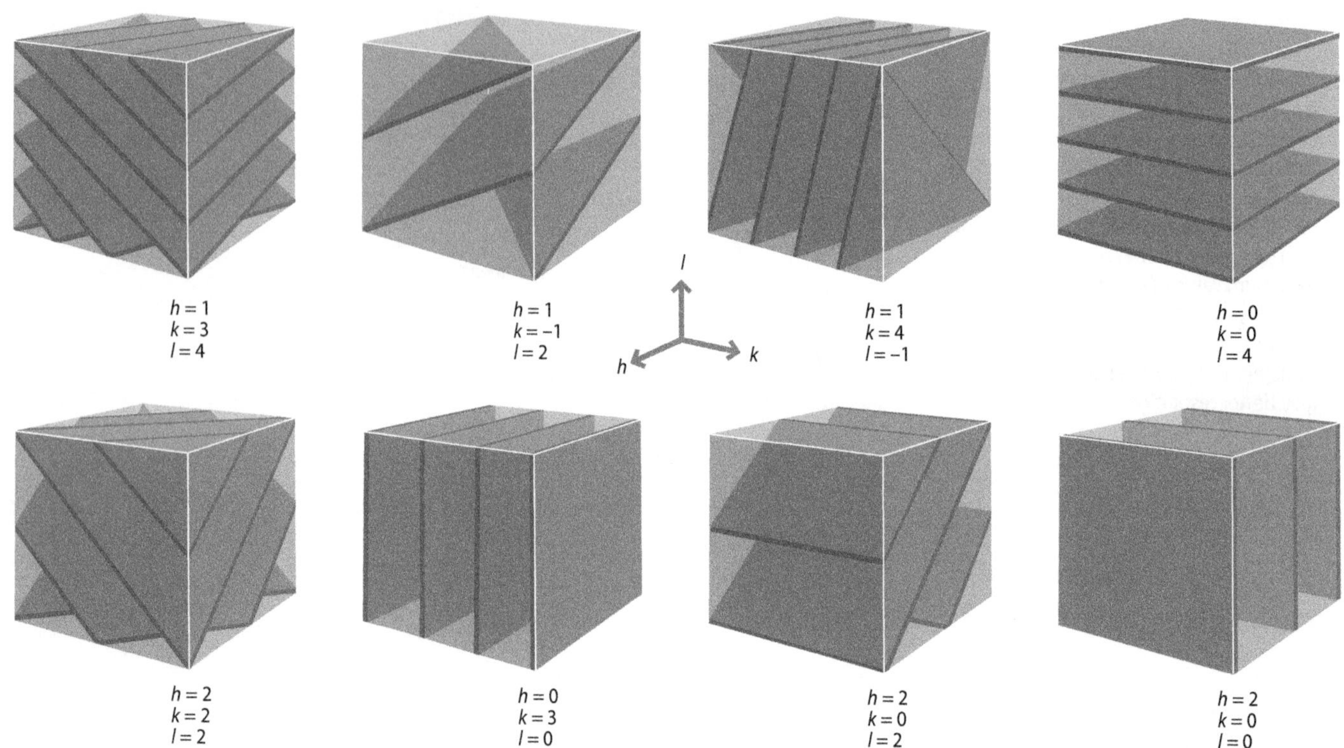

FIGURE 11.35
Crystal lattice planes are defined by three integers *h*, *k*, and *l*, which define how many intersections are made with each of the axes in the unit cell. (Adapted from M. F. Perutz, *Sci. Am.* 211: 64–76, 1964. With permission from the National Academy of Sciences.)

showed that this can be satisfied very simply if $n\lambda = 2d\sin\theta$, where λ is the wavelength of the X-rays, d is the spacing between planes, and θ is the angle between the incident beam and the plane. This is exactly the same principle as for light being diffracted by a grating.

There are many of these crystal planes, which can be constructed by dividing up the unit cell into an integer number of divisions in each of the x, y, and z directions and drawing planes between the divisions, as shown in **Figure 11.35**. The consequence is that when a crystal is irradiated by X-rays, constructive interference occurs in very precise directions, constituting a *diffraction pattern* (**Figure 11.36**) that contains distinct spots arranged in a regular

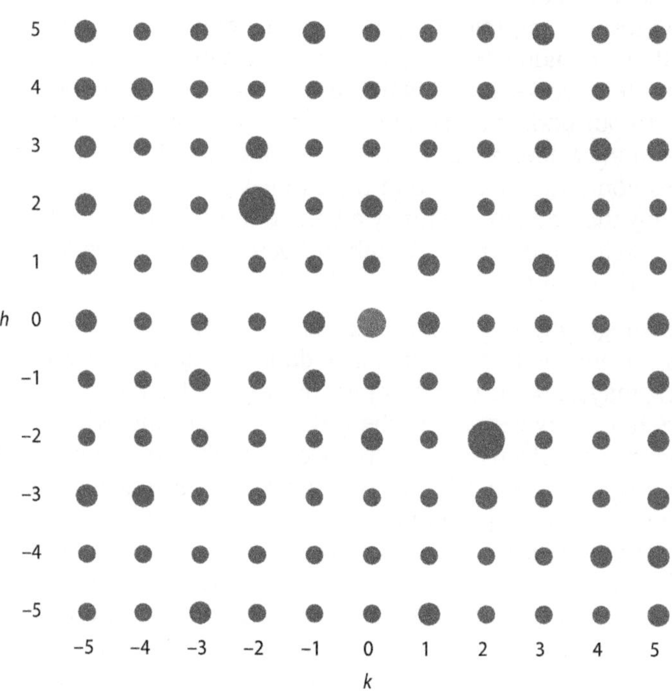

FIGURE 11.36
A diffraction pattern. Diffraction spots are found in a regular array, reflected off the different planes. They can therefore be indexed by the *hkl* integers, as indicated. This pattern has two-fold symmetry, implying that the structure that it came from must also have two-fold symmetry.

pattern. The diffraction pattern changes as the incident angle of the X-ray beam changes, which means that in order to observe all possible spots it is necessary to collect diffraction patterns from all possible orientations of the crystal. In practice this is done by collecting data while rotating the crystal by a small amount, for example 1°, and then collecting a second data set for the next 1° rotation, and so on (**Figure 11.37**). Crystal symmetry often means that for example only a 90° range need be covered, because the diffraction pattern has two-fold symmetry. The positions of the spots are defined by the dimensions of the unit cell, but the intensities of the spots in the diffraction pattern depend on how many atoms lie (strictly, how much electron density lies) along the planes: more atoms within that plane produce a more intense spot. The intensities therefore carry the structural information. In fact, each spot carries information on the entire structure.

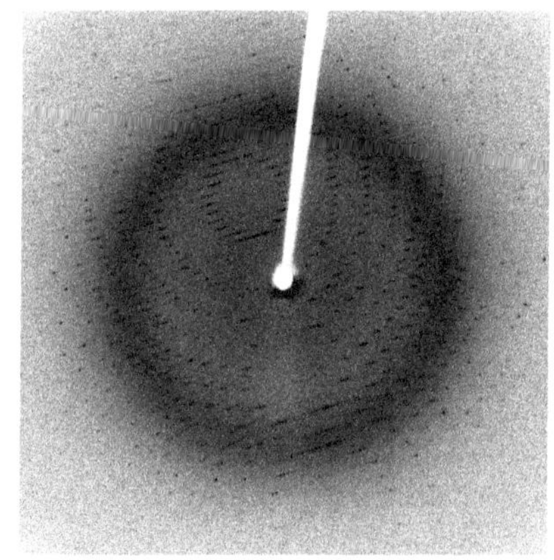

FIGURE 11.37
An actual protein diffraction pattern. The center of the image (white) corresponds to undiffracted X-rays. As we move outward from here, the average intensity of the diffracted spots gets weaker, corresponding to greater resolution: the resolution of the data therefore relates to how far out from the center there are useful data. The banded appearance arises from the fact that this image only comes from 1° of rotation of the crystal, and is therefore only a small fraction of the total number of diffraction spots measurable.

The Bragg equation given above shows that for a fixed wavelength λ, the measurement of a short distance d requires a large angle of incidence θ; in other words, the incident beam needs to be scattered by a large angle. This means that to obtain good *resolution*, we need to be able to measure diffraction intensities out to the edge of the diffraction pattern (see Figure 11.37).

For a protein of about 150 residues, one might expect to measure roughly 15,000 unique reflections for a resolution of around 1.8 Å, or 50,000 reflections for a resolution of around 1.2 Å. A protein of this size has about 900 unique heavy atoms and thus needs roughly 2700 data points to define it. There should therefore be enough information in a diffraction pattern for a well-determined structure. As the resolution gets worse (roughly worse than 2.5 Å), the amount of information available decreases, and modeling of the structure into the density becomes more dependent on the software used. A comparison with NMR data is instructive: X-ray produces very roughly 10–20 times more data than NMR, and moreover the data are in general more precisely determined. It is therefore little surprise that X-ray structures are in general of considerably better quality than NMR structures. This is of course not to say that NMR structures are not useful. NMR structures can be very accurate locally, but their long-range accuracy is necessarily low because they are constructed mainly 'from the bottom up,' using short-range distance information. By contrast, X-ray structures are constructed 'from the top down,' the low-resolution data producing the general topology, and the detailed placement of atoms requiring high-resolution data. In addition, X-ray structures require a non-physiological crystallization step. However, comparison of X-ray structures with NMR structures, and with biochemical data, suggests that in almost all cases the X-ray structure is a very accurate depiction of the structure in solution.

11.4.3 The phase problem in X-ray diffraction

Each spot in a diffracted X-ray pattern has both an amplitude and a phase, and to calculate the structure we need to know both. However, it is not possible to measure the phase, which implies that half the information is missing. The problem is somewhat analogous to trying to draw a three-dimensional map given only the distances between objects but not their angles. As long as we have a rough idea of the angles, we may be able to use this information (possibly together with other information such as bond lengths) as a basis to obtain an approximate map, from which we can refine the map further. Similarly, to solve an X-ray structure we need at least some idea of the phases (as many as possible), and then we can work out the rest.

The amplitude can be measured easily. However, there is no **direct (*11.12)** way to measure the phase. It is therefore necessary to perturb the phase in some small systematic way and repeat the diffraction measurement; by comparing the two diffraction patterns one can work out approximate values for

***11.12 Direct methods**
For small molecules, one can use geometrical relationships between different planes to compare intensities, and hence make some deductions about relative phases. This is called the direct method, and it does not require any additional information; it is, however, much simpler if the molecule is relatively small, because in this case there are effectively fewer simultaneous equations to solve. It is routinely used for organic crystals and has been used for very small protein crystals, but it is too computer-intensive for routine use in normal proteins. Herbert Hauptmann and Jerome Karle won the Nobel Prize for this in 1985.

***11.13 Max Perutz**
Max Perutz (1915–2002) (**Figure 11.13.1**) was born in Austria; he came to Cambridge, UK, in 1936 after completing his first university degree, and stayed there for the rest of his life. During World War II he worked on a project to build aircraft carriers from ice strengthened by wood fibers. His main achievement, for which he won the Nobel Prize in 1962 with John Kendrew, was to work out how to use X-rays to obtain protein structures, in particular how to obtain information on the phases of the signals by comparing diffraction patterns with and without heavy metal ions present. He used this method to work out the structure of hemoglobin, and then showed how the allosteric mechanism works, as well as the Bohr effect and sickle-cell anemia, and continued to research until his death, for example developing drugs for increasing oxygen delivery to tumors, and developing a model for amyloid formation in polyglutamine-rich proteins. He was appointed head of the Medical Research Council Laboratory Unit for Molecular Biology in 1947 (which at that time consisted of two people: Perutz and Kendrew), moving on to become Chairman of the Laboratory of Molecular Biology in 1962, an institution that has produced no less than 12 Nobel Prize winners (Perutz, Kendrew, Watson, Crick, Sanger, Klug, Milstein, Köhler, Walker, Brenner, Sulston, and Horvitz). He became a talented science writer, including the highly recommended book Is Science Necessary?.

FIGURE 11.13.1
Max Perutz. (From M.F. Perutz, *Annu. Rev. Physiol.* 52:1–25, 1990. With permission from Annual Reviews.)

at least some of the phases, which is generally enough to work out the rest. As implied above, the process is not unlike solving a logic problem: starting from a correct logical deduction one can fill in missing gaps and eventually complete the whole problem. Of course, if the initial deduction was wrong, then the initial error will propagate through the solution and give something completely wrong. This can happen in crystal structures: if the initial phases are incorrect, the result can be a structure that looks roughly right but contains errors. Fortunately such problems are usually easily detected.

The most common way to solve the phase problem is based on the observation that heavy atoms (that is, atoms significantly heavier than carbon, oxygen, or nitrogen) show marked differences in their ability to scatter X-rays at certain wavelengths (the absorption edge). This means that a small change in the wavelength of the X-ray beam close to this edge gives changes to the scattering by these atoms, and hence to the phases. Thus, by comparing three to five diffraction patterns obtained at different wavelengths, one can work out approximate values for the phases. This method is known as *multiwavelength anomalous dispersion* or *MAD*, and the heavy atom is most commonly selenium, obtained by growing the protein in a medium enriched with selenomethionine, which incorporates selenomethionine into the protein in place of methionine. It is usual to use a methionine auxotroph to ensure good incorporation levels. As a general rule, one selenium atom provides enough data to phase about 15 kDa of protein. It is also possible to use the native (sulfur-containing) protein and obtain the phases by single-wavelength anomalous dispersion from the sulfur atoms (SAD or S-SAD).

This method is recent, because it requires synchrotron radiation, as described below. The original method invented by **Max Perutz (*11.13)** was to incorporate heavy metals into the protein crystal, which again alter the intensities of the diffracted spots, a method called *isomorphous replacement*. It requires at least two different heavy-atom derivatives, and it also requires that the structures do not change when the heavy atoms are soaked in. One can also use a model structure, for example a structure built from lower-resolution data or the structure of a homolog, as a starting point to obtain phases. In using this method (called *molecular replacement*) one has to be particularly careful not to bias the resulting structure by the initial structure.

11.4.4 Structure, electron density, and resolution

X-rays are diffracted by electrons; consequently, the resultant X-ray structure is actually an electron density map, and not directly a structure. The structure has to be derived by fitting atoms into the electron density. Because hydrogens have very low electron density, they are usually not visible. The fitting tends to be done iteratively, checking against the electron density at each stage. For example, one can model a poly-Ala chain first and subsequently fit the side chains, based on the known sequence. If there are gaps in the sequence (as there often are, because some regions of the protein may be disordered and so not visible in the electron density map), then fitting the side chains requires working out where in the sequence you are.

The ease and accuracy with which this can be done depends markedly on the resolution of the data (**Figure 11.38**). Bond lengths are around 1.5 Å. Consequently, in a structure with a resolution of 2.5 Å or worse, it is difficult to resolve individual atoms, and (depending on the quality of the data) it can be hard to distinguish a backbone C=O from an N. This means that structures with a resolution of 2.5 Å or worse run the risk of having peptide planes inverted or misinterpreted, and it is very often difficult to follow side chains accurately. Most structures can now be determined to higher resolution than this and are more or less at atomic resolution. As the resolution improves further, the

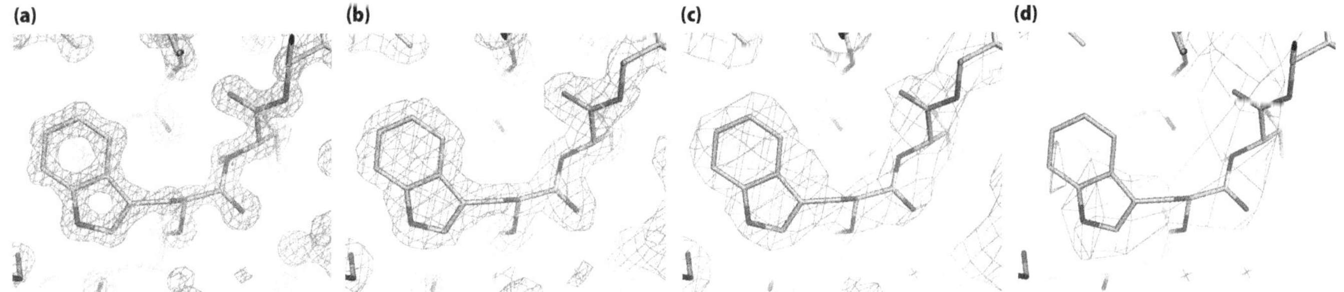

FIGURE 11.38
As the resolution improves, the ease with which individual atoms can be fitted into the electron density also improves. The resolutions are (a) 1.2 Å, (b) 2.0 Å, (c) 3.0 Å, and (d) 6 Å. At 3 Å, features such as aromatic side chains can be seen clearly, but the backbone is hard to distinguish. Note that the carbonyl group and backbone nitrogen cannot be distinguished clearly, so it is not obvious from the electron density which direction the chain is going in. At 2 Å, the backbone can be traced clearly. At 1.2 Å, individual atoms can be seen clearly. The mesh is drawn at contours of equal electron density, the green mesh being approximately twice the density of the brown mesh. The red crosses mark electron density peaks assigned to water. The protein is streptococcal Protein G B1 domain. (Courtesy of P. Artymiuk, University of Sheffield.)

definition improves, and at 1.2 Å it is sometimes possible to see protons. As a general rule, anything better than about 1.8 Å is currently considered 'high resolution.'

11.4.5 Measures of quality: *R* factor and *B* factor

There is a one-to-one correspondence between a protein structure and its diffraction pattern: if the structure is known, the diffraction pattern can be exactly calculated from it. (The reverse is also true, but only if you know the phases.) This is done using a Fourier transform (*11.8). This provides a good check of the correctness of a structure, because the back-calculated diffraction pattern can be compared with the experimentally observed pattern. The fractional difference between the two is known as the *R* factor, and for a good structure it is generally lower than 20%, depending on the resolution. A general rule of thumb is that the *R* factor might be expected to be about 1/10 of the resolution of the data.

However, the *R* factor alone does not provide a very reliable measure of quality, because after atoms have been fitted into the electron density, crystal structures are always *refined*: that is, atomic positions are moved around to produce the best fit between the model and the experimental data. It is thus not possible from the *R* factor alone to tell the difference between a genuinely good structure and one produced by over-refining a bad structure. The solution to this problem is to randomly select a small proportion of the experimental data, for example 10%, and not use this in the refinement. Calculation of the *R* factor is then done using these data, and the number so calculated is called R_{free}. If the refinement is a genuine improvement, then R_{free} should also improve, and should be similar to *R*. A good structure will typically have an *R* factor of less than 20%, and an R_{free} within about 5% of *R*.

Any real crystal is not perfect. It will have crystal defects. These can be long-range defects, in that the crystal does not extend infinitely and perfectly in all directions, or they can be local defects, in that some molecules in the lattice may have differences in their conformation. It is also very common to find that some parts of the protein are less well ordered than others. Typically, the termini and some surface loops may be less ordered or even completely disordered. These types of disorder are collectively called static disorder, meaning that it arises from molecules locked into different positions. There will also be dynamic disorder: thermal motions of the protein molecules will lead to small variations in position from one molecule to the next, which again will make the sample less crystalline. All this disorder makes the diffraction less precise, thus leading to some blurring of the diffraction pattern. When transformed to a structure, it makes the electron density smear out to a greater or smaller extent.

This is modeled by giving each atom a three-dimensional distribution rather than a single point location; it is called the temperature or *B* factor, which approximately represents the area within which the atom may be expected to be found. As noted above, it arises from disorder in the crystal. In low-resolution

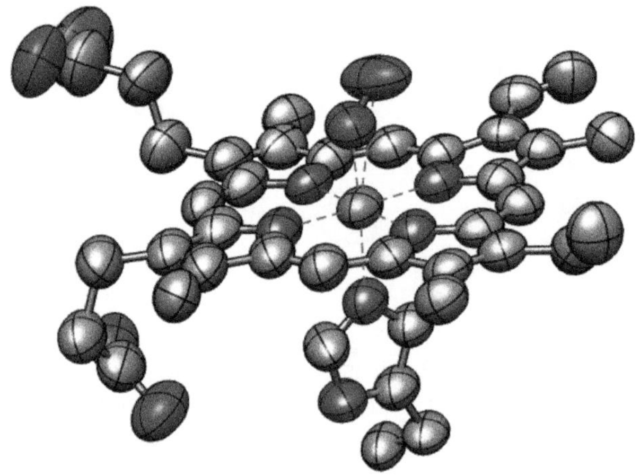

FIGURE 11.39
The heme ring (plus neighboring atoms) from oxymyoglobin (PDB file 1a6m), showing anisotropic *B* factors. The ellipsoids show the 50% probability positions of each atom; wobbling within the plane of the heme is of much larger amplitude than motions out of the plane. (This figure was created with UCSF Chimera http://www.rbvi.ucsf.edu/chimera/ [28], and is provided courtesy of UCSF.)

structures, in which the quantity of experimental data is limited, it would be typical to restrain the *B* factor during the refinement, to prevent over-refinement (that is, *B* factors changing to compensate for inadequacies in the data, leading to a structure that looks more precise than it is). However, in high-resolution structures, the *B* factor gives an accurate representation of local disorder; thus, for example, one tends to find the *B* factor increasing at the ends of long surface-exposed side chains, and in long loops. It is even possible to find whole domains missing from the density because they are too disordered (see, for example, Figure 8.8d). In very high-resolution structures, there are enough data to allow the *B* factor to be refined to give anisotropic disorder; that is, to show how the local disorder is distributed in three dimensions (**Figure 11.39**). An unusually high *B* factor may be due to disorder, but it could also be a warning that there is some kind of problem with the structure. The *B* factor thus provides another measure of the quality of a crystal structure.

11.4.6 Solvent and other molecules in protein crystals

In a protein crystal, roughly 50% of the volume is occupied by water molecules. This is a natural consequence of the irregular and roughly spherical shape of proteins, which means that inevitably there are significant gaps between one molecule and the next. As the structure is refined, one expects to see regions of electron density appearing that cannot correspond to protein. These can usually be assigned to water molecules. A medium- to high-resolution structure will generally contain roughly the same number of water molecules as it does amino acid residues. These waters tend to be located close to the protein surface, often in hydrophilic pockets, in places where they can form hydrogen bonds to hydrophilic groups on the protein or to other well-defined water molecules. They usually occupy positions that allow them to form hydrogen bonds with good geometry.

Water molecules will only be observed if they occupy the same position in all molecules in the lattice. If molecules are present but are in different places in different molecules, then only an average electron density is observed. This is the case farther from the protein surface and next to hydrophobic parts of the surface. During crystal structure refinement, there will inevitably be errors in the phases determined for the diffraction spots. When converted to an electron density map by Fourier transformation, the errors in the phases will lead to errors in the electron density. For example, what should be featureless regions of solvent will have apparent peaks and troughs of electron density. It is therefore usual to apply *solvent flattening* to the refinement calculation, assigning a uniform average electron density to these regions. When this modified electron density map is Fourier transformed back to a diffraction pattern, the phases of *all* the spots are improved; one can then combine the experimentally determined intensities with these improved phases and obtain a better electron density map, and so on iteratively.

It frequently happens that during the refinement, extra density is observed that cannot be assigned to water or to protein. Sometimes these are single peaks of density that can be assigned to metal ions. Metals have different hydrogen-bonding geometries that are characteristic of the metal; often these allow the metal to be identified. Sometimes the density clearly corresponds to additional molecules. Often these are molecules added to the crystallization solution to aid crystallization, but sometimes they can be ligands. It is not uncommon to find that a protein that was expressed, purified, and crystallized will have bound tightly enough to a ligand to be still retaining it in the crystal. This can be very helpful in defining how the protein works, because the ligand is very often functionally important.

In other cases, the researcher may well want to find out where and how a protein binds to a particular ligand such a substrate, inhibitor, or cofactor. There are two options. Often it is possible to add the ligand to the buffer surrounding the crystal. Protein crystals usually have large water-filled channels between molecules, providing enough space for ligands to diffuse into the crystal and bind. However, ligand binding in the correct position and orientation will typically require some rearrangement of the protein, which may not be consistent with the crystalline packing arrangement. It may therefore be that the ligand does not bind in its functional position; or else it may be that in binding, it disrupts the crystal contacts between one molecule and the next and the crystal 'cracks' and dissolves. The other option is to crystallize the protein together with its ligand. As discussed next, crystallization is far from a predictable science, and it is a matter of luck whether the complex will crystallize, even though the protein on its own may have done so.

11.4.7 The practicalities of protein X-ray diffraction

The major stumbling block in protein crystallography is crystallization. The area of contact between one molecule and its neighbor in the crystal is often rather small, meaning that the crystallization conditions have to be exactly right to encourage crystallization. In all crystallizations, the basic principle is that one wants to gradually alter the solution conditions to produce a slightly **supersaturated solution**. The most common method for doing this is called the *hanging drop method* (**Figure 11.40**). A solution of protein in suitable precipitant is placed in a small drop (for example, 1–2 μl) hanging under a microscope coverslip. This is sealed over a solution of higher osmotic strength. Over the course of several days or weeks, water will evaporate from the drop and condense in the precipitant so as to equalize the osmotic strength, slowly increasing the concentration of protein and precipitant in the drop. The drop can be periodically examined under a microscope to see whether crystals have formed.

Because it is almost impossible to predict the conditions necessary for crystallization, it is common to place protein solutions into a wide range of different conditions, for example with different precipitants (often short-chain alcohols or polyethylene glycols of different polymer lengths), buffers, salts, metal ions, and pH. There are several screens commercially available containing combinations that have previously been found to work well. The larger labs have robots for preparing these solutions; and in the very large labs, even the examination for crystals can be done automatically.

It is typical to find that most solutions produce nothing but precipitation, while some solutions produce poor-quality crystals. Further trials can then be conducted by subtly varying promising-looking conditions, possibly using very small crystals as seeds to encourage crystallization.

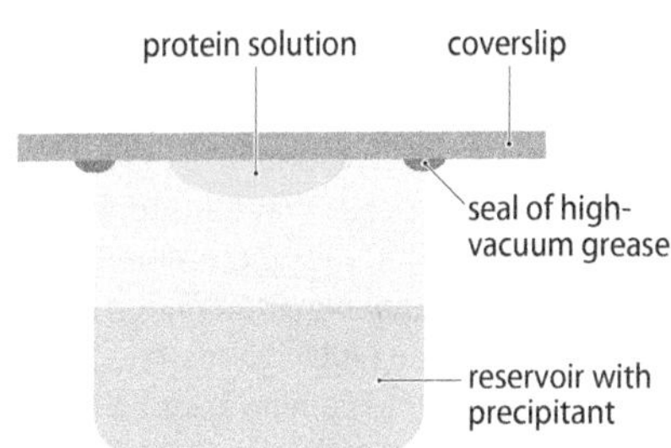

FIGURE 11.40
The hanging drop method for protein crystallization. See the text for details.

A protein crystal is easily damaged, because of the high solvent content and small intermolecular contacts. Crystals therefore have to remain in their buffers; otherwise they will dry out and break. Traditionally this was done by keeping the crystal and its mother liquor in a capillary tube. However, most crystallography is now performed at very low temperatures, the crystals being cooled by a jet of cold nitrogen gas at about −180°C. There are two reasons for this. One is that the thermal motions are smaller and hence the structure is better defined. However, the more important reason is that X-rays have very high energy and damage the crystal. Thus, crystals in an X-ray beam have a limited lifetime. It is possible to replace a crystal, but this is undesirable because no two crystals are identical, and some rather complicated analysis is required to scale and merge together the data from two crystals. At low temperature, damage is reduced. The normal procedure is therefore to pick up the crystal in a small drop of buffer using a loop of wire, and then cool it very rapidly by plunging it into a bath of liquid nitrogen. The rapid cooling is important because one does not want to allow crystalline ice to form, as this reduces the quality of the data; the aim is to produce 'vitreous' or disordered water. It is also common to add cryoprotectants such as glycerol to the solution to prevent the formation of ice crystals.

Traditionally, X-rays can be generated by firing high-energy electrons at a metal surface, often copper. This generates considerable heat, and the copper needs to be cooled by water, often as a 'rotating anode.' Many crystallography labs have this kind of equipment. However, the X-rays generated in this way are of relatively low intensity, and the wavelength is fixed, meaning that it is not possible to carry out MAD phasing. Therefore most crystal structures are now obtained in a *synchrotron*. These are large storage rings (up to 1 km in diameter) of high-energy particles under high vacuum. The particles generate X-rays when they are accelerated; that is, when they turn a corner. This means that X-ray beam lines are situated at corners, or even better at undulations or wiggles in the beam, placed deliberately so as to produce high-energy X-ray beams. Synchrotron sources are two orders of magnitude more intense than rotating anodes, and their wavelengths can be tuned to allow MAD (or SAD) phasing. The shorter-wavelength X-rays from a synchrotron have the added benefit that they produce (relatively) less damage to the crystal. Synchrotrons also allow much smaller crystals to be used (typically less than 100 μm in each direction), simplifying the problem of crystallization.

The diffraction patterns are detected by an *area detector*, a very sensitive solid-state charge-coupled device (incidentally the topic on which half of the Nobel Prize in Physics 2009 was based), providing rapid data collection. In recent years, considerable work has been done on speeding up data collection and data processing, with the result that from a good crystal it is often possible to calculate the structure of a small protein within an hour or so or collecting the data.

11.4.8 Structures of membrane proteins

Membrane proteins (*11.14) remain a challenge for crystallography, for several reasons. First, they are difficult to express; this is particularly true for eukaryotic membrane proteins, whose expression levels in bacteria can be low and unpredictable. If the expressed protein is located in the bacterial cell membrane, this can rapidly kill or disable the bacterium, giving low yield, whereas if it is expressed in the cytoplasm, it will be expressed in inclusion bodies and require resolubilization and refolding. Second, they are difficult to purify. Purification generally requires extraction from the membrane with detergents, and there are as yet no clear general methods for identifying a suitable detergent in which the protein remains stable. Membrane protein structure may require specific interactions with cellular lipids, which are likely to be

***11.14 Membrane proteins**

There is little in this book about membrane proteins. This is not because they are unimportant: as is frequently noted, roughly 30% of eukaryotic proteins contain transmembrane regions, and roughly half of the cell membrane by weight consists of protein. Rather, it is because they are difficult to study, and therefore far less is known about them. A good example of this is the membrane proteins of the red blood cell, a particularly simple and well-studied cell type. The two main membrane proteins in red blood cells are band 3 and glycophorin. Band 3 is an anion transporter that transports bicarbonate (HCO_3^-) and chloride across the membrane. Glycophorin is a very small and highly glycosylated protein of only 131 amino acids and has a single transmembrane helix. Remarkably, its function is still not really clear.

disrupted during the solubilization. This also implies that detergent extraction runs the risk of unfolding the protein, possibly irreversibly. It is noted in folding of membrane proteins (*7.11) that multiprotein membrane complexes often seem to require assembly in the correct order, with the help of specific chaperones. Because such conditions are unlikely to be present during membrane protein extraction, there is a high risk of irreversible loss of structure. Third, they are difficult to crystallize. Crystallization requires regular contacts between one molecule and its neighbors. The transmembrane parts of membrane proteins are surrounded by detergent (**Figure 11.41**), which means that most of the surface of a solubilized membrane protein is featureless and does not form unique interactions. Hence, their crystallization is particularly difficult and relies on the external segments. New tricks are being developed in all these areas, but they are slow and difficult, and membrane proteins look likely to be difficult for some time to come. For example, the crystal structure of the β_2-adrenergic receptor was solved by inserting the gene for T4 lysozyme into one of the intracellular loops, replacing a disordered loop by a larger folded protein, and thus removing some disorder and simultaneously presenting a larger surface for crystallization [6]. The protein had also been subject to a large number of random mutations, selecting ones that gave improved thermal stability and presumably reduced mobility.

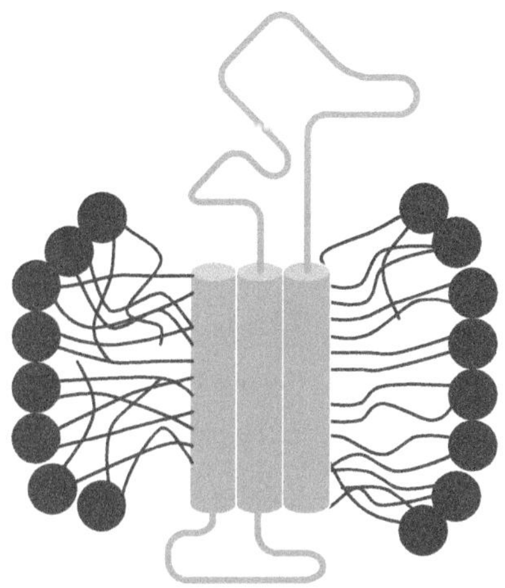

FIGURE 11.41
Membrane proteins are solubilized by detergent, and the transmembrane regions are therefore surrounded by detergent micelles. This makes it very difficult to achieve packing into a crystalline lattice, because the contacts between proteins are almost entirely nonspecific lipid–lipid or protein–lipid contacts.

11.4.9 Fiber diffraction

X-ray crystallography requires crystals: ordered three-dimensional arrays of proteins. It is possible to obtain diffraction patterns from fibers, if they are aligned in parallel bundles. Obviously the amount of information is significantly decreased, and it is usually possible only to determine the pitch of the helix (the repeat distance) and the angle of climb, which can be used to calculate the number of residues per repeat. This is clearly not enough information for a complete structure, implying that fiber diffraction requires the fitting of models to generate an atomic-resolution structure. The original information both on α helices and on the DNA double helix came from fiber diffraction data.

11.4.10 Neutron diffraction

Neutrons are small enough to have wave properties and can be diffracted by atomic nuclei, giving rise to diffraction patterns with resolution in the range 2–4 Å. The extent to which different atoms diffract X-rays is dependent on their atomic mass, implying that hydrogens usually cannot be seen. However, neutron scattering factors are quite different. In particular, hydrogen scatters neutrons by a similar amount to carbon, oxygen, and sulfur, while deuterium also scatters by a similar amount but in the opposite direction. This means that it is possible to determine the positions of hydrogen atoms by neutron diffraction, or to make the crystal invisible to neutrons by an appropriate ratio of ^2H to ^1H. The drawback with neutron diffraction is that one needs an intense source of neutrons, which is not easily available. Neutrons are also more damaging than X-rays. In addition, it is very difficult to obtain phases because, unlike X-ray diffraction, there are no real 'heavy' atoms (that is, atoms that scatter more strongly than carbon, nitrogen, and oxygen), and so it is usual to solve the X-ray structure first. Neutron diffraction is therefore not common.

11.4.11 Electron diffraction

By contrast with neutrons, electrons are readily scattered, and it is easy to generate high-energy electron beams by accelerating them in a high voltage (100 kV or greater). Electron beams can be focused by electromagnetic lenses, in an analogous way to light, although the lenses are of poorer quality as yet.

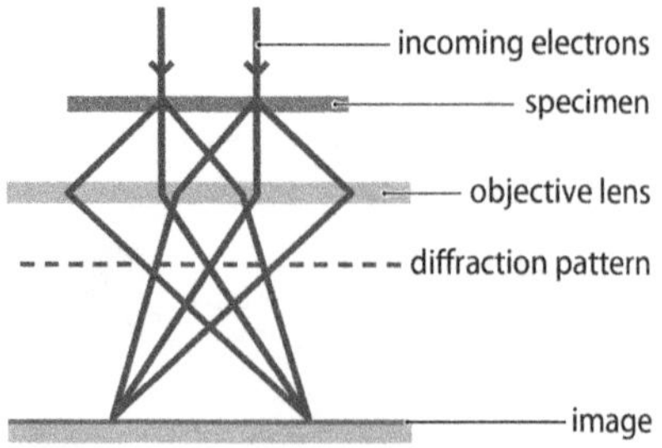

FIGURE 11.42
An electron microscope can be used to create either a diffraction pattern or an image, the latter of which can be used to provide phases to solve the diffraction pattern.

This means that electrons can be used both for microscopy and for diffraction. We discuss electron *microscopy* in the next section; here, we discuss electron *diffraction*. Electrons at 100 kV have a wavelength of 0.04 Å and are therefore well suited to the solution of protein structures. A major point in favor of electron diffraction, as compared with X-ray diffraction, is that the focused beam can be used either to generate a diffraction pattern or to create an image, depending on where the detector is placed (**Figure 11.42**). The diffraction pattern is similar to an X-ray diffraction pattern, and has the same phase problem, but the image *does* still contain the phase information and can be used to extract the phases (albeit at low resolution). Thus, the phase problem is much less of an issue in electron diffraction.

As described above, X-ray diffraction can be used to study the structure of membrane proteins, provided that the proteins can be induced to crystallize in a three-dimensional lattice. In some cases, three-dimensional crystals have proved elusive, but the protein can be crystallized into a two-dimensional lattice, usually one protein molecule thick. In such cases, electron diffraction can be used to investigate the structure. Electrons are much more strongly scattered from material than are X-rays (by a factor of about 10^4). This means that the sample has to be very thin (less than about 50 nm); otherwise too few electrons will be collected, and electrons will be scattered by several atoms, leading to major problems in analysis. This is a major disadvantage for electron diffraction studies on three-dimensional crystals but is well suited to two-dimensional crystals: indeed, a small number of proteins occur naturally at such high density that they form crystalline arrays in native membranes. One such is bacteriorhodopsin, a light-sensing protein from purple bacteria, which pumps a proton across the membrane in response to light and has been studied extensively.

A two-dimensional crystal is only a little thicker than the thickness of the membrane, approximately 40–50 Å, and the electron beam is therefore diffracted by a far smaller number of molecules than is an X-ray beam from a three-dimensional crystal. This means that the signal-to-noise ratio is very small, necessitating the use of multiple datasets and extensive signal processing even to see a signal. As in X-ray crystallography, samples are flash-frozen and studied at low temperature to decrease crystal damage (a major problem in such studies, even though electrons are less damaging than X-rays) and increase the crystalline order.

If the two-dimensional crystal is held in a plane perpendicular to the electron beam, the resulting structure is a *projection map* of the protein, in which all the depth information is lost, and one has effectively a sum of the electric potential through the depth of the membrane (**Figure 11.43**). In the figure, helices can be seen end-on as circular patches of high density, and it is clear that bacteriorhodopsin adopts a seven-helix transmembrane structure. Lower-intensity circular patches can sometimes be interpreted as tilted helices. Low-resolution projection maps (of the order of 7 Å) are relatively easy to obtain; they show the possible locations of helices and the overall molecular shape but little else. An increase in resolution requires a large increase in the number of samples

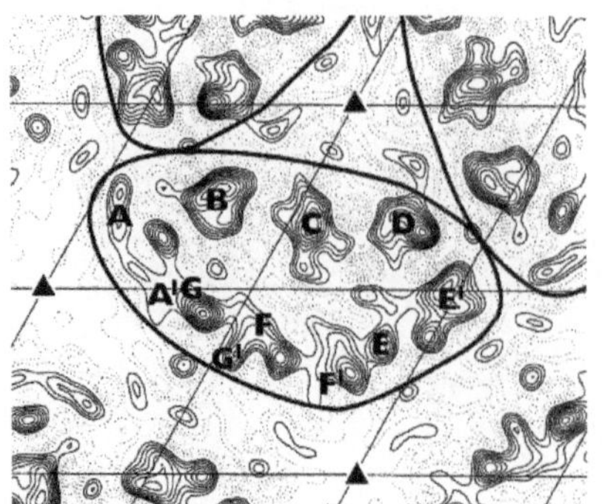

FIGURE 11.43
A projection map at 3.5 Å of bacteriorhodopsin. The seven helices are indicated. Helices A, E, F, and G are all quite strongly tilted out of the vertical, and hence their density is weaker and smeared out into two peaks with weaker intensity between. By contrast, helices B, C, and D are almost perpendicular to the membrane plane and therefore appear as single, more intense, peaks. (From P.A. Bullough and R. Henderson, *J. Mol. Biol.* 286: 1663–1671, 1999. With permission from Elsevier.)

studied. To obtain three-dimensional information, it is necessary to tilt the membrane at a series of angles (equivalent to the need to rotate crystals for X-ray structures): high resolution in the transmembrane direction requires a high angle of tilt. This becomes technically very difficult, not least to create a two-dimensional crystal flat enough that useful data can be obtained, and to prevent the sample from moving during the measurement. It is not surprising that very few high-resolution structures have been obtained by electron diffraction, one of the few being bacteriorhodopsin, which has been refined to 3.5 Å resolution [7]. In all such structures, the resolution in the plane of the membrane is much better than the resolution perpendicular to the membrane.

11.5 MICROSCOPY

11.5.1 Cryo-electron microscopy

Electron diffraction from two-dimensional crystals is likely to remain a very specialized and difficult technique. A somewhat easier method (although still far from routine!) is *single-particle cryo-electron microscopy*: the imaging of single macromolecular assemblies (about a few hundred kilodaltons or greater) [8]. This technique is very complementary to other structural techniques, and is particularly useful for the study of large complexes. Essentially, the electron beam is being used in exactly the same way as a light beam in a light microscope, but one with far shorter wavelength; it can therefore image to much greater resolution without running into problems with diffraction. The resolution is limited by the very small numerical aperture (*11.11) achievable in the microscope (a consequence of the incomplete focusing achieved by the lenses), although the main problem is usually achieving sufficient contrast.

There are two ways in which single-particle images can be studied. The simpler method is to make a dilute solution of the complex, add heavy metal ions, spot it onto a carbon grid, and dry it down (**Figure 11.44**). This leaves a metal-coated complex sitting on the grid. Not only does the metal improve the contrast, it also conducts away both electric charge and heat energy, and reduces the amount of damage. However, covering the protein with a metallic surface also obscures details of the structure and thus limits the resolution to 30 Å at best. It therefore cannot be used for high-resolution work. There is also a risk that the complex may become distorted during the drying process.

A higher resolution can be obtained by keeping the complex in solution and placing it on a grid in a very thin layer. The solution is then flash-frozen in liquid ethane to leave the proteins held in vitreous ice. Images are then collected, which are less distorted and at higher resolution than those from metal-coated samples, but have weak contrast and therefore a poor signal-to-noise ratio. This implies that it is necessary to use extensive signal averaging and statistical methods to obtain structural information. A very large number of images is collected of individual complexes. These images will most probably be of the

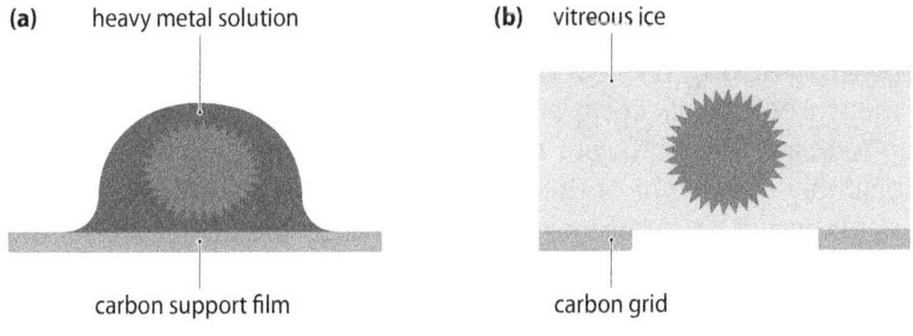

FIGURE 11.44
Two methods of preparing samples for single-particle electron microscopy imaging. (a) Negative stain. The particle is deposited onto a carbon film in a solution containing heavy metals, and dried down. The heavy metals absorb electrons strongly, so the protein is visible as an absence of heavy metals, hence the name negative stain. The drying process tends to distort the particle somewhat. (b) Cryo-electron microscopy. The sample is rapidly frozen in a thin film. Ideally, the sample is placed on a carbon grid so that the particle is over a hole in the grid. The contrast available from this method is much less good, but the sample is less distorted.

complex in a wide range of orientations, depending on how it happens to be sitting within the film. The images are therefore grouped into classes, hopefully representing different orientations of the complex. The classes are then averaged to give what ideally is a set of high-contrast high-resolution images of different orientations. Reconstruction methods are then used to produce three-dimensional structures. It is worth noting that an incorrect grouping of images into classes introduces misleading information into the structure and can give rise to completely misleading detail in the structure; it is as yet too early to say how widespread this problem is.

The simplest objects to image are ones with very high symmetry. Not only do such objects have higher redundancy (that is, more copies of the same basic unit per structure), but reliable alignment of the images also tends to be easier. The highest-resolution single-particle images are obtained for virus particles with icosahedral symmetry. Less symmetrical particles require a larger number of images. For example, to produce a high-resolution structure of the ribosome, it was necessary to collect more than 70,000 images [9]. This produced a structure with a resolution of about 10 Å, enough to identify possible protein domains, as well as the major and minor grooves of RNA helices. In most cases, the maximum resolution attainable is more like 20 Å.

A growing application of cryo-electron microscopy images is to provide a low-resolution framework for a protein complex, into which high-resolution crystal structures can be placed. The combined effect is to give an atomic-level model of very large protein assemblies. Chapter 9 describes several examples of such applications. Similar methods have been used for example to model actomyosin fibers. In principle it is possible to collect images of complexes 'in motion,' by freezing them a short time after application of a stimulus. This has been done successfully, although it requires considerable effort to control it sufficiently well to obtain useful and reliable images. Probably the best example is the acetylcholine-activated state of the nicotinic acetylcholine receptor, which has a lifetime of only 10 ms; it was studied by spraying acetylcholine onto a grid containing the receptor as it was in the process of being plunged into an ethane bath [10]. In complexes with mobile domains, such as the pyruvate dehydrogenase complex (Chapter 10), it is possible (at the cost of obtaining a very large number of images) to obtain structural information on sample heterogeneity, implying that one can get an idea of how mobile the domains are and where they are.

11.5.2 Atomic force microscopy (AFM)

We saw above that the resolution attainable in microscopy is determined by the wavelength of the radiation used. AFM (and the related technique *scanning tunneling microscopy* or STM) gets around this barrier by not using radiation at all.

The key part of an AFM is a very small and light cantilevered arm, at the end of which is a sharp point or probe, which is held in very close proximity to the sample (**Figure 11.45**). The sample is deposited onto a flat surface (often a sheet of silica cleaved across a fracture plane), which sits on a piezoelectric block. By varying the voltage across this block, it can be caused to move, with a resolution of less than 1 nm. As the sample moves, the tip of the AFM moves up and down, tracking the features of the sample. The height of the AFM tip can be measured by reflecting a laser off the back of the arm. The AFM can be set up to respond to different features of the sample. Most obviously, it can measure simply the height of the sample, but it can also respond for example to electrical charge or conductivity. By attaching specific proteins or ligands to the probe tip, AFM can identify specific binding partners. Thus, as the tip is scanned across the sample, it maps out the height (or charge or conductivity)

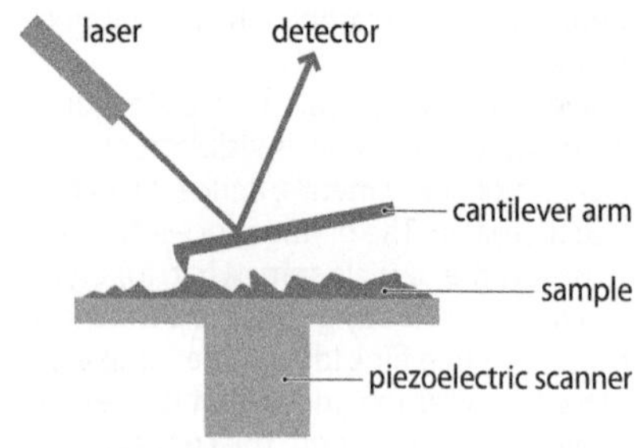

FIGURE 11.45
The experimental setup for AFM.

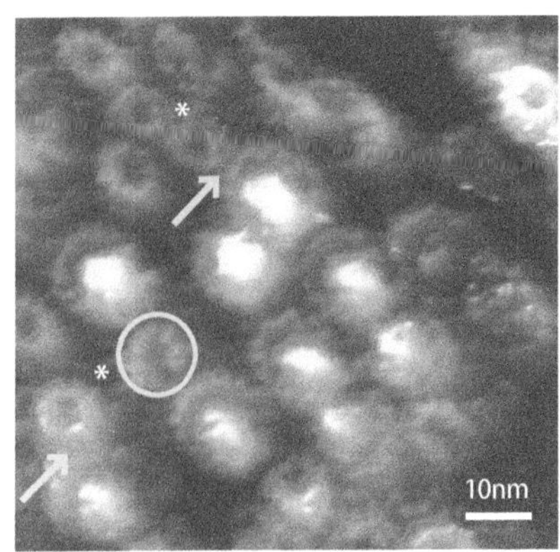

FIGURE 11.46
An AFM image of the bacterial light-harvesting (LH) complex from *Rhodobacter sphaeroides*. The height is coded by brightness. This complex consists of two different circular components. LH2 is a ring of nine αβ pairs, in which each α and β protein contains a single transmembrane helix and binds a bacteriochlorophyll, used to trap incident light. LH1 is a larger ring that also contains αβ pairs attached to bacteriochlorophyll, which surrounds the reaction center. LH1 rings generally occur in pairs. Light is absorbed by LH2 or LH1 rings and passed to the reaction center. Here, the LH1 rings with their reaction center are the bright objects, protruding an average of 37 Å above the membrane. The darker rings are the LH2 rings, which can be clearly seen to have nine subunits (asterisks). The green arrows show contact points between LH2 and LH1 rings. The green circle shows an LH2 ring surrounded on both sides by an LH1–RC complex. (From S. Bahatyrova, R. N. Frese, C. A. Siebert et al., *Nature* 430: 1058–1062, 2004. With permission from Macmillan Publishers Ltd.)

of the sample. A big advantage of AFM over electron microscopy is that the sample can remain hydrated, and there are therefore likely to be fewer artifacts caused by drying out of the sample.

Biological samples are likely to get damaged by dragging a tip across them. For these samples, it is more common to use the AFM in 'tapping' mode, in which the arm oscillates up and down, and the movement of the sample occurs while the arm is not in contact with the sample.

An example of an AFM image is shown in **Figure 11.46**. This shows the bacterial light-harvesting complex in remarkable detail, almost as good as that obtained by electron diffraction, even though the images are not averaged.

Essentially the same apparatus can be used not to *measure* a force between tip and sample, but to *generate* a force. If the tip is pressed down onto a protein, it will sometimes stick to it. When it is pulled away, it will provide a force to pull the protein away from the surface. When a graph is plotted of force against distance, one can see the force increasing until at some point the protein pulls away and the force goes back to zero (**Figure 11.47**). This can be used for example to measure intramolecular adhesive forces within a protein and to study the effect of ligand binding [11]. When the technique is applied to modular proteins, one obtains graphs such as that shown in **Figure 11.48**, which shows individual modules unfolding one by one as the protein is pulled away from the surface.

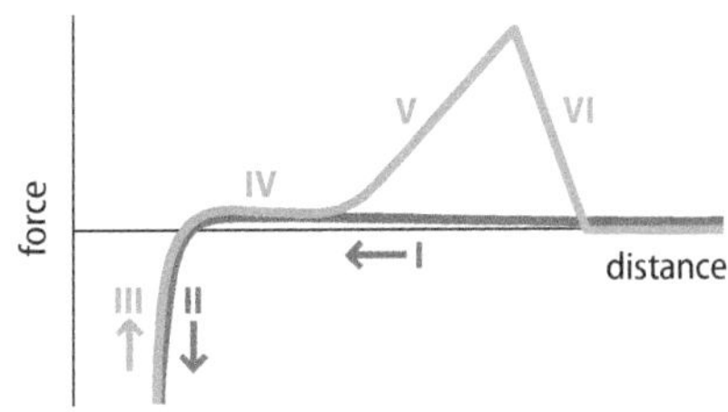

FIGURE 11.47
An AFM trace of a protein being pulled away from a surface. Stages I and II (brown) show the tip approaching the surface, and stages III–VI (cyan) show it coming away. Stage I: no interaction between tip and surface. Stages II and III: repulsive force between tip and surface. Stage IV: attractive force due to surface adhesion. Stage V: entropic force due to stretching and uncoiling a single protein chain. Stage VI: detachment of the protein. (Adapted from E. Jöbstl, J. R. Howse, J. P. A. Fairclough and M. P. Williamson, *J. Agric. Food Chem.* 54: 4077–4081, 2006. With permission from the American Chemical Society.)

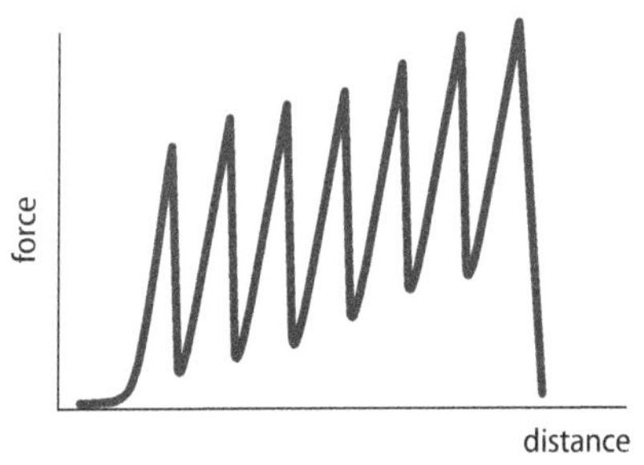

FIGURE 11.48
An AFM trace for a modular protein containing seven repeats being pulled away from a surface. The tip is (by chance) attached to one end of the protein, while the protein is attached to the surface at the other. Each module can be pulled apart sequentially. When the protein finally detaches from the surface, the force goes to zero. (Adapted from data obtained for the muscle protein titin from H. Li, A. F. Oberhauser, S. B. Fowler, J. Clarke and J. M. Fernandez, *Proc. Natl. Acad. Sci. USA* 97: 6527–6531, 2000. With permission from the National Academy of Sciences, USA.)

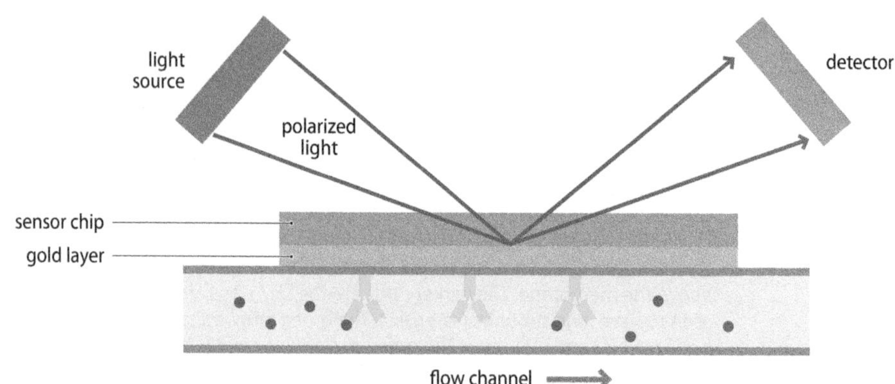

FIGURE 11.49
The experimental setup for SPR. The sensor chip has a highly reflective layer on it such as gold, and polarized light is reflected from it. A protein is attached to the bottom of the gold layer, and a solution of ligand is pumped past it. The binding of ligand causes a change in the mass distribution attached to the gold layer; this affects the evanescent wave and leads to a change in the angle of reflection, which can be detected by a suitably placed detector.

11.6 METHODS FOR STUDYING INTERACTIONS

11.6.1 Surface plasmon resonance (SPR)

SPR is measured with apparatus as shown in **Figure 11.49**. Light is reflected from a gold-coated surface, to the back of which one can attach a protein or ligand. When light is reflected off any surface, there is a sudden change in direction of the light. Wave theory does not allow a completely 'sudden' change, and the reflection in fact sets up a standing wave on the back of the reflective surface, which is called an *evanescent wave*. This decays away within about one-third of a wavelength from the surface: however, because the wavelength of light is around 500 nm, the evanescent wave extends a significant distance into the protein-coated solution on the back of the reflective surface. The evanescent wave is sensitive to the mass distribution through which it passes, implying that a change in mass (caused for example by the binding of a ligand to a protein-coated surface) affects the evanescent wave, and therefore affects the reflected light. In fact it causes a change in both the angle and the intensity of the reflected light. These changes are small but measurable.

In the most common experimental setup, a polysaccharide layer is coated onto the gold surface, to which the protein or ligand can be readily attached. One can then pump solutions along the surface. Typically, one first washes with buffer, and then adds a ligand to the solution. If there is binding, the evanescent wave is perturbed and the reflected light changes, resulting in a signal (**Figure 11.50**). The ligand can then be washed off again.

The resulting curves can be analyzed to give on- and off-rates, and hence binding affinities. The analysis is complicated for several reasons. The apparent on- and off-rates may depend on flow rate and various geometrical details such as how the protein is attached to the polysaccharide layer. There may be non-specific binding, and changes in the signal due merely to flow of the ligand past the surface. It is therefore not easy to obtain accurate affinities from SPR, although it is useful for comparing affinities of related ligands.

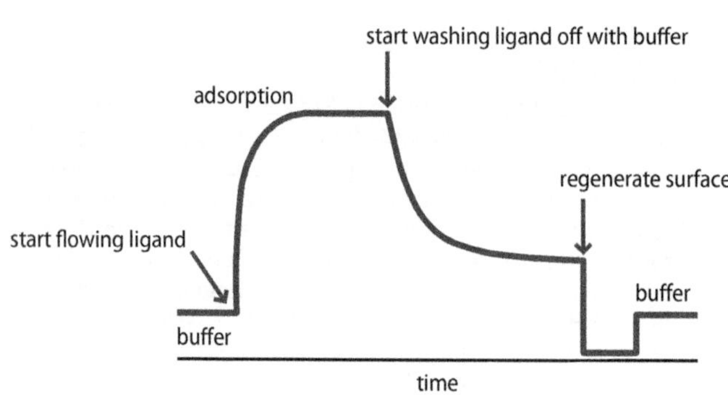

FIGURE 11.50
A typical SPR trace. The horizontal axis is time (a few minutes, left to right) and the vertical axis is an arbitrary scale. Once a baseline has been established, a ligand is washed over the gold surface and binds at a rate dependent on the on-rate. After the ligand has been passed over the surface for some time, it can be washed off again by buffer, at a rate dependent on the off-rate. The ratio of the off-rate to the on-rate gives the affinity.

11.6.2 Isothermal titration calorimetry (ITC)

ITC measures the heat given off or absorbed when two molecules bind. The apparatus is shown in **Figure 11.51**, and is (at least in concept) very simple. There are two identical cells in a carefully controlled thermal bath, about 1 ml in volume. Usually the reference cell contains buffer and the sample cell contains a solution of protein in exactly the same buffer. A ligand is titrated into the sample cell, in a series of steps with gaps between them to allow the temperature to settle down. This will either give off heat energy or absorb heat energy, depending on whether the binding is exothermic or endothermic. Sensitive thermocouples measure the difference in temperature between the two cells, and regulate a heater so as to maintain the same temperature in both. The changes in heat are very small, typically being measured in $\mu cal\, s^{-1}$; the experimenter must take great care to ensure that all solutions are identical and clean. A typical result is shown in **Figure 11.52**. Here, the addition of ligand is exothermic; as ligand is added, the binding site becomes saturated so that the heat evolved decreases during the titration. The magnitude of the signals gives the change in enthalpy on binding, while the shape of the curve gives the free energy, as well as the stoichiometry.

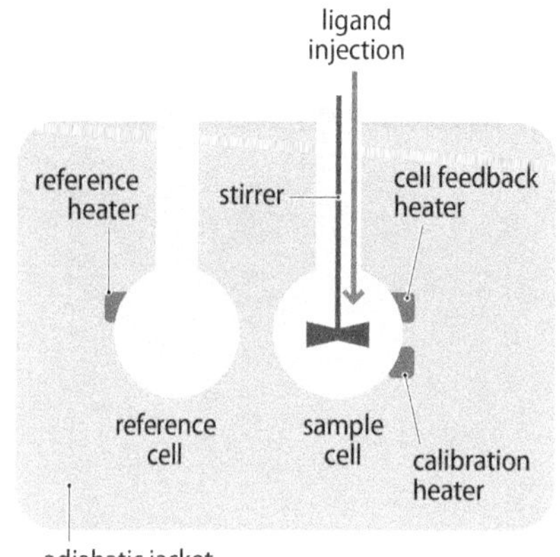

FIGURE 11.51
Experimental apparatus for isothermal titration calorimetry (ITC).

The technique is thus a powerful way of measuring the thermodynamics of binding events. As discussed in Chapter 1, there is a temptation to overinterpret these thermodynamic results, but this does not diminish the value of the technique!

11.6.3 The Scatchard plot: an object lesson

The classic method for analyzing binding, for example of antibody to antigen, is the Scatchard plot. Particularly in the early days of biochemistry, a simple analytical tool for binding assays was to use radioactively labeled ligand. Binding can be detected by mixing ligand with antibody, and subsequently precipitating the antibody by immunoprecipitation (*4.7). Bound ligand can then be measured by counting radioactivity in the precipitate (for example, after centrifuging or filtering the solution); free ligand is what is left in the supernatant. Alternatively, if binding of ligand to protein causes a change in the UV spectrum of the protein, one can calculate the proportion of protein binding sites that are occupied by ligand, by measuring the change in absorbance on addition of a known amount of ligand, and dividing it by the change in absorbance when the protein is saturated with ligand. Either of these methods gives the ratio r, which is the average number of molecules of ligand bound per molecule of protein. This number depends on the concentration of ligand, [L], and on the affinity of protein for ligand, as measured by the dissociation constant K_d.

It is of considerable interest to know how many ligand-binding sites there are on the protein. Let us call this n. Addition of ligand to protein gives a saturation curve, which has exactly the same hyperbolic shape as an enzyme kinetic saturation curve, and has the equation

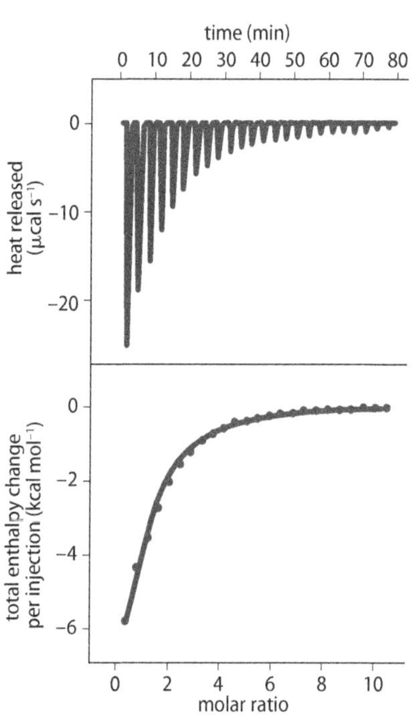

FIGURE 11.52
A typical ITC result. A solution of a family 29 carbohydrate-binding module was titrated with a solution of glucomannan, using a series of additions. The top trace shows the heat input required to keep the temperature the same in both chambers, which decreases with each addition because the fraction of bound protein increases and the fraction of bound protein decreases as more ligand is added. These traces are integrated to give the total enthalpy change per injection (bottom trace), which can be fitted to obtain the free energy of binding. (Courtesy of Harry Gilbert, University of Georgia.)

$$r = \frac{n[L]}{K_d + [L]}$$

which is drawn in **Figure 11.53a**. This curve has the same problem as a standard enzyme kinetics curve: it is difficult to extract values from it just by looking at it. It was therefore proposed by Scatchard to convert this equation to a linear equation:

$$\frac{r}{[L]} = \frac{n}{K_d} - \frac{r}{K_d}$$

Plotting $r/[L]$ against r gives the graph drawn in Figure 11.53b. This has the major advantage that it is a straight line if there is only one type of binding site on the protein. The gradient of the line is $-1/K_d$, and the intercept on the x axis is n, the number of ligand-binding sites on the protein. If the ligand binds with different affinities at different sites, a curved Scatchard plot is produced, as shown in Figure 11.53c.

This method has been taught to generations of biochemistry students, and is still very widely taught today. It is, however, a very misleading procedure, and as described above *should never be used* as a way of calculating K_d or n.

The reason is that the conversion involved changing the data from values that one can understand intuitively into values that have a much less obvious meaning. One can look at Figure 11.53a and spot problems, whereas one cannot so easily spot problems with the presentation in Figure 11.53b. The problem is all the more serious when the fitting is done semi-automatically by a computer program or spreadsheet, when the naive user will tend to trust in whatever output the fitting yields. Two examples illustrate the problem. The first is shown in Figure 11.53b. The 'correct' answer is shown by the solid line, whereas the dashed line is the result you would get just by doing a simple straight-line fit to the data. The two differ significantly both in K_d (the slope) and in n (the x intercept). The second example is shown in **Figure 11.54**. Panel (a) shows a typical saturation curve (that is, the actual experimental data): the dashed line is identical except for a background signal that has 5% of the intensity of the bound intensity. In panel (b) the effect of the background is to lead to a population of apparently very tight binding, plus a second population of weaker binding. This is, however, completely an artifact of the background.

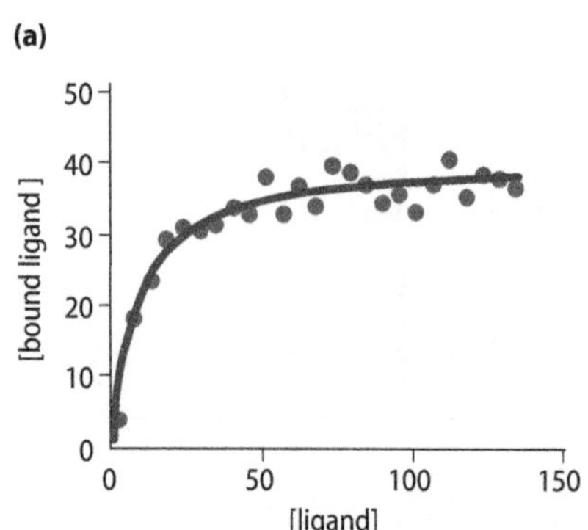

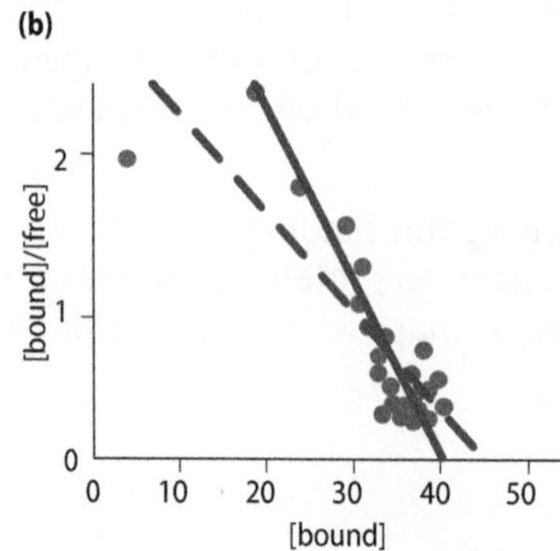

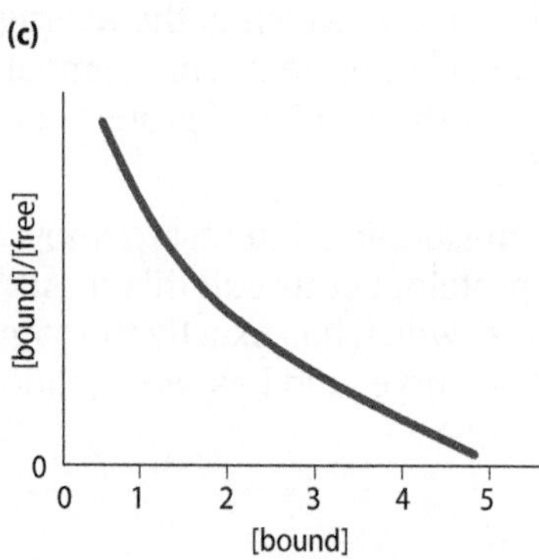

FIGURE 11.53
The Scatchard plot for analyzing protein–ligand binding. (a) Experimental data, showing typical (noisy) experimental data of bound ligand versus total ligand. The solid curve is the best nonlinear fit to the experimental data. (b) The same data converted to a Scatchard plot. The solid line is again the correct nonlinear fit, while the dashed line is a linear fit to the data and is for example what you would get from using a spreadsheet to calculate the line of best fit. The dissociation constant is –(gradient), and the number of equivalent binding sites on the protein is the intercept on the x axis: both have considerable error in the linear fit. (c) Scatchard plot for two binding sites. The steeper initial slope corresponds to a tight binding to a small number of sites, while the shallower later slope corresponds to a weaker binding at more sites. (With acknowledgement to GraphPad Software, Inc.)

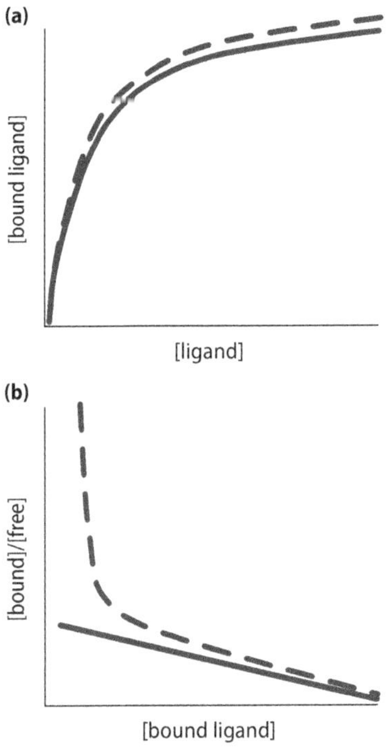

FIGURE 11.54
One of many problems with the Scatchard plot. If the data have a vertical offset because of a small amount of background binding (a), the result when transformed to a Scatchard analysis (b) is an apparent strong binding site. (With acknowledgement to Dmitriy Yuryev.)

The Scatchard plot was proposed in an era when computers were not readily available, and so data had to be converted into a form that could be plotted on a linear graph. Today, computers can easily undertake nonlinear regression on the original experimental data, which is how the data *should* be analyzed. It is instructive to convert the data into a Scatchard format to look for interesting features within the data, but this should never be used for analysis.

Exactly the same can be said for the analysis of enzyme kinetics data. The conversion of the hyperbolic enzyme kinetics curve to a linear Lineweaver–Burk plot leads to all sorts of problems, and there is now no need for it because fitting programs can easily fit the experimental data directly.

11.7 MASS SPECTROMETRY

The idea of mass spectrometry is extremely simple: it measures the mass of molecules. More exactly, it measures the mass-to-charge ratio, m/z. It does this by generating an ion in the gas phase. The ion is then accelerated by a voltage, and its mass can be measured in several ways. The most common of these is called *time of flight* (TOF): as the name implies, the time taken to move a fixed distance is measured, which is proportional to m/z (**Figure 11.55**). When properly calibrated, mass spectrometry can measure masses with an accuracy of 0.01%, or one mass unit in 10kDa. Mass spectrometers need to maintain a high vacuum so that ions are not deflected as they travel along the spectrometer.

Mass spectrometry can be applied to a wide range of molecules. In organic chemistry, molecules can be forced into the gas phase by bombarding them with electrons. These simultaneously kick molecules off a surface into the gas phase and provide them with a charge. However, the energy required to do this to biological macromolecules causes them to disintegrate long before they can be observed, and gentler methods are needed. The most common is *matrix-assisted laser desorption ionization* (MALDI). The macromolecule is prepared by mixing it into a suitable matrix, typically a small organic microcrystalline sample. A very brief irradiation of the matrix with a laser at an appropriate wavelength excites the matrix and effectively causes it to boil, taking with it neighboring proteins. These will have a charge, depending on the **p*I*** of the protein and their pH when added to the matrix, and this is used to accelerate them into the mass spectrometer, which can be run in negative-ion or positive-ion mode as required. Alternatively, a solution of a protein is

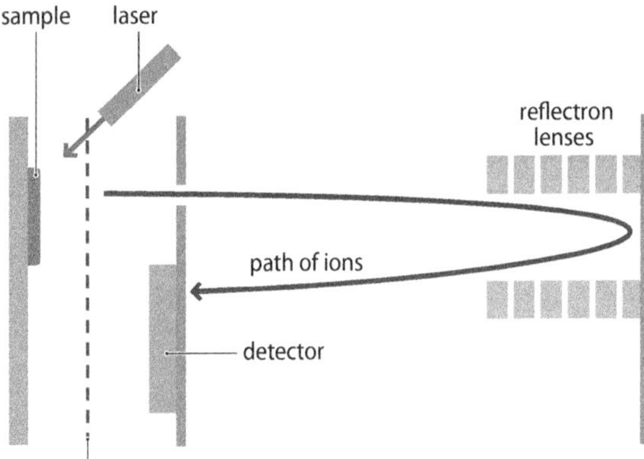

FIGURE 11.55
A MALDI-TOF mass spectrometer. The reflectron lenses are electromagnetic coils designed to focus and reflect charged particles.

sprayed into the mass spectrometer through a nozzle at high voltage: this is *electrospray ionization* or ESI. As the solution comes out of the nozzle, the solvent evaporates rapidly leaving a charged protein behind. This technique can be used directly attached to a chromatographic separation, giving rise to one of the popular 'hyphenated techniques,' liquid chromatography–mass spectrometry (LC-MS).

An important feature of MS is that it is very sensitive: almost every ion that passes down the spectrometer is detected. The efficiency of ionization is relatively low, but this still means that in favorable cases, spectra of good quality can be obtained from nanogram (sub-picomole) amounts or less. This allows MS to be used as a detection technique where sample amount is limited. In particular, it can detect proteins without requiring overexpression of the protein; it can also detect proteins separated on an SDS-PAGE gel.

In most instances, the macromolecule will be multiply charged. This is a great advantage, because it allows a greater precision in the measurement of its m/z ratio. A spectrometer designed to measure m/z in the range 0–5000 can achieve greater precision than one designed to work in the range 0–50,000, and clearly a multiply charged protein has higher z and therefore smaller m/z. A 50 kDa protein produces a series of signals in the mass spectrometer with increasing charge. Thus, $[M + H]^+$, $[M + 2H]^{2+}$, $[M + 3H]^{3+}$, $[M + 4H]^{4+}$, and so on, will have m/z 50001, 25001, 16668, 12501, and so on, and one can observe signals from a multiply charged protein at relatively low m/z. It is straightforward (on a computer) to locate the series of ions from multiply charged species and work out the original mass. There is also a complication from the fact that just over 1% of carbons have a mass of 13 rather than 12: one must therefore apply a correction to obtain the 'true' mass. It is also very common to observe $[M + Na]^+$ and other adducts, implying that salt-free solutions are preferred.

Several amino acids have identical or almost identical mass. Thus, Ile and Leu are identical in mass; Asp and Asn differ by only one mass unit; (Val + Thr) has the same mass as (Leu + Ser); and so on. And of course the mass of a peptide does not tell you its sequence. The exact mass of a protein is thus by no means a unique identifier of the protein sequence. It is in any case not usually possible to determine the *exact* mass; in addition, many proteins carry covalent modifications, which increases the mass beyond what is expected from the sequence. Nevertheless, it is sometimes possible to identify a protein only from its mass, particularly when some details are already known, such as host organism, location of the protein (cytoplasmic, nuclear, etc.), or p*I*, as discussed below.

A more reliable way of identifying the protein is to cut it up into several smaller and predictable fragments. Often, trypsin is used, which cuts the protein after Lys and Arg residues. Because these are rather common residues, the result is a fairly large number of peptides with a range of sizes. Peptides can be identified more reliably than proteins, because the masses are lower and therefore the accuracy of mass measurement is higher. Software linked to the spectrometer searches through the genome for peptides with masses matching those obtained. Because each protein produces many peptides, identification of several possible peptides from the same protein provides a much more reliable identification of the parent protein than a simple molecular ion. This method is known as *peptide mass fingerprinting*.

This method lends itself well to proteomics. One can for example run SDS-PAGE (*11.1) gels of a mixture of proteins, cut out an interesting band, digest it with trypsin while it is still in the gel, and then analyze it by MS, thereby providing a rapid and clean identification of unknown proteins (unless there are several proteins within the same band).

As an alternative to proteolysis, other methods are also being developed. There are a range of techniques that permit the selection of a particular parent ion on the basis of its m/z, followed by fragmentation of the ion to produce a range of fragments. In particular, proteins tend to fragment in the same way at each amino acid in the sequence, giving rise to a series of ions separated by the masses of the amino acids. This provides a direct way of sequencing the protein, or at least of obtaining a partial sequence. One such method effectively has two mass spectrometers back-to-back and is called tandem MS or MS-MS, another hyphenated technique.

Mass spectrometry is in principle a good way of identifying covalent modifications of proteins, particularly when combined with the peptide digestion methods described above. Rapid improvements are being made, for example in the identification of phosphorylation sites.

MS is not a very quantitative technique. The intensity of an ion bears little relationship to its abundance in the original sample. However, quantitative information can be obtained by spiking the sample with known amounts of isotopically labeled sample, a method finding considerable application in systems biology and proteomics.

Using gentle methods of ionization, protein complexes can be induced to stay together in the mass spectrometer. In this way one can use mass spectrometry as a technique for analyzing protein interactions, by seeing what components fly together. The organization of the proteins within the yeast exosome (Chapter 9) was worked out in this way [12], and similar techniques can be used to find out how many monomers there are in a homo-oligomer, or how the composition of molecular machines changes with growth conditions. A new development in MS is ion-mobility MS [13], which can provide information on the cross-sectional area of protein complexes, and thus on the architecture. There is now a wide range of types of mass spectrometer, each with its own range of applications, and new developments arise continuously.

11.8 HIGH-THROUGHPUT METHODS

11.8.1 Proteomic analysis

High-throughput (that is, rapid and automated) methods are used to obtain information on a genome-wide scale. The key point about high-throughput methods is that they should be fast and (if possible) comprehensive. *Transcriptomics* is an ideal subject for high-throughput methods. Its aim is to measure how much of each mRNA transcript is present, by using a chip dotted with a set of complementary DNA sequences. It is ideal because one sequence of RNA behaves much like another, and they can all be separated and handled in the same way. Therefore one can expect that the ratio of different mRNAs present in the original sample will be maintained throughout the various purification stages of the experiment, and hence that the method should be reasonably quantitative.

This is, however, not true for **proteomics (*11.15)**: the analysis of all the proteins present in a sample. This is because different proteins have different physical properties and behave very differently. It is not possible to find conditions under which all proteins can be isolated with equal efficiency. At best one could hope to find conditions under which a given class of proteins (for example cytosolic or membrane proteins) might be analyzed together; but even this is so far a rather distant hope. Another major problem for proteomics, much more severe than for transcriptomics, is that different proteins can be present at very different concentrations in the cell, varying in eukaryotes roughly from one copy to one million copies per cell.

***11.15 Proteomics**

The word proteomics is used to mean different things. Usually it means the quantitative analysis of all the proteins present in a sample (or in a species), but it could also include the determination of their function and role. An early definition in 1994 defined it as "the study of proteins, how they're modified, when and where they're expressed, how they're involved in metabolic pathways and how they interact with one another" [29]. It certainly implies a characterization of post-translational modifications as well as other covalent modifications such as alternative splicing. It is frequently observed that proteins in a 2D gel (identified by using antibodies in a Western blot, for example) show up as a series of horizontal spots rather than a single spot, very probably indicating different amounts of phosphorylation. As discussed in this chapter, we are still some way from a reliable proteomic characterization, because different proteins have very different physical properties, implying that it is not easy to purify all proteins to the same extent. Major techniques in proteomics research include 2D gels, mass spectrometry, immunology, and other methods for identifying proteins and their covalent modifications, and high-throughput techniques for studying protein interactions. A major difficulty in proteomics is the enormous range of protein concentrations present in the cell. Many of the most interesting proteins are present at only a few molecules per cell, whereas the most abundant proteins are present at concentrations of more than 500,000 per cell in yeast [30].

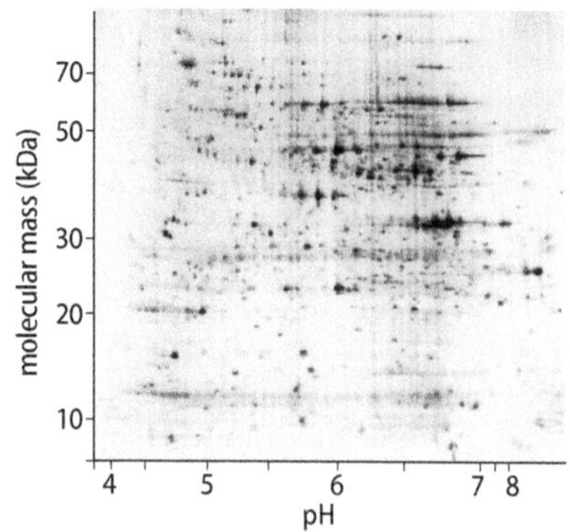

FIGURE 11.56
A two-dimensional protein gel. Protein phosphorylation alters the p*I* of a protein without significantly altering its molecular mass. Therefore if a protein is present in several different phosphorylation states, it will appear as two or more spots in a horizontal row. The figure shows several proteins with this appearance.

The aim of proteomics is to analyze what proteins are present in a cell or cellular compartment under a certain set of growth conditions; and very often to compare the protein content as growth conditions or genotype are changed. This is usually done by using *2D gel electrophoresis* (**Figure 11.56**). A protein mixture is spotted onto one corner of a two-dimensional gel, and separated first by **isoelectric focusing**. This separates proteins according to their **p*I***. One can select different ranges of p*I*, giving different separations. The gel is then rotated by 90°, and run as an SDS-PAGE (*11.1) gel, which separates according to size. The result can then be stained, for example by silver salts (a more sensitive stain than Coomassie brilliant blue). The spots can be probed by a variety of techniques, but normally by mass spectrometry. It is possible to place the gel directly into a MALDI-TOF spectrometer, although it is more common to cut out spots, perform trypsin digestions, and then analyze the peptide fragments.

The technique only picks up the proteins that are present in high abundance, and is bad at separating membrane proteins, which normally do not run onto the gel at all. A brief calculation shows that it must be true that only a relatively small proportion of all the proteins present can be visualized on a 2D gel. A good 2D gel can resolve at best only a few hundred proteins. The human genome has about 25,000 proteins, of which perhaps 10,000 may be expressed by any single cell. However, **alternative splicing** and post-translation modifications mean that there may be up to 100,000 different proteins present in each cell. (Alternative splicing increases the number of different proteins twofold or threefold, whereas post-translational modification increases it fourfold or fivefold.) We can therefore see no more than 1% of the proteins actually present. For membrane proteins other approaches can be used, such as proteolytic digestion before separation, by which one may hope to identify the extramembrane regions. The method is, however, being improved continuously, and is an important element in the complete analysis of cellular contents.

It is interesting to note that the abundance of different proteins derived from proteomics bears only an approximate relationship to that predicted from transcriptomics: an upregulated mRNA transcript does not necessarily result in an upregulated protein expression. It remains unclear how far this is genuine and how far it is a result of noise and experimental difficulties. This is a major question still to be answered. To some extent it can be answered by a **Western blot (*11.16)** analysis of interesting proteins, which is much more quantitative. There is certainly a better correlation between Western blots and transcription levels than there is between proteomic abundance and transcription level. Protein interactions can also be studied with **pull-down assays**.

11.8.2 Protein:protein interactions—yeast two-hybrid screens

A key goal of proteomics is to identify protein interaction networks. There are two major techniques that have been used for this purpose: two-hybrid screens and TAP-tagging. They have different aims, in that two-hybrid screens specifically pull out two-protein interactions between one target protein and any other, whereas TAP-tagging aims at finding all the partners that bind to a particular target within a complex. There has been a general feeling that two-hybrid screens suffer from an unduly high number of false positives (that is, they identify an interaction *in vitro* when no such interaction actually occurs

***11.16 Western blotting**

Western blotting involves the transfer of proteins from an SDS-PAGE gel to a membrane by placing the gel and membrane together and applying a voltage (**Figure 11.16.1**). The membrane is then washed with an antibody, which allows the detection of specific proteins. Specific antigens on a gel can thus be identified and quantitated in the presence of many other proteins.

FIGURE 11.16.1
Western blotting.

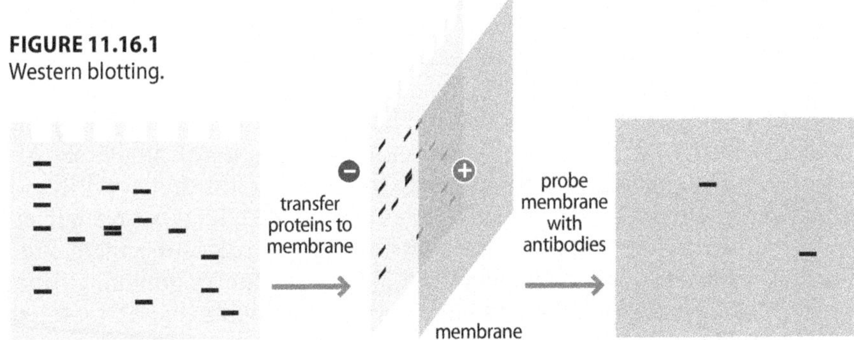

in vivo). As discussed in some detail in Chapter 9, this is possibly true, but it is also true that proteins do make many weak interactions, *in vivo* as well as *in vitro*, and it is so far unclear whether these have any functional relevance or are merely 'artifacts.'

The two-hybrid screen (**Figure 11.57**) is a simple and rapid method of finding the interaction partners of a target protein [14, 15]. The screen uses a *reporter gene*, whose activation is easily detected. A popular such gene is the *E. coli lacZ* gene, whose activation is detected because when grown in a medium containing 5-bromo-4-chloro-3-indolyl-β-D-galactoside (X-Gal), the cells turn blue. The gene is activated only when a neighboring region of DNA known as the upstream activation sequence (UAS) binds to a protein containing a transcription activation domain. In unmodified cells, this protein consists of two domains: one whose function is to bind to DNA, and a second which is the transcription activation domain. Protein:protein interactions can then be identified by splitting this protein into two parts, as shown in Figure 11.57. The DNA-binding domain is attached to a putative dimerization domain, and the transcription activation domain is attached in different clones to a library of putative dimerization partners. Activation of the reporter gene is observed only when the two dimerization domains interact, and so position the transcription activation domain in the correct location upstream of the reporter gene, giving rise to blue colonies.

11.8.3 Protein:protein interactions—TAP-tagging

TAP (tandem affinity purification)-tagging is a clever way of purifying low-abundance proteins and pulling them out of the cell still attached to any binding partners they may have [16]. It is an extension of the standard protein purification tag such as the histidine tag, in which a His_6 sequence is added to one end of the protein. This sequence binds to metals such as nickel, whereas normal proteins do not, and therefore the protein may be purified in one step by use of a nickel affinity column. In TAP-tagging, *two* tags are used sequentially (**Figure 11.58**). Typically, the C terminus of the protein is tagged with a sequence consisting of a calmodulin-binding peptide, followed by a cleavage site recognized by tobacco etch virus (TEV) protease, followed by Protein A. Protein A binds to IgG. The tagged protein is therefore purified first on IgG beads, after which the Protein A sequence is cleaved off by TEV protease. The protein is then purified on calmodulin beads in the presence of calcium, which is required for the correct folding of calmodulin. It is then eluted from

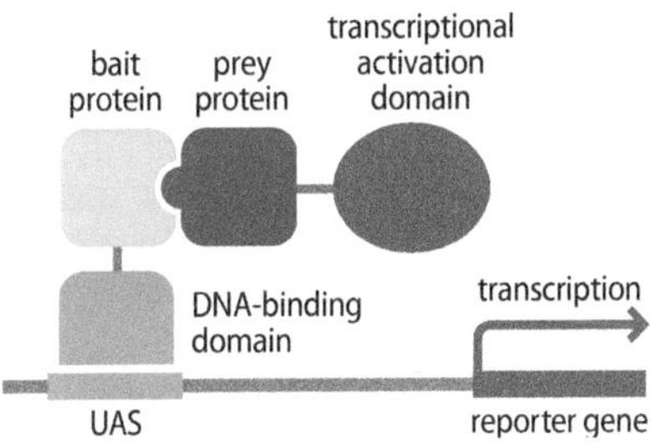

FIGURE 11.57
The two-hybrid method for identifying dimerization partners.

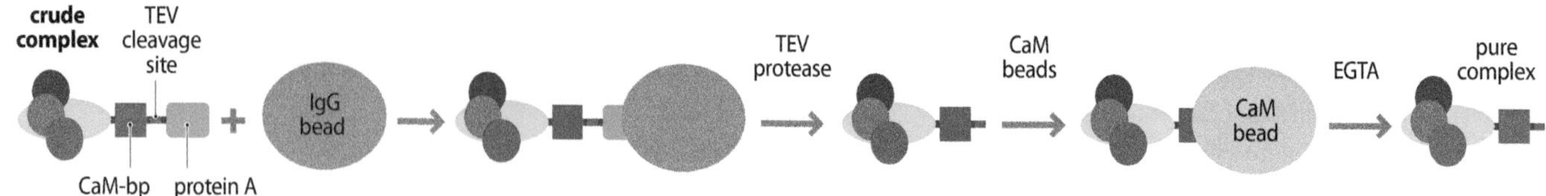

FIGURE 11.58
The TAP-tag method for isolating binding partners of a tagged protein. For good coverage of the interactome, one would want to attach TAP-tags to all proteins, and use both N and C termini in case the C terminus is involved in interactions, in which case placement of the tag there would interfere. CaM, calmodulin; CaM-bp, calmodulin-binding peptide.

the column with ethylene glycol tetraacetate (EGTA), which removes the calcium. The second affinity step serves as a second independent purification and also removes the TEV protease. The method gives a reproducibly very high purification factor, allowing the preparation of essentially completely pure protein from the cell without overexpression. The proteins that copurify with the tagged protein are usually identified by running the complex on a gel, cutting out bands corresponding to the partner proteins, digesting them in the gel with a protease such as trypsin, and identifying the fragments by mass spectrometry. The observation of several trypsin fragments that all match the same full-length protein is a reliable way of identifying the parent protein. The method is commonly used to isolate and characterize complexes, because the mild treatments allow complexes to be purified intact in roughly stoichiometric ratios. It should, however, be added that the necessary washing steps probably do dissociate weak complexes (see Problem N3 in Chapter 9). As discussed in Chapter 9, many important complexes (such as the putative metabolons) are weak enough that they cannot be observed using TAP-tagging. TAP-tagging also misses other complexes that one might expect to see and are presumably too weak to purify intact, such as kinase–substrate complexes and TRAF–TRAD complexes downstream of the TNF receptor.

In this context we should also mention protein microarrays, which are also used to identify proteins and their interactions. These typically consist of an array of antibodies attached to a surface such as a microscope slide, which can be used to capture and subsequently analyze protein mixtures. This technology suffers from the same problems as 2D gels, namely differential stability and solubility of proteins, and again is probably most useful in comparing different conditions, for which the results are more quantitative and reliable.

11.9 COMPUTATIONAL METHODS

11.9.1 Bioinformatics

In the final part of this chapter, we turn to computational methods that are used to study proteins. There is an enormous range of such methods, and we have not attempted to cover them all, merely those that are of most importance and are mentioned elsewhere in this book. Computational methods can be quick and relatively simple, in which case they can often be run on remote servers; at the other extreme they require supercomputers and specialist users. We first describe the simpler methods, starting with the very important field known broadly as **bioinformatics (*11.17)**.

***11.17 Bioinformatics**
There are many definitions of bioinformatics, but the most generally accepted would be the use of computational methods to investigate biological problems. Its origins are in protein sequence alignment, but it has expanded enormously, as discussed in Section 11.9.1. The latest and potentially most exciting developments of bioinformatics are in the field of systems biology (Section 11.9.3): the modeling of complete systems. In general it is probably true to say that we do not have complete enough data on most systems to allow us to model them usefully, but no doubt this will change.

One of the most obvious questions to ask about a protein is whether its sequence is similar to any other known sequence. This is a simple question, and if the sequence is very similar to another one, the answer is also very simple. It starts to get more complicated once the sequences start to become very different. There are several programs for performing this analysis, of which the most popular is NCBI's BLAST. Like other programs, it outputs a list of hits in order, together with a likelihood score (an E value). This represents how likely it is that the hit could have arisen by chance. Such programs can also identify motifs and use them for matching.

One of the problems is how to handle insertions and deletions (indels). Two homologous proteins may well have indels, and one needs to allow for them. Indels can be of almost any length, although short ones are more common; and they essentially always occur in loops rather than in regular secondary structure. Different programs have different ways of dealing with indels. As noted in Chapter 2, proteins are organized into domains, and one would expect that homologous sequences would fit into the same pattern: good matching of sequences within the domain boundaries, but a very much reduced success rate if the sequence runs over a domain boundary. In searching for homologies, trials with different length sequences are therefore a good idea. However, it is difficult to find matches to short stretches of sequence, because similar sequences could easily arise by chance. The same is true for low-complexity sequences.

One can now conduct much more sophisticated searches, using multiple alignments, which make it easier to identify weaker patterns of similarity. The program PSI-BLAST is an example of one such program.

Sequence comparisons are very useful. For one thing, they show which residues are **conserved**. We can then investigate *why* they are conserved. As discussed in Chapter 1, they may be structural (glycine, proline, cystine, large hydrophobic) or functional (active-site residues, interaction sites); in either case, we need to understand their role, for example by **site-directed mutagenesis (*11.18)**.

Bioinformatics is mainly concerned with linear sequences, for example protein sequences. Typical applications discussed elsewhere in the book are the prediction of secondary structure, the use of hydropathy plots to identify transmembrane regions, the identification of targeting sequences, and sequence comparisons aimed at working out evolutionary relationships. Large genomic centers usually have bioinformatics groups to try to make sense of the enormous amounts of data coming out of them, and also to organize the data into databases in a form that can be usefully analyzed. A major problem in genomic studies is **annotation**: making guesses as to the function of a protein from its sequence. A close match to a protein with known function is usually an excellent guide to function, as discussed in Chapter 1, but there are exceptions. A particular problem is that some annotations ascribe a function to a multidomain protein, when different domains within the protein perform different functions. The annotation can then get carried over to homologous proteins using the 'wrong' domain. There are a small proportion of annotations that are wrong in this way, and hunting them down is a far from trivial task.

Bioinformatics can also work on three-dimensional structures. A major class of such programs seeks to predict three-dimensional structure from sequence; these programs are discussed further below. This is a very important goal, because there is at least 100 times more sequence data than structure, and the gap is widening. It seems unlikely that we shall ever be able to determine the structures of all known proteins experimentally. There are programs to compare three-dimensional structures and identify binding sites, and to predict protein:protein interactions. There are also many programs for calculating pK values, docking proteins, calculating electrostatic potentials, and so on.

11.9.2 Dynamics simulations

This class of programs tends to be much more demanding of computer time, and requires experts to run them. They vary widely in the amount of detail that is included in the calculation. The most obvious are **molecular dynamics** programs. In this method, the dynamics of a protein is simulated using Newtonian mechanics. Each atom is assigned a mass and a partial charge, and all the forces between atoms are modeled, such as coulombic attractions

***11.18 Site-directed mutagenesis**

Site-directed mutagenesis (SDM) is the change of one amino acid in a polypeptide to another by altering the gene coding for it. SDM is relatively straightforward to do using PCR, by introducing mutations into primers at the desired positions. PCR can introduce errors, and it is always advisable to resequence a mutated plasmid. It is, however, not nearly so easy to interpret the mutation (see, for example, Problem 1 of this chapter). Mutation of residues that have an important structural role often results in a misfolded protein, and therefore low expression; mutation of catalytic residues results in an inactive enzyme. However, as discussed in Chapter 5, binding and catalysis are closely interlinked, and so mutation of an enzyme's 'binding-site' residue often results in changes to V_{max}, whereas changes to an 'active-site' residue often result in changes to K_m. Mutations often lead to structural rearrangement of the protein, implying that thermodynamic measurements are particularly difficult to interpret, and usually require a structure of the mutated protein. For these reasons, it is often most useful to make relatively small mutations, which are less likely to lead to major structural change, such as Glu to Gln or Asp; Phe to Tyr; Leu to Val. It is worth noting that mutations to charged residues will affect the pI of the protein, which could well affect both its purification and its stability.

and repulsions, van der Waals attractions and repulsions, bond vibrations and bending, angle bending, hydrogen bonding, and sometimes dihedral angle changes. The forces are based on physical parameters observed in small molecules, but the ultimate test of a dynamics program is whether it can reproduce experimental observations: the parameters are therefore changed in a rather empirical way to produce structures and dynamics that reproduce observations. The set of forces is described collectively as a **force field**. Once the molecule and all its forces have been set up, the atoms are given random velocities, consistent with the Boltzmann distribution (*5.2) at the desired temperature. Each force produces accelerations, which change the velocities. The atoms are then allowed to move for a very short time, after which everything is calculated all over again. The typical step size is of the order of 1 fs (10^{-15} second), and simulations often run for hundreds of picoseconds or more, so a lot of calculations are required. The CPU time required for the calculation depends markedly on the number of atoms in the simulation, and on the details of the force field. A particular problem in such calculations is what to do about water molecules. Water is of great importance in determining the dynamics of a protein. However, the inclusion of enough solvent to cover the protein sufficiently that one need not worry about what happens at the 'edge' of the water bath requires many thousands of additional atoms and greatly increases the scale of the calculation. There are a variety of 'continuum' and other models that try to avoid including too many solvent molecules.

A good test of the correctness of a force field is whether it can reproduce the experimental structure. If, for example, one takes a high-resolution X-ray crystal structure, which is as close to the 'true' structure as one can get, and minimizes its energy in a molecular dynamics force field, the resultant structure will be about 1 Å root mean squared deviation away from the crystal structure. This suggests that this may be about as accurate as force fields can get at the moment.

The restrained molecular dynamics programs used to calculate NMR structures are a cut-down version of full molecular dynamics, designed merely as a minimizer and not as a realistic calculation of dynamics; they were described in more detail earlier in this chapter. They also use *simulated annealing*; that is, a gradual lowering of the temperature of the system to find the global energy minimum.

Some of the most interesting protein motions occur on timescales longer than microseconds. As discussed in Chapter 6, these include induced-fit/conformational selection motions and most of the dynamics associated with enzyme recognition of substrate, transition state, or product. This is several orders of magnitude too long for conventional molecular dynamics. There are several ways of investigating such motions. One of the oldest, and still very useful, is to identify *normal modes* of motion. The vibrations possible for the static structure are analyzed to look for correlations in movement between one part of the protein and another. The largest-scale correlations form the low-energy normal modes, which are also assumed to represent slow dynamics, for reasons outlined in Chapter 6.

Another technique for investigating slower dynamics is *Monte Carlo* methods. In this method (which is applicable to a wide range of problems, not just dynamics), the system is given a fairly large change, and the energy of the resulting structure is calculated. If the energy is lower than the starting energy, the new structure is accepted as a possible solution. If the energy is higher, the structure may be accepted, with a probability given by the Boltzmann probability at the temperature of the simulation (hence the name Monte Carlo, based on the casinos to be found there). This is a good method for searching large regions of conformational space efficiently, because it does not need to find its way across potentially large energy barriers. Genetic algorithms (*1.21) can also be used for the same purpose.

Protein folding also occurs on long timescales, of the order of seconds. Calculations in this area fall into two categories. Some are aimed at investigating the principles of protein folding, and tend to use very simplified models of proteins, with the protein folding onto a regular lattice, using one point per amino acid residue. Others aim to predict a protein's structure. These use a large range of methods, but in general they use algorithms for folding an extended chain based on modeling of local and long-range interactions in the protein; the program Rosetta, mentioned in Chapter 1, is currently one of the most successful of such programs (as judged for example by its success in the regular 'contests' to find the most successful protein structure predictor, which go under the name CASP). Clearly, the prediction of protein structure is a key goal of structural biology. For proteins with a reasonable degree of sequence similarity to known structures, the success rate is good, but there is still a long way to go before we can claim truly to be able to predict structures, or to understand how to do so.

In all calculations of protein stability, for structure, dynamics, or folding, the variable normally computed is the enthalpy. Calculations of entropy are much harder, because they require a counting of relative frequencies of different conformations and therefore require a much wider coverage of the conformational energy surface. For this reason, calculation of free energy remains rather inaccurate.

An accurate calculation of protein structure and dynamics requires the use of quantum mechanics. Such calculations are particularly needed when calculating how reactions happen, because calculations of bond making and breaking clearly require some analysis of electron orbitals. These calculations are orders of magnitude harder and slower, and a complete quantum mechanical description of a whole protein is still far too costly to attempt. Typically, some hybrid approach is used, where the bond and its immediate vicinity are calculated using quantum mechanics, but more distant regions are calculated using faster and less accurate methods.

It is worth making again a point mentioned earlier in this chapter, and also in Chapter 6: the large majority of experimental approaches (for example to the measurement of protein structure, dynamics [including enzyme catalysis], or folding) can only observe ensemble averages, and provide little or no information on the spread of individual molecules. Almost the only way of investigating what individual molecules do is to use computational methods. They thus provide an important, and increasingly reliable, complement to experiment.

11.9.3 Systems biology

Systems biology is not a technique, nor indeed is it a set of techniques. Rather, it forms a way of thinking about biological systems, which is entirely in keeping with the organization of this book. It suggests that a biological system is more than just the sum of its parts, but that the different parts interact to produce a novel or 'emergent' property that was not there in the individual parts. The study of systems biology therefore consists of two tightly interlocked elements: characterization of the biological components present, and modeling how these components interact. The characterization and the modeling should be detailed enough for them to be able to produce a truly predictive model, in which for example the effect of altering the function of one of the components can be investigated.

This kind of study is still in its infancy. One of the most successful studies is a model of the heart [17]. The model itself consists of muscle cells, organized into a three-dimensional structure and containing nerve termini, ion channels, ion pumps, and so on; and when 'turned on' it predicts a beating heart, which can be used to model the effect of disease states (damage to a channel, for example) or drugs. Construction of such a model requires close collaboration between biologist and modeler, and at least in some cases it may provide the next step up in understanding what is required if we are truly to comprehend how proteins work.

11.10 SUMMARY

This chapter has provided a brief overview of the main techniques used in studying protein structure and function. We have not covered molecular biology in any detail. The major structural technique is X-ray crystallography, which is becoming faster and more automated, but it still requires human checking if errors are to be avoided. NMR is also a useful technique for structure determination, although its more important contribution is in characterizing dynamics. Electron diffraction is much slower and is only useful for membrane proteins; however, electron microscopy and AFM can be used to image individual large protein assemblies, and therefore provide an important way of understanding less well-defined complexes, of the kind discussed in Chapter 9. A range of spectroscopic techniques are available, which provide information on specific functional groups or chromophores and so are less general in their application. Fluorescence is particularly useful because it can identify the intracellular location of groups of molecules, measure distances between fluorophores in real time, and even image single molecules.

Mass spectrometry is becoming a very important technique with a wide variety of applications, including identifying and sequencing proteins, and even characterizing complexes. It can be linked to proteomics methods for the high-throughput separation and identification of proteins. High-throughput methods remain challenging for proteins because of their wide range of physical properties, but they are an important development goal because of their genome-wide information content.

Computational methods are a vital complement to experimental methods. In particular, they can describe properties at the level of individual molecules that are observable only as averages in experiments. They also provide a range of tools for comparing proteins, and analyzing their sequence and structure. A promising new field is systems biology, which seeks to integrate different kinds of information to show how the system has properties that are different from those of its parts.

11.11 FURTHER READING

NMR

Many textbooks go into far too much detail. For the more 'chemical' aspects, I recommend Modern NMR Techniques of Chemistry Research, by Derome [18], or its successor, High-resolution NMR Techniques in Organic Chemistry, by Claridge [19]; or Modern NMR Spectroscopy: a Guide for Chemists, by Sanders and Hunter [20]. For details on proteins, the best is still NMR of Proteins and Nucleic Acids, by Wüthrich [21]. I should also mention The Nuclear Overhauser Effect in Structural and Conformational Analysis, by Neuhaus and Williamson [22], which in addition to the NOE also covers basics of spins and relaxation, and gives a very clear description of 2D NMR.

X-ray crystallography

Biomolecular Crystallography, by Rupp [23], is an excellent recent textbook with as clear explanations as I have come across.

Also recommended is Crystallography Made Crystal Clear, by Rhodes [24].

Bioinformatics

Introduction to Protein Science, by A. M. Lesk [25], provides an excellent way in to the subject. It also provides a good coverage of proteomics and systems biology.

For most of the other techniques, manufacturers' websites are often a good place to find details. A good place to start with molecular biology is Molecular Cell Biology, by H. Lodish et al. [26].

11.12 PROBLEMS

Hints for some of these problems can be found on the book's website.

1. T4 lysozyme is a hydrolytic enzyme produced by the bacteriophage T4. Residue 96 in the enzyme is arginine, which is in a helix. The hydrocarbon part of the arginine side chain (that is, Cβ, Cγ, Cδ) is buried and contacts hydrophobic residues. The positively charged end is partly exposed on the surface and makes two hydrogen bonds to main-chain carbonyls. This residue was replaced by all 19 other possible amino acids, and the results are listed in Table 11.3 [27].

TABLE 11.3 Stability of Arg96 mutants of T4 lysozyme

Protein	ΔT_m (°C)	$\Delta\Delta H$ (kJ mol^{-1})	$\Delta\Delta G$ (kJ mol^{-1})
Wild type (Arg 96)	0.0	0.0	0.0
R96K	−0.2	+8	0.0
R96Q	−1.4	+29	−1.3
R96A	−5.1	−80	−8.4
R96V	−6.4	−96	−10.0
R96S	−7.0	−75	−10.9
R96E	−7.0	−13	−10.5
R96G	−7.1	−67	−10.9
R96M	−7.1	−92	−11.3
R96T	−7.6	−50	−11.7
R96C	−7.7	−151	−12.1
R96I	−7.9	−80	−12.1
R96N	−8.0	−46	−12.6
R96H	−8.3	−63	−13.0
R96L	−8.6	−92	−13.4
R96D	−9.5	−50	−14.6
R96F	−11.5	−117	−17.6
R96W	−12.8	−172	−18.8
R96Y	−13.2	−155	−19.7
R96P	−15.5	−138	−23.0

Changes in free energy are for unfolding at pH 5.35. ΔT_m is the change in melting temperature relative to the wild type (66.6 °C at this pH), $\Delta\Delta H$ is the change in enthalpy of unfolding, and $\Delta\Delta G$ is the change in free energy. (Data from B. H. M. Mooers, W. A. Baase, J. W. Wray, and B. W. Matthews, *Protein Sci.* 18: 871–880, 2009. With permission from John Wiley & Sons.)

What can you conclude about the determinants of protein stability? In particular, how important are (a) the hydrophobic side chain, (b) the volume of the side chain, and (c) suitable charge:charge interactions with the end of the side chain?

2. Explain why the number of high-resolution structures of membrane proteins is so small.

3. What technique might you use (and why) to detect the following post-translational modifications: (a) relative amounts of phosphorylation at one of three possible serines; (b) methylation or acetylation of specific lysines in the N-terminal tail of a histone; (c) attachment of one or more ubiquitin molecules; (d) glycosylation of a secreted protein; (e) sulfation of a tyrosine. Mass spectrometry has an accuracy of about 0.01% in the measured mass. This means that mass spectroscopy can be used for most of these modifications: it is, however, not the best or most common for all of them. For example, if phosphorylation is possible at three possible serines, how can you tell which of them is phosphorylated?

4. Why do most NMR studies of proteins now use ^{13}C,^{15}N-labeled protein? Why do studies of large proteins also use ^{2}H-labeled protein? What is the problem with using ^{2}H labels in particular for structure calculation?

5. Many NMR spectrometers now used for protein work are fitted with cold probes, also known as cryoprobes. What are they and why are they useful?

6. (a) Explain the difference between accuracy and precision in protein structure determination. In comparing X-ray and NMR structures, which usually has the better precision? And which has the better accuracy? (b) What measures are used to test the correctness of crystal structures? And NMR structures?

7. Crystal structures are usually obtained from solutions with very high ionic strength, and at very low temperature; in addition, they are obtained from regular crystalline arrays of proteins. Is there any evidence that this makes the structures non-physiological?

8. (a) What is the difference between BLAST and PSI-BLAST? (b) Run a BLAST search on *Rhodobacter sphaeroides* protein PufX (you will need to search in a protein database such as Expasy UniprotKB to find the sequence first, and then run BLAST, for example from the NCBI website). What do the E values mean? Which of these sequences do you think represents a genuine homolog? Does this allow you to say what the function of the protein is?

11.13 NUMERICAL PROBLEMS

N1. The energy E of a transition is related to its frequency ν by the relation $E = h\nu$, where h is Planck's constant (6.6×10^{-34} J s). The Boltzmann distribution $N_1/N_2 = e^{-\Delta G/kT}$ shows the population ratio between two states, N_1 and N_2, differing in energy by ΔG, where T is the absolute temperature and k is Boltzmann's constant (1.4×10^{-23} J K^{-1}). Use these two equations to calculate the excess population of ^{1}H nuclei in the lower spin state at 500 MHz. How much better is this ratio at 900 MHz? These numbers are important because the magnitude of the magnetization vector is proportional to the ratio in population difference, so we want this number to be as large as possible.

N2. Using Table 11.1, predict the molar extinction coefficient of mature (that is, post-secretion) hen egg-white lysozyme (you will need to find the sequence first!). Do the same with a Web server (for example within Expasy). Finally, use the Web to find the actual value. Are the differences significant? A solution of lysozyme has an absorbance of 0.308 in a cell with a path length of 1 cm. What is its concentration?

N3. Using the information in *11.11, calculate the numerical aperture of an oil-immersion lens with incident angle α of 60°. What is the maximum resolution available from such a lens?

N4. A scorpion toxin (64 residues, molecular mass 7250 Da) crystallizes in a rectangular crystal with dimensions $45.94 \times 40.68 \times 29.93\,Å^3$. There is one protein molecule in the unit cell. What is the concentration of protein in the cell, in $mg\,ml^{-1}$ and in mM? How does this compare with the typical concentration of proteins in a prokaryotic cell?

N5. The typical density of a protein in a crystal is roughly $1.4\ g\,ml^{-1}$. Use this value to calculate the expected volume of one molecule of the protein. Compare this with the volume of the unit cell in question N4 and calculate how much space is not occupied by water. At 20°C, the density of water is $0.998\,g\,ml^{-1}$. How many molecules of water would you expect to be in the unit cell? In the crystal structure, only 89 molecules of water were identified. Where are the rest?

11.14 REFERENCES

1. S Brenner (2004) Interview. *Discover* 25(4).
2. A Kornberg (2003) Ten commandments of enzymology, amended. *Trends Biochem. Sci.* 28:515–517.
3. S Gräslund, P Nordlund, J Weigelt et al. (2008) Protein production and purification. *Nat. Methods* 5:135–146.
4. Y Shen, O Lange, F Delaglio et al. (2008) Consistent blind protein structure generation from NMR chemical shift data. *Proc. Natl. Acad. Sci. USA* 105:4685–4690.
5. MP Williamson & CJ Craven (2009) Automated protein structure calculation from NMR data. *J. Biomol. NMR* 43:131–143.
6. V Cherezov, DM Rosenbaum, MA Hanson et al. (2007) High-resolution crystal structure of an engineered human b_2-adrenergic G protein-coupled receptor. *Science* 318:1258–1265.
7. N Grigorieff, TA Ceska, KH Downing et al. (1996) Electron-crystallographic refinement of the structure of bacteriorhodopsin. *J. Mol. Biol.* 259:393–421.
8. HR Saibil (2000) Macromolecular structure determination by cryo-electron microscopy. *Acta Cryst. D* 56:1215–1222.
9. J Frank (2001) Cryo-electron microscopy as an investigative tool: the ribosome as an example. *BioEssays* 23:725–732.
10. J Berriman & N Unwin (1994) Analysis of transient structures by cryomicroscopy combined with rapid mixing of spray droplets. *Ultramicroscopy* 56:241–252.
11. E Jöbstl, JR Howse, JPA Fairclough & MP Williamson (2006) Noncovalent cross-linking of casein by epigallocatechin gallate characterized by single molecule force microscopy. *J. Agric. Food Chem.* 54:4077–4081.
12. CV Robinson, A Sali & W Baumeister (2007) The molecular sociology of the cell. *Nature* 450:973–982.
13. G von Helden, T Wyttenbach & MT Bowers (1995) Conformation of macromolecules in the gas phase: use of matrix-assisted laser desorption methods in ion chromatography. *Science* 267:1483–1485.
14. S Fields & R Sternglanz (1994) The two-hybrid system: an assay for protein–protein interactions. *Trends Genet.* 10:286–292.
15. E Warbrick (1997) Two's company, three's a crowd: the yeast two hybrid system for mapping molecular interactions. *Structure* 5:13–17.
16. A Dziembowski & B Séraphin (2003) Recent developments in the analysis of protein complexes. *FEBS Lett.* 556:1–6.
17. D Noble (2006) Systems biology and the heart. *Biosystems* 83:75–80.
18. AE Derome (1987) Modern NMR Techniques of Chemistry Research. Oxford: Pergamon Press.
19. TDW Claridge (2009) High-resolution NMR Techniques in Organic Chemistry, 2nd ed. Oxford: Elsevier.
20. JKM Sanders & BK Hunter (1993) Modern NMR Spectroscopy: a Guide for Chemists, 2nd ed. Oxford: Oxford University Press.
21. K Wüthrich (1996) NMR of Proteins and Nucleic Acids, 2nd ed. New York: Wiley/Blackwell.
22. D Neuhaus & MP Williamson (2000) The Nuclear Overhauser Effect in Structural and Conformational Analysis, 2nd ed. New York: Wiley-VCH.
23. B Rupp (2009) Biomolecular Crystallography: Principles, Practice, and Application to Structural Biology. New York: Garland Science.
24. G Rhodes (2006) Crystallography Made Crystal Clear, 3rd ed. Burlington, MA: Academic Press.

25. AM Lesk (2010) Introduction to Protein Science: Architecture, Function, and Genomics, 2nd ed. Oxford: Oxford University Press.

26. H Lodish, A Berk, CA Kaiser et al. (2007) Molecular Cell Biology, 6th ed. New York: WH Freeman.

27. BHM Mooers, WA Baase, JW Wray & BW Matthews (2009) Contributions of all 20 amino acids at site 96 to the stability and structure of T4 lysozyme. *Prot. Sci.* 18:871–880.

28. EF Pettersen, TD Goddard, CC Huang et al. (2004) UCSF Chimera—a visualization system for exploratory research and analysis. *J. Comput. Chem.* 25:1605–1612.

29. MR Wilkins (1994) Conference on 2D Electrophoresis: From Protein Maps to Genomes; Siena, Italy.

30. J Norbeck & A Blomberg (1997) Two-dimensional electrophoretic separation of yeast proteins using a non-linear wide range (pH 3-10) immobilized pH gradient in the first dimension: reproducibility and evidence for isoelectric focusing of alkaline (pI > 7) proteins. *Cell* 13:1519–1534.

Glossary

Action potential
A change in electric potential across a cell membrane that moves along the cell to transmit a nerve impulse.

Activation energy
The free energy required to go from the lowest energy point on a pathway to the highest.

Active site
The region on an enzyme where the substrate binds and catalysis occurs. More loosely, also used to mean a functionally important surface region on a protein such as the site of protein:protein interaction.

Activity and concentration
The concentration of a substance in water is just the amount of substance divided by the volume. The *activity* is the *effective* concentration, for example when calculating reaction rates. For an ideal solution, which typically means a very dilute one, the *activity* of a compound is equal to its concentration. As the concentration increases, the activity can become substantially different, either bigger or smaller.

Alignment
The comparison of two protein (or DNA) sequences, indicating matches between them.

Allosteric
Allosteric regulation is the regulation of an enzyme by binding an effector molecule at a site other than the protein's active site (from the Greek for 'different structure'). (This is the definition used in this book; others are used elsewhere.)

Alternative splicing
A mechanism of RNA splicing in which introns are cut out and the exons joined together in different ways so as to give mature messenger RNA coding for different protein sequences.

Amino acid
In principle, any organic compound that has an amine group (-NH_3^+) and a carboxylic acid (-CO_2^-). In practice, it is normally assumed to mean an α-amino acid; that is, one in which the amine group is attached to the same carbon as the carboxylic acid.

Amphipathic
The property of having both hydrophobic and hydrophilic parts.

Amyloid
A non-native protein conformation in which the protein forms fibers composed of β strands that run in a direction perpendicular to the direction of the fiber. It occurs in several diseases such as Alzheimer's disease and bovine spongiform encephalopathy (BSE). The word amyloid means 'starch-like': the plaques formed by the amyloid fibers stain with iodine, as does starch.

Anabolism
Metabolic process of building up molecules from smaller building blocks.

Analogous
Two enzymes are *analogous* if they perform the same biological function but are structurally unrelated.

Annotation
The labeling of genes, proteins, and protein domains by their known or predicted function.

Anterior
The head end of a body or cell.

Antibody
A protein that forms part of the vertebrate immune system. Its function is to recognize foreign molecules (**antigens**) and bind to them.

Antigen
The molecule that is recognized by an **antibody**.

Apoptosis
Generally described as programmed cell death; a regulated and controlled process by which a cell destroys itself. It should be contrasted with *necrosis*, which is an uncontrolled death.

Autoinhibition
A protein that is inactive because there is an intramolecular interaction that blocks or inhibits the active site is said to be autoinhibited.

Auxotroph
A bacterial strain that has been engineered to require a particular nutrient in order to grow.

Axon
The long part of a nerve cell (neuron) that is used for transmitting nerve impulses to other cells.

Barnase
A ribonuclease from *Bacillus amyloliquefaciens*.

Beer–Lambert law
The absorbance of a solution A is given by $A = \epsilon ct$, where c is the concentration of the solute, t is its thickness, and ϵ is the molar extinction coefficient. Normally c is measured in moles litre^{-1} and t is in cm (1 cm), and therefore ϵ is in litre mole^{-1} cm^{-1}.

β barrel
A barrel-shaped protein structure formed by a series of β strands.

β hairpin
A peptide chain that has folded back on itself to make a simple two-stranded β sheet.

β-strand
A section of polypeptide chain that is in the extended or β region of the Ramachandran plot, and which hydrogen bonds to one or more neighboring strands to form a β sheet.

Bioinformatics
Computational methods applied to biological systems.

C terminus
The end of a polypeptide chain that has a free carboxylate group; that is, the right-hand end as normally written. Note the distinction between the adjective *terminal* and the noun *terminus*.

Caspase
A **c**ysteine protease that cleaves after **asp**artate. The result forms a signal that leads to apoptosis or programmed cell death.

Catabolism
The metabolic process of breaking down complicated molecules into smaller metabolic intermediates that can be recycled or used for energy generation.

Catalytic triad
The three amino acids that make up the key **active site** residues of serine proteases: Ser, His, and Asp.

Chaperone
A protein that shields an unfolded protein and prevents it from aggregating. Some chaperones ('chaperonins') help unfolded proteins to refold. Most chaperone function requires ATP.

Chemical shift
The frequency of an NMR signal, usually defined in ppm as $[(\nu - \nu_{ref})/\nu_{ref}] \times 10^6$, where ν is the NMR frequency and ν_{ref} is the frequency of a reference signal such as DSS.

Chimeric
A protein containing modules originating from different sources. From the *chimera*, a mythological creature having the body of a lion, a snake's head as a tail, and a goat's head.

Chloroplast
A membrane-enclosed organelle found in plants, where photosynthesis occurs.

Chromophore
A molecule that absorbs ultraviolet radiation. These are typically aromatic or conjugated (that is, containing alternating double and single bonds).

Codon
A three-nucleotide (triplet) messenger RNA sequence that codes for an amino acid.

Codon usage
The genetic code is universal (except for slight variations in mitochondria, in which up to 4 of the 64 possible codons have a different meaning, depending on species). Most amino acids are coded for by several codons, and the frequency at which a particular codon occurs varies markedly across species. In particular, codon usage in prokaryotes is different from that in eukaryotes, implying that an attempt to express a eukaryotic gene in a bacterium may fail or become very inefficient if the codon usage is too different.

Coiled coil
A simple structure formed by two α helices winding around each other.

Conformational selection
A model to explain enzyme function (and protein binding) that is an extension of the **induced-fit** model, discussed in Chapter 6. It says that a protein is in equilibrium between several states (at least one energy well characteristic of the free protein and one energy well characteristic of the bound protein), and that binding of a ligand *selects* the alternative conformation. Thus, the induced-fit model says that ligand binding induces a conformational change, whereas the conformational selection model says that that the conformational change happens before ligand binding.

Conserved/conservation
When comparing several protein sequences, a *conserved* residue is one that is not altered from one sequence to the next, or is almost always the same but is occasionally replaced by a chemically similar one (*highly conserved*). A *conservative* mutation changes one amino acid to a chemically similar one (for example, Phe to Leu, Thr to Ser, Glu to Asp), thus usually conserving the protein function.

Convergent evolution
A rather rare occurrence, in which a similar structural or functional feature in a protein is evolved independently more than once.

Correlated motion
The movement of two atoms is correlated when there is a statistical correlation between the position of one and the position of the other, as a function of time: for example, when they are moving in the same direction at the same rate.

Correlation time
A measure of the time it takes a molecule to rotate; roughly equal to the time it takes to rotate by one radian ($360/2\pi°$).

Cryptic genes
A gene that was once presumably functional but is no longer.

Curly arrows
Curved arrows indicating the movement of electrons during a chemical reaction.

Cytokinesis
The division of a growing cell.

Cytoplasmic streaming
A process used largely by plant cells in which an actin/myosin system leads to a circulation of organelles (and thus the cytoplasm) around the cell.

D-amino acid
An α-amino acid with D-chirality at the α carbon. See L-amino acid.

Deconvolution
A mathematical procedure in which a function is expressed as a sum of other known functions. Thus, for example, if reference CD spectra are known for α helix, β sheet, and random coil, any given spectrum can be expressed as a sum of these, and the relative amounts of each can be determined. Obviously the deconvolution procedure is only as good as the reference spectra, which is the problem, because small changes in the reference can have large effects on the fitting.

Denaturant
A chemical that destabilizes a protein and (at sufficient concentration) causes it to unfold. Typical examples include urea and guanidinium chloride.

Dielectric constant
The proportion by which a medium shields charges, compared with vacuum.

Dihedral angle
The angle defined by four atoms.

Dipole
Something with two poles, like a magnet (a magnetic dipole: north and south poles) or a molecule that has positive and negative charges at opposite ends (an electric dipole).

Dipole moment
A measure of the strength of a **dipole**, which for an electric dipole is equal to $q\mathbf{r}$, where q is the charge and $\mathbf{r}$ is the vector connecting the charges. A dipole moment therefore has both a magnitude and a direction.

Dissociation constant
For a complex AB splitting into A and B, defined as (concentration of A) × (concentration of B) / (concentration of AB). In this case it has units of concentration, and is roughly the concentration at which the complex is half dissociated. See **Binding and dissociation constants and free energy (*5.5)**.

Divergent evolution
A very common mechanism, in which the gene for an initial protein is duplicated, and the original and copy then diverge over the course of evolutionary time. Structure is retained more than function, and much more than sequence.

Domain
A domain is a polypeptide chain (or part of one) that can independently fold into a stable compact tertiary structure or fold.

Effective concentration
The ratio between the affinity of A for B when they are covalently connected and their affinity when they are not. This ratio has units of concentration, and is the concentration of B needed to give the same fraction of bound A as is found in the A-B fusion.

Electronegative
An atom or group that is good at attracting bonding electrons toward it. Examples include N and O.

Electron microscopy
A technique, described in Chapter 11, in which electrons are used instead of light to image very small objects.

Electrophile
A positively charged atom, which is attacked by, and forms a bond to, a nucleophile.

Embellishment
A word used to describe the tendency of evolution to tinker with a simple system and make it more complex by adding on extra features, for example to make it more specific.

Encounter complex
The initial complex formed when two molecules collide in solution. The molecules will typically be close together but not necessarily in the correct conformation or orientation to bind properly. The encounter complex must either dissociate or rearrange to form a functional complex. This is discussed more fully in Chapters 4 and 6.

Endoplasmic reticulum (ER)
The endoplasmic reticulum is a membrane system found within eukaryotic cells. It forms an interconnected system that is continuous with the nuclear membrane that is used for expression of membrane proteins, post-transcriptional modification and secretion, as well as the synthesis of glycogen and steroids. The volume inside the ER is known as the lumen. It forms approximately half of the total membrane of the cell.

Endosome
A membrane-enclosed vesicle that is formed from the cell membrane and transported toward the center of the cell, as a way of recycling or ingesting proteins or extracellular matter.

Energy landscape
A diagram indicating (usually in a very schematic way) the variation in energy of a protein with conformation. Discussed in Chapter 6.

Entropy/enthalpy compensation
The commonly observed tendency in a series of related measurements (for example binding interactions, dissolution, effects of mutations on stability or binding) for the free energy to remain roughly constant but with large compensatory changes in entropy and enthalpy. Discussed in more detail in Chapter 1.

Exchange rate of amide protons
The rate at which a backbone amide proton (HN) exchanges to a deuteron (^{2}H isotope, or D) after the protein has been placed in D_2O. It can be measured by NMR experiments with a time resolution of a few minutes. It can also be observed by mass spectroscopy on a faster timescale, but site-specific information is more difficult. The exchange rate is sensitive to local secondary structure and burial of the amide.

Expression vector
A DNA sequence that has been created to make the cloning and expression of proteins simple. There are many such vectors commercially available. They typically have an origin of replication (so that multiple copies of the plasmid are made), strong promoters (so that one copy of the vector will make multiple copies of the protein), multiple cloning sites (restriction enzyme cleavage sites, so that cutting of the plasmid and insertion of the desired gene is simple), and antibiotic resistance genes as selection markers.

Fibrous proteins
Proteins that form fibers. As discussed in Chapter 1, some fibers, such as those of actin and tubulin, are continually being remodeled. Although they are indeed fibers, they are made up of globular proteins. However, true fibrous proteins are those of permanent fibers such as keratin, collagen, and silk, which form long and usually cross-linked fibers.

Fluorophore
A chemical that is capable of fluorescence.

Fold
A fold is a three-dimensional arrangement or topology of secondary structure elements.

Force field
The set of forces used by a molecular dynamics program. In principle the forces are all derived from measurable properties, but in practice it is common to modify, add, or replace some, to make the function of the program match better to experimental data.

Frameshift mutation
A frameshift mutation, also known as a reading frame shift, is an insertion or deletion of nucleotides from a gene. Because the genetic code is based on triplets, the insertion or deletion of a number of nucleotides that is not a multiple of three leads to a shift in the reading frame. Usually this produces a nonfunctional or truncated product. Occasionally (as in the PGK/TIM system described in Section 9.4.5) it leads to two genes becoming fused together and transcribed as one polypeptide.

Free energy
The energy input needed to get from one side of a reaction to the other. See **Binding and dissociation constants and free energy (*5.5)**.

Fusion protein
A genetic construct in which a protein to be expressed is linked as a continuous polypeptide with a second protein. Typically the second protein will be used either to aid in purification (for example, glutathione S-transferase or GST, which can be purified using a glutathione column) or in expression or targeting (for example, a protein that targets the protein to the periplasm). It is often possible to construct the linker so that it can be cleaved by a specific protease.

Gel filtration
A chromatographic separation based on size. The resin through which the solution is passed contains channels or pores into which small molecules can fit, but large ones cannot. Therefore small molecules migrate more slowly through the resin, and emerge from the column later. There are a large number of commercially available resins, matched to the size range to be separated and the flow rate desired.

Gene duplication
The copying of a gene to leave two copies of the gene in the genome.

Gene sharing
A rather rare event in which one protein performs two different functions. Also known as **moonlighting**.

General acid
An amino acid residue that donates a proton during the course of a reaction and thereby catalyzes it.

General base
An amino acid residue that removes a proton during the course of a reaction and thereby catalyzes it.

Genetic algorithm
A computer method for optimizing complex problems based on the same principles as natural selection (see Section 1.4.12).

Genotype
The sequence of the DNA in an organism. This of course gives rise to the **phenotype**.

Golgi apparatus
A membrane-enclosed eukaryotic organelle in which proteins and lipids produced and modified within the **endoplasmic reticulum** are further modified and sorted ready for transport to their final destination.

Green fluorescent protein
A protein originally from a jellyfish that is naturally fluorescent. The fluorophore is composed of three amino acids that react together spontaneously, which means that **heterologous** expression yields a fluorescent protein without the need to add external cofactors.

GTPase-activating protein (GAP)
A protein that stimulates the hydrolysis of GTP to GDP bound to a Ras-like small GTPase, and thereby switches it off.

Guanine nucleotide exchange factor (GEF)
A protein that stimulates the release of GDP from a small GTPase protein, and therefore allows activation by binding to the more abundant GTP.

Gyromagnetic ratio
The ratio between the nuclear magnetic dipole moment and its angular momentum. The gyromagnetic ratio is proportional to the energy difference between α and β spin states and therefore also to the NMR observation frequency.

Heat capacity
A measure of the heat energy required to increase the temperature of an object. It usually has the symbol C_p, meaning the heat capacity at constant pressure. It can be measured by measuring the enthalpy at different temperatures, because $C_p = \partial H/\partial T$.

Helical wheel
A pictorial representation of the positions of the side chains in an α helix, usually used to identify hydrophobic or hydrophilic surfaces (Figure 1.38).

Helix–coil transition
The transition between a helical conformation and a **random coil** conformation. This is an example of a cooperative transition.

Henderson-Hasselbalch equation
$\text{pH} = \text{p}K_a + \log([\text{base}]/[\text{acid}])$, defining the relationship between pH, pK_a and the concentrations of protonated and unprotonated forms of a buffer.

Heterologous
Expression of a protein in a host organism different from that from which the protein originated.

Heteronuclear
An NMR experiment involving nuclei of more than one type, for example ^{1}H and ^{15}N.

Histidine tag
A histidine or His tag is a sequence of about six histidines that is engineered close to the terminus of a protein. It is useful for purification because it binds to metal ions such as nickel and can therefore be used to purify the protein using column chromatography.

Histone
The main protein component of chromatin. There are six classes of histone in humans, which act as a core around which DNA is wrapped (forming *nucleosomes*) and also as clamps to hold the nucleosomal DNA in place.

Homologous
Two proteins are homologous if they are related by evolution; that is, if they share a common ancestor or have at some point diverged from each other. This necessarily implies that they have some sequence similarity, but in distantly homologous proteins this relationship could be obscured. Most proteins that have sequence similarity are homologous, but not necessarily all of them.

Homologous recombination
The exchange of DNA between two double-stranded DNA molecules, by base pairing between identical or closely similar sequences in the two molecules.

Horizontal transfer
The transfer of genetic material (such as DNA) from one organism to another other than by inheritance. This is extremely common in bacteria, in which DNA can be physically passed between bacteria.

Housekeeping enzyme
An enzyme that is required in every cell and performs an essential task such as basic metabolism. It tends to be long-lived and relatively unregulated, with a concentration and activity that remain unchanged over time.

Hydropathy plot
A graph of amino acid hydrophobicity against sequence, used to determine the location of transmembrane sequences.

Hydrophilic
Literally *water-loving*: chemicals that have good affinity for water. These typically contain polar or hydrogen-bonding groups such as OH or NH, or charged groups. The hydrophilic amino acids are generally taken to be Asp, Asn, His, Lys, Glu, Gln, Arg, Ser, Thr, Trp, and Tyr. Cys has both hydrophilic and hydrophobic character.

Hydrophobic
Literally *water-hating*: chemicals that have no affinity for water and therefore tend to aggregate together away from water in aqueous solution. These typically have side chains that are hydrocarbons or aromatic rings. The hydrophobic amino acids are generally taken to be Ala, Phe, Gly (sometimes omitted from this list because it is so small and therefore not strongly hydrophobic), Ile, Leu, Met, Pro, and Val.

Immunoprecipitation
The precipitation of a protein antigen from solution by a specific antibody. Particularly useful for co-immunoprecipitation of interacting partners. See Section 4.2.1.

Inclusion body
When a bacterium is challenged to express a non-native protein (such as overexpression of a plasmid-borne gene, or expression of viral genes) it sometimes reacts by expressing the protein in inclusion bodies, which contain solid protein aggregates. In overexpression, the fraction of protein found within inclusion bodies can be decreased by expression at lower temperature. Inclusion bodies can sometimes be solubilized with denaturants, followed by gradual dialysis.

Induced fit
The theory proposed by Koshland in 1958, that an enzyme structure changes on binding to substrate. Chapter 6 discusses the conformational selection model, which is effectively an update on induced fit.

Insertion/deletion
When comparing aligned protein sequences, it is often observed that they differ because of the insertion or deletion of several amino acids at one point. Structurally, this almost always corresponds to the lengthening or shortening of a loop, or even the addition or removal of an extra domain. Insertions and deletions therefore hardly ever occur in regular secondary structure and are good indicators of loops.

Internal duplication
The duplication and subsequent evolution of a section of a gene coding for a protein, resulting in a protein that has two homologous sequences with similar structure.

Isoelectric focusing
A technique for purifying and characterizing proteins according to their ***pI***. Proteins are mixed with suitable buffers and run on a gel with the use of a voltage applied across the gel. They migrate to a position determined by their p*I*. This is the first stage in two-dimensional gel electrophoresis.

Isothermal titration calorimetry
A technique for measuring the heat change when adding one solution to another. Typically a solution of a ligand is injected stepwise into a solution of its receptor. The equipment consists of two identical jacketed cells (Figure 11.51) that are maintained at the same temperature: the power required to do this is the experimental result. The results are normally plotted as a graph of heat against time (Figure 11.52), from which the overall enthalpy of binding can be calculated, after making suitable corrections for the heat of dilution of the ligand. If the affinity falls into a suitable range and the enthalpy change is large enough, this curve can also be fitted to give the affinity of binding.

Karplus curve
A formula representing the relationship between a three-bond coupling constant and the dihedral angle. It usually

takes the form $^3J = A\cos^2\theta + B\cos\theta + C$ (or equivalently $A'\cos2\theta + B'\cos\theta + C$), where A, B, and C are derived from fitting to experimental measurements. It was first suggested by Martin Karplus in 1959.

kDa
Protein sizes are normally measured in kilodaltons (kDa). The mass of ^1H is 1 dalton, ^{12}C is 12 daltons etc. The average mass of one amino acid when linked together by peptide bonds to make a protein is 110 daltons. Therefore a protein of 100 residues has an approximate mass of $100 \times 110 = 11{,}000$ daltons or 11 kDa.

Kinase
An enzyme that adds a phosphate group, usually derived from ATP.

Last universal common ancestor (LUCA)
The original organism or community of organisms from which prokaryotes, eukaryotes, and archaea diverged, and therefore the base of the genetic tree.

Le Chatelier's principle
This states: "If a chemical system at equilibrium experiences a change in concentration, temperature, or total pressure, the equilibrium will shift in order to minimize that change." It is more an empirical observation than a law.

Lipid raft
The standard 'fluid mosaic' model of lipid membranes says that diffusion of lipids within the membrane is rapid. However, many lines of evidence now suggest that there are some regions of the membrane that have a different composition from the rest. They tend to be enriched in cholesterol and sphingolipids and have been called lipid rafts or more recently membrane rafts. They are suggested to be thicker and more rigid than normal membranes. They also attract and/or bind to some proteins, and repel others. They have been suggested for example to be key sites for cytoskeletal attachment and signaling. Despite considerable controversy over almost every aspect of lipid rafts, the evidence is fairly convincing.

Lock and key model
This model was suggested by **Emil Fischer (*5.11)** in 1894. It proposes that enzymes catalyze reactions by acting like a lock into which the substrate fits neatly. The idea was used to explain how it is that an enzyme can catalyze the hydrolysis of one stereoisomer but not another. It has been superseded by the induced-fit model (Chapter 6).

Low-complexity protein
A protein or protein section whose sequence contains only a small number of amino acids in a quasi-regular repeat. Many, but not all, of these form natively unfolded proteins, as exemplified by $(\text{Gly-Pro-Hypro})_n$ in collagen, and Ala_n and $(\text{Gly-Ala})_n$ in silks, which form fibers.

Meiosis
The way in which gametes (egg and sperm cells) are produced, in which a diploid cell (that is, one containing two copies of each chromosome, such as are found in animals) undergoes two cycles of cell division but only one round of DNA duplication, producing four haploid cells each containing only one copy of each chromosome.

Metabolon
A word formed from *metabolite* and *operon*, meaning a collection of metabolic enzymes from a metabolic pathway that are physically associated into a complex. See Chapter 9.

Metastable
A protein conformation that is stable under some circumstances but can be changed to a different conformation by a perturbation of the conditions.

Metazoan
A multicellular animal.

Minimal medium
Most bacterial growths are performed in a 'rich' medium such as the popular LB medium, which contains tryptone (the product of digestion of casein by trypsin) and yeast extract, a complicated mixture of vitamins, amino acids, lipids, and so on. For the production of isotopically labeled protein for NMR, LB medium cannot be used because the unlabeled molecules present would lead to poor levels of isotope incorporation. Therefore a chemically defined minimal medium is used, which contains for example glucose as the only carbon source and ammonium chloride as the only nitrogen source, together with salts and a small quantity of vitamins.

Mitochondrion
A membrane-enclosed eukaryotic organelle that produces most of the energy in the cell.

Mitogen
A chemical (typically a protein such as a growth factor) that stimulates cell proliferation, including cancer.

Mitosis
The division of a eukaryotic nucleus into two, involving condensation into chromatin and separation of the two halves. This results in the formation of two daughter cells with the same amount of DNA as the mother cell.

Mitotic spindle
A structure formed transiently during mitosis, which runs across the center of a dividing cell and has an important function in separating chromosomes evenly between the two daughter cells.

Module
A module is a highly conserved sequence that is observed in different contexts in multidomain proteins.

Molar extinction coefficient
The **Beer–Lambert law** says that the absorbance of a solution A is given by $A = \epsilon ct$, where c is the concentration of the solute, t is its thickness, and ϵ is the molar extinction coefficient. It is thus a characteristic of the molecule in question, and can be predicted reasonably well for a protein from the amino acid sequence, as discussed in Section 11.2.2.

Molecular dynamics
Simulation of the motion of a molecule or molecular system by using Newtonian mechanics.

Molten globule
This is a term that is most often used to describe proteins in mildly denaturing solution conditions, for example low

pH or highish denaturant, or apo-proteins (proteins without their cofactors). The name evokes a structure that has folded structure (fat globules) swimming around in a hydrated solution. Such proteins have native-like secondary structure but no well-formed tertiary structure, a hydrated interior, a noncooperative folding transition, and a rather larger radius than the folded protein. Because similar properties are also observed for protein folding intermediates, these are often described as being molten globules.

Moonlighting
A protein having two different functions, often where the 'main' activity is the clearly 'evolved' function, with a clearly defined active site, and the second function typically involves a different part of the protein surface and is an example of evolution tinkering with a random event to develop it. Tables 1.5 and 1.6 present some examples of moonlighting functions. Also known as **gene sharing**.

Mosaic
A word used to mean two different but related things:
1. A mosaic protein is one that consists of domains or modules brought in from different sources.
2. A mosaic or **patchwork** metabolic pathway is one in which the constituent enzymes are evolutionarily and structurally unrelated to each other.

Motif
Confusingly, the word motif is used in two different ways:
1. A **structure motif** is a grouping of elements of regular secondary structure. Examples include antiparallel helix bundle, helix–turn–helix, calcium-binding EF hand, coiled coil, β hairpin, Greek key, and β-α-β motifs. It is identical in meaning to supersecondary structure.
2. A **sequence motif** is a sequence of amino acids that is characteristic of a particular protein **fold**. The fold is therefore typically identified from the motif. Examples include GXGXXG, characteristic of the nucleotide-binding **Rossmann fold**; CXXCH, characteristic of class I soluble *c*-type cytochromes; and the sequence DEADx$_n$SAT, characteristic of the DEAD-box family of RNA helicases.
This book therefore makes clear whether it refers to a structure or a sequence motif.

N terminus
The end of a polypeptide chain that has a free amino group, namely the start or left-hand end as normally written. Note the distinction between the adjective *terminal* and the noun *terminus*.

Natively unstructured protein
A protein, or a region of a protein, that under physiological conditions has no defined secondary or tertiary structure. See Chapter 4 for more discussion.

Native structure
The structure of a protein in its native (that is, physiologically active) state.

Natural selection
The term used by Charles Darwin to describe the mechanism by which evolution occurs; sometimes but less helpfully also called *survival of the fittest*. As used in modern biology, it indicates that a genetic change will be selected for if it increases the number of surviving progeny of an organism.

Nonorthologous gene displacement
The occurrence in one organism of a gene coding for a specific function, which in another organism is coded for by an evolutionarily unrelated gene.

Nonribosomal synthesis
Protein biosynthesis that occurs using a series of specific enzymes in a large multienzyme complex (Chapter 10), by contrast with the standard means of biosynthesis using ribosomes.

Nuclear export signal
The opposite of a nuclear localization sequence: a sequence that directs a protein out of the nucleus into the cytoplasm.

Nuclear localization sequence (signal)
A short peptide sequence rich in positively charged lysines and arginines, which can be at any exposed part of the protein, the archetypal sequence being KKKRK from the SV40 virus T-antigen, and provides a signal to direct the protein to the nucleus.

Nucleophile
An electron-rich group that attacks an electrophile.

Oncogene
A gene whose expression causes cancer.

Order parameter
In NMR, a number taking values between 0 and 1 that describes the fraction of motion (typically of an N–H bond) that is due to overall tumbling, the rest being due to local internal motions. It has the symbol S^2.

ORF (open reading frame)
A DNA sequence that is presumed to code for a protein.

Orthologs
An ortholog is a homolog that was created by a speciation event; two orthologs fulfill the same role in evolutionarily related organisms because they were derived from the same ancestral protein.

Overexpression
The production of a protein at much higher levels than would normally be produced, by putting it under the control of a strong promoter.

Paralogs
A paralog is a homolog that was created by gene duplication.

Patchwork
Assembly of enzymes into pathways by duplication and co-option of enzymes from different metabolic pathways; named by analogy with the blankets and tablecloths made by sewing together cloth from different sources.

PCR
Polymerase chain reaction is a technique for increasing the amount of DNA, or amplifying it.

Peptide bond
The bond between the C=O of one amino acid and NH of the next. It is approximately flat and is almost always *trans*; that is, with the O and H facing in opposite directions (Figures 1.3 and 1.5).

Peptidylprolyl isomerase
An enzyme that catalyzes the exchange between *cis* and *trans* peptide bonds at the peptide bond preceding proline.

Periplasm
Bacteria are often grouped into two classes, depending on whether they can be stained violet with a dye discovered by Gram. Gram-negative bacteria (such as *Escherichia coli*) have an inner cytoplasmic membrane and an outer membrane, whereas Gram-positive bacteria have a cell membrane and a cell wall. The space between them (in both cases) is called the periplasm.

Phenotype
An observable characteristic of an organism: appearance, behavior, development, and so on. Arises as a consequence of the genome and is therefore to be contrasted with **genotype**.

Phosphatase
An enzyme that removes a covalently attached phosphate (and therefore the opposite of a **kinase**).

p*I*
The p*I* or isoelectric point is the pH at which a protein has net zero charge. It is therefore usually the pH at which proteins are least soluble. Its value can be calculated reasonably accurately from the sequence. Because two-dimensional protein gels, as used in proteomics analyses, have p*I* as one of the two dimensions (usually the horizontal one, obtained by isoelectric focusing, the other being size), it is an important parameter.

pK_a
The pH at which a titratable group is half ionized.

Plasmid
A small circular piece of DNA that replicates independently of the genome. It can therefore be used for cloning and expression, by carrying the gene of interest. It generally also contains a gene coding for resistance to some antibiotic, to provide a selectable marker to prevent the plasmid from getting lost from the bacterial host.

Polyproline II helix
A structure typically formed by proline polymers, with a three-residue repeat. See Section 4.2.6. This is an extended structure, also found in collagen fibers.

Posterior
The tail end of a cell or body.

Processive
A processive enzyme moves along a polymeric substrate carrying out the same reaction multiple times without dissociating: examples include DNA polymerase and cellulase.

Promiscuous
An enzyme that can catalyze two different reactions.

Pseudogene
A pseudogene is a DNA sequence that looks like a gene but does not express a protein.

Pseudo-twofold symmetry
A protein gene that is duplicated and fused creates a protein with two identical halves. Over the course of time, the two halves diverge, creating a protein with *pseudo* (not exact) two-fold symmetry.

Psychrophile
A psychrophilic organism grows preferentially at low temperature. Such organisms are typically found in cold oceans.

Pull-down assay
An immunological technique for identifying protein interactions. See **Immunoprecipitation (*4.7)**.

Pulse sequence
In NMR, the sequence of pulses and delays used to produce an NMR spectrum.

Quantum yield
Defined as the ratio of the number of photons emitted to the number absorbed. Anything with a quantum yield greater than 0.1 could be considered a reasonably strongly fluorescent molecule.

Racemization
Equilibration of a **chiral (*1.2)** molecule to make a 1:1 mixture of D and L isomers. For an amino acid this can be done by removing the proton on the α-carbon and replacing it randomly on either face.

Radius of gyration
Defined as the root mean square distance of the atoms from the center of gravity; effectively it is the radius of a uniform sphere that has the same mass and density as the protein, and is therefore the 'average' radius. It is often abbreviated as r_g. By contrast, the hydrodynamic radius r_h (also called the Stokes radius) is the radius of a uniform sphere that tumbles and diffuses at the same rate as the protein, and is larger than the radius of gyration roughly by the size of a single layer of water molecules.

Ramachandran plot
A diagram plotting the protein backbone dihedral angles ϕ and ψ for amino acids.

Random coil
A peptide in random coil conformation has its residues populating the (ϕ,ψ) map randomly according to their relative free energies. It therefore spends rather more time in the β-sheet region than in the α-helix region, and so is fairly extended.

Relaxation (in NMR)
Any process by which a set of nuclear spins return toward their equilibrium distribution.

Repeat expansion
It is surprisingly common to observe that trinucleotide sequences in coding DNA can get duplicated, leading to repeated amino acids in the translated protein. This may occur by some kind of unequal crossing or **homologous recombination** event. Often the sequence expanded is CAG, leading to polyglutamine stretches. This can cause a number of diseases, in which the severity of the disease is correlated with the number of glutamines.

Restriction enzyme
A nuclease that cuts DNA at a specific short sequence (four to six bases). Many restriction enzymes cut the two strands

in slightly different places, to leave short overhangs. They are used very widely in molecular biology for manipulating DNA sequences.

Ribosome
The large molecular machine composed of ribosomal RNA plus protein that translates messenger RNA into protein.

RING finger
A RING (Really Interesting New Gene) finger is a specialized zinc finger with two zinc ions and ligands C_3HC_4 (C = Cys, H = His), which binds both to ubiquitylation enzymes and to their substrates and hence acts as a ligase.

Rossmann fold
A protein **fold**, originally described by Michael Rossmann, containing a characteristic GXGXXG **sequence motif**, which functions to bind to nucleic acids.

Salt bridge
An electrostatic attraction between two charged groups in a protein, between Asp, Glu, or the C terminus and Lys, Arg, His, or the N terminus.

Scaffold protein
A protein that acts as a bridge between two other proteins, bringing them into close proximity so that they bind or react together.

Scale-free
In a scale-free network, the number of interactions made follows a power law, so that the frequency of one node having k links, $f(k)$, is proportional to k^{-n}, where n is usually between 2 and 3. Thus, most nodes have only one or two links, but a small number have a very large number of links. Such a structure has been observed for a very large range of networks including road maps, internet links, and protein interaction networks, which are said to be robust against small perturbations. It is called scale-free because the network looks similar when viewed on different scales.

Screening
The solvation of a charge by water or ions leads to an effective decrease in the charge, described by the word screening. See **Electrostatic screening (*4.4)**.

SDS-PAGE (sodium dodecyl sulfate polyacrylamide gel electrophoresis)
A gel is made from polyacrylamide into which is incorporated sodium dodecyl sulfate (SDS). When a voltage is placed across the gel, proteins move down the gel by a distance dependent on their size. The method thus forms a simple and reliable way to separate proteins by size.

Sequence motif
A sequence of amino acids that is characteristic of a particular protein **fold**. The fold is therefore typically identified from the motif. Examples include GXGXXG, characteristic of the nucleotide-binding **Rossmann fold**; CXXCH, characteristic of class I soluble *c*-type cytochromes; and the sequence DEADx$_n$SAT, characteristic of the DEAD-box family of RNA helicases.

Serine protease
Proteases are enzymes that digest proteins by hydrolyzing the peptide bond between amino acids. There are several classes of proteases, of which the largest group are called serine proteases, because the amino acid that performs the nucleophilic attack on the peptide bond is a serine. They always contain a **catalytic triad**, in which the serine is aided by a neighboring histidine and aspartate.

Signal recognition particle
A ribonucleoprotein (RNA–protein complex) that recognizes proteins coming off the **ribosome** that should be targeted to membranes. It halts translation until the ribosome is docked onto a receptor on the membrane, after which it dissociates.

Signal sequence
A short sequence, usually at the N terminus of a protein, that directs it to a particular cellular location.

Specificity constant
The ratio k_{cat}/K_m. This determines how specific an enzyme is for a particular substrate, and is also the apparent **second-order rate constant (*5.17)** for a bimolecular reaction (for example, an enzyme plus a single substrate).

Structure motif
A grouping of elements of regular secondary structure. Examples include antiparallel helix bundle, helix–turn–helix, calcium-binding EF hand, coiled coil, β hairpin, Greek key, and β-α-β motifs. It is identical in meaning to **supersecondary structure**.

Superfold
A fold found, with small modifications, in several related structures.

Supersaturated solution
A solution in which the concentration of the solute is greater than its solubility. In these circumstances, the thermodynamic preference is for the solute to either precipitate or crystallize; kinetic factors may mean that either or both of these processes may be slow.

Supersecondary structure
A grouping of elements of regular secondary structure. Examples include antiparallel helix bundle, helix–turn–helix, calcium-binding EF hand, coiled coil, β hairpin, Greek key, and β-α-β motifs. It is identical in meaning to **structure motif**.

Synapse
The junction between a nerve cell and another cell, such as a muscle cell.

TAP-tag
TAP (tandem affinity purification) tagging is a way of purifying low-abundance proteins by attaching two affinity tags that can be cleaved off sequentially. See Chapter 11.

TCA
Tricarboxylic acid. The *TCA cycle*, also known as the Krebs cycle, is the major metabolic route for using pyruvate to provide metabolic energy via its conversion into the tricarboxylic acid citrate.

Termination codon
Also known as a stop codon: one of the three mRNA **codons** that code not for an amino acid but for the end

of the polypeptide chain. These are UAG (*amber*), UAA (*ochre*) and UGA (*umber*), the last of which can also code for selenocysteine.

Thermophile
A thermophilic organism that grows preferentially at high temperature. Such organisms are typically found in hot springs.

Three-dimensional domain swapping
The exchange of two equivalent domains or secondary structure elements within a dimer to create an alternative dimeric structure in which the constituent monomers are interwoven, and the dimer is therefore much more strongly bound together.

Transition state
The part of a reaction with the highest free energy. By definition, the transition state (TS) is therefore very short-lived and cannot be isolated.

Triple resonance experiment
An NMR experiment involving pulses and magnetization transfer on three different nuclei, commonly ^{1}H, ^{13}C, and ^{15}N.

Two-hybrid screen
A technique used for detecting interacting pairs of protein domains, described in Chapter 11.

Ubiquitin
A small protein found widely in eukaryotic cells (hence the name). Its main function is to label proteins for degradation by covalent attachment ('ubiquitylation'), which targets them to the **proteasome** for degradation.

Vector
An agent used to carry genetic material. Very often it takes the form of a bacterial **plasmid**.

Western blot
A technique in which proteins are run on an SDS-PAGE or two-dimensional gel, and then moved from the gel onto a membrane by applying a voltage perpendicular to the gel. The membrane can then be washed with an antibody to a particular protein, allowing the identification of the protein on the membrane. This technique was developed from an earlier one in which nucleic acids were identified by hybridization to a probe in a similar manner, a technique invented by Ed Southern and called the Southern blot.

Zwitterion
A compound carrying both a positive and a negative charge.

Index

Note: Page numbers followed by F indicate figures, those followed by T indicate tables and those followed by * indicate blue-shaded information boxes.

B

C

N

O

P

T